WA 1227120 9

D0302723

LUMINESCENCE BIOTECHNOLOGY

Instruments and Applications

LUMINESCENCE BIOTECHNOLOGY

Instruments and Applications

Edited by
Knox Van Dyke
Christopher Van Dyke
Karen Woodfork

Boca Raton London New York Washington, D.C.

Library of Congress Cataloging-in-Publication Data

Luminescence biotechnology : instruments and applications / edited by Knox Van Dyke, Christopher Van Dyke, Karen Woodfork.
p. cm.
Includes bibliographical references and index.
ISBN 0-8493-0719-8 (alk. paper)
1. Luminescent probes. 2. Luminescence immunoassay. 3. Luminescence spectroscopy. I. Van Dyke, Knox, 1939- II. Van Dyke, Christopher. III. Woodfork, Karen.

TP248.25.L85 .L86 2001
660.6′028--dc21 2001043471

Direct all inquiries to CRC Press LLC, 2000 N.W. Corporate Blvd., Boca Raton, Florida 33431.

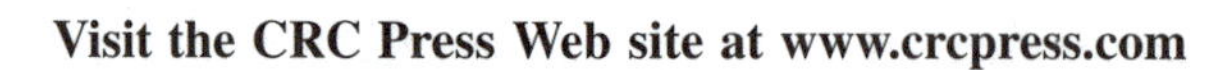

International Standard Book Number 0-8493-0719-8
Library of Congress Card Number 2001043471
Printed in the United States of America 1 2 3 4 5 6 7 8 9 0
Printed on acid-free paper

Preface

Luminescence Biotechnology: Instruments and Applications was written to record, in a single compendium, the tremendous progress in bioluminescence and chemiluminescence in recent years. The field has progressed in many ways, particularly in the development of new probes, enhancers, and instrumentation, as well as new methods to study genetics and oxidative stress.

More than 25 years ago, I told my students that luminescence would develop exponentially when new chemical probes that were specific and highly luminescent became available. Unknown to me at the time, Dr. Irena Bronstein had begun developing highly luminescent adamantyl derivatives and, consequently, the basis for the company Tropix (now Applied Biosystems). This company jump-started luminescence technology, providing the capital to move into new areas — such as molecular biology and high-throughput screening — that have transformed the way we search for new drugs. Dr. Bronstein's luminescence probes made the dream of replacing radioactivity assays a reality. Now highly sensitive assays can be done without the danger, environmental hazards, and/or regulations associated with radioactivity.

Another recent development is enhancers that boost the luminescence output from a probe. Originally reported by Larry Kricka et al., this concept has been applied to a variety of luminescent chemistry. Dr. Kricka found that the blue light generated from luminol and a reaction with hydrogen peroxide/horseradish was enhanced more than 1000-fold by *p*-iodophenol. Later, Dr. Bronstein developed a variety of extremely useful enhancers applicable to adamantyl-dioxetane chemistry (discussed in Chapter 1). We report that the Stratagene enhancer can stimulate luminescence from macrophages, possibly linked to peroxynitrite production.

New instrumentation has enabled luminescence to become both more accurate and highly efficient. Thirty years ago, we measured luminescence from single samples, plotted data by hand, and compared samples by cutting out the area under the curve and weighing it. Later, we automated sample changing using the belt-driven assays associated with liquid scintillation counters. At that time, assaying in real time analysis such as kinetic assessment and area under the curve was inconceivable. New digital assay systems are readily computerized (see Chapters 41 through 45). High-throughput assays and instrumentation allows hundreds of thousands of assays to be done in a day (see Chapter 44). Some high-throughput systems use photodiode array technology rather than photomultiplier tubes, which allows a measurement to occur in an *x-y* array quickly and easily. The development of highly luminescent chemicals allows the use of the photodiode array, which digitizes in real time from multiple points simultaneously.

The late Marlene Deluca pioneered research into gene expression with luminescence and, with the help of her associates at the University of California at San Diego, sequenced the firefly luciferase gene. This made it possible to insert the luciferase next to genes whose products were difficult to measure. Activation of the luciferase gene and the production of luciferase, thus "reports" activity of the target gene. This stimulated the use of other luminescent mechanisms for study of gene activation (e.g., renilla from sea pansy), which allows for dual labeling mechanisms. Such luminescent systems have become tremendously useful tools to study gene expression in plants, mammals, nonmammals, parasites, and bacteria. Many instances of the use of luminescence to report gene expression exist today.

The concept of oxidative stress that was originally developed by Helmut Seis drives a new, clear effort to understand the mechanisms by which a great variety of diseases cause damage in the human body. Almost all diseases cause an oxidative imbalance so that the ratio of oxidant to antioxidants is greater than one. As the oxidants are produced to fight disease, they destroy our antioxidants, which leads to the damage of DNA, RNA, and protein. Luminescence can be used to follow the oxidative activity of the body and the cells and to quantitate the oxidative damage to genes. In a series of chapters in this book we outline our work in this regard. To protect our genes from the ravages of oxidative mutation is to prevent the pressures that cause vascular and neurodegeneration as well as cancer. Once we learn how to control or negate oxidative stresses, we have a better chance of longer, more productive lives; I believe my studies using luminescence have contributed to that end. I predict that these studies will help end chronic diseases.

I thank all of the authors of this book for contributing their efforts. The international flavor of this book clearly indicates that scientists from many nations have realized the importance of this effort.

Knox Van Dyke

About the Editors

Knox Van Dyke, Ph.D., is a professor in the Department of Biochemistry and Molecular Pharmacology at West Virginia University, Morgantown. He obtained his A.B. in chemistry in 1961 from Knox College, Galesburg, Illinois and his Ph.D. in endocrinological biochemistry in 1966 from St. Louis University, under Professor Philip A. Katzman in the Edward A. Doisy Department of Biochemistry.

After completion of his graduate studies, Dr. Van Dyke joined Dr. Leroy Saxe at West Virginia University in the development of novel drug screening systems for malaria under a U.S. Army contract. The culmination of this work led to the discovery of mefloquine as an important antimalarial drug against *Plasmodium falciparum* and the discovery of the fundamental difference between purine and pyrimidine metabolism in malarial parasites. A purine metabolite, hypoxanthine, was found to be a key endogenous purine-nucleic-acid precursor for malarial parasites and it forms the basis of present screening systems developed against the deadly *P. falciparum* human malarial parasites.

As a part of the drug screening system for malarial parasites, Dr. Van Dyke developed the first automated system for the measurement of adenosine triphosphate (ATP) by luciferase light production which measured drug toxicity to the erthyrocyte. This led to the development of the Aminco Chem-Glo Photometer using a flowing stream. A collaboration with Dr. Madhu Manandhar produced the first research into the phenomenon of "delayed luminescence," which allows the use of almost any purine and pyrimidine nucleotide to produce light from *Photinus pyralis*.

In 1975, luminol-dependent chemiluminescence (CL) was developed in the laboratory of Dr. Van Dyke. Studies using this system examined chronic granulomatous disease, neonatal sepsis, and neutropenic states and were completed by former students Dr. Michael Trush, Dr. Mark Wilson, and Dr. Paul Stevens. The first published work with alveolar macrophages producing cellular chemiluminescence was accomplished in this era.

In the 1980s, Dr. Van Dyke published numerous studies applying cellular chemiluminescence to various disease states, including cancer, arthritis, black lung, and environmental problems, in collaboration with Dr. Vincent Castranova and Dr. David Peden.

Dr. Van Dyke has worldwide patents in malaria, cancer, and HIV chemotherapy. Additional patents are pending in reperfusion injury, vasodilation, and neurogenerative diseases. A multiple drug resistance inhibitor for cancer is in third-stage clinical trial and should be available clinically in 1 to 2 years. This may be the first drug available that inhibits major resistance mechanisms for more than ten different cancer drugs.

Presently, Dr. Van Dyke is developing new nutritional substances capable of blocking oxidant-based mutation. Some of the basis for this research is outlined in this book. This work was done with the collaboration of Paul McConnell, Candace Ogle, and Dr. Mark Reasor.

In addition to his duties as a professor of biochemistry and molecular pharmacology, Dr. Van Dyke maintains an active research laboratory fundamentally interested in cellular CL and CL immunoassay with current studies exploring the interplay of oxidants and antioxidants in disease formation and suppression. He is a member (past or present) of the following societies: American Chemical Society, National Association for the Advancement of Science, Sigma Xi, Society of Pharmacology and Experimental Therapeutics, American Society of Photobiology, Who's Who in The Eastern United States, and Who's Who in the Frontiers of Science.

Dr. Van Dyke is the editor of, and a contributing author to, several CRC Press publications including *Bioluminescence and Chemiluminescence: Instruments and Applications* (1985), *Cellular*

Chemiluminescence (1987), and *Luminescence Immunoassay and Molecular Applications* (1990). He is the author of seven chapters in *Modern Pharmacology*, 6th edition, edited by C. R. Craig and R. E. Stitzel, to be published in 2002.

Recently, Dr. Van Dyke believes he has found the basis for chronic diseases, e.g., Alzheimer's disease, Parkinson's disease, multiple sclerosis, amyotrophic lateral sclerosis (ALS, Lou Gehrig's disease) and others. This work was done with luminometry and some of the evidence lies in the pages of this book. In addition, Dr. Van Dyke believes the diseases can be corrected using a simple supplement to the diet. Let us hope he is correct.

Christopher Van Dyke, B.S.E.E., is a lecturer in the Department of Biochemistry and Molecular Pharmacology at West Virginia University, Morgantown. He obtained his B.S.E.E. in electrical engineering from West Virginia University in 1987.

During high school and college, Mr. Van Dyke worked on many projects in Dr. Knox Van Dyke's laboratory including writing computer programs, culturing *P. falciparum*, and performing luminescent and radioimmunoassays. Following graduation, he worked for a small biotechnology company on many different projects, including the initial development of an anticancer drug and an imaging device. This was followed by employment in the laboratory of Dr. Knox Van Dyke exploring the toxicity of inhaled silica. He then worked at the National Institute of Occupational Health and Safety, creating animal exposure systems to explore pulmonary mechanics and the toxicity of inhaled substances.

In his current position, Mr. Van Dyke develops Web sites and other electronic resources for medical education.

Karen Woodfork, Ph.D., is an adjunct associate professor in the Department of Biochemistry and Molecular Pharmacology at West Virginia University, Morgantown.

Dr. Woodfork received her B.A. in physics from Rice University, Houston, Texas, in 1987 and her Ph.D. in pharmacology from West Virginia University in 1993. Her graduate studies in the laboratory of Dr. Jeannine Strobl examined the role of potassium channels in the proliferation of breast cancer cells. She received postdoctoral training in Dr. James Garrison's laboratory in the Department of Pharmacology at the University of Virginia, where she investigated the effects of G protein lipid modifications on receptor-mediated signal transduction. As an assistant professor at Washington and Jefferson College in Washington, Pennsylvania, she authored laboratory manuals and designed curricula for undergraduate courses in cell biology, biochemistry, and general biology. Following a National Research Council Fellowship at the National Institute for Occupational Safety and Health, she joined the Department of Biochemistry and Molecular Pharmacology at West Virginia University, where she teaches graduate- and undergraduate-level courses, writes grants, and designs educational software for medical students. She is a contributing author to *Modern Pharmacology,* 6th edition (to be published in 2002).

Contributors

Maciej Adamczyk, Ph.D.
Department of Chemistry
Diagnostics Division
Abbott Laboratories
Abbott Park, Illinois, U.S.A.

Neena Agarwal, B.S.
Department of Biochemistry
and Molecular Pharmacology
West Virginia University School of Medicine
Morgantown, West Virginia, U.S.A.

Franz Allerberger, M.D.
Department of Bacteriology
Innsbruck University Hospital
Innsbruck, Austria

Gang An, Ph.D.
GlaxoSmithKline
Research Triangle Park,
North Carolina, U.S.A.

James Anthony, Ph.D.
Digene Corporation
Gaithersburg, Maryland, U.S.A.

Yoshio Aramaki, Ph.D.
Pharmaceutical Research Division
Takeda Chemical Industries, Ltd.
Osaka, Japan

Mario Baraldini, M.D., Ph.D.
Institute of Chemical Sciences
University of Bologna
Bologna, Italy

Fritz Berthold, Ph.D.
Berthold Technologies GMBH & Co.
Bad Wildbad, Germany

Peggy Biser, Ph.D.
Department of Chemistry
Frostburg State University
Frostburg, Maryland, U.S.A.

Deborah M. Boldt-Houle, Ph.D.
Applied Biosystems
Bedford, Massachusetts, U.S.A.

Berthold Breitkopf, Ph.D.
Berthold Technologies GMBH & Co.
Bad Wildbad, Germany

Reinhold Brettschneider, Ph.D.
Centre for Applied Plant Molecular Biology
Institute for General Botany
Hamburg, Germany

Judith Britz, Ph.D.
Cylex, Inc.
Columbia, Maryland, U.S.A.

Irena Bronstein, Ph.D.
Applied Biosystems
Bedford, Massachusetts, U.S.A.

Tom Brotcke, M.B.A.
Pierce Chemical Company
Rockford, Illinois, U.S.A.

Stacy M. Burns, Ph.D.
Division of Neonatal
and Developmental Medicine
Department of Pediatrics
Stanford University Medical Center
Stanford, California, U.S.A.

Mireille Caron, B.Sc.
BioSignal Packard, Inc.
Montreal, Quebec, Canada

Kelly Carter-Allen, B.S.
Packard BioScience
Downers Grove, IL, U.S.A.

Vince Castranova, Ph.D.
The Health Effects Laboratory Division
National Institute for Occupational Safety
and Health
Morgantown, West Virginia, U.S.A.

Marie Charrel, Ph.D.
Department of Immunology
Royal Free University College Medical School
Windeyer Institute of Medical Sciences
London, United Kingdom

Fei Chen, M.D., Ph.D.
The Health Effects Laboratory Division
National Institute for Occupational Safety
and Health
Morgantown, West Virginia, U.S.A.

Anthony C. Chiulli, B.S.
Applied Biosystems
Bedford, Massachusetts, U.S.A.

Mark A. Christenson, Ph.D.
Life Sciences Division
Roper Scientific
Trenton, New Jersey, U.S.A.

Milan Číž, Ph.D.
Institute of Biophysics
Academy of Science of the Czech Republic
Brno, Czech Republic

Carrie Clothier, B.S.
Pierce Chemical Company
Rockford, Illinois, U.S.A.

Christopher H. Contag, Ph.D.
Division of Neonatal
and Developmental Medicine
Department of Pediatrics
Stanford University Medical Center
Stanford, California, U.S.A.

Pamela R. Contag, Ph.D.
Xenogen Corporation
Alameda, California
and
Division of Neonatal
and Developmental Medicine
Department of Pediatrics
Stanford University Medical Center
Stanford, California, U.S.A.

Sylvia Daunert, Pharm.D., Ph.D.
Department of Chemistry
University of Kentucky
Lexington, Kentucky, U.S.A.

Peter J. Delves, Ph.D.
Department of Immunology
Royal Free University College Medical School
Windeyer Institute of Medical Sciences
London, United Kingdom

Sapna K. Deo, Ph.D.
Department of Chemistry
University of Kentucky
Lexington, Kentucky, U.S.A.

Brian J. D'Eon, B.S.
Applied Biosystems
Bedford, Massachusetts, U.S.A.

Manfred P. Dierich, M.D., Prof.
Department of Hygiene
Innsbruck University Hospital
Innsbruck, Austria

Pierre Dionne, Ph.D.
BioSignal Packard, Inc.
Montreal, Quebec, Canada

Thomas Eberl, M.D.
Department of Transplant Surgery
Innsbruck University Hospital
Innsbruck, Austria

Kelli Feather-Henigan, B.S.
Pierce Chemical Company
Rockford, Illinois, U.S.A.

Matthias Fladung, Ph.D.
Federal Research Centre for
Forestry and Forest Products
Institute for Forest Genetics and
Forest Tree Breeding
Grosshansdorf, Germany

William Fleming, Ph.D.
Department of Physiology
and Pharmacology
West Virginia University School
of Medicine
Morgantown, West Virginia, U.S.A.

John L. A. Fordham, Ph.D.
Department of Physics and Astronomy
University College London
London, United Kingdom

Kevin P. Francis, Ph.D.
Xenogen Corporation
Alameda, California, U.S.A.

Laura Frost, Ph.D.
Department of Immunology and Cell Biology
West Virginia University School
of Medicine
Morgantown, West Virginia, U.S.A.

Dale L. Godson, Ph.D.
Veterinary Infectious Disease Organization
Saskatoon, Saskatchewan, Canada

Susantha M. Gomis, Ph.D.
Veterinary Infectious Disease Organization
Saskatoon, Saskatchewan, Canada

Jürgen Gräßler, M.D.
Department of Internal Medicine
Carl Gustav Carus Medical School
Technical University of Dresden
Dresden, Germany

Christopher Gruber, Ph.D.
Division of Neonatal and Developmental
Medicine
Department of Pediatrics
Stanford University Medical Center
Stanford, California, U.S.A.

Massimo Guardigli, Ph.D.
Department of Pharmaceutical Sciences
University of Bologna
Bologna, Italy

Patrick C. Hallenbeck, Ph.D.
Department of Microbiology
and Immunology
University of Montreal
Montreal, Quebec, Canada

Michael A. Harvey, Ph.D.
Schleicher & Schleicher, Inc.
Keene, New Hampshire, U.S.A.

Paul Hengster, Ph.D.
Department of Transplant Surgery
Innsbruck University Hospital
Innsbruck, Austria

Kimberly K. Hines, M.S.
Pierce Chemical Company
Rockford, Illinois, U.S.A.

Ineabel Horneij, M.B.A.
Pierce Chemical Company
Rockford, Illinois, U.S.A.

Shuntaro Hosaka, Ph.D.
Department of Applied Chemistry
Faculty of Engineering
Tokyo Institute of Polytechnics
Kanagawa, Japan

Benoit Houle, Ph.D.
BioSignal Packard, Inc.
Montreal, Quebec, Canada

Yuzo Ichimori, Ph.D.
Pharmaceutical Research Division
Takeda Chemical Industries, Ltd.
Osaka, Japan

Isuke Imada, Ph.D.
Department of Biochemistry
and Molecular Pathology
Osaka City University Medical
School
Osaka, Japan

Masayasu Inoue, M.D., Ph.D.
Department of Biochemistry
and Molecular Pathology
Osaka City University
Medical School
Osaka, Japan

Muhammad Iqbal, Ph.D.
Division of Animal
and Veterinary Sciences
West Virginia University
Morgantown, West Virginia, U.S.A.

John C. Jackson, Ph.D.
Pierce Chemical Company
Rockford, Illinois, U.S.A.

Danny Joh, Ph.D.
Xenogen Corporation
Alameda, California, U.S.A.

Carl Hirschie Johnson, Ph.D.
Department of Biological Sciences
Vanderbilt University
Nashville, Tennessee, U.S.A.

Erik Joly, Ph.D.
BioSignal Packard, Inc.
Montreal, Quebec, Canada

Akihito Kanauchi, Ph.D.
Department of Biological Sciences
Vanderbilt University
Nashville, Tennessee, U.S.A.

Hillar Klandorf, Ph.D.
Division of Animal and Veterinary Sciences
West Virginia University
and
Pathology and Physiology Research Branch
National Institute for Occupational Safety and Health
Morgantown, West Virginia, U.S.A.

Jian-Qiang Kong, M.D., Ph.D.
Department of Pharmacology
Brody School of Medicine
East Carolina University
Greenville, North Carolina

Steffi Kopprasch, Ph.D.
Department of Internal Medicine
Carl Gustav Carus Medical School
Technical University of Dresden
Dresden, Germany

Richard Kowalski, Ph.D.
Cylex, Inc.
Columbia, Maryland, U.S.A.

Vivian Kuhlenkamp
Federal Research Centre for Forestry and Forest Products
Institute for Forest Genetics and Forest Tree Breeding
Grosshansdorf, Germany

Marialuise Kunc, B.S.
Department of Transplant Surgery
Innsbruck University Hospital
Innsbruck, Austria

Anne Labonté, B.Sc.
BioSignal Packard, Inc.
Montreal, Quebec, Canada

Byeong-ha Lee, Ph.D.
Department of Plant Sciences
University of Arizona
Tucson, Arizona, U.S.A.

Jennifer C. Lewis, Ph.D.
Department of Chemistry
University of Kentucky
Lexington, Kentucky, U.S.A.

Li Li
Packard Instrument Company
Meriden, Connecticut, U.S.A.

Yunbo Li, M.D., Ph.D.
Department of Pharmaceutical Sciences
College of Pharmacy
St. John's University
Jamaica, New York, U.S.A.

Bill Lipton, M.S.
Pierce Chemical Company
Rockford, Illinois, U.S.A.

Betty Liu, B.S.
Applied Biosystems
Bedford, Massachusetts, U.S.A.

Antonín Lojek, Ph.D.
Institute of Biophysics
Academy of Science of the Czech Republic
Brno, Czech Republic

Attila Lorincz, Ph.D.
Digene Corporation
Gaithersburg, Maryland, U.S.A.

Torben Lund, Ph.D.
Department of Immunology
Royal Free University College Medical School
Windeyer Institute of Medical Sciences
London, United Kingdom

Masako Maeda, Ph.D.
School of Pharmaceutical Sciences
Showa University
Tokyo, Japan

Raimund Margreiter, Ph.D.
Department of Transplant Surgery
Innsbruck University Hospital
Innsbruck, Austria

Walter Mark, Ph.D.
Department of Transplant Surgery
Innsbruck University Hospital
Innsbruck, Austria

Phillip G. Mattingly, Ph.D.
Department of Chemistry
Abbott Laboratories
Abbott Park, Illinois, U.S.A.

Paul McConnell, M.S.
Department of Genetics
West Virginia University School of Medicine
Morgantown, West Virginia, U.S.A.

Luc Menard, Ph.D.
BioSignal Packard, Inc.
Montreal, Quebec, Canada

Gary F. Merrill, Ph.D.
Department of Cell Biology
and Neuroscience
Rutgers University
Piscataway, New Jersey, U.S.A.

Greg Milosevich, B.S., B.A.
Alpha Innotech Corporation
San Leandro, California, U.S.A.

Monica Musiani, M.D., Ph.D.
Department of Clinical
and Experimental Medicine
Division of Microbiology
University of Bologna
Bologna, Italy

Candace L. Ogle, B.S.
Department of Pharmacology
and Toxicology
West Virginia University School
of Medicine
Morgantown, West Virginia, U.S.A.

Corinne E. M. Olesen, Ph.D.
Applied Biosystems
Bedford, Massachusetts, U.S.A.

Breck O. Parker, Ph.D.
Schleicher & Schleicher, Inc.
Keene, New Hampshire, U.S.A.

Patrizia Pasini, Ph.D.
Department of Pharmaceutical Sciences
University of Bologna
Bologna, Italy

Jens Pietzsch, Ph.D.
Institute and Policlinic of Clinical Metabolic
Research
Carl Gustav Carus Medical School
Technical University of Dresden
Dresden, Germany

David W. Piston, Ph.D.
Department of Molecular Physiology
and Biophysics
Vanderbilt University
Nashville, Tennessee, U.S.A.

Nino Porakishvili, Ph.D.
Department of Immunology
Royal Free University College
Medical School
Windeyer Institute of Medical Sciences
London, United Kingdom

Andrew A. Potter, Ph.D.
Veterinary Infectious Disease Organization
Saskatoon, Saskatchewan, Canada
and Canadian Bacterial Diseases Network
Health Sciences Center
University of Calgary
Calgary, Alberta, Canada

Nader Pourmand, Ph.D.
Stanford Genome Technology Center
Stanford University
Palo Alto, California, U.S.A.

Dinesh S. Rathore, M.S.
Division of Animal
and Veterinary Sciences
West Virginia University
Morgantown, West Virginia, U.S.A.

Mark J. Reasor, Ph.D.
Department of Pathology
West Virginia Medical Center
Morgantown, West Virginia, U.S.A.

Annadi Ram Reddy, Ph.D.
Department of Chemistry
Nizam College
Osmania University
Hyderabad, India

Aldo Roda, Ph.D.
Department of Pharmaceutical Sciences
University of Bologna
Bologna, Italy

Ivan M. Roitt, Ph.D.
Department of Immunology
Royal Free University College
Medical School
Windeyer Institute of Medical Sciences
London, United Kingdom

Mostafa Ronaghi, Ph.D.
Stanford Genome Technology Center
Stanford University
Palo Alto, California, U.S.A.

Meir Sacks, Ph.D.
Department of Pharmacology
and Toxicology
West Virginia Medical Center
Morgantown, West Virginia, U.S.A.

Eisuke F. Sato, Ph.D.
Department of Biochemistry and
Molecular Pathology
Osaka City University Medical School
Osaka, Japan

Xianglin Shi, Ph.D.
Pathology and Physiology
Research Branch
National Institute for Occupational Safety
and Health
Morgantown, West Virginia, U.S.A.

Melvin F. Simoyi, M.S.
Division of Animal and
Veterinary Sciences
West Virginia University
Morgantown, West Virginia, U.S.A.

Peter Sottong, M.S.
Cylex, Inc.
Columbia, Maryland, U.S.A.

Wolfgang Steurer, M.D.
Department of Transplant Surgery
Innsbruck University Hospital
Innsbruck, Austria

Becky Stevenson, Ph.D.
Department of Plant Sciences
University of Arizona
Tucson, Arizona, U.S.A.

Viktor Stolc, Ph.D.
NASA Ames Research Center
Moffett Field, California, U.S.A.

Beth A. Strachan, Ph.D.
Pierce Chemical Company
Rockford, Illinois, U.S.A.

Thankiah Sudhaharan, Ph.D.
Department of Chemistry
Nizam College
Osmania University
Hyderabad, India

Lee A. Sylvers, Ph.D.
Pierce Chemical Company
Rockford, Illinois, U.S.A.

David Taylor, Ph.D.
Department of Pharmacology
Brody School of Medicine
East Carolina University
Greenville, North Carolina, U.S.A.

Michael Taylor, Ph.D.
Department of Pharmacology and Toxicology
West Virginia University School of Medicine
Morgantown, West Virginia, U.S.A.

Sean C. Taylor, Ph.D.
BioSignal Packard, Inc.
Montreal, Quebec, Canada

John L. Tonkinson, Ph.D.
Schleicher & Schleicher, Inc.
Keene, New Hampshire, U.S.A.

Michael A. Trush, Ph.D.
Department of Environmental Health Sciences
Bloomberg School of Public Health
Johns Hopkins University
Baltimore, Maryland, U.S.A.

Takafumi Uchida, Ph.D.
Department of Pathology
Institute of Development, Aging, and Cancer
Tohoku University
Sendai, Japan

Val Vallyathan, Ph.D.
The Health Effects Laboratory Division
National Institute for Occupational Safety and Health
Morgantown, West Virginia, U.S.A.

Christopher Van Dyke, B.S.E.E.
Department of Pharmacology and Toxicology
West Virginia Medical Center
Morgantown, West Virginia, U.S.A.

Knox Van Dyke, Ph.D.
Department of Pharmacology and Toxicology
West Virginia Medical Center
Morgantown, West Virginia, U.S.A.

Robert W. Veltri, Ph.D.
UroCor, Inc.
Oklahoma City, Oklahoma, U.S.A.

John C. Voyta, Ph.D.
Applied Biosystems
Bedford, Massachusetts, U.S.A.

David von Schack, Ph.D.
CLONTECH Laboratories, Inc.
Palo Alto, California, U.S.A.

Mark Walter, Ph.D.
Department of Transplant Surgery
Innsbruck University Hospital
Innsbruck, Austria

Karen Woodfork, Ph.D.
Department of Pharmacology and Toxicology
West Virginia Medical Center
Morgantown, West Virginia, U.S.A.

Yao Xu, Ph.D.
Department of Biological Sciences
Vanderbilt University
Nashville, Tennessee, U.S.A.

Alexander F. Yakunin, Ph.D.
Department of Microbiology and Immunology
University of Montreal
Montreal, Quebec, Canada

Yu-Xin Yan, Ph.D.
Applied Biosystems
Bedford, Massachusetts, U.S.A.

Jian-Kang Zhu, Ph.D.
Department of Plant Sciences
University of Arizona
Tucson, Arizona, U.S.A.

Birgit Ziegenhagen, Ph.D.
Federal Research Centre for Forestry and Forest Products
Institute for Forest Genetics and Forest Tree Breeding
Grosshansdorf, Germany

Table of Contents

Section I

Introduction to Luminescence Assay

1 Light Probes

Knox Van Dyke and Karen Woodfork

CONTENTS

0-8493-0719-8/02/$0.00+$1.50

1.1 LUMINESCENCE IN THE LABORATORY

The most useful assay methods are those that produce a low-background signal and offer a wide range of detection, typically 10^6-fold. Initially, radioactivity was used for this purpose; however, its purchase price and disposal costs, as well as its associated personal and environmental hazards, are prohibitively high. In addition, the safety training and maintenance of government-mandated records require extensive staff time and effort. These factors combine to make radioactive methods an alternative, but not the primary choice of detection technique. Luminescence detection, in contrast, is at least as sensitive as isotopic methods, is generally less expensive, and lacks the associated safety risks. Luminescence detection of a direct-labeled DNA probe is the most sensitive method of detection, eliminates interference from indirect labeling, is stable and reliable, and utilizes reagents with an extended shelf life (Table 1.1).[1]

1.2 GENERAL PRINCIPLES OF LUMINESCENCE

1.2.1 Luminescence Reactions

When chemicals react, energy in the form of heat is either released or absorbed (Figure 1.1). In certain reactions, the initial reactants reach a higher energy (excited) state, then release light as they decay to ground state (Figure 1.2). Luminescence is often described as "cold light" because no external energy needs to be added to produce the light and little heat is generated by the reaction. In contrast, fluorescence requires the addition of light energy to push the reactants into an excited state. Phosphorescence, like fluorescence, requires the addition of energy to excite the reactant chemicals, but phosphorescence produces light for minutes or hours after the excitation energy has been removed.

Luminescent reactions can be classified according to the source of the compounds that generate light. Bioluminescence refers to the release of light when the original source of the chemicals is biological. Chemiluminescence refers to light that originates from the reaction of synthetic chemicals. Luminescent reactions can also be classified by the method used to achieve the excited state (Table 1.2). Luminol (3-aminophthalhydrazide) is an example of a chemical that can participate in many types of luminescence reactions. It can produce light in solution if it is exposed to γ rays, high frequency sonic waves, frictional forces, or peroxides.

Flash-type luminescence involves a rapidly developing kinetic reaction (Figure 1.3). In contrast, glow-type luminescence involves sustained equilibrium kinetic reactions. Each type of luminescence reaction has distinct advantages and disadvantages. Generally, flash reactions are more difficult to reproduce consistently. If the injection of chemicals is too rapid, frictional forces can excite the

TABLE 1.1
Comparative Sensitivity of Detection Methods

Luminescence	10^{-19} *M*
Radioisotope	10^{-18} *M*
Fluorescence	10^{-12} *M*
Absorbance	10^{-9} *M*
Latex agglutination	10^{-5} *M*

TABLE 1.2
Classification of Luminescence Reactions

Reaction Type	Method Used to Activate Probe to the Excited State
Chemiluminescence	Oxidative reactions
Triboluminescence	Frictional stress
Lyoluminescence	X or γ rays
Sonoluminescence	Sound waves
Electroluminescence	Electrical energy, terbium/yttrium chemistry, laser light

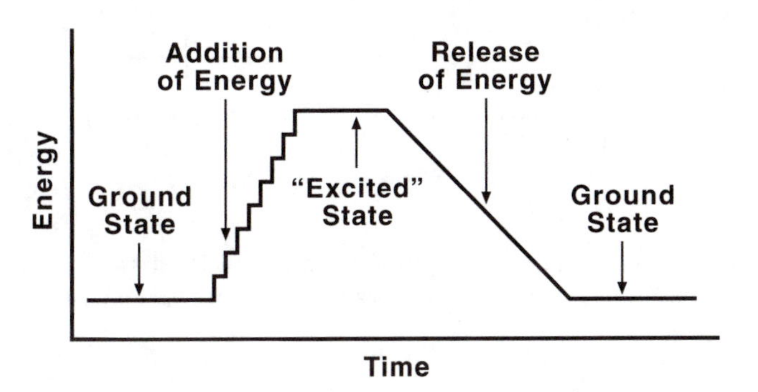

FIGURE 1.1 The change in energy of a hypothetical reaction with time, displaying the ascent to excited state and the return to ground state.

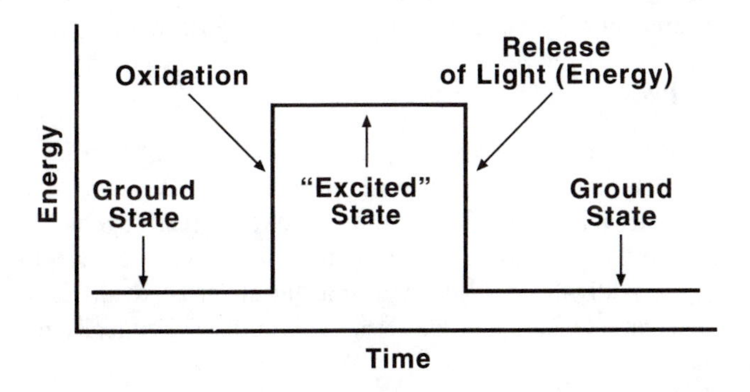

FIGURE 1.2 The change in energy with time of a luminescent reaction. Note that the ascent to the excited state is driven by oxidation. The return to ground state produces light or luminescence.

TABLE 1.3
Comparison of Flash and Glow Luminescence Systems

Flash	Glow
Rapid assay	Injection not necessary
Multiple samples	Mixing not difficult
High signal to noise (S/N) ratio	Samples can be preloaded, read sequentially
Less temperature dependence	Signal adjustable at different enzyme levels
Linear response	Read time maximized
Wide dynamic range	High signal levels; additives helpful
Signal depends on fast kinetics	Easy-to-read large numbers of samples
Injection independent of injectee	Temperature dependent

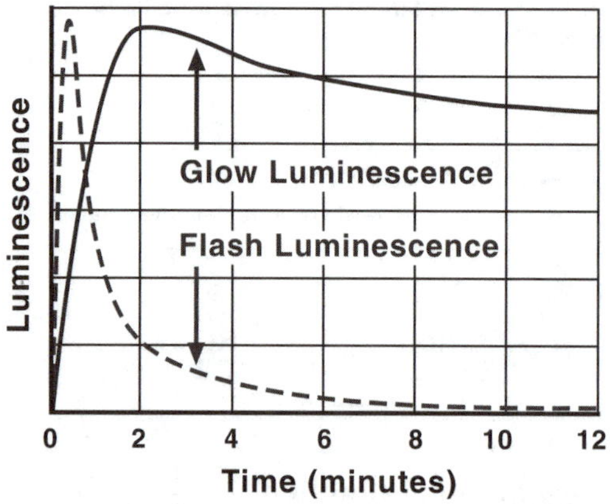

FIGURE 1.3 Kinetics of flash- and glow-type luminescence reactions. The flash reaction occurs quickly, peaks, and dissipates quickly. The glow reaction is slower to start but maintains light production for a longer time.

probe and produce an injection artifact. In addition, flash reactions require rapid and thorough mixing, which depends upon variables such as the force, angle, and volume of injection; the geometrical relationship of the injector to the container; and the speed and completeness of mixing relative to the kinetics of reaction. To minimize these variables it is necessary to use pressure-independent mechanical injection or continuous-flow systems with coiled-tubing cuvettes. From a practical standpoint, the slower glow-type reactions are much easier to perform and tend to be more reproducible. A comparison of flash and glow reactions is shown in Table 1.3.

1.2.2 Luminescent Probes

Luminescent probes are compounds that, when placed in solution with attacking chemicals, readily reach excited states, which then release light while decaying to the ground state. Light-producing reactions must release sufficient energy to cause the electronic rearrangement of the probe to the excited state. The attacking chemicals produced by these reactions are generally highly energetic free radicals or strongly oxidizing peroxides that transfer electrons to the luminescent probe and push it into the excited state. Once the probe releases light, it is chemically rearranged and is not able to emit light again.

The color of light emitted depends upon the identity of the luminescent probe and the amount of energy that is released by the reaction. In general, more energetic reactions produce light of shorter wavelengths (e.g., blue), whereas less energetic reactions produce light of longer wavelengths (green, red). Some luminescent probes emit light in the near infrared and therefore require

TABLE 1.4
Major Classes of Luminescent Probes

Probe	Wavelength(s) Emitted (nm)
1. Adamantane-dioxetanes	Enhancer dependent
2. Acridinium compounds	430–435
3. Luminol compounds	425
4. Lucigenin	500
5. Firefly luciferin	565
6. Photoproteins	470
7. Active oxalates	Fluorescent dependent
8. Hydrazides and Schiff bases	540
9. Electrochemiluminescent probes	620
10. Luminescent oxygen channeling probes	520–620

a less energetic reaction than probes that produce light in the visible range. Other probes produce ultraviolet light and require an even higher energy reaction to produce light.

There are ten major classes of luminescent probes. Table 1.4 summarizes the major features of these classes.

1.2.3 Enhancers of Luminescence

Enhancers are chemicals that, when added to a luminescent reaction, make the excited state easier to attain, prolong its duration, or increase its efficiency so that more light is produced. Various enhancers can amplify light production from a few fold to over 1000-fold. Enhancers may act by a process similar to that of liquid scintillation counting. In liquid scintillation counting, the energy of a radioactive particle is transferred, via the solvent, to a fluorescent compound whose spectrum overlaps that of a second fluorescent substance. The second fluorescent compound then becomes excited and emits light at a higher wavelength, which is detected by a photomultiplier tube. In the case of luminescence, a chemical reaction causes a luminescent probe to produce light. This light energy is then transferred to the fluorescent enhancer via the solvent. The enhancer is raised to an excited state and emits light at a wavelength that overlaps the emission spectrum of the luminescent probe.

1.3 MECHANISMS OF LIGHT PRODUCTION BY LUMINESCENT PROBES

Two major mechanisms are believed to be responsible for light emission by luminescent probes. The most common of these involves the production of intermediate dioxetanes and dioxetanones. AMPPD, AMPGD, acridinium esters, and firefly luciferin produce light in this manner.[2,3] A second mechanism involves a direct attack by peroxides on a luminescent chemical. Here, the luminescent chemical produces an intermediate in an excited state that emits light as the spent chemical returns to ground state. In some of these reactions, electron transfer or charge transfer may occur.[4] The specific mechanisms involved in the production of light by the major classes of luminescent probes are described in detail below.

1.3.1 Adamantane-Dioxetanes

The adamantane-dioxetanes, also known as the 1,2 enzymatically activated dioxetanes, were originally synthesized by Bronstein and are available from Applied Biosystems (Foster City, CA). Two of the most common adamantane-dioxetanes are AMPPD—3-(2′-spiroadamantane)-4-methoxy-4-(3″-phoshoryloxy)

TABLE 1.5
Enhancers of Adamantane-Dioxetane Luminescence

Additive	Emission Max (nm)	Relative Light Intensity
None	477	1
Sapphire I,II	461	20–100
Emerald I,II	542	100–1000
Ruby™	620	50–300

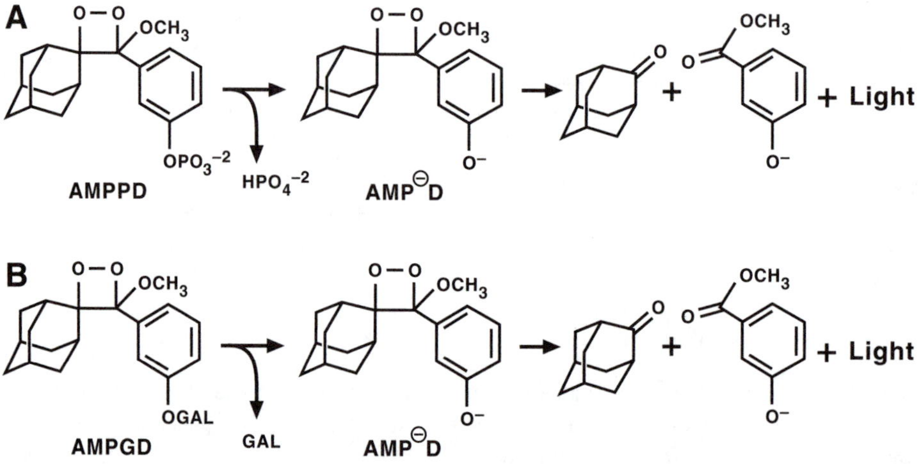

FIGURE 1.4 Chemical reactions of the adamantyl derivatives AMPPD and AMPGD. (A) AP dephosphorylates AMPPD. The AMPD anion then splits into the adamantone and a phenolate ester, which is in an excited state producing light. (B) AMPGD reacts with *β*-galactosidase, splitting off galactoside and leaving the AMPD anion, which rearranges to produce adamantone and a phenolate ester in an excited state producing light.

phenyl-1,2-dioxetane disodium salt—and AMPGD—3-(4-methoxyspiro[1,2 dioxetane-3,2′-tricyclo [3.3.1.1,3,7]decan]-4-yl)phenyl-*β*-*d*-galactopyranoside). AMPPD and AMPGD are enzymatic substrates that are cleaved by alkaline phosphatase (AP) or *β*-galactosidase, respectively, to produce a common adamantane anion phenolate intermediate. This phenolate then splits into the meta-oxy-benzoate anion, which is responsible for luminescence, and adamantone (Figure 1.4). The unenhanced light from this system is emitted at a wavelength of 477 nm (blue). Substrates of this type produce a very intensive glow-type luminescence, which can be easily detected, usually with a minimum background. In addition, fluorescent enhancers can be added that can boost the signal dramatically.

The mechanism by which AMPPD produces light involves dephosphorylation by AP to form a metastable chlorophenolate dioxetane anion intermediate that emits light at 466 nm. The intermediate does not decompose readily and emits a light almost immediately, although the maximal signal may occur within 1 to 9 min. AMPPD produces a steady-state emission of light in 20 min. Light is emitted for prolonged periods of time (hours to days). This allows for either immediate or delayed assay. Using X-ray film, a weak signal can be detected over long exposure times and retain quantitation. Signal-to-noise ratios can be optimized by varying exposure times.

The light generated by the adamantane-dioxetanes can be significantly amplified using enhancers;[5] see Table 1.5. Emission spectra can be preselected to ensure that the emitted light is well matched to the detection system. The Sapphire™ and Emerald™ II reagents produce high signal-to-noise ratio

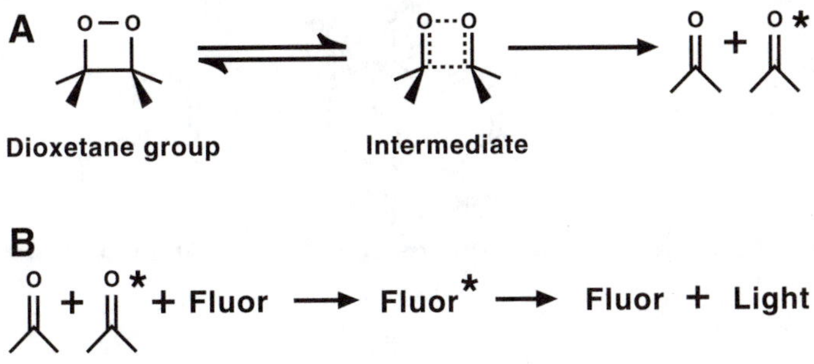

FIGURE 1.5 An intermediate dioxetane (cyclic peroxide) forms in many light-producing reactions. (A) The dioxetane rearranges to an intermediate that forms aldehydes, one of which is in an excited state. (B) When the excited state aldehyde reacts with a highly fluorescent compound, energy is transferred to the fluorescent compound, which produces luminescence.

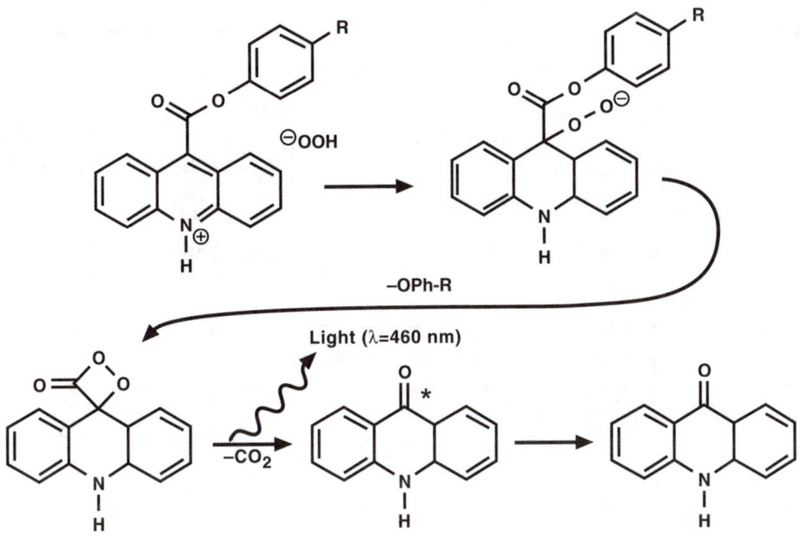

FIGURE 1.6 An acridinium ester reacts in the presence of a basic peroxide. This produces a new peroxide that forms a dioxetanone, which rearranges to produce an acridone in an excited state. Light is produced and molecules return to the ground state.

compared with their Sapphire and Emerald I counterparts. All these enhancers are available in an enhancer variety pack (Applied Biosystems) to facilitate the optimization of a light reaction.

1.3.2 Acridinium Compounds

The mechanism by which acridinium derivatives and esters produce luminescence was among the first to be understood. When oxidized by hydrogen peroxide under alkaline conditions, acridinium compounds form dioxetane, a four-member cyclic peroxide (Figure 1.5).[2,6] A brilliant blue (435-nm) flash-type luminescence is produced (Figure 1.6).

1.3.3 Luminol Compounds

Luminol (3-aminophthalhydrazide) and its derivatives produce blue light (425 nm) in the presence of horseradish peroxidase (HRP) and its substrate hydrogen peroxide. A product of this enzymatic reaction then oxidizes luminol to diazoquinone, which reacts with the hydrogen peroxide anion to

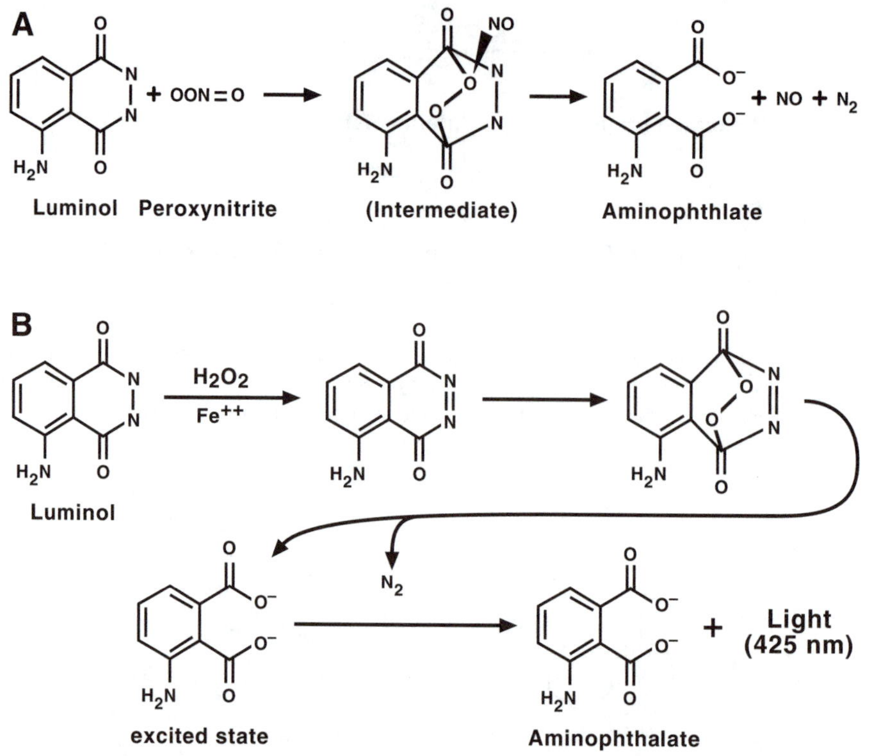

FIGURE 1.7 Two reactions with luminol to produce blue light at 425 nm. (A) Luminol reacts with peroxynitrite to produce a transient intermediate that rearranges to produce spent phthalate and light. (B) Luminol is oxidized using Fenton chemistry of hydrogen peroxide and ferrous iron to an intermediate that rearranges to give light and eventually spent product.

form an α-hydroxy-hydroperoxide intermediate that rapidly decomposes to the aminophthalate product with the emission of light (Figure 1.7).[7,8] It was initially believed that luminol produces light by reacting with the oxygen-based free radicals that are produced in this oxidative enzymatic reaction.[9,10] However, it appears that the process may be far more complex.

Although the actual compound that oxidizes luminol has never been isolated, there has been much research into its identity. Brestel[11] showed that hypochlorite (bleach) reacts with hydrogen peroxide at physiological pH to form a product that reacts with luminol to generate vigorous chemiluminescence. In these experiments, no peroxidase is present, suggesting that the compound that oxidizes luminol is a product of the reaction between hypochlorite and hydrogen peroxide. It is known that these two compounds react to produce singlet oxygen, which emits red light; however, this light is produced transiently, in the presence or absence of luminol, and has little to do with blue luminol luminescence. A follow-up of Brestel's original study showed that hydrogen peroxide is necessary for the chemiluminescent oxidation of luminol by hypochlorite.[12] Brestel suggested that hypochlorite and hydrogen peroxide might produce a new compound, possibly $OOCl^-$, which could react with luminol to produce blue light. Peroxy-hypochlorite ($OOClO^-$) might also serve this function. The reaction of luminol with hydrogen peroxide produces no light and the reaction of luminol hypochlorite produces minimal light; however, the combination of hydrogen peroxide and hypochlorite with luminol produces 100-fold more light than the individual compounds.[11] The small amount of light produced by hypochlorite and luminol alone may result from the presence of a minute amount of a long-lived peroxide generated by the industrial bleach-making process of bubbling chlorine gas through a solution of sodium hydroxide.

Peroxynitrite is also able to oxidize luminol under some circumstances. The reaction by which Kupffer cells (*in situ* liver macrophages) produce light from luminol can be inhibited by superoxide dismutase and an inhibitor of nitric oxide synthesis.[13] This suggests that an unusual peroxide consisting of nitric oxide and superoxide oxidizes luminol under these circumstances (Figure 1.7A). Peroxynitrite is a good candidate; it is 1000-fold more potent than hydrogen peroxide as an oxidant.[14] The effect of pH on the oxidation of luminol also suggests that peroxynitrite or some other very potent oxidant is the species that oxidizes luminol. At basic pH of 11, luminol is easily oxidized by weak oxidants such as hydrogen peroxide.[7] At pH values between 7.0 and 8.5, luminol is hardly oxidized by hydrogen peroxide. In cellular luminescence, holding pH around 7.4 is important to avoid metabolic alkalosis leading to cell death.

In reactions involving HRP, another oxidant species may react with luminol to produce light. As stated previously, hydrogen peroxide alone does not produce significant luminol luminescence. When HRP is added to hydrogen peroxide, light is emitted in measurable quantities. HRP contains heme iron, which is responsible for a portion of the catalysis. A free radical may be produced by a reaction between the heme iron and hydrogen peroxide (Figure 1.7B). This radical could then autoreact to form a more vigorous peroxide such as –OOOO– which might readily oxidize luminol to produce the characteristic blue light. A number of enhancers have been used to amplify the luminescence of luminol. Luciferin can stimulate the luminescence of luminol fivefold.[15] Para-iodophenol can enhance luminol luminescence 1000-fold.[16]

1.3.4 Lucigenin

Lucigenin (bis-*N*-methylacridinium nitrate) reacts with superoxide at or around neutral pH to produce yellow-green light (Figure 1.8). In this reaction lucigenin accepts an electron from superoxide. The lucigenin radical then reacts directly with superoxide to form an excited state *N*-methyl acridone in one half of the molecule, which then decays ground state and releases light. At alkaline

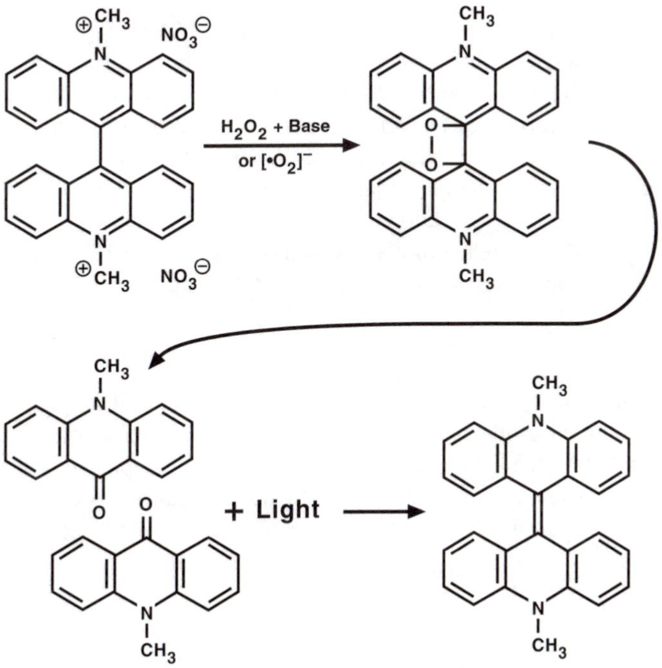

CH_3 $NO_3^{\ominus}$ N N CH_3 $NO_3^{\ominus}$

H_2O_2 + Base

or $[\bullet O_2]^-$

CH_3 N O O N CH_3

CH_3 N O O N CH_3 + Light

CH_3 N N CH_3

FIGURE 1.8 Luminescent reaction of lucigenin. Yellow-green light is produced in the presence of base and hydrogen peroxide or superoxide anion.

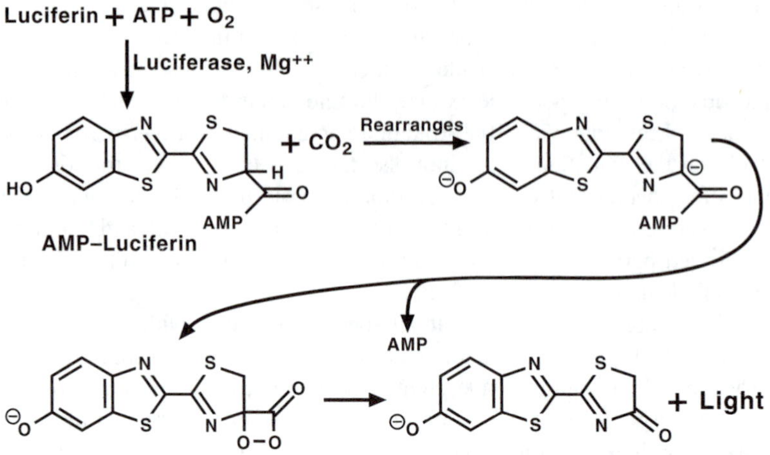

FIGURE 1.9 The reaction of firefly luciferase with ATP and oxygen in the presence of magnesium ion. An AMP-luciferin intermediate and carbon dioxide are produced. The intermediate rearranges to a dioxetane and AMP. The dioxetane produces an excited state with production of yellow-green light.

pH, lucigenin reacts directly with hydrogen peroxide and produces light for many minutes. At or around neutral pH, lucigenin does not react OCl^-, $OOClO^-$, singlet oxygen, or hydrogen peroxide. Therefore, it is not a substrate for myeloperoxidase or other haloperoxidases. If used properly, lucigenin can be a fairly selective indicator of superoxide generation. As such, it can be used as an intracellular indicator of superoxide generated by mitochondria (see Chapter 21).

1.3.5 Luciferase and Luciferin

Firefly luciferase reacts with luciferin, adenosine triphosphate (ATP), oxygen, and magnesium ions at pH 8 to produce yellow-green luminescence at 562 nm (Figure 1.9). In this reaction, luciferin is attacked by oxygen in the thiazoline ring carbon adjacent to the carboxyl to form a dianion. The dianion is oxidized, then rearranges to form a dioxetanone. The dioxetanone loses oxygen and forms an oxyluciferin excited state. The formation of oxyluciferin is thought to involve electron transfer from the dianion. The quantum efficiency of this reaction is about 33% and among the most efficient light-emitting reactions known.

Luciferin can be added to the reaction of hydrogen peroxide and HRP in the presence of luminol to produce an enhancing luminescent effect, which is synergistic.[15]

1.3.6 Photoproteins

The best studied of the photoproteins is aequorin, a protein found in jellyfish. The chromophore of aequorin is a substituted imidazolopyrazinone. A variety of photoproteins use the same or a similar chromophore.[17] The aequorin complex is composed of the apoaequorin protein, molecular oxygen, and the luminophore coelenterazine.[18,19] When the aequorin complex binds to three Ca^{2+} ions, coelenterazine is oxidized to coelenteramide, resulting in the release of carbon dioxide and blue light of ~466 nm.[20]

1.3.7 Active Oxalates

Light production by the active oxalates involves a reaction between an organically soluble peroxide, oxalate ester (TCPO), and flurophore (e.g., fluorescein) (Figure 1.10).[21] The mechanism involves an attack by the peroxide on the oxalate ester in the presence of the fluorophore in an organically soluble medium. Electron transfer takes place and structural rearrangement of the monoperoxyoxalic

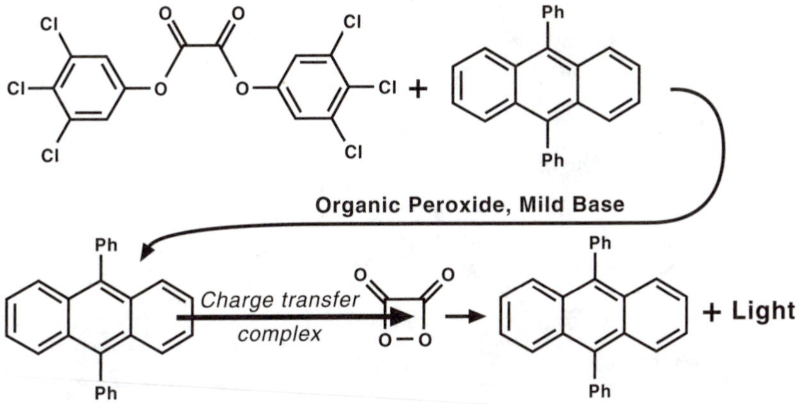

FIGURE 1.10 Oxalic acid ester (a TCPO-type compound) reacts with an organic peroxide or hydrogen peroxide in the presence of a base. A dioxetane intermediate then transfers charge to the fluorescent compound causing it to produce light at its fluorescence emission wavelength.

acid intermediate produces the excited state diphenylanthracene. Electron-withdrawing groups such as trichloro, dinitro carboxyester, or fluorine on the benzene rings attached to the oxalic portion of the structure produce extremely efficient luminescence. The color of light produced depends on the wavelength generated by the fluorescent compound; it is possible to produce blue, yellow, green, or red light with different compounds. Since the fluorophore is responsible for the final color of the light, it is plausible that the reaction between the oxalate ester and the peroxide produces an initial excited state and that the fluorophore becomes excited by the light released by the oxalate. The active oxalates produce cold light for many hours at room temperature and for months when the reaction is fast-frozen.

1.3.8 Hydrazides and Schiff Bases

Schiff bases are formed when aldehydes react with primary amines. Therefore, this reaction could be used to detect primary amines by luminescence. The hydrazides react with bases and could be used in their detection.

1.3.9 Electrochemiluminescent Probes

The ruthenium trisbipyridyl group, in the presence of tripropylamine, can be oxidized to produce light at an electrode.[22,23] The excited label then returns to ground state and can be used for another cycle (Figure 1.11).

1.3.10 Luminescent Oxygen Channeling Probes

Luminescent oxygen channeling uses a laser to trigger the production of singlet oxygen, which then initiates the luminescence of an olefin.[24]

1.4 SPECIFIC LUMINESCENT PROBES

1.4.1 Adamantane-Dioxetanes

The dioxetane probes are used in a wide variety of assays. In one of the most frequently used methods, an unlabeled primary antibody is used to detect a protein or polymer of interest. An enzyme-labeled secondary antibody then binds the primary antibody. A luminescent probe that is a substrate for the labeling enzyme is added, and the amount of light produced is measured. Another common

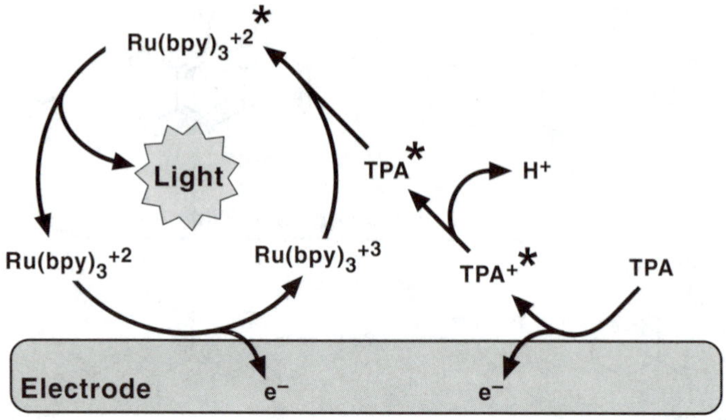

FIGURE 1.11 Electroluminescence used by Origen in its automatic analyzers. The system uses a ruthenium metal chelate in the presence of tripropylamine, which produces luminescence by a change in electrical potential.

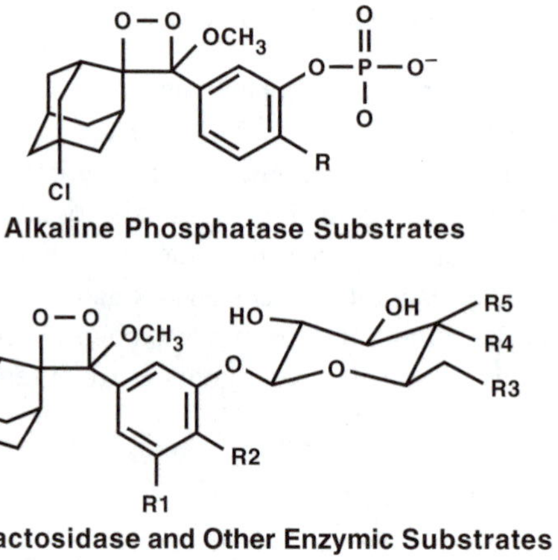

FIGURE 1.12 AMPPD and AMPGD and other structural derivatives for different enzymes.

AP Substrates

Substrate	R
CSPD	H
CDP-*Star*	Cl

***β*-Galactosidase and Other Enzymic Substrates**

Name	R_1	R_2	R_3	R_4	R_5	Substrate
Galacton®	H	H	OH	H	OH	*β*-Galactosidase
Galacton-Plus®	Cl	H	OH	H	OH	*β*-Galactosidase
Galacton-*Star*®	H	Cl	OH	H	OH	*β*-Galactosidase
Glucon®	H	H	OH	OH	H	*β*-Glucosidase
Glucuron®	H	H	OONa	OH	H	*β*-Glucuronidase

method uses a biotinylated antibody, which is reacted with streptavidin-labeled enzyme to initiate the luminescent reaction.

AMPPD is activated to produce light following dephosphorylation by AP. The two most widely used AMPPD substrates are CSPD®, the disodium salt of AMPPD, and CDP-*Star*® (Applied Biosystems) (Figure 1.12). CDP-*Star* produces a five- to tenfold higher signal than CSPD and produces light more rapidly. AMPPD has been used in conjunction with AP labeling and is applied in a variety of immunoassays and as a label of DNA and RNA probes. Because AMPPD can penetrate cells, it can be used to follow small numbers of cells or aggregates and is used in flow cytometry. The AP label is used with 1,2-dioxetane substrates in the following technologies: Access® (Beckman Coulter, Fullerton, CA), IMMULITE® and IMMULITE® 2000 (Diagnostics Products, Los Angeles, CA), and Lumipulse 1200 (Fujirebio, Fairfield, NJ).

Galacton®, Galacton-Plus®, and Galacton-*Star*® (Applied Biosystems) are AMPGD substrates for *β*-galactosidase, which are useful in the femtogram range (Figure 1.12). These substrates allow *β*-galactosidase to be used as a reporter gene or as a label for immunoassays; as few as 10^3 molecules of *β*-galactosidase can be measured under optimal conditions using enhancers and accelerators. Galacton-Plus offers prolonged light emission kinetics relative to Galacton. Galacton-*Star* has luminescent properties similar to Galacton but can be used in membrane applications. The Galacton-*Star* assay system is 1000-fold more sensitive than colorimetric assays and generates constant light emission for over 1 h, allowing for automated high-throughput screening of samples. All three systems can be used in single-tube format, but Galacton-Plus and Galacton-*Star* are preferred for microplate and scintillation counter applications.

Glucuron® (Applied Biosystems) is a dioxetane-producing chemiluminescent substrate for *Escherichia coli* *β*-glucuronidase and other enzymes that hydrolyze conjugated glucuronic acid substrates (see Figure 1.12). This system is used in reporter gene assays, environmental testing, biomedical research, clinical evaluation, toxicology, and pharmaceutical screening.

Glucon™ (Applied Biosystems) is a dioxetane-producing chemiluminescent substrate of *β*-glucosidase, which is used for the rapid identification of enterococci or streptococci. *β*-Glucosidase has been cloned and sequenced and can therefore be used as a reporter gene (see Figure 1.12).

1.4.2 Acridinium Esters

The molecular size of the acridinium esters is similar to that of an iodine atom. This makes them a good substitute for ^{125}I in both basic and clinical research (see Chapter 5). Because the original acridinium phenyl esters were unstable in storage, the molecules were modified to include a bulky sulfonyl-activated amide-leaving group instead of phenolate.

Abbott Laboratories (Abbott Park, IL) developed the *N*-sulfonylacridinium-9-carboxamides for immunoassays in its Prism and Architect instruments. Abbott has also developed a new thiol-specific acridinium probe that labels protein and nucleic acids.[25] These compounds react with *N*-acetyl cysteine and are specific for thiolated proteins or nucleic acids. The same group has developed acridinium hydroxylamines as carbonyl-reactive luminescent labels. Acridinium hydroxylamines are used to label various substrates containing aldehydes and ketones such as cortisol, estrone, norethinodrone, 6-oxoestradiol, progesterone, and digitoxin.

The Flashlight™ system (Assay Designs, Ann Arbor, MI), utilizes an *N*-hydroxysuccinimide-derivatized acridinium ester that is singly triggered and stable in aqueous solution. This compound produces blue light (465 nm) and can be used for the labeling of peptides, proteins, and nucleic acids (Figure 1.13).

Lumigen (Southfield, MI) has developed a family of acridinium-based substrates that efficiently produce blue light (430 nm). Among the more commonly used are PS-3 and APS-5. PS-3 is activated

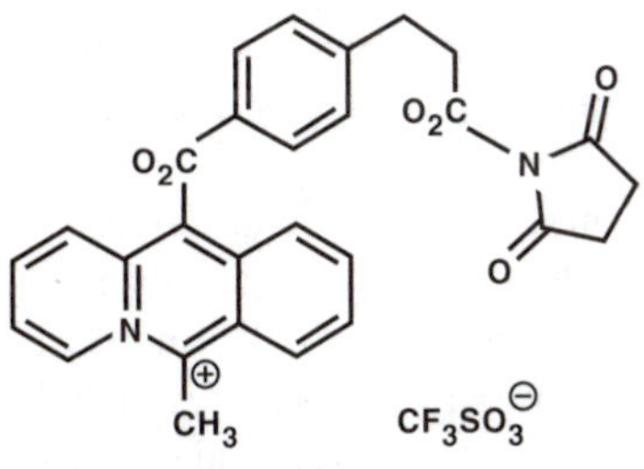

Acridinium C_2 NHS Ester

FIGURE 1.13 Structural formula for acridinium ester available from Lumigen and used in the Flashlight™ system.

by reaction with HRP and hydrogen peroxide; APS-5 is activated by AP. Both these compounds form dioxetanone intermediates that produce chemiluminescence. The decomposition of the intermediates produces *N*-methylacridone, which can be detected using fluorimetric techniques. PS-3-based enhanced chemiluminescence systems produce between 4- and 20-fold more light than the luminol-based systems. Lumigen PS-3 is used in the ECL-Plus™ system (Amersham-Pharmacia Biotechnology, Piscataway, NJ) for the detection of proteins. APS-5 formulations exhibit a linear response over five to six orders of magnitude with low backgrounds and high sensitivity (10^{-21} *M*). The Lumi-Phos™ WB system (Pierce, Rockford, IL) utilizes Lumigen PPD with AP. The Duo-LuX™ substrate (Vector Laboratories, Burlingame, CA) is a form of APS-5 that can be used with either AP or HRP and is detected by either luminescence or fluorescence.

A number of acridinium probes have been developed for the measurement of various biomolecules and enzyme activities. Steijger et al.[26] have developed an acridinium sulfonylamide for the determination of carboxylic acids by high-pressure liquid chromatography. Waldrop et al.[27] have synthesized a 9-acridinecarbonylimidazole for the measurement of hydrogen peroxide, glucose oxidase, AP, hydrolases, and dehydrogenases. Unlike most acridine luminescent compounds, this imidazole derivative is a ketone. The detection limit of these enzymes is in the 1 to 10 attomole range.

Acridinium or sulfonamide labels are used in the following systems: ACS-180, ACS-180 plus, and ADVIA® Centaur™ (Bayer Diagnostics, Tarrytown, NY); Advantage® (Nichols Institute, San Juan Capistrano, CA); Prism and Architect (Abbott Diagnostics), and PACE-2® (Gen-Probe, San Diego, CA).

1.4.3 Luminol Compounds

Luminol (Figure 1.14) is an inexpensive and versatile probe that is widely used in immunoassays, molecular biology, measurement of oxidative reactions, and forensic pathology. Systems using luminol with para-iodophenol as an enhancer have been patented and used in various assays such as the Amersham-Pharmacia ECL™ systems for Northern, Southern, and Western blots. The SuperSignal™ (Pierce) systems, also used for the detection of nucleic acid and antibody probes, use luminol with sodium phenothiazine 10-yl propane sulfonate as an enhancer (see Chapter 10). The SuperSignal systems are similar in sensitivity to radioactive labeling and have a higher signal-to-noise ratio. Enhanced luminol is used in the VITROS® ECi instrument sold by Ortho-Clinical Diagnostics (Rochester, NY).

Luminol

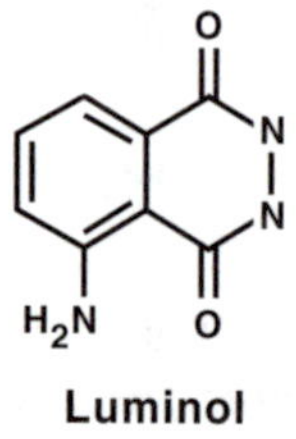

FIGURE 1.14 Structural formula for luminol.

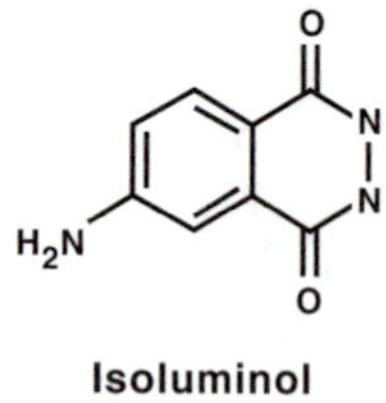

Isoluminol (Figure 1.15) does not produce luminescence as efficiently as luminol; however, it is sometimes used in place of luminol as a conjugate in various immunoassays.[28] Isoluminol may be preferable for use in phagocytic cells such as neutrophils because it does not inhibit the formation of superoxide.[29,30] At high concentrations luminol can inhibit endogenous cellular reactions. Isoluminol does not cross the plasma membrane and therefore cannot interfere with cellular reactions.

Isoluminol

FIGURE 1.15 Structural formula for isoluminol.

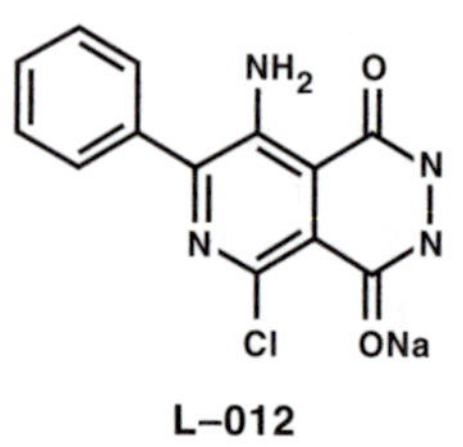

L-012 (Figure 1.16) is much more expensive than luminol but produces several-hundred-fold more luminescence in cellular assays with macrophages or neutrophils. In addition, it is much more water soluble than luminol. It reacts with a variety of peroxides including hydrogen peroxide, peroxynitrite, peroxyhypochlorite, and superoxide at or around neutral pH.[31] Although L-012 is less selective in its reactions with peroxides than is luminol, it is useful in assays where maximal light production is desired.[32]

L–012

FIGURE 1.16 Structural formula for L-012.

1.4.4 Lucigenin

Lucigenin or bis-*n*-methyl acridinium nitrate is a biacridine compound that has been used to measure superoxide. It can penetrate cells and has been used to measure superoxide in cellular mitochondria or in isolated mitochondria.[33] There has been criticism of the superoxide–lucigenin method; however, these experiments have used extremely high amounts of lucigenin that itself can generate superoxide and confound the results.[34]

1.4.5 Luciferase and Luciferin

Firefly luciferase is frequently used as a reporter gene to measure the activity of gene-regulatory sequences. In addition, it can be linked to other genes whose product may be difficult to measure (e.g., growth hormone) to produce a chimeric protein that produces light as an end point of gene activity. Luciferase is also used for pyrosequencing DNA, for eukaryotic cell viability assays, and to quantify viable bacteria in the food production industry.

Luciferin from *Cypridinia* is chemically different from firefly luciferin but forms a similar dioxetanone dianion intermediate after reaction with oxygen, which continues to the excited state with concomitant emission of light.[35]

De Silva et al.[36] have synthesized luciferin and dimethyl luciferin substrates that are exocyclic enol phosphate derivatives of aromatic esters and thioesters. These compounds are substrates of alkaline phosphatase and produce red light. A sensitivity of 0.01 attomoles has been achieved.

1.4.6 Photoproteins

Aequorin has been widely used for the measurement and visualization of intracellular calcium. Aequorin can be used for the measurement of Ca^{2+} concentrations from ~0.1 to >100 μM.[37] Recombinant aequorin can be loaded into individual cells using microinjection or cell permeabilization techniques. Alternatively, the aequorin gene can be used as a transgene in cell lines or whole organisms. A recombinant aequorin complex is available from Molecular Probes (Eugene, OR). Aequorin also forms the luminescent basis of the CHEMICON (Temecula, CA) kits for various cytokines.

1.4.7 Active Oxalates

Rauhut and associates[38] at American Cyanamid have used the reaction of certain oxalic esters (oxalyl chloride) in the presence of peroxides and one of many different colored, organically soluble fluorescers to produce a variety of luminescent products. They proposed a monoperoxyoxalic acid intermediate that reacts with a variety of fluorescent compounds to produce a rainbow of colored light from blue to red (see Figure 1.10). The oxalates are used in many commercial products, such as light sticks, toys, and jewelry.

1.4.8 Hydrazides and Schiff's Bases

Several hydrazides including 7-dimethylamino-naphthalene-1,2-dicarboxylic acid hydrazide (NH), 7-aminobutyl-ethylamino-naphthalene-1,2-dicarboxylic acid hydrazide (A), and 7-(isothiocyanato-butyl)-ethylamino-naphthalene-1,2-dicarboxylic acid hydrazide (I) are available from Assay Designs. Naphthalene hydrazide (NH) can be used to assay acetylcholine. Acetylcholine is reacted with acetylcholinesterase followed by choline oxidase to produce hydrogen peroxide. HRP is used to break down this hydrogen peroxide to produce an oxidant that reacts with NH and generates light.

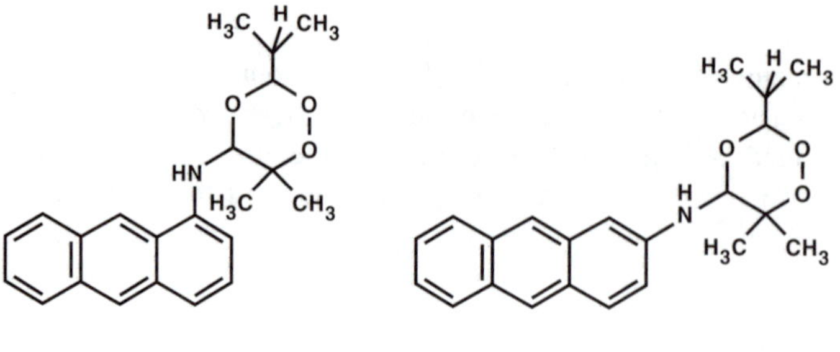

FIGURE 1.17 Structural formula of anthryl Schiff base 1 (a) and 2 (b) from Lumigen.

NH and I are very stable and can be used in the development of sensitive immunoassays where robustness is important.

Anthryl Schiff base 1 and 2 (Figure 1.17) can be used in luminescence assays and are available from Assay Designs. The 2-anthryl Schiff base has a quantum yield of 48%. The mechanism comes from an excited singlet state and produces yellow light at 540 nm. Because the reaction occurs when the Schiff base is attacked by a strong base, it can be used to detect strong bases.

1.5 LUMINESCENCE ASSAYS

Thousands of luminescent assays are currently in use; examples of different types are shown in Table 1.6. A number of the more commonly used luminescent assays are discussed in more detail below.

1.5.1 High-Throughput Screening

High-throughput screening (HTS) systems that utilize luminometers and robotics are important tools in the search for the next generation of receptor-selective drugs. With these systems it is possible to screen thousands of chemicals each day through chemical assays, immunoassays, and assays of receptor, enzyme, and reporter gene activity.[39] Many HTS instruments can be used for fluorescence as well as luminescence and fluorimetry.[40]

The NorthStar™ HTS system (Applied Biosystems) can screen 100,000 compounds per day (see Chapter 44). This system could be used for the polymorphic screening of drugs to best fit an individual's specific biochemical and genetic parameters. Applied Biosystems has combined its luminescence technology with the Digene Hybrid Capture® Technology (described in Section 1.7) to produce Xpress-Screen®, which can detect a two- to threefold change in rare messenger RNA transcripts present at 0.0017 attomoles (1000 copies/well).

BioSignal Packard (Montreal, Quebec, Canada) has developed the AlphaScreen™ system (see Chapter 42), which measures biological interaction as a function of proximity. In this system, two potentially interacting substances are conjugated to two different types of beads: a photosensitizer bead that generates singlet-state oxygen and an acceptor bead that contains chemiluminescent groups. A light signal is generated only if the two beads are in close proximity (250 nm) by virtue of a biological interaction between the conjugated substances. This technology can be used in assays of kinases, proteases, helicases, protein–protein interactions, immunoassays, and G protein–coupled receptor function.

Bioluminescence Resonance Energy Transfer ($BRET^2$, BioSignal Packard) is a technology designed to measure interactions between two proteins in a cell-based assay. Here, genetic fusion proteins are made by combining one protein of interest with *Renilla* luciferase and another protein

TABLE 1.6
Examples of Luminescent Assays

	Chemiluminescence	
8-Oxoguanine	High-throughput screening	Phagocytosis
Allergens	Immunoassays	Prostrate specific antigen
Antioxidant load	Luminescent O_2 channeling	Protein quantitation
Cytotoxicity	Membrane lipid oxidation	Quantitative PCR
Cigoxin	Nitric oxide	Redox assay
Dot blots	Nucleic acid probe	Thyrotropin
Enhancer development	Nucleic acid quantitation	TNF-α
Enzyme activity	Oxidative stress	TSA 100/membrane
Ferritin	p53	TSH
hCG	Peroxynitrite	Virus and virus antibodies
	Bioluminescence	
Antibiotic sensitivity	Enterovirus	Oxidative burst
ATP bacteria assay	Estradiol	Oxidative products
Atrial naturetic peptide	Genetic testing	Perfused organ assay
Bacteriophage sensitivity	GIMA antigen	Phagocytosis
BRMA antigen	Growth factors	PL4 antigen
CA 125	Growth hormone	Reperfusion inhibitor
CEA antigen	Luteinizing hormone	Reporter gene assays
Cyclosporin		T-cell activation
Cytotoxicity	Mutagenesis	Toxicity testing
Enhanced luminol/lucigenin	Nitric oxide	Tumor chemosensitivity
Enhancer development	OMMA antigen	
	Electroluminescence	
High-throughput screening immunoassays (40+ available)		

of interest with green fluorescent protein. When the two proteins interact, blue light energy from luciferase is transferred to the green fluorescent protein and green light is produced. The ratio of green to blue light increases as the magnitude of the interaction increases.

1.5.2 Immunoassays

Immunoassays are a routine use of luminescence, representing 10 to 30% of assays performed in some clinical laboratories. These techniques usually involve probing a protein of interest with a primary antibody that is then reacted with a secondary antibody. Light is produced when a luminescent probe is acted upon by the enzyme, which is bound, either directly or indirectly, to the secondary antibody. Although conventional procedures utilizing Schiff's base or aldehyde attack can be used to produce labeled antibodies and other proteins, genetically engineered fusion conjugates consisting of a luminescent protein linked to a protein of interest produce higher enzymatic activity.[41] Fusion proteins useful in immunoassays include HRP or AP linked to the biotin-binding protein streptavidin; aqueorin fused with IgG heavy chain or bacterial protein A; and luciferases linked to protein A, protein G, or streptavidin. There are over 1500 assays listed as luminescence immunoassays from the Web site of the National Library of Congress.

The recently developed ORIGEN® (IGEN International, Gaithersburg, MD) method of immunoassay utilizes electroluminescence in an assay that is selective, robust, and exhibits low background. In this system, a magnet is used to capture an analyte that is linked between antibodies bound to a magnetic bead and antibodies carrying the ruthenium trisbipyridyl label.[22,23] The amount of analyte is assayed by measuring the amount of light produced by the ruthenium trisbipyridyl

TABLE 1.7
Advantages of the ORI-TAG System

Attribute	ORI-TAG	Advantage
Size, mol. wt.	<700 a.m.u	High loading capacity, small effect on biological activity
Stability	months–years	Long shelf life
Solubilty	Aqueous, org.	Compatible with all assay media
Functional	Hydrophilic	Resists sticking, low nonspecific binding
Reactivity	Low	Compatible with a variety of chemistries
Specificity	High	Light emission specific to electrochemistry, high S/N

at an electrode (see Figure 1.11). The ORI-TAG system has the advantages shown in Table 1.7. More than 5000 ORIGEN-based assay systems are currently in use, including enzyme activity assays, binding assays, immunoassays, and nucleic acid probe assays. Electroluminescent immunoassays have also been used with the Elecsys® 1010 and 2010 instruments from Roche Diagnostics (Indianapolis, IN).

The luminescent oxygen channeling immunoassay (LOCI) is another novel immunoassay technique. Here, luminescence is produced in an interaction between olefin-coated latex beads and beads coated with a dye such as phthalocyanine or naphthalocyanine. Analyte-specific bead pairs form in a manner of a latex agglutination assay. When laser light (680 nm) reaches a dye-coated particle, singlet oxygen is produced and a short burst of light is released. This causes the olefin on the second bead to release a longer burst of light, which is measured. An advantage of LOCI is that a separation of bound and free analyte is not necessary. One can measure small molecules (e.g., homocysteine) and large polymers (e.g., HIV RNA) in serum or whole blood.[42]

1.5.3 DNA and RNA Blots

A number of kits are available for the detection of membrane-bound DNA and RNA. The mechanisms, detection limits, and emission of the several commonly used kits are shown in Table 1.8. The most frequently used probes for DNA and RNA blots are the adamantane-dioxetanes, luminol compounds, and acridinium esters.

1.5.3.1 Adamantane-Dioxetanes

CDP-*Star* and CSPD can be used in the detection of DNA or RNA bound to nylon membranes. Direct detection of nucleic acids involves the use of an AP-labeled DNA probe. The dioxetane substrate is dephosphorylated by AP, resulting in the production of luminescent bands at the sites of the bound probe. Because the dephosphorylated dioxetanes have a strong hydrophobic attraction for the nylon membrane, distinct bands are formed. Picogram or smaller quantities of DNA can be imaged on X-ray or instant film using exposure times ranging from minutes to several hours. This method offers high sensitivity and low background; a single copy of a gene can be measured in 0.25 μg of DNA. Indirect detection of nucleic acids can also be performed using CDP-*Star* or CSPD. Here, biotin- or fluorescein-labeled probes bind specifically to conjugates consisting of streptavidin–AP or antifluorescein antibody–AP, respectively. CDP-*Star* or CSPD is then dephosphorylated by the AP, resulting in the production of luminescent bands. Detection of other labels such as dioxigenin or DNP can also be accomplished with the appropriate antibody–AP conjugates. Indirect detection methods are often of lower sensitivity than direct methods. In addition to Southern and Northern blots, dioxetanes can be used in differential display, plaque hybridization, gel shift assays, RNAse protection assays, and the detection of RNA binding proteins.

TABLE 1.8
Examples of Detection Systems for DNA and RNA Blots

Kit	Manufacturer	Mechanism	Detection Limit	Emission Stability
AlkPhos Direct™	Amersham Pharmacia Biotech	Direct AP-labeled probe, CDP-*Star* substrate	60 fg	5 days
Gene Images™	Amersham Pharmacia Biotech	Fluorescein-11-dUTP-labeled probe, AP–antifluorescein conjugate, CDP-*Star* substrate	50 fg	5 days
North2South®	Pierce	Biotin-labeled probe, streptavidin–HRP conjugate, enhanced luminol substrate	fg	6 h
North2South® Direct	Pierce	Direct HRP-labeled probe, enhanced luminol substrate	fg	6 h
ECL Direct™	Amersham Pharmacia Biotech	Direct HRP-labeled probe, enhanced luminol substrate	500 fg	1–2 h
ECL 3′ Oligolabelling™	Amersham Pharmacia Biotech	Fluorescein-labeled probe, HRP–antifluorescein Ab conjugate, enhanced luminol substrate	500 fg	1–2 h
ECL Plus™	Amersham Pharmacia Biotech	Biotin-labeled probe, HRP–streptavidin conjugate, PS-3 substrate	50 fg	24 h
UltraSNAP™	Vector	Biotin-labeled probe, AP–streptavidin conjugate, Duo-LuX substrate	fg	8 h
Southern-*Star*™	Applied Biosystems	Biotin- or fluorescein-labeled probe, AP–streptavidin or antifluorescein Ab conjugate, CDP-*Star* substrate	<100 fg	Days
Southern-Light™	Applied Biosystems	Biotin- or fluorescein-labeled probe, AP–streptavidin or antifluorescein Ab conjugate, CSPD substrate	<1 pg	Days
ChemiGlow™	Alpha Innotech	Enhanced luminol substrate	500 fg	1–2 h

fg = femtogram; pg = picogram.

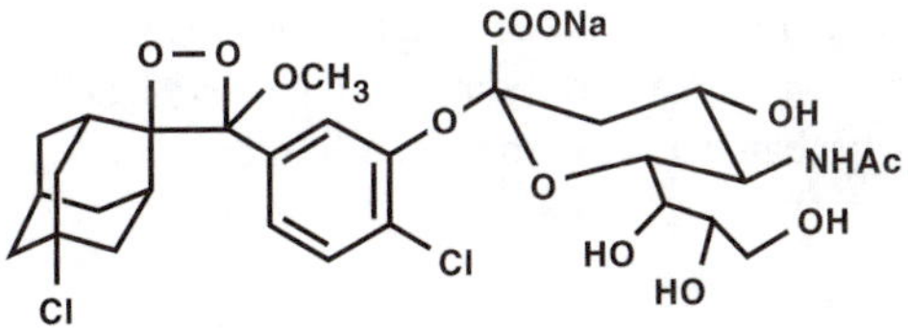

FIGURE 1.18 Structural formula for the NA-*Star* substrate for neuraminidase assay.

Tropix has developed a new adamantane derivative called NA-*Star*™, which is used to measure viral neuraminidase (Figure 1.18). This work has resulted in a new drug for treatment and prevention of type A and B influenza called oseltamavir (Tamiflu®). The drug binds sialic acid of the host cells and inhibits the viral neuraminidase from binding and reinfecting new host cells.

1.5.3.2 Luminol Compounds

Enhanced luminol is the most commonly used system for the detection of nucleic acids in Northern blots, Southern blots, plaque hybridizations, and gel shift assays. Direct-labeling kits allow one to produce probes that are covalently cross-linked to HRP, hybridized with nucleic acids, then detected

TABLE 1.9
Examples of Detection Systems for Western Blots

Kit	Manufacturer	Mechanism	Detection Limit	Emission Stability
Western-Light™	Applied Biosystems	AP-labeled 2° Ab, CPSD substrate	High fg	Hours–days
Western-*Star*™	Applied Biosystems	AP-labeled 2° Ab, CDP-*Star* substrate	Mid fg	Hours–days
SuperSignal® West Pico	Pierce	HRP-labeled 2° Ab, enhanced luminol substrate	Low pg	6–8 h
SuperSignal® West Femto	Pierce	Enhanced luminol	Low fg	8 h
SuperSignal® West Dura	Pierce	Enhanced luminol	Mid fg	24 h
ECL™	Amersham Pharmacia	HRP-labeled 2° Ab, enhanced luminol	<1 pg	1 h
ECL Plus™	Amersham Pharmacia	HRP-labeled 2° Ab, PS-3 substrate	<100 fg	1–24 h
LumiPhos™ WB	Pierce	PPD substrate	Low pg	Hours
VECTASTAIN® ABC-AmP	Vector	AP- or HRP-labeled 2° Ab, Duo-LuX substrate	1 pg	8 h
ChemiGlow™	Alpha Innotech	HRP-labeled 2° Ab, enhanced luminol substrate	fg	1–2 h

fg = femtogram; pg = picogram.

with enhanced luminol. Indirect detection systems for enhanced luminol utilize biotin-labeled probes and a streptavidin–HRP conjugate or fluorescein-11-dUTP-labeled probes and a HRP–antifluorescein antibody conjugate. An enhanced luminol system (ChemiGlow™, Alpha Innotech, San Leandro, CA) has been developed specifically for use with charge-coupled device (CCD) imaging devices. CCD digital cameras provide a wider dynamic range (0 to 4) than is possible with film (0 to 1.8).

1.5.3.3 Acridinium Esters

Acridinium compounds are also used to detect nucleic acids on membranes. Lumigen PS-3 and Duo-LuX™ substrates have been used in various direct and indirect labeling systems. These compounds often exhibit higher sensitivity than classical enhanced luminol.

1.5.4 Western Blots

Many systems are available for the detection of proteins bound to nitrocellulose or PVDF membranes (Table 1.9). Here, the protein of interest binds to a primary antibody that is in turn bound by an HRP- or AP-labeled secondary antibody. A chemiluminescent substrate is then activated by the enzyme and used to detect the bands to which the antibodies are bound.

1.5.5 Cell Proliferation, Cytotoxicity, Genotoxicity, and Activation Assays

1.5.5.1 Cell Proliferation and Cytotoxicity

Intracellular ATP is closely regulated in living cells and is rapidly degraded following cell death. Measurement of intracellular ATP is an accurate way to quantify the number of live cells in a culture. The ViaLight™ kit from LumiTech (Nottingham, U.K.) uses luciferase luminescence to measure ATP and can detect ten or fewer viable cells. The assay is rapid, easily automated, and works with adherent and nonadherent mammalian cells. A high-throughput screening system is also available.

1.5.5.2 Apoptosis and Necrosis

States of apoptosis and necrosis can be distinguished by changes in the ratio of cellular ADP: ATP. LumiTech has recently developed an apoptosis assay (ApoGlow™) that uses luciferase luminescence to measure changes in the ratio of cellular ADP:ATP. This technique can detect 10% cell death in 100 cells. It is performed in 96-well plates and is easily adapted to automation. The system is linear over five orders of magnitude and works well with many types of cultured mammalian cells and cell lines.[44]

1.5.5.3 Genotoxicity

The Vitotox™ genotoxicity assay (Thermo Labsystems, Franklin, MA) is a rapid genotoxicity screen similar in principle to the Ames assay. Unlike the Ames test, which detects mutations in a small number of genes involved in histidine synthesis, the Vitotox system measures damage to the entire bacterial genome. The Vitotox system also claims a higher sensitivity and a shorter assay time (4 h) than the Ames test. In the Vitotox assay, samples with or without S9 metabolic activator are mixed with two recombinant *Salmonella typhimurium* strains. These strains contain different reporter genes, which utilize the bacterial SOS response to measure either genotoxicity or cytotoxicity. The genotoxicity reporter plasmid contains the luciferase gene under control of the recN promoter. The recN promoter is repressed under normal conditions; however, DNA damage causes the RecA protein to initiate a cascade that turns on the recN promotor and, in turn, luciferase expression. Light emission due to luciferase activity is a function of the genotoxicity of the test compound. A parallel control for cytotoxicity is performed with the other test strain to ensure that the genotoxicity light signal is not the result of cell death or nonspecific enhancement of luminescence by the test compound. Thermo Labsystems Luminoscan reader automatically runs the experiment and interprets the data. This assay is available in 96- and 384-well plate formats.

1.5.5.4 Chemosensitivity

Cree et al.[45] have developed a luminescence-based assay for chemotherapeutic drug sensitivity. Tumor cells from individual patients are grown in the presence of different chemotherapeutic drugs; then the cell number is measured by a luminescent assay of ATP. The resulting chemosensitivity profile correlates with clinical outcome 70 to 80% of the time. A phase-III clinical trial comparing assay-directed drug selection with physician preference is currently in progress.

1.5.5.5 T-Cell Activation

Conventional methods for the measurement of T-cell activation typically require days of incubation and the use of radioisotopes. A novel assay system (Luminetics™, Cylex, Columbia, MD) uses luciferase-based measurement of intracellular ATP to determine T-cell activation within 4 to 24 h. In the Luminetics system, T cells are stimulated with an antigen or mitogen, then separated into T-cell subsets by monoclonal antibodies attached to magnetic particles. Intracellular ATP is measured using luciferase; increased levels of ATP indicate activated T-cell populations. This technology has clinical applications in the diagnosis and treatment of AIDS, autoimmune disorders, cancer, and organ transplantation. By using monoclonal antibodies with different lymphocyte surface marker specificities, the activation of different B-lymphocyte subsets can be measured. The Luminetics assays may be applied to mouse and possibly other animal species where antibodies exist for specific cell surface epitopes (see Chapter 24).

1.5.6 Assay of Gene Activity and Recombinant Proteins

1.5.6.1 Reporter Gene Assays

Luminescent reporter gene assays are used extensively in molecular biology. These constructs allow scientists to follow genes that do not produce an easily measured end product and to monitor the activity of gene regulatory sequences. Major reporter genes used in luminescent reporter gene assays include firefly luciferase, *Renilla* luciferase from marine bacteria, aqueorin, obelin, and alkaline phosphatase. Dual reporter genes produce light of different wavelengths and allow one to follow two reactions simultaneously. Such genes can include a combination of firefly luciferase, β-glucosidase, β-glucuronidase, or alkaline phosphatase. A combination of β-galactosidase and alkaline phosphatase is also used in dual detection.

1.5.6.2 Epitope Tagging

Epitope tagging is a powerful tool for the detection and purification of proteins expressed in *E. coli*, *Drosophila*, baculovirus, yeast, and mammalian systems. Luminescence has recently been combined with this method to facilitate the detection of tagged proteins. Epitope tags such as FLAG are sequences attached to the N or C terminus of the expressed protein. These tags are unlikely to interfere with structure or function of the protein and are detectable by monoclonal antibodies. Using mammalian expression systems, the yields are low and detection of proteins using established methods can prove difficult. Luminescent detection using AP-labeled antibodies against the anti-FLAG monoclonals and CPSD with Sapphire enhancer can greatly improve sensitivity. Tagging may also be done with HRP with detection via an enhanced luminol system.

1.5.7 DNA Pyrosequencing

Pyrosequencing is a DNA-sequencing technique that does not require primers, labeled nucleotides, or electrophoresis. In this method, one of the four deoxynucleotide triphosphates (dNTP) is added to a DNA template. If the base is complementary to the template strand, DNA polymerase catalyzes its addition to a growing complementary strand. Each incorporation event causes the release of pyrophosphate, which is then converted to ATP by ATP sulfurylase. This ATP is detected by a luciferase–luciferin system. The amount of pyprophosphate released and the magnitude of the light peak generated are directly proportional to the number of each dNTP incorporated into the complementary strand. Only one dNTP is added at a time; between additions, apyrase is used to degrade unincorporated dNTPs and excess ATP. A modified coupled enzyme pyrosequencing system has been used to sequence over 100 bases.[46] This pyrosequencing system requires no separation on gels and is being used to measure single polynucleotide polymorphisms at a rapid rate—50,000/day/instrument (see Appendix A, Pyrosequencing AB (Sweden); for an actual discussion of the method, see Chapter 17).

1.5.8 Quantitative Polymerase Chain Reaction

1.5.8.1 CSPD

Luminescence can be used in the measurement of small amounts of DNA or RNA by quantitative polymerase chain reaction (PCR).[47] In this process, TAQ polymerase along with specific primers is used to amplify a DNA sequence. RNA can be similarly amplified after cDNA is produced by reverse transcriptase. If the number of amplification cycles is selected so that substrate concentration is not limiting and product DNA accumulation is exponential, a proportional relationship exists between the amount of original DNA added and the quantity of product DNA formed. Luminescent PCR detection systems such as PCR-Light™ (Applied Biosystems) utilize one biotinylated primer and one unlabeled primer. The PCR product is captured on a streptavidin-coated microplate or bead; then a

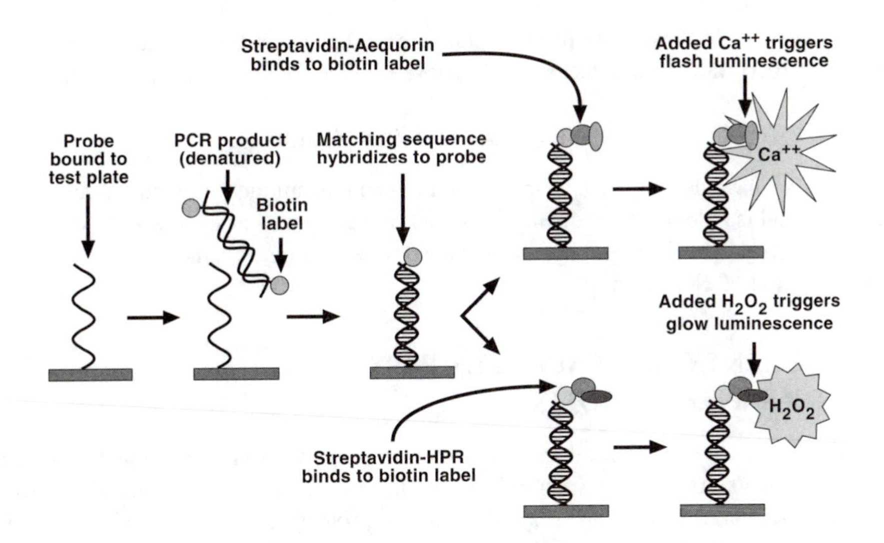

FIGURE 1.19 The aequorin-based OligoDetect assay. Hybridization with a biotin-labeled probe is used to measure a variety of genes.

fluorescein-labeled oligonucleotide probe is hybridized to the captured DNA. An antifluorescein AP conjugate, CSPD chemiluminescent substrate, and Sapphire-II enhancer are used to generate light. There is a linear relationship between light produced and femtomoles of PCR product formed over a range of 0.1 to 1000 fmol. A similar sandwich hybridization procedure may be performed by capturing an unlabeled PCR product with a bound probe then hybridizing it with an AP-labeled reporter probe.

1.5.8.2 Aequorin

Luminescent measurement of reverse transcription PCR (RT-PCR) products with aequorin is capable of detecting these amplicons with a 30- to 60-fold higher sensitivity than radioactive methods.[48] CHEMICON has recently developed a kit for cytokines, etc., which utilizes this technique. In the OligoDetect™ system, biotinylated primers are used to generate biotin-labeled PCR products from cDNA (Figure 1.19). These PCR products are then denatured and added to a probe-coated plate. Streptavidin-labeled aequorin (AquaLite™) is used to detect the bound, biotinylated PCR product. This system is capable of detecting 10^7 copies of DNA and has a dynamic range of 10^6-fold. The CHEMICON XpressPack mRNA expression analysis kits utilize the OligoDetect system along with specific primer sets for the detection of various cytokine, tumor marker, apoptosis marker, and housekeeping mRNAs.

1.5.9 Biosensors

Biomet sensors have been used to quantify copper, zinc, cadmium, chromium(VI), lead, nickel, and mercury. *Alcaligenes eutrophus* were genetically altered to produce luminescence in presence of heavy metals by fusing promoters involved in regulation of heavy metal resistance with the luxCDABE genes of *Vibrio fischeri* (see Chapter 40). Individual sensors were used to follow soil contamination and remediation. A luminescent bacterial sensor sensitive to 300 pg/ml of tetracycline has been produced.[49]

1.5.10 Luminescent Microarray Assay

A new concept in luminescent microarray assay is the use of microchip technology. This is covered in detail in Chapter 45 by Berthold et al., who are pioneering in this technology. Recently, OriGene developed the SmartArray™ DNA microarray, which utilizes luminescent techniques to find genes that interact at each step of a regulatory process. Chips are available for sets of nuclear hormone receptors, homeobox/b-zip/HLH transcription factors, and tissue-specific/inducible transcription factors.

1.5.11 Gas-Phase Luminescence to Measure Nitric Oxide

Maurer and Fung[50] have developed a gas-phase luminescent technique for the measurement of nitric oxide. This method is able to detect quantities of NO in headspace gas as low as 11 pmol. Because of its high sensitivity, speed, simplicity, and nondestructive nature, this method is ideal for kinetic studies of inhibition of NO synthase.

1.6 LUMINESCENT CHEMICALS AS THERAPY AND THERAPEUTIC PROBES

Because luminescent chemicals produce light when reacting with strong oxidants, they are in a real sense antioxidants. As such, they should be used in a concentration that does not interfere with a given assay. Since antioxidants are very important in protecting the body against damage done by highly toxic oxidants, it is not surprising that light probes can be used as antioxidant drugs and as probes of oxidative-based disease.

Ideally, a therapeutic luminescent chemical should react with the strong oxidant where it is appropriate and interact with vital body processes as little as possible. Unfortunately, most luminescent probes have never been tested against strong oxidants in a whole-animal model. Recently, Dubuisson[51] reported that coelenterazine and methyl coelenterazine are inhibitors of tert-butyl hydroperoxide-induced oxidative stress. Luminescence labeling has been used to follow K-ras mutations as a measure of pancreatic cancer growth using a DNA probe labeled with an acridinium ester. Tyrosinase mRNA in melanoma has been measured by RT-PCR using electroluminescence.[52]

1.7 CLINICALLY RELEVANT LUMINESCENT ASSAYS

Luminescence has become a mainstay of clinical and biomedical assays (see Table 1.7). This is the result of the development of new and brighter luminescence chemistry and the electroluminescent/magnetic immunoassay. If one searches Medline using the search term "luminescence immunoassay," over 1250 citations are noted with more than 400 dated between 1997 and early 2001. This illustrates the tremendous increase in the development of these assays. Advances in microplate luminometer technology using stacked microplates, multiple well formats, and multiple assay modes (luminescence, fluorescence, etc.) have made assays faster and more reliable.

The development of high-throughput screening systems like NorthStar HTS (Applied Biosystems) has made processing a large number of samples quick and easy. Development and use of combinatorial chemistry and luminous tag strategy has opened a new dimension in the integration of pharmacology and toxicology with combinatorial chemistry.[53] These types of studies should stimulate the development of high-throughput assays of membrane, receptor, and gene mechanisms necessary for the development of new highly selective drugs.

Recently, Digene Corporation (Gaithersburg, MD) has developed Hybrid Capture Technology for the luminescent detection of a number of bloodborne pathogens. Hybrid Capture greatly magnifies the product so that the luminescence assay is highly sensitive and specific for even a mutant strain of the disease. In this system, viral or bacterial DNA is hybridized with specific RNA probes. These DNA–RNA hybrids then bind to anti-DNA–RNA hybrid antibodies coated on a microwell plate (Figure 1.20). AP-labeled secondary antibodies then bind to the DNA–RNA hybrids in sandwich fashion. An amplified luminescence signal is produced with a dioxetane substrate. At present, the Hybrid Capture system is used to detect human papilloma virus, cytomegalovirus, hepatitis B, HIV, *Chlamydia trachomatis*, and *Neisseria gonorrhoeae*. A test for herpes simplex virus is now available.

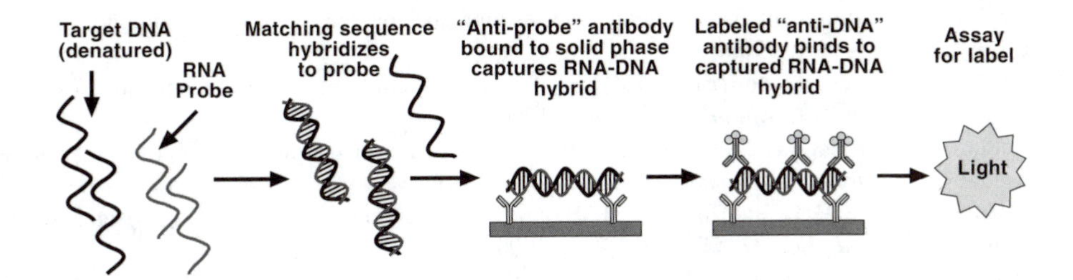

FIGURE 1.20 Hybrid Capture technology. The DNA of a pathogen is hybridized with specific RNA probes. These DNA–RNA hybrids then bind to antibodies coated on a microwell plate. AP-labeled secondary antibodies then bind to the DNA–RNA hybrids in sandwich fashion. A dioxetane substrate is used to produce luminescence.

A comprehensive review of the use of Southern blot hybridization has recently been published.[54] Another review discusses the advantages of luminescence detection and includes detailed information on the following probes: AP, HRP, digoxin-antidigoxin–AP, fluorescein-probe/antifluorescein–HRP, BrdU-mouse anti-BrdU/goat anti-mouse IgG-HRP, biotin/avidin–AP, and biotin/streptavidin–HRP.[55] These probes are used in assays for Lyme disease, HIV, human papilloma virus, human parvovirus, hepatitis C virus, respiratory synctial virus, Norwalk-like virus, Lassa virus, *Babesia, Mycobacterium* species, and *Clostridium difficile*. Genes that have been assayed using luminescent techniques include interleukin-2 receptor, Philadelphia chromosome for chronic myelogenous leukemia, T-cell rearrangement in T-cell lymphomas, Prader-Willi/Angelman syndromes, and trinucleotide repeats such as Fragile X syndrome, Huntington's disease, and spinocerebellar ataxia type I, proto-oncogenes, immunoglobulins, HLA markers, and peptide hormones.

1.8 CONCLUSIONS

Luminescence has clearly found a niche in many assays linked to molecular biology and medical science. Disposal of luminescent materials is without the financial, ecological, safety, and record-keeping problems of radioactivity. But the main advantage is that the light produced with luminescent chemistries can allow the researcher to produce better and clearer analyses of signals. In addition, the quantitation from luminescence is more rapid than using X-ray development for blots and competition assays. Luminescence in its sundry forms has developed rapidly because of its high sensitivity and broad range of detection, the quickness and ease of assay, the ability to reprobe experiments, and low cost.

REFERENCES

1. Advertising Flyer, Gen Probe Incorporated Chemiluminescence, The Ideal Detection System, Gen Probe Incorporated, San Diego, CA, 1999.
2. McCapra, F. and Beheshti, I., Selected chemical reactions that produce light, in *Bioluminescence and Chemiluminescence: Instruments and Applications,* Van Dyke, K., Ed., CRC Press, Boca Raton, FL, 1985, 9–42.
3. McCapra, F., Chemical generation of excited states: the basis of chemiluminescence and bioluminescence, *Methods Enzymol.,* 305C, 3–46, 2000.
4. Adam, W., Matsumoto, M., and Trofimov, A., Hydrogen bonding effects on the fluorescence versus electron-transfer-initiated chemiluminescence spectra of the m-oxybenzoate ion derived from a bicyclic diroxetane, *J. Org. Chem.,* 65, 2078–2082, 2000.
5. Bronstein, I., Edwards, B., and Voyta, J. C., 1,2-Dioxetanes: novel chemiluminescent enzyme substrates—application to immunoassays, *J. Biolumin. Chemilumin.,* 4, 99–111, 1989.

6. Weeks, I., McCapra, F. et al., Acidinium ester as high specific activity labels in immunoassay, *Clin. Chem.,* 29, 1474–1479, 1983.
7. Merenyi, G., Lind, J., and Eriksen, T. E., Luminol chemiluminescence: chemistry, excitation emitter, J. *Biolumin. Chemilumin.,* 5, 53–65, 1990.
8. Bottu, G., The effect of buffers and chelators on the reaction of luminol with Fenton's reagent near neutral pH, *J. Biolumin. Chemilumin.,* 6, 147–151, 1991.
9. Allen, R. C., Dale, D. C., and Taylor, F. B., Blood phagocyte luminescence: gauging systemic immune activation, *Methods Enzymol.,* 305, 591–632, 2000.
10. Van Dyke, K., Introduction to bioluminescence and chemiluminescence, in *Bioluminescence and Chemiluminescence: Instruments and Application*, Van Dyke, K., Ed., CRC Press, Boca Raton, FL, 1985, 1–8.
11. Brestel, E., Mechanisms of cellular chemiluminescence, in *Cellular Chemiluminescence,* Vol. I, Van Dyke, K. and Castranova, V., Eds., CRC Press, Boca Raton, FL, 1987, 93–104.
12. Arnhold, J., Mueller, S., Arnold, K., and Grisson, E., Chemiluminescence intensities and operation of luminol oxidating by sodium hydrochlorite in the presence of hydrogen peroxide, *J. Biolumin. Chemilumin.,* 6, 189–192, 1991.
13. Wang, J. F., Komarov, P., Sies, H., and de Groot, H., Contribution of nitric oxide synthase to luminol-dependent chemilumilnescence generated by phorbol-ester-activated Kupffer cells, *Biochem. J.,* 279, 311–314, 1991.
14. Radi, R., Cosgrove, P. T., Beckman, J. S., and Freeman, B. A., Peroxynitrite-induced luminol chemiluminescence, *Biochem. J.,* 290, 51–57, 1993.
15. Thorpe, G. H., Whitehead, T. P., et al., Enhancement of the horseradish peroxidase-catalyzed chemiluminescence oxidation of cyclic diacyl hydrazide by 6-hydroxybenzothiazides, *Anal. Biochem.,* 145, 96–100, 1985.
16. Thorpe, G. H., Kricka, L. J., Moseley, S. B., and Whitehead, T. P., Phenols are enhancers of the chemiluminescent horseradish peroxidase-luminol-hydrogen-peroxide reaction: application in luminescence-monitored enzyme immunoassays, *Clin. Chem.,* 31, 1335–1341, 1985.
17. Brini, M. et al., Transfected aequorin in the measurement of cytosolic Ca^{2+} concentration ($[Ca^{2+}]C$). A critical evaluation, *J. Biol. Chem.,* 270, 9896–9903, 1995.
18. Head, J. F., Inouye, S., Teranishi, K., and Shimomura, O., The crystal structure of the photoprotein aequorin at 2.3 A resolution, *Nature,* 405(6784), 372–376, 2000.
19. Jones, K., Hibbert, F., and Keenan, M., Glowing jellyfish, luminescence and a molecule called coelenterazine, *Trends Biotechnol.,* 17(12), 477–481, 1999.
20. Ohmiya, Y. and Hirano, T., Shining the light: the mechanism of the bioluminescence reaction of calcium-binding photoproteins, *Chem. Biol.,* 3(5), 337–347, 1996.
21. Rauhut, M. M., *Acc. Chem.,* 2, 80, 1969.
22. Hoyle, N. R., Eckert, B., and Kraiss, S., Electrochemiluminescence: leading edge technology for automated immunoassay, *Analyte Detection Clin. Chem.,* 42, 1576–1578, 1996.
23. Blackburn, G. F. et al., Electrochemiluminescence detection for development of immunoassays and DNA probe assays for clinical diagnostics, *Clin. Chem.,* 37, 1534–1539, 1991.
24. Ullman, E. F. et al., Luminescent oxygen channeling assay (LOCI™) sensitive, broadly applicable homogenous immunoassay method, *Clin. Chem.,* 42, 1518–1526, 1996.
25. Adamcyzk, M. et al., Linker modulation of the chemiluminescent signal from *N*-(10)-(3-sulfopropyl)-*N*-sulfonylacridinium-9 carboxamido tracers, *Bioconjug. Chem.,* 11, 714–724, 2000.
26. Steijgen, O. M. et al., An acridinium sulphonylamide as a new chemiluminescent label for the determination of carboxylic acids in liquid chromatography, *J. Biolumin. Chemilumin.,* 13, 31–40, 1998.
27. Waldrop, A. A., Fellers, J., and Vary, C. P., Chemiluminescent determination of hydrogen peroxide with 9-acridine carbonylimidazol and use in measurement of glucose oxidase and alkaline phosphatase, *Luminescence,* 15, 169–182, 2000.
28. Kohen, F., DeBoever, J., and Kim, J. B., Recent advances in chemiluminescence-based immunoassays for steroid hormones, *J. Steroid Biochem.,* 27, 71–79, 1987.
29. Lundqvist, H. and Dahlgren, C., Isoluminol-enhanced chemiluminescence: a sensitive method to study the release of superoxide anion from human neutrophils, *Free Radical Biol. Med.,* 20, 785–792, 1990.
30. Dahlgren, C. and Karlsson, A., Respiratory burst in human neutrophils, *J. Immunol. Methods,* 17, 3–14, 1999.

31. Imada, I. et al., Analysis of reactive oxygen species generated from neutrophils using a chemiluminescence probe L-012, *Anal. Biochem.,* 271, 53–58, 1999.
32. Sohn, H. Y. et al., Sensitive superoxide detection in vascular cells by new chemiluminescence dye L-012, *J. Vasc. Res.,* 36, 456–464, 1999.
33. Li, Y., Stansbury, K. H., Zhu, H., and Trush, M. A., Biochemical characterization of lucigenin (bis-*n*-methyl acridinium) as a chemiluminescent probe for detection of intramitochondrial superoxide anion radical production, *Biochem. Biophys. Res. Commun.,* 262, 80–87, 1999.
34. Liochev, S. I. and Fridovich, I., Lucigenin (bis-*n*-methyl acridinium) as a mediator of superoxide anion production, *Arch. Biochem. Biophys.,* 337, 115, 1997.
35. Swada, H., Japanese patent 5,286,976, 1993.
36. De Silva, R. et al., Novel red chemiluminescent substrates for alkaline phosphatase, *Luminescence,* 15, 205, 2000.
37. Blinks, J. R., Detection of Ca^{2++} with photoproteins, in *Bioluminescence and Chemiluminescence: Instruments and Applications*, Vol. II, Van Dyke, K., Ed., CRC Press, Boca Raton, FL, 1985, 185–226.
38. Rauhut, M. M. et al., *Photochem. Photobiol.,* 4, 1097, 1965.
39. Lazo, J. S. and Wipf, P., Combinatorial chemistry and contemporary pharmacology, *J. Pharmacol. Exp. Ther.,* 293, 705–709, 2000.
40. Cortese, J., High throughput screening, *Scientist,* 14, 18–21, 2000.
41. Kricka, L. J., Voyta, J. C., and Bronstein, I., Chemiluminescent methods for detecting and quantitating enzyme activity, *Methods Enzymol.,* 305; *Bioluminescence and Chemiluminescence,* 26, 370–389, 2000.
42. Patel, R. et al., Quantification of DNA using luminescent oxygen channeling assay, *Clin. Chem.,* 46, 1471–1477, 2000.
43. Tropix, Luminescence Catalog, 2001–2002.
44. Bradbury, D. A., Measurement of the ADP:ATP ratio in human leukaemic cell lines can be used as an indicator of cell viability, necrosis and apoptosis, *J. Immunol. Methods,* 20, 79–92, 2000.
45. Cree, I. A. et al., Development of an ATP-based chemosensitivity assay, *Luminescence,* 15(4), 204, 2000.
46. Nyren, P., Pyrosequencing—a new method for fast DNA sequencing, *Luminescence*, 15(4), 219, 2000.
47. Martin, C. S., Voyta, J. C., and Bronstein, I., Quantitative polymerase chain reaction and solid phase capture nucleic acid detection, *Methods Enzymol.,* 305, 466–476, 2000.
48. Actor, K. et al., A FLASH-type bioluminescent immunoassay that is more sensitive than radioimaging: quantitative detection of cytokine cDNA in activated and resting human cells, *J. Immunol. Methods,* 211, 65–77, 1998.
49. Borresmans, B. et al., Biosensors for the detection of heavy metals, genotoxic compounds and antiobiotics, *Luminescence,* 15, 203, 2000.
50. Maurer, T. S. and Fung, H.-L., Evaluation of nitric oxide synthase activity and inhibition kinetics by chemiluminescence, *Nitric Oxide Biol. Chem.,* 4, 372–378, 2000.
51. Dubuisson, M. L. N., Antioxidant properties of natural coelenterazine and synthetic methyl coelenterazine in rat hepatocytes subjected to tert-butyl hydroperoxide induced oxidative stress, *Biochem. Pharmacol.,* 60, 471–478, 2000.
52. O'Connell, C. D. et al., Detection of tyrosinase mRNA in melanoma by reverse transcription-PCR and electrochemiluminescence, *Clin. Chem.,* 44, 1161–1169, 1998.
53. Tenner, K. S. and O'Kane, D. J., Clinical application of Southern blot hybridization with chemiluminescence detection, *Methods Enzymol.,* 305, 450–466, 2000.
54. Bronstein, I. et al., Detection of DNA in Southern blot with chemiluminescence, *Methods Enzymol.,* 217, 398–413, 1993.
55. Olesen, C. E. et al., Novel methods for chemiluminescence detection of reporter genes, *Methods Enzymol.,* 326(2), 175–202, 2000.
56. Turner, G. K., Measurement of light from chemical or biochemical reactions, in *Bioluminescence and Chemiluminescence, Instruments and Applications,* Vol. I, Van Dyke, K., Ed., CRC Press, Boca Raton, FL, 1985, 43–78.
57. Clontech Laboratories, Inc., 1020 East Meadow Circle, Palo Alto, CA, 94303, USA; 1-800-662-2566.
58. Spring, K., Use of CCD cameras and imaging devices, *BioTechniques,* 29, 70–76, 2000.
59. Ellis, R. J. and Wright, A. G., Optimal use of photomultipliers for chemiluminescence applications, *Luminescence,* 14, 11–18, 1999.

2 Instrumentation for the Measurement of Luminescence

Knox Van Dyke and Karen Woodfork

CONTENTS

2.1 INTRODUCTION

Two types of instruments are used for the measurement of luminescence reactions: luminometers containing photomultiplier tubes and charge-coupled device (CCD) cameras. Luminometers are generally used for reactions that are performed in test tubes or multiwell plates; CCD devices are most frequently used for imaging large areas such as blots, gels, or plates. This chapter discusses the operational principles of each type of instrument, the types of measurements most appropriate for each, and some general principles of luminescence measurement regardless of the instrumentation.

2.2 PHOTOMULTIPLIER TUBES

2.2.1 Theory of Operation

The photomultiplier is a vacuum tube that converts light energy into an electrical signal that can be manipulated. A photomultiplier can measure between 10^3 and 10^{10} photons over an entire range of 10^7. A photomultiplier tube consists of a photocathode, an anode, and a series of dynodes that accelerate electrons using high voltage (Figure 2.1). The photocathode and dynodes are coated with

0-8493-0719-8/02/$0.00+$1.50

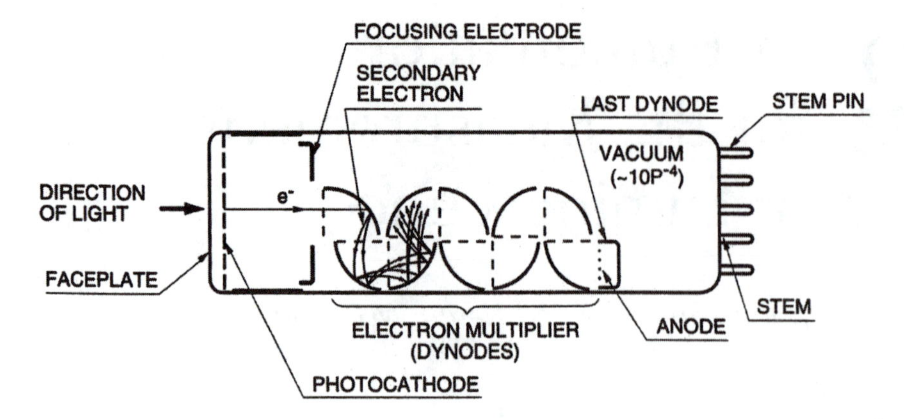

FIGURE 2.1 Construction of a photomultiplier tube. Light that enters a photomultiplier tube is detected and produces an electrical current, which can be amplified. This occurs by the following processes: (1) Light passes through the input window (face plate). (2) Light strikes the photocathode and photoelectrons are emitted in a vacuum (external photoelectric effect). (3) Photoelectrons are accelerated and focused by the focusing electrode onto the first dynode where they are multiplied by means of secondary electron emission. This secondary emission is repeated at each of the successive dynodes. A single photon can be amplified a million times! (4) The multiplied secondary electrons emitted from the last dynode are collected by the anode. (5) The electrons produce a current that is proportional to the original amount of light entering the photomultiplier tube. The current is amplified and analyzed as counts per minute. (Courtesy of Hamamatsu Corporation.)

bi- or trialkaline earth substances (e.g., potassium/cesium/antimony, silver/magnesium/cesium), which emit electrons when attacked by photons. The electrons produced when photons strike the photocathode are then accelerated by a positive potential difference into the first dynode. The impact of these electrons results in the production of multiple secondary electrons that are then focused onto the second dynode. Additional secondary electrons are produced at the second dynode. This process is repeated through a series of dynodes, each of which produces additional electrons and amplifies the original signal. A final collecting anode accumulates the electrons. In a photomultiplier tube, a single photoelectron can produce a million secondary-type electrons. The magnitude of this multiplication is controlled by adjusting the positive voltage applied to the tube.

All photomultiplier tubes produce an electrical signal unrelated to that from incoming light. This is called dark current, dark counts, or background. It differs from noise in that noise affects both the light signal and the background. Background can be reduced by cooling the tube. Most modern bialkali or trialkali tubes have low background at room temperature. However, if the ambient temperature rises above 22°C the background can become very significant.

A major advantage of photomultiplier tubes is that low-energy noise can be mostly eliminated using pulse height discrimination (Figure 2.2). The main drawback of the photomultiplier tube, other than cost of producing uniform tubes, is that it is difficult to miniaturize. Therefore, photomultiplier tubes cannot be used for counting a two-dimensional array of light-producing points simultaneously.

2.2.2 Types of Photomultiplier Tubes

There are two major types of photomultiplier tubes: the venetian-blind-type EMI (Figure 2.3) and the box and grid Hamamatsu (Figure 2.4).

The photocathode quantum efficiency is defined as the ratio of released electrons to incident photons. New photomultiplier tubes can have an efficiency of over 90%, whereas older tubes were less than 40% efficient. A 13-dynode photomultiplier tube can produce electron multiplication of 50 million-fold.

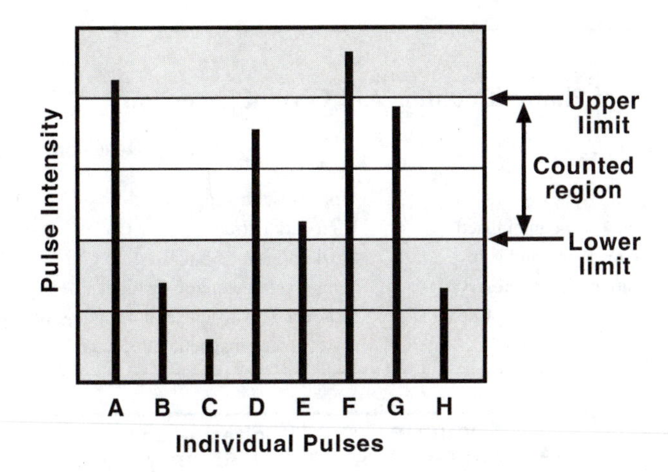

FIGURE 2.2 Pulse height discrimination utilizes electronic gating of pulses based on height analysis.

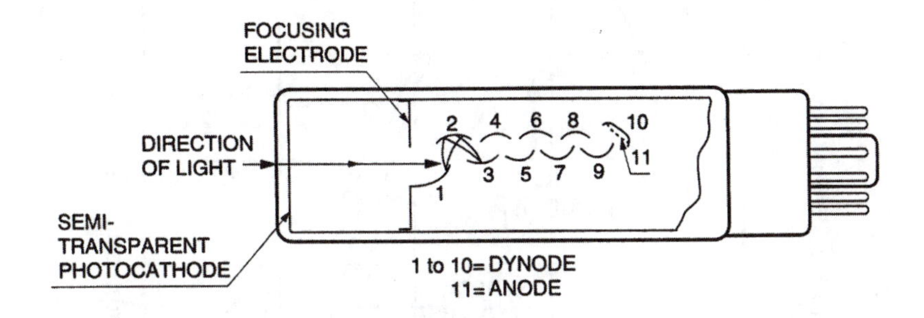

FIGURE 2.3 Venetian-blind-type photomultiplier. (Courtesy of Hamamatsu Corporation.)

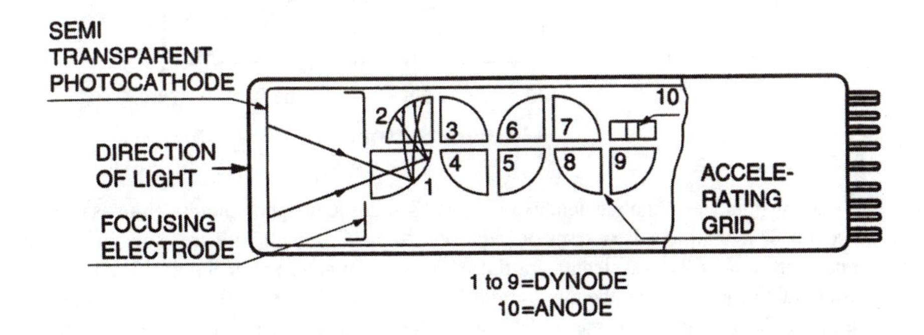

FIGURE 2.4 Box and grid-type photomultiplier. (Courtesy of Hamamatsu Corporation.)

Luminometers that use photomultiplier tubes can be classified as either current detecting or photon counting. Current-detecting luminometers integrate the charge signals from the photomultiplier tube with a direct current (DC) amplifier. Photon-counting luminometers use a fast pulse amplifier and count only those pulses with an amplitude above a certain threshold value. Current-detecting instruments are rugged but less sensitive to small amounts of light. Photon-counting instruments are more efficient and exhibit superior signal recovery over the entire range of the photomultiplier tube. A comparison of photon counting and current detection is shown in Table 2.1. For a detailed discussion of the photomultiplier tubes in chemiluminescence and bioluminescence, see Ellis and Wright.[1]

TABLE 2.1
Comparison of Photon Counting and Current Detection

Photon Counting	Current Detection
Less noise in gain	More noise in gain
Low dark current, can be minimized	High dark current
Larger dynamic range at fixed gain	Smaller dynamic range
Temporal structure of signal preserved	Temporal structure of signal can be lost
Less sensitive to various types of noise	More sensitive to temperature, aging, voltage stability, rate effects, magnetic effects, and microphonics

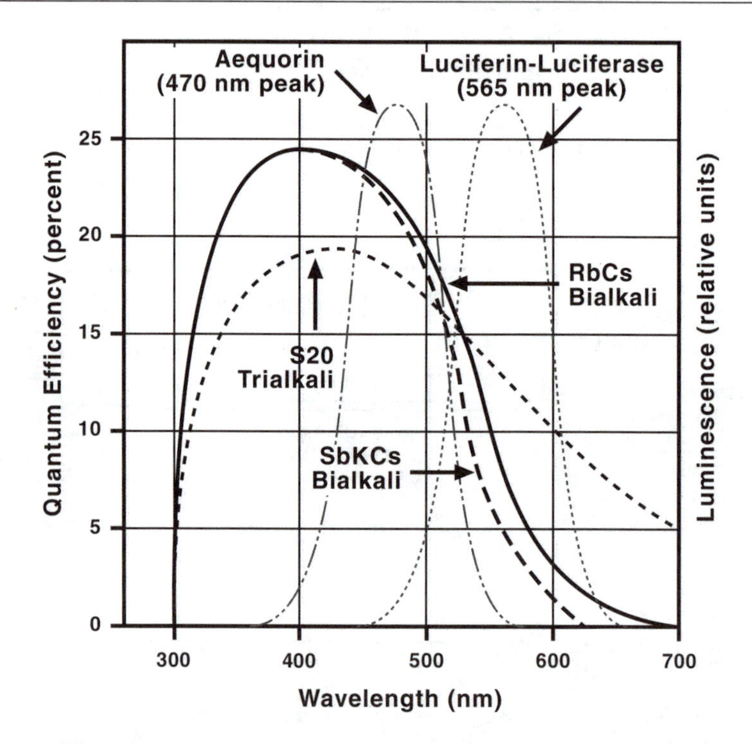

FIGURE 2.5 A graph of the quantum efficiency of various bi- and trialkali photomultiplier tubes at wavelengths from 300 to 700 nm. The luminescence curves of aequorin and luciferin/luciferase are shown to describe tubes that might be most efficient at the wavelengths of those emitters. Luminol peaks about 425 nm; it is clear that a rubidium/cesium tube would provide the highest efficiency. The S value relates to sensitivity of detection. An S20 value is a high efficiency tube, whereas an S13 tube is less sensitive because it has a lower S value.

2.2.3 Spectrum Matching of Photomultiplier Tubes

For measurements of highest sensitivity, the spectrum of the photomultiplier tube should be matched to the emission spectrum of the luminescent reaction. The relationship between quantum efficiency and wavelength for the different types of photomultiplier tubes is shown in Figure 2.5. Light produced by luminol (425 nm) should be detected with a trialkali/sensitivity 20 photomultiplier tube. Aequorin luminescence at 470 nm is best measured with an antimony/potassium/cesium bialkali tube. Firefly luciferin/luciferase light, maximal at 565, is most closely matched with a rubidium/cesium bialkali tube.

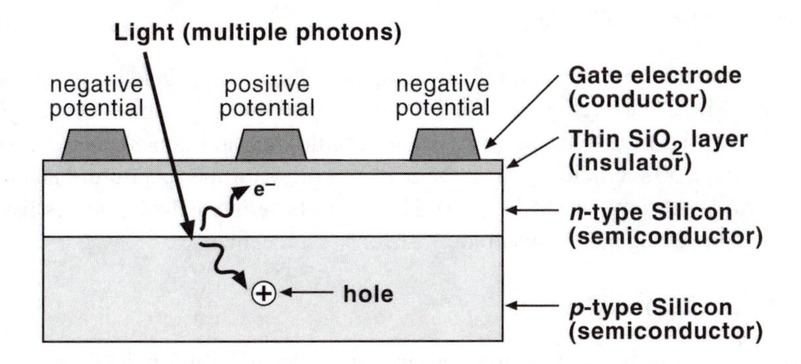

FIGURE 2.6 A cross section of a typical pixel of a CCD is shown. Light interacts with a semiconductor, which produces an electron–hole pair via the photoelectric effect. This charge is collected and eventually sensed as a voltage and digitized to produce an image that can be viewed and analyzed.

2.3 CHARGE-COUPLED DEVICES

2.3.1 Theory of Operation

A CCD camera contains a two-dimensional array of photodiodes (pixels), each of which is a single detector. Charge is generated and accumulated when photons strike the doped silicon layer of a detector. The accumulated charge is periodically read as a voltage value for each pixel (Figure 2.6). Longer exposure times can be used to detect faint signals. Excellent resolution is possible with CCD detectors because they are composed of millions of pixels measuring only micrometers. However, CCD cameras are orders of magnitude less sensitive than photon-counting luminometers. They are often used with high-output luminescence reactions such as enhanced luminol or dioxetane chemistry. The spectral range of CCD detectors is from visible to near-infrared wavelengths (400 to 1100 nm) and is greater than that of photomultiplier tubes. If a converter is used, the range of a CCD camera can be extended into ultraviolet (200 to 400 nm).

A camera using larger detectors is more sensitive because it has more area to gather photons, but also has lower resolution. The resolution of a CCD camera is related to the size and total number of pixels. Adequate resolution requires two to three resolvable units per sampling area. To mimic the resolution of a light microscope (approximately 22 μm), adequate resolution would be provided by an array of pixels measuring 11×11 μm; however, improved resolution would be obtained using 7×7 μm pixels. When the image is projected onto the CCD and adjusted for proper sampling, a larger number of pixels only increases the field of view and not the resolution. Multiple scans may be necessary to produce a clear image when using a less-sensitive detector.

The quantum efficiency of CCD cameras in the blue-green range is approximately 70%. Back-illuminated CCDs employed in slow-scanning cameras have a 90% efficiency. The dynamic range of a CCD device is relatively low (about 10^4) compared with photomultiplier tube (10^7). The use of small sensors in a CCD can enhance resolution but limit dynamic range. For a complete discussion of the dynamic range of CCD cameras, see Spring.[2]

2.3.2 Color CCD Cameras

Two types of color CCD cameras are commonly used in scientific applications. Single-sensor CCD cameras utilize a filter wheel or liquid crystal tunable filter to generate a sequence of colors. Beam-splitting cameras use a prism to separate colors and trim filters to allow each sensor to produce the correct color. Color cameras are less sensitive than single-color instruments because the additional beam-splitting or wavelength selection causes some loss of the light signal.

2.3.3 Intensified Digital Cameras

Intensified cameras are used for very low light imaging, for ratio imaging, and for the study of dynamic events. In this type of CCD camera (e.g., The Berthold Night Owl®, Wildbad, Germany), the photocathode is very close to a microchannel plate electron multiplier and phosphorescent output screen. The photocathode has a 50% efficiency in the blue-green portion of the spectrum. The intensifier gain is adjustable over a large range up to ~80,000. However, these intensified cameras have a reduced dynamic range and lower signal-to-noise ratio than slow-scan cameras.

2.3.4 Slow-Scan Digital Cameras

Slow-scan cameras generally produce the best image in lower light levels. These CCD cameras have control over readout rate, pixel size, and field of view. They offer at least two rates of readout so that speed can be decreased to reduce noise. Pixel size can be increased by a process known as binning in which the charge from a cluster of pixels is treated as if it came from a single large detector. This is particularly useful when light levels are low; however, spatial resolution is sacrificed for sensitivity of detection. Many slow-scan cameras allow a selected part of an image to be displayed and the remainder discarded. A camera that allows adjustment of the field of view and the rate of framing is adaptable to a wider range of circumstances than a fixed framing rate camera. This is particularly useful for the detection of gels and blots.

Cooled slow-scan CCD instruments have high sensitivity and image quality and broad dynamic range. These characteristics make such instruments well suited to a glow assay, which emits light consistently over a range of minutes. For assays using flash-type reactions, an intensified CCD detector with fast image detection is a better choice.

2.3.5 Backlighting

The sensitivity of CCD devices can be improved by backlighting, a technique in which light illuminating the nondetector side (*p*-type silicon) decreases the energy needed for a photon to generate charge on the detector. Detectors utilizing backlighting are more difficult to make and therefore more expensive than standard CCDs. A backlit CCD camera with a phosphorescent screen is most useful for rapid scan of a large area.

2.3.6 Software for Image Analysis

The imaging software accompanying many CCD devices controls settings of the camera such as exposure time, focus, storage, and display of the image. This software subtracts background, adjusts image brightness and contrast, and corrects for distortion problems and differences in the sensitivity of pixels. Images can be added and subtracted or converted to false color. Software can manipulate data for quantification and analysis of data from blots and gels. Imaging software varies considerably from one manufacturer to another. To ensure that the software is useful to the individual investigator, it needs to be used under actual experimental conditions in the laboratory. In general, one should look for software that is designed as an integral part of the system rather than as an afterthought.

2.4 GENERAL PRINCIPLES OF LUMINESCENCE DETECTION

1. Minimize the distance between the sample and detector

When measuring luminescence with any type of instrument, it is important to have the detector as close as possible to the light-emitting sample. This is because light intensity varies inversely with the square of the distance from the emitting source. Even when a lens is used to focus the source of light, it is important to have the lens as close as possible to the source of the light because of possible light losses in the lens and the aforementioned light intensity/distance relationship.

2. Use reflectance to capture available light efficiently

Under optimal conditions, emitted light can be thought of as a point source in the middle of a silvered sphere that reflects the light back into the detector from every possible angle. This relationship is described as 4π geometry. To take advantage of this, the best luminometers have sample chambers with geometrically efficient mirroring to reflect light back to the photomultiplier.

3. Maximize the signal-to-noise ratio

When using either a photomultiplier tube or a CCD mechanism, it is important to maintain the maximum signal-to-noise ratio. Noise results from events that create an electrical signal unrelated to the light being measured. Sources of noise for a photomultiplier tube include magnetic fields, heat, electromagnetic radiation, and spurious light not emanating from the sample. An advantage of photon counting is that output from tube noise can be subtracted from the signal using electrical discrimination based on energy levels. For CCD devices, noise results mainly from the effect of heat on pixel electrons and noise inherent in the camera electronics. Heat increases the amount of statistical noise that results from the random arrival of charges on the surface of the sensor. Camera noise is additive with statistical noise.

There are several ways to increase the signal-to-noise ratio. Complete darkness should be maintained in the detection chamber. In addition, spurious current from tubes or CCD devices must be kept to a minimum. Cooling can greatly improve the signal-to-noise ratio and is essential for slow-scan CCD cameras. Photomultiplier tubes tend to be much noisier as they approach the red region. In this situation, cooling is used to reduce the noise. Several systems are commonly used to cool luminescence instrumentation. Peltier elements are the least expensive, and they are capable of cooling to –30°C. Refrigeration is a more expensive option, with one-stage units able to reach –40°C and two-stage units able to reach –80°C. Liquid nitrogen–cooled devices are most expensive and are able to reach the lowest temperatures. A practical hint is to run luminescent reactions in a cool, semidarkened room. Reagent tubes should be kept in the dark to keep tube luminescence to a minimum prior to assay.

4. Mix the sample well for maximal reproducibility

In flash-type luminescence reactions, light production occurs within seconds and is highly dependent on the completeness of mixing prior to production of the flash (Figure 2.7). In glow reactions the light can be generated over minutes to hours and mixing speed is less crucial. In many cases, the chemistry of a flash reaction can be converted to a more prolonged glow. An example of this is the luciferase/ATP

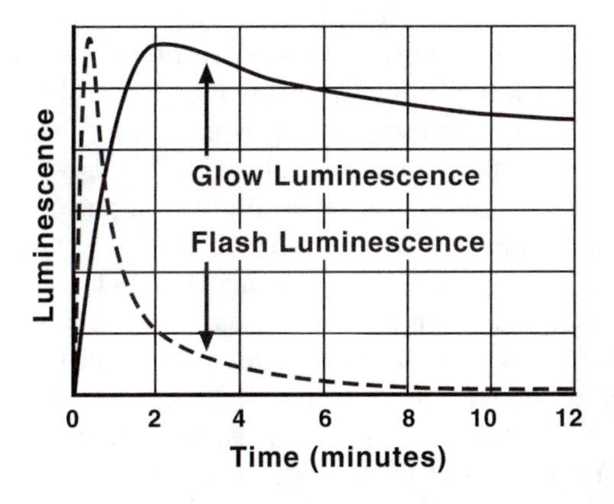

FIGURE 2.7 Luminescence can be separated into two major types based upon kinetics. A flash occurs and terminates quickly. A glow reaction occurs more slowly, and the light is sustained for longer periods of time.

reaction, where either a flash or glow reaction can be accomplished. Packard Instruments and Promega are among the companies that utilize this flexibility in their assays.

5. Be aware of the cost-to-benefit ratio

The photomultiplier luminometer operating as a photon counter is most cost-effective for the measurement of low light levels. Because the CCD camera uses a two-dimensional array, it is best for area or blot measurements, although expensive. Processing of samples is very rapid and allows for high-throughput screening of samples. With CCD devices, the cost per sample is dramatically reduced because of efficient processing. This is an important consideration in pharmaceutical, toxicological, and genetic screening (see discussion in Chapter 44).

6. Use of luminometers to handle a large number of samples

Plate luminometers are very efficient for handling large numbers of individual samples. There are many plate luminometers on the market, which vary with respect to their flexibility and reproducibility. Some instruments have an injector for flash or burst luminescent reactions. Others have no provision for flash reactions and measure only continuous light in each well. Such luminometers can only be used for light reactions that are stable over the period of time required to read all relevant wells. Otherwise, the intensity of the wells can change before the entire plate is read, resulting in highly variable data. Plate luminometers are inherently compromised because they do not measure light as efficiently as luminometers that utilize test tubes. Furthermore, they have a much-reduced dynamic range which prevents their use in either very intense light or low-light applications.

Several variations on the plate luminometer allow for increased flexibility in measurement. A strip luminometer (the Berthold MPL-2®) reads samples in a single strip of wells and allows the user to run a fairly large number of samples while still retaining precise control over the reaction conditions. Packard Instruments has developed a plate luminometer in which stacked plates can be loaded at one time (TOP COUNT®). Such an instrument is useful for high-throughput screening assays. Packard Instruments has also developed a dual-wavelength luminometer that can read two wavelengths simultaneously (Fusion Universal® Microplate Reader).

There are several issues to consider when using a plate luminometer. One is that multiwell plates vary in size from manufacturer to another. The luminometer should have flexible programming to allow variation in plate size, number of wells per plate, and positioning of wells on the plate. It is particularly convenient to have a luminometer that can handle plates containing between 6 and over 1000 wells. Another factor to take into consideration is the style of multiwell plate to use. Because plate luminometers must read the well from either its bottom or top, wells should have opaque sides to prevent errors resulting from luminescent "cross talk" between the wells. Plates containing wells with clear bottoms and either black or white sides are useful for luminescence measurements. White plates give higher counts than black plates because the sides reflect light. Ideally, the best plate would have silvered sides and a clear bottom. In this situation there would be maximal internal reflectance from the walls with little cross talk between wells.

A large number of samples can be read efficiently with a photomultiplier tube luminometer using a system in which multiwell plates are stacked and read robotically. Such instruments include the Packard TopCount® or Fusion® and the EG&G Wallac/Perkin Elmer LS55®. However, there are also various instruments that hold samples in individual containers on a belt. Berthold produces instruments of this type with and without injectors (AutoLumat®). The LKB luminometers (WALLAC 1420 Victor®) use a turret- or tray-loading system in which samples are preloaded and programmed for assay. The most recently developed systems for handling large numbers of samples employ CCD cameras. Here, a large number of samples are imaged simultaneously in a single cassette. A clear advantage of these CCD devices is that multiple scans can be added together to create a final sample reading. With digital image analysis, real-time calculations on large numbers of samples can be performed easily and efficiently. It is important to note that CCD devices are 1000 to 10,000 less sensitive at detection of light than photomultiplier tube luminometers; therefore, a higher intensity of light from individual samples is needed.

A simple device that can be purchased or can be built by a researcher is a Polaroid camera used to photograph a multiwell plate. Quantitation of the photograph can be done by densitometry. With these devices it is important not to reach saturation of the image.

ACKNOWLEDGMENT

We would like to thank Hamamatsu Corporation for permission to use Figures 2.1, 2.3, and 2.4, which were reproduced from Reference 3. This extensive reference is quite useful for a wide variety of photomultiplier applications.

REFERENCES

1. Ellis, R. J. and Wright, A. G., *Luminescence,* 14, 11–18, 1999.
2. Spring, K., *Biotechniques,* 29, 70–76, 2000.
3. *Photomultiplier Tubes: Basics and Applications,* 2nd ed., Hamamatsu Corporation, 1999.

3 Calculations to Quantify Luminescent Measurements

Christopher Van Dyke

CONTENTS

3.1 QUANTIFYING LUMINESCENT MEASUREMENTS

Luminescence measurements are commonly produced as a series of data points taken at various reaction times. The investigator is then faced with the task of reducing such data to a single number so that the treatments or reaction conditions can be statistically compared. Three basic approaches to this task are described below. Not all approaches are useful for all types of measurements.

In the following discussion, the assumption is made that measurement data for each sample have been collected so that each time of measurement (e.g., independent variable or abscissa) is paired with a measured quantity of light (dependent variable or ordinate). Thus, each data point is represented by a number pair in the form of (x, y) with the x value representing time and y representing the measured luminescence at time x.

3.2 PEAK HEIGHT

Peak height is the simplest method of quantifying luminescence data. Peak height may be expressed as either the measurement (y value) or the time of the measurement (x value). The latter is less often encountered, but may be used where the goal of the experiment is to explore reaction kinetics. Peak height can be used to describe either very fast reactions (flash luminescence) or very stable reactions (glow luminescence). In the case of glow luminescence, peak height is a misnomer since a "peak" does not really exist. Once the reaction is known to follow glow kinetics, a single measurement may be taken, which is, in effect, a peak height.

Several conditions must be met to maximize the consistency of flash luminescence peak measurements. It is important to duplicate conditions that affect reaction kinetics precisely. These include factors such as sample mixing, temperature, and pH. It is also important to assure the actual peak is captured. Two situations may prevent this from occurring. First, beginning measurement after the start of the reaction may cause the peak to be missed. When the first point measured is larger than the second, the actual peak was probably not captured. Second, a relatively slow

0-8493-0719-8/02/$0.00+$1.50

measurement rate may miss the peak if it falls between two sample periods. To check for this condition, construct an x–y plot of the data. If the peak appears as a sharp point rather than a rounded curve, the actual peak may have been missed. If possible, the best solution is to measure more often. If this is not possible, quantifying by integration (see below) may minimize this error.

3.3 RATE OF REACTION

The rate of a reaction describes the creation of luminescence over a period of time and is most commonly used in the investigation of reaction kinetics. This quantity is calculated as the slope of the line between two selected measurements. The measurements most commonly selected are the first measurement and the peak height. It is therefore important to observe the aforementioned recommendations for maximizing peak consistency.

Calculation of slope

$$m = \frac{y_{\text{peak}} - y_{\text{initial}}}{x_{\text{peak}} - x_{\text{initial}}}$$

where

m = slope (rate of reaction)
y_{peak} = measurement at peak
y_{initial} = initial measurement
x_{peak} = time at peak measurement
x_{initial} = time at initial measurement

3.4 AREA UNDER THE CURVE

Calculation of the area under the curve of a light-producing assay measures the total light production over a selected period of time and is sometimes referred to as integration. Integration removes reaction kinetics from consideration so that variables such as inconsistent mixing and noise are effectively averaged out. The software supplied with many instruments performs this calculation. If such capacity is lacking, several methods for calculating the integrated area are presented below.

Two important considerations are the range of integration and background. The range or time period of interest must be selected to achieve meaningful values. Initially, the measurement may be taken for a long period of time to allow the full reaction to be viewed. The total time of the observation may then be reduced. Any significant background or nonspecific signal will increase the area under the curve and may disguise important differences. Hence, measurement and subtraction of the background signal are highly recommended.

Four different methods for determining the area under the curve are presented here. Simpson's rule is generally the most accurate but requires a fixed sampling interval and an odd number of data points. The trapezoidal rule is more flexible and easier to implement than Simpson's rule, but is generally 1 to 5% less accurate. Two different versions of the trapezoidal rule are shown: the standard one, which requires a fixed interval between samples, and a variation, which permits an arbitrary sample interval. Finally, a physical method is discussed.

3.4.1 SIMPSON'S RULE

Simpson's rule is a mathematical technique for integration that uses a series of parabolas (e.g., quadratic equations) to estimate the shape of the curve. Note that this requires an odd number of observations spaced at a fixed interval. In the mathematical notation below, the integration ranges from data point 0 to data point n.

TABLE 3.1
Example Data

Sample Number	Elapsed Time (s)	Luminescence Detected (light units)
0	0	6
1	15	71
2	30	316
3	45	171
4	60	76
5	75	45
6	90	28
7	105	17
8	120	14

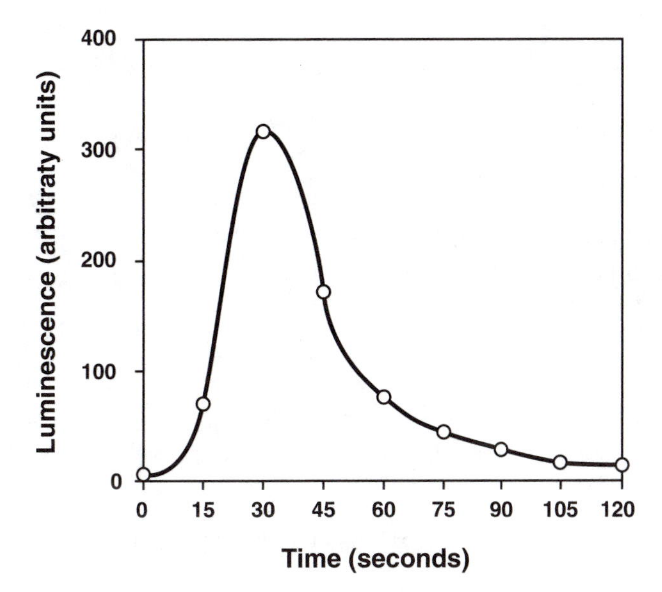

FIGURE 3.1 The data in Table 3.1 plotted. These data are used for the example calculations.

Simpson's rule

$$\text{Area} \approx \frac{\Delta t}{3}\left(y_0 + 2\sum_{i=1}^{n-1} y_{2i} + 4\sum_{i=1}^{n} y_{(2i-1)} + y_n \right)$$

where

Δt = the sampling interval (constant)
y_i = measurement number i (index starting at zero)
n = the total number of samples (must be an odd number)

Example 1

The area under the curve for the data in Table 3.1 is calculated via Simpson's rule. The data is graphed in Figure 3.1. In this case, the sampling interval Δt = 15 s and n = 8. Thus,

$$\text{Area} \approx \frac{15}{3}[6 + 2(316 + 76 + 28) + 4(71 + 171 + 45 + 17) + 14]$$

or

$$\text{Area} \approx 10{,}380\ (\text{arbitrary light units})(\text{s})$$

3.4.2 Trapezoidal Rule

The trapezoidal rule is another mathematical technique for integration. In this method, adjacent data points are joined with a straight line; then trapezoids are formed by connecting the points to the x-axis with vertical lines. The resultant areas are summed to approximate the area under the curve. Generally, this method is less accurate than the Simpson's rule, although the accuracy is dependent on the curve shape.

Two different versions are presented here. The simpler case requires a constant sampling interval, whereas the second allows for a variable sample interval. The area under the curve from data points 0 to n is approximated by:

Trapezoidal rule

$$\text{Area} \approx \frac{\Delta t}{2}\left(y_0 + 2\sum_{i=1}^{n-1} y_i + y_n\right)$$

where

Δt = the sampling interval (constant)
y_i = measurement number i (index starting at zero)
n = the total number of samples

Example 2

The area under the curve for the data in Table 3.1 is calculated via the trapezoidal rule.

Again, the sampling interval $\Delta t = 15$ s and $n = 8$. Thus,

$$\text{Area} \approx \frac{15}{2}[6 + 2(71 + 316 + 171 + 76 + 45 + 28 + 17) + 14]$$

or

$$\text{Area} \approx 11{,}010\ (\text{arbitrary light units})(\text{s})$$

3.4.3 Trapezoidal Rule (Variable Sample Interval)

Some practical laboratory situations do not allow for a fixed time between samples. For such a situation, the following modification of the trapezoidal rule is presented. The area under the curve from data points 0 to n is approximated by:

Modified trapezoidal rule

$$\text{Area} \approx \frac{1}{2}\left(-x_0 y_0 + \sum_{i=1}^{n-1} x_i y_{i-1} - \sum_{i=1}^{n-1} x_{i-1} y_i + x_n y_n\right)$$

where

x_i = time at measurement i (index starting at zero)
y_i = measurement number i (index starting at zero)
n = the total number of samples

Example 3

While the data in Table 3.1 has a fixed sample interval, the area under the curve is calculated via the modified trapezoidal rule as an illustration of the method.

$$\sum_{i=1}^{n-1} x_i y_{i-1} = 15(6) + 30(71) + 45(316) + 60(171) + 75(76) + 90(45) + 105(17) + 120(17)$$

$$\sum_{i=1}^{n-1} x_i y_{i-1} = 41{,}430$$

$$\sum_{i=1}^{n-1} x_{i-1} y_i = 0(71) + 15(316) + 30(171) + 45(76) + 60(45) + 75(28) + 90(17) + 105(14)$$

$$\sum_{i=1}^{n-1} x_{i-1} y_i = 21{,}090$$

Substituting,

$$\text{Area} \approx \frac{1}{2}[-0(6) + 41430 - 21090 + 120(14)]$$

or

$$\text{Area} \approx 11{,}010 \text{ (arbitrary light units)(s)}$$

3.4.4 Estimation by Mass

This is the simplest—and most time-consuming—method for estimating the area under the curve. Regions of interest are cut from a data plot and weighed. Given a paper of uniform thickness, an accurate analytical balance, and a fixed scale, the mass of the cutouts will be proportional to the area under the curve. Note that the quality of estimation is largely dependent on the accuracy of the cut, so larger graphs will minimize errors. This alternative is useful if data exist only in a paper form or when an alternative method for verification of software calculations is needed.

4 A Low-Budget Luminometer for Sensitive Chemiluminescent Immunoassays*

Nino Porakishvili, John L.A. Fordham, Marie Charrel, Peter J. Delves, Torben Lund, and Ivan M. Roitt

CONTENTS

4.1 INTRODUCTION

The enzyme-linked immunosorbent assay (ELISA) for antigens and antibodies with one of the reagents bound to a solid phase is in widespread use. The final readout of the enzyme product is most commonly made spectrophotometrically using a chromogenic substrate in 96-well microtiter plates, but there is now a general movement toward the development of chemiluminescent rather than colorimetric probes because of low backgrounds, large linear concentration ranges, and sensitivities that may be 10- to 100-fold greater.[1,2]

One drawback for laboratories with modest budgets is the relatively high cost of luminometers. This chapter describes an instrument that can be simply constructed in the typical university workshop, based on a moderately inexpensive charge-coupled device (CCD) camera housed in a light-tight box with a single lens. The camera control and acquisition software can be installed in a personal computer of modest performance. Acquired data in the 96 wells are then analyzed after transfer of the images to a custom-built software package that corrects for errors introduced by the lens.

As a further development, the use of a CCD camera for quantitative imaging allows detection of multiple assays carried out on an array of very small dots of the ligand-binding reagent bound to the bottom of a microwell; this leads to economy in the use of reagents and test sample and

* This chapter is a concise version of a paper by Porakishvili, N. et al. with the same title appearing in *J. Immunol. Methods,* 234, 35–42, © 2000. Reproduced by permission of the publisher, Elsevier Science.

0-8493-0719-8/02/$0.00+$1.50

permits direct comparison and standardization of reactivity of the sample against several minidots simultaneously.

4.2 MATERIALS AND METHODS

4.2.1 Titration of Rabbit Polyclonal Anti-hCG Sera

Rabbits were primed and boosted twice at 6-week intervals with 100 μg/ml of human chorionic gonadotropin (hCG) expressed in a baculovirus system and conjugated with tetanus toxoid (TT). Blood samples were collected and the sera separated for parallel chemiluminescent ELISA (CHELISA) and ELISA studies. Black (Corning Costar, High Wycombe, Bucks, U.K.) and white (Nunc, Roskilde, Denmark) 96-well microtiter plates were coated with 50 μl/well of 1 μg/ml recombinant hCG (Sigma-Aldrich, Poole, Dorset, U.K.) in carbonate-bicarbonate buffer (CBB), pH 9.6, overnight at 4°C. The plates were washed three times in PBS–0.05% Tween-20 (PBS-T) and blocked with 200 μl/well of 2% dried skimmed milk powder in CBB overnight at 4°C. Following three further washes in PBS-T, 50 μl of dilutions of the rabbit antiserum or 1 μg/ml monoclonal anti-hCG (OT3A; a gift from Dr. W. Stevens, Organon Technika, Netherlands) in PBS-T-2% bovine serum albumin (BSA) were added to the wells. The plates were incubated at 37°C for 2 h and washed three times in PBS-T. Then, 50 μl of 1/1000 goat anti-rabbit IgG horseradish peroxidase (HRP) conjugate (Sigma) in PBS-T-BSA were added and the plates incubated at 37°C for 2 h, followed by washing three times in PBS-T. Then, 100 μl of tetramethyl benzidine (TMB) (Sigma) for ELISA and Power Signal Luminol/Enhancer (Pierce, Warrington, U.K.) for CHELISA were added to each well.

Absorbance values at 450 nm were read in a Multiskan spectrophotometer (Labsystems, Helsinki, Finland) after 10 min of incubation for the ELISA, or in the luminometer with the number of the photons emitted from each well recorded after different periods (10 s to 5 min) for the CHELISA.

4.2.2 Production of "Nanodots" (for CHELISA)

Microarrays of antigen spots of volumes as small as 20 nl or less can be laid down on the bottom of a microtiter well using robotic inkjet techniques (Biodot Ltd, Huntingdon, Cambridge, U.K.). Thereafter, the protocol was identical with that for normal coating of an entire well with antigen. The results of CHELISA assays carried out on these spots were read with the plate in the upper position in the luminometer.

4.2.3 The Luminometer Detector

A schematic diagram of the luminometer is shown in Figure 4.1. A 96-well microtiter plate was placed in a sample holder within a light-tight chamber coated with matt black paint. The emission from the spots under investigation was focused by a high-quality, f1.8, Nikon lens onto a commercially available CCD camera manufactured by the Santa Barbara Instrumentation Group (Santa Barbara, CA). This camera, type ST6,* has built-in thermoelectric cooling to allow long integration times on faint samples and a software package for setup and operation. The temperature during operation of the luminometer was set at –35°C. Two focus positions are available; the lower position allows imaging of all 96 wells, whereas the upper position allows more-detailed study of just 4 wells

* Although the ST6 camera was utilized in these experiments, two other cameras from the same suppliers are potential alternatives: the ST7 of comparable price, has a smaller imaging area but more pixels and would be usable in high-resolution applications such as imaging within single wells, whereas the ST9, which is slightly more expensive, has a larger imaging area which means that less demagnification is required when imaging a 96-well plate.

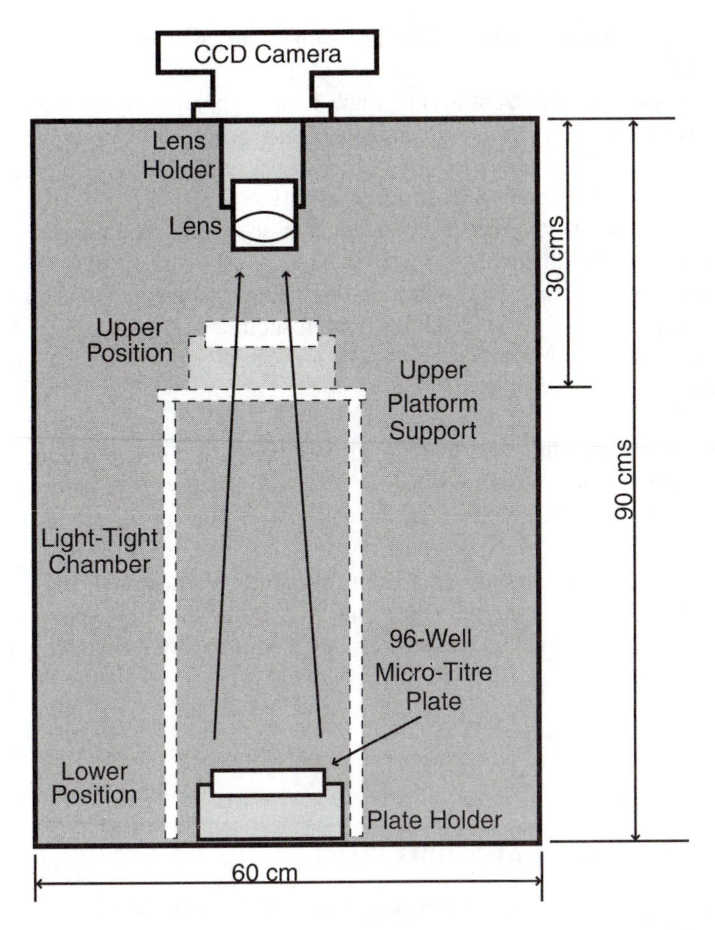

FIGURE 4.1 Diagram of the luminometer. (From Porakishvili, N. et al., *J. Immunol. Methods,* 234, 35, 2000. With permission from Elsevier Science.)

in a 2 × 2 array. Black microtiter plates are used to prevent transmission of emission between wells. Details of the mechanical design employed at University College London can be obtained by contacting the authors.

Prior to operation of the luminometer, the imaging system must be calibrated. In particular, optical aberrations (blooming) and vignetting introduced by the lens must be quantified. This calibration needs to be carried out only once as the two positions are fixed and the plate holder provides accurate positioning. For the calibration a beta light that provides a constant light output at 530 nm is employed. In all 96 separate exposures on this light are then carried out with it positioned in all of the wells. The resultant data set provides the optical efficiency variations between wells due to vignetting. The well counts in all subsequent analyses are then modified during data reduction to account for this artifact.

Images for analysis were acquired using the software package provided with the CCD camera. Thermal buildup in the CCD during each exposure was automatically subtracted; a bias of 100 was then added to prevent generation of negative numbers by noise.

The images when using the lower position were then reduced using the following procedure:

1. The raw counts in the center of each well were found using a custom-built software package available at University College London. A grid of 96 windows was placed over

the image and centered on the wells. The sum counts within each window were then automatically produced.
2. By using the same software routine, the counts in the webbing between wells was obtained, which provided data on image contamination due to optical aberrations (blooming) in the lens.
3. The two data sets were transported to Microsoft Excel.
4. The bias of 100 was then removed from each data set, with account taken of the window size used for count extraction.
5. The contamination in each well was found by taking the average of the four plate web positions surrounding each well. This was then subtracted from the well data set.
6. Finally, the data set was normalized to account for efficiency variations using the calibration data.

It is intended in the near future to include all these tasks in a single routine, the input being the raw image and the output being the normalized counts for each well. A more accurate blooming subtraction would also be included that uses the calibration data set for providing contamination levels.

For the upper focus position, only data from four wells are acquired, these centered on the optical axis and hence suffering minimal vignetting. At present, data from spots within these wells are analyzed using the data reduction package available with the CCD camera, this automatically removing any bias introduced.

4.3 RESULTS

4.3.1 ANTIBODY TITRATION IN MICROTITER PLATES

As an illustrative example of the equipment in use as an ELISA microplate reader, a rabbit antiserum to hCG was titrated out in parallel with a monoclonal anti-hCG in a microtiter plate (Figure 4.2). The rabbit antiserum gave a clearly positive result at a dilution of 1 in 10^7 and a weak positive at 1 in 5×10^7. The *murine* monoclonal antibody titrated out to 1 in 5×10^6. The background values obtained in the absence of serum were negligible.

The sensitivity of the CHELISA was compared with that obtained using conventional ELISA with a chromogenic substrate and a spectrophotometric readout. The titration curves are shown in Figure 4.3, which plots the results for increasing dilutions of a polyclonal rabbit anti-hCG and the prebleed serum. The greater sensitivity of the chemiluminescent assay is immediately apparent. At a serum dilution of 1 in 5×10^5, the ratio of values for antiserum relative to prebleed control is 50:1 for CHELISA compared with 5:1 for the values in the conventional colorimetric assay; the ratios at a dilution of 1 in 10^6 are 12.5:1 and 1.2:1, respectively. Indeed, at a dilution of 1 in 10^6, the conventional assay is barely distinguishable from the control. Another feature is the impressive dynamic range of the CHELISA, which in the example shown runs from 3000 photons/well/min in the control diluted 1 in 10^4 to 10×10^6 photons/well/min in the polyclonal antiserum. Comparable figures for the conventional ELISA are 0.13 for the absorbance in the control to 3.0 in the antiserum assay. In both assays, replicates were reasonably tight; the coefficient of variation for CHELISA for the clearly positive dilutions from 1 in 10^4 to 1 in 5×10^6 was 9.5% and that for conventional ELISA for the assay values of dilutions from 1 in 10^4 to 1 in 10^6 was 7.7%.

Human sera from patients with systemic lupus erythematosus were titrated for antibodies to DNA and gave comparable results in CHELISA and ELISA tests although, as expected, the chemiluminescent assay showed greater sensitivity (data not given).

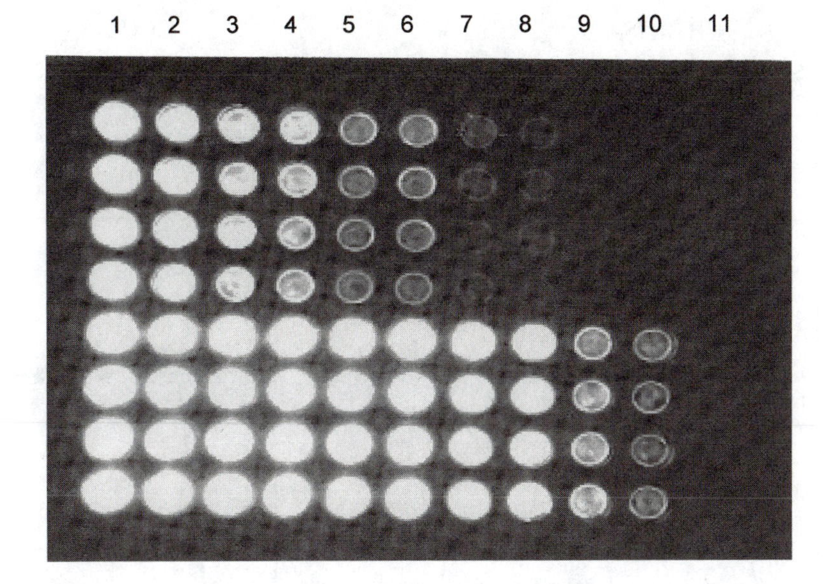

FIGURE 4.2 Luminometer image of microtiter plate showing CHELISA tests on progressive dilutions from 1 in 10^3 (column 1) to 1 in 5×10^7 (column 10) of a mouse monoclonal antibody (top four rows) and a rabbit polyclonal antiserum (bottom four rows) to hCG. All wells except those in the negative column (11) were coated with 1 μg/ml hCG antigen. (From Porakishvili, N. et al., *J. Immunol. Methods,* 234, 35, 2000. With permission from Elsevier Science.)

4.3.2 CHELISA Tests on Nanodots of Antigen

Volumes of hCG ranging from 200 to 20 nl were deposited robotically on the bottom of microtiter plate wells and stained with anti-hCG using the same conditions employed for staining manually prepared spots except that, for logistic reasons, the plates were dried in a convection oven at 40°C for 5 min and treated with the blocking solution the next morning. Clear results with discrete dots of light were obtained down to the lowest size array. The largest dots of 200 nl coating volume gave a mean light detection (±SD) of 23.6 ± 3.7 photons/pixel/s (Figure 4.4a). The background value given by the uncoated surface of the well was 2.4 ± 0.01 photons/pixel/s. The microheterogeneity within each of these spots was determined by dividing each spot into six equal areas using the analytic matrix setting; this gave an intraspot coefficient of variation of 4.9%. The mean light detected from an array of six minidots obtained by coating with volumes of 100 nl was 18.0 ± 2.2 photons/pixel/s (Figure 4.4b). The mean figures for coating volumes of 50 and 20 nl were 20.0 ± 2.0 (Figure 4.4c) and 15.2 ± 1.1 photons/pixel/s (Figure 4.4d), respectively.

4.4 DISCUSSION

The objective was to produce a sensitive ELISA reader capable of achieving the economies of miniaturization, and affordable by laboratories with low to modest budgets. An apparatus based on chemiluminescence was a fairly obvious choice in view of the low detection limits, the large dynamic range with linear signal response, and the short assay times, all of which contribute to a slow general movement away from spectrophotometric readout systems.

The authors were considerably influenced by the experience of one author (J.L.A.F.) in designing detection systems incorporating CCD cameras to record low-intensity pinpoints of light emitted by stars.[3,4] It was decided to base the instrument around an economically priced (£2000), off-the-shelf

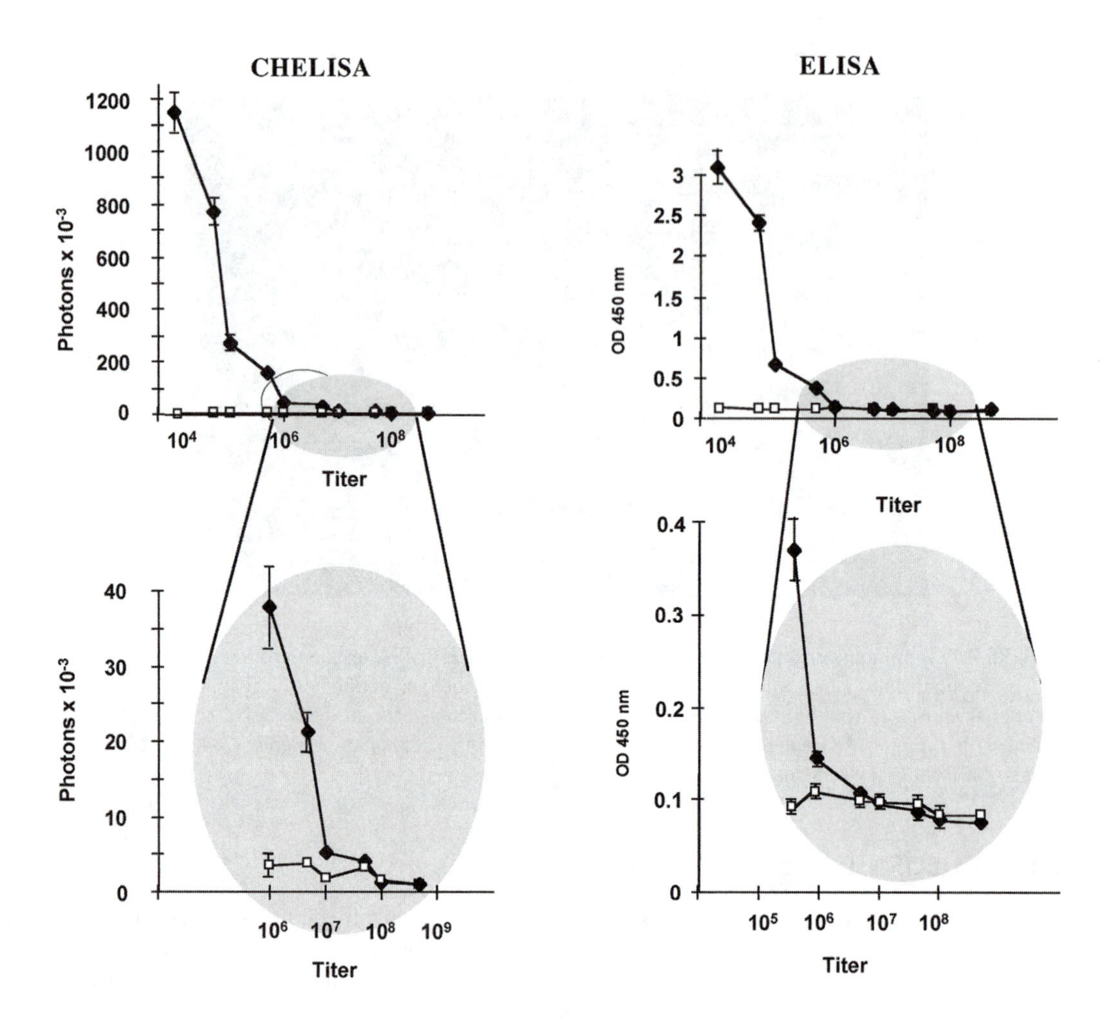

FIGURE 4.3 Comparison of the CHELISA and colorimetric ELISA tests for titration of a polyclonal rabbit antiserum to hCG on hCG-coated microtiter plates. ◆, antiserum; □, prebleed serum. (From Porakishvili, N. et al., *J. Immunol. Methods,* 234, 35, 2000. With permission from Elsevier Science.)

CCD camera, which could simultaneously record light emission from a whole optical field such as the 96 wells in a conventional microtiter plate, with no need for the serial *xy* scanning required with photomultiplier tube devices. Installing the camera with a lens in a light-tight box with a plate holder should be well within the competence of a relatively unsophisticated workshop and should cost for less than an extra £1000.

The software provided by the manufacturers allows visualization of the image through digital readout using a relatively undemanding personal computer as specified by the manufacturer's (Santa Barbara Instrumentation Group) data sheet. Thus, photon emission by each part of the chemiluminescent image can be quantified, and a program is available to permit each well or portion of a well to be framed within a single element of an adjustable matrix. The software has also been extended to correct for lens aberrations.

This chapter has illustrated how readily the simultaneous imaging of an entire microtiter plate can be demonstrated and has compared the performance of the chemiluminescent (CHELISA) system with a parallel assay for antihuman gonadotropin with a rabbit polyclonal antiserum using a conventional colorimetric ELISA. The chemiluminescent system showed a tenfold increase in

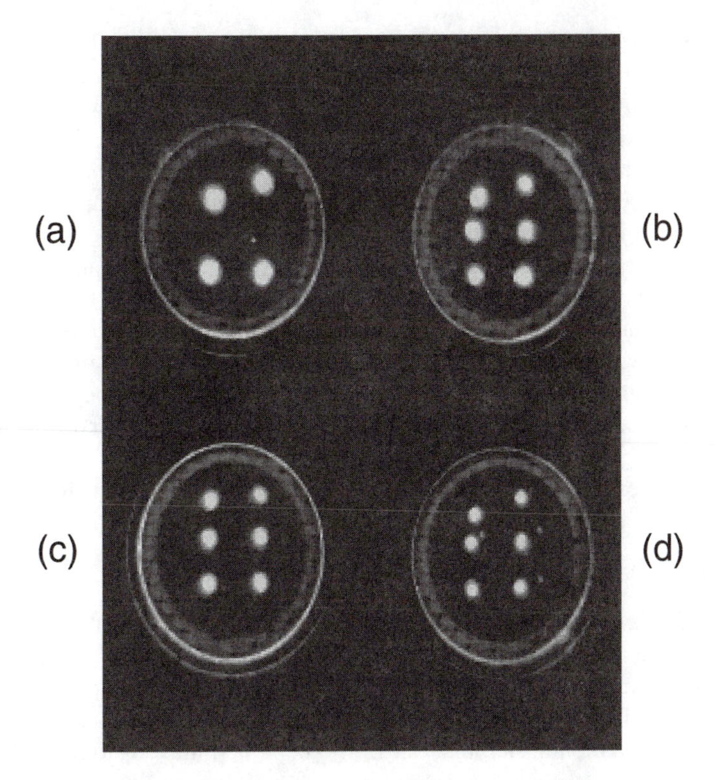

FIGURE 4.4 Rabbit anti-hCG (diluted 1/1000, a dilution previously shown to give 50% of maximum binding to hCG-coated wells) assayed on nanodots formed by coating the bottom of a microtiter well with hCG drops of volume (a) 200 nl, (b) 100 nl, (c) 50 nl, and (d) 20 nl. Antibody binding was revealed by enhanced luminol chemiluminescence using an HRP conjugated anti-rabbit serum. (From Porakishvili, N. et al., *J. Immunol. Methods,* 234, 35, 2000. With permission from Elsevier Science.)

sensitivity, a much better signal-to-background ratio, and a much wider dynamic range of readout, which enables both high and low antibody concentrations to be measured. In contrast, the logarithmically based optical densities of the color reaction are not linear, with greater deviation at higher values. Furthermore, the colorimetric readout depends on a reduction from 100% light transmission in the negative control, whereas the CHELISA starts from a very low background photon detection, with the majority of the signal generated by thermal noise in the camera, which is grossly reduced by Peltier cooling.

As might be anticipated, the CHELISA system was shown to be effective with a number of different antigens and could detect human, rabbit, and mouse antibodies.

Miniaturization can be readily achieved by using different format microtiter plates with much smaller wells, but the ability of a CCD camera to record images over a whole field leads to a most important application, namely, one involving assays of antibodies on minidot arrays of antigens.

Although using a crude manual method, the authors were able to place three discrete spots of antigen on the bottom of a single microtiter well, but significant improvement really requires robotic spotting. Although only a single exploratory study, we encouraging results were obtained with an array of six identical spots of antigen using ink-jet printing technology in which as little as 20 nl of coating antigen was deposited. Reproducibility was good considering that the antigen was coated directly onto the untreated plastic without any attempt to optimize the process; the average coefficient of variation was 11.4%, but was lower the smaller the volume used to form the nanodot. Much larger numbers of assays are possible using the full area of the well and delivering substantially smaller volumes.

Analyte assays using large arrays of ligand binder, in the present study antigen, deposited as minidots is a growing and exciting development[5] for many reasons. It allows *multiple tests* to be carried out on a single small fluid sample such as a finger prick of blood (or even mouse cerebrospinal fluid). It is extremely economical in plastic ware and in solid-phase reagents since a few micrograms of purified or recombinant antigen can produce thousands of minidots. One only requires enough developing reagents for a single rather than for multiple tests. Interestingly, duplicate or even triplicate assays can be included in the same array to give greater confidence in the accuracy of the result and a quality control check on the minidot deposition process. There is an obvious economy in operator time and reduction in chance of technical error. Reading time is shortened since there is no need for serial measurements. The economics of the system make it feasible to afford the inclusion of duplicate wells with a chaotropic agent to estimate the avidity of the antiserum for the various antigen minidots if required, or with a readout for different antibody classes.

Potential applications in the immunoassay field are almost limitless and extend to all ligand-binding assays including, of course, those involving DNA arrays. Whether a more sensitive instrument will need to be developed to cope with assays on ultrasmall dots remains to be seen.

ACKNOWLEDGMENTS

We thank Mr. C. Flack of Biodot Ltd. for making an array of hCG nanodots. We are grateful to the Wellcome Trust for a Senior Travelling Fellowship to Professor N. Porakishvili to enable her to carry out the research for this project at University College London. Support was also provided by research funds within University College London and by the Cleveland Immunological and General Trust. The original publisher, Elsevier B.V., kindly permitted us to reproduce the material presented here.

REFERENCES

1. Kricka, L. J., Chemiluminescent and bioluminescent techniques, *Clin. Chem.,* 37, 1472, 1991.
2. Vesterberg, O., A new luminometer for sensitive quantification by chemiluminescence of specific proteins in microtiter plates and on blot membranes, *J. Biochem. Biophys. Methods,* 30, 301, 1995.
3. Fordham, J. L. A., The use of CCDs in photon counting detectors, in *Proc. ESO/OHP Workshop.* The optimization of the use of CCD detectors, in *Astronomy,* Baluteau, J.-P. and D'Odorico, S., Eds., ESO, 1986, 309.
4. Fordham, J. L. A. et al., The MIC photon counting detector. Electron Image Tubes and Image Intensifiers II, *SPIE,* 1449, 1991.
5. Ekins, R. P. and Chu, F., Multianalyte microspot immunoassay. The microanalytical "compact disk" of the future, *Ann. Biol. Clin.,* 50, 337, 1992.

APPENDIX A

Functionality of Commercially Available Instruments for Luminescence

Knox Van Dyke

Details of Instruments

Instrument Assay — **General Characteristics**

1. Junior (Berthold)

Instrument Assay		General Characteristics
Sample type	Tube	Handheld portable
Luminescence type	Glow only	Battery operated
Reporter gene assay (RGA)	Yes	Low priced, high value
Dual-label RGA	Yes	
ATP measurements	Yes	
Enzymes assays	Yes	
Luminescence immunoassay	Fair	
DNA probe assays	Fair	
Cellular luminescence	Fair	
Redox luminescence	Yes	
BRET assays	No	

2. Minilumat (Berthold)

Instrument Assay		General Characteristics
Sample type	Tube	Operates on battery/mains
Luminescence type	Glow only	Sensitive, excellent dynamic range
Reporter gene assay (RGA)	Yes	Easy protocol entry
Dual-label RGA	Yes	Flexible software
ATP measurements	Yes	Operates with printer
Enzymes assays	Yes	
Luminescence immunoassay	Yes	
DNA probe assays	Yes	
Cellular luminescence	Fair	
Redox luminescence	Yes	
BRET assays	No	

3. Lumat (Berthold)

Instrument Assay		General Characteristics
Sample type	Tube	Two variable-volume injectors
Luminescence type	Flash/glow	Sensitive
Reporter gene assay (RGA)	Yes	Versatile
Dual-label RGA	Yes	Low to medium sample output

0-8493-0719-8/02/$0.00+$1.50

Instrument Assay		General Characteristics
ATP measurements	Yes	
Enzymes assays	Yes	
Luminescence immunoassay	Yes	
DNA probe assays	Yes	
Cellular luminescence	Yes	
Redox luminescence	Yes	
BRET assays	No	
4. Flash'n'glow (Berthold)		
Sample type	30 tube sampler	Three variable-volume injectors
Luminescence type	Flash/glow	Flexible Windows PC software
Reporter gene assay (RGA)	Yes	Remote control and data evaluator
Dual-label RGA	Yes	Internal printer for stand-alone assay
ATP measurements	Yes	
Enzymes assays	Yes	
Luminescence immunoassay	Yes	
DNA probe assays	Yes	
Cellular luminescence	Fair	
Redox luminescence	Yes	
BRET assays	No	
5. AutoLumat (Berthold)		
Sample type	180 tube sampler	Automatic sample changer for up to 180 tubes
Luminescence type	Flash/glow	Three variable-volume injectors possible
Reporter gene assay (RGA)	Yes	Heating: optional
Dual-label RGA	Yes	TubeMaster Windows Software (PC)
ATP measurements	Yes	
Enzymes assays	Yes	With heating, cellular luminescence is good
Luminescence immunoassay	Yes	
DNA probe assays	Yes	
Cellular luminescence	Yes	
Redox luminescence	Yes	
BRET assays	No	
6. FB-17 Microplate luminometer (Berthold)		
Sample type	96/384 wells	Extended capability vs. LB 96 V
Luminescence type	Flash/glow	Multiple filter sets—multiple wavelengths
Reporter gene assay (RGA)	Yes	Wavelengths changed quickly, useful for simultaneous dual wavelengths: BRET
Dual-label RGA	Yes	Dynamic range extended × 100 at high signal level—with intensity recalculation
ATP measurements	Yes	Easy switchover from 96/384 wells
Enzymes assays	Yes	Automatic position readjustment with different microplates, high efficiency
Luminescence immunoassay	Yes	Low cross talk between wells
DNA probe	Yes	Orbital and linear shaking
Cellular luminescence	Fair	Robot access; optimized sample presentation
Redox luminescence	Yes	Appropriate software
BRET assays	Yes	Bar code reader with sample processing
7. FB-12 (Zylux–Berthold)		
Sample type	Tubes, vials, dishes, microfuge tubes	Stand-alone instrument or computer-driven with printer; high sensitivity, excellent sample versatility
Luminescence type	Glow assays	Low-cost instrument with high versatility

Instrument Assay		General Characteristics
Reporter gene assay (RGA)	Yes	Can be maintained in incubator at 33.5°C
Dual-label RGA	Yes	For cellular assays
ATP measurements	Yes	35-mm dishes with adherent cells can be assayed at 33.5°C
Enzymes assays	Yes	Lightweight unit with versatile assay capability
Luminescence immunoassay	Yes	Excellent, adaptable software
DNA probe assays	Yes	Good sensitivity 300–650 nm
Cellular luminescence	Yes	Six-decade range
Redox luminescence	Yes	
BRET assays	No	
	8. Sirius (Berthold)	
Sample type	Tubes, dishes, vials, microfuge tubes	Similar to the FB-12 but with two possible injectors; excellent sample versatility; a stand-alone instrument or linked to Windows PC computer/printer
Luminescence type	Flash/glow	Can be placed into incubator at 33.5°C
Reporter gene assay (RGA)	Yes	For cellular assays
Dual-label RGA	Yes	Lightweight unit with injectors for flash assays; has its own external printer for stand-alone assays
ATP measurements	Yes	Good sensitivity 300–650 nm
Enzymes assays	Yes	Six-decade range
Luminescence immunoassay	Yes	
DNA probe assays	Yes	
Cellular luminescence	Yes	
Redox luminescence	Yes	
BRET assays	No	
	9. MPL-1 (Xylux–Berthold)	
Sample type	96 well/strip	Heated 96-well or strip-well luminometer
Luminescence type	Glow only	Cycles in 120–200 s
Reporter gene assay (RGA)	Yes	Good sensitivity 300–650 nm
Dual-label RGA	Yes	Six-decade range
ATP measurements	Yes	
Enzymes assays	Yes	
Luminescence immunoassay	Yes	
DNA probe assays	Yes	
Cellular luminescence	Yes	
Redox luminescence	Yes	
BRET assays	No	
	10. MPL-2 (Xylux–Berthold)	
Sample type	96 well/strip	Seems to be an updated version of the MPL-1
Luminescence type	Glow	Scan single well access in 96-well pattern
Reporter gene assay (RGA)	Yes	Heated
Dual-label RGA	Yes	Good sensitivity from 300–650 nm
ATP measurements	Yes	
Enzymes assays	Yes	
Luminescence immunoassay	Yes	
DNA probe assays	Yes	
Cellular luminescence	Yes	
Redox luminescence	Yes	
BRET assays	No	

Instrument Assay		General Characteristics
	11. Orion Microplate (Berthold)	
Sample type	96, 384, strip	Easy-to-use, sensitive plate luminometer with injection and heating as an option
Luminescence type	Flash/glow	Good for all flash and glow assays
Reporter gene assay (RGA)	Yes	Except Bret, which requires two wavelengths simultaneously
Dual-label RGA	Yes	Up to four injection ports
ATP measurements	Yes	Excellent sensitivity and broad range of wavelengths 300–650 nm
Enzymes assays	Yes	
Luminescence immunoassay	Yes	
DNA probe assays	Yes	
Cellular luminescence	Yes	
Redox luminescence	Yes	
BRET assays	No	
	12. Vega (Xylux–Berthold)	
Sample type	Tubes	Miniature battery-operated tube luminometer
Luminescence type	Glow	Spectral range 300–650 nm
Reporter gene assay (RGA)	Yes	Six-decade dynamic range
Dual-label RGA	Yes	Photon counter
ATP measurements	Yes	1–991 s per run
Enzymes assays	Yes	Compatible with Windows operating system
Luminescence immunoassay	Yes	
DNA probe assays	Yes	
Cellular luminescence	No	
Redox luminescence	Yes	
BRET assays	No	
	13. NightOWL™ (Berthold)	
Sample type	2D, 3D gels and samples	Versatile CCD imaging system to measure an array of light or fluorescence measuring ultralow light
Luminescence type	Usually this instrument would measure the glow, color, or fluorescence of sample	Microscopic and macroscopic samples can be analyzed; blots, dots, microarrays, petri dishes, gels, etc. can be measured for luminescence, color, or fluorescence
Reporter gene assay (RGA)	Yes	System can actually scan for hours to detect even the faintest signal
Dual-label RGA	Yes	Operates under complete computerized control using WinLight software; light-tight housing containing cooled-Peltier CCD camera to −70°C
ATP measurements	Yes	The CCD camera can be adapted to a fluorescent microscope
Enzymes assays	Yes	Camera is mounted on a movable tract, which is motor driven for adjustment of magnification and brings the camera close to the source of light or fluorescence
Luminescence immunoassay	Yes	State-of-the-art imaging system, which comes with the standard CCD camera or one that uses the higher-efficiency back-thinned chip that is illuminated from the rear for 80% quantum efficiency
DNA probe assays	Yes	Genetically altered plants that produce luminescent signals can be visualized
Cellular luminescence	Yes	WinLight™ application software from Media Cybernetics is used
Redox luminescence	Yes	Gel scan 3D and Gelscript software is utilized and application software from Phoretix is used for 1D and 2D gels
RNA, protein blots or gels	Yes	

Instrument Assay		General Characteristics
14. FluorChem™ (Alpha Innotech Corp.)		
Sample type	2D, 3D gels	Cooled-Peltier CCD camera to measure luminescence, fluorescence, color with multi-image light box; Alpha Innotech's top of the line system
Luminescence type	Glow	System does a good job with a variety of different gels and blots; its sensitivity would not be as good as the NightOWL system but its costs are $20–30,000 less
Reporter gene assay (RGA)	Yes	There have been major improvements in the software recently supplied by Alpha Innotech; uses Alpha Ease™ software using 36-bit architecture
Dual-label RGA	Yes	Selects different colors for fluorescence using a filter wheel
ATP measurements	Yes	1.3 million pixel resolution
Enzymes assays	Yes	Spatial resolution of 30–180 μm
Luminescence immunoassay	Yes	Lens zoom 12.5–75 mm
DNA probe assays	Yes	Works well with a variety of light and fluorescent probes
Cellular luminescence	No	Chroma light illumination
Redox luminescence	Yes	Can be used for multiplexing, multicolor arrays, colony counting, TLC, etc.
RNA, protein blots or gels	Yes	MovieMode™-capture multiplexed images
15. ChemiImager™ (Alpha Innotech)		
Sample type	2D, 3D gels	ChemiNova CCD camera and Multimage II Light Cabinet
Luminescence type	Glow	Zoom lens (6×)
Reporter gene assay (RGA)		Not as sensitive as the FluroChem but it is very good for a variety of blots and gels and colonies; can be used for luminescence, fluorescence, or color
Dual-label RGA	Yes	Uses 32-bit ChemiImager AlphaEase architecture software; has MovieMode for luminescent and kinetic imaging; transilluminator with 260–395 nm with ultraviolet or Chromalight™; has thermal and color printing
ATP measurements	Yes	
Enzymes assays	Yes	
Luminescence immunoassay	Yes	
DNA probe assays	Yes	
Cellular luminescence	No	
Redox luminescence	No	
RNA, protein blots or gels	Yes	
16. LAS-1000 (Fuji Photo Film USA)		
Sample type	2D, 3D gels, blots, membranes	Fuji Cooled (–30°C) CCD camera at 1.3 M pixels; used for luminescent, fluorescent analysis; used with intelligent light box
Luminescence type	Glow, fluorescence	Hardware PC/Mac compatible; operating system Windows 95
Reporter gene assay (RGA)	Yes	
Dual-label RGA	Yes	
ATP measurements	Yes	
Enzymes assays	Yes	
Luminescence immunoassay	Yes	
DNA probe assays	Yes	
Cellular luminescence	No	
Redox luminescence	No	
Northern, Western, Southern blots	Yes	

Instrument Assay		General Characteristics
17. ImageStation 44CF (Kodak Digital Science)		
Sample type	2D, 3D gels, blots, membranes	Cooled CCD (756 × 582 pixels) imaging sensing device; 6× zoom lens; closed optical path image; simple chamber design with few parts
Luminescence type	Glow, fluorescence	1.3 million pixel CCD chip at −30°C by electronic cooling; linearity –4 orders; used with intelligent light box
Reporter gene assay (RGA)	Yes	Software for PC/Mac Windows 95
Dual-label RGA	Yes	Kodak Digital Science Software ID Analysis
ATP measurements	Yes	
Enzymes assays	Yes	
Luminescence immunoassay	Yes	
DNA probe assays	Yes	
Cellular luminescence	No	
Redox luminescence	No	
Gels and blots—Western, Southern, Northern	Yes	
18. MicroMax 1300 (Roper Scientific)		
Sample type	2D, 3D gels, blots, luminescent plants	Liquid nitrogen–cooled CCD camera, 512 × 512 pixels with 13 × 13 μm individual pixel spatial resolution; one side of chip has pixels and other side has silver mask; after exposed region collects signal, a fast electronic shutter shifts the charge to an unexposed region; as the second exposure begins, the camera reads out the first, which allows user to collect data quickly and continuously; this cooled backlit camera is 90% efficient; state-of-the-art imaging with very low background
Luminescence type	Glow and fluorescence	Measure plants in several petri dishes simultaneously
Reporter gene assay (RGA)	Yes	Outstanding luminescence measurements
Dual-label RGA	Yes	
ATP measurements	Yes	A high-performance, back-illuminated CCD device for low-light applications; combines high quantum efficiency across the whole visible spectrum, low readout noise and low binning noise; available in a thermoelectric or liquid nitrogen–cooled format; has an imaging array of 1340 × 1300 and 20 × 20 μm pixels; a large-format back-illuminated camera applicable to low-light condition
Enzymes assays	Yes	
Luminescence immunoassay	Yes	
DNA probe assays	Yes	
Cellular luminescence	Yes	
Redox luminescence	Yes	
Gels, blots—Western, Southern, Northern	Yes	
19. Origen® Analyzer (Igen International, Inc.)		
Sample type		
Luminescence type	Electroluminescence	Electroluminescence analyzer; uses a ruthenium metal chelate in presence of tripropylamine, which produces luminescence that is activated by a change in electrical potential; more than 40 assays available for the system

Instrument Assay		General Characteristics
Reporter gene assay (RGA)	Yes	Includes sandwich, direct, and competitive assays to be used in immunoassays, nucleic acid assays, receptor ligand binding and measurements of bacteria, and viruses; tiny beads (paramagnetic) act as a solid phase enabling fast reaction kinetics to occur; the bead complex is channeled through a flow cell and captured at the surface of an electrode by magnetic field; when voltage is applied, luminescence occurs and is assayed; both clinical and veterinary assays are available
Dual-label RGA	Yes	
ATP measurements	—	
Enzymes assays	Yes	
Luminescence immunoassay	Yes	
DNA probe assays	Yes	
Cellular luminescence	No	
Redox luminescence	No	
DNA, RNA, protein assays	Yes	
20. Architect i2000 (Abbott Diagnostics)		
Sample type	Glow	A completely automated immunoassay instrument for clinical diagnostics
Luminescence type	Acridinium-derivative luminescence	Uses Chemiflex luminesence detection technology based on an acridinium luminescent rection (see Chapter 5)
Reporter gene assay (RGA)	Yes	
Dual-label RGA		Sample processing is immediate, random, or continuous
ATP measurements	No	250 assays are done per minimodule and 250/multimodule; extensive use of bar codes is made for sample identity; tracking of over 50,000 samples on computer; the first assay is done in 28 min and 200 tests per module can be done per hour
Enzymes assays	Yes	
Luminescence immunoassay	Yes	
DNA probe assays	Yes	
Cellular luminescence	No	
Redox luminescence	No	
21. Lumicount (Packard Instruments)		
Sample type	Variety of plates 6–384 wells	Microplate luminometer with state-of-the-art electronics
Luminescence type	Glow	Maximum counts = 140,000
Reporter gene assay (RGA)	Yes	Can do fairly high throughput assays with reactions that produce a steady glow
Dual-label RGA	Yes	
ATP measurements	Glow type	Five-decade dynamic range; read time per well can be from 0.1 to 25.5 s
Enzymes assays	Yes	When instrument is set for 0.5 s read, it takes 60 s to scan plate
Luminescence immunoassay	Yes	This instrument has provisions for shaking and heating from 8°C above ambient to 45°C
DNA probe assays	Yes	Uses Window-based software; data can be exported to a variety of file formats, e.g., Excel and Lotus; a comprehensive data analysis with curve-fitting and SMART™ spline, etc.
Cellular luminescence	No	
Redox luminescence	No	
BRET assays	No	

Instrument Assay		General Characteristics
	22. TopCount® NXT™ (Packard Instruments)	
Sample type	24-, 96-, 384-well plates	Automated luminescence and scintillation counter using stacked plates; single photon counting or time-resolved scintillation counting; can do 15,360 assays unattended and can do scintillation counting without added cocktail using Flashplates; can read 50,000 assays per day; TR-LSC technology provides maximum accuracy; light does not travel between wells; back reflected within each well
Luminescence type	Glow	Built-in IBM-compatible computer/Pentium II
Reporter gene assay (RGA)	Yes	Stacking system can use 20 of the 384- or 96-well plates or 15 of the 24-well plates
Dual-label RGA	Yes	The stacker is automatic for loading, counting, and repeat cycle
ATP measurements	Glow	Temperature control from 19 to 35°C
Enzymes assays	Yes	Hologram ODBC built-in database
Luminescence immunoassay	Yes	Automatic bar code reader for sample identification
DNA probe assays	Yes	Used with LucLite™ and LucLite™ Plus glow luciferase reporter gene assays
Cellular luminescence	Yes	Used with CytoLite luminescent cell proliferation assay or cytotoxicity assay
Redox luminescence	Yes	Used with ATPLite™ M for ATP measurements without injectors
BRET assays	No	The TopCount system can have as many as six detectors that can assay six different wells simultaneously; does cell proliferation, receptor binding, and proximity assays, constant quanta luminescence, and LucLite gene expression assays, etc.; can be combined with multiprobe automated liquid handling system; set up for single-button, walk-away operation
	23. Fusion™ HT (Packard Instruments)	
Sample type	Microplates from 6–1536 wells for fluorescence; 6, 12, 24, 48, 96, 384 wells for luminescence	Universal microplate analyzer measures top or bottom luminescence, fluorescence (intensity or time-resolved or fluorescence energy transfer) or absorbance; nine read modes deliver unmatched versatility; external 40-plate stacker; uses a quartz-halogen or flash xenon light source for 240–700 nm wavelength coverage; can be used in high-throughput or slower-throughput mode by setting the time to read sample
Luminescence type	Glow or flash	Measures FRET (fluorescence energy transfer), TRF (time-resolved fluorescence), HTRF (homogeneous time-resolved fluorescence), enhanced flash luminescence, glow luminescence
Reporter gene assay (RGA)	Yes	Charge integration circuitry extends linear dynamic range; light source normalization with photodiode monitors light source fluctuation
Dual-label RGA	Yes	Temperature controlled from ambient 5°C to ambient 20°C
ATP measurements	Glow	Bar code reader for sample identification
Enzymes assays	Yes	Real-time display function
Luminescence immunoassay	Yes	Data export feature
DNA probe assays	Yes	On-screen sample mapping
Cellular luminescence	Yes	HTRF reagents, LucLite™ and LucLite plus™, ATP-Lite™ M, CytoLite luminescent cell proliferation assay
Redox luminescence	Yes	Not used for BRET analysis (bioluminesence resonance energy transfer)

Instrument Assay		General Characteristics
Multimode BRET™ technology	Yes	BRET technology for whole-cell functional assays; proximity interaction based on renilla luciferase and green fluorescent protein large assembly of vectors for easy assay assembly; Deep Blue C™ new coelenterazine for a larger Stokes shift (large difference between excitation and emission wavelengths); BRET technology is used to measure β arrestin functional ligand binding for orphan receptor screening, target validation, and high-throughput screening; also measures receptor–dimer interaction; homodimer–heterodimer in membranes and whole cells
24. Fusion™ αHT (Packard Instruments)		
Sample type	Flash or glow assay, reads two different wavelengths; 6-, 12-, 24-, 48-, 96-, 384-well microplates	A universal microplate reader, which is similar to the Fusion HT but with the added feature of doing Alpha Screen technology, which uses labeled acceptor and donor beads for highly sensitive homogeneous assays that are easily miniaturized
Luminescence type	Yes	Monitors two luminescent wavelengths simultaneously and therefore can be used for Alpha Screen technology
Reporter gene assay (RGA)	Yes	There is also a Fusion α™ instrument that has the multiple modes of assay but has more optional features than the alpha high-throughput system discussed above, which has all modes of assay as standard; instrument can be used for Alpha Screen technology without modification
Dual-label RGA	Yes	
ATP measurements	Yes	
Enzymes assays	Yes	
Luminescence immunoassay	Yes	
DNA probe assays	Yes	
Cellular luminescence	Yes	
Redox luminescence	Yes	
BRET and ALPHA assays	Yes	
25. DML 2000™ System (Digene)		
Sample type	This is a PM tube–based signal-amplified microplate luminometer	The purpose of this instrument is to measure viral RNA or DNA; designed to provide walk-away simplicity; custom configured using Windows 95 platform in a variety of languages
	These assays can also be tube-based Glow assay	Assays are human papilloma virus, *Chlamydia* trachomatis and *Neisseria* gonorrheae, cytomegalovirus hepatitis type B, and human immunodeficiency virus Herpes simplex will be available; Hybrid Capture technology is a DNA or RNA signal amplification assay using a luminescence end point
Luminescence type	Glow	
Reporter gene assay (RGA)	Yes	When these assays are tube based a standard luminometer should be sufficient
Dual-label RGA	Yes	
ATP measurements	Yes	See explanation for Hybrid technology in Chapter 9
Enzymes assays	Yes	
Luminescence immunoassay	Yes	
DNA probe assays	Yes	
Cellular luminescence	Yes	
Redox luminescence	Yes	
BRET assays	No	

Instrument Assay		General Characteristics
26. TD-20/20 (Turner Designs)		
Sample type	Tube type, multiple sizes from microfuge tubes to scintillation vials	Updated model of original Model 2E; uses a ten-stage photomultiplier but assay current output not photon discrimination Rugged instrument but not as sensitive to low light levels as a photon counter; comes with autoinjector for flash assays, which is user independent; can be used as a stand-alone instrument with a printer or connected to computer via an RS-232 connection
Luminescence type	Flash or glow	
Reporter gene assay (RGA)	Yes	
Dual-label RGA	Yes	
ATP measurements	Yes	
Enzymes assays	Yes	
Luminescence immunoassay	Yes	
DNA probe assays	Yes	
Cellular luminescence	No	
Redox luminescence	Yes	
BRET assays	No	
27. Reporter™ Microplate Luminometer (Turner Designs)		
Sample type	96 well	This is an unusually small instrument used for glow luminescence assays; spectral range is 300–650; linked to a Windows 95-based computer, which controls its machine functions; no temperature control (warming) for this unit; runs at ambient (room) temperature
Luminescence type	Glow	
Reporter gene assay (RGA)	Yes	
Dual-label RGA	Yes	
ATP measurements	Yes, glow	
Enzymes assays	Yes	
Luminescence immunoassay	Yes	
DNA probe assays	Yes	
Cellular luminescence	No	
Redox luminescence	Yes	
BRET assays	No	
28. LS 55 (PerkinElmer)		
Sample type	1-cm cells or tubes	A luminescence spectrometer, which can measure the wavelengths of a luminescence, fluorescence, or phosphorescence; has a xenon lamp for excitation purposes, which is not needed when doing luminescence measurements; temperature controlled; uses FL WinLab™ software for controlling the instrument
Luminescence type	Glow	
Reporter gene assay (RGA)	Yes	
Dual-label RGA	Yes	
ATP measurements	Yes	
Enzymes assays	Yes	
Luminescence immunoassay	Yes	
DNA probe assays	Yes	
Cellular luminescence	Yes	
Redox luminescence	Yes	
BRET assays	Yes	

Instrument Assay		General Characteristics
	29. GENios Multi-Label Reader (Tecan)	
Sample type	6–384 plate reader	Multipurpose instrument for absorbance, fluorescence, and luminescence; can measure absorbance from 230–1000 nm; uses Excel (Xfluor) or Magellan software; reads luminescence from 400–700 nm; fluorescence assay is from 340–700 nm; can read wells from top or bottom; has temperature control and warms to 42°C; equipped with shaking and appropriate filters for fluorescence
Luminescence type	Glow	
Reporter gene assay (RGA)	Yes	
Dual-label RGA	Yes	
ATP measurements	Yes	
Enzymes assays	Yes	
Luminescence immunoassay	Yes	
DNA probe assays	Yes	
Cellular luminescence	Yes	
Redox luminescence	Yes	
BRET assays	Yes	
	30. Lumistar Galaxy (BMG Technologies)	
Sample type	Multiple plate formats, 384 wells	Microplate luminometer with photon counting and extended dynamic range, three injectors access a well simultaneously; fast reading 23 s for 96 wells and 66 s for 384 wells; top and bottom reading; incorporated into major robotic systems via DDE interface and fast loading transport system; incubation up to 60°C. Windows-based software using Excel spreadsheet for data reduction; Galaxy software is an option; raw data accessed through pull-down menus
Luminescence type	Flash and glow	
Reporter gene assay (RGA)	Yes	
Dual-label RGA	Yes	
ATP measurements	Yes	
Enzymes assays	Yes	
Luminescence immunoassay	Yes	
DNA probe assays	Yes	
Cellular luminescence	Yes	
Redox luminescence	Yes	
BRET assays	No	
	31. FLUOstar Galaxy (BMG Technologies)	
Sample type	Multiple plate formats, 384 wells	Similar to Lumistar but with added fluorescence and absorbance capability; has dual excitation and dual emission capability and multicolor detection and automatic gain adjustment; can be linked with robotic capability and can do FRET, BRET, and Alpha assay (see LUMIstar for other comments); real-time graphics; a very versatile instrument
Luminescence type	Flash and glow	
Reporter gene assay (RGA)	Yes	
Dual-label RGA	Yes	
ATP measurements	Yes	
Enzymes assays	Yes	
Luminescence immunoassay	Yes	
DNA probe assays	Yes	
Cellular luminescence	Yes	

Instrument Assay		General Characteristics
Redox luminescence	Yes	
BRET assays	Yes	
32. PSQ™96 (Pyrosequencing AB)		
Sample type	384-well plates	Luminescence plate reader built for pyrosequencing DNA; system was originally developed to sequence single-stranded DNA but has been developed to sequence double-stranded DNA as well
Luminescence type	Glow	Unit can be linked to robotics and score up to 100,000 SNP (single nucleotide polymorphisms); there are no gels so that sequencing is more time efficient (see Chapter 17)
Reporter gene assay (RGA)	No	Comes complete with kits to do the luciferase assay used in pyrosequencing and software is built into analyze the relevant seqencing data
Dual-label RGA	No	
ATP measurements	Yes	
Enzymes assays	No	
Luminescence immunoassay	No	
DNA probe assays	Yes	
Cellular luminescence	No	
Redox luminescence	No	
BRET assays	No	
33. Victor 2™ (Wallac)		
Sample type	1536-well plates	Plate-type instrument that measures absorbance, fluorescence (all types including polarization), and luminescence; instrument can handle 60 stacked plates robotically, which makes it valuable for drug discovery
Luminescence type	Glow	Uses Windows 95 or 98 programming; all models come with scanning, shaking, and kinetic modules; preprogrammed modules for LANCE™ homogeneous assay applications; reagent dispensers and heating are available; this is a compact unit and features an amazing versatility in assay types
Reporter gene assay (RGA)	Yes	
Dual-label RGA	Yes	
ATP measurements	Yes	
Enzymes assays	Yes	
Luminescence immunoassay	Yes	
DNA probe assays	Yes	
Cellular luminescence	Yes	
Redox luminescence	Yes	
BRET assays	Yes	
34. Fluorosc Ascent FL™ (Thermo Labsystems)		
Sample type	1–864 well plates	Plate luminometer with fluorescence assay as well; direct fiberless optics from above or below wells; temperature control and orbital shaking; three injectors are available; unit is driven by Ascent software
Luminescence type	Flash or glow	
Reporter gene assay (RGA)	Yes	
Dual-label RGA	Yes	
ATP measurements	Yes	
Enzymes assays	Yes	
Luminescence immunoassay	Yes	
DNA probe assays	Yes	

Instrument Assay		General Characteristics
Cellular luminescence	Yes	
Redox luminescence	Yes	
BRET assays	Yes	
35. NorthStar™ HTS Workstation (Applied Biosystems–PerkinElmer)		
Sample type	96–1536 wells	High-throughput CCD luminometer used for fast drug research for chemical libraries of compounds; can run 500,000 assays per day and can handle a variety of different cell-based and molecular targets; the user saves time and money with such an efficient screening process; a complete integrated system for a variety of different assays; particularly set up for miniaturization of assays to save liquid handling and excessive reagent costs; system allows walk-away robotic automation; can be used with both flash and glow luminescence; CCD is linked to a camera lens system with an optical collimator with integrated Fresnel field lens, an arrangement that produces an excellent signal-to-noise ratio for assays (see Chapter 44)
Luminescence type	Flash or glow	System has bar code identification to match samples and data files; since Applied Biosystems has pioneered dioxetane luminescent chemistry, the NorthStar system features is luminescence chemistries and adaptations with other systems as well, e.g., Xpress-Screen technology for MRNA signal amplification or ICAST™ protein–protein interaction technology; in addition Galacto-Light™, GL-plus, Gal-Screen™, or dual light reporter gene assays can be done on this system; Luc Screen™ and Phospha-Light™ reporter gene analysis can be used as well; even cAMP-Screen, which is a luminescence assay for cyclic AMP, can be used for screening; system can measure aequorin assays as it has injection capabilities
Reporter gene assay (RGA)	Yes	
Dual-label RGA	Yes	
ATP measurements	Yes	
Enzymes assays	Yes	
Luminescence immunoassay	Yes	
DNA probe assays	Yes	
Cellular luminescence	Yes	
Redox luminescence	Yes	
BRET assays	No	
36. GeneGnome™ (Syngene)		
Sample type	Gels or arrays	Automated imaging station for luminescence detection for Western, Northern, and Southern blots; CCD camera distinguishes 65,356 shades of gray
Luminescence type	Glow	
Reporter gene assay (RGA)	Yes	
Dual-label RGA	Yes	
ATP measurements	Yes	
Enzymes assays	Yes	
Luminescence immunoassay	Yes	
DNA probe assays	Yes	
Cellular luminescence	No	
Redox luminescence	No	
BRET assays	No	

APPENDIX B

Survey of Commercially Available Instruments for Luminescence

Christopher Van Dyke

The following table details available instrumentation for luminescence known by the author at press time. Because of the nature of the marketplace and advances in technology, this information is subject to change.

Unless otherwise stated, addresses and telephone numbers are assumed to be in the United States. All prices are in U.S. dollars.

Company and Address Telephone Number	Instrument (Price) Detector	Web Site E-mail Address	Functions and Features
Advanced American Biotechnology & Imaging 1166 E. Valencia Dr., #6C Fullerton, CA 92831 714-870-0290	Chemi-16 ($9,995) CCD cooled 16 bit, 1.21 M (1260 × 960) pixels	www.aabi.com/Chem16/10site1.htm aab@ix.netcom.com	Scanning
	Core Facility ($12,950) CCD	aabi.com/systems/systems.html aab@ix.netcom.com	Fluorescence, scanning
Alpha Innotech 14743 Catalina St. San Leandro, CA 94577 800-795-5556	FluorChem ($49,000–$55,000) CCD cooled 16 bit, 1.35 M (1312 × 1032) pixels	www.alphainnotech.com/bio/lsmaster2.html info@aicemail.com	Absorbance, fluorescence, scanning
	ChemiImager ($18,000–$24,000) CCD cooled Multiple options including: 12 bit, 1 M pixels; 8 bit, 380 K pixels	www.alphainnotech.com/included/chemi.html info@aicemail.com	Scanning
Amersham Pharmacia Biotech 800 Centennial Ave. Piscataway, NJ 08855 800-526-3593	ImageMaster VDS-CL CCD cooled 480 K (800 × 600) pixels	www.apbiotech.com/stiboasp/showmodule.asp?nmoduleid=163722	Fluorescence, scanning

0-8493-0719-8/02/$0.00+$1.50

Company and Address Telephone Number	Instrument (Price) Detector	Web Site E-mail Address	Functions and Features
	Storm ($46,000–$70,000) Scanning laser, storage phosphor 16 bit, 3.76 M (2150 × 1750) pixels	http://www.apbiotech.com/ stiboasp/showmodule. asp?nmoduleid= 163723	Fluorescence, radioisotope
	Typhoon 8600 ($85,000) Scanning laser, storage phosphor 16 bit, 3.76 M (2150 × 1750) pixels	www.apbiotech.com/ stiboasp/showmodule. asp?nmoduleid= 163721	Fluorescence, radioisotope
Bio-Rad Laboratories 2000 Alfred Nobel Drive Hercules, CA 94547 800-424-6723	Molecular Imager FX System ($64,900) Scanning laser, storage phosphor	www.bio-rad.com/	Fluorescence, radioisotope
	Fluor-S MultiImager System ($29,950) CCD cooled 12 bit, 1.38 M (1340 × 1032) pixels	www.bio-rad.com/	Colormetry, fluorescence, scanning
Bio-Tek Instruments Highland Park, Box 998 Winooski, VT 05404 888-451-5171	FL600 PMT	www.biotek.com/ labcsr@biotek.com	Microtiter plates, absorbance, fluorescence, incubation
	FLx800 PMT	www.biotek.com/ labcsr@biotek.com	Microtiter plates, absorbance, fluorescence, incubation
Bioscan 4590 MacArthur Blvd., NW Washington, D.C., 20007 202-338-0974	LUMI-One	www.bioscan.com/ plumione.htm sales@bioscan.com	Tubes, battery operated
	LUMI-SCINT PMT	www.bioscan.com/ plumiscint.htm sales@bioscan.com	Tubes, battery operated, liquid scintillation
	LUMI/96 PMT	www.bioscan.com/ plumi96.htm sales@bioscan.com	Microtiter plates
BMG Labtechnologies Hanns-Martin-Schleyer-Str. 10 D-77656 Offenburg, Germany	FLUOstar Galaxy PMT (290–700 nm)	www.bmg-labtechnologies. com/fluostar_galaxy. htm usa@bmg-labtechnologies.com	Microtiter plates, sample changer (optional), absorbance, fluorescence, two injectors, incubation
	LUMIstar Galaxy PMT (290–700 nm)	www.bmg-labtechnologies.com/ lumistar_galaxy.htm usa@bmg-labtechnologies.com	Microtiter plates, sample changer (optional), absorbance, fluorescence, three injectors, incubation

Company and Address Telephone Number	Instrument (Price) Detector	Web Site E-mail Address	Functions and Features
Dynex 14340 Sullyfield Circle Chantilly, VA, 20151-1683 800-336-4543	MLX ($15,600–$25,950) PMT	www.dynextechnologies.com/16_lum_shl.htm info@dynextechnologies.com	Microtiter plates, three injectors (optional), incubation
Eastman Kodak Co. (Distributed by NEN NEN Life Science Products, Inc. 549 Albany St. Boston, MA 02118 800-551-2121)	Image Station 440CF ($29,100) CCD cooled 12 bit, 438 K (752 × 582) pixels	www.kodak.com/ (www.NEN.com/products/imgstation/about.htm)	Absorbance, fluorescence, scanning, images from below sample
Fuji Medical Systems U.S.A., Inc. 419 West Avenue Stamford, CT 06902 800-431-1850	LAS-1000 ($48,950) CCD cooled 14 bit, 1.28 M (1384 × 922) pixels	www.fujimed.com/science/LAS1000.html	Absorbance, fluorescence, scanning
Gen-Probe 10210 Genetic Center Dr. San Diego, CA 92121 800-523-5001 ext.8080	Leader 50	www.gen-probe.com/leader50.html	Tube, fixed-volume injector
	Leader 450i	www.gen-probe.com/leader450i.html	—
Labsystems Labsystems Oy Sorvaajankatu 15 P.O. Box 208 FIN-00811 Helsinki, Finland +358-9-329-100	Luminoskan Ascent PMT	www.labsystems.fi/products/luminometric/lumino_index.htm labsystems.oy@thermobio.com labsystems.us@thermobio.com	Microtiter plates, three injectors (one standard), incubation
	Luminoskan FL PMT (185–680 nm)	http://www.labsystems.fi/products/fluorometric/fluoro_fl_index.htm labsystems.oy@thermobio.com labsystems.us@thermobio.com	Microtiter plates, up to three injectors, incubation (optional), sample changer (optional)
	Luminoskan TL, TL Plus PMT	www.labsystems.fi/products/luminometric/lumino_tl_index.htm labsystems.oy@thermobio.com labsystems.us@thermobio.com	Tube, one (TL) or two (TL Plus) manual injectors (optional)
LJL BioSystems 404 Tasman Drive Sunnyvale, CA 94089 888-611-4555	Analyst AD	www.ljlbio.com/products/instruments/analytad.htm sales@ljlbio.com	Microtiter plates, absorbance, fluorescence
	Analyst HT	www.ljlbio.com/products/instruments/analytht.htm sales@ljlbio.com	Microtiter plates, high-throughput, absorbance, fluorescence

Company and Address Telephone Number	Instrument (Price) Detector	Web Site E-mail Address	Functions and Features
	Acquest Ultra-HTS System	www.ljlbio.com/products/instruments/acquest.htm sales@ljlbio.com	Microtiter plates, high-throughput, absorbance, fluorescence
Mirai Bio (Hitachi Genetic Systems) 1201 Harbor Bay Parkway Suite 150 Alameda, CA 94502 800-624-6176	CCDBIO 8 CCD cooled (8 bit)	www.miraibio.com/	Absorbance, fluorescence, scanning
	CCDBIO 16C CCD cooled (16 bit, 1.34 M (1300 × 1030) pixels)	www.miraibio.com/	Fluorescence, scanning
Molecular Devices 1131 Orleans Avenue Sunnyvale, CA 94089 800-635-5577	SPECTRAmax GEMINI XS PMT	www.moleculardevices.com/pages/gemini.html nfo@moldev.com	Microtiter plates, incubation
Nucleotech 1400 Fashion Island Blvd. Suite 510 San Mateo, CA 94404 888-869-4080	NucleoVision 920 CCD cooled 374 K (768 × 494) pixels	www.nucleotech.com/920sheet.html info@nucleotech.com	Absorbance, fluorescence, scanning
Packard 800 Research Parkway Meriden, CT 06450 800-323-1891	Fusion	www.packardinst.com/	Microtiter plates, sample changer, colormetry, fluorescence, incubation
	LumiCount PMT	www.packardinst.com/	Microtiter plates, incubation
	TopCount NXT PMT	www.packardinst.com/	Microtiter plates, sample changer, liquid scintillation
PerkinElmer Life Sciences 761 Main Avenue Norwalk, CT 06859 800-762-400	Berthold MiniLumat LB 9506 PMT (380–630 nm)	www.wallac.fi/catalog/lb9506.htm	Tube, portable, optional battery
	Berthold junior PMT (380–630 nm)	www.wallac.fi/catalog/juntech.htm	Tube, portable, battery operated
	Berthold LB 9507 PMT (380–630 nm)	www.wallac.fi/catalog/lb9507.htm	Tube, up to two volume-variable injectors
	Berthold LB 955 "flash'n'glow" PMT (370–620 nm)	www.wallac.fi/catalog/lb955.htm	Tube, 30 sample changer, up to three injectors
	Berthold LB 953 PMT (370–620 nm)	www.wallac.fi/catalog/lb953.htm	Tube, 180 sample changer, up to three fixed-volume injectors

Company and Address Telephone Number	Instrument (Price) Detector	Web Site E-mail Address	Functions and Features
	Berthold NightOWL ($60,000) CCD cooled 217 K (578 × 375) pixels or backlit 262 K (512 × 512) pixels	lifesciences.perkinelmer.com/products/catalog/products/prod626.asp	Microtiter plates/others, fluorescence
	Wallac ViewLux™ ultraHTS CCD cooled	www.wallac.fi/catalog/vl.htm	Microtiter plates, high-throughput, absorbance, fluorescence
	Wallac 1420 VICTOR^2 V PMT Photodiodes 340–700 nm, optionally up to 850 nm	www.wallac.fi/victor/index.htm	Microtiter plates/others, sample changer (optional), absorbance, fluorescence, up to four injectors (optional), incubation (optional)
	Berthold LB 96V Microlumat Plus PMT	www.wallac.fi/catalog/Lb96v.htm	Microtiter plates, up to three variable-volume injectors, sample temperature control
	Wallac MicroBeta® TriLux PMT	www.wallac.fi/catalog/42.htm	Liquid scintillation counter
	Wallac MicroBeta® Jet PMT	www.wallac.fi/catalog/189.htm	Microtiter plates and other formats, liquid scintillation, up to six injectors (optional, depends on model)
Stratagene 11011 North Torrey Pines Rd. La Jolla, CA 92037 800-894-1304	Eagle Eye® II ($18,995) CCD cooled 374 K (768 × 494) pixels	www.stratagene.com/instruments/eagle_eye.htm	Colormetry, fluorescence, scanning
Sygene 97H Monocacy Blvd. Frederick, MD 21701 877-435-3627	GeneGnome CCD cooled 16 bit, 343 K (694 × 494) pixels	www.syngene.com/genegnomeshort.asp ussales@syngene.com	Fluorescence (option), scanning (option)
	ChemiGenius ($29,995) CCD cooled 16 bit, 343 K (694 × 494) pixels	www.syngene.com/chemigenius.asp ussales@syngene.com	Absorbance, fluorescence, scanning
	ChemiGenius Plus ($41,450) CCD cooled 16 bit, 1.34 M (1300 × 1030) pixels	www.syngene.com/chemishort.asp ussales@syngene.com	Absorbance, fluorescence, scanning
	MultiGenius ($23,950) CCD cooled (8 bit)	www.syngene.com/multigenius.asp ussales@syngene.com	Fluorescence
Tecan P.O. Box 13953 Research Triangle Park, NC 27709 919-361-5208	Ultra 230–850 nm	www.tecan.com/ info@tecan-us.com	Microtiter plates, fluorescence, absorbance

Company and Address Telephone Number	Instrument (Price) Detector	Web Site E-mail Address	Functions and Features
Tropix, Inc. (Applied Biosystems) 47 Wiggins Avenue Bedford, MA 01730 (800) 542-2369	TR717 380–630 nm	www.appliedbiosystems.com/products/productdetail.cfm?id=70 tropix@pebio.com	Microtiter plates, up to two injectors, incubation, high-throughput (optional)
	NorthStar CCD cooled (16 bit)	www.appliedbiosystems.com/products/productdetail.cfm?id=65	Microtiter plates, high-throughput, injectors (optional), temperature control, sample changer (optional)
Turner Designs 845 W. Maude Ave. Sunnyvale, CA 94086 404-749-0994	TD-20/20 PMT (300–650 nm)	www.turnerdesigns.com/instruments/998_2020.htm sales@turnerdesigns.com	Tube or 35-mm culture dishes, temperature control (optional), up to two injectors (optional)
	The Reporter PMT (300–650 nm)	www.turnerdesigns.com/instruments/998_2020.htm sales@turnerdesigns.com	Microtiter plates
UVP, Inc. 2066 W. 11th St. Upland, CA 91786 (800) 452-6788	Biochemi CCD cooled (–40°C) 16 bit, 1.31 M (1280 × 1024) pixels	www.uvp.com/chemi/html/chemi.html	Fluorescence, scanning
	Optichemi CCD cooled (–75°C) 16 bit (14 acq), 1.31 M (1280 × 1024) pixels	www.uvp.com/chemi/html/chemi.html	Fluorescence, scanning
Zylux Corporation 1742 Henry G. Lane St. Maryville, TN 37801 865-379-6016	MPL2 PMT (300–650 nm)	http://www.zylux.com/researchproducts/mpl2_specifications.htm	Fluorescence, scanning
	FB15 (Sirius) PMT (370–630 nm)	www.zylux.com/fb15_specifications.htm Bbloomf@aol.com	Tubes and petri dishes, up to three injectors
	FB12 PMT (370–630 nm)	http://www.zylux.com/researchproducts/fb12_specifications.htm Bbloomf@aol.com	Tubes and petri dishes
	MPL3 (Orion) PMT (300–650 nm)	http://www.zylux.com/researchproducts/mpl3_specifications.htm Bbloomf@aol.com	Microtiter plates
	FB14 (Vega) PMT (300–650 nm)	http://www.zylux.com/researchproducts/fb14_specifications.htm Bbloomf@aol.com	Tube, portable, battery powered

Section II

Light Probes/Clinical Assay

5 Chemiluminescent *N*-Sulfonylacridinium-9-Carboxamides and Their Application in Clinical Assays

Maciej Adamczyk and Phillip G. Mattingly

CONTENTS

5.1 INTRODUCTION

> ***Bi·o·tech·nol·o·gy*** (bi′o tek näl′o je) n. [Gr. < bios, life] [Gr. technologia, systematic treatment: see TECHNIC & LOGY] a set of powerful tools that employ living organisms (or part of organisms) to make or modify products, improve plants or animals, or develop microorganisms for specific uses.
>
> —*National Science and Technology Council, July 1995*

The medical diagnostic industry has been both innovator and recipient of products born of modern biotechnology. With contributions from fields as far ranging as synthetic and physical chemistry, immunology, molecular biology, engineering, material and computer science, the industry has delivered powerful tools to diagnose rapidly, and economically, a variety of maladies and to monitor therapies to combat those conditions. The use of immunoassay, and most recently, nucleic acid hybridization assay technologies has been crucial to the *specificity* of these tools, while luminescent detection has provided the required *sensitivity* and *economy*. This chapter details the state of the art of medical diagnostics technology at Abbott Laboratories (Abbott Park, IL), with particular emphasis on the development of a new class of luminescent labels: the *N*-sulfonylacridinium-9-carboxamides and their use on the Abbott Laboratories Prism® and Architect™ platforms.

0-8493-0719-8/02/$0.00+$1.50

The current level of sophistication has taken years of research and development. Along the way, detection technologies have evolved considerably. The genesis of the industry may be traced to Yalow and Berson's demonstration of radioimmunoassay (RIA)[1] for the determination of insulin in human plasma. As the name of the technique indicates, the radioisotopes were used for detecting the specific binding interaction between an antibody and the corresponding antigen. The assay consisted of a competition between radiolabeled antigen and unlabeled antigen for the binding sites on the specific antibody. The antigen–antibody complex was separated from unbound antigen, and the amount of radiolabeled antigen present was determined by scintillation counting. The unknown concentration of an antigen in a patient sample could be determined by comparison with a standard curve constructed from known concentrations of the antigen. Thus, an industry was born.

As an adjunct to its radiopharmaceutical business, Abbott Laboratories was among the first to commercialize radioimmunoassays (RIAs) with the introduction of AusRIA®, a test for serum hepatitis, in 1972.[2] A year later, the company formed the Abbott Diagnostics Division (ADD) and has been an industry leader ever since. There were advantages to this initially marketed biotechnology: radioisotope detection provided the gold standard for assay sensitivity; the labeled antigen and unlabeled antigen may only differ isotopically, so there is no perturbation of binding with the antibody. For the end user and manufacturer, these advantages were largely outweighed by the inherent disadvantages of handling radioisotopes (i.e., contamination, disposal, shelf life, and conjugate stability).

At Abbott, and elsewhere in the industry, a new wave of research to replace RIA with friendlier detection technologies was initiated. By 1979, analyzers based on enzyme immunoassays (EIA) were being marketed (e.g., the Abbott Laboratories Quantum®). The format of the Quantum assays was heterogeneous, employing a quarter-inch polystyrene bead solid phase. The system had a low degree of automation, and the horseradish peroxidase (HRPO) enzyme label provided a simple colorimetric readout.

Various luminescence detection technologies—fluorescence, fluorescence polarization (FPIA), time-resolved fluorescence (TRF), and chemiluminescence—and automation improvements were also pursued. The early 1980s saw the introduction of analyzers like the Abbott Laboratories TDx (automated FPIA), IMx (automated, microparticle-enhanced EIA), and Vision (automated analyzer for the doctor's office).

The late 1980s brought a dramatic shift in the medical diagnostics industry and the biotechnology products that it provided. Managed care directed the market away from widely distributed testing (i.e., physician's office, small hospital, and clinical laboratories) to large, centralized facilities. Low cost and high throughput became more critical requirements to satisfy. At the same time, advances in other areas of medicine brought requirements for greater assay sensitivity.[3–6]

Chemiluminescence, as the most sensitive non-isotopic detection technology for diagnostic assays and one of the most easily automated, seemed to be a prime candidate for development. Within that technology, several options were available. Chemists at Miles Laboratories[7] had demonstrated the utility of the phthalhydrazide class of chemiluminescent compounds (luminol, isoluminol, and their analogues) in protein binding assays for both low-molecular-weight analytes and for macromolecules. Even though the sensitivity of the assays was very good, the low aqueous solubility of the compounds was a major drawback. The necessity of using a peroxidase enzyme to catalyze the release of light from these compounds made the assays no less complicated to run than other EIAs.

Chemiluminescent acridinium salts (**1**, in Figure 5.1) were first alluded to as a better alternative to the phthalhydrazide labels by Woodhead et al.[8] and later fully disclosed in a report that used phenyl N^{10}-methyl-acridinium-9-carboxylate fluorosulfate (**1**, R = CH_3, X = OAr) in an immunoassay for α-fetoprotein.[9]

As shown in Figure 5.1, the chemiluminescent reaction of acridinium salts proceeds by the addition of hydrogen peroxide anion to the electron-deficient 9-position of the acridinium nucleus (**2**) subsequently forming a tetrahedral dioxetane intermediate (**4**). Recent calculations[10] have indicated

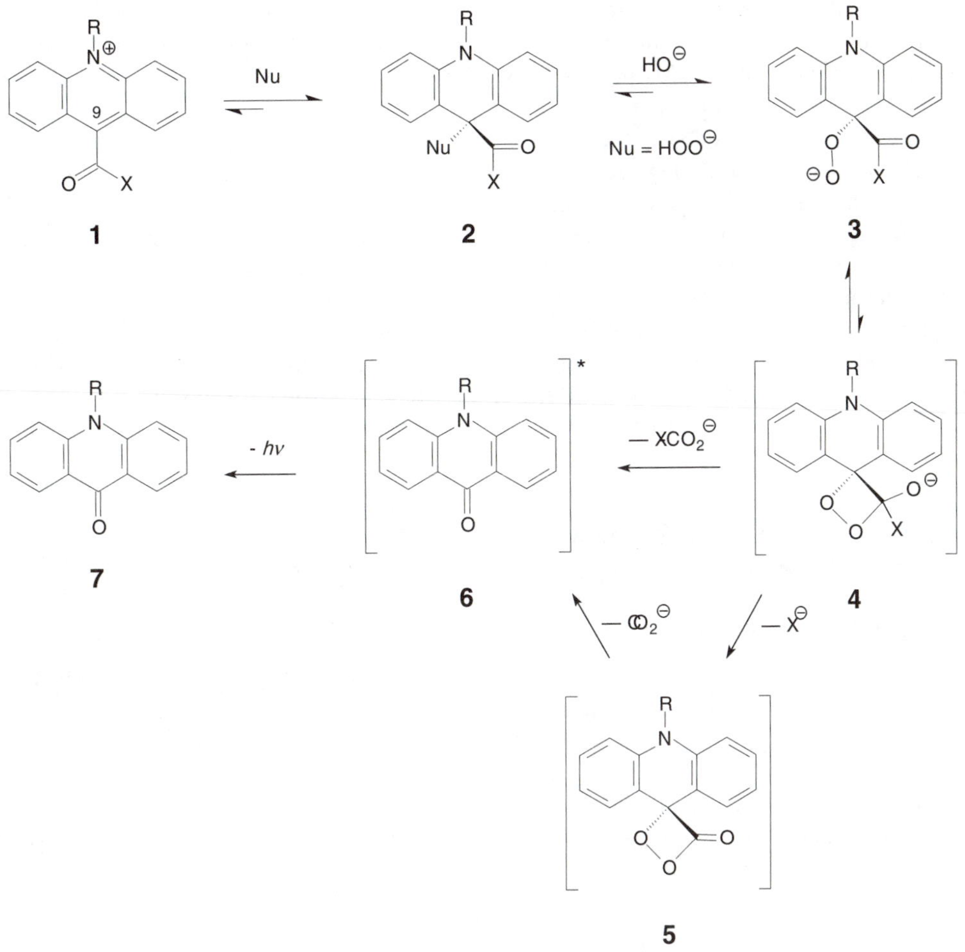

FIGURE 5.1 Proposed reaction mechanism of acridinium chemiluminescent salts. (From Adamczyk, M. et al., *Tetrahedron,* 55, 10, 899, 1999. With permission.)

that **4** should give rise to the excited acridone **6** directly, without the intermediacy of the dioxetanone **5**.[9] On return to the ground state, **6** emits light that can be measured. The advantages over the aryl hydrazides included higher quantum yield, better aqueous solubility, and no requirement for an enzymatic catalyst.

To those who examined the technology more closely, however, it soon became apparent that the original acridinium phenyl esters were not suitable for commercialization because of their inherent instability[11] on storage in the aqueous buffers used for immunoassays. Their use was further complicated by the facile formation of an inactive pseudobase form (**2**, Nu = OH).[9] These were disappointing discoveries, but the research continued with the goal of rationally designing an acridinium chemiluminophore that circumvented these limitations.

The fact that modifying the X-leaving group could have a great effect on the chemiluminescence of acridinium salts was well known. The original report of Rauhut et al.[12] demonstrated the high chemiluminescence efficiency of 9-chlorocarbonyl-10-methyl-acridinium salts (**1**, X = Cl). Later studies illustrated that any acridinium-9-carboxylate derivative with an X-leaving group (**1**, X = OAr or OAlk) that had a pK_a lower than that of hydrogen peroxide (~12) efficiently produced light.[13,14] The trend indicated that the efficiency rose as the pK_a decreased, i.e., as X became a better leaving group. Conversely, derivatives with even poorer leaving groups, e.g., carboxamides (**1**, X = NH_2)

were considered nonchemiluminescent.[15] Preparation of a *more stable* acridinium chemiluminescent label by modifying the leaving group ability of X seemed to be at odds with having an *efficient* label.

One approach to solving this problem has been to increase the steric bulk on the phenolate leaving group and thus slow the detrimental hydrolysis side reaction that occurs in aqueous buffers.[16] The answer we,[17] and later others,[18,19] arrived at was to use a sulfonyl-activated amide-leaving group in place of the phenolate-leaving group. Contrary to the earlier literature reports,[15] amides were suitable leaving groups if properly activated.

The *N*-sulfonylacridinium-9-carboxamides [**1**, X = $N(R')SO_2R''$] have been successfully introduced commercially on the Abbott Laboratories Prism and Architect chemiluminescent immunoassay platforms, satisfying the initial goals. As a class, these chemiluminescent acridinium labels have proved to be very versatile. The peripheral substituents (**1**, R, R′, and R″) are easily manipulated to control the kinetics of the light output,[20,21] modify the solubility properties of the label,[22,23] or introduce a variety of functional groups useful for bioconjugation.[24] The following discussion focuses on the preparation and use of *N*-sulfonylacridinium-9-carboxamides in clinically useful assays.

5.2 PREPARATION AND CHARACTERIZATION OF *N*-SULFONYLACRIDINIUM-9-CARBOXAMIDES

5.2.1 THE N^{10}-METHYL SERIES

At the outset in the development of the *N*-sulfonylacridinium-9-carboxamide labels, the proper substituents on the sulfonamide-leaving group had to be chosen to arrive at a species that demonstrated both shelf stability and the kinetics suitable for high-throughput diagnostic testing.

A series of compounds were prepared according to Scheme 5.1.[21] The potassium salt of sulfonamide **10** was acylated with 9-chlorocarbonyl-acridinium hydrochloride[12] to give the *N*-sulfonylacridine-9-carboxamide (**11**). Unlike the aryl acridine-9-carboxylate esters, the *N*-sulfonylcarboxamides

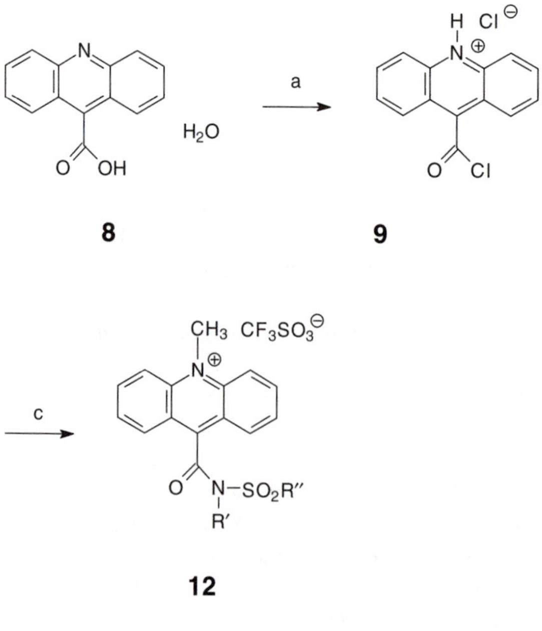

SCHEME 5.1 (a) $SOCl_2$, reflux, 4 h;[12] (b) $R'NHSO_2R''$ (**10**), KO-*t*-Bu or KH, THF, 0°C to room temperature; (c) $CF_3SO_3CH_3$, CH_2Cl_2, or C_6H_6, 0°C to room temperature. (From Mattingly, P. G., *J. Biolumin. Chemilumin.*, 6, 107, 1991. With permission.)

TABLE 5.1
N^{10}-Methyl-*N*-Sulfonylacridinium-9-Carboxamide Salts Ranked by Chemiluminescence Lifetime

Compound 12	R′	R″
a	C_6H_6	CF_3
b	C_6H_6	$o\text{-}NO_2C_6H_4$
c	C_6H_6	$p\text{-}BrC_6H_4$
d	$i\text{-}C_3H_7$	CF_3
e	$n\text{-}C_4H_9$	$p\text{-}NO_2C_6H_4$
f	$i\text{-}C_3H_7$	$o\text{-}NO_2C_6H_4$
g	C_6H_6	$p\text{-}CH_3C_6H_4$
h	$n\text{-}C_4H_9$	$o\text{-}NO_2C_6H_4$
i	$n\text{-}C_4H_9$	$o,p\text{-}(NO_2)_2C_6H_4$
j	$n\text{-}C_4H_9$	$p\text{-}BrC_6H_4$
k	$i\text{-}C_3H_7$	$p\text{-}BrC_6H_4$
l	$n\text{-}C_4H_9$	$p\text{-}CH_3C_6H_4$
m	$i\text{-}C_3H_7$	$p\text{-}CH_3C_6H_4$
n	$n\text{-}C_4H_9$	$2,4,6\text{-}(i\text{-}C_3H_7)_3C_6H_4$
o	$n\text{-}C_4H_9$	$2,4,6\text{-}(CH_3)_3C_6H_4$

Source: Mattingly, P. G., *J. Biolumin. Chemilumin.*, 6, 107, 1991. With permission.

existed as a mixture of rotamers with a coalescence temperature of 120°C on the ^{1}H NMR time scale. The mixed rotamers were then methylated using methyl triflate to give the desired *N*-sulfonylacridinium-9-carboxamides **12a–o** (see Table 5.1). In most cases, the product precipitated from the reaction mixture and required no further purification. On evaluation of the chemiluminescent properties of the compounds, it was noted that there was little difference in light yield through the series. There was a noticeable difference in the kinetics of the light generation (Table 5.1), however. The duration of chemiluminescence ranged from less than 1 s for compound **12a** to greater than 50 s for compound **12o**. Compound **12l** (R′ = *n*-butyl, R″ = $p\text{-}CH_3C_6H_4$) displayed a chemiluminescence emission for 6 s and had excellent stability in aqueous buffer. This combination of substituents was judged to be optimal for further development into a labeling reagent for use in clinical assays.

Scheme 5.2 depicts the preparation of a series of N^{10}-methyl-*N*-tosyl-acridinium-9-carboxamides **14a–g** bearing alkylcarboxy arms for linking the acridinium label to analytes of clinical interest. The preparation of **14c** was typical:[22]

To a 500-mL oven-dried three-neck round-bottomed flask was added potassium hydride dispersion (9.1 g, 35 wt%, 79.4 mmol) and the hydride was washed with hexanes (45 ml × 2), then was replaced by anhydrous THF (100 ml). The mixture was cooled in an ice bath. To the potassium hydride suspension was added methyl *N*-tosyl-4-aminobutyrate[23] (12.3 g, 45.3 mmol) in anhydrous THF (50 ml) over 30 to 45 min. The internal temperature was monitored using an immersed thermometer during the addition, and the temperature was kept between 2 and 3°C. The reaction mixture was stirred for additional 45 min under nitrogen. Acridine-9-carbonylchloride hydrochloride **9** (11.0 g, 39.6 mmol) was added in one portion with a rinse of anhydrous THF (50 ml). The ice bath was removed and the reaction mixture was stirred at room temperature for 5 h under nitrogen. The reaction mixture was diluted with ethyl acetate (200 ml), filtered through Celite (50 g) and the Celite was washed with ethyl acetate (200 ml × 3). The combined filtrate was washed with brine (200 ml × 2), dried over anhydrous sodium sulfate, filtered, and evaporated *in vacuo*. The crude material was separated by column chromatography (600 g silica

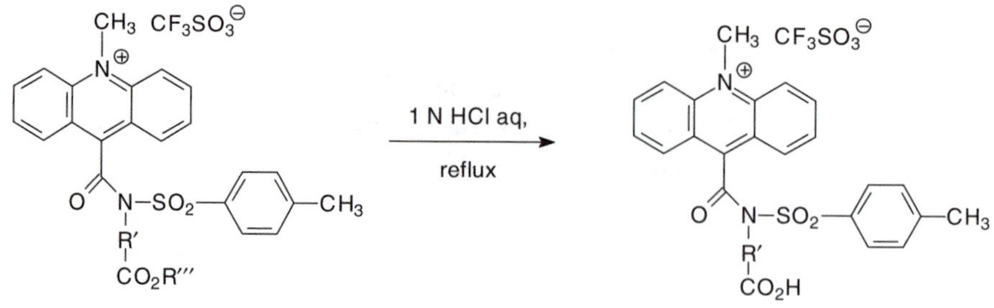

	R′	R‴
a	$-(CH_2)_2-$	$-CH_3$
b	$-(CH_2)_2-$	$-C(CH_3)_3$
c	$-(CH_2)_3-$	$-CH_3$
d	$-CH_2(4\text{-}C_6H_{10})-$	$-CH_3$
e	$-(CH_2)_4-$	$-CH_2C_6H_5$
f	$-(CH_2)_5-$	$-CH_2C_6H_5$
g	$-(4\text{-}C_6H_4)CH_2-$	$-CH_2C_6H_5$

SCHEME 5.2

gel, 30% ethyl acetate/hexanes). Fractions were combined and evaporated *in vacuo* to afford 13.71 g (73%) of pure product **11** (R′ = $-(CH_2)_3CO_2CH_3$, R″ = $p\text{-}C_6H_4CH_3$).

Compound **11** (R′ = $-(CH_2)_3CO_2CH_3$, R″ = $-p\text{-}C_6H_4CH_3$) (1.9 g, 4 mmol) was dissolved in dry dichloromethane (20 ml), then stirred with methyl trifluoromethane sulfonate (1 ml) under nitrogen overnight. The solution was then concentrated *in vacuo* and purified by flash chromatography on silica gel to remove traces of unquarternized material. The resulting methyl ester **13c** was heated at reflux in 1 *N* aq HCl for 2 h. After cooling and then lyophilization of the aq solution, **14c** was obtained in 89% yield.

It was generally found that some ester exchange (R‴) occurred during the methylation reaction with methyl triflate, even when a *t*-butyl ester was used (**13c**). Since all the esters hydrolyzed under the aq HCl conditions, this was not limiting and compounds **14a–g** were all efficiently prepared. All of the compounds showed good aqueous solubility and stability. However, lengthening the alkyl linker (R′) did have an effect on the hydrophilicity of the compounds. The longer the alkyl chain, the less hydrophilic was the label. These differences, although subtle, were enough to encourage us to explore the preparation of acridinium labels with even more hydrophilic character.

5.2.2 Sulfopropylated Acridinium-9-Carboxamide Salts

Sulfopropylation has long been a means of increasing the hydrophilicity of surfaces,[25,26] polymers,[27,28] dyes,[29] and even some proteins.[30,31] The most straightforward way of introducing the sulfopropyl group into a chemiluminescent *N*-sulfonylacridinium-9-carboxamide is shown in Scheme 5.3 and detailed below:[22,23]

A mixture of compound **11** (R′ = $-(CH_2)_3CO_2CH_3$, R″ = $-p\text{-}C_6H_4CH_3$) (9.0 g, 18.9 mmol) and 1,3-propane sultone (25.0 g, 205 mmol) in a 250 ml round-bottomed flask covered with aluminum foil was heated at 125°C in an oil bath for 5 h under nitrogen. The reaction mixture was then cooled to room temperature, dissolved in a small amount of methanol, added to silica gel (100 g), and evaporated *in vacuo*. The compound adsorbed on silica gel was loaded onto a flash column (600 g silica gel). The column was eluted with dichloromethane (500 ml), 5% methanol in dichloromethane (3.5 l) and 15%

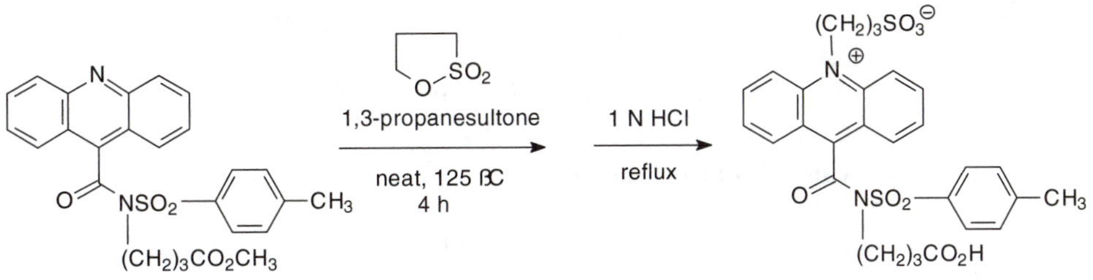

SCHEME 5.3 (From Adamczyk, M. et al., *J. Org. Chem.*, 63, 5636, 1998. With permission.)

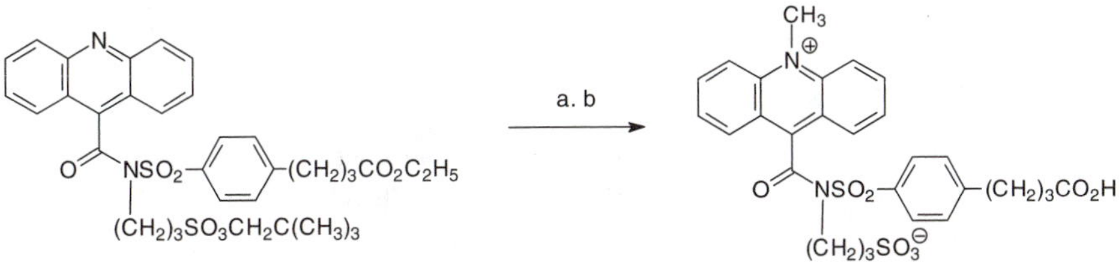

SCHEME 5.4 (a) $CF_3SO_3CH_3$, CH_2Cl_2, ambient, 7 days; (b) 1 *N* aq HCl, reflux.

SCHEME 5.5 (a) $CF_3SO_3(CH_2)_3SO_3CH_2(CH_3)_3$, 2,6-di-*t*-butyl-4-methylpyridine, ambient, CH_2Cl_2, 7 days; (b) 1 *N* aq HCl, reflux.

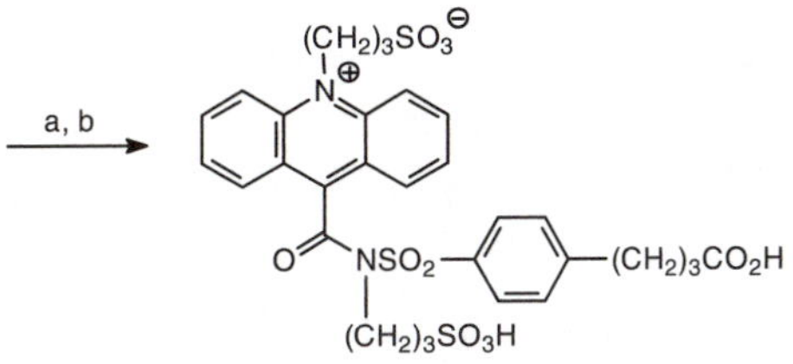

methanol in dichloromethane (6 l). The fractions that showed strong fluorescent yellow spots on TLC (silica gel, 15% methanol in dichloromethane) were combined and evaporated *in vacuo*. The resulting mixture of esters (15.2 g) was heated to reflux in aq 1 *N* HCl (500 ml) for 4.25 h under nitrogen, then cooled at room temperature for 16 h. During this time, a yellow precipitate formed. The mixture was cooled in ice for 45 min and the precipitate was collected by filtration, washed with acetonitrile (40 ml), and dried *in vacuo* to afford pure **15** (8.56 g, 78%).

Alternatively, the sulfopropyl group could be incorporated at an earlier step in the synthesis following the protocol shown in Scheme 5.4. *N*-Sulfonylacridine-9-carboxamide **16**[22,23] was simply methylated using methyl triflate as before. In this case, however, the reaction was much slower, taking 7 days to complete. Hydrolysis afforded the N^{10}-methyl-*N*-(3-sulfopropyl)-*N*-sulfonylacridinium-9-carboxamide **17**, which is isomeric with acridinium **15**.

From the same intermediate **16**, a bis-sulfopropylated acridinium salt, **18**, was prepared (Scheme 5.5). The N^{10}-sulfopropylation did not proceed with 1,3-propane sultone, but was successful using the more reactive neopentyl 3-triflyloxypropanesulfonate.[22,23]

TABLE 5.2
N^{10}-(3-Sulfopropyl)-*N*-Sulfonylacridinium-9-Carboxamide Salts Ranked by Chemiluminescence Lifetime

	R′	R″
a	$-(CH_2)_3CO_2H$	*p*-$CH_3C_6H_4$
b	$-(CH_2)_3CO_2H$	*p*-$CH_3OC_6H_4$
c	$-(CH_2)_3CO_2Hi$	*o*-$CH_3C_6H_4$
d	*p*-$CH_3OC_6H_4$	$-(CH_2)_3CO_2H$
e	*i*-C_3H_7	*p*-$C_6H_4(CH_2)_3CO_2H$
f	$-(CH_2)_3CO_2H$	*i*-C_3H_7
g	*o,p*-$(CH_3O)_2C_6H_3$	$-(CH_2)_3CO_2H$
h	$-(CH_2)_3CO_2H$	2,4,6-$(CH_3)_3C_6H_2$

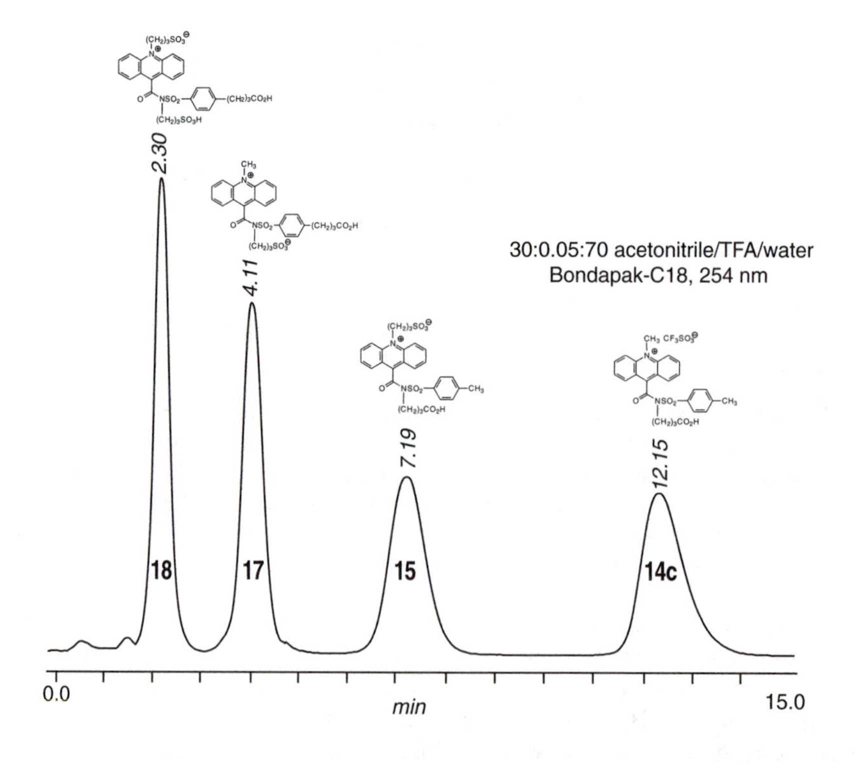

FIGURE 5.2 Comparison of the hydrophilicity of acridinium sulfonamide salts. (From Mattingly, P. G. et al., *10th Int. Symp. Biolumin. Chemilumin.*, Bologna, Italy, September 1998, Roda, A. et al., Eds., John Wiley & Sons, Chichester, U.K., 1999. With permission.)

The effect of sulfopropylation on the hydrophilicity of *N*-sulfonylacridinium-9-carboxamides was illustrated by the relative retention time of compounds **14c**, **15**, **17**, and **18** (Figure 5.2) under the conditions of reversed-phase high-performance liquid chromatography (HPLC).[22] The theory is that the retention time is inversely proportional to the hydrophilicity of the analyte. The retention time for N^{10}-sulfopropylated acridinium salt **15** was roughly half that of the N^{10}-methyl analogue **14c**, while addition of a second sulfopropyl group (**18**) reduced the retention time one third further. An explanation for the nearly twofold difference in the retention time of the isomeric compounds **15** and **17** is not clear at this time.

Sulfopropylation had minimal effect on the chemiluminescence profile of the acridinium compounds (Figure 5.3), while the overall light yield remained constant. The same trends in kinetics that were noted in the N^{10}-methyl series (Table 5.1) were maintained in a similar series of N^{10}-(3-sulfopropyl) acridinium compounds (Table 5.2).[20] There was an overall 20-fold decrease in the rate of chemiluminescence between compound **15** (entry **a**) and the mesitylene-substituted compound (entry **h**).

5.2.3 Detection of Reaction Intermediates

One undesirable characteristic of the acridinium phenyl ester class of labels had been the unproductive formation of the pseudobase (Figure 5.1, **2**, Nu = OH) at neutral and basic pH. Observations of the N^{10}-sulfonylacridinium-9-carboxamides have indicated that these compounds are much less

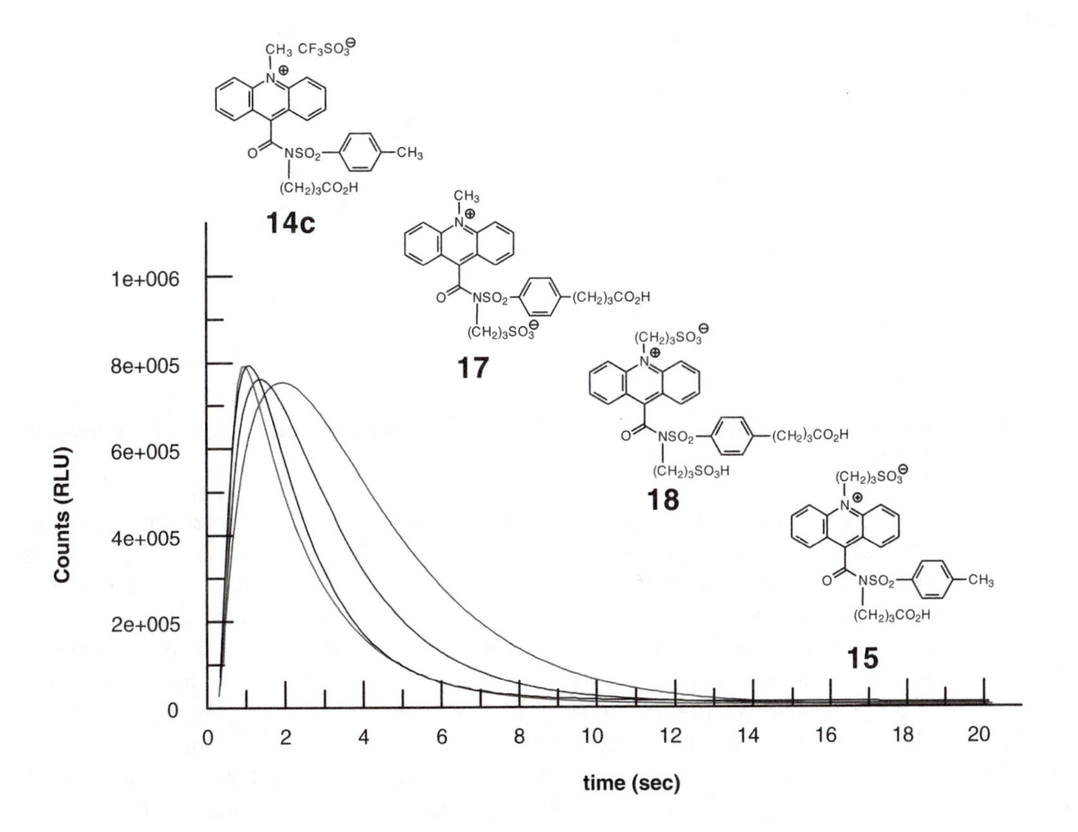

FIGURE 5.3 Chemiluminescence profiles of acridinium sulfonamide salts.

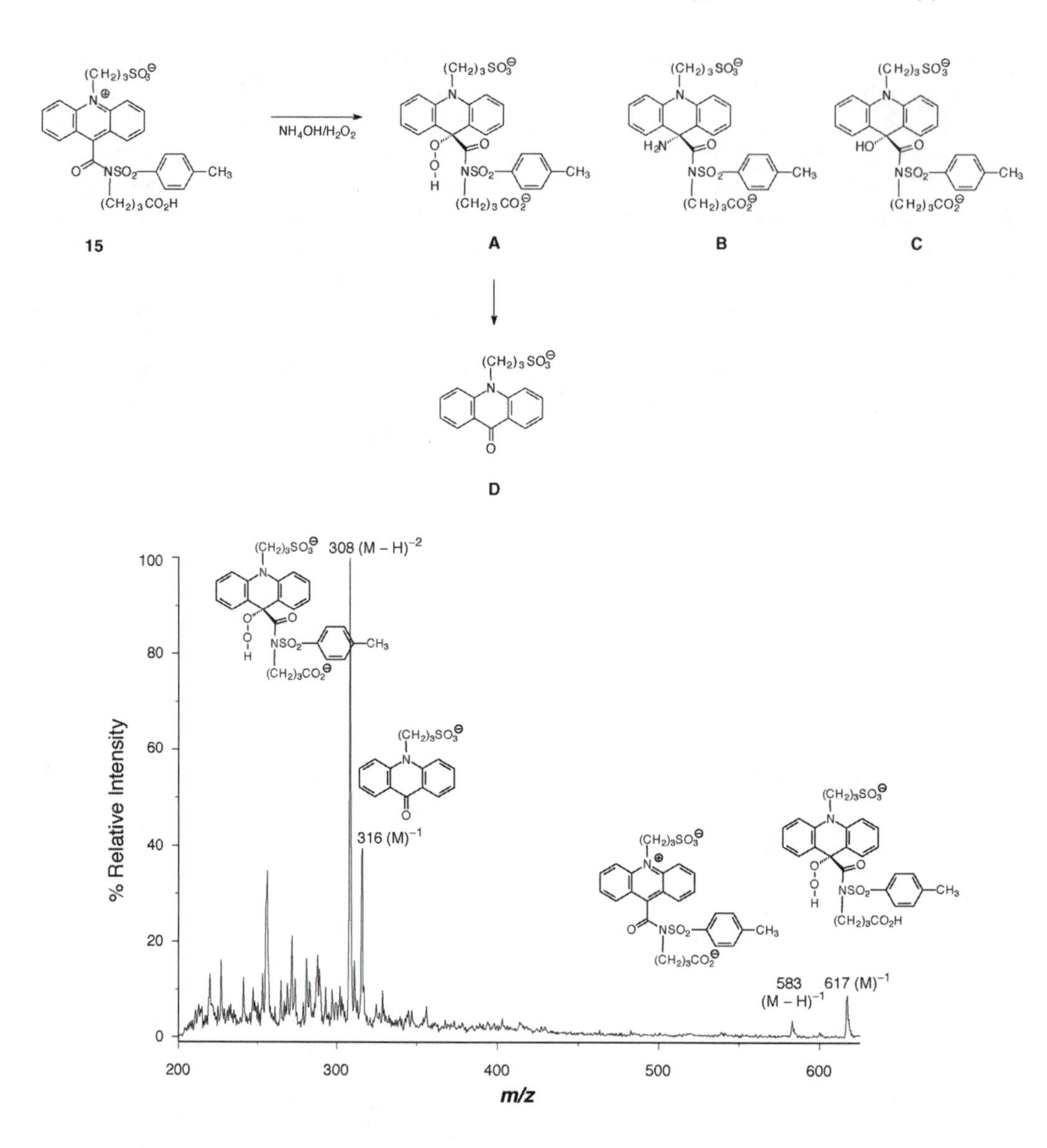

FIGURE 5.4 ESI-MS of acridinium peroxide adducts. (From Adamczyk, M. et al., *Eur. Mass Spectrosc.,* 4, 121, 1998. With permission.)

susceptible to this phenomenon. We recently demonstrated this difference between the acridinium phenyl esters and N^{10}-sulfonylacridinium-9-carboxamides by flow-injection electrospray mass spectrometry.[32] In those experiments an aqueous solution of the N^{10}-sulfonylacridinium-9-carboxamide **15** was introduced into the mass spectrometer along with a trigger solution containing hydrogen peroxide with ammonium hydroxide as the base. We observed ions that corresponded to the hydrogen peroxide adduct **A** (Figure 5.4) and the acridone end product **D**, but not the ammonia adduct **B** or the pseudobase **C**. In the absence of hydrogen peroxide only the ammonia adduct **B** was seen (Figure 5.5), and, when the trigger solution was removed completely, only the molecular ion and fragments from **15** were observed (Figure 5.6). Under the same conditions, an acridinium phenyl ester produced only ions related to the pseudobase, regardless of the presence of ammonia and hydrogen peroxide (Figure 5.7).

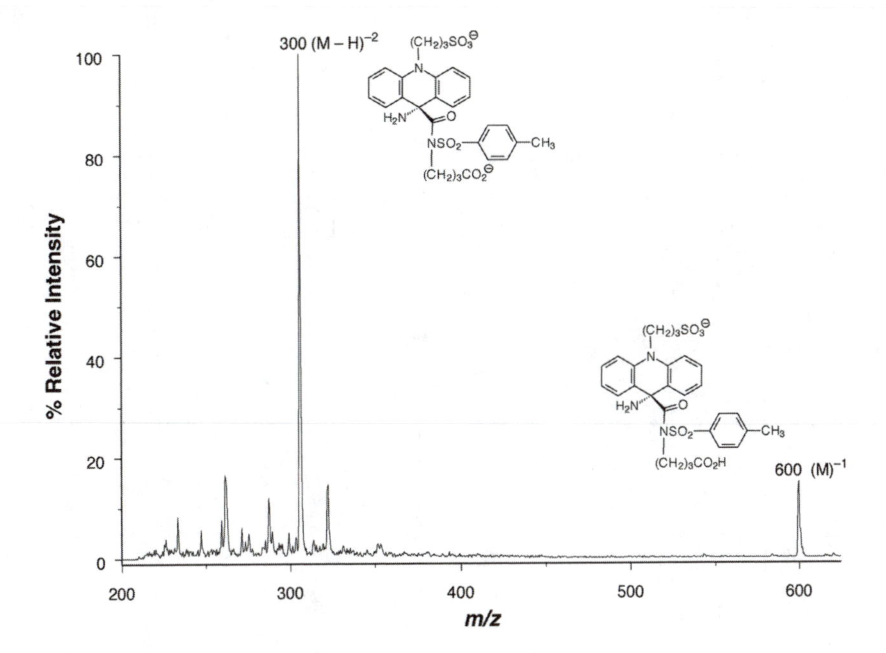

FIGURE 5.5 ESI-MS of acridinium ammonia adducts. (From Adamczyk, M. et al., *Eur. Mass Spectrosc.*, 4, 121, 1998. With permission.)

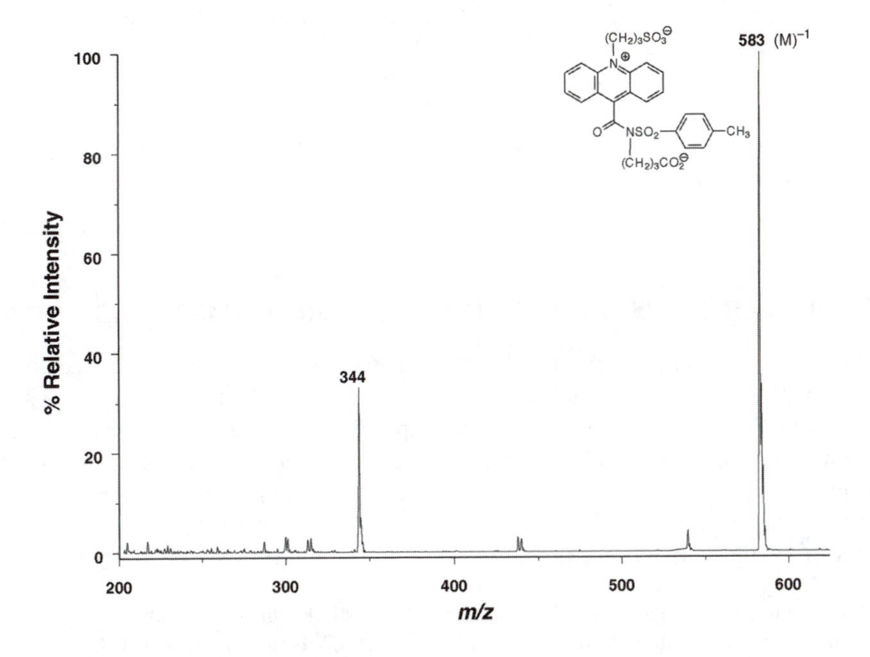

FIGURE 5.6 ESI-MS of salt (acridinium salt). (From Adamczyk, M. et al., *Eur. Mass Spectrosc.*, 4, 121, 1998. With permission.)

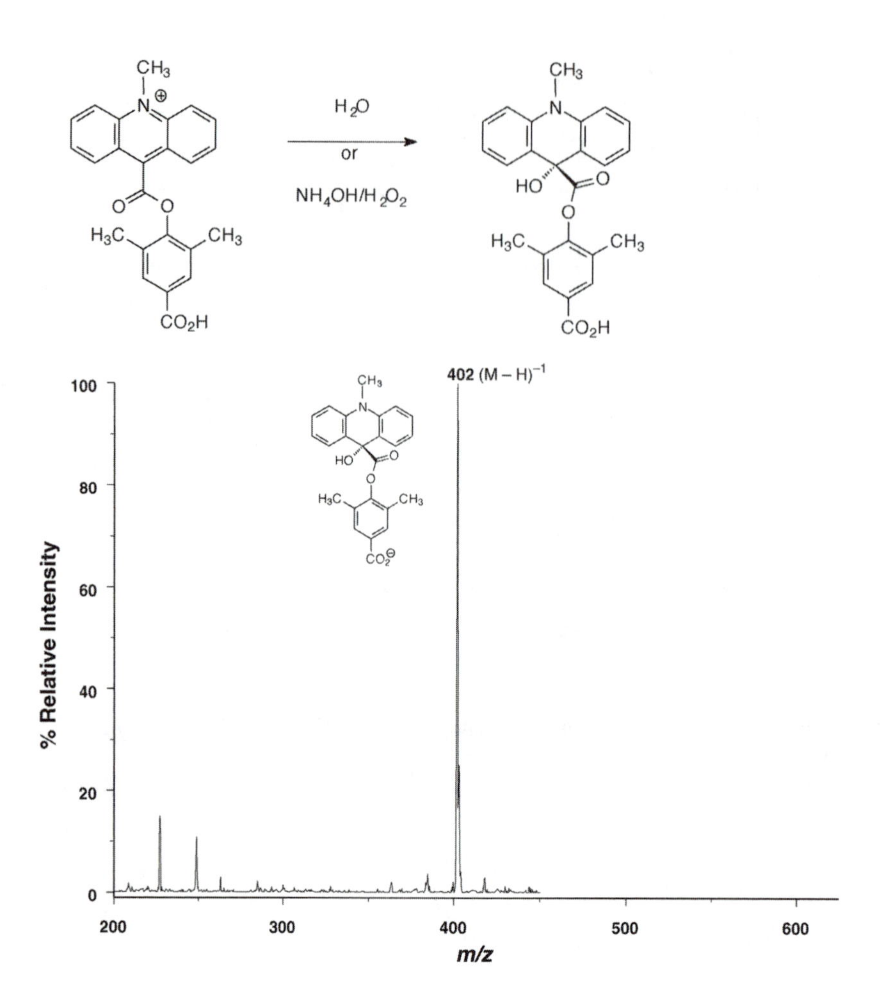

FIGURE 5.7 ESI-MS of acridinium phenyl ester adducts. (From Adamczyk, M. et al., *Eur. Mass Spectrosc.*, 4, 121, 1998. With permission.)

5.3 CONJUGATION USING ACRIDINIUM-9-CARBOXAMIDE SALTS

The N^{10}-sulfonylacridinium-9-carboxamide salts thus far described have been utilized in a variety of assay formats encountered in clinical diagnostics. Most of those formats require that the acridinium compound be conjugated to a component specific for the particular assay. The carboxylic acid–bearing compounds **14**, **15**, **17**, and **18** can be activated *in situ* and coupled to amino-substituted components in organic solvents, buffered solution, or mixtures of both. Often we have found that preparation of pure active esters[33] to be most advantageous (e.g., **19a–d**, Scheme 5.6). The preparation of **19b**[34] is illustrative of the process.

> Acridinium acid **15**[23] (8 g, 13.68 mmol) was suspended in DMF (80 ml) and pyridine (11 ml). A solution of *N*-trifluoroacetoxysuccinimide (29 g, 137.38 mmol) in DMF (80 ml) was added. All the solids gradually dissolved, after which a yellow precipitate formed. After stirring for 24 h in the dark, the precipitate was collected by filtration and washed sequentially with ethyl acetate (200 ml), cold ethanol (60 ml), and again with ethyl acetate (2 × 200 ml). Further drying *in vacuo* afforded the desired active ester **19b** (9.5 g, 99%).

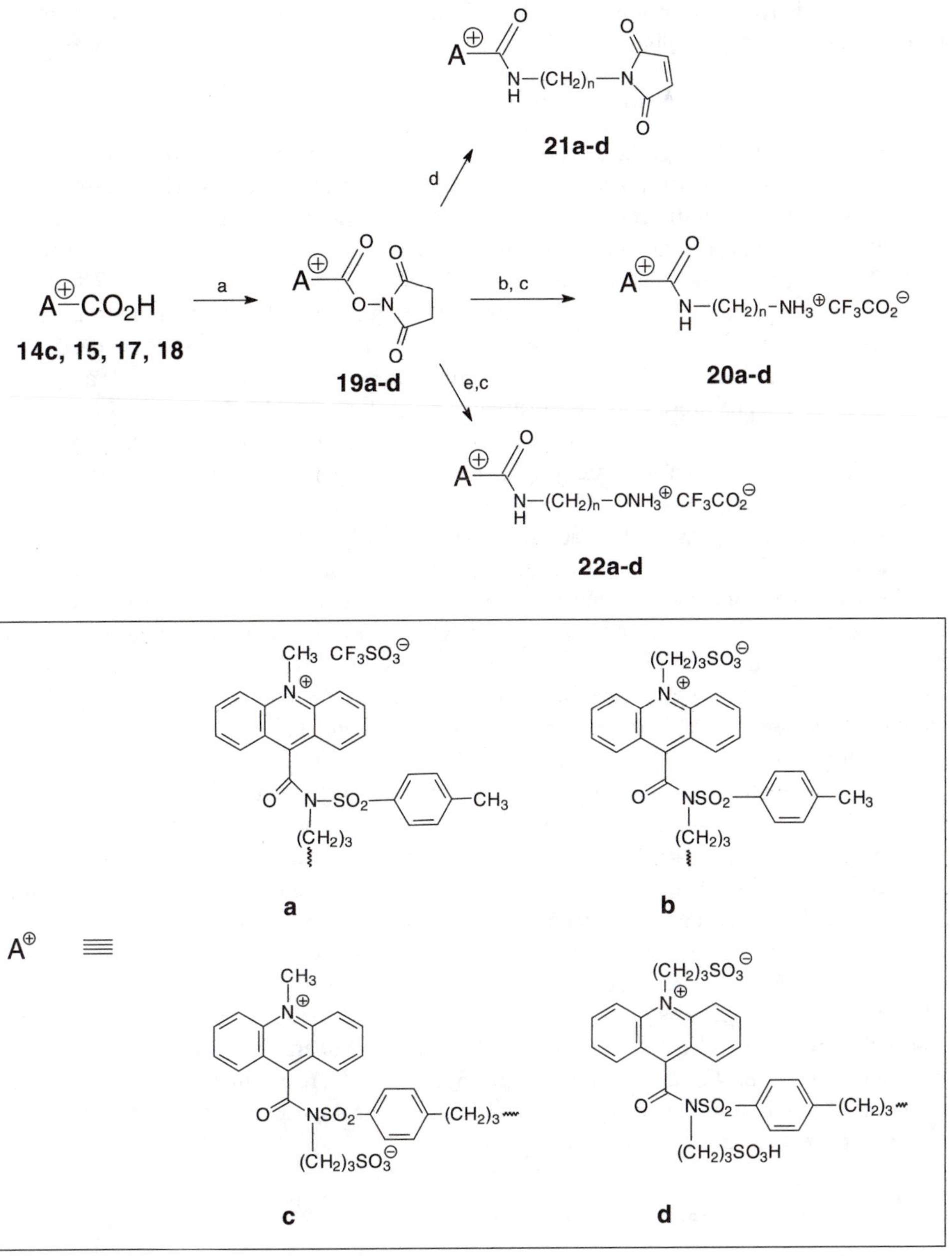

SCHEME 5.6 (a) *N*-Succinimidyltrifluoroaceate, pyridine; (b) *t*-BOCNH$(CH_2)_n$$NH_2$, DMF; (c) TFA, CH_2Cl_2; (d) *N*-aminoalkylmaleimide, DMF, Et_3N; (e) ethyl *O*-aminoalkoxyacetimidate, DMF; (f) aq 3 *N* HCl.

Active esters such as **19b**, while useful in direct conjugation of clinically important substrates, are also convenient intermediates in the preparation of amino-substituted N^{10}-sulfonylacridinium-9-carboxamide salts (**20a–d**). The active ester is coupled to a mono *N*-*t*-butoxycarbonyl-protected diamine,[35] then deprotected with trifluoroacetic acid (TFA). The resulting TFA salts are stable powders that can be coupled conveniently to carboxylic acid containing substrates. Similarly, the active esters can be used to acylate aminoalkyl maleimides[36–38] to give the thiol reactive labels **21a–d**. Recently,[24] we have utilized the active esters to prepare N^{10}-sulfonylacridinium-9-carboxamide reagents that react directly with aldehydes and ketones. The acridinium hydroxylamines[24] (AHA, **22a–d**) are prepared much like the aminoalkyl analogues, by the acylation of *O*-aminoalkyl-*N*-protected hydroxylamines.[39]

5.3.1 Low-Molecular-Weight Analytes

There are many clinically significant analytes that have molecular weights of 1500 Da or less and are often referred to as small molecule or low-molecular-weight analytes. Their conjugates with a detectable label are generally referred to as tracers. Low-molecular-weight analytes include most therapeutic or abused drugs and many endogenous hormones. The preparation of N^{10}-sulfonylacridinium-9-carboxamide conjugates with such analytes is illustrated in Schemes 5.7 to 5.11. Most often an N^{10}-sulfonylacridinium-9-carboxamide active ester is reacted with an amine-bearing analogue of the analyte in an organic solvent in the presence of a tertiary amine base (Scheme 5.7). When using an active ester such as **19b**, there is no need to protect hydroxyl or carboxylic acid functional groups that may also be present in the substrate. This is illustrated by the conversion of the thyroid hormone analogues **23a–e**,[40,41] the estradiol steroid hormone analogues **25** to **29**,[42–44] the antibiotic vancomycin **24**,[45] and the cardiotonic drug analogue digoxigenin **30**[46] into the respective chemiluminescent tracers **31** to **38**.

In some cases it is possible to react the N^{10}-sulfonylacridinium-9-carboxamide directly by the *in situ* activation of its free carboxylic acid in the presence of an analyte. A peptide bond between a simple, monofunctional amino-containing substrate and the label can be formed with a variety of coupling reagents.[47] This approach has been successful even with some very complex multifunctional substrates. Acridinium salt **15** was used to label the *N*-methylleucine nitrogen of vancomycin (Scheme 5.8).[45] In contrast, the reaction of the isolated *N*-hydroxysuccinimide active ester **19b** with vancomycin (Scheme 5.7) led to the *N*-vancosylamine conjugate.

When the analyte or its analogue contains a free carboxylic acid group, an N^{10}-sulfonylacridinium-9-carboxamide bearing an amino group can be used to form the tracer. For example, 7-carboxyalkyl estradiol **40** and aminoalkyl N^{10}-sulfonylacridinium-9-carboxamide **20b** (Scheme 5.9) react in the presence of dicyclohexylcarbodiimide and *N*-hydroxybenztriazole to give the estradiol tracer **41**.

In all the preceding examples the N^{10}-sulfonylacridinium-9-carboxamide tracers are isolated by reverse-phase HPLC eluting with aqueous acetonitrile containing trifluoroacetic acid as the modifier. One could envision acid-sensitive analytes that might not be amenable to such treatment. Further, there could be advantages of speed and economy if such purification could be avoided entirely. We have addressed these possibilities by demonstrating the feasibility of preparing and using solid-supported active esters of an N^{10}-sulfonylacridinium-9-carboxamides (Scheme 5.10).[48] The solid-supported active ester **42** was prepared from **15** and a NHS resin, then reacted with aminoethyl phenytoin **43** as the limiting reagent. The resulting tracer **44** was isolated in excellent yield and purity without the need for purification.

5.3.2 High-Molecular-Weight Analytes

Proteins (antigens, antibodies) and nucleic acids are clinically relevant analytes themselves in many cases, or they are the labeled components in the diagnostic assay. In either case, the concerns and strategies for the preparation of chemiluminescent N^{10}-sulfonylacridinium-9-carboxamide conjugates of these materials are somewhat different from that of small molecule tracers. Small molecule tracers can be purified to homogeneity and characterized to establish their structure. Conjugates of high-molecular-weight analytes are usually heterogeneous mixtures whose structure cannot be determined with the same degree of certainty or ease. There is a degree of flexibility in designing a small molecule tracer; functional groups can be differentially blocked; new functional groups can be introduced; total synthesis is always an option. This is seldom the case with high-molecular- weight substrates. One has to deal with the native reactive groups that are present, based primarily on the selectivity of labeling reagent alone. The molar concentration of many high-molecular-weight analytes is lower than that of

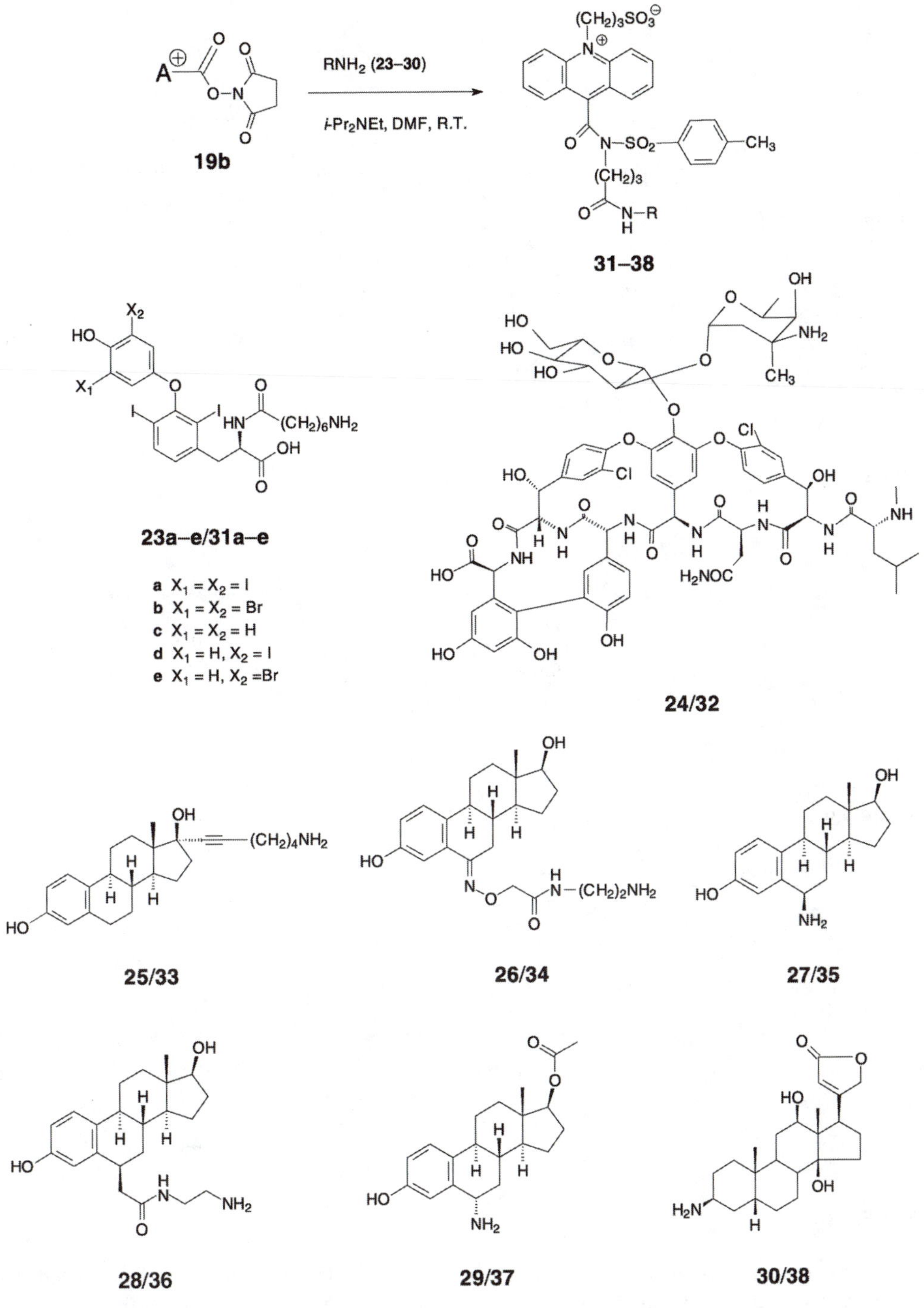

19b

RNH_2 (23–30)

i-Pr_2NEt, DMF, R.T.

31–38

23a–e/31a–e

a $X_1 = X_2 = I$
b $X_1 = X_2 = Br$
c $X_1 = X_2 = H$
d $X_1 = H, X_2 = I$
e $X_1 = H, X_2 = Br$

24/32

25/33

26/34

27/35

28/36

29/37

30/38

SCHEME 5.7

small molecule analytes. Thus, in assays for high-molecular-weight analytes, sensitivity becomes a larger concern. Unlike most tracers, high-molecular-weight conjugates usually must contain more than one equivalent of the label to reach the desired level of sensitivity. However, the level of substitution by the label and the site of labeling can have a profound effect on the activity of the conjugate.[49]

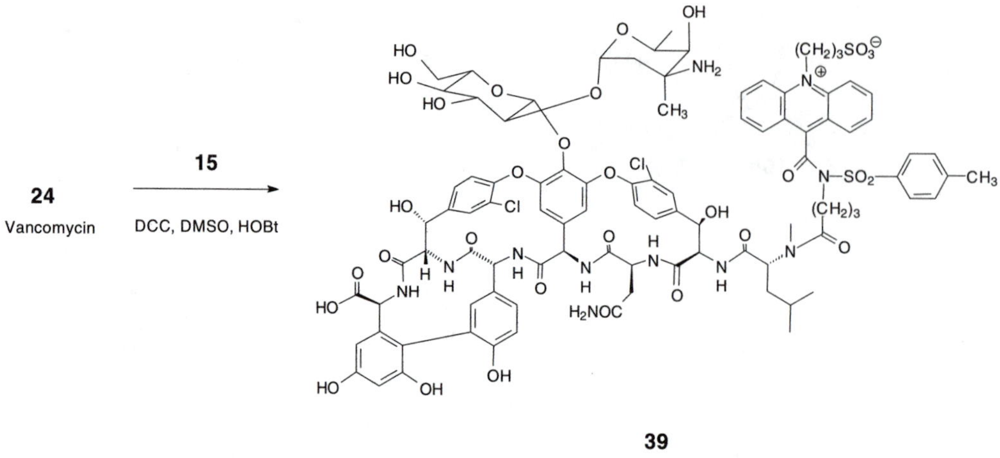

24 Vancomycin → 15, DCC, DMSO, HOBt → 39

SCHEME 5.8

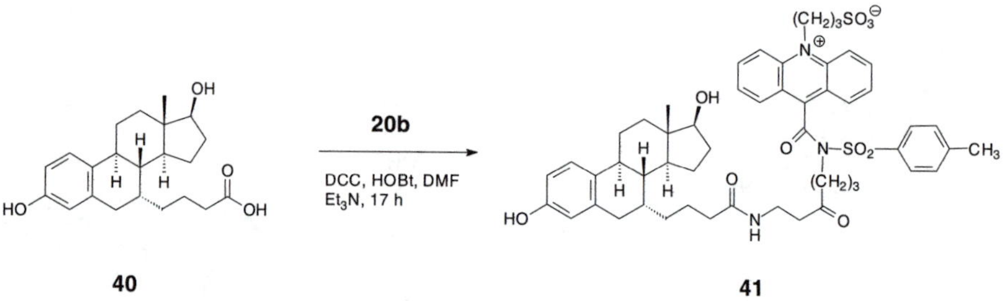

40 → 20b, DCC, HOBt, DMF, Et_3N, 17 h → 41

SCHEME 5.9

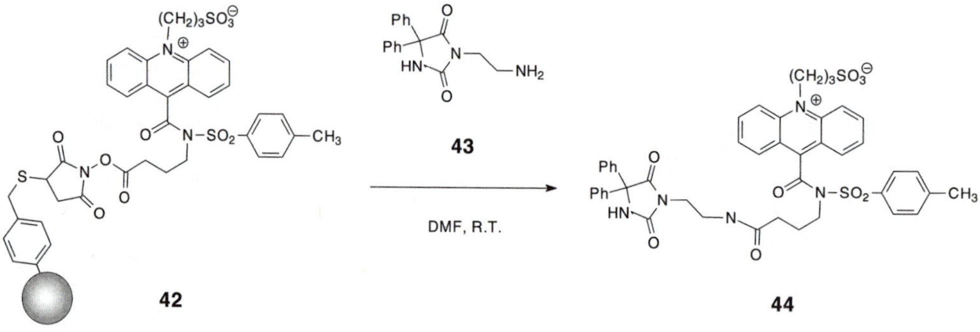

42 → 43, DMF, R.T. → 44

SCHEME 5.10

Proteins contain reactive amines, carboxylic acids, and thiols. Often they are glycosylated. The carbohydrate residues are susceptible to oxidation and provide a ready source of reactive aldehydes.

The available amino groups on proteins may be acylated by active esters, such as **19b**, (Scheme 5.11) to give the corresponding chemiluminescent conjugate (**46**).

Since this method of conjugation to amines is random, the higher the incorporation of the label, the greater is the likelihood that an essential amino acid residue will be modified to the detriment of the specific binding interaction of the assay. In cases where this is a problem, the introduction of more N^{10}-sulfonylacridinium-9-carboxamide residues per reactive site on the protein can be a solution. Scheme 5.12 illustrates the preparation of a chemiluminescent active ester that contains

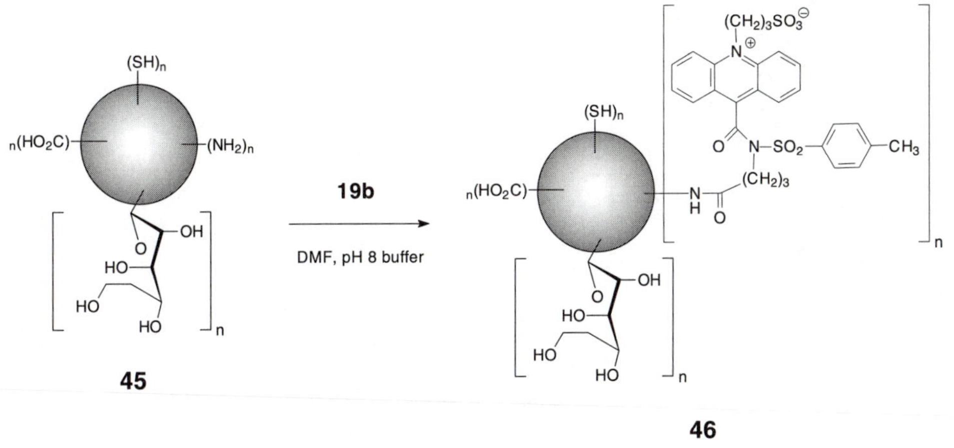

45 → **19b**, DMF, pH 8 buffer → **46**

SCHEME 5.11

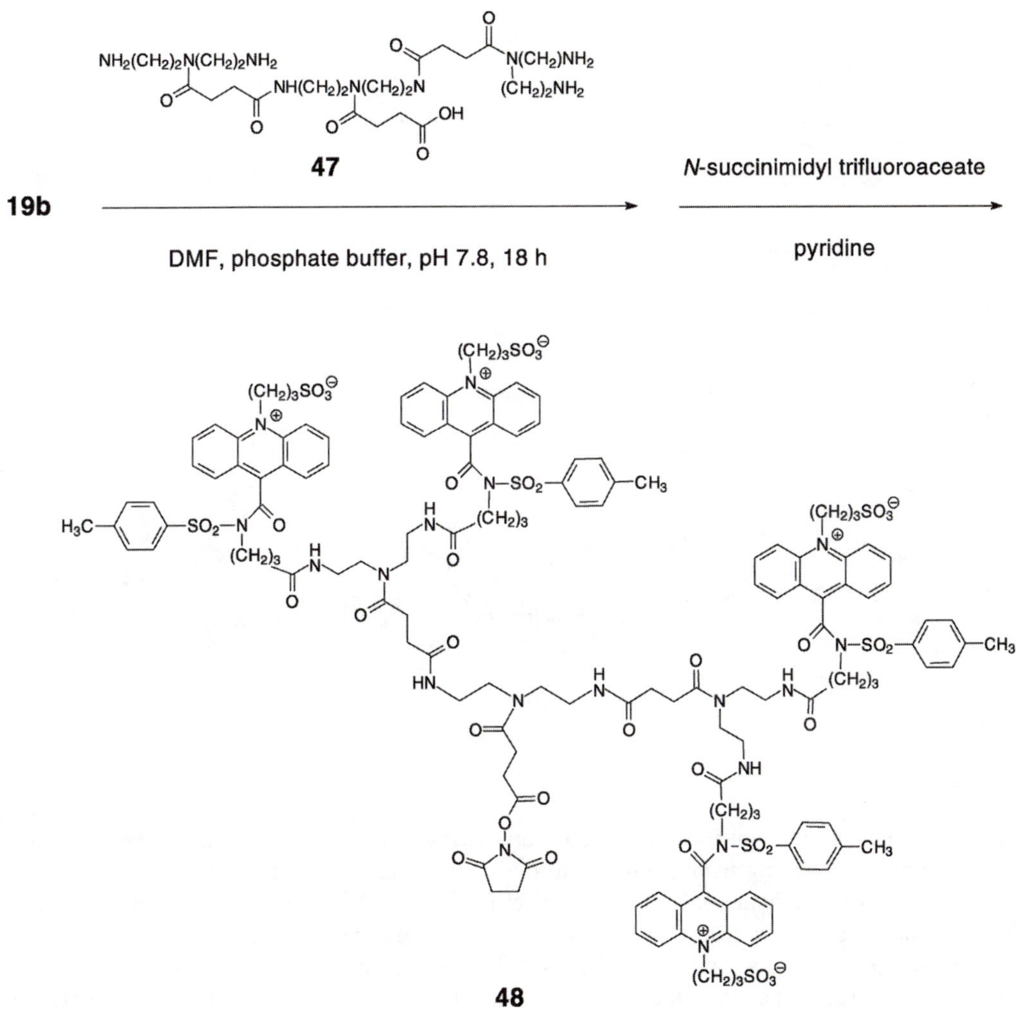

19b → **47**, DMF, phosphate buffer, pH 7.8, 18 h → *N*-succinimidyl trifluoroaceate, pyridine → **48**

SCHEME 5.12

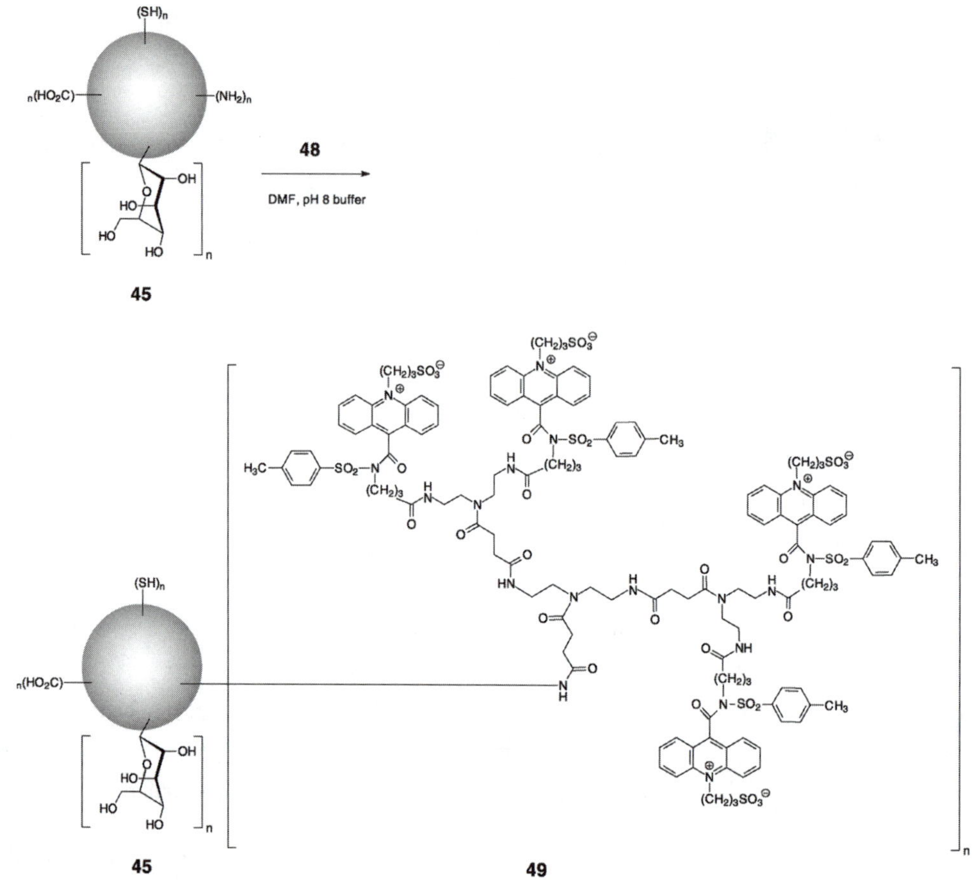

SCHEME 5.13

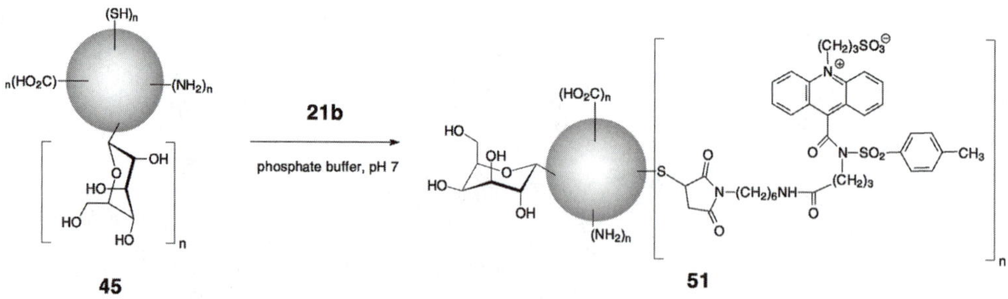

SCHEME 5.14

four N^{10}-sulfonylacridinium-9-carboxamide units. Arborol Tracermers[34] of this type react with protein amino groups in a manner similar to the monomeric active ester **19b** (Scheme 5.11). At the same level of incorporation (n) the Tracermer conjugate **49** (Scheme 5.13) produces a chemiluminescent signal greater than four times that of the monomer conjugate **46**.

Proteins contain many fewer thiol groups than amino groups. Many of the thiol groups may be present as disulfides and are thus unreactive until they are reduced. This offers the possibility of greater selectivity in the conjugation reaction. N^{10}-Sulfonylacridinium-9-carboxamides bearing a maleimide substituent (**21b**) react selectively with the thiol groups on proteins (Scheme 5.14)

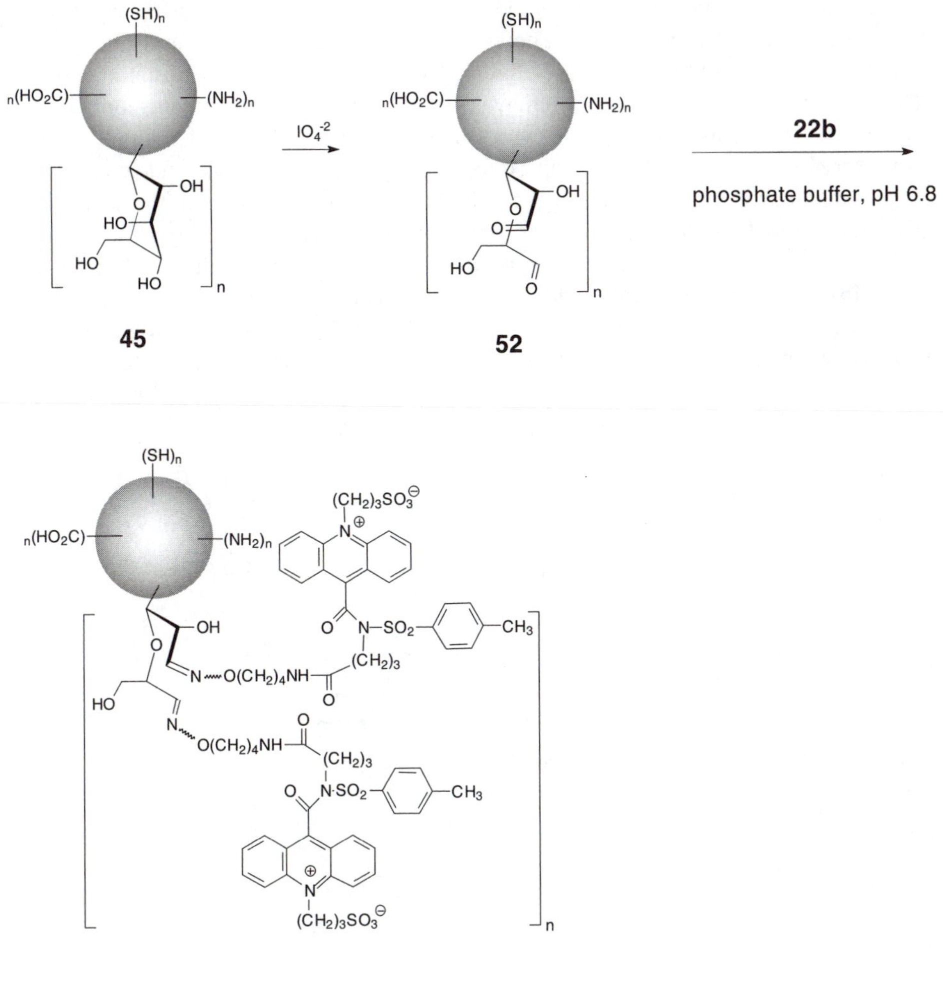

SCHEME 5.15

under neutral to slightly acidic pH to give the corresponding conjugate (**51**). At higher pH and extended reaction times, amines will also add to the maleimide.

On oxidation with periodate, the carbohydrate present on proteins (antibodies, in particular) gives rise to aldehydes[49] (**52**, Scheme 5.15), which selectively react with *O*-(N^{10}-sulfonylacridinium-9-carboxamide) hydroxylamines[24] (**22b**) to give a chemiluminescent conjugate with a stable oxime linkage. In the case of antibodies, most of the glycosylation is on the Fc portion of the molecule, distant from the site of binding to the antigen.

The N^{10}-sulfonylacridinium-9-carboxamide-conjugates regardless of the method of preparation must be rigorously purified to separate unconjugated N^{10}-sulfonylacridinium-9-carboxamide active ester **19b**, **49**, **21b**, or **22b**, or their equally chemiluminescent by-products. The ratio of the chemiluminescent label to protein can be determined by ultraviolet difference spectroscopy or mass spectrometry. Electrospray ionization is the bioanalytical mass spectrometry technique of choice, allowing the determination both of conjugate composition and the level of residual unconjugated label. A further degree of characterization can be achieved by digesting the protein and mapping the fragments.

Nucleic acids are another class of high-molecular-weight analytes that have been labeled with N^{10}-sulfonylacridinium-9-carboxamides. The small oligonucleotide probes used in hybridization assays are usually prepared by solid-phase oligonucleotide synthesis using phosphoramidite chemistry. The resulting oligonucleotides are removed from the solid support by nucleophilic cleavage of an ester linker. These conditions are somewhat harsh for the N^{10}-sulfonylacridinium-9-carboxamide labels and therefore they are most often introduced postsynthetically on oligonucleotides that bear a reactive 5′-amine or thiol, or a similarly modified nucleobase elsewhere in the sequence, using the active ester or maleimide derivatives, respectively, that were already discussed.

Nucleic acid can also be directly labeled with N^{10}-sulfonylacridinium-9-carboxamides (Scheme 5.16).[50] N^{10}-sulfonylacridinium-9-carboxamide carboxylic acid **14a** was activated with EEDQ to form the mixed anhydride that reacts with the N^4-amino group of cytosine bases. Alternatively, an aminoalkyl linker can be interposed between the chemiluminescent label and the

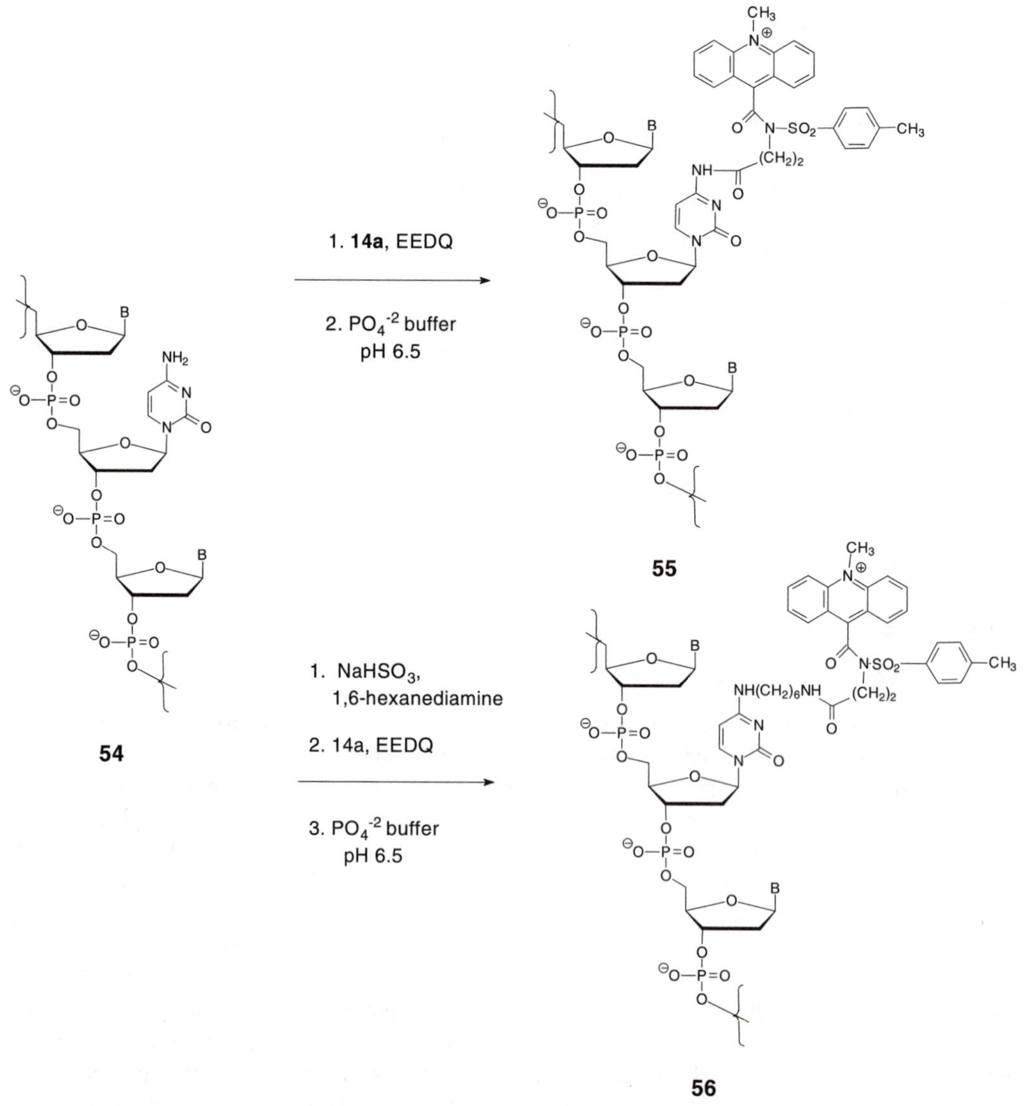

1. **14a**, EEDQ
2. PO_4^{-2} buffer pH 6.5

1. $NaHSO_3$, 1,6-hexanediamine
2. 14a, EEDQ
3. PO_4^{-2} buffer pH 6.5

54 **55** **56**

SCHEME 5.16

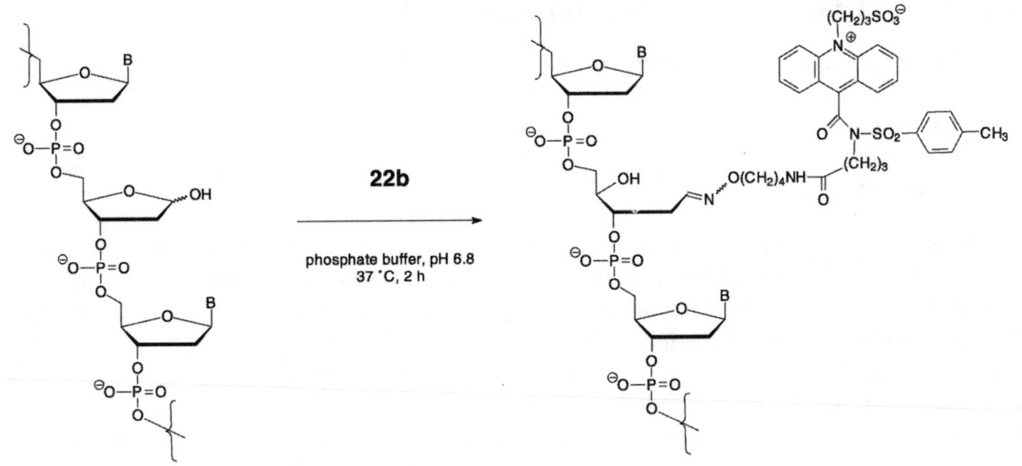

22b

phosphate buffer, pH 6.8
37 °C, 2 h

57

58

SCHEME 5.17

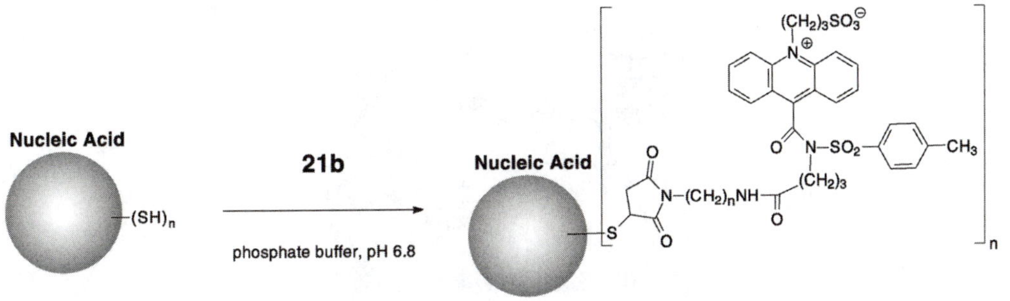

Nucleic Acid

21b

phosphate buffer, pH 6.8

59

60

SCHEME 5.18

nucleic acid by the sulfite-catalyzed exchange of the N^4-amino group of cytosine residues[51] in the sequence with 1,6-hexanediamine, followed by acylation with the activated label. Oligonucleotide probes with an acridinylated poly-dC tail were also prepared by these methods. Care had to be taken to choose a hybridization sequence that was low in dC or had been prepared with an unreactive dC equivalent. Alternatively, nucleic acids have been depurinated under acid conditions. The resulting deoxyribosyl aldehyde abasic sites (**57**, Scheme 5.17) react readily with *O*-(N^{10}-sulfonylacridinium-9-carboxamide) hydroxylamines (**22b**) affording the corresponding oxime conjugates (**58**).[24] Abasic sites can be incorporated into synthetic oligonucleotides[52–54] as well, making this a simple and versatile approach to label nucleic acid probes. Recently, commercial kits for the introduction of thiol groups onto nucleic acids[55] have provided yet another means of preparing N^{10}-sulfonylacridinium-9-carboxamide conjugates. Thiolated nucleic acids react similarly to thiol-bearing proteins with the N^{10}-sulfonylacridinium-9-carboxamide maleimide **21b** (Scheme 5.18).

When the intended use is in a diagnostic assay, the principal limitation of any of the random approaches for the labeling of nucleic acids is that the hybridization-binding region is susceptible to undesirable modification. On the other hand, nucleic acids can contain millions of bases, and in those cases hundreds of labels can be introduced before the effect is seen.

5.4 ABBOTT CHEMILUMINESCENCE ANALYZERS

Abbott Laboratories currently markets two *in vitro* diagnostic chemiluminescence-based analyzers that employ N^{10}-sulfonylacridinium-9-carboxamide-labeled tracers and conjugates: the Prism and the Architect *i*2000™ (Figure 5.8).

5.4.1 Abbott Prism

The Abbott Prism is a high-throughput analyzer designed specifically for the screening of donated blood[56–60] for hepatitis surface antigen (HbsAg) and antibodies to hepatitis B core antigen (HBc),[61,62] hepatitis C virus (HCV),[59] human immunodeficiency virus types I and II (HIV-1/HIV-2), and human T-cell lympotrophic virus type I and II (HTLV-I/HTLV-II). The system[63] is totally automated and designed for 24-h operation with the capacity to generate up to 900 test results per hour. The throughput is achieved by the simultaneous operation of as many as six independently operating channels. Each channel is divided into nine segments or stations where reagents, diluents, wash solutions, and, finally, the chemiluminescent triggering solution are dispensed. The reagents and solutions necessary for the immunoassay are stored within the analyzer at ambient temperature or under refrigeration as required. The operator can load up to 280 bar-coded patient samples and

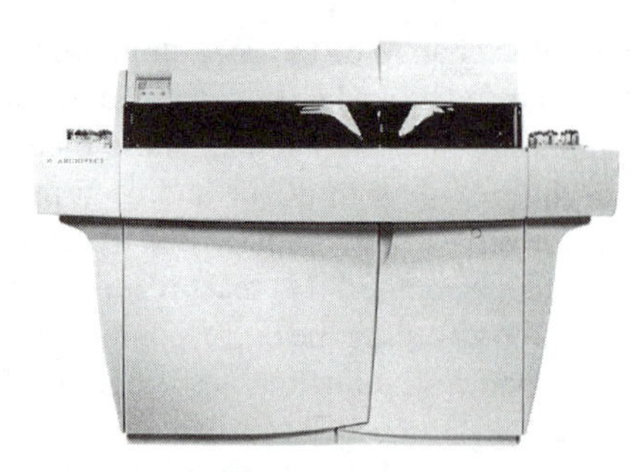

FIGURE 5.8 Abbott Laboratories Prism and Architect chemiluminescence-based analyzers. (Courtesy of Abbott Laboratories.)

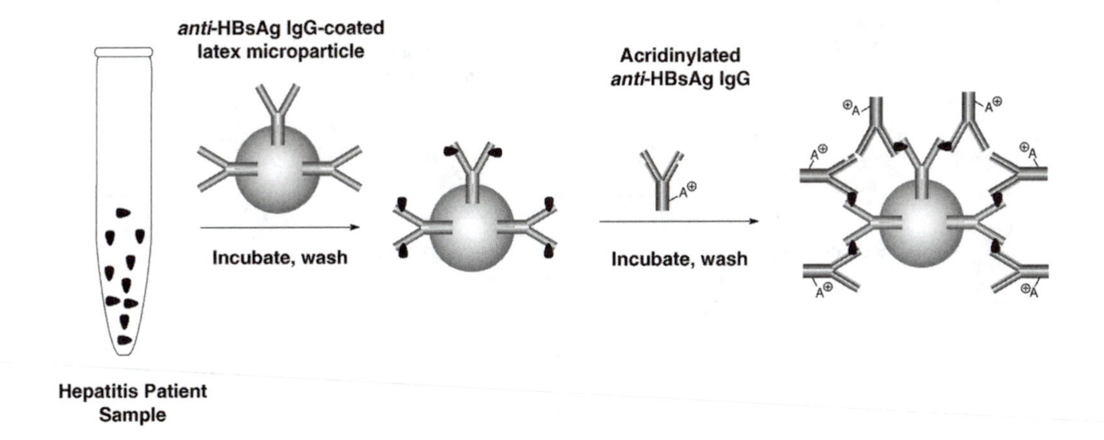

FIGURE 5.9 Prism two-step antigen sandwich assay format.

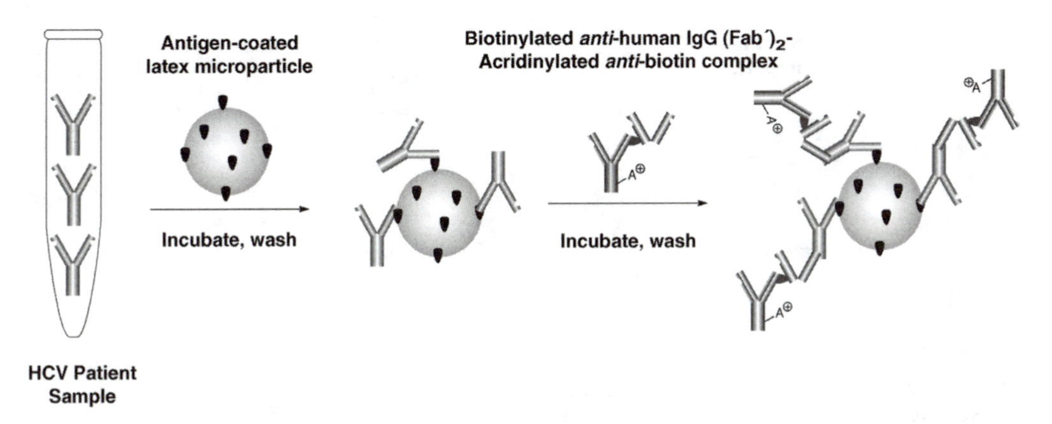

FIGURE 5.10 Prism two-step antibody sandwich assay format.

controls at a time. The analyzer then automatically loads a reaction tray onto the appropriate channel, pipettes the sample, and proceeds with the immunoassay. The chemiluminescent reaction is initiated at the final station, and the signal is detected by a fiber-optic-connected photomultiplier assembly[64,65] before the tray is deposited in a waste disposal container.

Two- and three-step formats are employed in the immunoassays (Figures 5.9 through 5.12). Each assay format utilizes a latex microparticle as the capture phase and an N^{10}-sulfonylacridinium-9-carboxamide-labeled conjugate as the detection reagent. The simplest assay format is the two-step sandwich assay used for the detection of hepatitis surface antigen (HbsAg) (Figure 5.9). The instrument then pipettes the samples into the incubation wells of the reaction tray,[66] moves to the next station, and dispenses the latex microparticle reagent that has been coated with *anti*-HbsAg IgG. The sample is incubated, then washed into the reaction well. At the bottom of the reaction well is a glass fiber matrix that serves as a filter to separate bound from free antigen. An N^{10}-sulfonylacridinium-9-carboxamide-anti-HbsAg conjugate is then added. The reaction mixture is incubated, then washed before proceeding to the chemiluminescence detection station.

The Prism HCV assay (Figure 5.10) also uses a two-step assay format, but differs from the HbsAg assay, in that antibodies to the virus in the patient sample are detected. In this case the latex microparticles are coated with recombinant viral antigen and the detection reagent is an N^{10}-sulfonylacridinium-9-carboxamide-*anti*-human IgG complex. The assay for antibodies to hepatitis core antigen (HBcAg) is also a two-step format (Figure 5.11), but it is a competitive/blocking assay. The latex microparticle,

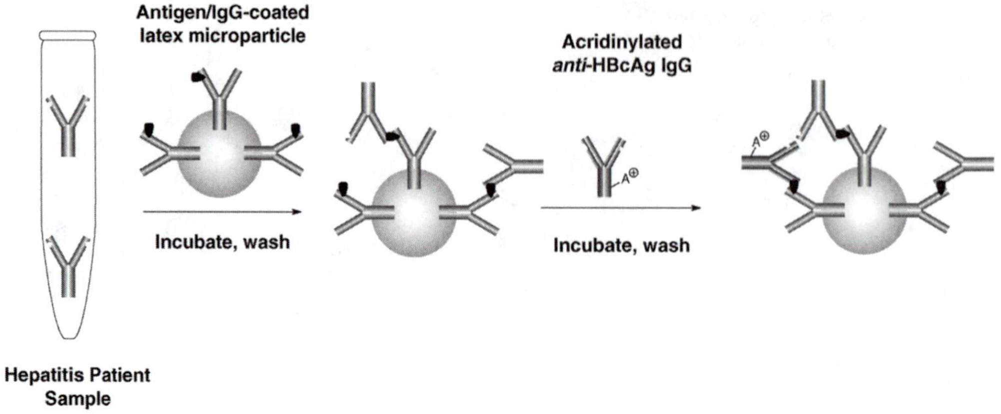

Antigen/IgG-coated latex microparticle

Acridinylated *anti*-HBcAg IgG

Incubate, wash

Incubate, wash

Hepatitis Patient Sample

FIGURE 5.11 Prism two-step competitive blocking assay format.

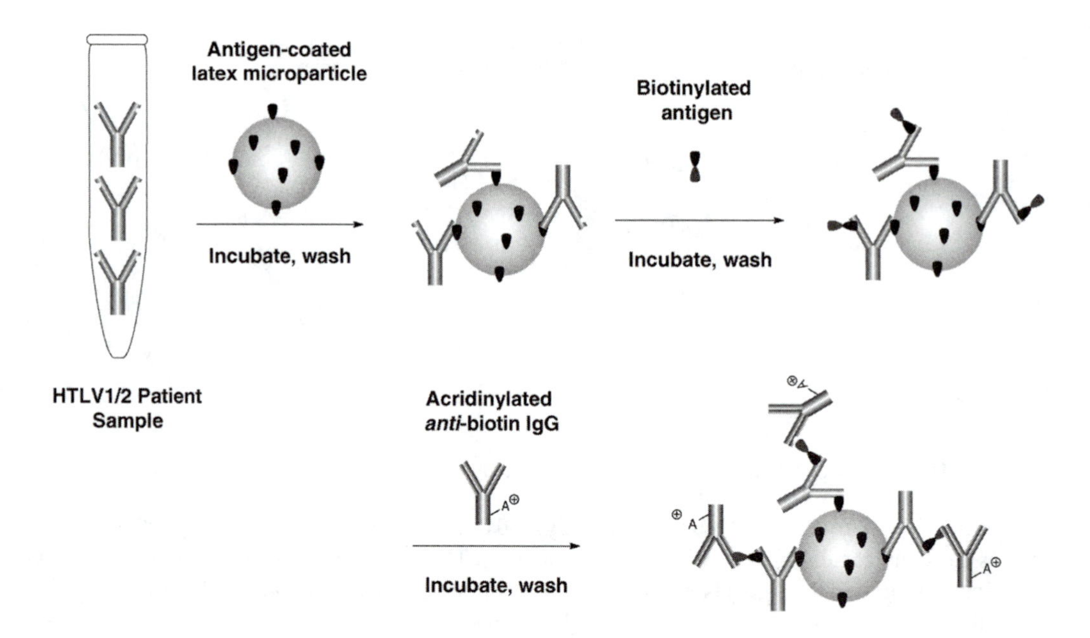

FIGURE 5.12 Prism three-step antibody sandwich assay format.

coated with *anti*-HBcAg IgG and antigen, is incubated with the patient sample to form a sandwich. The remaining microparticle-bound *anti*-HBcAg IgG–antigen complex is reacted with a second *anti*-HBcAg IgG, labeled with an N^{10}-sulfonylacridinium-9-carboxamide to form a sandwich.

The three-step format used in the detection of antibodies to HTLV-1/2 is illustrated in Figure 5.12. Antigen-coated microparticles are incubated with patient sample, and then with biotinylated antigen. In a third step, N^{10}-sulfonylacridinium-9-carboxamide-labeled *anti*-biotin is added to complete the sandwich assay.

5.4.2 Abbott Architect *i*2000

The Abbott Architect *i*2000 (Figure 5.8) is a fully automated chemiluminescence-based analyzer[67] for both high-[68–77] and low-molecular-weight analytes.[70–72,76,78] The system consists of a PC-based system control center, a sample handler for up to 125 samples, and a processing module that dispenses sample, calibrators, controls, and assay reagents in addition to detecting the chemiluminescent signal. Like the Prism, the Architect *i*2000 has on-board refrigeration of reagents, where necessary, and waste disposal. The system may be loaded with reagents for 25 different assays; random and continuous access is available in addition to priority processing of samples. In this configuration as many as 200 tests results per hour are possible. At present, the system may be configured with an additional processing module that increases the capacity to 250 samples and 400 results per hour with up to 50 different assays. Since instrument has been recently introduced, the menu of assays is continually expanding (see http://Abbottdiagnostics.com/ for the latest listing).

In contrast to the Prism, the Abbott Architect *i*2000 employs paramagnetic microparticles as the solid support. The particles are coated with a capturing molecule specific to the assay. This can be an antigen, hapten, antibody, or viral particle depending upon the analyte. The one- or two-step chemiluminescent assay is performed in individual disposable reaction vessels that are automatically loaded into the processing module carousel. In the one-step format, the sample, paramagnetic microparticles, and N^{10}-sulfonylacridinium-9-carboxamide conjugate or tracer are added to the reaction vessel and incubated. During washing steps to remove unbound materials, the paramagnetic microparticles binding the immune complex are held to the side of the reaction vessel by a magnet while the liquid is removed by pipette. A solution containing acid and hydrogen peroxide is then added and a background chemiluminescence signal is recorded. Finally, a base activator is added and the chemiluminescence is initiated and read by a photomultiplier tube for a defined time. The two-step assay interposes a incubation between the addition of the microparticles and the N^{10}-sulfonylacridinium-9-carboxamide conjugate or tracer. For those assays that require a pretreatment step, two reaction vessels are typically used. The first receives the sample and the pretreatment reagent. After a period of incubation, the treated sample is transferred to a new reaction vessel and the assay proceeds using the one- or two-step protocol.

5.5 CONCLUSION

The N^{10}-sulfonylacridinium-9-carboxamides have been developed as stable and efficient chemiluminescent labeling reagents for the production of tracers and conjugates for medical diagnositic testing. The physical properties and ease of use of these labels far surpass any other labeling reagents. They have served as the essential ingredient in the successful launch of Abbott Laboratories' newest and most highly automated immunoassay analyzers. However, in many ways we have only touched the surface in our exploration of their utility. In some assays it may be clinically desirable to detect fewer than 1000 molecules of the analyte, so enhancing sensitivity to that level is still a worthwhile goal. All the assay formats presented here have been heterogeneous; homogeneous assay formats would also be desirable. A detailed theoretical or kinetic analysis of the chemiluminescent reaction of N^{10}-sulfonylacridinium-9-carboxamides has yet to be presented that explains the observed differences between them and the acridinium phenyl esters.

It is our belief that the N^{10}-sulfonylacridinium-9-carboxamides will prove equally successful in the development of nucleic acid–based and clinical chemistry assays as they have in immunoassays.

REFERENCES

1. Yalow, R. S. and Berson, S. A., Assay of plasma insulin in human subjects by immunological methods, *Nature,* 184, 1648, 1959.
2. Pratt, W. D., *The Abbott Almanac: 100 Years of Commitment to Quality Health Care,* The Benjamin Company, Elmsford, NY, 1987, 224.
3. Burtis, C. A., Advanced technology and its impact on the clinical laboratory, *Clin. Chem.,* 33, 352, 1987.
4. Forsman, R. W., Why is the laboratory an afterthought for managed care organizations? *Clin. Chem.,* 42, 813, 1996.

5. Koch, D. D., How to evaluate and implement new technologies in an era of managed care and cost containment, *Clin. Chem.,* 42, 797, 1996.
6. Logue, J., Federal reimbursement to laboratories, *Clin. Chem.,* 42, 817, 1996.
7. Schroeder, H. R., Vogelhut, P. O., Carrico, R. J., Boguslaski, R. C., and Buckler, R. T., Competitive protein binding assay for biotin monitored by chemiluminescence, *Anal. Chem.,* 48, 1933, 1976.
8. Woodhead, J. S., Simpson, J. S. A., Campbell, A. K., Ryall, M. E. T., Hart, R., and McCapra, F., Practical immunoassay—the present state of the art. Chemiluminescence immunoassay, *Anal. Proc.* (London), 18, 102, 1981.
9. Woodhead, J. S., Weeks, I., Beheshti, I., Campbell, A. K., and McCapra, F., Acrdinium esters as high-specific-activity labels in immunoassay, *Clin. Chem.,* 29, 1474, 1983.
10. Rak, J., Skurski, P., and Blazejowski, J., Toward an understanding of the chemiluminescence accompanying the reaction of 9-carboxy-10-methylacridinium phenyl ester with hydrogen peroxide, *J. Org. Chem.,* 64, 3002–3008, 1999.
11. Hart, R. C. and Taaffe, L. R., The use of acridinium ester-labelled streptavidin in immunoassays, *J. Immunol. Methods,* 101, 91, 1987.
12. Rauhut, M. M., Sheehan, D., Clarke, R. A., Roberts, B. G., and Semsel, A. M., Chemiluminescence from the reaction of 9-chlorocarbonyl-10-methylacridinium chloride with aqueous hydrogen peroxide, *J. Org. Chem.,* 30, 3587–3592, 1965.
13. McCapra, F., Chemical mechanisms in bioluminescence, *Acc. Chem. Res.,* 9, 201, 1976.
14. McCapra, F., The chemistry of bioluminescence, *Proc. R. Soc. London, Ser. B,* 215, 247, 1982.
15. McCapra, F., Chemiluminescence of organic compounds, *Prog. Org. Chem.,* 8, 231, 1973.
16. Law, S. J., Miller, T., Piran, U., Klukas, C., Chang, S., and Unger, J., Novel poly-substituted aryl acridinium esters and their use in immunoassay, *J. Biolumin. Chemilumin.,* 4, 88, 1989.
17. Mattingly, P. G. and Bennett, L. G., Chemiluminescent Acridinium Salts, U.S. Patent Appl. 921,979, 1986.
18. Molz, P., Skrzipczyk, H. J., Lübbers, H., Strecker, H., Schnorr, G. E., and Kinkel, T., New Chemiluminescent 9-Carboxy-Acridinium Compounds and Their Conjugates with Biologically Interesting Substances, Used in Luminescent Immunoassays, German Patent Appl. DE3628573, 1988.
19. Zomer, G. and Stavenuiter, J. F. C., New Acridinium Compounds Useful as Chemiluminogenic Labels for Both Heterogeneous and Homogeneous Immunoassays, European Patent Appl. 324,202, 1989.
20. Adamczyk, M., Chen, Y.-Y., Mattingly, P. G., Moore, J. A., and Shreder, K., Modulation of the chemiluminescent signal from N^{10}-(3-sulfopropyl)-*N*-sulfonylacridinium-9-carboxamides, *Tetrahedron,* 55, 10899, 1999.
21. Mattingly, P. G., Chemiluminescent 10-methylacridinum-9-(*N*-sulfonylcarboxamide) salts. Synthesis and kinetics of light emission, *J. Biolumin. Chemilumin.,* 6, 107, 1991.
22. Adamczyk, M., Mattingly, P. G., Chen, Y.-Y., and Pan, Y., Sulfopropylated chemiluminescent *N*-sulfonylacridinium-9-carboxamide salts, in *Bioluminescence and Chemiluminescence. Perspectives for the 21st Century. Proceedings of the 10th International Symposium on Bioluminescence and Chemiluminescence,* Bologna, Italy, September 1998, A. Roda, M. Passagli, L. J. Kricka, and P. E. Stanley, Eds., John Wiley & Sons, Chichester, 1999, 37.
23. Adamczyk, M., Chen, Y.-Y., Mattingly, P. G., and Pan, Y., Neopentyl 3-triflyloxypropanesulfonate. A reactive sulfopropylation reagent for the preparation of chemiluminescent labels, *J. Org. Chem.,* 63, 5636, 1998.
24. Adamczyk, M., Mattingly, P. G., Moore, J. A., and Pan, Y., Synthesis of a chemiluminescent acridinium hydroxylamine (AHA) for the direct detection of abasic sites in DNA, *Org. Lett.,* 1, 771, 1999.
25. Schmitt, K. D., Surfactant-mediated phase transfer as an alternative to propanesultone alkylation. Formation of a new class of zwitterionic surfactants, *J. Org. Chem.,* 60, 5474, 1995.
26. Gautun, O. R., Carlsen, P. H. J., Maldal, T., Vikane, O., and Gilje, E., Selective synthesis of aliphatic ethylene glycol sulfonate surfactants, *Acta Chem. Scand.,* 50, 170, 1996.
27. Ikenoue, Y., Saida, Y., Kira, M., Tomozawa, H., Yahima, H., and Kobayashi, M., A facile preparation of a self-doped conducting polymer, *J. Chem. Soc. Chem. Commun.,* 1694, 1990.
28. Grosius, P. and Gallot, Y., Synthesis of polyelectrolytes by sulfopropylation of poly(*p*-bromostyrene) chains, *C. R. Acad. Sci., Ser. C,* 271, 487, 1970.
29. Flanagan, J. H., Khan, S. H., Menchen, S., Soper, S. A., and Hammer, R. P., Functionalized tricarbocyanine dyes as near-infrared fluorescent-probes for biomolecules, *Bioconjugate Chem.,* 8, 751, 1997.

30. Ruegg, U. T. and Rudinger, J., Alkylation of cysteine thiols with 1,3-propane sultone, *Methods Enzymol.,* 47, 116, 1977.
31. Ruegg, U. T. and Rudinger, J., Reaction of cysteine thiol groups with 1,3-propane sultone. *S*-3-Sulfopropyl as a modifying group for protein chemistry, *Int. J. Peptide Protein Res.,* 6, 447, 1974.
32. Adamczyk, M., Fishpaugh, J. R., Gebler, J. C., Mattingly, P. G., and Shreder, K., Detection of reaction intermediates by flow injection electrospray ionization mass spectrometry: reaction of chemiluminescent *N*-sulfonylacridinium-9-carboxamides with hydrogen peroxide, *Eur. Mass Spectrosc.,* 4, 121, 1998.
33. Adamczyk, M., Gebler, J. C., and Mattingly, P. G., Characterization of protein-hapten conjugates. 2. Electrospray mass spectrometry of bovine serum albumin-hapten conjugates, *Bioconjugate Chem.,* 7, 475–481, 1996.
34. Adamczyk, M., Fishpaugh, J., Mattingly, P. G., and Shreder, K., Tracermer signal generators: an arborescent approach to the incorporation of multiple chemiluminescent labels, *Bioorg. Med. Chem. Lett.,* 8, 3595, 1998.
35. Mattingly, P. G., Mono-protected diamines. N^{α}-(tert-Butoxycarbonyl)-α,ω-alkanediamine hydrochlorides from amino alcohols, *Synthesis,* 366, 1990.
36. Huber, E., Klein, C., and Betz, H.-G., Preparation of Aminoalkylmaleimides and Their Conjugates with Haptens and Antigens, German Patent Appl. DE3919915A1, 1990.
37. Dean, R. T., One Vial Method for Labeling Protein/Linker Conjugates with Technetium-99m, U.S. Patent 5,180,816, 1993.
38. Chorev, M., Caulfield, M. P., Roubini, E., McKee, R. L., Gibbons, S. W., Leu, C. T., Levy, J. J., and Rosenblatt, M., A novel, mild, specific and indirect maleimido-based radioiodolabeling method. Radiolabeling of analogs derived from parathyroid hormone (PTH) and PTH-related protein (PTHrP), *Int. J. Peptide Protein Res.,* 40, 445, 1992.
39. Khomutov, A. R., Vepsalainen, J. J., Shvetsov, A. S., Hyvonen, T., Keinanen, T. A., Pustobaev, V. N., Eloranta, T. O., and Khomutov, R. M., Synthesis of hydroxylamine analogs of polyamines, *Tetrahedron,* 52, 13751, 1996.
40. Moore, J. A., Adamczyk, M., Mattingly, P. G., and Pan, Y., Determination of solution binding affinities of an anti-T4 Fab fragment for a library of thyroxine analogs and tracers using biacore surface plasmon resonance, in *Book of Abstracts, 214th ACS National Meeting,* American Chemical Society, Washington, D.C., 1997, ORGN114.
41. Adamczyk, M., Johnson, D. D., Mattingly, P. G., Moore, J. A., and Pan, Y., Immunoassay reagents for thyroid testing. 3, Determination of the solution binding affinities of a T-4 monoclonal-antibody Fab fragment a library of thyroxine analogs using surface-plasmon resonance, *Bioconjugate Chem.,* 9, 23, 1998.
42. Adamczyk, M., Mattingly, P. G., and Reddy, R. E., The synthesis of 6β-amino-estradiol and its biotin, acridinium, and fluorescein conjugates, *Steroids,* 63, 130, 1998.
43. Adamczyk, M., Chen, Y.-Y., Moore, J. A., and Mattingly, P. G., Estradiol-mimetic probes. Preparation of 17α-(6-aminohexynyl)estradiol biotin, fluorescein and acridinium conjugates, *Bioorg. Med. Chem. Lett.,* 8, 1281, 1998.
44. Reddy, R. E., Adamczyk, M., and Mattingly, P. G., A convenient stereoselective synthesis of 6α and 6β-aminoestradiol. Preparation of estradiol probes, in *Book of Abstracts, 214th ACS National Meeting,* American Chemical Society, Washington, D.C., 1997, ORGN213.
45. Adamczyk, M., Grote, J., Moore, J. A., Rege, S. D., and Yu, Z., Structure-binding relationships for the interaction between a vancomycin monoclonal antibody Fab fragment and a library of vancomycin analogues and tracers, *Bioconjugate Chem.,* 10, 176, 1999.
46. Adamczyk, M. and Grote, J., Efficient synthesis of 3-aminodigoxigenin and 3-aminodigitoxigenin probes, *Bioorg. Med. Chem. Lett.,* 9, 771, 1999.
47. Bodanszky, M. and Bodanszky, A., *The Practice of Peptide Synthesis,* Springer-Verlag, New York, 1984, 284.
48. Adamczyk, M., Fishpaugh, J. R., and Mattingly, P. G., Resin-supported labeling reagents, *Bioorg. Med. Chem. Lett.,* 9, 217, 1999.
49. Abraham, R., Moller, D., Gabel, D., Senter, P., Hellström, I., and Hellström, K. E., The influence of periodate oxidation on monoclonal antibody avidity and immunoreactivity, *J. Immunol. Methods,* 144, 77, 1991.
50. Fino, J. R., Codacovi, L., Chan, C., and Mattingly, P. G., unpublished data, 1988.

51. Nitta, N., Kuge, O., Yui, S., Tsugawa, A., Negishi, K., and Hayatsu, H., A new reaction useful for chemical cross-linking between nucleic acids and proteins, *FEBS Lett.,* 194, 1984.
52. Laayoun, A., Decout, J.-L., Defrancq, E., and Lhomme, J., Hydrolysis of oligodeoxyribonucleotides containing 8-substituted purine nucleosides. A new route for preparing abasic oligodeoxyribonucleotides, *Tetrahedron Lett.,* 35, 4991, 1994.
53. Peoc'h, D., Meyer, A., and Imbach, J. L., Efficient chemical synthesis of oligodeoxynucleotides containing a true abasic site, *Tetrahedron Lett.,* 32, 207, 1991.
54. Vasseur, J. J., Peoc'h, D., Rayner, B., and Imbach, J. L., Derivatization of oligonucleotides through abasic site formation, *Nucleosides Nucleotides,* 10, 107, 1991.
55. Daniel, S. G., Westling, M. E., Moss, M. S., and Kanagy, B. D., FastTag nucleic acid labeling system: a versatile method for incorporating haptens, fluorochromes and affinity ligands into DNA, RNA and oligonucleotides, *BioTechniques,* 24, 484, 1998.
56. Bonini, A., Mackowiak, J., Kotlinski, S., Abunimen, N., and Keirans, W., Results assurance in automated blood bank virology testing through design and system integretion, *Vox Sanguinis,* 70(Suppl. 2), 40, 1996.
57. Hughes, W., Clinical evaluation of the Abbott Prism, *Vox Sanguinis*, 70(Suppl. 2), 40, 1996.
58. Kay, D. J., Seed, C. R., and Cobain, T. J., The Abbott Prism viral screening system: the Western Australian experience, *Abstr. General Meeting of the Am. Soc. for Microbiol.,* 97, 585, 1997.
59. Leete, J., Hojvat, S., Hu, R., Makela, R., Guidinger, P., Hughes, W., and Smilovici, W., Multi-center evaluation of an automated assay for the detection of antibodies to hepatitis C virus, *Vox Sanguinis,* 70(Suppl. 2), 40, 1996.
60. Sekiguchi, S., Kato, T., and Ikeda, H., Evaluation of Prism HBs antigen assay for blood donor screening, *Vox Sanguinis,* 70(Suppl. 2), 40, 1996.
61. Wolf-Rogers, J., Weare, J. A., Rice, K., Robertson, E. F., Guidinger, P., Khalil, O. S., and Madsen, G., A chemiluminescent, microparticle-membrane capture immunoassay for the detection of antibody to hepatitis B core antigen, *J. Immunol. Methods,* 133, 191, 1990.
62. Weare, J. A., Robertson, E. F., Madsen, G., Hu, R., and Decker, R. H., Improvement in the specificity of assays for detection of antibody to hepatitis B core antigen, *J. Clin. Microbiol.,* 29, 600, 1991.
63. Khalil, O. S., Zurek, T. F., Tryba, J., Hanna, C. F., Hollar, R., Pepe, C., Genger, K., Brentz, C., Murphy, B., Abbunimeh, N., Carver, R., Harder, P., Coleman, C., Roberston, G., and Wolf-Rogers, J., Abbott Prism: a multichannel heterogeneous chemiluminescence analyzer, *Clin. Chem.,* 37, 1540, 1991.
64. Khalil, O. S., Zurek, T. F., Pepe, C., Genger, K., Huff, D. G., Coleman, C., Hanna, C., Hu, R., Mackowiack, J., and Bennett, L., Detection apparatus for multiple heterogeneous chemiluminescence immunoassay configurations, *Anal. Biochem.,* 196, 61, 1991.
65. Khalil, O. S., Mattingly, P. G., Genger, K., Mackowiak, J., Butler, J., Pepe, C., Zurek, T. F., and Abunimeh, N., Automated chemiluminescence immunoassay measurements, *Proc. Ultrasensitive Lab. Diag.,* 1895, 28, 1993.
66. Khalil, O. S., Hanna, C. F., Huff, D., Zurek, T. F., Murphy, B., Pepe, C., and Genger, K., Reaction tray and noncontact transfer method for heterogeneous chemiluminescence immunoassays, *Clin. Chem.,* 37, 1612, 1991.
67. Li, D. J., Sokoll, L. J., and Chan, D. W., Automated chemiluminescent immunoassay analyzers, *J. Clin. Ligand Assay,* 21, 377, 1998.
68. Danna, A., Figard, S., Ramp, J., Groskopf, W., Kawamoto, T., Vannest, R., Abano, D., George, S., Wikstrom, K., Kumagai, Y., and Shaw, N., Automated chemiluminescent paramagnetic microparticle immunoassays for AFP, CEA, CA19-9, PSA and free PSA on the Abbott Architect i2000 instrument, *Tumor Biol.,* 18(Suppl. 2), P-26, 1997.
69. Chan, D. W., Sokoll, L. J., Jones, K. A., Partin, A. W., and Ramp, J. M., Performance of free and total PSA on the Abbott Architect i2000 automated immunoassay system, *Clin. Chem.,* 44, A46, 1998.
70. Anawis, M., Fico, R., Jeng, K. Y., Quinn, F., Ahmed, A., Berman, M., Brotherton, D., Engstrom, E., Finley, D., Hansen, J., Kaplan, M., Kapsalis, A., Klein, C., Lach, A., Leonard, B., McCarrier, J., Mooney, M., Nelson, M., Scopp, R., Vorwald, M., Wong, M., Grenier, F., and Trimpe, K., Performance of automated chemiluminescent microparticle immunoassays for TSH, free T4, total T4, free T3 and total T3 on the Abbott i2000 system, *Clin. Chem.,* 43, 301, 1997.

71. Black, W., Aoys, E., Demedina, M., Jaffe, K., Kuhns, M., Lisnic, A., Prostko, I., Stewart, S., Xu, L., Schiff, E. R., and Spronk, A. M., Performance of an automated chemiluminescent paramagnetic microparticle immunoassay for the detection of antibodies to HIV-1, HIV-2 and HIV-1 group-o on the Abbott i2000 system, *Clin. Chem.,* 43, 616, 1997.
72. Blevins, L. M., Baugher, W., Drake, C. J., Nolan, C. H., Bicok, B., Davidson, C. L., Eng, K., Khalil, G., Kondic, K. A., Kramer, C. E., Levy, A., McInerney, C. C., Milovanovic, M. D., Pacenti, D., Pestel, C. D., Picking, J. M., Spring, T. G., Wang, P., and Trimpe, K. L., Performance of automated chemiluminescent paramagnetic microparticle immunoassays for β-HCG, LH, FSH and prolactin on the Abbott i2000 system, *Clin. Chem.,* 43, 302, 1997.
73. Kanehiro, M., Kawamoto, T., Groskopf, W., George, S., Sai, K., Wikstrom, K., Kumagai, Y., and Shaw, N., Performance of automated chemiluminescent paramagnetic microparticle immunoassays for α-fetoprotein, carcinoembryonic antigen and carbohydrate antigen-19-9 on the Abbott i2000 system, *Clin. Chem.,* 43, 533, 1997.
74. Lipka, J., Chiba, T., Kitamura, M., Vickstrom, R., Wiesner, D., Yamada, K., Krishnan, K., Kuhns, M., Schneck, A., Demedina, M., Schiff, E. R., and Spronk, A. M., Performance of automated chemiluminescent paramagnetic microparticle immunoassays for Hbsag, Hbsab, and HbCab on the Abbott i2000 system, *Clin. Chem.,* 43, 618, 1997.
75. Moklee, E., Adamczyk, J., Christensen, M., Shipchandler, M., Westerberg, D., Yao, H., and Beggs, M., Performance of an automated chemiluminescent paramagnetic microparticle immunoassay for glycated hemoglobin assay on the Abbott i2000 system, *Clin. Chem.,* 43, 183, 1997.
76. Pennington, C., Rudnick, S., Simondsen, R., Dubler, R., Jeanblanc, N., Joysumpoa, J., Kern, L., Palafox, M., Preisig, J., Rupani, H., Sanchez, B., Seekins, K., Sullivan, M., and Beggs, M., Performance of automated chemiluminescent paramagnetic microparticle immunoassays for ferritin, vitamin-B12, and folate on the Abbott i2000 system, *Clin. Chem.,* 43, 791, 1997.
77. Rufo, G., Callear, K., Demedina, M., Johnson, I., Kuhns, M., Lagedrost, J., Markese, J., Noonan, G., Schiff, E. R., and Spronk, A. M., Performance of an automated chemiluminescent paramagnetic microparticle immunoassay for hepatitis-c virus (HCV) on the Abbott i2000 system, *Clin. Chem.,* 43, 617, 1997.
78. Bouma, S., Worobec, S., Baker, A., Dubler, R., Frias, E., Ginsburg, S., Hehmann, M., Ishman, S., Gardiner, M., Hsu, S., Pacenti, D., Pascucci, T., Patel, C., Piechura, J., Schueller, M., Sheu, M., Wendland, D., and Wang, P., Performance of automated chemiluminescent paramagnetic microparticle immunoassays for estradiol, progesterone, and testosterone on the Abbott i2000 system, *Clin. Chem.,* 43, 295, 1997.

6 Bioluminescence Detection of Proteolytic Bond Cleavage by Using Recombinant Aequorin*

Sapna K. Deo, Jennifer C. Lewis, and Sylvia Daunert

CONTENTS

6.1 INTRODUCTION

Proteases catalyze the cleavage of amide bonds of proteins producing small oligopeptides or free amino acids, and, thus, these enzymes play a critical role in various cell processes.[1,2] The actions of certain proteases within the cell are important in that they are involved in metabolic digestion, complement activation, fertilization, and the production of peptide hormones.[3] For example, serine peptidases are essential for coagulation and fibrinolysis in blood plasma.[4] Therefore, considerable interest has been placed on the study of proteases and on the detection of peptide bond cleavage.

Processing of polyproteins not only occurs in plant and animal cells, but is also crucial for viral assembly and replication. Viruses encode for their own specific proteolytic enzymes, and do not rely on host cell proteinases.[2] The human immunodeficiency virus type 1 (HIV-1) protease

* This chapter is reprinted from Deo, S. K. et al., *Anal. Biochem.*, 281, 87–94, © 2000 by Academic Press. Reprinted by permission of the publisher.

0-8493-0719-8/02/$0.00+$1.50

encoded by HIV plays a key role in the development of AIDS, and has been extensively studied.[5,6] The protease has been identified as a prime target for the design of inhibitors to be used as potential treatment for the AIDS virus.[7] Given the importance of this protease, it was chosen as a model for the development of the bioluminescent system for detection of peptide bond cleavage that employs a mutant of the photoprotein aequorin.

Aequorin is a naturally bioluminescent protein originally isolated from the jellyfish *Aequorea victoria*. It has been used extensively as a calcium indicator[8] and, more recently, as a highly sensitive quantitative label in analytical assay systems.[9–11] Aequorin exists as a stable complex of the apoprotein, a chromophoric unit (coelenterazine), and molecular oxygen. Upon addition of Ca^{2+}, aequorin undergoes a change in conformation, which leads to the oxidation of coelenterazine to coelenteramide, with release of CO_2 and light (λ_{max} ~ 469 nm). The flash-type emission of light occurs as a single-turnover event lasting less than 5 s. Unlike fluorescence, the excited state is produced through a chemical reaction providing a bioluminescence signal that has virtually no background, allowing it to be detected down to the attomole range.[9]

Several methods and commercially available systems have been developed to measure and determine the kinetic parameters of enzymatic bond-breaking events. In earlier studies, bioluminescent protein bacterial luciferase was used as a direct substrate in assay for proteases.[12] However, to date, none of the methods mentioned above has employed a genetically engineered bioluminescent fusion protein that can function as a highly sensitive direct means of detecting hydrolysis of amide bonds by proteases. The bioluminescence detection system reported here was developed by constructing a fusion protein between a chosen recognition sequence for the HIV-1 protease and a mutant of aequorin.

The mutant of aequorin employed contained no cysteine residues in the apoprotein structure. The recognition site selected for the cleavage was S-E-N-Y-P-I-V, which corresponds to an optimum natural substrate for the HIV-1 protease located on the *gag-pol* polyprotein.[13,14] Specifically, the peptide bond between the amino acid residue tyrosine and proline is cleaved by the protease. The spacers were introduced before and after the recognition site present within the fusion protein to limit the possibility of steric hindrance, and to produce a more accessible cleavage site for the protease. The fusion protein was biotinylated through a unique cysteine residue introduced at the N terminus, and then site-specifically immobilized onto a neutravidin-coated 96-well microtiter plate. Upon cleavage by the HIV-1 protease, the aequorin label is released into the liquid phase, thus reducing the bioluminescent signal measured from the solid phase as shown in Figure 6.1. The decrease in the bioluminescence signal is correlated to the activity of the protease. The assay was also employed to determine inhibition constants of three HIV-1 protease inhibitors.

6.2 EXPERIMENTAL

6.2.1 MATERIALS

Tris(hydroxymethyl)amino methane (Tris) free base, ethylenediaminetetraacetic acid (EDTA) sodium salt, dithiothreitol (DTT), agar, glucose, sodium dodecyl sulfate (SDS), sodium phosphate, and all other reagents were purchased from Sigma (St. Louis, MO). HQH, quaternized polyethyleneimine anionic exchanger, was purchased from Perseptive Biosystems (Cambridge, MA). Polyethylene oxide (PEO) maleimide-activated biotin, reactibind 96-well neutravidin-coated white polystyrene microtiter plates, biotin-free bovine serum albumin (BSA), and the polyacrylamide 6000 desalting column were purchased from Pierce (Rockford, IL). Microlite 2 polystyrene microtiter plates were purchased from Dynex (Chantilly, VA). Bradford protein assay kit was purchased from Bio-Rad Laboratories (Hercules, CA). Coelenterazine was purchased from Biosynth International (Naperville, IL). Luria Bertani (LB) broth, *Eco*R I and *Sal* I restriction enzymes, *Taq* and *Pfu* polymerases, and all other molecular biology reagents were purchased from Gibco-BRL (Gaithersburg, MD). Recombinant HIV-1 protease, acetyl pepstatin, and Ac-Leu-Val-phenylalaninal were purchased from Bachem

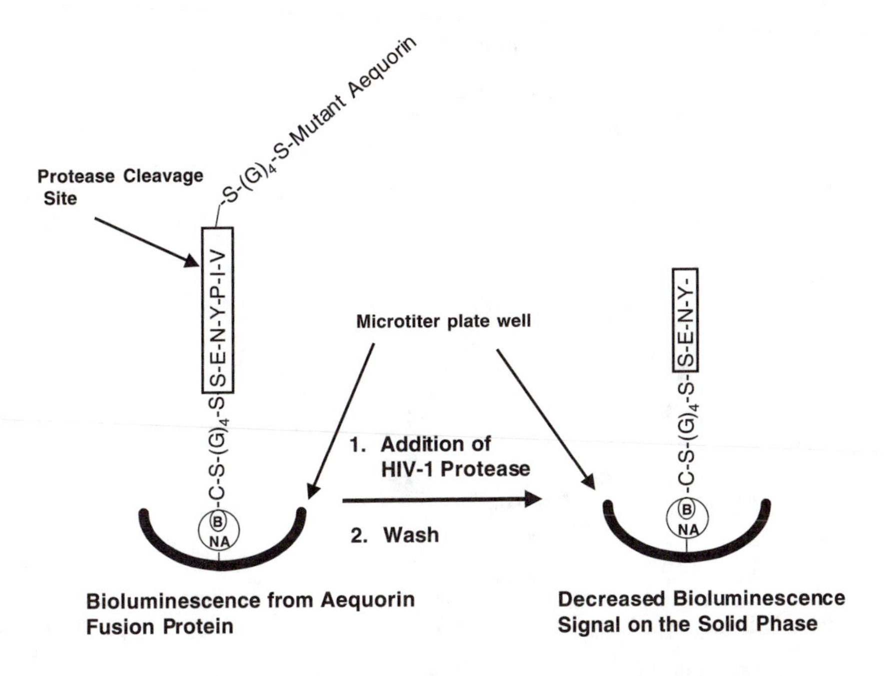

FIGURE 6.1 Schematic of the assay design using the aequorin fusion protein in which B represents biotin and NA represents neutravidin immobilized on the wells. Amino acids are represented with their standard one-letter code. (From Deo, S. K. et al., *Anal. Chem.*, 281, 87, 2000. With permission.)

Biosciences (King of Prussia, PA). Pepstatin A was purchased from CALBIOCHEM (San Diego, CA). All the primers used for the polymerase chain reaction (PCR) were purchased from Operon Technologies (Alameda, CA). All solutions were prepared using deionized (Milli-Q Water Purification System, Millipore, Bedford, MA) distilled water. All chemicals were reagent grade or better and were used as received.

6.2.2 Apparatus

PCR was performed on a Perkin Elmer Gene Amp PCR System 2400 (Norwalk, CT). Ion-exchange chromatography was performed using a BioCAD-SPRINT perfusion chromatography system by Perseptive Biosystems. Fractions containing the aequorin fusion protein were lyophilized using a VirTis Bench Top 3 freeze dryer (Gardiner, NY). The purity of the fusion protein was verified by sodium dodecyl sulfate-polyacrylamide gel electrophoresis (SDS-PAGE) on a Phast System (Pharmacia Biotech, Uppsala, Sweden). Bioluminescence measurements were made on a MLX microtiter plate luminometer from Dynex, using a 100-μl fixed-volume injector. All luminescence intensities reported are the average of a minimum of three replicates and have been corrected for the contribution of the blank. All molecular biology procedures were performed using standard protocols.[15]

6.2.3 Methods

6.2.3.1 Preparation and Isolation of the Fusion Protein

The gene of a mutant apoaequorin in which all three cysteines of the native aequorin were replaced by serines was obtained by PCR from plasmid pSD110. To obtain the gene construct that codes for the HIV-1 recognition site–mutant apoaequorin fusion protein, the following four primers were employed:

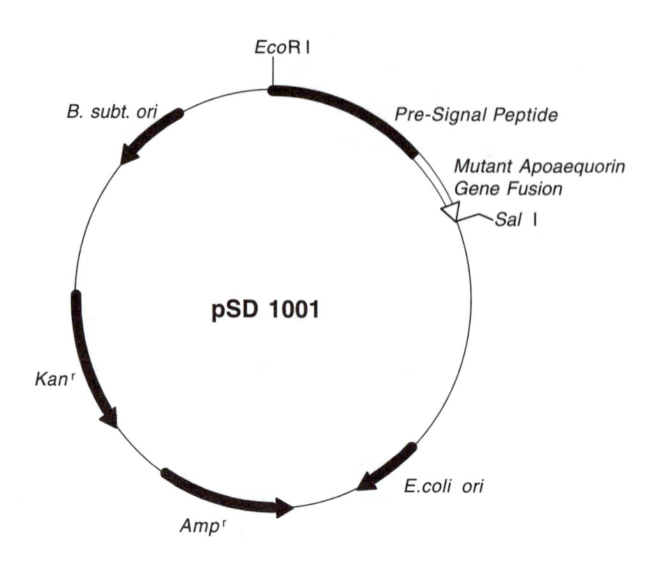

FIGURE 6.2 Schematic representation of the plasmid pSD1001 containing the oligonucleotide sequences of the pre-signal peptide and the gene of fusion protein. (From Deo, S. K. et al., *Anal. Chem.*, 281, 87, 2000. With permission.)

1. TCTGAAAACTATCCTATAGTTTCTGGAGGAGGAGGATCTGTGAAATTGACGTCA-GAC
2. GCGGCGGCGGTCGACTTAGGGGACAGCTCCGTA
3. GCCCAGGCGGCAGGGAAATGTTCTGGAGGAGGAGGATCTTCTGAAAACTATC-CTATA
4. TAATACGACTCACTATAGGG

Primers 1 and 2 were used to amplify a DNA fragment containing the HIV-1 protease recognition site followed by a six amino acid spacer of two serines and four glycine residues, and then the mutant apoaequorin. The second DNA fragment amplified with primers 3 and 2 contained the unique cysteine residue, another spacer, and the overlap region of "pre" signal sequence from the *Bacillus* expression system. The "pre" signal peptide facilitates the secretion of the protein into the medium. The two DNA fragments were then used in an overlap extension PCR performed to obtain the desired DNA construct of the fusion protein with the "pre" signal peptide using the outside primers (1 and 4) by following the standard protocol.[16] This construct was digested with the restriction enzymes *Eco*R I and *Sal* I. Next, the fragment was ligated into the pSbt vector (obtained from L. G. Bachas, University of Kentucky),[17] which had also been digested with the same two enzymes, to yield the pSD1001 vector (Figure 6.2). DNA sequencing was performed at the Macromolecular Center (University of Kentucky) to confirm the presence of the gene for HIV-1 protease recognition site–apoaequorin fusion protein. Bacteria (*B. subtilis*) were transformed with the pSD1001 vector, and then cultured to express the fusion protein. Specifically, bacteria were grown in 250 ml of LB broth containing kanamycin (30 μg/ml) for 16 to 18 h. The culture was centrifuged at 5930 × *g* at 10°C for 30 min to collect the cells. The supernatant containing the fusion protein was removed, and the pH was adjusted to 4.2 by adding concentrated acetic acid to precipitate the fusion protein. This mixture was then centrifuged at 12100 × *g* at 4°C for 30 min, and the precipitate containing the fusion protein was obtained.

6.2.3.2 Purification of the Fusion Protein

The precipitate containing the fusion protein was resuspended in 30 m*M* Tris-HCl, pH 7.0, containing 2 m*M* EDTA and 1 m*M* DTT. The final concentration of DTT was adjusted to 10 m*M* and the protein was mixed with slow stirring for 15 h at 4°C. Then, the protein was purified by perfusion chromatography using a HQH anion-exchange column. A salt gradient from 0.0 to 0.5 *M* NaCl was employed to elute the protein, which was obtained between 0.2 to 0.25 *M* NaCl. Fractions containing the photoprotein were pooled and glucose was added to a final concentration of 30 m*M*. The protein was then lyophilized and resuspended in deionized distilled water. Excess salt and DTT were removed by using a polyacrylamide 6000 desalting column. The purity of the fusion protein was verified by SDS-PAGE using 12.5% polyacrylamide gels, which were developed by silver staining (Pharmacia Biotech). The protein concentration was determined by using the Bradford protein assay, with BSA as the standard.

6.2.3.3 Biotinylation of the Fusion Protein

A fivefold molar excess of PEO maleimide-activated biotin was added to a solution of the fusion protein in a buffer containing 30 m*M* sodium phosphate, 2 m*M* EDTA, and 100 m*M* NaCl at pH 7.2. The reaction mixture was stirred for 4 h at room temperature, and then stirred for 15 h at 4°C. After the conjugation reaction was completed, a fivefold molar excess of coelenterazine was added to the apoaequorin fusion protein, and this mixture was stirred for 15 h at 4°C. Excess of the conjugation reagent and coelenterazine was removed from the protein by employing the polyacrylamide 6000 desalting column pre-equilibrated with a buffer containing 30 m*M* Tris-HCl, 2 m*M* EDTA, 150 m*M* NaCl at pH 7.3.

6.2.3.4 Bioluminescence Emission Study

A volume of 100 μl of the biotinylated fusion protein at a concentration of 4.4×10^{-8} *M* was added to a Microlite 2 polystyrene microtiter plate, and the bioluminescence was measured by injecting 100 μl of luminescence triggering buffer (100 m*M* Tris-HCl, 100 m*M* $CaCl_2$, pH 7.5). The bioluminescence signal was collected at 0.1-s intervals over a 5-s time period.

6.2.3.5 Binder-Saturation Study of Biotinylated Fusion Protein

A stock solution of the biotinylated fusion protein at a concentration of 4.4×10^{-6} *M* was serially diluted with the dilution buffer (30 m*M* Tris-HCl, 2 m*M* EDTA, 150 m*M* NaCl, pH 7.3, containing 0.1 mg/ml biotin-free BSA). A volume of 300 μl of varying concentrations of the fusion protein was immobilized onto a neutravidin-coated microtiter plate by incubating the protein in the wells of the plate for 3 h with shaking at 150 rpm on a rotary shaker. The unbound fusion protein was removed by washing three times with wash buffer (30 m*M* Tris-HCl, 2 m*M* EDTA, 150 m*M* NaCl, pH 7.3, containing 0.1 mg/ml biotin-free BSA). A volume of 100 μl of luminescence-triggering buffer was injected into each microtiter well, and the bioluminescence emitted was collected over a 5-s time period.

6.2.3.6 Temperature Stability Study

A volume of 300 μl of the biotinylated fusion protein of concentration 4.4×10^{-8} *M* was immobilized to the neutravidin-coated wells as described above. Next, the plate containing immobilized fusion protein was incubated with 300 μl of the assay buffer (50 m*M* sodium acetate, 0.15 *M* NaCl, 2 m*M* EDTA, pH 5.8, containing 0.1 mg/ml BSA and 10% glycerol), either at 37°C or at 30°C for time periods ranging from 10 to 120 min. At the end of each time period, the wells were washed with the assay buffer three times and bioluminescence signal was measured.

6.2.3.7 Dose–Response Curve for the HIV-1 Protease

To the plates containing 300 μl of the biotinylated fusion protein of concentration 4.4×10^{-8} *M* immobilized to the neutravidin-coated wells, 300 μl of HIV-1 protease ranging in concentration from 5×10^{-7} *M* to 1×10^{-12} *M* in assay buffer was added, and the plate was incubated at 30°C for 30 min. The neutravidin-coated wells of the microtiter plate were then washed with the assay buffer three times and the bioluminescence was measured on the solid phase.

6.2.3.8 Determination of Inhibitation Constant (IC_{50}) for Pepstatin A

Following immobilization of the fusion protein (300 μl of concentration 4.4×10^{-8} *M*), 150 μl of pepstatin A ranging in concentration from 0.01 to 100 μM in assay buffer was added to the plate followed by addition of 3×10^{-13} moles of HIV-1 protease in assay buffer. The plate was then incubated for 30 min at 30°C prior to a washing step (three times with the assay buffer) that was followed by measurement of the bioluminescence intensity.

6.2.3.9 Determination of Inhibition Constants for Acetyl Pepstatin and Ac-Leu-Val-Phenylalaninal

To the plates containing 300 μl of the biotinylated fusion protein of concentration 4.4×10^{-8} *M* immobilized, a volume of 150 μl of acetyl pepstatin was added at concentrations ranging from 10 to 60 n*M* in assay buffer. The immobilized protein with the acetyl pepstatin was then incubated with 3×10^{-13} mol of HIV-1 protease in assay buffer for time periods of 2 to 30 min at 30°C. After each time period, the neutravidin-coated wells of the microtiter plate were washed with the assay buffer three times and the bioluminescence intensity was measured. This bioluminescence intensity was then related to the concentration of the fusion protein remaining immobilized after the cleavage by the protease employing a calibration curve. A calibration curve was generated by immobilizing several concentrations of the fusion protein and determining the corresponding light intensity (data not shown). The concentration of inhibitor vs. time was plotted, and the velocity was determined from the slope of the line. A Dixon plot of $1/v$ vs. inhibitor concentration was then constructed. The inhibition constant for Ac-Leu-Val-phenylalaninal was determined by following the same procedure as described above for the inhibitor acetyl pepstatin.

6.3 RESULTS AND DISCUSSION

Several strategies have been reported for the detection of amide bond cleavage by HIV-1 protease, including high-performance liquid chromatography (HPLC), spectrophotometric analysis, fluorescence resonance energy transfer (FRET) assay, and ELISA with chromogenic substrate.[18–21] Difficulties with spectrophotometric assays arise from the poor solubility of the peptide substrates, as well as the occurrence of only small changes in the maximum absorption wavelength of the products that results in a high background for the measured signal.[18] The HPLC-based assays, some of which are commercially available, are time-consuming and not adaptable to automation.[19] The detection limit for the commercially available FRET assay is in the nanomolar range and the method is adaptable to automation. The drawback of this assay is that the substrate for the protease is chemically conjugated to the fluorophore, which often leads to the production of nonreproducible heterogeneous conjugates. In addition, the substrate peptide employed is prepared synthetically and, hence, is not as cost effective as when prepared by genetic means.[20] In the modified ELISA, absorbance was used as a mode of detection, and is not a sensitive method.[21] An ideal assay for the detection of amide bond cleavage and for use in high-throughput screening of potential drug inhibitors of the HIV-1 protease would be rapid, easy to perform, and sensitive. In this regard, the use of a bioluminescence protein label such as aequorin, which can be detected at attomole levels, should be well suited for the purpose.

To produce the fusion protein between the natural HIV-1 protease substrate and the mutant apoaequorin employed in the system, a gene encoding for both DNA sequences was constructed by using PCR. The sequence for the fusion protein was then ligated into the expression vector pSbt to obtain the plasmid pSD1001 (see Figure 6.2). In this plasmid design, a six amino acid spacer was introduced between the HIV-1 protease recognition site and the apoaequorin gene. The length of the spacer arm chosen was based on previous work performed in the authors' laboratory where a spacer of the same length had been introduced in the preparation of a fusion protein between aequorin and a single chain anti-salmonella antibody.[22] This fusion protein was employed in the development of an immunoassay for the salmonella antigen. In another study, an immunoassay for an octapeptide was developed using an octapeptide–aequorin fusion protein without any spacer.[9] Although the binder in this case was an antibody and, thus, a much larger size protein than the HIV-1 protease, no steric hindrance was observed. These studies led to the belief that the length of the spacer, six amino acids, was adequate for the development of an assay for HIV-1 protease.

Although aequorin has been successfully expressed in *Escherichia coli* in the authors' laboratory, in this study it was expressed in *B. subtilis*.[9] These cells were chosen as the "pre"-signal peptide of the protein subtilisin from *B. subtilis* provides ease of purification by allowing the protein to be excreted out of the cell into the media. *Bacillus subtilis* cells were transformed with the plasmid pSD1001. The expressed protein was purified using perfusion anion-exchange chromatography. The purity and concentration of the protein were determined using SDS-PAGE and the Bradford assay, respectively. The yield of the purified protein was 21 mg/l of culture.

The purified fusion protein was biotinylated using a maleimide-activated biotin that reacts selectively with the unique sulfhydryl group of the fusion protein. Following the biotinylation reaction, the aequorin fusion protein was generated by addition of an excess amount of coelenterazine. Then, the biotinylated fusion protein was immobilized onto a neutravidin-coated 96-well microtiter plate. A bioluminescence emission study was performed to characterize the protein in terms of its luminescence properties. The fusion protein demonstrated the same flash-type emission characteristics of the native protein with 95% of the total light emitted during a 5-s period. Thus, a 5-s integration time was chosen for all remaining experiments. The half-life of the bioluminescence decay of the fusion protein was determined to be 0.68 s, which is comparable to 0.60 s, obtained for the native aequorin.[9] The difference in half-life could be attributed to the presence of the recognition site for HIV-1 protease, which is seven amino acids in length, flanked by two spacers of six amino acids each, and a biotin molecule.

To determine the amount of fusion protein containing the HIV-1 protease substrate to be used in subsequent experiments, a binder-saturation study was performed. A typical curve depicting the saturation of binding sites in the wells of the plate by the biotinylated fusion protein is presented in Figure 6.3. This figure shows that incubating 300 μl of a 4.4×10^{-8} M solution of the fusion protein per well for the immobilization onto the microtiter plate saturates the available binding sites for biotin on the neutravidin-coated wells. Thus, for the rest of the studies, a concentration of the protein of 4.4×10^{-8} M was employed.

Although the HIV-1 protease is active at both 30 and 37°C, a considerable loss in activity of aequorin above 30°C has been reported in the literature.[20,23] Therefore, the immobilized biotinylated fusion protein was tested for its stability in terms of retaining its bioluminescent properties at 30 and 37°C as shown in Figure 6.4. It was observed that the fusion protein retained its bioluminescent activity at 30°C for about 40 min, and lost 30% of its bioluminescent activity after 50 min. After a 60-min incubation period, it was observed that there is considerable loss (up to 80%) in the bioluminescence emission of the fusion protein. At 37°C, the loss of activity occurred at a much faster rate. In this case, there was a 30% loss of the bioluminescence emission of the protein after 30 min and complete loss of activity after 50 min. Consequently, 30°C was chosen as the optimum temperature and 30 min was selected as the incubation time for the development of the assay. The immobilized fusion protein is stable and active at 4°C for at least 3 weeks.

By using the fixed amount of HIV-1 protease substrate determined from binder-saturation curve, a dose–response curve was generated for the HIV-1 protease as shown in Figure 6.5. The study was performed by incubating various concentrations of the HIV-1 protease at 30°C for 30 min on the microtiter plate. It was demonstrated that the minimum concentration of protease required to yield a

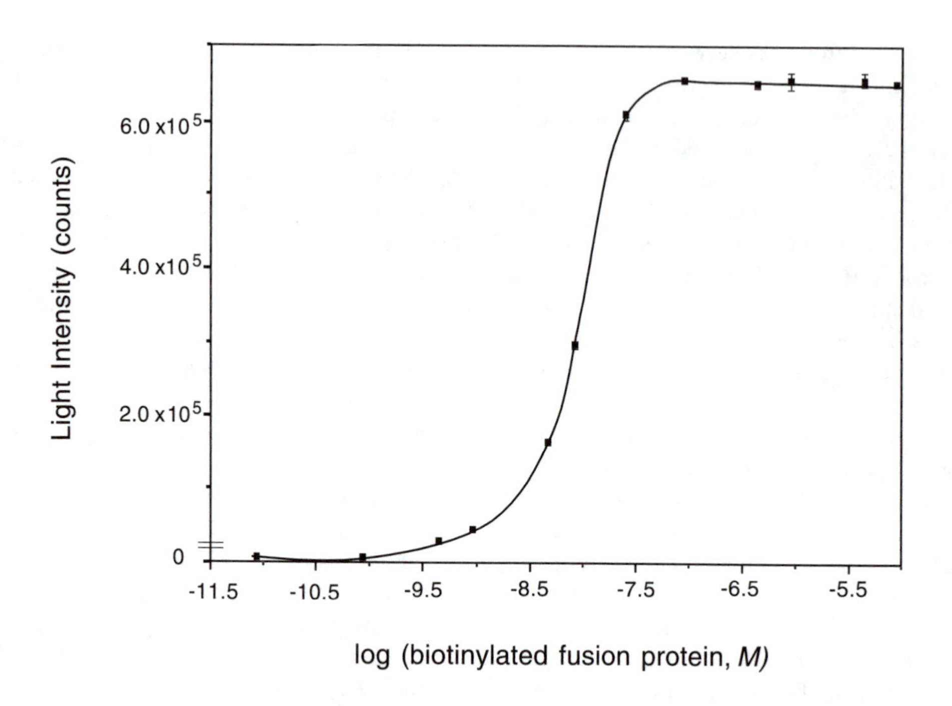

FIGURE 6.3 Curve for the saturation of the binding sites for biotin on the neutravidin-coated wells obtained by immobilizing biotinylated fusion protein. Data are average ±1 SD ($n = 3$). Some error bars are obstructed by the symbols for the points. (From Deo, S. K. et al., *Anal. Chem.,* 281, 87, 2000. With permission.)

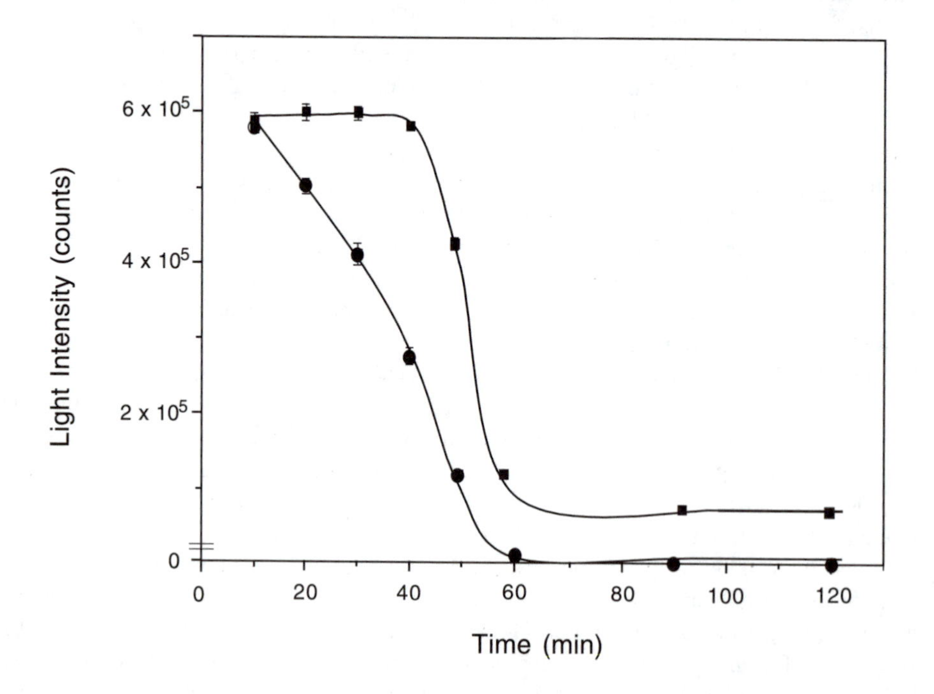

FIGURE 6.4 Temperature–stability curve obtained by incubating immobilized fusion protein either at 30°C (■) or at 37°C (●) for various time periods. Data are the average ±1 SD ($n = 3$). Some error bars are obstructed by the symbols for the points. (From Deo, S. K. et al., *Anal. Chem.,* 281, 87, 2000. With permission.)

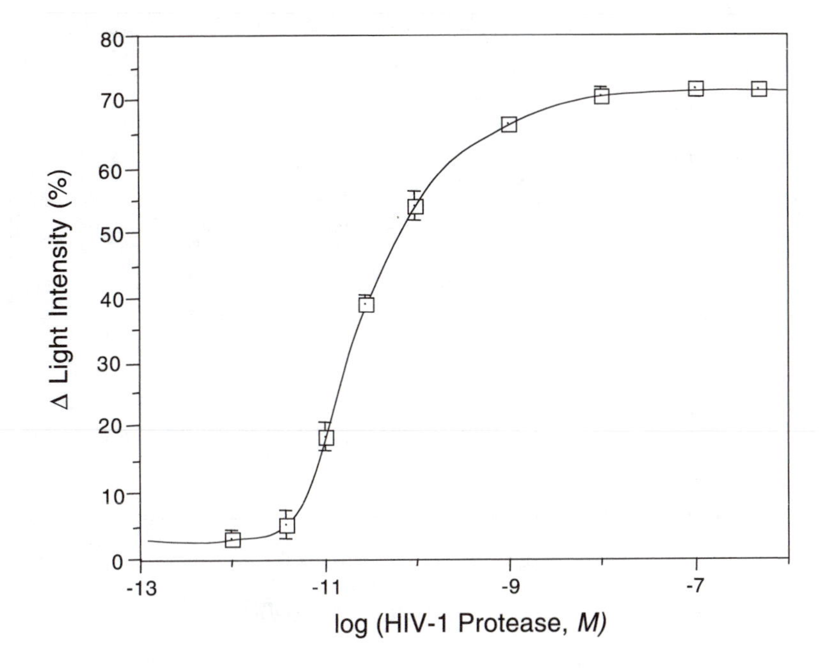

FIGURE 6.5 Dose–response curve for the HIV-1 protease generated by incubating varying concentrations of HIV-1 protease with the immobilized biotinylated fusion protein for 30 min at 30°C. The decrease in the light intensity on the solid phase was measured after a washing step. Data are the average ±1 SD (n = 3). (From Deo, S. K. et al., *Anal. Chem.*, 281, 87, 2000. With permission.)

maximum decrease of 75% in bioluminescence signal after bond cleavage was 1×10^{-8} *M*. There was no significant decrease in the bioluminescence signal with concentrations above 1×10^{-8} *M*. At concentrations between 1×10^{-8} *M* and 1×10^{-11} *M* of the protease, a decrease in the bioluminescence signal ranging from 60 to 70% was observed. Therefore, a concentration of protease of 1×10^{-9} *M* was selected for the remaining studies, which yields an almost 70% decrease in the signal.

To evaluate the bioluminescence system developed for assessing the activity of inhibitors of the HIV-1 protease, inhibition constants for three inhibitors of the HIV-1 protease were determined. The IC_{50} of the competitive inhibitor, pepstatin A was determined employing this assay as shown in Figure 6.6. Almost 100% inhibition of the protease activity, or, in other words, 100% retention of the bioluminescence activity of the immobilized fusion protein, was obtained when a concentration of 90 μM of pepstatin A was used in the presence of 1×10^{-9} *M* HIV-1 protease. The IC_{50} value obtained was 0.4 μM, which is comparable to the literature-reported IC_{50} value of 0.1 to 10 μM for pepstatin A.[24,25]

A Dixon plot ($1/v$ vs. inhibitor concentration) for a competitive inhibitor, acetyl pepstatin, is presented in Figure 6.7A. The equation of the line for competitive inhibitor is $1/v = K_m\,[\mathrm{I}]/V_{max}K_i\,[\mathrm{S}] + (1 + K_m/[\mathrm{S}])/V_{max}$. The inhibition constant (k_i) for acetyl pepstatin was found to be 40 n*M* and is obtained by substituting $v = V_{max}$ in the equation for the Dixon plot. The V_{max} value was determined by constructing a standard Lineweaver–Burk double-reciprocal plot (plot not shown). Values reported in the literature for the k_i of acetyl pepstatin range from 17 n*M* to 1 μM, depending on the assay conditions and the substrates used for the assay.[26] A Dixon plot was also constructed for the inhibitor Ac-Leu-Val-phenylalaninal as shown in Figure 6.7B. The inhibition constant for Ac-Leu-Val-phenylalaninal, a noncompetitive inhibitor of HIV-1 protease, was found to be 0.147 μM. This value was determined by assuming $1/v = 0$ in the equation for noncompetitive inhibition, which is given by $1/v = (1 + K_m/[\mathrm{S}])\,[\mathrm{I}]/V_{max}K_i + (1 + K_m/[\mathrm{S}])/V_{max}$. The literature reported K_i for this inhibitor is 0.9 μM, which was determined by employing solid-phase enzyme immunoassay

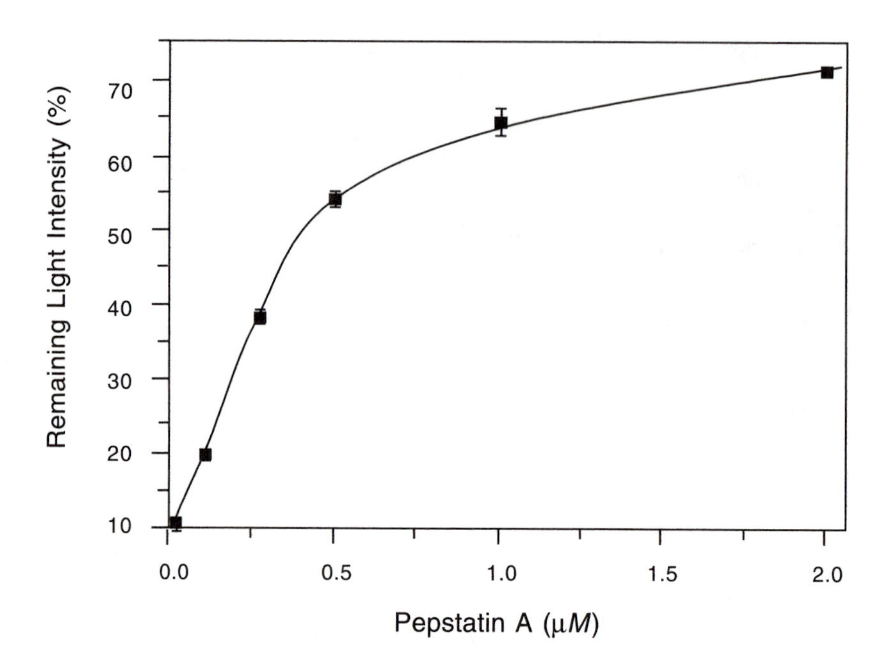

FIGURE 6.6 Inhibition of HIV-1 protease by pepstatin A. The curve for the inhibitor was generated by incubating immobilized fusion protein with a solution of pepstatin A and HIV-1 protease for 30 min at 30°C. Remaining light intensity was measured on the solid phase after a washing step. Data points beyond a concentration of 2×10^{-8} M of pepstatin A are not shown. Data are the average ±1 SD ($n = 3$). Some error bars are obstructed by the symbols for the points. (From Deo, S. K. et al., *Anal. Chem.*, 281, 87, 2000. With permission.)

with a chromogenic substrate.[27] The difference in the values of inhibition constant may be due to the differing assay conditions and the label employed.

In summary, the authors have demonstrated that a bioluminescence-based strategy can be employed in the sensitive detection of peptide bond cleavage by proteases. The assay designed can be readily adapted to automation, and consists of a single incubation step followed by washing, allowing for the direct assay of proteolytic bond cleavage. Another advantage of this method is that the labeled substrate for the protease is genetically engineered, which allows for the production of a highly reproducible labeled substrate in unlimited quantities. Since the label used is a protein, the system does not suffer from any solubility problems under the aqueous conditions of the assay, which can be a drawback when using certain synthetic labels.[20] The detection limit of 1×10^{-11} M obtained for the HIV-1 protease using this assay is two orders of magnitude better than that of currently available methods, which have detection limits in the nanomolar range. A linear range of two to three orders of magnitude is achieved employing this assay. The amounts of HIV-1 protease and aequorin fusion protein substrate employed for the assay were significantly lower than the amounts used in other reported assays. For example, the amount of substrate employed (4.4×10^{-8} M) was two orders of magnitude less in comparison with that used in the commercially available FRET assay from Bachem Biosciences (10.7×10^{-6} M).[20] Also demonstrated is that the assay developed is useful in evaluating the activity of competitive, as well as noncompetitive, inhibitors of the protease.

In conclusion, the method employed here to detect protease activity is quite versatile in that different recognition sites for other physiologically and pharmacologically important proteases could be incorporated into the assay design through the preparation of other bioluminescent fusion proteins. Moreover, the system may find applications in the high-throughput screening of biopharmaceutical drugs that are potential inhibitors of a target protease.

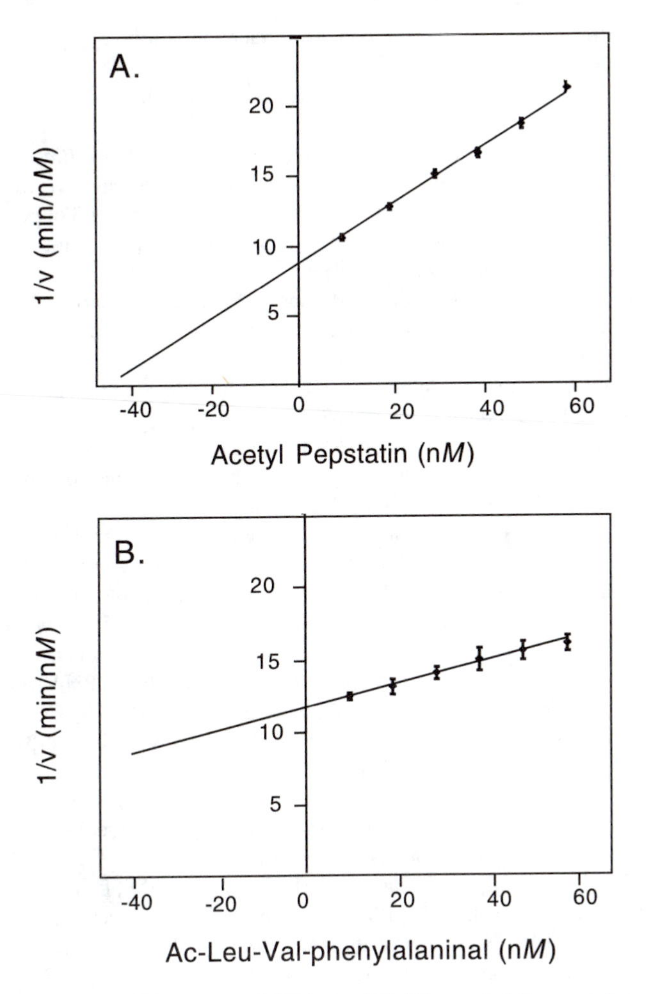

FIGURE 6.7 Dixon plots for HIV-1 protease inhibitors. (A) Dixon plot for acetyl pepstatin generated by incubating immobilized fusion protein with acetyl pepstatin and HIV-1 protease for 2 to 30 min. (B) Dixon plot for Ac-Leu-Val-phenylalaninal was generated under the same conditions as in A. Data are the average ±1 SD ($n = 3$). Some error bars are obstructed by the symbols for the points. (From Deo, S. K. et al., *Anal. Chem.*, 281, 87, 2000. With permission.)

ACKNOWLEDGMENTS

This work was supported by grants from the National Institutes of Health (GM47915) and the Department of Energy (DE-FG05-95ER62010) to S.D. S.D. is a Cottrell Scholar and a Lilly Faculty Awardee. S.K.D thanks the University of Kentucky for an RCTF Fellowship in Biological Chemistry and J.C.L. thanks the National Science Foundation and the University of Kentucky for an NSF-IGERT Fellowship.

REFERENCES

1. Martin, M. T, Angeles, T. S., Sugasawara, R., Aman, N. I., Napper, A. D., Darlsey, M. J., Sanchez, R. I., Booth, P., and Titmas, R. C., Antibody-catalyzed hydrolysis of an unsubstituted amide, *J. Am. Chem. Soc.*, 116, 6508, 1994.

2. Jentsch, S. and Schlenker, S., Selective protein degradation: a journey's end within the proteasome, *Cell,* 82, 881, 1995.
3. Kay, J. and Dunn, B. M., Viral proteinases: weakness in strength, *Biochim. Biophys. Acta,* 1048, 1, 1990.
4. Rawlings, N. D. and Barrett, A. J., Evolutionary families of peptidases, *Biochem. J.,* 290, 205, 1993.
5. Barrie, K. A., Perez, E. E., Lamers, S. L., Farmerie, W. G., and Dunn, B. M., Natural variation in HIV-1 protease, Gag p7 and p6, and protease cleavage sites within Gag/Pol polyproteins: amino acid substitutions in the absence of protease inhibitors in mothers and children infected by human immunodeficiency virus type 1, *Virology,* 219, 407, 1996.
6. Venturini, A., Lopez-Ortiz, F., Alvarez, J. M., and Gonzalez, J., Theoretical proposal of a catalytic mechanism for the HIV-1 protease involving an enzyme-bound tetrahedral intermediate, *J. Am. Chem. Soc.,* 120, 1110, 1998.
7. Wlodawer, A. and Vondrasek, J., Inhibitors of HIV-1 protease: a major success of structure-assisted drug design, *Annu. Rev. Biophys. Biomol. Struct.,* 27, 249, 1998.
8. George, C. H., Kendall, J. M., Campbell, A. K., and Evans, W. H., Connexin-aequorin chimerae report cytoplasmic calcium environments along trafficking pathways leading to gap junction biogenesis in living COS-7 cells, *J. Biol. Chem.,* 273, 29,822, 1998.
9. Ramanathan, S., Lewis, J. C., Kindy, M. S., and Daunert, S., Heterogeneous bioluminescence binding assay for an octapeptide using recombinant aequorin, *Anal. Chim. Acta,* 369, 181, 1998.
10. Crofcheck, C. L., Grosvenor, A. L., Anderson, K., Lumpp, J. K., Scott, D., and Daunert, S., Detecting biomolecules in picoliter vials using aequorin bioluminescence, *Anal. Chem.,* 69, 4768, 1997.
11. Galvan, B. and Christopoulos, T. K., Bioluminescence hybridization assays using recombinant aequorin: application to the detection of prostate-specific antigen mRNA, *Anal. Chem.,* 20, 3545, 1996.
12. Billich, S., Knoop, M., Hansen, J., Strop, P., Sedlacek, J., Mertz, R., and Moelling, K., Synthetic peptides as substrates and inhibitors of human immune deficiency virus-1 protease, *J. Biol. Chem.,* 263, 17,905, 1988.
13. Njus, D., Baldwin, T., and Hastings, J. W., A sensitive assay for proteolytic enzymes using bacterial luciferase as a substrate, *Anal. Biochem.,* 61, 280, 1974.
14. Darke, P. L., Nutt, R. F., Brady, S. F., Garsku, V. M., Ciccarone, T. M., Leu, C., Luma, P. K., Freidinger, R. M., Veber, D. F., and Sigal, I. S., HIV-1 protease specificity of peptide cleavage is sufficient for processing of Gag and Pol polyproteins, *Biochem. Biophys. Res. Commun.,* 156, 297, 1988.
15. Maniatis, T., Fritsch, D. F., and Sambrook, J., Eds., *Molecular Cloning: A Laboratory Manual,* Cold Spring Harbor Laboratory Press, Cold Spring Harbor, NY, 1989.
16. Horton, R. M., Cai, Z., Ho, S. N., and Pease, L. R., Gene splicing by overlap extension: tailor-made genes using the polymerase chain reaction, *Biotechniques,* 8, 528, 1990.
17. Huang, W., Wang, J., Bhattacharyya, D., and Bachas, L. G., Improving the activity of immobilized subtilisin by site-specific attachment to surfaces, *Anal. Chem.,* 69, 4601, 1997.
18. Nashed, N. T., Louis, J. M., Sayer, J. M., Wondrak, E. M., Mora, P. T., Oroszlan, S., and Jerina, D. M., Continuous spectrophotometric assay for retroviral proteases of HIV-1 and AMV, *Biochem. Biophys. Res. Commun.,* 163, 1079, 1989.
19. Tomaszek, T. A., Jr., Maggard, V. W., Bryan, H. G., Moore, M. L., and Meek, T. D., Chromophoric peptide substrates for the spectrophotometric assay of HIV-1 protease, *Biochem. Biophys. Res. Commun.,* 168, 274, 1990.
20. Matayoshi, E. D., Wang, G. T., Kraft, G. A., and Erickson, J., Novel fluorogenic substrates for assaying retroviral proteases by resonance energy transfer, *Science,* 247, 954, 1990.
21. Fournot, S., Roquet, F., Salhi, S. L., Seyer, R., Valverde, V., Masson, J. M., Jouin, P., Pau, B., Nicolas, M., and Hanin, V., Development and standardization of an immuno-quantified solid phase assay for HIV-1 aspartyl protease activity and its application to the evaluation of inhibitors, *Anal. Chem.,* 69, 1746, 1997.
22. Wang, J., Ensor, C. M., Dubuc, G. J., Narang, S. A., and Daunert, S., Fusion proteins of single-chain antibody and photoproteins for assay development in the detection of salmonella antigen, presented at *218th National ACS Meeting Spring,* 1999.
23. Inouye, S., Sakaki, Y., Goto, T., and Tsuji, F., Expression of apoaequorin complementary DNA in *Escherichia coli, Biochemistry,* 25, 8425, 1986.

24. Buttner, J., Dornmair, K., and Schramn, H. J., Screening of inhibitors of HIV-1 protease using an *Escherichia coli* cell assay, *Biochem. Biophys. Res. Commun.*, 233, 36, 1997.
25. Stella, S., Saddler, G., Sarubbi, E., Colombo, L., Stefanelli, S., Denaro, M., and Selva, E., Isolation of α-MAPI from fermentation broths during a screening program for HIV-1 protease inhibitors, *J. Antibiotics*, 44, 1019, 1991.
26. Richards, A., Roberts, R., Dunn, B., Graves, M., and Kay, J., Effective blocking of HIV-1 proteinase activity by characteristic inhibitors of aspartic protinases, *FEBS Lett.*, 247, 113, 1989.
27. Sarubbi, E., Seneci, P. F., Angelastro, M. R., Peet, N. P., Denaro, M., and Islam, K., Peptide aldehydes as inhibitors of HIV protease, *FEBS Lett.*, 319, 253, 1993.

7 Development of New Label Enzymes for Bioluminescent Enzyme Immunoassay

Masako Maeda

CONTENTS

7.1 INTRODUCTION

Bioluminescence is widely used in assays that require high sensitivity.[1] The use of these techniques has been facilitated by the expression of various luciferase and phosphoprotein genes in *Escherichia coli* or *Saccharomyces cerevisiae*.[2–4] Recombinant firefly luciferase is commercially available and has been used for enzyme immunoassay (EIA). However, firefly luciferase loses much of its activity when treated with the chemical cross-linking reagents used in production of conjugates. To address this problem, we have established two highly sensitive bioluminescent EIAs (BLEIAs) in which thermostable acetate kinase (AK) is used to label the conjugate.[5] We have utilized biotinylated AK in our EIA systems to assay for human chorionic gonadotrophin (hCG), murine interleukin-6 (mIL-6), and other clinically important substances.[6,7] In addition, we have used biotinylated mutants of thermostable firefly luciferase (*Luciola lateralis*), which emit different wavelengths of light,[8,9] to develop a simultaneous BLEIA of pepsinogen I and II (PGI and PGII) in serum.[10]

7.2 BLEIA USING AK AS A LABEL ENZYME

We have developed the first sensitive BLEIA that utilizes AK-labeled antibody or antigen coupled with the firefly luciferase–luciferin system. AK is an enzyme that converts ADP and acetylphosphate to ATP and acetate (Figure 7.1). AK activity can thus be determined by using luciferin and recombinant firefly luciferase to measure the ATP that is generated. The AK–luciferase system differs from conventional chemiluminescence immunoassays in that it involves no decay of light effect.

End point and rate assays of AK activity were performed (Figure 7.2). A comparison of these methods is shown in Table 7.1. Both standard curves are linear between 10^{-20} and 10^{-16} mol/assay. The end point assay is somewhat more sensitive than the rate assay; however, the rate assay is much faster and easier in that only one reagent is added to initiate the reaction. The sensitivity of this method is almost the same as that of the chemiluminescence assay of alkaline phosphatase using AMPPD and is higher than the bioluminescence assay of alkaline phosphatase using D-luciferin-*O*-phosphate as the substrate.[11,12]

0-8493-0719-8/02/$0.00+$1.50

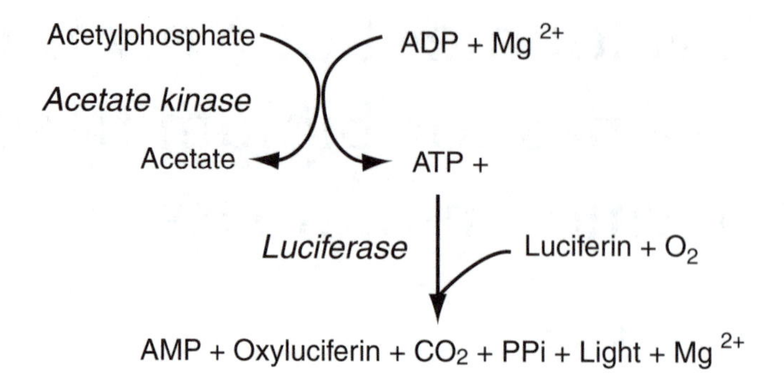

FIGURE 7.1 Principle of bioluminescent assay for acetate kinase.

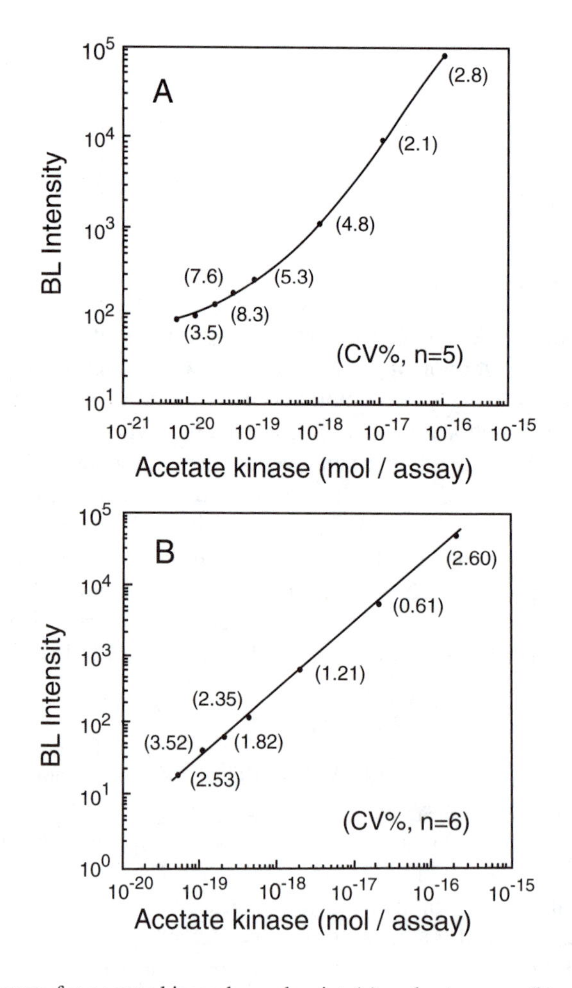

FIGURE 7.2 Standard curve for acetate kinase by end point (a) and rate assay (b).

We constructed BLEIAs for the measurement of 17-α-hydroxyprogesterone (17-OHP), human thyroid-stimulating hormone (TSH), and anti-rabbit IgG Fab′-rabbit IgG. In these systems, an AK-labeled hapten was used to compete for antibody binding with the unlabeled hapten. A solid phase coated by secondary antibody, anti-TSH, or anti-rabbit IgG allowed the removal of potential interference by serum components. Conditions such as buffer pH and composition, reaction time

TABLE 7.1
Comparison of End Point and Rate Assays of Acetate Kinase

Assay Method	Sensitivity (mol/assay)	Reaction or Lag Time (min)	Total Assay Time (min/60 samples)
End point assay	1.4×10^{-20}	60	90
Rate assay	6.5×10^{-20}	15	16

TABLE 7.2
Comparison of 17-OHP, TSH, and Rabbit IgG BLEIAs

Assay	Measurable Range	Detection Limit
17-OHP	0.1–50 pg/assay	0.1 pg (300 amol)/assay
TSH	0.006–45 μIU/ml	—
Rabbit IgG	0.01–25 ng/assay	12.5 pg (83 amol)/assay

TABLE 7.3
Reproducibility of the TSH BLEIA

TSH Concentration (μIU/ml)	CV% ($n = 6$)
0	1.72
0.006	4.06
0.013	3.99
0.025	2.51
0.05	3.35
0.1	1.79
1.0	3.99
10	7.14

and temperature, and concentration of the AK-labeled hapten, antibody, and AK substrates were optimized for maximal sensitivity and specificity.

The standard curves for the BLEIAs of 17-OHP, TSH, and rabbit IgG are shown in Figure 7.3. A comparison of the measurable range and detection limits of each assay is shown in Table 7.2.

The AK-linked BLEIA is highly sensitive and consistent. For example, the TSH BLEIA is approximately 25 times more sensitive than the radioimmunoassay (RIA). It is extremely reproducible (Table 7.3) and shows a high degree of correlation with a commercially available fluorometric EIA (AIA1200, Tosoh Corporation, Tokyo) (Figure 7.4). In addition, seven samples that were below the detection limit of the conventional fluorimetric EIA were measured by the BLEIA. Thus, this assay will be useful for the detection of low concentrations of TSH seen clinically in conditions such as hyperthyroidism resulting from Grave's disease or thyroid tumor.

To produce a universal reagent for the AK-linked BLEIA system, we prepared biotinylated AK. Biotin (vitamin H) binds with high affinity to streptavidin (SA), a protein produced by *Streptomyces avidinii*. Because of the excellent sensitivity provided by this high affinity binding, the biotin–streptavidin system has been used extensively in immunoassays. The biotin–streptavidin system is also advantageous because it allows for the standardization of sandwich-type immunoassays in which only one labeled substance may be used for the detection of multiple antigens. Biotinylated polycolonal

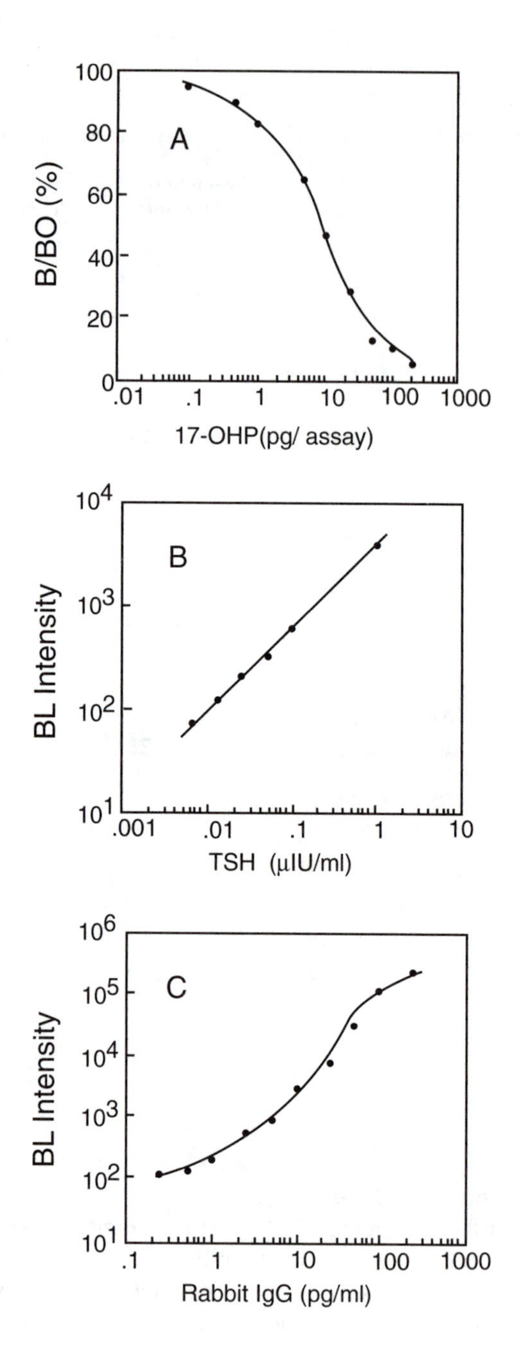

FIGURE 7.3 Standard curves for 17-OHP (a), TSH (b), and rabbit IgG (c).

or monoclonal antibodies can easily be prepared under mild conditions using *N*-hydroxysuccinimide ester or maleimide derivatives of biotin.

We used the biotin–streptavidin system to develop BLEIAs for TSH and hCG. The protocol for the SA-based BLEIA of hCG and the standard curve are shown in Figure 7.5. The biotinylation of AK resulted in minimal loss of enzyme activity. The measurable range of this assay was 0.003 to 20 mIU hCG/ml. There was a high degree of correlation between the BLEIA and a commercially available, time-resolved fluorescent immunoassay (DELFIA, Pharmacia, Turku, Finland) (Figure 7.6). In addition, the BLEIA was capable of measuring 11 samples that were below the detection limit of the DELFIA.

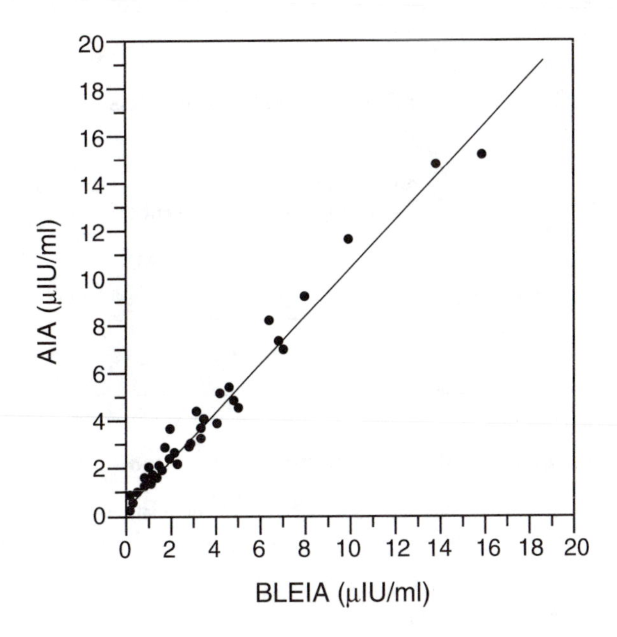

FIGURE 7.4 Correlation between BLEIA and AIA.

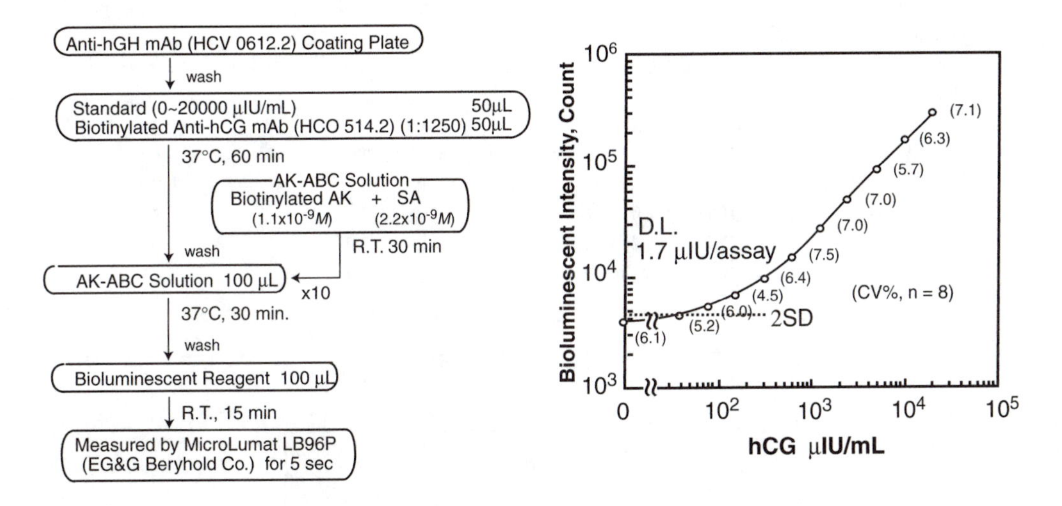

FIGURE 7.5 Assay procedure and standard curve for hCG.

We further utilized the SA-based BLEIA in the measurement of mIL-6, a system that requires extreme sensitivity. Recently, Takaki et al.[13] reported that plasma concentrations of the cytokine IL-6 are increased by noninflammatory stresses such as immobilization. Therefore, we developed a BLEIA to measure mIL-6 levels in various mouse tissues. The mIL-6 BLEIA was sixfold more sensitive than a colorimetric EIA and had a measurable range of 7.8 to 4000 pg/ml and a detection limit of 6.6 pg/ml. It was able to detect accurately mIL-6 in mouse plasma and tissue extracts spiked with mIL-6 (Table 7.4). We used this BLEIA to measure mIL-6 in the tissues of normal control mice, mice stressed by immobilization for 1 h followed by a 1-h rest, and mice treated with the inflammatory stimulus lipopolysaccharide (LPS, 1 mg/kg injected i.p. 2 h prior to sacrifice). The IL-6 levels of the LPS-treated mice were generally significantly higher than those of the control group, as would be expected (Table 7.5). The stressed mice displayed higher levels of mIL-6 in the liver and plasma and lower IL-6 levels in the brain and kidney than did the control group.

TABLE 7.4
Recovery of mIL-6 from Plasma and Tissue Extract

	Amount Added to Sample (pg/ml)	Percent Recovery
Plasma	31, 125, 500	85.2 ± 17.6 (n = 6)
Tissue extract	0, 125, 500, 2000	84.3 ± 15.3 (n = 12)

TABLE 7.5
Tissue Concentrations of mIL-6 in LPS-Treated, Stressed, and Normal Mice

Tissue	Control	Immobilization-Stressed	LPS-Treated
Brain	7.25 ± 0.8	5.14 ± 1.2	14.06 ± 1.5
Heart	5.47 ± 2.8	3.41 ± 0.9	N.D.
Lung	4.98 ± 1.2	4.58 ± 1.4	N.D.
Stomach	9.88 ± 6.8	6.20 ± 2.0	N.D.
Duodenum	<0.08	0.70 ± 1.2	N.D.
Jejunum	2.56 ± 2.8	1.78 ± 1.6	N.D.
Ileum	4.36 ± 0.9	2.74 ± 0.8	N.D.
Colon	1.89 ± 0.4	2.15 ± 0.2	N.D.
Mesentery	50.94 ± 21.7	56.56 ± 33.8	160.59 ± 75.5
Liver	146.37 ± 17.2	200.77 ± 27.2	165.69 ± 12.4
Spleen	2.53 ± 0.5	4.79 ± 4.1	269.76 ± 105.8
Kidney	105.03 ± 7.0	78.64 ± 36.3	370.35 ± 63.4
Muscle	3.24 ± 1.2	4.47 ± 3.5	N.D.
Plasma	Below detection limit	148.8 ± 53.2	125.6 ± 73.7

N.D. = not determined.

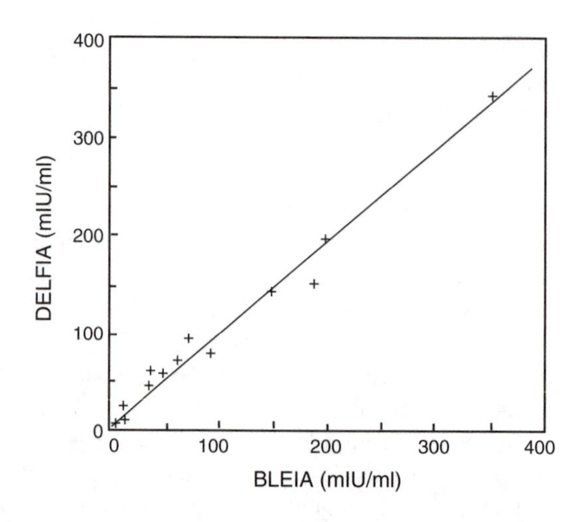

FIGURE 7.6 Correlation between BLEIA and DELFIA.

TABLE 7.6
Tissue Concentrations of PACAP in SD Rats (ng/g wet tissue)

Tissue/Brain Region	With Perfusion	Without Perfusion
Hypothalamus	207.00 ± 41.13	427.3 ± 92.7
Cortex	11.06 ± 0.225	42.08 ± 3.37
Hippocampus	13.72 ± 6.19	61.56 ± 8.54
Anterior pituitary	8.56	—
Posterior pituitary	46.45	—
Lung	1.67 ± 0.61	3.83 ± 0.26
Atrium	1.75 ± 0.38	4.90 ± 0.99
Liver	1.53 ± 0.63	5.82 ± 1.08
Spleen	3.04 ± 0.87	9.90 ± 1.88
Pancreas	1.19 ± 0.47	9.71 ± 0.87
Stomach	5.11 ± 1.53	12.88 ± 2.22
Duodenum	5.88 ± 1.39	19.84 ± 2.33
Jejunum	4.12 ± 0.17	23.13 ±5.57
Ileum	2.82 ± 1.27	23.48 ± 4.96
Colon	2.88 ± 1.10	24.63 ± 6.01
Kidney	1.54 ± 0.71	6.79 ± 2.9
Adrenal gland	4.75 ± 2.30	16.70 ± 2.68
Testis	51.32 ± 5.46	66.15 ± 9.75
Epididymis	1.20 ± 0.17	13.00 ± 2.75
Ovary	1.53 ± 0.22	17.42 ± 5.8

We also used the SA-based BLEIA to detect extremely low concentrations of pituitary adenylate cyclase activating polypeptide 38 (PACAP38). PACAP38 is a peptide hormone that was originally isolated from ovine hypothalamus. Our BLEIA used biotinylated PACAP38 as a labeled antigen and biotinylated AK-SA bound to a solid phase as a detector. The measurable range of PACAP38 in this assay was 62 to 16,000 pg/ml. We used this BLEIA to measure PACAP38 concentrations in the tissues of perfused or nonperfused SD strain rats (Table 7.6). The highest concentration of PACAP38 was found in the hypothalamus; however, other brain areas also showed high concentrations. The testis contained the highest level of PACAP38 found in the peripheral tissues. Perfusion decreased the amount of PACAP38 seen in all tissues, suggesting that PACAP38 in the blood accounts for this decrease. The plasma levels of PACAP38 measured in SD rats and humans were 265.5 ± 82.6 pg/ml ($n = 6$) and 596.4 ± 180.6 pg/ml ($n = 40$).

7.3 BLEIA USING BIOTINYLATED FIREFLY LUCIFERASE AS A LABEL

Recently, Tatsumi et al.[14] produced biotinylated luciferase consisting of *L. lateralis* luciferase fused to either an artificial biotin acceptor peptide or the carboxyl-terminal 87 residues of *E. coli* biotin carboxyl carrier protein. We utilized this method to produce biotinylated luciferase mutants that emit light of different colors.[8] We then employed these mutants in a simultaneous BLEIA of PGI and PGII in serum.[10]

PGI and PGII are inactive zymogens of pepsins, the proteolytic enzymes found in gastric secretions. Serum levels of PGI and PGII and the ratio of PGI/PGII are used to evaluate gastric atrophy, gastric ulcer, and gastric acidity. Because lesions with a great degree of gastric atrophy are associated with a high incidence of stomach cancer, the determination of serum PGI and PGII may be useful in the screening of populations at high risk of stomach cancer. A simultaneous assay of serum PGI and PGII would be a cost-effective and labor-saving method of performing this screen.

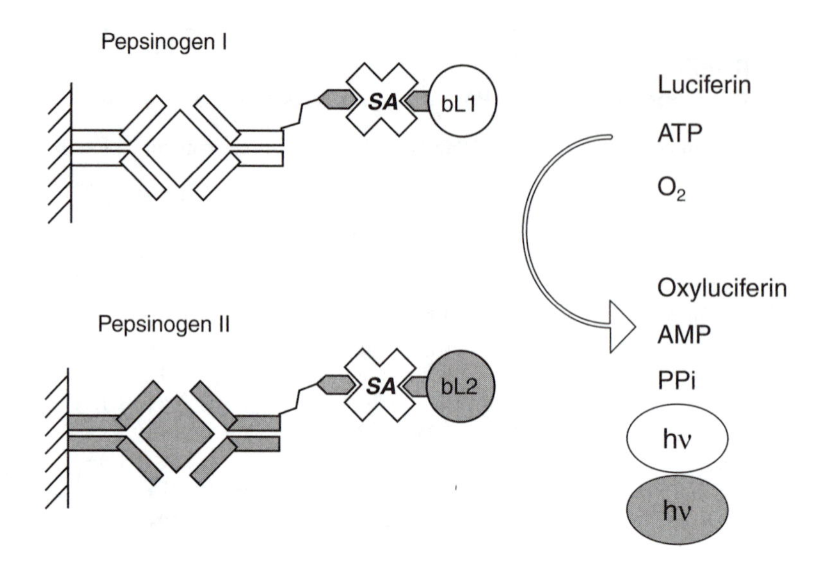

FIGURE 7.7 Principle of simultaneous assay of pepsinogens.

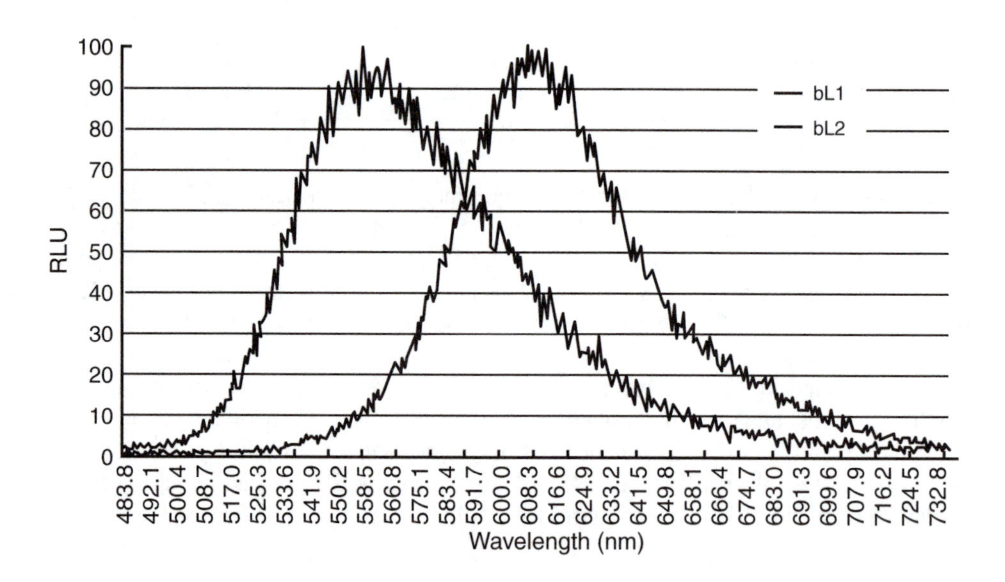

FIGURE 7.8 Analysis of spectra for two kinds of luciferase (bL1 and bL2).

The principle of the simultaneous BLEIA of PGI and PGII is shown in Figure 7.7. Magnetic particles coated with anti-PGI and anti-PGII antibodies were used to capture serum PGI and PGII. A triplex of biotinylated anti-PGI and anti-PGII antibodies, streptavidin, and biotinylated luciferase mutants was then added. The particles were washed to remove the unbound triplex. The bound enzyme activities of the two luciferases were measured at $\lambda_{max} = 607$ nm (PGI) and $\lambda_{max} = 559$ nm (PGII).

The emission spectra of the two luciferases did not interfere with each other's luminescence intensity over the concentration range used (Figure 7.8). The calibration range of PGI was 2 to 200 ng/ml and of PGII was 1 to 100 ng/ml. Both assays are highly reproducible; the values obtained by the simultaneous assay correlate well with the values obtained by single assay (Table 7.7).

TABLE 7.7
Simultaneous BLEIA of PGI and PGII in Serum

	CV (%)	Comparison with Single Assay
PGI	3.4–10.2	y (simultaneous assay) = 0.995x (ELISA) + 0.84, r = 0.948
PGII	4.0–8.6	y (ELISA) = 1.055x (simultaneous assay) + 1.47, r = 0.974

Thus, the simultaneous BLEIA of PGI and PGII has many features that would make it a useful clinical tool.

7.4 CONCLUSION

We have developed two types of highly sensitive bioluminescent assays in which firefly luciferase was used to determine the amount of ATP produced by an AK label bound to an antigen or antibody. These BLEIAs were as much as 25 times more sensitive than standard RIAs and do not exhibit the decay of light effect seen in conventional chemiluminescence immunoassays. We utilized AK in combination with the biotin–streptavidin system to measure extremely small amounts of mIL-6 and PACAP38 accurately without concentration of samples or any other pretreatment. We then developed a system for the simultaneous detection of two antigens using two complexes of biotinlyated antibody–streptavidin–biotinylated luciferase that emitted two different wavelengths of light. This system was used in a clinical application for the detection of PGI and PGII in serum. Further investigations are under way to determine whether multiple analytes in the same sample can be analyzed by BLEIA.

REFERENCES

1. Miska, W. and Geiger, R., Luciferin derivatives in bioluminescence-enhanced enzyme immunoassays, *J. Biolumin. Chemilumin.*, 1, 119–128, 1989.
2. Tatsumi, H., Kajiyama, N., and Nakano, E., Molecular cloning and expression in *Escherichia coli* of a cDNA clone encoding luciferase of a firefly, *Luciola lateralis, Biochim. Biophys. Acta,* 1131(2), 161–165, 1992.
3. Thompson, E. M., Nagata, S., and Tsuji, F. I., Cloning and expression of cDNA for the luciferase from the marine ostracod *Vargula hilgendorfii, Proc. Natl. Acad. Sci. U.S.A.,* 86(17), 6567–6571, 1989.
4. Inouye, S., Noguchi, M., Sakaki, Y., Takagi, Y., Miyata, T., Iwanaga, S., Miyata, T., and Tsuji, F. I., Cloning and sequence analysis of cDNA for the luminescent protein aequorin, *Proc. Natl. Acad. Sci. U.S.A.,* 82(10), 3154–3158, 1985.
5. Maeda, M., Ikeda, H., Tsuji, A., Murakami, S., Ito, S., and Kamada, S., Development of a new label enzyme for bioluminescent EIA, *Anal. Lett.,* 28, 383–394, 1995.
6. Murakami, S., Nakajima, M., Sekine, H., Maeda, M., and Tsuji, A., Streptavidin-biotin based bioluminescent EIA using biotinylated AK and recombinant firefly luciferase, *Anal. Lett.,* 29, 969–980, 1996.
7. Murakami, S., Ito, K., Goto, T., Kamada, S., and Maeda, M., Bioluminescent EIA using thermostable mutant luciferase and AK as a labeled enzyme, *Anal. Chim. Acta,* 361, 19–26, 1998.
8. Fukuda, S., Tatsumi, H., and Maeda, M., Bioluminescent enzyme immunoassay with biotinylated firefly luciferase, *J. Clin. Ligand Assay,* 21, 358–362, 1998.
9. Kajiyama, N. and Nakano, E., Isolation and characterization of mutants of firefly luciferase which produce different colors of light, *Protein Eng.,* 4(6), 691–693, 1991.
10. Ohkuma, H., Abe, K., Kosaka, Y., and Maeda, M., Simultaneous assay of pepsinogen I and pepsinogen II in serum by bioluminescent EIA using two kinds of *Luciola lateralis* luciferase, *Anal. Chim. Acta,* 395(3), 265–272, 1999.

11. Schaap, A. P., Sandison, M., and Handley, R. S., Chemical and enxymatic triggering of 1,2-dioxetane, *Tetrahedron Lett.,* 28, 1159–1162, 1987.
12. Bronstein, I., Edwards, B., and Voyta, J. C., 1,2-Dioxetanes: novel chemiluminescent enzyme substrates. Applications to immunoassays, *J. Biolumin. Chemilumin.,* 4(1), 99–111, 1989.
13. Takaki, A., Huang, Q. H., Somogyvari-Vigh, A., and Arimura, A., Immobilization stress may increase plasma interleukin-6 via central and peripheral catecholamines, *Neuroimmunomodulation,* 1(6), 335–342, 1994.
14. Tatsumi, H., Fukuda, S., Kikuchi, M., and Koyama, Y., Construction of biotinylated firefly luciferases using biotin acceptor peptides, *Anal. Biochem.,* 243(1), 176–180, 1996.

8 Assay of Diversified Biomolecules with a Luminogenic Conjugate Substrate: 5-(5′Azoluciferinyl)-2,3-Dihydro-1,4-Phthalazinedione*

Thankiah Sudhaharan and Annadi Ram Reddy

CONTENTS

* This chapter is an expansion of a paper originally published in *Anal. Biochem.*, 271, 159–167, 1999. Reprinted with permission.

0-8493-0719-8/02/$0.00+$1.50

8.1 OVERVIEW

Horseradish peroxidase (HRP) catalyzes the oxidative chemiluminescent reaction of luminol, and firefly luciferase catalyzes the oxidation of firefly D-luciferin. A novel substrate, 5-(5′azoluciferinyl)-2,3-dihydro-1,4-phthalazinedione (ALPDO) that contains both luminol and luciferin functionalities triggers the activity of both the enzymes HRP and firefly luciferase in solution. It is synthesized by diazotization of luminol and its subsequent azo coupling with firefly luciferin. Nuclear magnetic resonance (NMR) spectral data show that the C5′ of benzothiazole of luciferin connects the diazophthalahydrazide. ALPDO has electronic absorption and fluorescence properties different from its precursor molecules. The chemiluminescence emission spectra of the conjugate substrate display biphotonic emission characteristics of azophthalatedianion and oxyluciferin. It has an optimum pH 8.0 for maximum activity with respect to HRP as well as luciferase. At pH 8.0 the bifunctional substrate has enhanced activity 12 times that of luminol, but has depressed activity 7 times that of the firefly luciferin–luciferase system. The specific enhancement of light emission from the cyclic hydrazide part of ALPDO helps in the sensitive assay of HRP to 2.0×10^{-13} *M*, whereas that of ATP is sensitive to 1.0×10^{-14} *M*. However, at higher than 3 m*M* [H_2O_2] the sensitivity underwent a reversal for HRP. The choline-liberated H_2O_2 oxidation of ALPDO produced sensitivity of assay of choline to 1.0×10^{-7} *M*. NADH diminished the chemiluminescence of the HRP–ALPDO–H_2O_2 system. The linear decrease in the chemiluminescent light output with NADH results in the indirect detection of NADH to 5×10^{-8} *M*. Addition of enhancers such as firefly luciferin and *p*-iodo phenol (PIP) to the HRP–ALPDO–H_2O_2 system enhances the light output.

8.2 INTRODUCTION

The recent advances in biotechnology, especially in the areas of immunology and DNA diagnosis, are basically the result of the development of the various analytical methods for the detection of target molecules. Although various techniques were introduced for molecular recognition, isotopic labeling remains the most widely exploited. Because of the requirement for specialized handling skills associated with use of isotopic probes during experiment and their subsequent disposal, a constant search is on for simpler alternatives. Among them, the chemiluminescent, bioluminescent, and fluorescent methods are considered to have the most potential for the detection and assay of various enzymes, cofactors, and their metabolites.[1] A number of chromogenic and fluorogenic substrates are in use at present for the determination of unmodified enzymes.[2] The sensitivity, speed, and specificity of the two enzymes firefly luciferase[3] and HRP,[4] catalyzing conversion of the chemical energy to light energy via the formation of an activated chemical structure whose decay produces the light are comparable with that of radiometric assays. Chemiluminescent systems based on HRP are widely used for immunoassays.[5–8] HRP catalyzes the H_2O_2 oxidation of luminol (5-amino-2,3-dihydro-1,4-phthalazinedione) (**I**, in Scheme 8.1) to produce a luminol radical.[9] This radical then forms an endoperoxide, which upon decomposition yields an electronically excited 3-aminophthalate dianion, emitting light on its return to the ground state (Equation 8.1).

$$\text{Luminol} \xrightarrow[\text{Alkaline pH}]{\text{HRP, } H_2O_2/\,^1O_2/\,^{\cdot}OOR} \text{3-aminophthalate dianion} + N_2 + h\nu \quad (8.1)$$

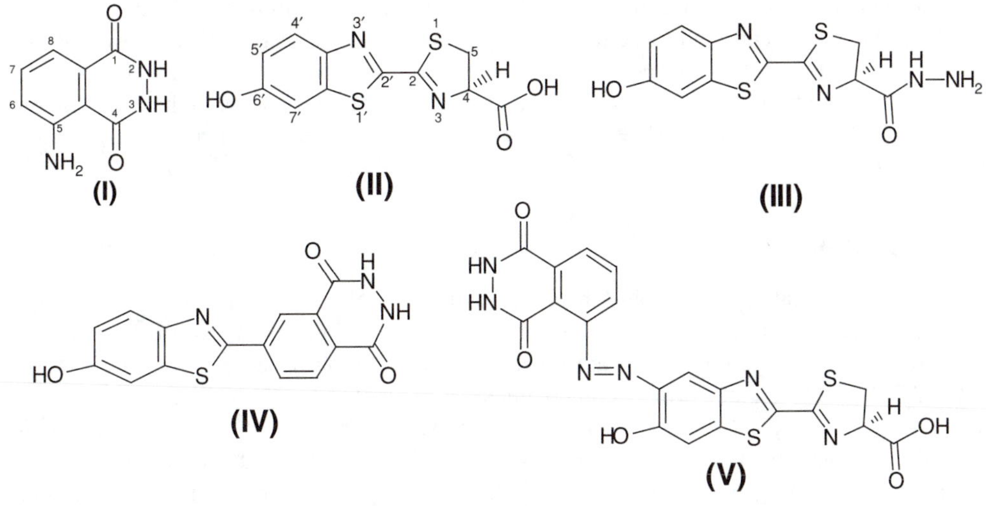

SCHEME 8.1 Structure of various chemiluminescent molecules.

Firefly luciferase catalyzes the oxidative decarboxylation of D-luciferin [(4,5-dihydro-2-(6′-hydroxybenzothiozolyl)-4-thiazole carboxylic acid)] (**II**) in the presence of ATP (Equations 8.2 and 8.3).

$$\text{LA} + \text{LN} + \text{ATP} \xrightleftharpoons{\text{Mg}^{2+}} \text{LA-LN-AMP} + \text{PP}_i \tag{8.2}$$

$$\text{LA-LN-AMP} + \text{O}_2 \longrightarrow \text{OLN} + \text{AMP} + \text{CO}_2 + h\nu \tag{8.3}$$

where LA is luciferase, LN is luciferin, and OLN is oxyluciferin. Firefly bioluminescence involves the incorporation of oxygen into luciferin to yield a dioxetanone intermediate,[10,11] which is spontaneously cleaved to form excited oxyluciferin. The excited oxyluciferin undergoes radiative decay to ground state emitting at ~565 nm, at neutral pH.[12] The firefly luciferin–luciferase reaction is the best method for rapid, sensitive measurement of ATP and other nucleotides by coupled bioluminescence,[13] providing a linear assay over five to six orders of magnitude down to 10^{-16} mol.

Several publications aimed at improving the efficiency and elucidating the chemiluminescent reaction mechanism of luminol and firefly luciferin,[14–17] respectively, have appeared in recent years. The chemiluminescent active functional moiety in luminol is the hydrazide group, whereas in luciferin the reactivity originates from the unsubstituted 5′ position of the thiazole ring. White and Rosewell[18] combined the above two active functional groups and synthesized dehydroluciferin monoacylhydrazide (**III**) and found that the chemiluminescence results from the excited carboxylate anion via an azadioxetane intermediate. Because of their poor solubility in aqueous medium and their low acidity, these luciferyl hydrazides have not found many applications in water-based chemiluminescent systems. On the other hand, Whitehead and co-workers[19] discovered peroxidase catalyzed enhanced luminescence of luminol severalfold in the presence of firefly luciferin and other 6-hydroxy benzothiazoles. Thus, the use of chemiluminescence enhancers such as firefly luciferin and PIP improved the detection limit of peroxidase, enabling the development of simple non-isotopic immunoassay systems. To explain the potential of benzothiazoles as chemiluminophores, Sasamoto et al.[20] designed a cyclic hydrazide, 4-(5′-hydroxybenzothiazolyl) phthalyl hydrazide (**IV**), a hybrid of luciferin and luminol. It is the first example of a substrate that has proved to serve as both a fluorogenic and a chemiluminogenic substrate. Similarly, Klel and Obreen[21] observed that diazoluminol, derived from 3-amino-L-tyrosine and luminol, has increased the luminescence of the latter by 5- to 20-fold. We have coupled luminol and firefly D-luciferin by diazo group and obtained a novel bifunctional substrate, 5-(5′-azoluciferinyl)-2,3-dihydro-1,4-phthalazinedione (ALPDO) (**V**), which can bind to both HRP and firefly luciferase. It is stable and water-soluble, the D-luciferin part is not racemized, and it has maximum activity at pH 8. Hence, we considered it worthwhile to employ ALPDO as an efficient chemiluminescent substrate to the detection of the two biologically active enzymes HRP and firefly luciferase and their related cofactors, coenzymes, and substrates.

8.3 MATERIALS

Firefly D-luciferin, luminol, and phosphatydylcholine (*p*-choline) were purchased from Sigma-Aldrich (USA) and HRP from Bangalore Genei Pvt. Ltd. (Bangalore, India). North American firefly luciferase was obtained from Amersham (U.K.), as an ATP bioluminescent assay RPN1630 kit. ATP, NADH, βNAD, and NMN were obtained from Boehringer Mannheim. H_2O_2, $NaNO_2$, *p*-iodophenol, choline, and HCl were obtained from E-Merck (India) as AR grade and used without any further purification. The original stock solutions of luciferin–luciferase were made in 50 m*M* tricine buffer containing 10 m*M* $MgCl_2$ and 1 m*M* EDTA at pH 8.0. Whereas the stock solution of ALPDO, luminol, and HRP was prepared in 0.1 *M* Tris-HCl, pH 8.0, and stored at −80°C, the working stock solutions were prepared on the day of use by diluting in the respective buffers.

8.4 PHYSICOCHEMICAL CHARACTERIZATION

8.4.1 Synthesis of ALPDO

Initially, 10 mg (0.033 m*M*) of luciferin and 6.3 mg (0.033 m*M*) of luminol were chilled in ice. Luminol was dissolved in minimum amount of water. Once the temperature was lower than 3°C, 100 μl of 1.2 *N* HCl was added, drop by drop, maintaining the temperature below 3°C. Meanwhile, 3.1 mg (0.039 m*M*) of $NaNO_2$ was dissolved in minimum amount of H_2O, chilled, and added to the precooled luminol solution without rise in temperature. The excess HNO_2 was tested with KI-starch paper and destroyed by urea solution. The diazonium salt obtained was kept in ice for about 15 min. About 4 mg (0.1 *M*) of NaOH was dissolved in minimum of H_2O and added to the ice-cooled luciferin. Finally, the chilled diazonium salt solution was added to the alkaline luciferin slowly with stirring. After complete addition, it was allowed to stand at low temperature for about 1 h. The precipitated ALPDO was separated by centrifugation and washed with chilled dilute HCl and water and vacuum-dried. It has an R_f value of 0.5 in $CHCl_3$:CH_3OH (90:10, v/v), whereas luminol and luciferin have an R_f of 0.1 and 0.15, respectively. The compound decomposed around 340°C. The elemental analysis and NMR data are given in Table 8.1.

TABLE 8.1
NMR and Elemental Analysis Data of ALPDO

	Elemental Analysis		
NMR[a] (in DMSO-d_6)	Element	Theoretical (%)	Experimental (%)
8.58 (s,1H), 8.03 (d,1H),	C	50.44	50.38
7.95 (d,1H), 7.46 (s,1H),	H	2.65	2.60
7.42 (s,1H), 7.12 (d,d,1H),	N	18.59	18.76
4.32 (s,2H,broad), 6.72	O	14.16	—
(t,1H), 3.63 to 3.75 (m,2H),	S	14.16	—
10.8 (offset region,1H)			

[a] d = doublet; s = singlet; t = triplet; d,d = double doublet; m = multiplet.

8.4.2 Spectral Studies

The ultraviolet–visual (UV-VIS) spectra of ALPDO were obtained using a Hitachi UV-VIS-150 Spectrophotometer in water. The corrected fluorescence spectra of ALPDO were obtained employing Hitachi Spectrofluorimeter 4010 maintaining 5 nm entrance and exit slit width. The concentration of ALPDO used was 1×10^{-6} *M* and the excitation wavelength was 360 nm. NMR spectrum was obtained from Bruker WH 300 spectrometer by dissolving 5 mg of ALPDO in DMSO-d_6. Elemental analysis data were obtained from Perkin Elmer Analyzer 240C. Bioluminescence data were obtained using LKB luminometer 1250.

Azobenzene has rich possibilities for various technological and fundamental studies. It is a photochemically reactive chromophore. Coulombic interaction between the two polar groups in the ALPDO may stabilize the structure and prevent its decomposition.[22] The new substrate is not only stable but also bioactive, and the stock solutions prepared display profound *in vitro* bioluminescence even after 2 months, when stored at –80°C.

8.4.3 Electronic Absorption, Fluorescence, and Chemiluminescence Properties of the ALPDO

The absorption spectrum of ALPDO in water (pH ~7.2) is shown in Figure 8.1A. It differs markedly from the absorption spectra of its corresponding precursor components. It exhibited three electronic transitions at 255, 288, and 360 nm in water. The longer- and shorter-wavelength absorption maxima have more intensity than the 288-nm absorption maximum and have $A_{360}/A_{288} = 2.66$ and $A_{255}/A_{288} = 3$. Following Griffiths,[23] the 288-nm absorption maxima can be attributed to the syn azo form and are probably stabilized by the intramolecular hydrogen bonding between 6′ hydroxyl hydrogen of luciferin and diazo nitrogen. The fluorescence spectrum of the bifunctional substrate in water is shown in Figure 8.1B. It can be noticed that when excited at 360 nm it gave a broad emission doublet maxima centering at 542 nm. The emission maxima obtained resemble the emission spectrum of luciferin, which exhibits an emission maximum at 536 nm in water. However, compared with luciferin, ALPDO is relatively less fluorescent. The 6-nm red shift in the fluorescence maximum and the decreased fluorescence quantum yield of ALPDO over luciferin are probably due to the intramolecular charge transfer from donor luciferyl part to the acceptor luminol moiety. The large Stokes shift (~175 nm) and also the absorption and emission bands not intersecting one another are clearly a benefits of ALPDO as a fluoroprobe. As expected in phenols, the fluorescence intensities in ALPDO display pH dependence. An increase in pH increases the fluorescence intensity, reaching a maximum at pH 8 to 9. Although the compound contains both the luminol and luciferin moieties,

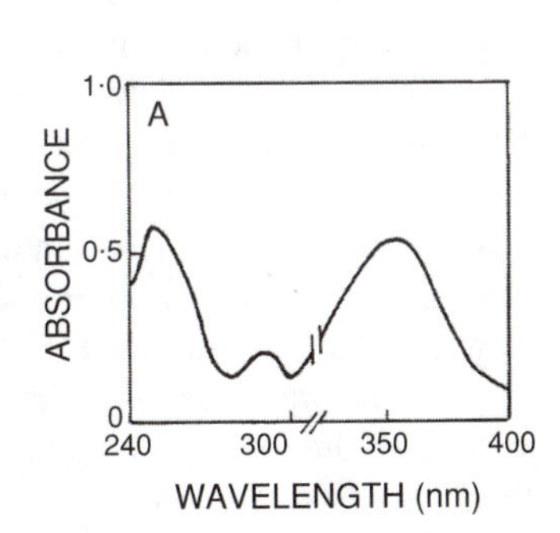

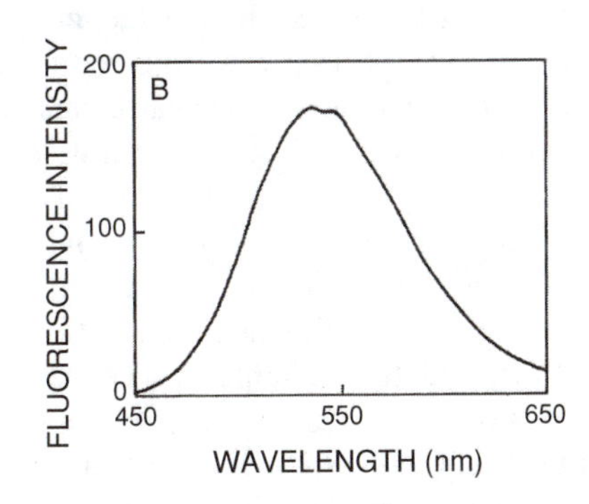

FIGURE 8.1 (A) Absorption spectrum of ALPDO (2×10^{-4} *M*) in H_2O (OD scaled down two times). (B) Fluorescence spectrum of ALPDO (1×10^{-6} *M*) in H_2O λ_{ex} at 360 nm.

TABLE 8.2
Absorption and Emission Maxima of Luciferin, Luminol, and ALPDO in H_2O

Compound	λ_{max} (nm)	$\lambda_{em.flu.}$[a] (nm)	$\lambda_{emi.CL}$[b] (nm)
D-Luciferin	260, 335	536	550
Luminol	305, 340	424	425
ALPDO	255, 288, 360	542	554, 448

[a] Fluorescence emission maxima.
[b] Chemiluminescence emission maxima.

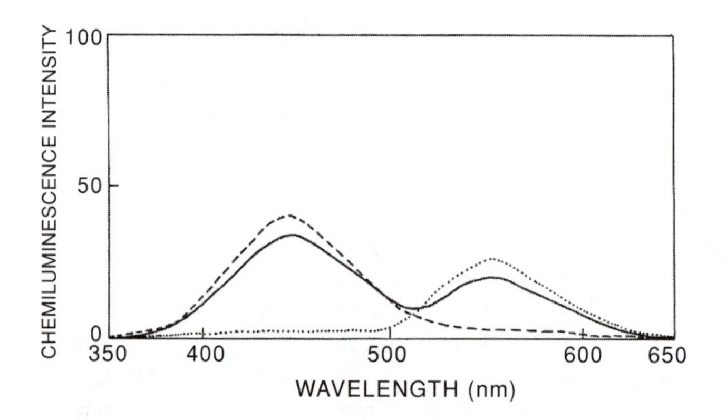

FIGURE 8.2 Chemiluminescence emission spectra of ALPDO (the noisy chemiluminescence emission spectra are converted into smooth curves) in 50 m*M* tricine buffer at pH 8.0 containing 10 m*M* $MgCl_2$ and 1 m*M* EDTA. (—) ALPDO (2.4×10^{-3} *M*) in the presence of $ATP^{Mg^{2+}}$ (20 p*M*) + luciferase (4.3×10^{-6} *M*) + HRP (5.6×10^{-6} *M*) + H_2O_2 (3×10^{-7} *M*); (···) ALPDO (2.4×10^{-3} *M*) in the presence of ATP Mg^{2+}(20 p*M*) + luciferase (4.3×10^{-6} *M*); (---) ALPDO (2.4×10^{-3} *M*) in the presence of HRP (5.6×10^{-6} *M*) + H_2O_2 (3×10^{-7} *M*).

it does not exhibit any wavelength-dependent emission spectra, revealing that the two moieties are in one plane with extensive delocalization of π electrons from one part of ALPDO to another taking place via the azo bridge.

The chemiluminescence emission spectrum of the substrate is shown in Figure 8.2. The chemiluminescence emission spectra of the substrate were obtained employing a Hitachi Spectrofluorimeter 4010 by closing the entrance slit and keeping the exit slit in energy mode wide open (15 nm). The chemiluminescence reaction was triggered for the spectrum by injecting either ATP (20 p*M*) or H_2O_2 (3×10^{-7} *M*) or both into the cuvette containing premixed solutions of enzymes (5.6×10^{-6} *M* HRP and luciferase 4.3×10^{-6} *M*) and substrate ALPDO (2.4×10^{-3} *M*) in 50 m*M* tricine buffer containing 10 m*M* $MgCl_2$ and 1 m*M* EDTA at pH 8.0. The signals emitted from chemiluminescence system were scanned immediately. It can be seen from the figure that when the reaction was initiated by injecting ATP into the equilibrated mixture of firefly luciferase–ALPDO, a luminescence maximum at 554 nm was observed; when the reaction was triggered by injecting H_2O_2 (3.0×10^{-7} *M*) to the ALPDO–HRP mixture, a luminescence maximum at 448 nm was observed. The chemiluminescence maxima of ALPDO are slightly different from the chemiluminescence maxima of firefly luciferin and luminol (Table 8.2), when obtained with their respective enzymes.

The difference in chemiluminescence maxima of the ALPDO can be attributed to the different light-emitting species, as shown in the Scheme 8.2. In the case of firefly luciferin–luciferase bioluminescence, the light-emitting species is excited oxyluciferin, whereas in luminol–HRP it is

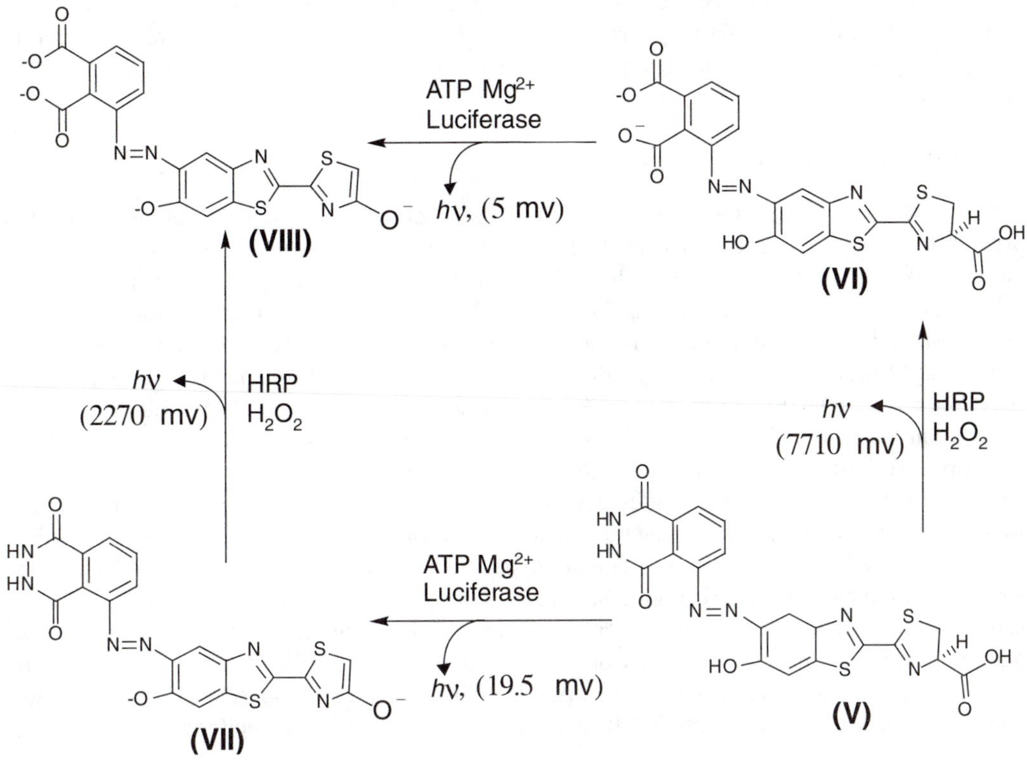

SCHEME 8.2

excited 3-aminophthalate dianion. When ALPDO reacts with H_2O_2 in presence of HRP, the light-emitting species is 3-(5′-luciferin) azophthalatedianion (**VI**), whereas when it reacted with ATP, the light-emitting species is 5′-(3′-luminol)-azo oxyluciferin (**VII**). Interestingly, the azo conjugate substrate exhibits two chemiluminescence emission maxima, one at 448 nm and the other at 554 nm in the presence of both luciferase and HRP enzymes when the chemiluminescent reaction was initiated with a mixture of ATP and H_2O_2. The biphotonic emission at two wavelengths that are marginally shifted from the emission maxima of their respective natural substrates reveals that ALPDO can bind to either one or both the enzymes. This is contrary to the fluorescence spectra, where only emission maxima characteristic of luciferin appeared.

The *in vitro* bioluminescence spectral intensity of azo phthalate dianion moiety far exceeds that of the bioluminescence intensity of azo oxyluciferin moiety, whatever the order of mixing of the reagents. A similar observation was made in the case of phthalahydrazide.[24] When phthalahydrazide is oxidized to excited phthalate anion, the latter is practically nonfluorescent. However, when coupled by a methylene bridge to either *N*-methylacridone or diphenylanthracene the chemiluminescence of phthalahydrazide increased substantially as a result of intramolecular energy transfer. The bioluminescent intensity of ALPDO at 554 nm is always less than the bioluminescent intensity of luciferin at 550 nm. It was noticed in the luminometer that once one of the moieties of the substrate is oxidized to the light-emitting part, the light emitted by its counterpart in the second stage, with the addition of H_2O_2/ATP, is always less than the light emitted in the first stage. For example, when the ALPDO first oxidizes by H_2O_2 in the presence of HRP, the light output given by the luminol moiety is 7710 mV. But, when the same reaction was carried out after the oxidation of luciferin moiety, the light output is only 2270 mV, under similar experimental conditions, after the blank correction (Scheme 8.2). Similarly, when the luciferin part of the bifunctional substrate was first oxidized, with the injection of ATP, the light output was 19.5 mV. The same reaction yielded only 5 mV when it occurred in the second stage. This is probably because ALPDO binds to both the enzymes sequentially, preferring HRP, and the partially oxidized substrate can bind to the first enzyme so that it can neither perfectly orient nor be completely available to the active site of the second enzyme.

8.4.4 Optimization of pH

Although HRP is stable in the pH range 4.2 to 12.0, its activity in the soluble state is maximal at pH 7.0 and greatly reduced under alkaline conditions. However, luminol chemiluminescence is most efficient at a pH between 10 and 13. This limits its use in biology. However, the optimum pH for maximum luciferase activity is 7.8 at 25°C. Hence, one of the important properties of the new substrate is its light-emitting ability with respect to the two native substrates, vis-à-vis its pH dependency. The activity of ALPDO at various pH ranges was evaluated by fixing H_2O_2 (6.0×10^{-5} *M*), HRP (5.6×10^{-8} *M*), ALPDO (8.5×10^{-5} *M*), luciferase (4.3×10^{-6} *M*), $ATP^{Mg^{2+}}$ (1.25×10^{-13} *M*), and varying buffers of different pH. The total reaction volume of 400 μl was reached by adding the respective buffers. The buffers used to obtain a pH between 7.0 and 12 are 0.1 *M* Tris HCl buffer and 0.1 *M* glycine/NaOH buffer. The mixing was done by an external mixer, and readings were taken after 5-s delay. The standard reactions at various pH ranges for luciferin and luminol were also carried out under identical conditions employing the same buffers. The activity of the analyte was plotted against pH. Figure 8.3 presents the activity of luciferase and HRP with their natural substrates and ALPDO against pH. Examination of the figure reveals two phenomena. First, the optimum pH for maximum activity of HRP shifted from pH 10, with luminol, to pH 8.0 with ALPDO, while there is no change in the optimum pH for the luciferase–ALPDO system, which retained the maximum activity at pH 8.0, as in the case of the luciferase–luciferin system. Second,

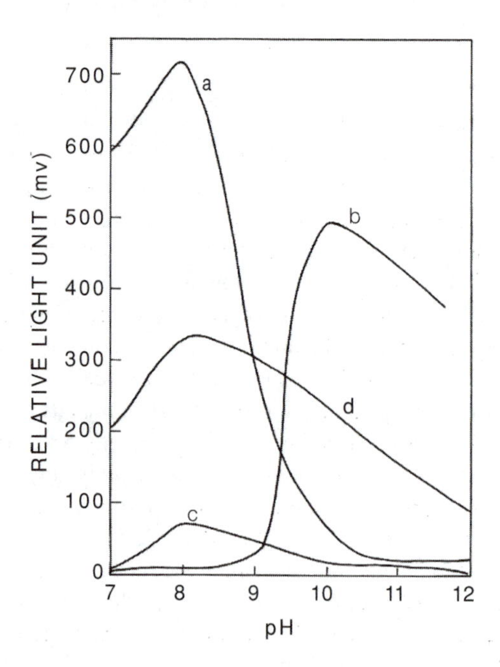

FIGURE 8.3 Optimization of pH for maximum activity of ALPDO in comparison with luminol and luciferin in varied (7 to 12) pH buffers. (a) ALPDO (8.5×10^{-5} *M*) + HRP (5.6×10^{-6} *M*) + H_2O_2 (6.0×10^{-6} *M*); (b) luminol (8.5×10^{-5} *M*) + HRP (5.6×10^{-6} *M*) + H_2O_2 (6.0×10^{-6} *M*); (c) ALPDO (1.77×10^{-4} *M*) + ATP (1.25×10^{-12} *M*) + luciferase (4.3×10^{-6} *M*); (d) luciferin (1.77×10^{-4} *M*) + ATP (1.25×10^{-12} *M*) + luciferase (4.3×10^{-6} *M*).

with the bifunctional substrate, the activity of HRP increased by 12 times and that of luciferase decreased by 7 times when compared with luminol and luciferin as substrates. This can be attributed to the intramolecular synergistic effect of luciferin on the light emission from the peroxidase-catalyzed oxidation of the luminol moiety. Hence, the diazo phthalazinedione moiety is more chemiluminescent than the luminol moiety. The results obtained reveal that when two different substrates are engineered into one entity, with a diazo bridge, the new synthon can still retain the chemiluminescence properties of the two components, although they are affected differently. Thus, two different enzymes can be assayed with a single substrate. After ascertaining the HRP- and luciferase-discriminating ability of ALPDO, we have evaluated the specificity factor, ϕ, of the substrate toward the two enzymes employing the relation:[25]

$$\phi = \frac{v_{HRP} X \,[\text{luciferase}]}{v_{luciferase} X \,[\text{HRP}]}$$

where v_{HRP} and $v_{luciferase}$ are the chemiluminescent light intensities obtained at equal concentrations of HRP and luciferase, under optimal conditions and assuming that the two enzymes behave well under steady-state conditions obeying Briggs–Haldane kinetics;[26] that is, substrate binding is faster than subsequent stages. The specificity factor obtained is 60,720 and this value shows that ALPDO has a profound preference for HRP over luciferase. The high specificity of ALPDO chemiluminescence toward HRP would allow a sensitive detection of the latter and its metabolites. The enhanced chemiluminescence of ALPDO with HRP at pH 8.0 has greater significance, as it is a chemiluminogenic probe. Thus, the biological pH compatible with maximum activity offers the provision to estimate a variety of luminol-related metabolites, cofactors, enzymes, etc., under physiological conditions as it has emission at longer wavelengths, a large Stokes shift, and requires a long time to reach peak height in chemiluminescence. The light yield of the coupled derivative was also linear over the range of concentrations.

8.5 BIOANALYTICAL APPLICATIONS

8.5.1 Chemiluminescence Assay of HRP

HRP is one of the most important and one of the smallest enzymes (mol. wt. ~44 kDa) used as a label. The reason for wide applicability of HRP as a label is that, in presence of a small amount of enhancers, the intensity of the light emission is increased by several orders of magnitude and background light emission from the luminol–peroxide reagent is greatly reduced, which leads to a dramatic increase in the signal: background ratio in chemiluminescence reactions. Chemiluminescent luminol and isoluminol are its known substrates, and are used to measure its activity. As ALPDO proved to be a better substrate for HRP than luminol, it is of profound significance if it can be employed to assay the activity of HRP following the same principle of HRP–luminol, i.e., under the constant but excess concentration of ALPDO and H_2O_2 ($\leq$3 m*M*). The HRP assay was carried out by keeping ALPDO (8.5×10^{-6} *M*) and H_2O_2 (6.0×10^{-5} *M*) at the optimum pH of 8.0 and varying the concentration of HRP. The reaction was initiated by the addition of HRP to the assay mix, and the reaction volume was kept at 400 μl using 0.1 *M* Tris HCl buffer. A control assay was also carried out using identical concentration of luminol as a standard for comparison with ALPDO. The amount of light produced by the photon-emitting final product showed linearity with the HRP concentration. Figure 8.4 shows that ALPDO exhibited linearity from 0.84 to 8.4×10^{-8} *M* of HRP. Owing to an improved signal-to-noise ratio and higher light output, the detection of HRP reached 2×10^{-13} *M* with 1.6×10^{-4} *M* ALPDO and 1.5×10^{-4} *M* H_2O_2, whereas it was detected to 1×10^{-13} *M* with enhanced chemiluminescence of luminol substrate.[27] The Michealis–Menton

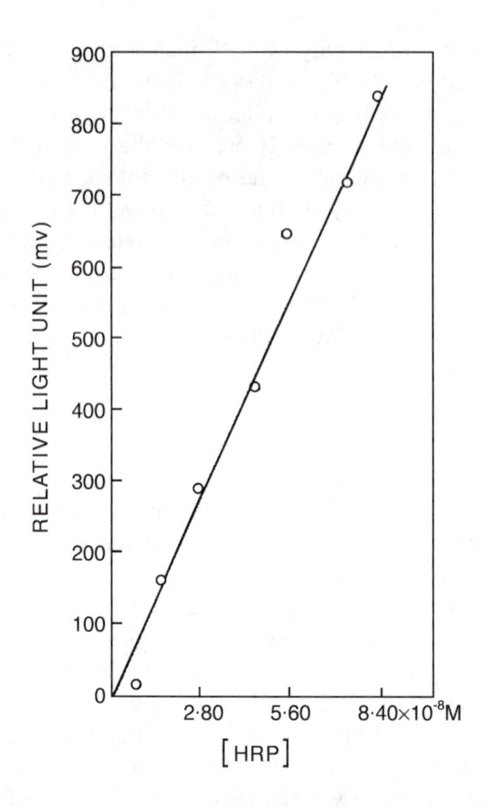

FIGURE 8.4 Estimation of HRP activity using ALPDO. ALPDO (8.5×10^{-5} *M*) and H_2O_2 (6.0×10^{-5} *M*) at pH 8.0 in 0.1 *M* Tris-HCl buffer.

constant K_m calculated for ALPDO is $1.2 \pm (0.2) \times 10^{-5}$ *M* for HRP at pH 8.0 in 0.1 *M* Tris-HCl buffer.

8.5.2 ATP Assay

Firefly luciferin–luciferase bioluminescence is a unique technique for accurate and sensitive assay of cellular ATP as well as ATP in solution.[28] As ALPDO is a 5′-azo phthalazine derivative of firefly luciferin and exhibits the activity of luciferin, it is considered worthwhile to estimate ATP based on its oxidative chemiluminescence reaction, catalyzed by firefly luciferase. The procedure adopted is similar to that of firefly luciferin–luciferase bioluminescence, where the single enzyme uses ATP and ALPDO and emits light. The ATP estimation was carried out by injecting various amounts of ATP concentration to the reaction mixture containing firefly luciferase (4.3×10^{-6} *M*) and ALPDO (1.77×10^{-4} *M*) in 50 m*M* tricine buffer containing 10 m*M* $MgCl_2$ and 1 m*M* EDTA at pH 8.0. The total reaction volume was 200 μl. A standard assay was carried out under identical conditions for luciferin for comparison. At very low ATP concentrations, the chemiluminescence decay is much slower as the potential quencher, oxyluciferin, forms at a slower rate. In Figure 8.5 a typical ATP standard curve employing ALPDO as a firefly luciferase substrate is shown. Addition of ATP to 20×10^{-12} *M* exhibited linearity in increase of the light output. The Michealis–Menton constant, K_m, estimated for ALPDO is $8.25 \pm (0.25) \times 10^{-5}$ *M* for luciferase at pH 8.0 in tricine buffer. From the above figure, it can be inferred that, although less sensitive than luciferin, ALPDO can be used as a substrate for firefly luciferase and can be employed to estimate ATP. By using ALPDO, it is possible to detect ATP in a femtomolar range.

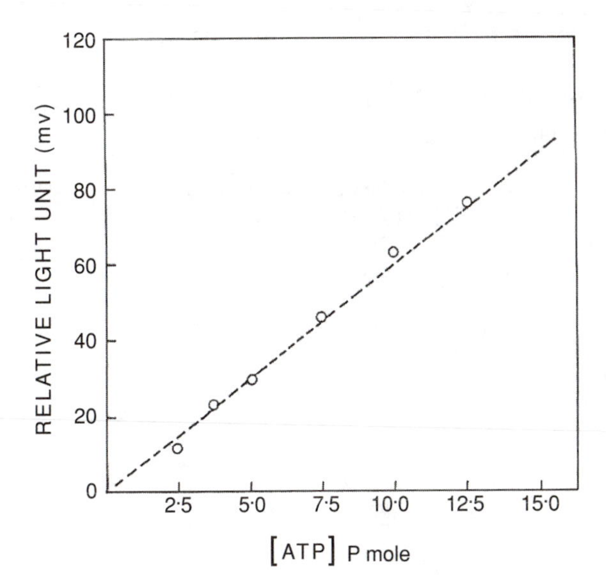

FIGURE 8.5 Estimation of ATP employing ALPDO–luciferase system. ALPDO (1.77×10^{-4} *M*) + luciferase (4.3×10^{-6} *M*) in 50 m*M* tricine buffer at pH 8.0 containing 10 m*M* $MgCl_2$ and 1 m*M* EDTA.

8.5.3 Effect of PIP

Light emission from luminol–H_2O_2–peroxidase systems can be greatly enhanced by the addition of substituted phenols, *p*-hydroxy cinnamic acid, benzothiazoles, and luciferin.[8] The underlying mechanism of this enhanced chemiluminescence is not completely understood. The most probable explanation is that these chemiluminescence enhancers can accelerate one or more of the oxidation steps to generate luminol radicals during the complex reaction pathway of the enzymatic oxidation of luminol.[29] A similar attempt to improve the HRP assay with ALPDO using PIP and firefly luciferin as chemiluminescence enhancers was made. The procedure involves the simultaneous use of PIP and ALPDO in the presence of oxidant, H_2O_2, and HRP at pH 8.0. The experiment was performed by adding various concentrations of luciferin–PIP to the reaction mixture containing ALPDO (8.5×10^{-5} *M*), HRP (5.6×10^{-10} *M*), and H_2O_2 (6.0×10^{-5} *M*) in 0.1 *M* Tris-HCl buffer at pH 8.0. The total reaction volume was maintained at 400 μl. A control experiment was also carried out using luminol as the standard. The total light emission in the chemiluminescent peroxidation of ALPDO in the presence of PIP is more than that in its absence, as shown in Table 8.3.

However, if the PIP concentration is more than 2.5×10^{-4} *M*, then it acts as a quencher. The reason for quenching of the HRP activity by PIP is not immediately known. The probable reason given for not enhancing the activity at higher concentration is that at pH 8.0 ionization of PIP to phenoxy iodophenol may not be occurring, thereby not mediating the electron transfer between the HRP and ALPDO. The effect of PIP on ALPDO–H_2O_2–HRP system has not yielded a satisfactory result although it was enhanced 6.5 times at $\leq 2.5 \times 10^{-4}$ *M* PIP concentration.

8.5.4 Effect of Luciferin

Whitehead and collaborators[7] made the surprising discovery that the addition of firefly luciferin to luminol–HRP–H_2O_2 not only enhanced the peroxidase catalyzed chemiluminescence severalfold but also reduced the chemical blank from luminol and H_2O_2. Table 8.4 shows the effect of luciferin on luminol–ALPDO with and without HRP. It can be observed from the table that luciferin enhances the ALPDO–H_2O_2–HRP chemiluminescence but not to an extent found with luminol–H_2O_2–HRP system. It is now clear that the molecules that provoke the light output from luminol when linked

TABLE 8.3
Comparison of Luminol, Luciferin, and ALPDO Activity Catalyzed by HRP and Luciferase in Their Respective Buffers at pH 8.0

Compound (*M*)	HRP (*M*)	Luciferase ($\times 10^{-6}$ *M*)	Time of Maximum Light Output	Light Emission (mV)	
				Background	With Enzyme
Luminol (8.5×10^{-5}) H_2O_2 (6×10^{-5})	5.6×10^{-8}	—	20 min	33	75
ALPDO (8.5×10^{-5}) H_2O_2 (6×10^{-5})	5.6×10^{-8}	—	55 s	29	920
Luciferin (1.77×10^{-4}) ATP (1.25×10^{-13})	—	4.3	5 s	—	290
ALPDO (1.77×10^{-5}) ATP (1.25×10^{-13})	—	4.3	5 s	—	30
Luminol (8.5×10^{-5}) PIP (2.5×10^{-4}) H_2O_2 (6×10^{-5})	5.6×10^{-10}	—	15 min	10	3400 (1.8)
ALPDO (8.5×10^{-5}) PIP (2.5×10^{-4}) H_2O_2 (6×10^{-5})	5.6×10^{-10}	—	13 s	6	52 (10.2)

Note: The values in the parentheses are before the addition of PIP and after background corrections.

TABLE 8.4
Effect of Chemiluminescence Enhancer Luciferin (4.46×10^{-3} *M*) on Luminol–ALPDO (1.25×10^{-4} *M*) and H_2O_2 (3.3×10^{-4} *M*) in the Presence and Absence of HRP (2.8×10^{-10} *M*), at pH 8.0

Substrate	Chemical Blank (mV)	Light Output (mV)
Luminol	10	44
Luminol + Luciferin	5	>10,000
Luminol + Luciferin + HRP	4	>10,000
ALPDO	6	28
ALPDO + Luciferin	3	801
ALPDO + Luciferin + HRP	2	8,195

to it by a covalent bond no longer act as efficient radical accelerators. For example, firefly luciferin enhanced the chemiluminescence of luminol more than 10,000 times, but when coupled covalently to luminol as in diazo phthalazinedione, the enhancement is less than 10,000 times. This can be explained by the presence of the amino group in luminol. Amines and phenols are good electron donors. Like transition metal ions, the free radicals formed from these are more stable. When luciferin–PIP is added to luminol during the chemiluminescence reaction, the latter can yield a stable radical more rapidly than in the absence of enhancers, resulting in enhanced chemiluminescence.[30] As ALPDO is devoid of amino substituent and in its place an electron-withdrawing diazo group is present, it is less prone to electron transfer mediators, such as luciferin and PIP. However, the diazo group can impart more light-emitting character than can an amino group and ALPDO can be irreversibly oxidized more efficiently than can luminol. Although amines are good electron donors, they are not fluorescent and, hence, ALPDO has more light yield than luminol. A similar observation was noticed by Yang and Yang[31] in the case of anthracene adducts, where electron-withdrawing groups at the bridge head enhanced the pericyclic chemiluminescence.

8.5.5 Assay of H_2O_2

The effect of H_2O_2 on the oxidation of ALPDO is studied following the same principle as that of luminol oxidation. However, unlike luminol, ALPDO bears no continuous linear relationship with H_2O_2 concentration but experiences a maximum and minimum activity. The H_2O_2 assay was carried out by fixing the [ALPDO] 50×10^{-6} *M* and HRP 4.3×10^{-13} *M* at pH 8.0, the total reaction volume was made 200 μl using Tris-HCl 0.1 *M* buffer. The reaction was initiated by the external additon of H_2O_2. The light output was measured, and the peak light intensity against [H_2O_2] was plotted. When the concentrations of HRP and ALPDO are kept constant and H_2O_2 concentration varied, the amount of light output obtained is represented in Figure 8.6. It can be seen from the figure that the activity of HRP is readily enhanced as the H_2O_2 concentration increases and attains a maximum at ~3 m*M* and subsequently undergoes a linear decrease in the activity. However, with the luminol–HRP system, a linear increase in the activity is noticed under identical conditions. The difference in the activity is probably because, in the case of luminol, the added H_2O_2 can act only at the hydrazide center, ultimately giving the carboxylate anion, but in case of ALPDO it could be acting at two centers, one at the hydrazide giving the carboxylate anion and another at —N=N— center. Azo compounds are known to react with peracids and peroxides to yield azoxy compounds. It is quite likely that the diazo group of substrate is unaffected to 3 m*M* but with more H_2O_2 is converted into azoxy

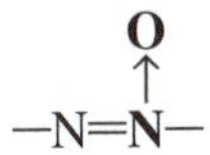

which reduces the activity of the luminol moiety. However, a closer examination of the figure reveals that at lower concentration of H_2O_2, ALPDO is more sensitive with less chemical blank and it can detect H_2O_2 to 3.0×10^{-8} *M*, with a maximum limit of 3 m*M*, whereas luminol can detect H_2O_2 to 3.0×10^{-7} *M* without any constraint on the upper limit of H_2O_2 concentration.

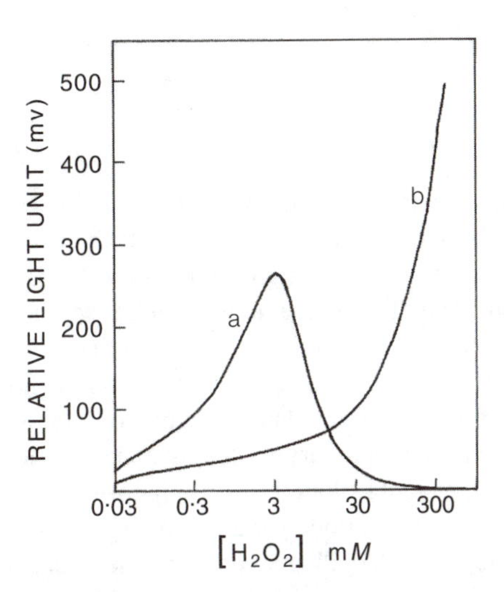

FIGURE 8.6 Estimation of H_2O_2 at pH 8.0. (a) In the presence of ALPDO (5×10^{-5} *M*) + HRP (4.5×10^{-13} *M*); (b) in the presence of luminol (5.0×10^{-5} *M*) + HRP (4.5×10^{-13} *M*).

8.5.6 Chemiluminescence Assay of Choline

p-Choline is one of the important phospholipids forming a lipid bilayer at the interfaces of two aqueous phases. Its breakdown assumes physiological significance as many hormones, neurotransmitters, growth factors, and phenol esters stimulate the release of choline and *p*-choline. It is observed that ALPDO is more sensitive than luminol at lower alkaline pH. By using the principle of Lucas and co-workers,[32–35] choline and *p*-choline can be determined in a single step using ALPDO–HRP chemiluminescence. The enzyme coupled reactions involved for the estimation of *p*-choline are

$$\text{p-Choline} + 3\,H_2O \xrightarrow[-\,H_3PO_4]{\text{Alkaline Phosphatase}} \text{diglyceride} + \underset{\text{Choline}}{HOCH_2{-}CH_2{-}N^{+}(CH_3)_3\,OH^{-}} \quad (8.4)$$

$$\underset{\text{Choline}}{HOCH_2{-}CH_2{-}N^{+}(CH_3)_3\,OH^{-}} + O_2 + H_2O \xrightarrow{\text{Choline oxidase}} \underset{\text{Betaine}}{{}^{-}OCH_2{-}CH_2{-}N^{+}(CH_3)_3} + 2H_2O_2 \quad (8.5)$$

$$\text{ALPDO} + H_2O_2 \xrightarrow{\text{HRP}} \text{ALP} + N_2 + h\nu \quad (8.6)$$

The liberated choline thus reacts as in Equations 8.4 and 8.5 to liberate H_2O_2, and the liberated H_2O_2 in reaction with ALPDO yields light (Equation 8.6). If the concentrations of all the reactants except *p*-choline are kept constant, then the light emitted by the reaction is directly proportional to the amount of *p*-choline present. This reaction also applies to choline. When all other reactants except choline are kept saturating and constant, as choline is increased there is an increase in light emission, thus making estimation of choline possible. Estimation of choline and *p*-choline was carried out by quantifying the H_2O_2 released. A standard reaction containing 2.5×10^{-5} *M* ALPDO, 2.5×10^{-8} *M* HRP, 0.5 U choline oxidase, and varying [choline] in the range of 5 to 25×10^{-6} *M* was carried out and the relative light units plotted against [choline]. A single procedure to measure choline released from *p*-choline by the action of alkaline phosphatase was performed by the coupled enzymatic reaction of exogenously added alkaline phosphatase, incubated at 37°C for about 30 min. Choline oxidase (0.5 U) and peroxidase (2.5 U) were added in the final step to trigger the chemiluminescent reaction of the oxidation of ALPDO, which occurred by the H_2O_2 derived from choline oxidation in the phosphate buffer saline (137 m*M* NaCl, 2.6 m*M* KCl, 8.1 m*M* Na_2HPO_4,

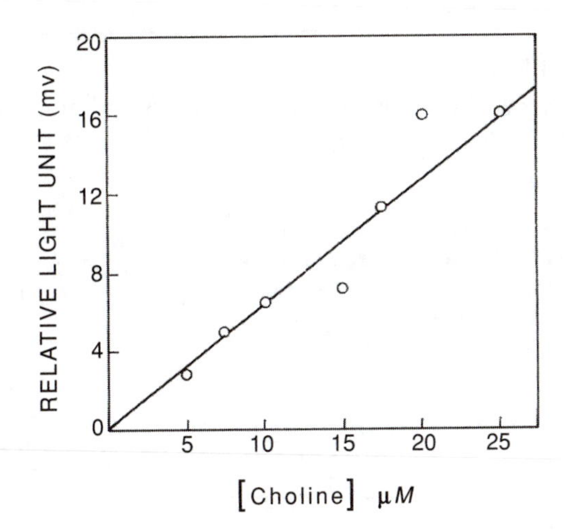

FIGURE 8.7 Estimation of choline in the presence of ALPDO (2.5×10^{-5} *M*) + HRP (2.5×10^{-8} *M*) + choline oxidase (0.5 U) at pH 7.4.

2 m*M* KH_2PO_4, pH 7.4, supplemented with 1.2 m*M* $CaCl_2$, 1.2 m*M* $MgCl_2$). The standard curve obtained by plotting the chemiluminescent light intensity against increasing choline concentration is provided in Figure 8.7. Note that Figure 8.7 shows that a good linearity exists between the chemiluminescent light intensity over a range of [choline], thereby providing an efficient non-isotopic technique to assay not only choline to 0.01 to 1 μm. However, when a similar technique is extended to the quantification of *p*-choline, the results obtained are nonlinear. The nonlinear relationship in *p*-choline may be due to the rate of coupled enzymatic reactions of added alkaline phosphatase and cholineoxidase, which may not create a linear relationship with that of the HRP catalyzed oxidation of ALPDO.

8.5.7 Quenching of ALPDO by NADH

Nicotinamide adenine dinucleotide (NAD) is a coenzyme and plays a vital role in several biological redox reactions. The presence of an antioxidant in the test sample depresses the luminescence of luminol, and this phenomenon is utilized for the assay of antioxidants.[36] As the ALPDO contains the luminol moiety, the effect of the reductogenic donor NADH on the luminescence of peroxidized ALPDO system was studied. The effect of NADH on ALPDO was investigated in the standard reaction of 59 m*M* pottasium phosphate buffer at pH 6.9, HRP (1×10^{-8} *M*), ALPDO (50×10^{-6} *M*), and varying [NADH]. The reaction was initiated by the addition of H_2O_2 (5×10^{-5} *M*). The light output was plotted against [NADH] to obtain a standard NADH assay curve. In Figure 8.8 the effect of NADH on ALPDO–HRP chemiluminescence is shown. It can be seen from the figure that NADH not only quenched the luminescence (Figure 8.8B), but also brought a delay in the time to reach maximum luminescence of the system (Figure 8.8A). However, all the reagents including NADH and HRP are kept constant but of ALPDO a linear increase in the chemiluminescent intensity with increase in [ALPDO] was noticed without affecting the chemiluminescence maximal time (data not shown). In Table 8.5 the data on the effect of reduced and oxidized pyridine nucleotides on the chemiluminescence of ALPDO–HRP system are shown.

It can be inferred that the chemiluminescence of ALPDO is decreased with the addition of βNAD^+ and NMN without change in the time of maximal activity. The quenching of the chemiluminescence of ALPDO by both the oxidized or reduced pyridine nucleotides may be because

TABLE 8.5
Effect of Pyridine Nucleotides on the Chemiluminescence of ALPDO (50×10^{-6} *M*), HRP (1×10^{-8} *M*), H_2O_2 (5.0×10^{-5} *M*) System

Pyridine Nucleotide for Maximal Activity ($\times 10^{-6}$ *M*)	% of Luminescence Quenched	Time for Maximal Activity (s)
	0	5
NADH (25)	12	13
NADH (50)	26	15
NADH (75)	40	20
NADH (100)	52	25
βNAD^+ (50)	33.3	5
NMN (50)	8.3	5

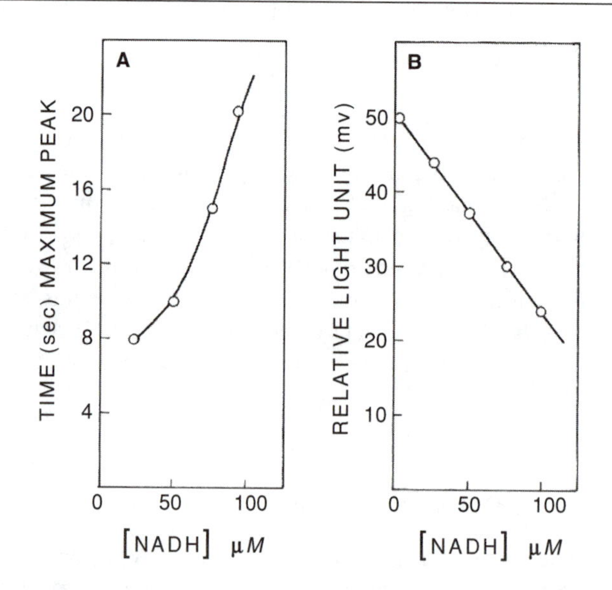

FIGURE 8.8 (A) Plot of delay in maximum light output against [NADH]. ALPDO (50.0×10^{-6} *M*) + HRP (1×10^{-8} *M*) + H_2O_2 (5.0×10^{-5} *M*) at pH 6.9. (B) Typical quenching plot of NADH with increasing [NADH]. ALPDO (50.0×10^{-6} *M*) + HRP (1×10^{-8} *M*) + H_2O_2 (5.0×10^{-5} *M*) at pH 6.9.

these reagents react with the intermediates involved[37] in the oxidation of ALPDO catalyzed by HRP as shown below:

$$H_2O_2 + \text{HRP} \longrightarrow \text{Compound } \mathbf{I}$$

$$\text{Compound } \mathbf{I} + X_2H_2 \longrightarrow \text{Compound } \mathbf{II} + X_2H^{\cdot}$$

$$\text{Compound } \mathbf{II} + X_2H_2 \longrightarrow \text{HRP} + X_2H^{\cdot}$$

$$X_2H^{\cdot} + H_2O_2 \longrightarrow \text{AP*}$$

$$\text{AP*} \longrightarrow \text{AP} + h\nu$$

where X_2H_2 is ALPDO, $XH^\bullet$ is ALPDO radical, and AP is luciferyl azopthalate dianion. It is most probable that $XH^\bullet$ may react with NADH to convert back to X_2H_2 or may react with compound **I**, thereby increasing the time for maximal activity by preventing the $XH^\bullet$ oxidation, which will eventually lead to decreased efficiency of the system. Similarly, βNAD^+ and NMN may react with redogenic intermediates to quench the chemiluminescence.

ACKNOWLEDGMENT

T.S. is employed at Jonaki, Board of Radiation and Isotope Technology, DAE. He would like to thank Dr. S. Gangadharan, Chief Executive, BRIT for granting permission to continue this research work and Shri B. Muralidharan, Officer-in-Charge, Jonaki, BRIT for continuous encouragement.

REFERENCES

1. Campbell, A. K., in *Chemiluminescence Principles and Applications in Biology and Medicine,* Ellis Harwood, Chichester, U.K., 1988.
2. Geiger, R. Hauber, R., and Misaka, W., *Mol. Cell. Probes,* 3, 309–328, 1989.
3. Conti, E., Franks, N. P., and Brick, P., *Structure,* 4, 287–298, 1996.
4. Lasagna, M., Vargas, V., Jameson, D. M., and Brunet, J. E., *Biochemistry,* 35, 973–979, 1996.
5. Yakunin, A. F. and Hallenbeck, P. C., *Anal. Biochem.,* 258, 146–149, 1998.
6. Arakawa, H., Maeda, M., and Tsuji, A., *Anal. Biochem.,* 192, 238–242, 1991.
7. Whitehead, T. P., Thorpe, G. H. G., Carter, T. J. N., Groucutt, C., and Kricka, L. J., *Nature,* 305, 158–159, 1983.
8. Thorpe, G. H. G. and Kricka, L. J., *Methods Enzymol.,* 133, 331–351, 1986.
9. Thorpe, G. H. G., Kricka, L. J., Moseley, S. R., and Whitehead, T. P., *Clin. Chem.,* 31, 1335–1341, 1985.
10. Koo, J. Y., Schmidt, S. P., and Schuster, G. B., *Proc. Natl. Acad. Sci. U.S.A.,* 75, 30–33, 1977.
11. Beck, S. and Koster, H., *Anal. Chem.,* 62, 2258–2270, 1990.
12. DeLuca, M. and McElroy, W. D., *Methods Enzymol.,* 57, 3–124, 1978.
13. Goodrich, G. A. and Burrell, H. R., *Anal. Biochem.,* 127, 395–404, 1982.
14. Thorpe, G. H. G. and Kricka, L. J., *J. Biolumin. Chemin.,* 3, 97–100, 1989.
15. Miska, W. and Geiger, R., *J. Clin. Chem. Clin. Biochem.,* 25, 23–30, 1987.
16. Ugarova, N. N., Vozny, Y. A., Kutuzova, G. D., and Dementieva, E. I., in *Bioluminescence Chemiluminascence, Current Status,* Stanley, P. E. and Kricka, L. J., Ed., Wiley, Chichester, 1991, 511.
17. Mayer, A. and Neuenhoffer, S., *Angew. Chem. Int. Ed. Engl.,* 33, 1044–1072, 1994.
18. White, E. H. and Rosewell, D. F., *Acc. Chem. Res.,* 3, 54–62, 1970.
19. Thorpe, G. H. G., Kricka, L. J., Gillespie, E., Moseley, S., Amess, R., Baggett, N., and Whitehead, T. P., *Anal. Biochem.,* 145, 96–100, 1985.
20. Sasamoto, K., Deng, G., Ushijima, T., Ohkura, Y., and Ueno, K., *Analyst,* 120 1709–1714, 1995.
21. Klel, J. L. and Obreen, G. J., U.S. Patent 5003050, 1991.
22. Liu, Z., Zhao, C., Tang, M., and Cai, S., *J. Phys. Chem.,* 100, 17,337–17,344, 1996.
23. Griffiths, J., *Chem. Soc. Rev.,* 1, 481–493, 1972.
24. Roberts, D. R. and White, E. H., *J. Am. Chem. Soc.,* 92, 4861–4867, 1978.
25. Thompson, J. E. and Jordan, D. B., *Anal. Biochem.,* 256, 7–13, 1998.
26. Briggs, G. E. and Haldane, J. B. S., *Biochem. J.,* 19, 338–339, 1925.
27. Kim, B. B., Pisarev, V. V., and Egorov, A. M., *Anal. Biochem.,* 199, 1–6, 1991.
28. Hastings, J. G. M., Wheat, P. F., and Oxiey, K. M. *Soc. Appl. Bacteriol. Tech. Ser.,* 26, 229–233, 1989.
29. Thorpe, G. H. G., Stott, R. A. W., Sankolli, G. M., Catty, D., Raykundalia, C., and Kricka, L. J., in *Bioluminescence and Chemiluminescence; New Perspectives,* Scholmerich, J., Andreesen, R., Kapp, A., Ernst, M., and Woods, W. G., Eds., Wiley, Chichester, 1987.
30. Gilbert, A. and Baggott, J., *Essentials of Molecular Photochemistry,* Blackwell Scientific Publications, London, 1991.
31. Yang, N. C. and Yang, X. Q., *J. Am. Chem. Soc.,* 109, 3804–3805, 1989.

32. Lucas, M., Sanchez-Margalet,V., Pedrera, C., and Bellido, M. L., *Anal. Biochem.,* 231, 19, 1995.
33. Bcckino, S. B., Blackmore, P. F., Wilson, P. B., and Exton, J. H., *J. Biol. Chem.,* 262, 15,309, 1987.
34. Irving, H. R. and Exton, J. H., *J. Biol. Chem.,* 262, 3440, 1987.
35. Kanaho, Y., Nakai, Y., Katoh, M., and Nozawa, Y., *J. Biol. Chem.,* 262, 12,492, 1993.
36. Thorpe, G. H. G. and Whitehead, T. P., British Patent Appl. 2,245,062, 1991.
37. Wong, J. K. and Salin, M. L., *Photochem. Photobiol.,* 33, 737, 1981.

9 Hybrid Capture: A System for Nucleic Acid Detection by Signal Amplification Technology

Attila Lorincz and James Anthony

CONTENTS

9.1 INTRODUCTION

The methodology and performance characteristics of the Hybrid Capture® (HC) test have been extensively described in the literature. The first-generation Hybrid Capture test, HC1, was replaced in 1996 by the more sensitive and reproducible second-generation HC2 test.[1–3] HC2 technology has been applied to the detection of a number of infectious bacterial and viral organisms, including human papillomavirus (HPV), cytomegalovirus (CMV), *Chlamydia trachomatis* (CT), *Neisseria gonorrhoeae* (GC), hepatitis B virus (HBV), and herpes simplex virus (HSV).[4–10] Several of the HC tests have been approved by the U.S. Food and Drug Administration, the most important of which is the HPV test for carcinogenic viral types. HC2 can detect one or more of 13 HPV types at the level of 1 pg/ml each, which corresponds to 5000 HPV genomes per test well. Most HPV testing is focused on the 13 key carcinogenic types—16, 18, 31, 33, 35, 39, 45, 51, 52, 56, 58, 59, and 68. In addition, HC2 technology has been applied to the detection of RNA. This HC2 RNA detection format, termed the Hybrid Capture Expression Analysis System (HC-EAS), has been used for transcript detection.

All HC technology operates on the principle of signal amplification and thus requires minimal specimen preparation. Interference from materials present in clinical specimens that may cause inhibition or negative results in other tests, such as target-amplification-based procedures, is not a concern with HC. Specimens for sexually transmitted disease (STD) testing are usually collected from brush samples or swabs and placed directly into a transport medium requiring no refrigeration. Blood viruses are usually tested from white cells (CMV), serum, or plasma (HBV). Specimens in appropriate transport media (STM, UCM) may be stored at ambient conditions for up to a week and may be tested immediately or after an additional storage period of 2 weeks or longer at 4°C.

0-8493-0719-8/02/$0.00+$1.50

HC tests can be performed from residual specimens used for liquid cytology examinations with no loss in sensitivity. We have developed a novel liquid cytology medium, the Universal Collection Medium (UCM), which has favorable characteristics for cytological and molecular testing, in particular the preservation of RNA in patient samples for up to 6 weeks at ambient temperature.

9.2 HYBRID CAPTURE 2

HC2 utilizes unlabeled RNA probes prepared by *in vitro* transcription to facilitate capture and detection of the target molecules (Figure 9.1). The current HC2 design uses a 96-well microplate format. Microplates containing immobilized antibody specific for RNA:DNA hybrids, in combination with target-specific RNA probes, are used to capture the target DNA. Following hybridization,

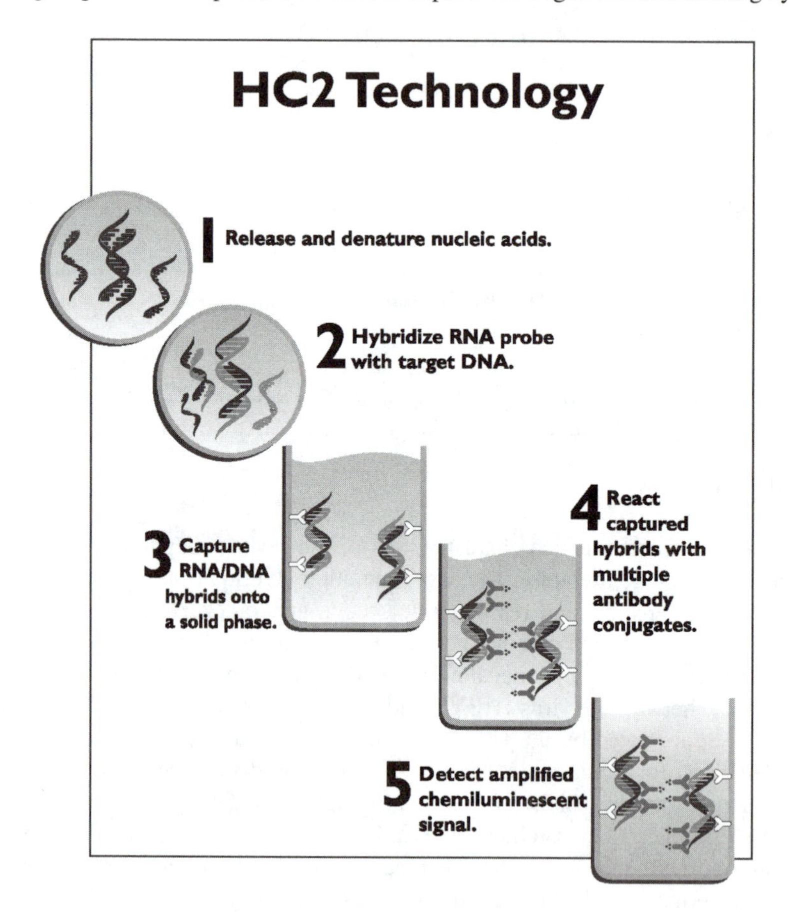

FIGURE 9.1 Essential steps of the HC2 test. Unlabeled full genomic RNA probes are hybridized in solution with denatured target DNA from specimens. Resulting RNA:DNA hybrids are captured on the surface of microplate wells by immobilized antibody that specifically recognizes RNA:DNA hybrids. Captured RNA:DNA hybrids are reacted with a second antibody conjugated to alkaline phosphatase. Detection is accomplished by addition of a chemiluminescent dioxetane-based substrate. As the substrate is cleaved by the bound alkaline phosphatase conjugate, light is emitted in a long-lived glow reaction and measured in relative light units (RLUs) using a microplate luminometer. Emitted light intensity is proportional to the amount of target DNA in the specimens and is usually expressed as a ratio of the signal to the positive control (RLU/PC). Alternatively, this ratio is termed RLU/CO, which is the ratio of the specimen RLU to the cutoff level of light based on the mean or a function of the mean of triplicate positive control specimens. In some cases, the data may be expressed as the ratio of the signal-to-noise, or S/N. Subtracting 1 from this value provides for a straight line through the cutoff.

FIGURE 9.2 Microplate luminometer (instrument on the left), computer, and other accessory equipment for the HC2 test.

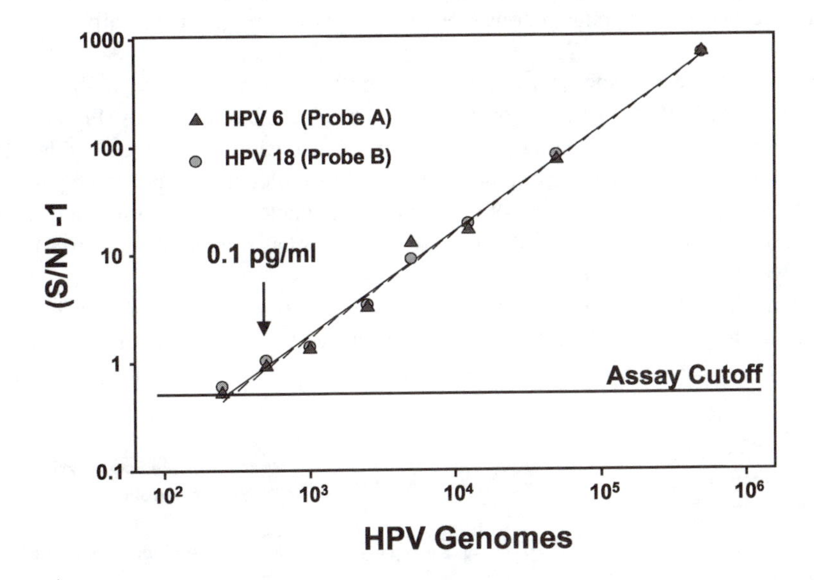

FIGURE 9.3 Luminometer readings for an HC2 HPV test. Various concentrations of HPV 6 or HPV 18 target DNAs were tested in the HC2 test using either a low risk HPV type cocktail (Probe A, types 6, 11, 42, 43, 44) or a high risk HPV cocktail (Probe B, types 16, 18, 31, 33, 35, 39, 45, 51, 52, 56, 58, 59, 68). The data are expressed as signal-to-noise ratios minus 1 (S/N – 1) as a function of the HPV genome equivalents per test well.

RNA:DNA hybrids are captured onto the plates and after the nonhybridized nucleic acid has been removed, an alkaline phosphatase-labeled anti-RNA:DNA monoclonal antibody is used to detect the immobilized hybrids. Incubation of the target-bound alkaline phosphatase antibody conjugate with the chemiluminescent substrate CDP Star (Applied Biosystems, Foster City, CA) produces light that is measured by a luminometer (Figure 9.2). The readings are then directly transferred into a software analysis program where the results are analyzed and the number of hybrids immobilized can be quantified (Figure 9.3).[1]

9.3 HYBRID CAPTURE 3

A new format of HC, Hybrid Capture 3 (HC3, Figure 9.4), has been developed specifically to address the need for a rapid, robust, quantitative, and sensitive method for the detection and discrimination of highly homologous nucleic acid targets. The HC3 format is a highly specific test for discriminating closely related nucleic acid targets, and HC3 can also be applied to the detection of a variety of mutations. The specificity of HC3 is achieved by using biotinylated DNA capture molecules directed to unique sequence regions within the desired target. By recognizing only unique sequence regions within closely related genomes, and by simultaneously using RNA probes that hybridize to other regions of these sequences to generate RNA:DNA hybrids (which provide the signal), HC3 provides both a highly selective and sensitive method for identifying closely related nucleic acid targets. The HC3 technique can be used to discriminate accurately between any highly homologous or nearly identical DNA targets.

HC3 virtually eliminates the issue of RNA probe cross-reactivity. With the HC3 format only the specificity of the biotinylated DNA capture probes is essential. Unless the cross-reactive DNA molecule hybridizes with the biotinylated capture oligonucleotides, it will be washed off the plate prior to the detection step regardless of whether or not the RNA probe hybridizes to the cross-reactive target present in the sample.

The assay format is indistinguishable from the HC2 test because the oligonucleotides and RNA probes can be premixed and added simultaneously. Another advantage of HC3 is that target capture is a function of hybridization to the biotinylated DNA capture probe, so there is less concern about background effects caused by endogenous DNA:RNA hybrids present in partially denatured clinical specimens.

To further enhance the specificity and to reduce the cross-reactivity of the HC3 assay, an additional component called blocker oligos was developed. These blocker oligos (Figure 9.5) are small, unlabeled oligos that are complementary to the biotin-labeled capture oligonucleotides. The capture and blocker oligonucleotides form a double-stranded DNA molecule that does not generate a signal. The blocker oligos are present in excess of the capture oligonucleotides and are designed to hybridize to the capture oligos at temperatures well below those used for target hybridization so as not to interfere with target hybridization. Optimal cross-reactivity reduction is achieved by using multiple small blocker oligonucleotides, typically two blocker oligonucleotides, specific for each capture oligonucleotide such that the sum of the blocker oligo sequences covers the entire or nearly the entire capture oligo sequence. The blocker oligonucleotides hybridize to the capture oligonucleotides at

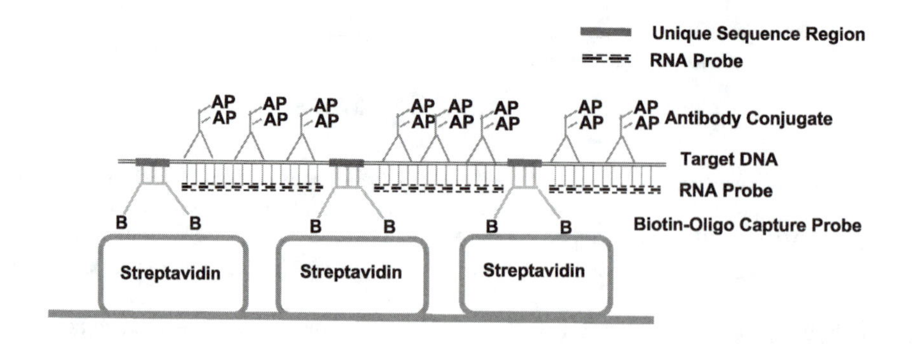

FIGURE 9.4 HC3 method. Clinical specimens containing DNA targets are denatured by heating in alkali as described previously for HC2.[1] Then, unlabeled RNA probes and small DNA oligonucleotides labeled with biotin are hybridized to the target. The oligos are chosen to hybridize in specific unique regions of the target to minimize or eliminate unwanted cross-reactivity. The hybrid complexes are then captured on streptavidin-coated plates, washed to remove unreacted molecules, and detected by supplying a dioxetane substrate as in the HC2 test.

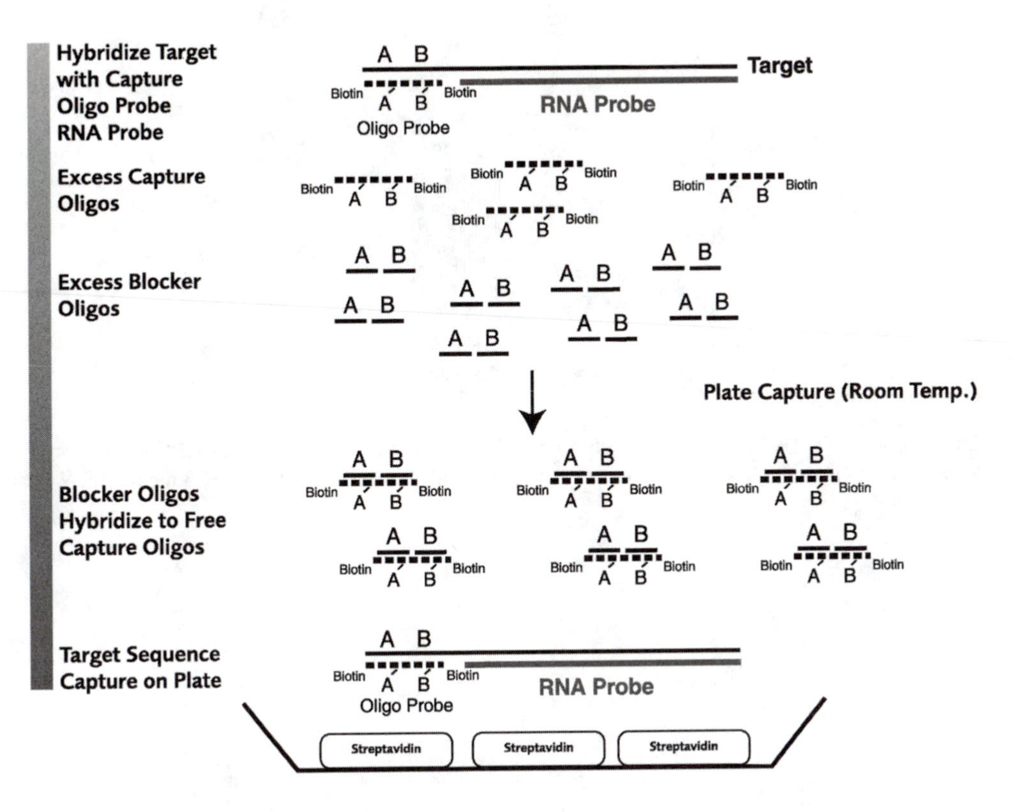

FIGURE 9.5 Blocker oligonucleotide method to reduce cross-reactivity: small unlabeled oligos complementary to the biotinylated capture probes reduce cross-reactive background during room-temperature plate capture.

temperatures below the hybridization temperature, particularly those encountered during room-temperature plate capture, and prevent cross-reactive hybridization from occurring.

HC3 has been applied to the specific typing and detection of HPV (Figure 9.6) and HSV (Figure 9.7). The HC3 format is capable of discriminating HPV 45 from the highly homologous HPV 18. Both HC2 and HC3 perform similarly in the detection of HPV 45, as the sensitivity of HPV 45 is comparable in the two formats. However, the cross-reactive detection of HPV 18 has been virtually eliminated in the HC3 format. HC3 also greatly increases the specificity of HSV typing. Both the HC2 and HC3 formats work well in detecting HSV-1, but HSV-2 cross-reactivity is reduced to background levels in the HC3 assay (see Figure 9.7).

The HPV HC3 format can be used to detect, differentiate, and accurately quantify HPV types from specimens containing a mixture of HPVs at various concentrations (see Figure 9.8). The detection levels of four HPV types at three different concentrations (5 pg/ml HPV 16, 50 pg/ml HPV 18, 1 pg/ml HPV 31, and 5 pg/ml HPV 56) when spiked as mixtures (target mix) into a single pooled negative clinical specimen were compared. The spiked specimen was then analyzed by the individual HPV HC3 typing assays for HPV 16, HPV 18, HPV 31, and HPV 56 and compared with HPV detection of the same types at the same concentrations in spiked specimens containing only a single HPV type. As a control, we also probed the specimen containing the HPV mix with an HPV 45 probe, an HPV type not present in the mix, to demonstrate the specificity of the assay. Using the HC3 test, we were able to accurately identify and quantify all the HPV types present in the mixture. No positive signal is evident with the HPV 45 probe, which was expected to produce negative reactivity.

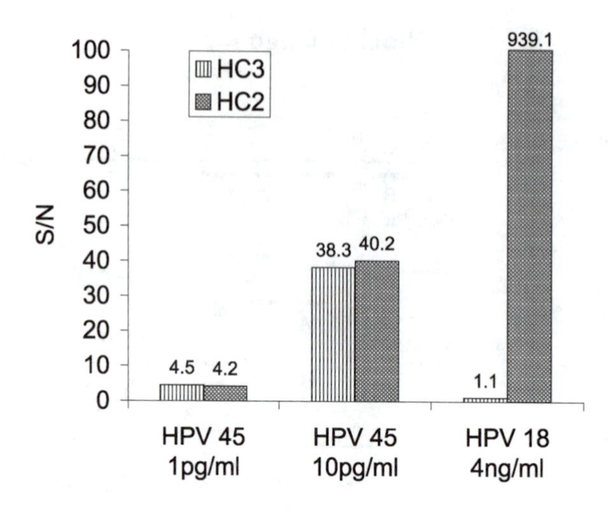

FIGURE 9.6 HC3 vs. HC2 type-specific detection of HPV 45. Sensitive detection of HPV 45 is maintained while cross-reactivity to highly homologous HPV 18 is virtually eliminated with the HC3 method. S/N = ratio of signal to noise.

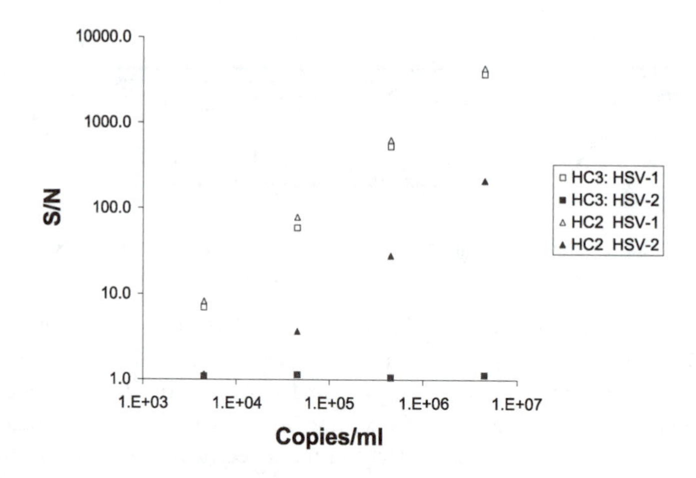

FIGURE 9.7 HC3 vs. HC2 type-specific detection of HSV-1. The experiment employed an HSV-1 probe with various HSV target DNAs. HSV-1 detection sensitivity is maintained in the HC3 format while cross-reactivity to highly homologous HSV-2 is vastly reduced with the HC3 method. S/N = ratio of signal to noise.

9.4 HYBRID CAPTURE EXPRESSION ANALYSIS SYSTEM

A sensitive method for RNA detection, the HC Expression Analysis System (HC-EAS) (Figure 9.9), is another adaptation of the HC technology. Direct mRNA measurements are an important tool for drug discovery and pharmaceutical screening. The HC-EAS method can detect RNA over a large range of concentrations directly from lysed cells without the need for complicated sample preparation or centrifugation. The HC-EAS method utilizes single-stranded biotinylated DNA probes that serve both capture and signal amplification functions. Cells or virions are lysed and the released RNA targets are hybridized with DNA probes and captured onto streptavidin-coated microplate wells. Following removal of unbound material, the wells are incubated with alkaline phosphatase–

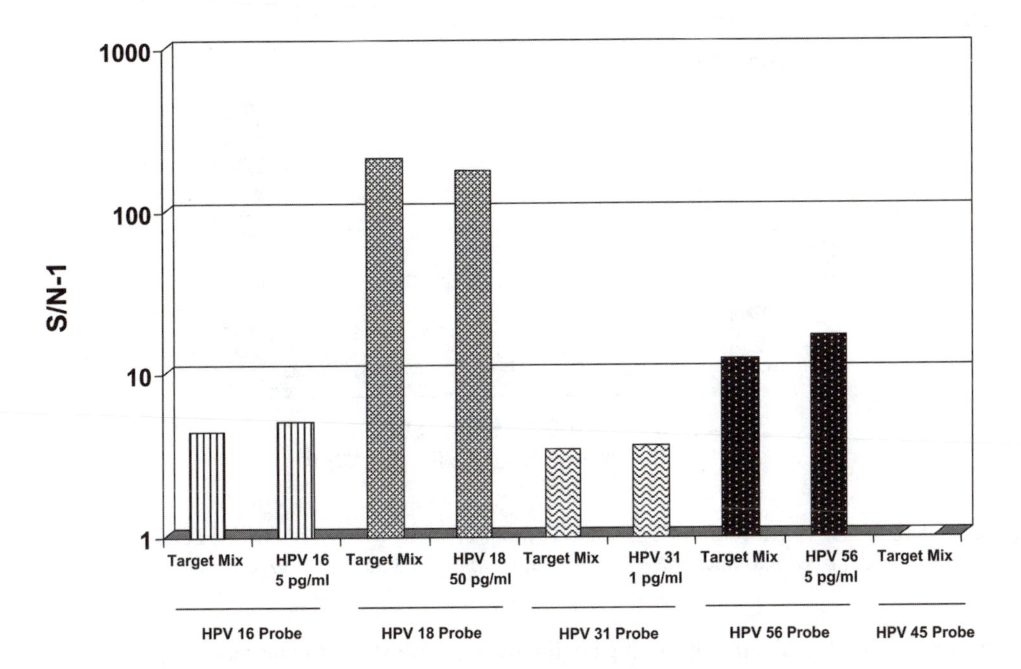

FIGURE 9.8 HC3 typing and detection from HPV target mixtures: HC3 detection of HPV 16, HPV 18, HPV 31, and HPV 56 from a mixed spiked negative clinical specimen (reading from the left: bars 1, 3, 5, and 7) and pure target samples (bars 2, 4, 6, and 8). Bar 9 is the HPV 45 probe negative control. The probes employed are shown under each pair of bars and under the last bar to the right.

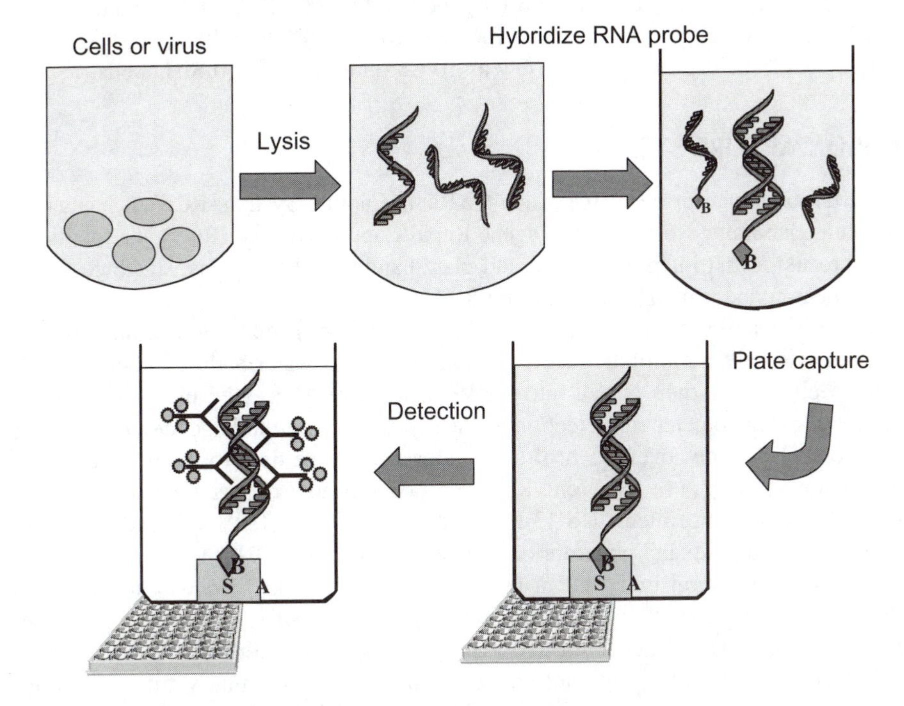

FIGURE 9.9 HC-EAS RNA detection method. Samples are lysed with detergent and protease, then the endogenous RNA is hybridized with a biotinylated single-stranded DNA probe. Following hybridization, hybrids are captured onto the wells of a streptavidin-coated microplate and then reacted with anti-RNA:DNA antibody conjugated to alkaline phosphatase. A chemiluminescent dioxetane substrate is added, and the light produced is read on a plate luminometer.

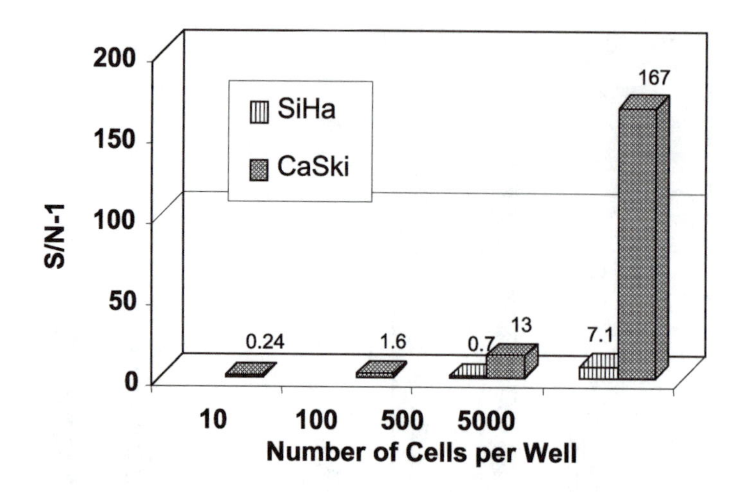

FIGURE 9.10 HC-EAS detection of E6-E7 HPV 16 mRNA directly from lysed SiHa and CaSki cells. Cultured cells were diluted, placed into microplate wells, lysed *in situ*, and directly analyzed for the presence of E6-E7 HPV 16 mRNA.

labeled anti-RNA:DNA antibody conjugate. Following wash steps, the target-bound conjugates are reacted with the chemiluminescent substrate CDP Star producing light that is measured by a luminometer. The readings are then analyzed and the number of hybrids immobilized can be quantified.

HC-EAS was used to detect E6-E7 HPV 16 mRNA directly from cell lysates (Figure 9.10). CaSki cells containing 600 copies of the E6-E7 gene[11] and SiHa cells containing one to two copies of the E6-E7[12] region were lysed and analyzed for the presence of E6-E7 mRNA by HC-EAS. HPV 16 E6-E7 mRNA could be detected from as few as 10 CaSki cells and 500 SiHa cells.

9.5 RAPID CAPTURE SYSTEM

An automated robotic platform for HC, called the Rapid Capture System (RCS), is being developed for high-volume laboratory testing. The Digene Rapid Capture System (RCS) is a robotic, 96-well microplate processor that integrates liquid and plate handling, incubations, shaking, and washing directly from bar-coded primary tubes (Figure 9.11).

Bulk denaturation of specimens is performed directly in the specimen collection tubes, utilizing a custom rack assembly, a multitube rack vortexer, and a 65°C water bath. Following this denaturation step, specimens are then placed into the RCS platform. Processed plates are transferred to the DML 2000™ luminometer for detection and analysis utilizing the Digene qualitative software. The RCS protocol provides over 3.5 h of continuous hands-free time to the user. The automated application allows a single user with one RCS to process up to 352 specimens (four microplates) in an 8-h shift and 704 specimens in a 13-h period. Currently, the HPV, CT, and GC assays have been adapted to this format, and plans are being made to automate HBV testing as well. Results for the CT/GC assays in several independent laboratories are good, and evaluations are proceeding. The RCS method for HC2 HPV testing has produced good "in-house" results on clinical specimens (Figure 9.12), but to date independent performance data are not available.

RCS provides for "walk-away" automation with bar-coded primary tube sampling (post-denatured specimen rack loading), integration of liquid and microplate handling, incubation, plate shaking, and plate washing. The specimens are bar-coded by specialized software for identification and then processed in the robotic platform. Currently, all processing steps occur on the platform, except for the specimen denaturation and plate reading steps.

FIGURE 9.11 Rapid Capture System (RCS).

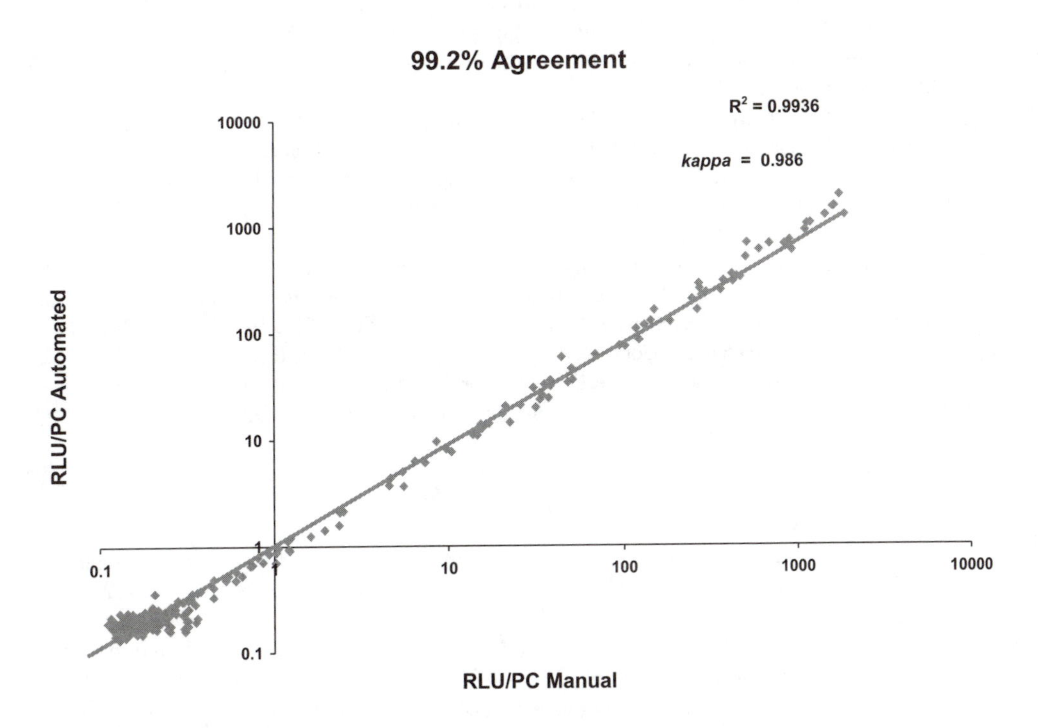

FIGURE 9.12 Scatterplot comparison of the manual HC2 and the automated RCS for HPV detection from a split panel of identical clinical specimens. The diagonal line passing through the origin shows excellent agreement between the two tests.

9.6 CONCLUSION

In conclusion, HC has evolved into a flexible and accurate method for the routine detection of viruses, bacteria, and genetic changes and for gene expression profiles. An HC robotic platform, the RCS, has been developed to allow automated high-volume laboratory testing. Invention of the HC3 technology has extended the useful range of applications to the detection and discrimination of closely related nucleic acid targets. HC-EAS allows for sensitive and rapid quantification of mRNA for expression profiling. HC is poised to be among the leading technologies used for disease detection at the molecular level for the next decade.

ACKNOWLEDGMENTS

We thank Iwona Mielzynska-Lohnas and James Lazar for the use of unpublished data and Katherine Mack for assistance in preparation of the manuscript.

REFERENCES

1. Lorincz, A., Hybrid Capture™ method for detection of human papillomavirus DNA in clinical specimens, *Pap. Rep.,* 7, 1–5, 1996.
2. Peyton, C. L., Schiffman, M., Lorincz, A. T., Hunt, W. C., Mielzynska, I., Bratti, C., Eaton, S., Hildesheim, A., Morera, L. A., Rodriguez, A. C., Sherman, M. E., and Wheeler, C. M., Comparison of PCR- and Hybrid Capture-based HPV detection systems using multiple cervical specimen collection strategies, *J. Clin. Microbiol.,* 36, 3248–3254, 1998.
3. Anthony, J. G., Linske-O'Connell, L., and Lorincz, A. T., Nucleic acid hybridization, in *Clinical Virology Manual,* S. Specter et al., Eds., ASM Press, Washington, D.C., 2001, 169–181.
4. Schiffman, M., Herrero, R., Hildesheim, A., Sherman, M.E., Bratti, M., Wacholder, S. et al., HPV DNA testing in cervical cancer screening: results from women in a high-risk province of Costa Rica, *JAMA,* 283, 87–93, 2000.
5. Wright, T. C., Jr., Denny, L., Kuhn, L., Pollack, A., and Lorincz, A., HPV DNA testing of self-collected vaginal samples compared with cytologic screening to detect cervical cancer, *JAMA,* 283, 81–86, 2000.
6. Tong, C. Y. W., Cuevas, L. E., Williams, H., and Bakran, A., Comparison of two commercial methods for measurement of cytomegalovirus load in blood samples after renal transplantation, *J. Clin. Microbiol.,* 38, 1209–1213, 2000.
7. Schachter, J., Hook, E.W., III, McCormack, W.M., Quinn, T.C., Chernesky, M., Chong, S., et al., Ability of the Digene Hybrid Capture II test to identify *Chlamydia trachomatis* and *Neisseria gonorrhoeae* in cervical specimens, *J. Clin. Microbiol.,* 37, 3668–3671, 1999.
8. Niesters, H. G. M., Krajden, M., Cork, L., de Medina, M., Hill, M., Fries, E., and Osterhaus, A. D. M. E., A multicenter study evaluation of the Digene Hybrid Capture II signal amplification technique for detection of hepatitis B virus DNA in serum samples and testing of EUROHEP standards, *J. Clin. Microbiol.,* 38, 2150–2155, 2000.
9. Pas, S. D., Fries, E., de Man, R. A., Osterhaus, A. D. M. E., and Niesters, H. G. M., Development of a quantitative real-time detection assay for hepatitis B virus DNA and comparison with two commercial assays, *J. Clin. Microbiol.,* 38, 2897–2901, 2000.
10. Cullen, A. P., Long, C. D., and Lorincz, A. T., Rapid detection and typing of herpes simplex virus DNA in clinical specimens by the Hybrid Capture II signal amplification probe test, *J. Clin. Microbiol.,* 35, 2275–2278, 1997.
11. Guerin-Reverchon, I., Chardonnet, Y., Chignol, M. C., and Thivolet, J., A comparison of methods for the detection of human papillomavirus DNA by *in situ* hybridization with biotinylated probes on human carcinoma cell lines: application to wart sections, *J. Immunol. Methods,* 123, 167–176, 1989.
12. Heiles, H. B. J., Genersch, E., Kessler, C., Neumann, R., and Eggers, H. J., *In situ* hybridization with digoxigenin-labeled DNA of human papillomaviruses (HPV16/18) in HeLa and SiHa cells, *BioTechniques,* 6, 978–981, 1988.

Section III

Molecular Biological Applications of Luminescence

10 Chemiluminescent Technology: Optimization of Western Blots and ELISAs Utilizing a Horseradish Peroxidase Label in Automated Systems

Kimberly K. Hines, Kelli Feather-Henigan, Li Li, Ineabel Horneij, Carrie Clothier, Greg Milosevich, Bill Lipton, and Tom Brotcke

CONTENTS

10.1 INTRODUCTION

Western blotting and ELISAs are two of the most common techniques used by researchers today to analyze protein samples for various applications. Traditionally, color-producing substrates and radioactivity were used to analyze results in both these techniques. With the continued concerns about radioactive waste and contamination, along with the insensitivity provided by colorimetric substrates, enhanced chemiluminescent methods have become a viable and often preferable alternative. The use of luminol-based systems for the detection of the horseradish peroxidase (HRP) label has become very popular. Several enhanced chemiluminescent substrates are commercially available that utilize a phenolic compound as the enhancer.[1] Other enhancers include various substituted phenols, naphthols, amines, and 6-hydroxbenzothiazole.[2,3] A more recent development utilizes an azine enhancer, sodium phenothiazine 10-yl-propane sulfonate (Figure 10.1).[4] Upon reaction with HRP in the presence of a hydrogen peroxide source, luminol in combination with this azine enhancer produces an intense light that can be captured either on film, by an imaging device, such as a charge-coupled device (CCD) camera, or by a luminometer. This azine enhancer-based system

0-8493-0719-8/02/$0.00+$1.50

is commercialized under the SuperSignal® Family name (Pierce Chemical Company, Rockford, IL).

The SuperSignal phenothiazine enhancer system has been optimized to meet different sensitivity requirements. The SuperSignal West Product Family was developed for Western blotting applications with varying degrees of assay sensitivity, light intensity, and duration. The SuperSignal ELISA Product Family was prepared for ELISA-based applications that require high signal-to-noise ratios with rapid kinetics at various temperatures.

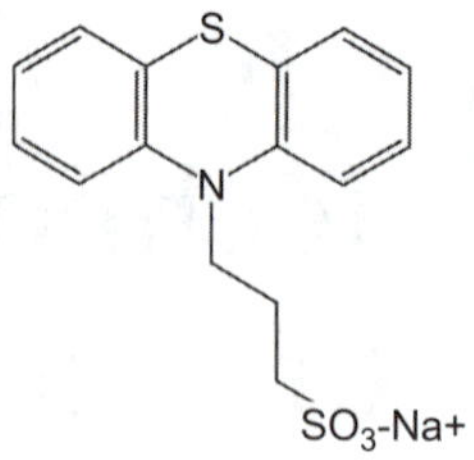

S
N
SO_3-Na+

FIGURE 10.1 Sodium phenothiazine 10-yl propane sulfonate, the enhancer utilized in the SuperSignal West and SuperSignal ELISA Chemiluminescent substrates.

To utilize the SuperSignal Family or any other sensitive chemiluminescent system, it is important to optimize the individual systems. Each step in the immunoassay should be evaluated to give the best assay performance and to enhance all of the characteristics inherent in the SuperSignal system.

10.2 WESTERN BLOT TECHNOLOGY: FROM SAMPLE TO IMAGE

To analyze protein samples, the proteins can be separated by gel electrophoresis and stained with various dyes such as silver or Coomassie® dye (ICI Americas, Wilmington, DE). Stained gels do not provide a significant amount of information about a specific protein, but instead provide a general view of all of the proteins that are present in the sample. To detect a specific protein or a specific antigen in a gel, immunochemical detection is often necessary unless a highly purified protein is analyzed. Originally, researchers tried to utilize antibodies directly on the gel but found that the procedure often resulted in the gels tearing and in poor diffusion into a gel that prevented consistent results. In 1979, Towbin et al.[5] described a method for the transfer of proteins out of a gel and onto a nitrocellulose membrane followed by immunodetection. This method was labeled "Western blotting" because it followed the success of "Southern blotting."[6] Southern blotting is the process of transferring DNA from agarose to a membrane followed by detection of DNA with specific probes.

Unpurified test samples typically contain several different proteins. By utilizing the Western blot technique through specific antibodies and the HRP label, the presence, relative quantity, and size of the antigen in a crude mixture can be detected. The protein solution of interest is collected and prepared in a sample buffer. The sample buffer typically contains sodium dodecyl sulfate (SDS) to unfold the proteins and to generate a constant anionic charge-to-mass ratio for the unraveled protein chains; glycerol to give the sample a higher density than the buffer allowing the protein sample to "sink" to the bottom of the well; a low-molecular-weight dye for the determination of the dye front that runs electrophoretically ahead of the sample; and a reducing agent such as Tris(2-carboxyethyl)phosphine (TCEP) or dithiothreitol (DTT), to break disulfide bonds in the protein. Once the sample has been prepared in the appropriate sample buffer, it is then added to a SDS-polyacrylamide gel (SDS-PAGE). The use of commercially available precast gels is highly recommended. Precast gels maintain consistent results from experiment to experiment and eliminate a potential source of variability.

The proteins are then separated electrophoretically by molecular size. Once the dye front has reached the bottom of the gel, the gel is removed to a transfer unit and the proteins are transferred electrophoretically to a membrane such as nitrocellulose or polyvinylidene difluoride (PVDF). Once the proteins have been transferred, the membranes are removed and the immunoperoxidase assay procedure begins.

Pierce Lane Marker Sample Buffer contains a bright pink hydrophobic tracking dye that transfers from the gel to nitrocellulose. By using this tracking dye, the dye front is easily monitored, calculations of molecular weight from the membrane can be performed, and a quality control check for the successful transfer to nitrocellulose is achieved. If a membrane must be cut into individual lanes, the sample buffer without protein can be added to the top of the gel at the end of electrophoresis. The hydrophobic dye should be allowed to electrophorese into the gel. After transfer, the membrane has pink markings for each of the lanes, allowing for easy cutting of the membrane.

Once the proteins have been transferred to the membrane, blocking of nonspecific sites is required. A blocking buffer is used to "block" the unreacted sites and should improve assay sensitivity and reduce background interference. "Blocking buffer" is a collective term for various additives that prevent nonspecific binding, but have no active part in the specific immunochemical reaction of the particular assay. Blocking buffers are generally available in two formats: homemade or commercial. Homemade blocking buffers are formulated by mixing the raw protein of choice with an appropriately buffered solution, on an as-needed basis. Commercially available blocking buffers are preformulated and are generally stable over a long period of time. Bovine serum albumin (BSA) and BLOTTO (nonfat dry milk) have traditionally been popular blocking reagents in immunoblots. Blocking buffers have been selected based on convenience, tradition, and literature references. For optimization of the blocking step for a particular immunoassay, empirical testing is essential. Many factors can influence nonspecific binding, including various protein-to-protein interactions that may be unique to a given immunoassay. The most important parameter when selecting a blocking buffer is the signal-to-noise ratio, measured as the signal obtained from a sample containing the target analyte, compared with that obtained from a sample without the target analyte.

To determine if there is a potential problem with cross-reactivity between the antibodies and the blocking buffer, it is recommended that a sample of the membrane of choice be blocked with different blocking buffers for 30 min at room temperature (RT). After blocking, the primary antibody should be added for 1 to 2 h or as the protocol recommends. Following addition and removal of the primary antibody, a wash step should be performed. The purpose of the wash step is to "rinse" away nonspecifically bound antibodies. Typical wash buffers include buffering salts such as mono- and dibasic phosphates to maintain the appropriate pH and a detergent such as 0.05% Tween®-20 (ICI Americas), to help break apart nonspecific interactions. After the wash step, the HRP-labeled secondary antibody should be added and incubated for 1 h or as the protocol recommends. The membrane should be washed again thoroughly, prior to the addition of the chemiluminescent substrate. The chemiluminescent substrate should by used according to the manufacturer's recommendations. After incubation in the chemiluminescent substrate is complete, the membrane should then be placed in a plastic sheet protector or plastic wrap and exposed to X-ray film. No signal should develop with these controlled experiments. If the results indicate that there is background noise produced at this point, the above should be repeated with the secondary antibody only, to determine which antibody is the source of cross-reactivity. Polyclonal secondary antibodies have a greater tendency to cross-react than monoclonal antibodies. A comparison of different blocking buffers should be done with the specific antibody system for optimal results. No signal should develop with these controlled experiments.

Alternatively, blocking buffers can "mask" the signal and decrease the signal-to-noise ratio of the samples. To test this phenomenon, different blocking buffers should be compared in different Western blotting samples. In the following example, various proteins were analyzed by Western blotting to determine the optimal blocking condition for nonspecific sites. Recombinant human cyclin B1, wild-type p53, and mouse fos baculovirus immunoblots were blocked with SuperBlock® Blocking Buffer (Pierce Chemical Company, Rockford, IL), milk, casein, or BSA, and the resulting blots were compared (Figure 10.2). The results indicate that there is no blocking reagent that is optimal for all systems. SuperBlock Blocking Buffer had the best overall consistently

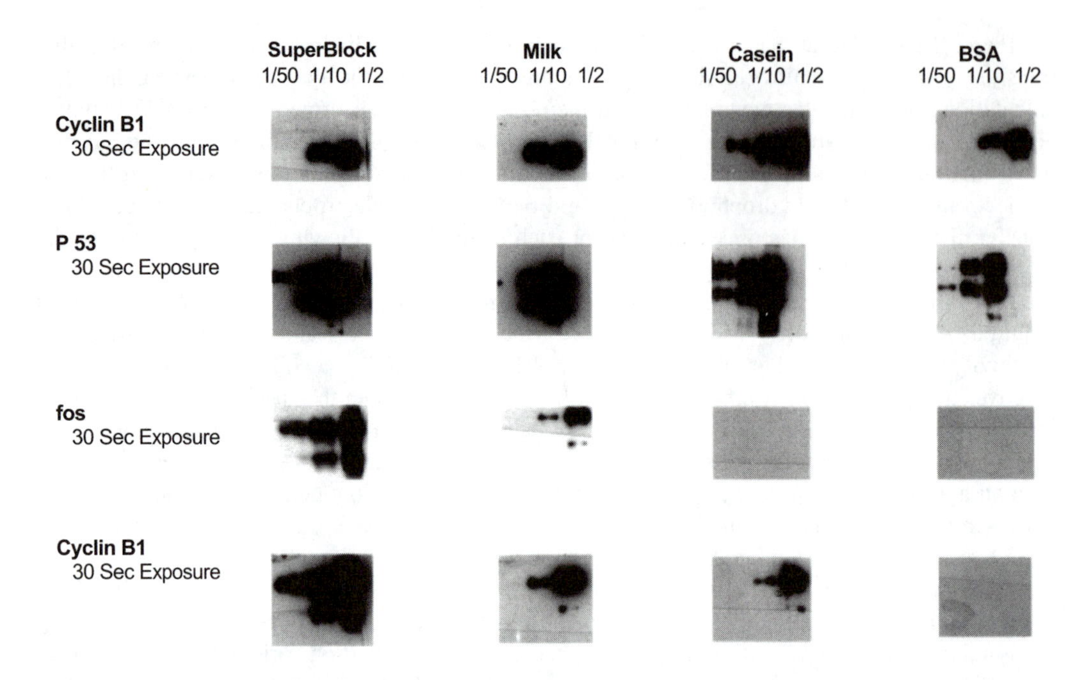

FIGURE 10.2 Recombinant human cyclin B1, wild-type p53, and mouse for baculovirus lysates (Pierce Chemical Company) were diluted in Reducing Sample Buffer (Pierce Chemical Company) at various concentrations. The samples were separated electrophoretically on a 12% SDS-PAGE gel (Novex). The proteins were transferred to nitrocellulose membrane and cut into strips. The membrane strips were blocked for 1 h at RT with shaking in different blocking buffers from Pierce Chemical. Tween-20 (0.05%) was added to all of the blocking buffers (Pierce Chemical Company). The membranes were then incubated with the appropriate primary antibody (Pharmingen) at 0.5 μg/ml prepared in the different blocking buffer solutions for 1 h at RT with shaking. Each membrane strip was washed with Tris-buffered saline followed by a 1-h incubation in 25 ng/ml dilution of HRP-labeled goat anti-mouse IgG (Pierce Chemical Company) prepared in the different blocking buffers. The membranes were washed with TBS. A working solution of SuperSignal West Pico Substrate (Pierce Chemical Company) was prepared and added to each membrane for 5 min. The membranes were removed and placed in sheet protectors before being exposed to film for 30 s and 5 min as indicated. The film was developed per the manufacturer's instructions (Pierce Chemical Company). The resulting blots were analyzed for signal-to-noise ratio and compared. (Courtesy of Pierce Chemical Company.)

high signal-to-noise results. The signal in p53 and fos after 5 min is actually too strong, and a recommendation would be to decrease either the antibody concentrations or the loading quantity of protein. Nonfat dry milk has good signal except with the fos system where the signal is slightly less than the SuperBlock Buffer. Casein appeared to be superior in the cyclin B1 system and worked well in the p53 system, but no signal was detected in the fos system without a longer exposure. BSA worked well in the cyclin B1 system but not in the other systems. BSA "masked" the signal in the fos system, and no signal was detected even with longer exposures. These blots demonstrate the importance of empirically testing the blocking buffer for each immunoassay system for both background noise and signal-to-noise ratios because blocking buffers may mask the system.

After the blocking buffer is optimized for the specific immunoassay, it is crucial to optimize the concentration of both the primary and HRP-labeled secondary antibody systems. Antibodies are host proteins produced in response to the presence of foreign molecules in the body. These foreign molecules are referred to as antigens. Antibodies, which are made primarily by plasma cells and the precursor B lymphocytes, circulate throughout the blood and lymph, where they bind to the specific antigen against which they were produced. Because this antigen–antibody reaction is specific, antibodies are important reagents for immunological research. The antibody that recognizes the antigen of interest is the primary antibody. The primary antibody is added directly to the blot and incubated from 1 h to overnight. An indirect two-step immunoassay method first described by Weller and Coons in 1954[7] is still the most common method for Western blotting. This method uses an HRP-labeled secondary antibody molecule for detection. After the primary antibody binds to the antigen, the HRP-conjugated secondary antibody is added, which binds to the primary antibody.

Increased sensitivity can be achieved by increasing the number of enzyme molecules bound to the antigen of interest. The multiple binding sites between the tetravalent avidin and biotin are ideal for this amplification. The avidin–biotin complex (ABC) was developed by Hsu and colleagues in 1981.[8–10] The primary antibody is incubated with the antigen of interest, followed by addition of a biotinylated secondary antibody. Biotinylated HRP (B-HRP) is preincubated with avidin, forming large complexes, which are then incubated with the biotinylated secondary antibody on the membrane. Optimization is required for preincubation of the ABC complex, and the concentration of biotinylated antibody as well as interaction with the blocking buffer. BLOTTO and other blocking reagents that contain endogenous biotin should be avoided to prevent cross-reactivity with the avidin–biotin system.

As a result of the increased signal intensity, sensitivity, and light emission kinetics of the SuperSignal West Substrate Family, assay conditions should be optimized when switching from colorimetric precipitating substrate systems and, possibly, when switching from other chemiluminescent substrates. The optimal antibody concentration to use with a given antigen is dependent on the specific antigen and antibody. The affinity/avidity of the antibody for the antigen and the activity of both the primary and secondary antibody will vary between systems. This, combined with the different sensitivity levels demonstrated by the different SuperSignal West Substrates, requires optimization of the antibody concentrations along with the exposure time to X-ray film or imaging systems. The ideal immunoblot is one without bands that are overexposed as a result of antigen or antibody concentrations are too high, or bands that are not seen because antigen or antibody concentrations are too low. The ideal blot does not have background and results in intense signal.

SuperSignal West Pico Substrate was the first generation of chemiluminescent substrates based on the azine enhancer system and produces a very intense signal with picogram-level detection of protein upon reaction with HRP. This two-component substrate contains concentrates of a stable luminol solution with an enhancer and of a stable peroxide solution. The working solution is stable for a minimum of 24 h at RT. The solutions can be used under both light and dark conditions. The SuperSignal West Pico Substrate Kit is stable for at least 1 year at RT. The light generated from this substrate has approximately 6 h of emission, allowing for multiple exposures against film or imaging devices. The recommended starting primary antibody dilution is a 1/1000 to 1/5000 dilution from a 1 mg/ml stock. For the HRP-labeled secondary antibody, the recommended starting dilution is 1/20,000 to 1/100,000 from a 1 mg/ml stock solution. To demonstrate how important the optimization of antibody concentration can be for the level of sensitivity allowed by chemiluminescence, several recombinant human wild-type p53 immunoblots were performed using various concentrations of primary and secondary antibodies (Figure 10.3). In Blot 1, the resulting blot was totally black because both the primary and HRP-labeled secondary antibody concentrations were too high. In Blot 2, the background is inconsistent but very dark, again a result of too much antibody. In Blots 3 and 4, the signal-to-noise ratio was much better because both the primary and secondary antibody concentrations were reduced. Typically, the manufacturer's recommended antibody concentration is based on results from colorimetric systems rather than from chemiluminescent systems. It is highly recommended to use the manufacturer's starting dilutions for any chemiluminescent substrate and to optimize from there.

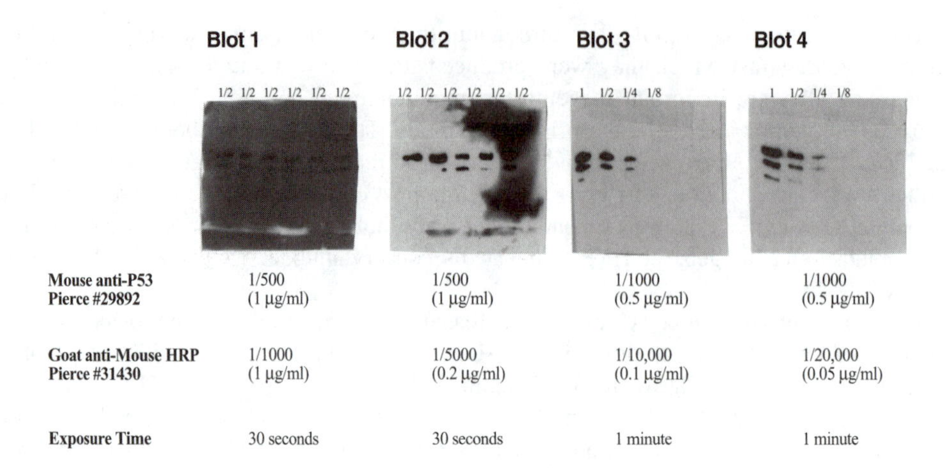

FIGURE 10.3 Recombinant human wild-Type p53 baculovirus lysate was separated electrophoretically and transferred to nitrocellulose membrane. The membrane was blocked with BSA (Pierce Chemical Company) and then incubated with various dilutions of mouse anti-human p53 starting at the manufacturer's recommended dilution (1/500 or 1 μg/ml). HRP-labeled goat anti-mouse was added at different concentrations and the signal was developed with SuperSignal West Pico Substrate. The exposure times against film were also varied. The resulting blots were compared. (Courtesy of Pierce Chemical Company.)

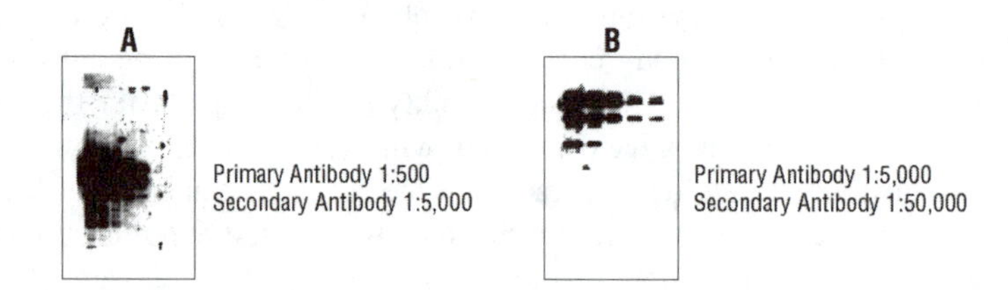

FIGURE 10.4 Recombinant human wild-type p53 baculovirus lysate was separated electrophoretically and transferred to nitrocellulose membrane. The blot was blocked with SuperBlock Blocking Buffer (Pierce Chemical Company) in Tris-buffered saline containing 0.05% Tween-20 followed by either a 1/500 dilution or a 1/5000 dilution of the primary antibody, mouse anti-p53. HRP-labeled goat anti-mouse was diluted to 1/5000 or 1/50,000 and added to the membrane. After washing off any excess HRP, the blot was incubated with SuperSignal West Dura Substrate (Pierce Chemical Company) and the blots were exposed to film for 1 min. (Courtesy of Pierce Chemical Company.)

SuperSignal West Dura Substrate was developed for researchers whose work requires both increased sensitivity and greater light emission duration. This substrate is ideal for use with cooled CCD instruments and other imaging devices. Femtogram levels of protein can be detected with this substrate. The substrate produces an immediate signal with light emission maintained over the next 24 h. This allows researchers to perform multiple exposures without rushing to darkrooms or imaging instruments. Because of the high sensitivity provided by this substrate, precious antibodies can be conserved. The recommended starting primary antibody dilution is 1/5000 to 1/50,000 from a 1 mg/ml solution. For the HRP-labeled secondary antibody, the recommended starting dilution is 1/50,000 to 1/250,000 from a 1 mg/ml stock. To demonstrate this antibody requirement, another recombinant human wild-type p53 baculovirus lysate immunoblot was prepared. The blot that was incubated with 1/500 primary and 1/5000 secondary demonstrates a blot that the antibody levels are too high (Figure 10.4A). The background is not excessively high, but the resulting bands are too intense and blur together, resulting in poor resolution. A large number of nonspecific protein

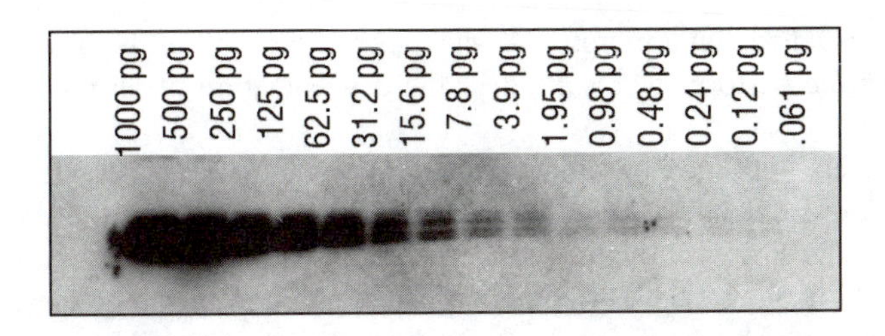

FIGURE 10.5 Serial dilutions of recombinant mouse IL-2 (Pharmingen) were prepared and separated electrophoretically on a 4 to 20% SDS-PAGE gel (Novex). The separated protein was then transferred electrophoretically to nitrocellulose membrane and blocked with SuperBlock Blocking Buffer in PBS containing 0.05% Tween-20. The ideal blot resulted when the primary antibody, rat anti-mouse IL-2 (Pharmingen), was prepared at a 1/5000 dilution in the blocking solution. The membrane was washed and then incubated in a 1/400,000 dilution of HRP-labeled goat anti-rat (Pierce Chemical Company). The membrane was washed again and then incubated with the working solution of SuperSignal West Femto Substrate (Pierce Chemical Company) and exposed to film for 1 min. (Courtesy of Pierce Chemical Company.)

bands are also visible. An optimal blot was achieved by using a 1/5000 primary antibody dilution and a 1/50,000 secondary antibody dilution (Figure 10.4B).

SuperSignal West Femto Substrate was designed to provide the ultimate in sensitivity to very low femtogram-level detection. This substrate enables the detection of very small quantities of antigen that normally could not be seen. At the same time, the substrate conserves precious antibodies by decreasing the amount of antibody required to detect the desired antigen. An extremely intense signal is produced immediately and persists for as long as 6 h. The recommended starting primary antibody dilution is 1/5000 to 1/100,000 from a 1 mg/ml solution. For the HRP-labeled secondary antibody, a 1/100,000 to 1/500,000 dilution from a 1 mg/ml solution is suggested for optimal blots. A recombinant mouse IL-2 immunoblot was performed using a 1/5000 dilution of the primary and a 1/400,000 dilution of the secondary and resulted in good signal-to-noise ratio with high sensitivity (Figure 10.5).

Traditionally, light generated by SuperSignal West Substrates and other enhanced chemiluminescent substrates have been captured and analyzed on X-ray film. An immunoblot is performed, reacted with the chemiluminescent substrate, placed in a plastic protector, and exposed to X-ray film. The film is processed in a darkroom by conventional methods or by a film processor. The film is stored easily in notebooks and is readily available for publication purposes. To analyze the data on film, a reflectance densitometer is required. The film is imaged on the densitometer and then the data are analyzed, resulting in the determination of relative intensity per data point. These are relative quantities but can allow for direct comparisons of samples from the same immunoblot experiment.

A more recent imaging development is the CCD camera. CCD cameras do not require a darkroom or the use of film, eliminating the need for film-developing chemicals or costly automatic film-developing equipment. CCD cameras have had many recent improvements in their sensor technology, making this a viable and often preferable method for detection of immunoblots. By cooling the CCD sensor through a thermoelectric device such as a single- or multistage Peltier device, there is a significant increase in the detection sensitivity. Although these CCD camera systems have more up-front costs compared with film development, there are many long-range advantages, including no need for darkroom or chemicals that require special disposal, an increase in the linear dynamic range (0 to 1.8 o.d. with film compared with 0 to 2.5 or greater o.d. with a CCD camera) and immediate quantitative data analysis. Once the Western blot has been processed with the chemiluminescent substrate, the blot is placed under the CCD camera in a light-tight environment and the CCD chip is exposed to the image over a specific exposure time. A digitized image is acquired that is visualized on a monitor. At this point, a printout of the blot can be generated or further analysis can be performed, resulting in the determination of an integrated density value (IDV) per data point.

The ChemiImager™ 4000 (Alpha Innotech Corp., San Leandro, CA) uses a CCD sensor that is cooled 60°C below ambient or –40°C. This reduction of temperature substantially reduces the dark current signal and noise, allowing for the necessary increase in exposure time without the interference of CCD image noise. By combining the enhanced signal and extended duration of light benefits demonstrated by the SuperSignal Products with the ChemiImager 4000, CCD analysis provides a flexible and easy-to-use alternative over film.

An IL-2 Western blot was performed and detected on both the ChemiImager 4000 and on X-ray film using SuperSignal West Pico Substrate. The blot was exposed to film for 1 min and for 5 min and to the CCD chip for 1, 5, and 15 min (Figure 10.6). After a 5-min exposure to film, 3 pg of IL-2 could be detected, whereas 100 pg of IL-2 were detected with a 15-min exposure with the CCD camera. An increase in exposure time allows for an increase in the capture of light and further analysis.

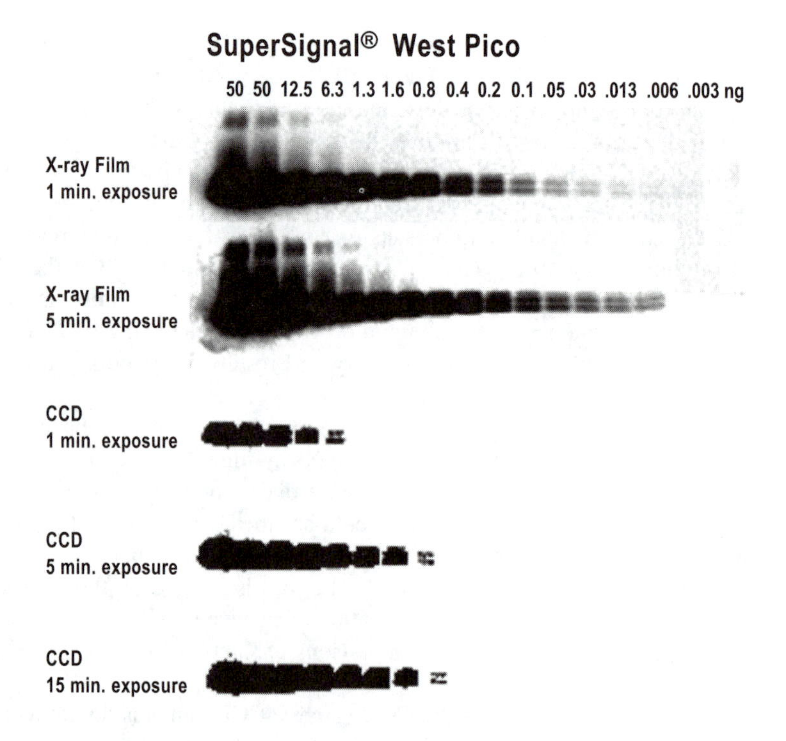

FIGURE 10.6 In the study, 50 ng of recombinant mouse IL-2 was serially diluted from 50 to 0.003 ng and electrophoresis was performed. The gels were transferred to nitrocellulose membranes and then blocked. After blocking, the membranes were incubated with 1 μg/ml dilution of the primary antibody, rat anti-mouse IL-2. The membranes were washed six times for 5 min with large volumes of PBS. Next, the membranes were incubated with a 20 ng/ml dilution of the secondary antibody, HRP-conjugated goat anti-rat IgG, and then washed again three times for 5 min each with PBS and three times each with PBS containing 0.05% Tween-20. The working solution of SuperSignal West Pico Substrate was prepared and added to the membranes for 5 min. The membranes were removed from the substrate and placed in plastic sheet protectors prior to exposure. Each membrane was exposed to X-ray film for 1 and 5 min, and 1-, 5-, and 15-min exposures at F1.6 were performed on the ChemiImager 4000 (Alpha Innotech). (From Feather-Henigan, K. et al., *Immunoblot Imaging with a Cooled CCD Camera and Chemiluminescent Substrator,* ISC Publishing. With permission.)

SuperSignal West Dura Substrate has both an extended light duration and a 25-fold increase in sensitivity over SuperSignal West Pico Substrate, to mid-femtogram levels. SuperSignal West Femto Substrate has the most intense signal and low femtogram detection. Because of these characteristics, the SuperSignal West Dura and Femto Products were analyzed in the ChemiImager 4000 and compared with X-ray film. Under the conditions of the blot tested (0.5 μg/ml of the primary antibody, rat anti-mouse Il-2 and 3.33 ng/ml of the secondary antibody, HRP-labeled goat anti-rat IgG), detection of purified Mouse IL-2 was rapidly captured with a 1-min exposure to film (Figure 10.7). After a 1-min exposure in the CCD camera, SuperSignal West Femto Substrate could detect 250 pg of IL-2, whereas there was no detection with SuperSignal West Dura Substrate. By increasing the exposure time to 5 min, SuperSignal West Femto Substrate could detect 62.5 pg of IL-2 and SuperSignal West Dura Substrate could detect 250 pg. SuperSignal West Femto Substrate could detect all seven lanes at least 15.6 pg of IL-2 after a 30-min exposure. SuperSignal West Dura Substrate could detect 62.5 pg of IL-2 (Figure 10.8). For each of these substrates, increasing the exposure time increases the amount of light captured and, as a result, increases the sensitivity (Figure 10.8). For detection on both X-ray film and the ChemiImager 4000, SuperSignal West Femto Substrate provides the maximum sensitivity. SuperSignal West Dura Substrate is recommended when multiple exposures are required.

The ChemiImager 4000 has a unique feature, the MovieMode, which allows for the automatic capture of several images with defined exposure times over an extended period of time. To demonstrate this feature, B-HRP was filtered through a nitrocellulose membrane at various concentrations and detected with SuperSignal West Dura Substrate. Exposure times varied from 1 to 15 min on the CCD camera, and the images were captured in the MovieMode at F1.6. With a 1-min exposure, signal was detected to 6.2 pg of B-HRP (Figure 10.9). With a 2-min exposure, 3.1 pg was detected. By exposing the blot for 4 min, 1.5 pg could be detected. With a 12-min exposure, 0.77 pg of B-HRP could be detected above background. The MovieMode allows for easy evaluation of an immunoblot requiring different exposure times with minimal hands-on time required. The resulting blots can then be analyzed for the best signal-to-noise results, as well as sensitivity limits.

To demonstrate the dynamic range of the ChemiImager 4000, the log of the IL-2 concentration was plotted against the log of the integrated density values for each concentration on a Western blot that was detected with SuperSignal West Dura Substrate. After a 30-min exposure to film, the linear dynamic range for this Western blot was at least 2.5 log orders in magnitude (Figure 10.10).

These data demonstrate that the sensitivity achieved on the CCD camera is not equivalent to X-ray film using equal exposure times. By increasing both the primary and secondary antibody concentrations, equal or better sensitivity can be achieved on the CCD camera without compromising signal-to-noise. A twofold increase of primary antibody, from 0.5 to 1 μg/ml, combined with a sixfold increase in the HRP-labeled goat anti-rat IgG secondary antibody, from 3.33 to 20 ng/ml, allowed similar detection on film and the CCD camera (Figure 10.11). The film detected 25 pg of IL-2 after a 30-s exposure and the CCD camera was able to detect 12 pg of IL-2 after a 30-min exposure.

Although the CCD camera is not as sensitive as X-ray film in direct comparisons, equal or better sensitivity can be achieved with longer exposure times as well as by increasing both the primary and secondary antibodies. The ChemiImager 4400 offers a Chemi-Nova-cooled camera that has similar resolution and dynamic range to the ChemiImager 4000 with many additional advantages. The ChemiImager 4400 camera (Alpha-Innotech Corp.,) has an improved CCD sensor, proprietary antireflective coating, larger pixel, and faster optics, which result in exposure times that are eight times faster than the ChemiImager 4000. A further improvement is the FluorChem™ 8000 CCD camera (Alpha-Innotech Corp.), which provides megapixel resolution and 16-bit files that result in a 0 to 4 o.d. dynamic range and provide up to 60 times faster exposure times than those of the ChemiImager 4000. Analysis of these new generations of CCD cameras with the SuperSignal West Products will be performed upon availability of the instrument.

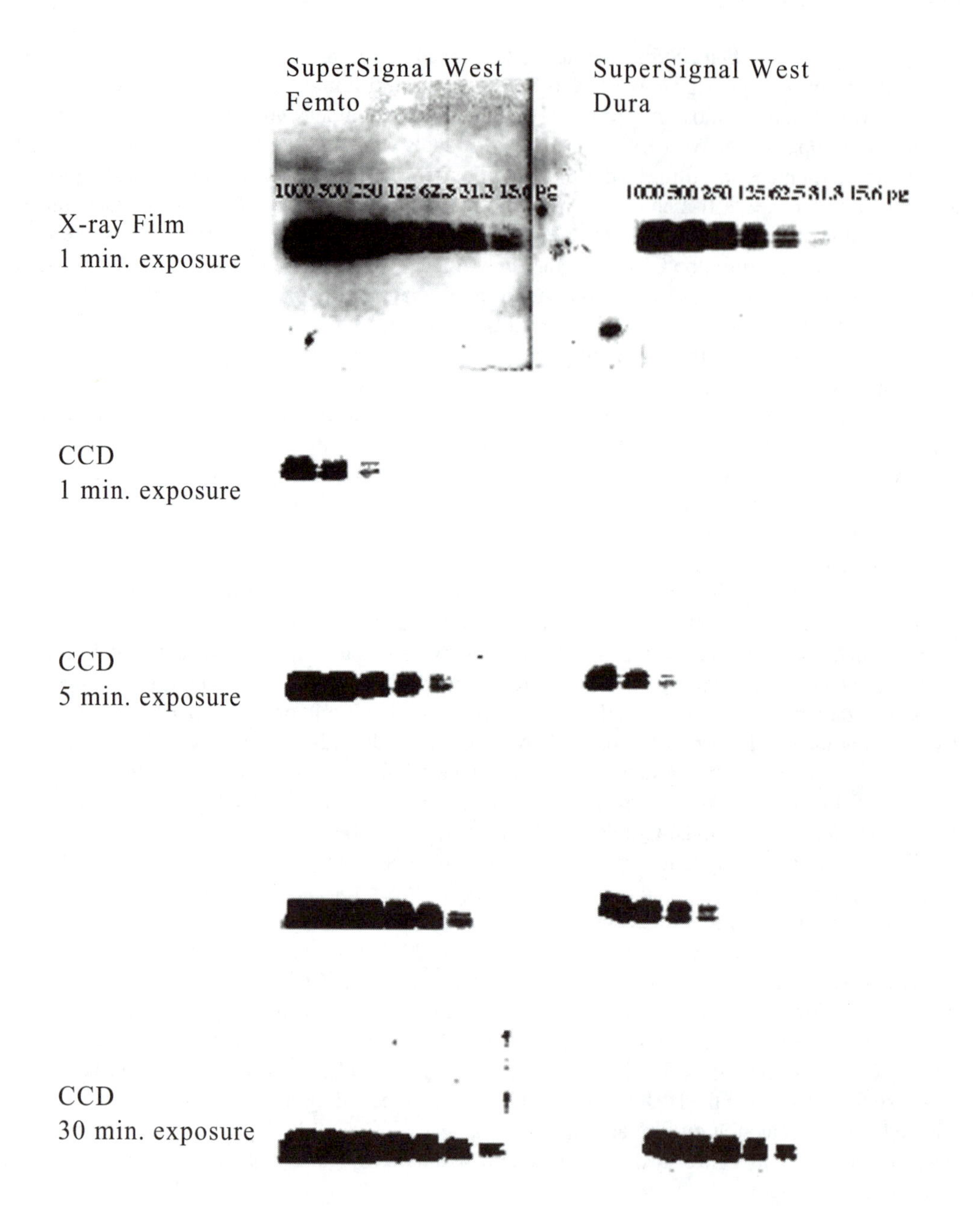

FIGURE 10.7 Recombinant mouse IL-2 was serially diluted from 1000 to 15.6 pg and electrophoresis was performed. The gels were transferred to nitrocellulose membranes and the membranes were blocked. After blocking, the membranes were incubated with a 0.05 μg/ml dilution of the primary antibody, rat anti-mouse IL-2. The membranes were washed six times for 5 min each with large volumes of PBS. The membranes were then incubated with a 3.33 ng/ml dilution of the secondary antibody, HRP-conjugated goat anti-rat IgG, and then washed again. Working solutions of SuperSignal West Femto Substrate and SuperSignal West Dura Substrate were prepared and added to the membranes for 5 min. The membranes were removed from the substrates and placed in plastic sheet protectors prior to exposure. The membranes were exposed to X-ray film for 1 min and 1-, 5-, 15-, and 30-min exposures at F1.6 were performed on the ChemiImager 4000. (From Feather-Henigan, K. et al., *Immunoblot Imaging with a Cooled CCD Camera and Chemiluminescent Substrator,* ISC Publishing. With permission.)

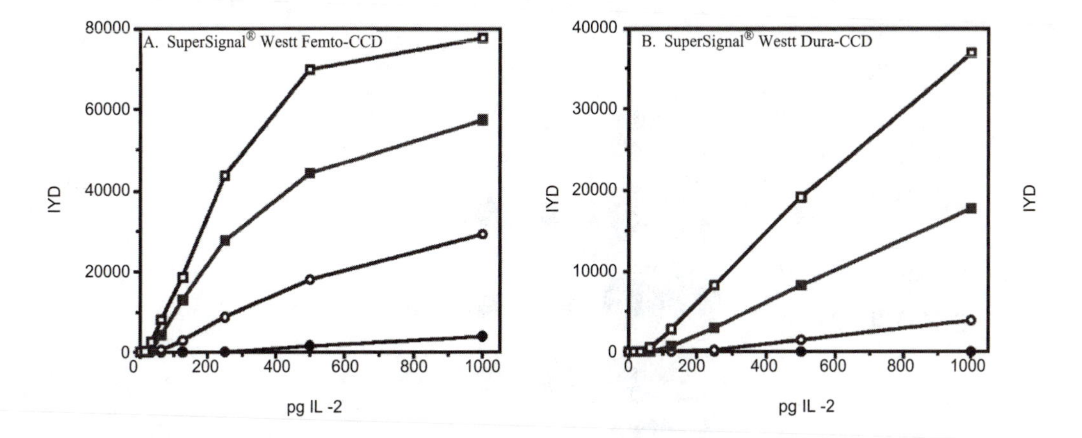

FIGURE 10.8 SuperSignal West Femto and SuperSignal West Dura Substrates were compared on film and the ChemiImager 4000. The concentration of IL-2 was plotted vs. the integrated density value (IDV). (From Feather-Henigan, K. et al., *Immunoblot Imaging with a Cooled CCD Camera and Chemiluminescent Substrator,* ISC Publishing. With permission.)

By combining the advantages of the SuperSignal West Substrates with digital imaging systems, chemiluminescent detection of Western blots is easily performed and analyzed. By using the CCD camera for analysis of Western blots, the processing time is significantly decreased as both exposure and analysis are done on the same piece of equipment. The multiple exposures often required for both CCD and film imaging can be done easily on an instrument such as the ChemiImager 4000 using the MovieMode or individual exposures. The SuperSignal West Product line provides a variety of sensitivity levels as well as light duration, allowing for detection of low levels of proteins that may require very long exposures. By combining the benefits of the SuperSignal West Products and CCD cameras, high-quality images can be achieved for a variety of blotting needs.

10.3 ELISA TECHNOLOGY: FROM MICROWELL PLATE TO DETECTION

The enzyme-linked immunosorbent assay (ELISA) was developed as an alternative to the traditional radioimmunoassay (RIA).[8] By using enzyme labels in place of radioisotopes in an ELISA, the problems associated with safety, disposal, expense, and short shelf life inherent in RIA reagents are eliminated. The ELISA technique provides a method to detect highly specific proteins with great sensitivity and relative ease. Several companies provide instruments that help automate part or all of the ELISA technique.

ELISA protocols involve the adsorption of antigens or antibodies onto a solid-phase support to allow separation of the immunologically bound material from the unbound material. This procedure depends on the reproducible adsorption of the reactant onto a solid phase. With the introduction of a polystyrene microwell plate in the early 1960s, the analysis of sample compounds by enzyme-linked antibodies was simplified. Crude or purified cellular lysates could be applied to the wells of the plates, and the compounds of interest were adsorbed to the polystyrene over a defined period of time. These compounds were then identified by the subsequent addition of the enzyme-linked antibodies and by the use of an identifiable substrate.

In recent years, the format of the microwell plates has changed significantly. The initial microwell plate was defined by an 8-well by 12-well matrix (96 total wells), but more recently has been modified to formats that offer 384, 1536, 3456, and even 9000 wells per plate. These plates, which offer the advantage of performing multiple assays per plate using numerous different compounds, have been

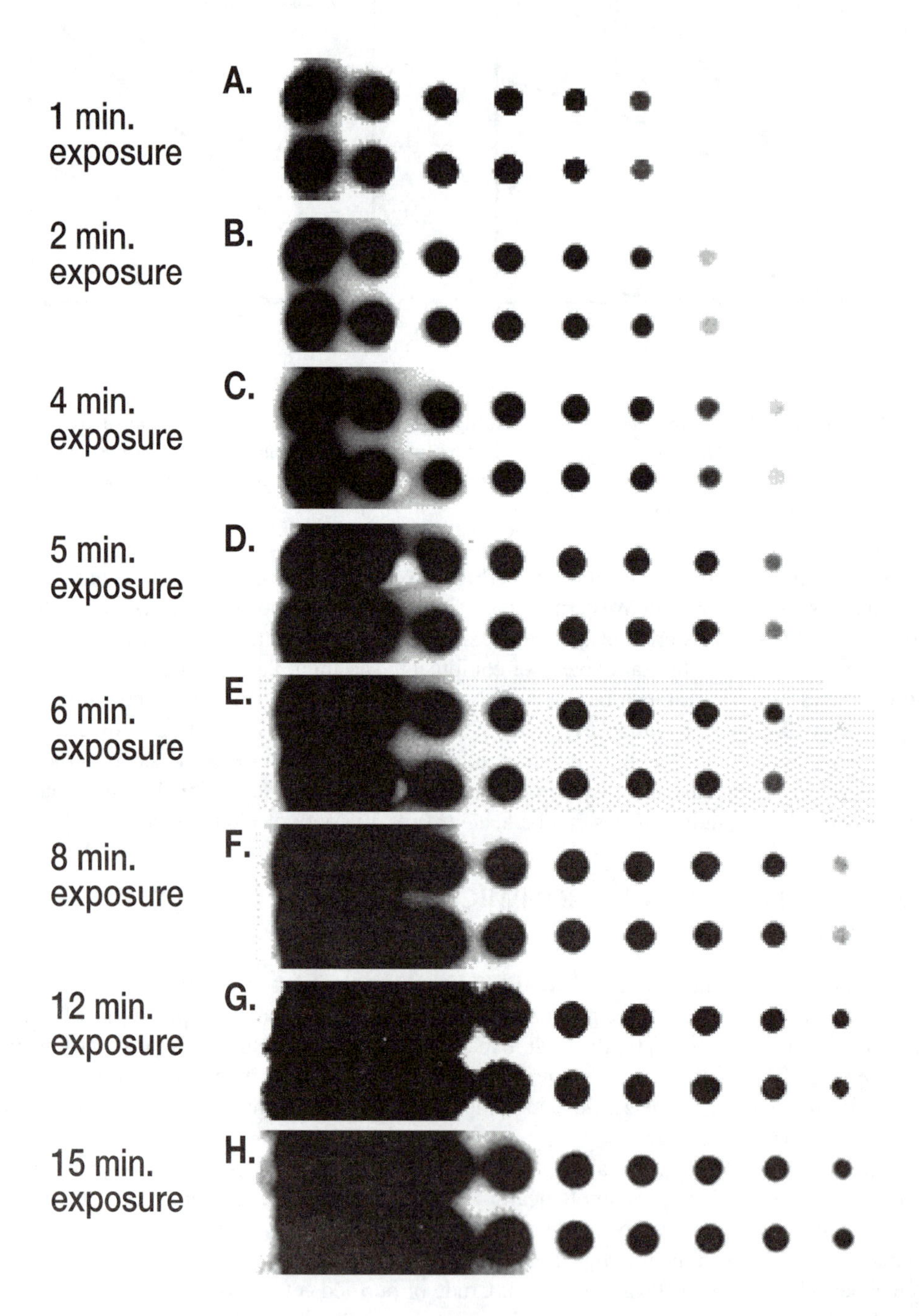

FIGURE 10.9 Biotinylated HRP was serially diluted from 1000 to 0.2 pg and a dot blot was performed. SuperSignal West Dura Substrate working solution was added to the membrane for 5 min. The membrane was removed from the substrate and placed in a plastic sheet protector prior to exposure to the CCD camera. Images were captured by the ChemiImager 4000 at various exposure times at F1.6 using the MovieMode. (From Feather-Henigan, K. et al., *Immunoblot Imaging with a Cooled CCD Camera and Chemiluminescent Substrator,* ISC Publishing. With permission.)

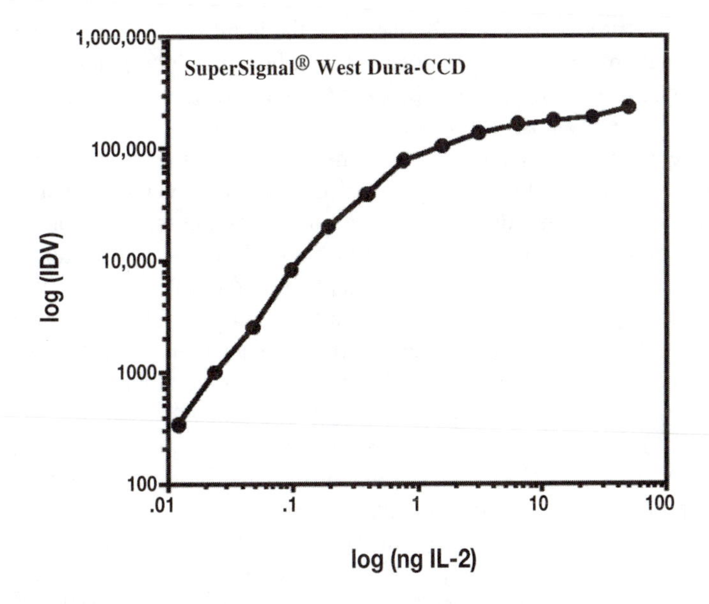

FIGURE 10.10 The log of the IL-2 concentration was plotted against the log of the IDV. (From Feather-Henigan, K. et al., *Immunoblot Imaging with a Cooled CCD Camera and Chemiluminescent Substrator*, ISC Publishing. With permission.)

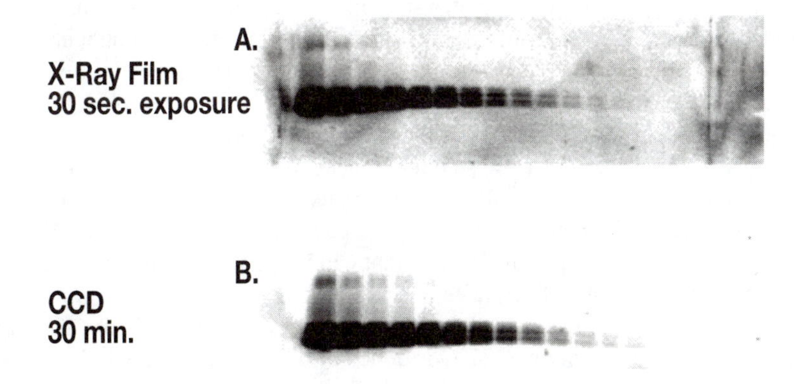

FIGURE 10.11 Recombinant mouse IL-2 was serially diluted from 50 to 0.003 ng and electrophoresis was performed. The gels were transferred to nitrocellulose membranes and the membranes were blocked. After blocking, the membranes were incubated with either a 0.5 μg/ml dilution (A) or a 1 μg/ml dilution (B) of the primary antibody, rat anti-mouse IL-2. The membranes were washed six times for 5 min each with large volumes of PBS. Next, the membranes were incubated with a 3.33 ng/ml dilution (A) or a 20 ng/ml dilution (B) of the secondary antibody, HRP-conjugated goat anti-rat IgG. The membranes were washed again six times for 5 min each with PBS (A) or three times for 5 min each with PBS and three times for 5 min each with PBS containing 0.05% Tween-20 (B). The working solution of SuperSignal West Dura Substrate was added to the membranes for 5 min. The membranes were removed from the substrate and placed in plastic sheet protectors prior to exposure. A 30-s exposure was done on film (A) and a 30-min exposure at F1.6 was performed with the ChemiImager 4000 (B). (From Feather-Henigan, K. et al., *Immunoblot Imaging with a Cooled CCD Camera and Chemiluminescent Substrator*, ISC Publishing. With permission.)

used extensively in the process of drug discovery. As the number of wells-per-plate increases and the amount of sample compound-per-well decreases, it becomes increasingly important to utilize detection systems that provide the greatest sensitivity. Fluorescent and chemiluminescent technologies have been employed extensively for this purpose.

To facilitate the binding of test compounds to the wells of the microwell plates, the polystyrene surface of the plate may be chemically or physically modified. Several "coated" microwell plates have been introduced for this purpose. These surface coatings and modifications include streptavidin, NeutrAvidin™ Biotin-Binding Protein (Pierce Chemical Company, Rockford, IL), glutathione, metal chelate, dextrin, Protein A and Protein G, and antibodies. The surface of these plates may also be "activated" for covalent binding to molecules of interest. These activated plates include maleic anhydride, maleimide, and amine-binding surfaces. In addition, recent technology has allowed for the covalent binding of any amine-containing molecule to the surface of either polystyrene or polypropylene plates.

In a traditional ELISA, the wells of a polystyrene microwell plate are coated with a specific antigen or antibody. For a typical protein, optimal adsorption to the microwell is achieved at 1 to 10 μg/ml in an appropriate coating buffer such as carbonate/bicarbonate buffer at pH 9.6. Different microwell plates may have different optimal coating conditions. Complete adsorption usually occurs within 1 to 2 h at RT or overnight at 4°C. Once the coating is complete, the excess coating reagent must be removed by washing with a wash buffer such as PBS containing 0.05% Tween-20.

Blocking of nonspecific sites in the microwell is important to eliminate any nonspecific binding of the immunoreagents used throughout the assay. There is a wide variety of blocking reagents that are commercially available, including serum, BSA, gelatin, dry milk, and casein. The best blocking reagent minimizes any background detection and enhances signal, increasing the signal-to-noise ratio. Empirical testing with the appropriate controls is essential.

SuperBlock Blocking Buffer from Pierce can be used to block the nonspecific sites quickly and can also be used to keep the coated plates stable for an extended period. The plates should be blocked with SuperBlock Blocking Buffer and stored at 4°C in a sealed container with desiccant. This allows the use of consistently precoated and preblocked plates for several different experiments over time.

Because an ELISA consists of a series of incubations with different reagents, wash steps must be performed to limit carryover from one step to the next. Wash procedures reduce background by removing unbound reagents from the well. Automatic plate washers are available to reduce hands-on assay time and to increase the reproducibility of the reaction.

There are several different types of ELISAs that can be performed utilizing either antigen- or antibody-coated microwell plates. For plates that are coated with a specific antibody, the sample containing the antigen of interest is added to the microwell and incubated for approximately 1 to 2 h at RT. Increasing the temperature to 37°C is done often, but caution should be taken to prevent hot spots and evaporation across the microwell plate. For best results, the antigen and all subsequent immunoreagents should be prepared in the blocking reagent containing 0.05% Tween-20. After the antigen step, a capture antibody is added to "sandwich" the antigen of interest. The antibody is incubated for 1 to 2 h at RT. This is followed by a wash step and then an enzyme-labeled secondary antibody directed against the capture antibody. The reaction time varies from 30 min to several hours. Alternatively, the capture antibody can be labeled directly with the enzyme. There are several commercially available kits that allow for the easy conjugation of an antibody. After the addition of the enzyme, it is necessary to perform an extensive wash step to remove any unbound reagent prior to detection with a substrate.

The optimal dilution of the immunoreagents is dependent on the affinity of the antigen and antibody, as well as the detection system of choice. Several colorimetric substrates are available that are ideal for quick analysis when high sensitivity is not required. With the introduction of chemiluminescent substrates and improvements in luminometers used to detect the signal, the sensitivity of ELISA assays has increased. Instruments such as the LumiCount® System (Packard Instruments,

Meriden, CT) are designed to offer exceptional sensitivity advantages for glow-type chemiluminescent detection. The LumiCount System provides a large linear detection range in both 96- and 384-well plate formats with similar sensitivities. The detection system utilizes a single photomultiplier tube (PMT) and digital photon integration (DPI) technology. The DPI electronics allow automatic selection of an optimized gain and PMT high voltage setting for attomolar sensitivity and an extended dynamic range for the specific assays.

By combining the Packard LumiCount System capabilities with the Pierce SuperSignal ELISA Substrate-based technology, rapid quantitative results can be achieved. SuperSignal ELISA Pico Substrate, upon reaction with HRP, results in immediate generation of light with low picogram-level detection. The signal output is enhanced over other luminol-based systems, resulting in high signal-to-noise ratios. The product has RT storage and a light duration of approximately 30 min. SuperSignal ELISA Femto Substrate results in even higher light output with femtogram-level sensitivity. This product also has increased low-end linearity ideal for ELISA assays requiring high sensitivity.

The capability of the LumiCount System to detect 96-well and 384-well plates is ideal for high-throughput screening applications. Dilution assays show that the assay sensitivity is not significantly affected by switching the assay from a 96-well to a 384-well plate (Figure 10.12). HRP-labeled anti-mouse IgG (whole molecule) was serially diluted and then added to the SuperSignal ELISA Femto Substrate working solution. The samples were then added to either the 96-well or 384-well OptiPlat™ (Packard Instruments), microwell plate. The plates were read on the LumiCount luminometer for 1 s/well read-length. The net relative light units (RLU) values were compared in a dose–response curve. The 96-well plate detected down to 90 fg of the HRP-labeled anti-goat IgG, whereas the 384-well plate system detected down to 45 fg. The minimal detection level is

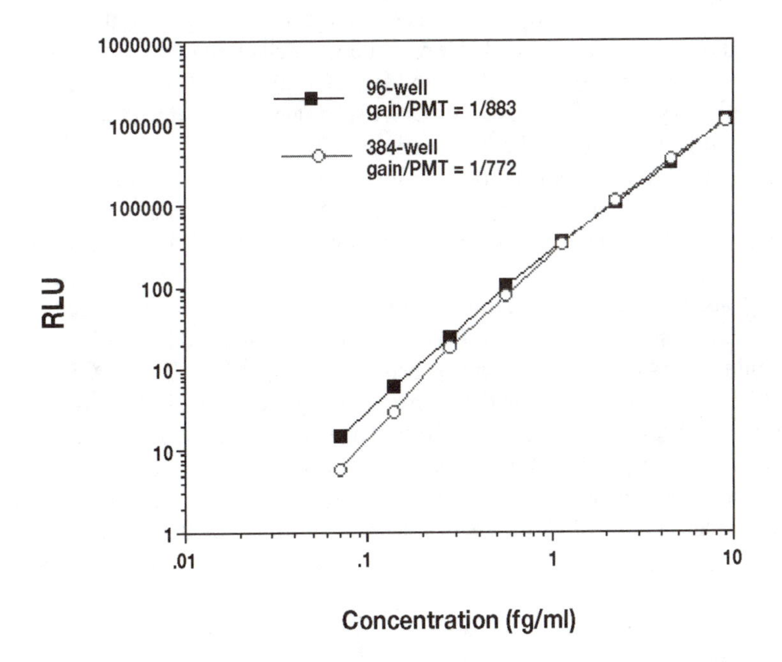

FIGURE 10.12 HRP-conjugated anti-mouse IgG (Sigma, St. Louis, MO) was serially diluted in 0.05 *M* carbonate/bicarbonate buffer, pH 9. The working solution of SuperSignal ELISA Femto Substrate was prepared and added to an OptiPlate 96-well or 384-well microplate. HRP-conjugated anti-mouse IgG (10 μl/well) was added to 100 μl of the substrate in the 96-well plate. In the 384-well plate, 5 μl of the HRP conjugate was added to 50 μl of the substrate. The net RLU were calculated at 1 s/well read-length. The net RLU was plotted against the concentration of the HRP-labeled antibody. (Courtesy of Packard Instrument Company.)

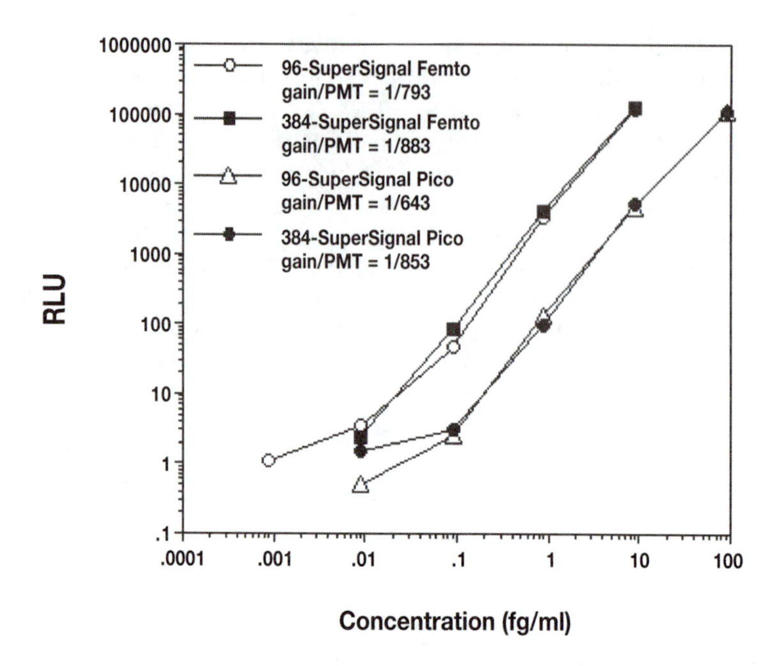

FIGURE 10.13 Minimal detection limit and low-end detection linear range analysis using serial dilutions of HRP-conjugated anti-mouse IgG and SuperSignal ELISA Femto and SuperSignal ELISA Pico Substrate in both a 96-well and 384-well OptiPlate microplate. (Courtesy of Packard Instrument Company.)

defined as the quantity of peroxidase at which its signal is above the signal of the reagent background plus two standard deviations. The extended low-end linear detection range with both of these assays was at least three orders of magnitude in peroxidase concentration (Figure 10.13).

The high sensitivity of the SuperSignal ELISA substrates can also lead to a reduction in assay time in comparison with a traditional colorimetric system. 3,3′,5,5′-Tetramethyl benzidine (TMB) has been the most common colorimetric substrate used in ELISA assays. To demonstrate the reduction of assay time, an interleukin-2 (IL-2) ELISA was performed using the manufacturer's recommended incubation and developed with TMB substrate. This ELISA was repeated using 25% of the original incubation periods. The assay time was reduced from 5.25 h with TMB to 1.35 h using SuperSignal ELISA Femto Substrate. Not only was the assay time reduced significantly but the sensitivity was eightfold greater than that of the TMB system (Table 10.1).

By combining the advantages of such substrates as SuperSignal ELISA Pico Substrate and SuperSignal ELISA Femto Substrate with optimized luminometers, ELISA assays can be performed rapidly, resulting in ultrasensitivity and versatility. The Packard LumiCount microwell luminometer offers high sensitivity in glow luminescence assays in both the 96- and 384-well formats. The virtually cross-talk-free optical design minimizes noise background, resulting in superior signal-to-noise ratios. SuperSignal ELISA Pico Substrate provides picogram detection with extended light duration, and SuperSignal ELISA Femto Substrate provides femtogram detection with superior low-end linearity.

10.4 CONCLUSION

Chemiluminescence is an ideal alternative to colorimetric and radioactive detection methods for both immunoblotting and ELISA-based assays. SuperSignal technology provides several advantages. By combining the effectiveness of the SuperSignal substrates with instrumentation such as the ChemiImager 4000 CCD camera and the LumiCount microwell luminometer, high-quality blots and ELISA assays can result with minimal but essential optimization.

TABLE 10.1
TMB vs. SuperSignal ELISA Femto Substrate

Substrate	Assay Time, min	Lowest Detectable Limit
TMB	315	1.05 pg IL-2
SuperSignal ELISA Femto Substrate	81	0.127 pg IL-2

Rat anti-mouse IL-2 was coated in a clear microwell plate (1 μg/ml) for 60 min at RT. The plate was blocked with SuperBlock Blocking Buffer and incubated with various concentrations of purified mouse IL-2 for 2 h at RT. The plate was washed followed by the addition of biotinylated rat anti-mouse IL-2 (1 μg/ml) (Pharmingen) for 1 h at RT. After washing, the plate was incubated with 1 μg/ml HRP-conjugated NeutrAvidin™ Biotin-Binding Protein (Pierce Chemical Company) for 30 min at RT. The plate was washed to remove any excess enzyme. TMB (Pierce Chemical) was added to the wells and incubated for 15 min prior to stopping with 1 *N* H_2SO_4. The absorbance was read at 450 nm on an ultraviolet microplate spectrophotometer (Molecular Dynamics). The same assay was repeated in a white microwell plate using 25% of the original incubation periods. The IL-2 was detected with SuperSignal ELISA Femto Substrate after a 1-min incubation. The dose–response curves for each of the substrates was generated. From the linear portion of each curve the lowest detectable limit was determined by the method of Rodbard.[12]

REFERENCES

1. Kricka, L. J., Thorpe, G. H. G., and Whitehead, T. P., U.S., Patent 4,598,044, 1986.
2. Kricka, L. J., Stot, R. F. A. W., and Thorpe, G. H., Enhanced chemiluminescence enzyme immunoassays, in *Complementary Immunoassays,* Collins, W. P., Ed., Wiley, Chichester, 1988, 169–179.
3. Thorpe, G. H. G. and Kricka, L. J., Enhanced chemiluminescent reactions catalyzed by horseradish peroxidase, *Methods Enzymol.,* 133, 331–354, 1986.
4. Davis, P. D., Feather-Henigan, K. D., and Hines, K. K., Assay of Peroxidase Activity, PCT patent application, pending.
5. Towbin, H., Staehelin, T., and Fordon, J., Electrophoretic transfer of proteins from polyacrylamide gels to nitrocellulose sheets: procedure and some applications, *Proc. Natl. Acad. Sci. U.S.A.,* 76, 4350–4354, 1979.
6. Southern, E. M., Detection of specific sequences among DNA fragments separated by gel electrophoresis, *J. Mol. Biol.,* 98, 503–517, 1975.
7. Weller, T. H. and Coons, A. H., Fluorescent antibody studies with agents of varicella and Herpes zoster propagated *in vitro, Proc. Soc. Exp. Biol. Med.,* 86, 789–794, 1954.
8. Hsu, S.-M., Raine, L., and Fanger, H., A comparative study of the peroxidase-antiperoxidase method and an avidin-biotin complex method for studying polypeptide hormones with radioimmunoassay antibodies, *Am. J. Clin. Pathol.,* 75, 734–738, 1981.
9. Hsu, S.-M., Raine, L., and Fanger, H., Use of antiavidin antibody and avidin-biotin-peroxidase complex in immunoperoxidase technics, *Am. J. Clin. Pathol.,* 75, 816, 1981.
10. Hsu, S.-M., Raine, L., and Fanger, H., Use of avidin-biotin-peroxidase complex (ABC) in immunoperoxidase techniques: a comparison between ABC and unlabeled antibody (PAP) procedure, *J. Histochem. Cytochem.,* 29, 577–580, 1981.
11. Engvall, E. and Perlmann, P., Enzyme-linked immunosorbent assay (ELISA). Quantitative assay of immunoglobulin G, *Immunochemistry,* 8, 871–875, 1971.
12. Rodbard, D., Statistical estimation of the minimal detectable concentration ("sensitivity") for radioligand assays, *Anal. Biochem.,* 90, 1–12, 1978.

11 A Luminol/Iodophenol Chemiluminescent Detection System for Western Immunoblots

Alexander F. Yakunin and Patrick C. Hallenbeck

CONTENTS

11.1 INTRODUCTION

Chemiluminescent (CL)-based procedures have recently become one of the most popular methods for the immunoblot detection of proteins and nucleic acids. CL has been explored as an detection method alternative to the use of radioisotopes because it eliminates the health hazard, cost, and disposal problems inherent in handling radioactive reagents. Furthermore, CL techniques are remarkably sensitive, require only short film exposure times, and produce sharply defined bands on immunoblots. At present, several chemiluminescent detection systems based on the use horseradish peroxidase (HRP),[1,2] alkaline phosphatase,[2,3] or glucose oxidase[2,4] have become available. HRP, an especially popular label for immunoassays, catalyzes the oxidation of a cyclic diacylhydrazide, luminol, by hydrogen peroxide to produce a luminol radical.[1,5] This radical then forms an endoperoxide that decomposes to yield an electronically excited 3-aminophtalate dianion, and the dianion emits light upon return to its ground state.[1,5] It has been demonstrated that light emission from luminol–H_2O_2–peroxidase systems can be greatly enhanced by the addition of 6-hydroxybenzothiazole derivatives or substituted phenols, which act as electron-transfer mediators between peroxidase and luminol.[1,5] In the presence of these enhancers the intensity of light emission may

0-8493-0719-8/02/$0.00+$1.50

be 1000-fold greater than that of the unenhanced reaction. The concept of enhanced CL catalyzed by peroxidase serves as the basis of several commercially available (Amersham, Boehringer, Pierce) detection kits for Western immunoblotting.

A previously published luminol–iodophenol detection system[6] was not very sensitive because it was not optimized. Since CL detection of Western blots is widely used, we decided to investigate various parameters that might affect the performance of a luminol–iodophenol-based detection system. In this report we describe a luminol–iodophenol detection system optimized for use in the detection and quantitative assay of proteins on Western immunoblots. The described system is simple, cheap, based upon commercially available chemicals, stable for years, sensitive, produces only a very low nonspecific background, and can be used for the quantitative analysis of proteins in crude bacterial extracts. A brief report of this system has been published.[7]

11.2 MATERIALS AND METHODS

11.2.1 Optimization of Developing Solution

Commercial HRP conjugate (goat anti-rabbit IgG–HRP conjugate, Bio-Rad) was diluted 10,000-fold with TBS buffer (0.15 *M* NaCl in 25 m*M* Tris-HCl buffer, pH 7.4), and 3-μl aliquots were loaded onto strips of nylon membrane (Biotrans+, ICN Biomedicals, Inc., Irvine, CA, 1.5 × 5 cm), which had been pre-equilibrated with TBS buffer by placing them on a piece of Whatman 3MM paper prewetted with TBS buffer. After complete adsorption of conjugate drops (~5 min), the nylon strips were incubated for 2 min in 10 ml of developing solution. To determine the optimal pH for CL reaction catalyzed by HRP conjugate adsorbed on a nylon membrane, the developing solution was buffered at different pHs with the following buffers (50 m*M* final concentration, prepared as described in Reference 8): MOPS-NaOH (3-(*N*-morpholino) propanesulphonic acid (ICN Biochemicals), pH 6.8; Na-phosphate, pH 7.4 and 8.0; Tris-HCl, pH 7.4 and 8.0; imidazole-NaOH, pH 8.0; bicine-NaOH (*N,N*-bis (2-hydroxymethyl) glycine (Fisher Scientific, Fair Lawn, NJ), pH 8.5 and 9.5; glycine-NaOH, pH 8.6, 9.0, 9.6, 10.0, 10.6; carbonate-bicarbonate, pH 9.2, 9.5, 10.0, 10.5, and 10.7; hydrazine-glycine, pH 9.5; CAPS-NaOH (cyclohexylaminopropane sulfonic acid (ICN Biomedicals), pH 10.2, 10.5, 11.0. Titration of optimal reagent concentrations was performed using the following stock solutions: luminol (5-amino-2,3-dihydro-1,4-phtalazine-dione; ICN Biochemicals), 10 m*M* solution in 50% dimethyl sulfoxide (Me_2SO); 4-iodophenol (Aldrich Chemical Company, Milwaukee, WI), 50 m*M* solution in 50% Me_2SO; H_2O_2 (Fisher Scientific), 30%, commercial preparation. After incubation, excess liquid was drained off, nylon strips were wrapped in Saran wrap film, and exposed to X-ray film (Fuji RX, Fisher Scientific) for 30 s. X-ray films were developed in a X-ray film autoprocessor (Mini-Med/90, AFP Imaging, AFP Imaging Corp, Elmsford, NY), scanned with a laser densitometer (Personal Densitometer, Molecular Dynamics, Sunnyvale, CA), and CL signals were quantitated using Image Quant™ software (Molecular Dynamics).

11.2.2 Antisera

The three antibodies used in this study were raised in rabbits to pyruvate carboxylase (subunit M_r 128,000), flavodoxin (M_r 21,500), and nitrogenase MoFe-protein (subunit M_r 55,000 and 59,500) isolated from the photosynthetic bacterium *Rhodobacter capsulatus*. Homogeneous preparations of these proteins were purified as previously described,[9–11] antisera were obtained by subcutaneous injections in rabbits, and prepared as described.[12]

11.2.3 Quantitative Immunoblot Analysis of Proteins

For this analysis, protein samples were prepared by mixing aliquots of homogeneous preparations or bacterial cultures with an equal volume of Laemmli sample buffer[13] followed by 1-min or 5-min (cell samples) incubation in boiling water. Variable amounts of protein (determined by the Bradford method[14]) (0.01 to 10 ng/well for homogenous preparations; 0.2 to 5 μg/well for cell samples) were separated on SDS-polyacrylamide gels (PAGE) (7.5% total acrylamide for pyruvate carboxylase; 12.5% for MoFe-protein; 15% for flavodoxin) according to Laemmli[13] on a Mini-PROTEAN® gel apparatus (Bio-Rad, Hercules, CA). Gel electrophoresis was conducted at 200 mA for 40 to 60 min. Following electrophoresis, a transfer "sandwich" was assembled as described by Towbin et al.,[15] and polypeptides were electophoretically transferred from the gel to a nylon membrane (Biotrans-PVDF, ICN) using a Mini Trans-Blot Electrophoretic Transfer Cell (Bio-Rad). Nylon membrane blotting was carried out in Laemmli electrode buffer (without SDS) at 50 to 70 mA overnight or in 10 m*M* CAPS-NaOH buffer (pH 11.0) at 200 mA for 1 h. When nitrocellulose (Trans-Blot® Transfer Medium, Bio-Rad) or polyvinylidene fluoride (PVDF, Immobilon P, Millipore) membranes were used as a transfer membrane, they were treated according to the manufacturer's protocol, and blotting was carried out in Laemmli electrode buffer (without SDS) containing 10% methanol (50 to 70 mA), overnight.

After transfer, membranes were washed in TBS buffer (0.15 *M* NaCl in 25 m*M* Tris-HCl, pH 7.4) for 5 min on a rocking platform. This and all subsequent incubations were performed at room temperature. Nonspecific binding was blocked by incubation of membranes in 50 ml of blocking solution (1% skim milk powder in TBS buffer) for 30 min followed by two 5-min washings in TBS buffer containing 0.05% Tween® 20 (Tween-TBS buffer). Membranes were then treated with a 1/5000 dilution of rabbit anti-MoFe-protein (or other protein) antiserum in Tween-TBS buffer containing 0.5% skim milk powder for 2 h. After two 5-min washings in Tween-TBS buffer, the membranes were incubated in 1/25,000 dilution of goat anti-rabbit IgG–HRP conjugate (Bio-Rad) in Tween-TBS buffer containing 0.5% skim milk powder for 2 h. After incubation, membranes were washed twice (5 min) in Tween-TBS buffer and once (5 min) in TBS buffer.

To visualize the immunoblotting results, membranes were incubated for 2 min in developing solution consisting of 50 m*M* glycine-NaOH buffer (pH 9.6), 0.2 m*M* luminol, 4 m*M* 4-iodophenol, and 17.6 m*M* H_2O_2 (10 ml solution per membrane 6 × 10 cm). The stock solution was prepared by the mixing of all components except H_2O_2 (luminol and 4-iodophenol were initially dissolved in Me_2SO) and kept in an amber bottle at room temperature. Immediately prior to use, H_2O_2 was added to the developing solution (20 μl of 30% H_2O_2 per 10 ml of developing solution). After 2-min incubation, excess developing solution was drained off, the membranes were wrapped in Saran wrap film, and exposed to X-ray film (Fuji RX, Fisher Scientific) for the indicated time (30 s to overnight). After development, the X-ray films were scanned with a laser densitometer (Personal Densitometer, Molecular Dynamics), and the processed images of CL signals were quantitated using Image Quant software (Molecular Dynamics). All data points represent the averages from at least four independent determinations, with standard deviations indicated by error bars.

11.3 RESULTS AND DISCUSSION

11.3.1 EFFECT OF PH ON SIGNAL INTENSITY

Although HRP is stable in the pH range of 4.5 to 12.0, its activity in the soluble state is maximal at pH 7.0 and greatly reduced under alkaline conditions.[16,17] However, luminol CL is most efficient at pHs above 9.0,[18] and this large difference in pH optima potentially significantly limits the application possibilities of peroxidase-catalyzed CL. The reported pH optimum (8.0 to 8.6) for luminol CL catalyzed by a peroxidase–antibody conjugate in solution[1] probably reflects a compromise, where peroxidase retains some activity and luminol CL is possible. During immunoblot analysis, the CL signal is produced by a peroxidase–antibody conjugate that is adsorbed (immobilized) on a porous membrane. Since immobilized proteins often demonstrate changed, usually increased, stability, we determined the effect of variation of pH (6.8 to 11.0) on luminol CL catalyzed by a peroxidase–conjugate adsorbed on a nylon membrane (Figure 11.1). In these experiments, diluted commercial

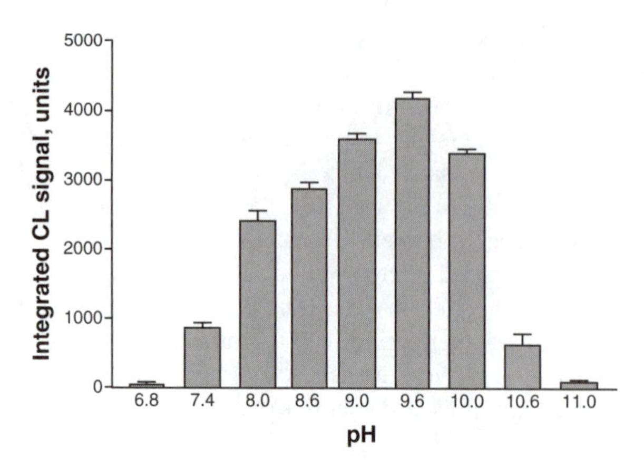

FIGURE 11.1 Effect of pH of the developing solution on the CL signal produced by HRP conjugate adsorbed on nylon membrane. Diluted HRP conjugate was adsorbed on strips of nylon membrane and incubated in developing solution buffered at different pHs (as described in Section 11.2). X-ray film was exposed for 30 s to the strips, developed, scanned with a laser densitometer, and the obtained signal intensity was plotted vs. the pH of the developing solution. The buffers (50 m*M* final concentration) used were MOPS-NaOH, pH 6.8; Na-phosphate, pH 7.4 and 8.0; glycine-NaOH, pH 8.6, 9.0, 9.6, 10.0, and 10.6; CAPS-NaOH, pH 11.0. (From Yakunin, A. F. and Hallenbeck, P. C., *Anal. Biochem.,* 258, 146, 1998. With permission.)

(Bio-Rad) peroxidase–antibody conjugate was adsorbed on strips of nylon membrane that were incubated in developing solutions at different pHs. Subsequent exposition to X-ray film demonstrated a significant signal over a broad pH range with maximal light emission at pH 9.6 (Figure 11.1). A similar pH profile was observed when *p*-coumaric acid was used as an enhancer instead of iodophenol (data not shown). This pH optimum is shifted to alkaline pH compared with the pH optimum of the same reaction catalyzed by soluble peroxidase, and is therefore closer to the pH optimum of luminol CL. Consequently, these results suggest that membrane-adsorbed peroxidase is more active at alkaline pHs than the soluble enzyme and retains significant activity under conditions that are optimal for luminol CL.

11.3.2 Effect of Different Buffers on the Signal Intensity

In general, the CL response to different buffers was as previously reported for the soluble system,[19] although we found that nylon-adsorbed HRP was partially active in bicine buffer. In addition, this HRP conjugate gave a similar signal intensity at pH 8.0 in sodium phosphate, Tris-HCl, or imidazole buffers. Under alkaline conditions (pH 9 to 11) the greatest light intensity was observed with glycine-NaOH and CAPS-Na buffers, whereas the signal was slightly reduced in carbonate buffer (data not shown). In subsequent work on the optimization of conditions for peroxidase-catalyzed luminol CL, we used Tris-HCl buffer (pH 8.0) and glycine-NaOH buffer (pH 10.0).

11.3.3 Effect of pH on Signal Duration

Signal duration was also greatly affected by the pH of the developing solution. After 1-h incubation, no signal was detected at pH 7.4 and 8.0; at pH 9.0 and 9.6 the signal was greatly reduced; at pH 10.0 the signal intensity was unchanged; and at pH 10.6 and 11.0 the signal intensity was increased about fourfold (Figure 11.2). At pH 10.6, and especially 11.0, light emission was significantly delayed, and although maximal intensity was lower than that at pH 9.6 or 10.0, the duration of CL was significantly longer. Thus, at pH 11.0 an appreciable signal can be detected even after 17-h incubation in developing solution.

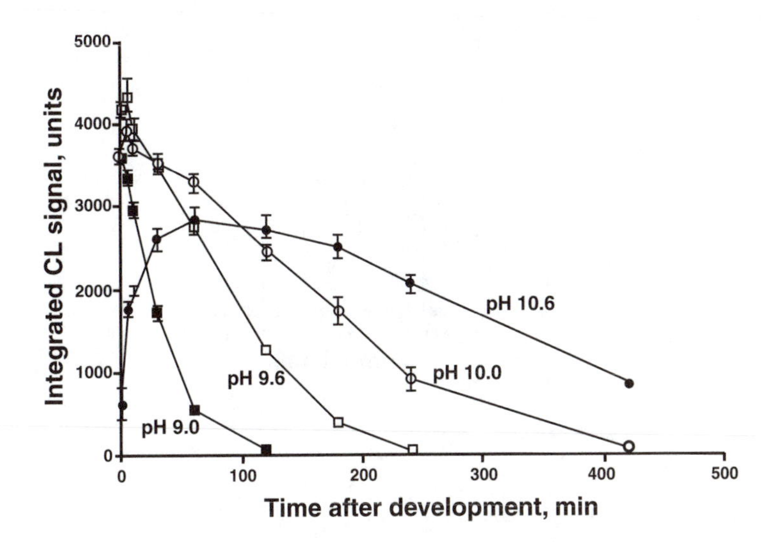

FIGURE 11.2 Effect of the pH of the developing solution on the duration of CL produced by HRP conjugate adsorbed on nylon membrane. Experimental conditions—as described in the legend to Figure 11.1. At the indicated times, X-ray films were exposed to the nylon strips for 30 s, developed, and scanned with the laser densitometer. (From Yakunin, A. F. and Hallenbeck, P. C., *Anal. Biochem.*, 258, 146, 1998. With permission.)

11.3.4 Optimal Concentrations of the Components and Stability of Developing Solutions

We also investigated the optimal concentrations of the components of the developing solution (luminol, 4-iodophenol, H_2O_2). As a starting point, we used the concentrations previously reported to be optimal for the peroxidase-catalyzed reaction in solution.[1] Using different buffer systems, the optimal concentrations were determined to be luminol—0.2 m*M*; 4-iodophenol—4.0 m*M*; H_2O_2—17.6 m*M* (Figure 11.3). Similar results were obtained when a standard protein (*R. capsulatus* MoFe-protein) was electrophoresed on SDS-PAGE, transferred onto a nylon membrane that was subsequently cut into several strips, and developed in different solutions. These values are about tenfold higher than those reported for the reaction catalyzed by peroxidase conjugate in solution.[1] This difference is presumably due to an increased stability of membrane-adsorbed peroxidase, which permits an increased reaction time. Based on these results, the optimized composition of the suggested developing solution is 0.5 m*M* luminol, 4 m*M* 4-iodophenol, and 50 m*M* glycine-NaOH buffer (pH 9.6). A significantly higher luminol concentration (9 m*M*) was used in the previously published system.[6] This may be an additional reason for the low sensitivity observed since our results demonstrated that luminol inhibits CL catalyzed by membrane-adsorbed HRP conjugate at concentrations higher than 0.5 m*M* (Figure 11.3).

The solutions can be prepared and stored at room temperature (in an amber bottle) for an indefinitely long period of time. Our experience has shown that this developing solution produces excellent results even after 3 years of storage at room temperature. H_2O_2 should be added to the developer immediately prior to use (17.6 m*M* final concentration). Additionally, this developing solution can be reused several times without significant effects on the quality of the development with the addition of some H_2O_2 to compensate for its degradation during previous use or storage.

11.3.5 Detection Limits

To determine approximately the detection limits of Western blotting with the luminol–iodophenol system as optimized here, we used nitrogenase MoFe-protein, an $\alpha_2\beta_2$ tetramer as standard protein. Decreasing amounts of MoFe-protein were electrophoretically separated on SDS-PAGE gels,

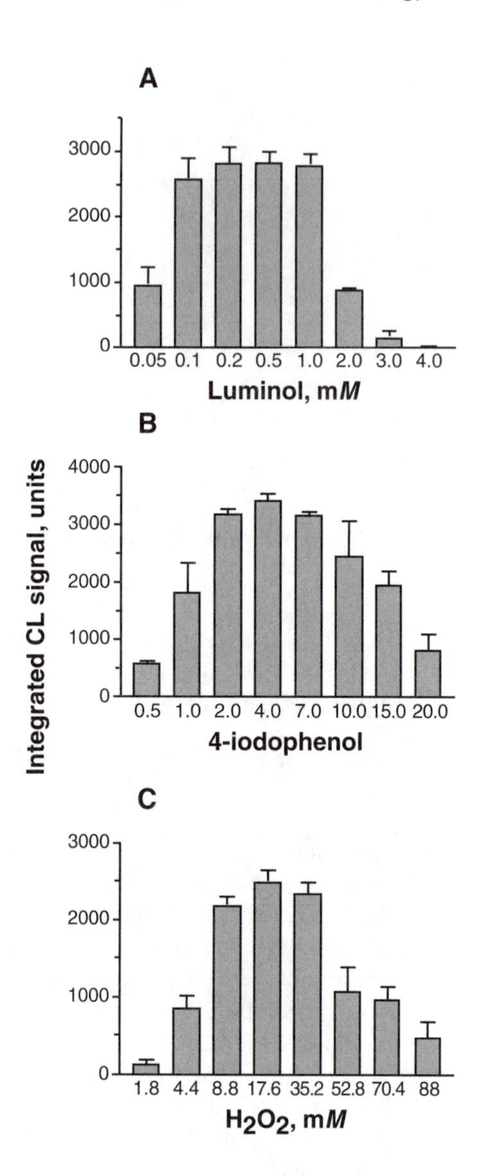

FIGURE 11.3 Optimization of reagent concentrations: CL signal as a function of (A) luminol concentration; (B) 4-iodophenol concentration; (C) H_2O_2 concentration. All developing solutions contained 50 m*M* glycine-NaOH buffer (pH 10), and (A) 4 m*M* 4-iodophenol, 17.6 m*M* H_2O_2; (B) 0.2 m*M* luminol, 17.6 m*M* H_2O_2; (C) 0.2 m*M* luminol, 4 m*M* 4-iodophenol. Other experimental conditions are as described in the caption to Figure 11.1. (From Yakunin, A. F. and Hallenbeck, P. C., *Anal. Biochem.*, 258, 146, 1998. With permission.)

electroblotted onto a nylon membrane, treated sequentially with anti-MoFe-protein antibody and Bio-Rad HRP conjugate, and subsequently visualized by incubation in the developing solution. As presented in Figure 11.4A, virtually no nonspecific background was observed with the use of optimized developer and exposition times with X-ray film between 1 min and 1 h. With a 1-min exposition time, 1 ng of MoFe-protein could be detected and 0.5 ng after a 1-h exposition. As little as 0.1 to 0.3 ng of MoFe-protein can be detected on immunoblots after overnight exposition with X-ray film with only a modest nonspecific background (Figure 11.4A). Higher sensitivity, detection limit (0.03 to 0.05 ng), was found with immunoblots of a monomer protein (*R. capsulatus* flavodoxin) or pyruvate carboxylase, which contains identical subunits (data not shown). Significantly lower nonspecific background after overnight exposure with X-ray film was observed when nitrocellulose

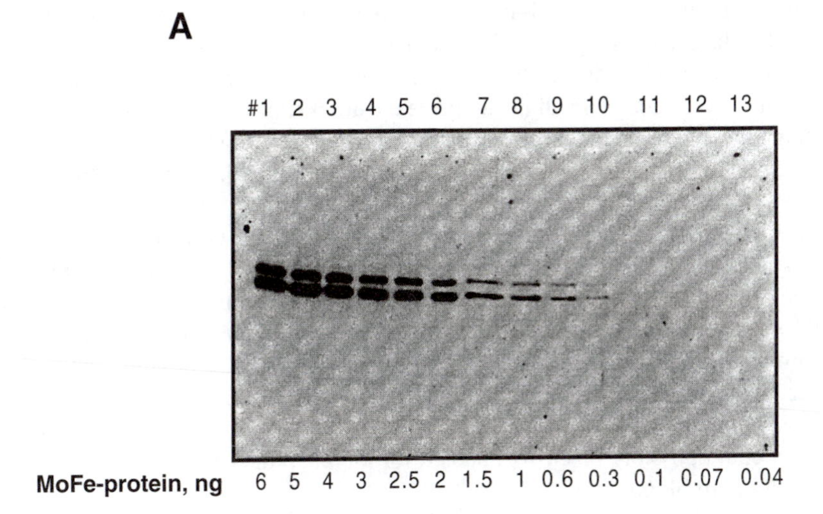

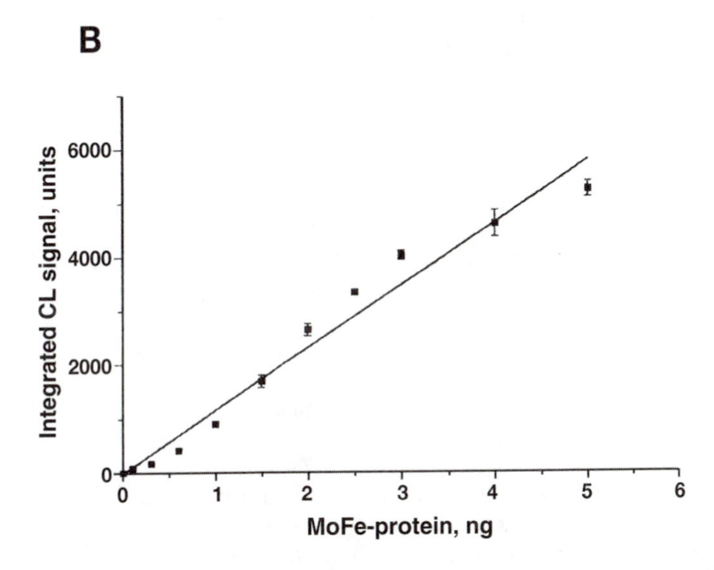

FIGURE 11.4 Quantitative immunoblot analysis of nitrogenase MoFe-protein from *R. capsulatus*. The indicated amounts of MoFe-protein were fractionated on a 12.5% SDS-PAGE, transferred to a nylon membrane, and probed with anti-MoFe-protein-specific antibody. The primary antibody was detected with HRP conjugate using as developing solution: 50 m*M* glycine-NaOH buffer (pH 9.6), 0.2 m*M* luminol, 4 m*M* 4-iodophenol, 17.6 m*M* H_2O_2. The resulting X-ray film was analyzed by densitometry, and the integrated CL signal was plotted vs. the amount of MoFe-protein that had been loaded. (A) MoFe-protein immunoblot (computer-generated image). (B) Immunoblot signal vs. the amount of MoFe-protein. (From Yakunin, A. F. and Hallenbeck, P. C., *Anal. Biochem.*, 258, 146, 1998. With permission.)

or PVDF membranes were used. However, the nitrocellulose membrane provided a significantly lower detection sensitivity than nylon membrane, presumably because of its lower affinity for protein. PVDF membranes demonstrated the same level of sensitivity as nylon membranes, but due to their higher hydrophobicity their treatment is more complicated.

By using the optimized luminol–iodophenol developer, a linear relationship between the amounts of protein loaded on the gel and the resulting luminescent signals was found (Figure 11.4B). The range of linearity can be adapted for different protein amounts by varying the exposition time with X-ray film (lower protein loading, longer exposition times; higher protein loading, shorter exposition times) to avoid the localized saturation of film by the light signal. We have applied the described developer for quantitative immunoblotting analysis of various bacterial proteins in whole-cell extracts. Modestly abundant proteins (0.1% of total cell protein) can be detected and quantified in samples from dilute bacterial cultures with gel loading of as little as 0.2 to 1.0 μg of total protein/well. Samples for analysis can be prepared by simple mixing of culture aliquots with SDS-loading buffer followed by a 5- to 10-min incubation in boiling water. For the detection and quantification of less abundant proteins (<0.1% of total cell protein), culture samples can be fixed and concentrated by precipitation with trichloroacetic acid and correspondingly higher protein loading on the gel.

Thus, here we have described a simple, stable CL reagent system that gives a very low background. This system should be of widespread utility in detection of immunoblots (Western blots). Furthermore, it could possibly be easily adapted for the nonradioactive detection of nucleic acids (Southern or Northern blots).

ACKNOWLEDGMENT

This research was supported in part by Grant OGP0036584 from the Natural Sciences and Engineering Research Council of Canada.

REFERENCES

1. Thorpe, G. H. G. and Kricka, L. J., Enhanced chemiluminescent reactions catalyzed by horseradish peroxidase, *Methods Enzymol.,* 133, 331, 1986.
2. Arakawa, H., Maeda, M., and Tsuji, A., Chemiluminescent assay of various enzymes using indoxyl derivatives as substrate and its applications to enzyme immunoassay and DNA probe assay, *Anal. Biochem.,* 199, 238, 1991.
3. Arakawa, H., Maeda, M., Tsuji, A., and Takahashi, T., Highly sensitive biotin-labelled hybridization probe, *Chem. Pharm. Bull.,* 37, 1831, 1989.
4. Arakawa, H., Maeda, M., and Tsuji, A., Chemilumunescence enzyme immunoassay for thyroxin with use of glucose oxidase and a bis(2,4,6-trichlorophenyl)oxalate-fluorescent dye, *Clin. Chem.,* 31, 430, 1985.
5. Thorpe, G. H. G., Kricka, L. J., Moseley, S. B., and Whitehead, T. P., Phenols as enhancers of the chemiluminescent horseradish peroxidase–luminol–hydrogen peroxide reaction: application in luminescence-monitored enzyme immunoassays, *Clin. Chem.,* 31, 1335, 1985.
6. Leong, M. M. L. and Fox, G. R., Enhancement of luminol-based immunoblot and Western blotting assays by iodophenol, *Anal. Biochem.,* 172, 145, 1988.
7. Yakunin, A. F. and Hallenbeck, P. C., A luminol/iodophenol chemiluminescent detection system for Western blots, *Anal. Biochem.,* 258, 146, 1998.
8. Stoll, V. S. and Blanchard, J. S., Buffers: principles and practice, *Methods Enzymol.,* 182, 24, 1990.
9. Hallenbeck, P. C., Meyer, C. M., and Vignais, P. M., Nitrogenase from the photosynthetic bacterium *Rhodopseudomonas capsulata*: purification and molecular properties, *J. Bacteriol.,* 149, 708, 1982.
10. Yakunin, A. F., Gennaro, G., and Hallenbeck, P. C., Purification and properties of a *nif*-specific flavodoxin from the photosynthetic bacterium *Rhodobacter capsulatus, J. Bacteriol.,* 175, 6775, 1993.
11. Yakunin, A. F. and Hallenbeck, P. C. Regulation of synthesis of pyruvate carboxylase in the photosynthetic bacterium *Rhodobacter capsulatus, J. Bacteriol.,* 179, 1460, 1997.
12. Garvey, J. S., Cremer, N. W., and Sussdorf, D. H., *Methods in Immunology,* 3rd ed., W.A. Benjamine, Inc., Reading, MA, 1977, chap. 1.
13. Laemmli, U. K. and Favre, M., Maturation of the head of bacteriophage T4. I. DNA packaging events, *J. Mol. Biol.,* 80, 575, 1973.

14. Bradford, M. M., A rapid and sensitive method for the quantitation of microgram quantities of protein utilizing the principle of protein-dye binding, *Anal. Biochem.,* 72, 248, 1976.
15. Towbin, H., Staehelin, T., and Gordon, T., Electrophoretic transfer of proteins from polyacrylamide gels to nitrocellulose sheets: procedure and some applications, *Proc. Natl. Acad. Sci. U.S.A.,* 76, 4350, 1979.
16. Luck, H., Peroxidase, in *Methods of Enzymatic Analysis,* Bergmeyer, H. U., Ed., Academic Press, New York, 1963, 895.
17. Hodgson, M. and Jones, P., Enhanced chemiluminescence in the peroxidase-luminol-H_2O_2 system: anomalous reactivity of enhancer phenols with enzyme intermediates, *J. Biolumin. Chemilumin.,* 3, 21, 1989.
18. Seitz, R., Chemiluminescence detection of enzymatically generated peroxide, *Methods Enzymol.,* 57, 445, 1978.
19. Schroeder, H. R., Boguslaski, R. C., Carrico, R. J., and Buckler, R. T., Monitoring specific protein-binding reactions with chemiluminescence, *Methods Enzymol.,* 57, 424, 1978.

12 Chemiluminescent Detection of Immobilized Nucleic Acids—From Southern Blots to Microarrays

John L. Tonkinson, Breck O. Parker, and Michael A. Harvey

CONTENTS

12.1 INTRODUCTION

Immobilization of DNA to solid and microporous surfaces has become a hallmark method in molecular biology. DNA can be immobilized to a variety of surfaces including nitrocellulose and nylon membranes, as well as to modified glass surfaces.[1,2] Interactions between nucleic acids and the immobilization surface can be either covalent or noncovalent, but must be irreversible. Perhaps the most important aspect of any immobilization methodology is that the nucleotide bases of the immobilized DNA must be available for Watson–Crick base pairing. This is necessary so that efficient hybridization to probe sequences can occur. In traditional methods such as Southern blotting, a sequence of DNA is used to probe immobilized DNA from cell or tissue extracts.[1] Recently, reverse hybridizations in the form of ordered arrays have become increasingly useful. In these methods, extracted DNA is used to probe a series of purified or synthetic sequences that have been robotically spotted and immobilized to a surface.[2]

Examination of DNA that has been immobilized to a surface requires an efficient and sensitive probing method. Traditionally, radiolabeling of probes has been the most common detection method. DNA can be labeled with radionuclides by nick-translation, reverse transcription, or end labeling with terminal transferase.[1] Alternatively, radioactive phosphates can be added to DNA probes by

0-8493-0719-8/02/$0.00+$1.50

enzymatic methods.[1] Specific activity of a radiolabeled probe can be determined easily by combining optical data for DNA concentration with number of counts from the isotope. Traditional beta emitters are useful in DNA hybridizations because they can excite silver grains in X-ray film or europium complexes in phosphor imagers. Semiquantitative data can then be obtained concerning the immobilized sequence of interest.

Despite the relative ease with which radioactive probes can be made and used, there are several disadvantages associated with them. One of the primary problems associated with radioisotopes is half-life. Two commonly used isotopes, ^{32}P and ^{33}P, have half lives of 14 and 25 days, respectively, rendering probes with these labels useless relatively quickly. Another concern with isotopes is the cost of disposal. Many institutions store short-lived species such as ^{32}P and ^{33}P in-house while they decay. This reduces the cost of disposal, but it requires dedicated and sometimes specially designed storage locations.

Chemiluminescent detection of immobilized DNA has been an option available to molecular biologists for almost 10 years.[3–5] Standard methods that enzymatically generate light were originally used in immunoassay applications, but have been adapted to the detection of immobilized nucleic acids and are gaining widespread popularity.[3,6] Chemiluminescent detection of DNA provides similar sensitivity and specificity as radioactive detection without the disadvantages of disposal and degradation. In addition, the signal can be captured on X-ray film, requiring no additional instrumentation compared with radiation.

The methods most commonly employed to generate light for detecting immobilized DNA are phosphate cleavage from protected 1,2-dioxetanes with alkaline phosphatase, and luminol excitation by peroxide radicals formed by horseradish peroxidase (HRP).[3,7–9] Initially, light-generating systems were problematic for detection of immobilized biomolecules due to the instability and short half-life of the excitable substrates. However, stable and efficient substrate reagents have been developed for each of the systems, increasing their quantum efficiency and stability. There are a variety of chemiluminescent detection methodologies currently available that vary according to the method used to label probe DNA with the appropriate enzyme. The user can choose between direct or indirect labeling, one- or two-step systems, and enzymatic or nonenzymatic methods of labeling.

Based on the number of available commercial products and the number of recent research articles where chemiluminescent detection has been used for novel applications, it is apparent that chemiluminescent detection of immobilized DNA is being used with increasing frequency.[10–14] Recently, a substantial amount of data has been published describing the use of chemiluminescent detection of DNA in bacterial and PCR-based arrays,[15–17] and one group has reported the ability to multiplex with chemiluminescent detection.[18] These advances in commercialization of chemiluminescent products and labeling systems, as well as their increased use in basic biological research, testify to the utility and efficiency of chemiluminescent detection of immobilized DNA. This chapter provides an overview of various chemiluminescent detection and labeling strategies, followed by some novel uses of chemiluminescence in the detection of immobilized DNA.

12.2 CHEMILUMINESCENT STRATEGIES

During the past decade, two methods of producing chemiluminescence have risen to the forefront as efficient and sensitive for the detection of immobilized nucleic acids. These are alkaline phosphatase cleavage of a phosphate protecting group on 1,2-dioxetanes to produce an unstable intermediate that emits light energy as it degrades, and an HRP-initiated reaction that oxidizes luminol. Both of these methods have been developed into commonly used commercial products for the detection of immobilized DNA.

12.2.1 Alkaline Phosphatase

The routine use of alkaline phosphatase–initiated chemiluminescent reactions was dependent on the development of a class of thermostable substrates. Schaap[7,8] described such a class of substituted 1,2-dioxetanes in 1988. Until then, alkaline phosphatase had been used primarily to generate colored end points in biological assays, a method lacking the sensitivity and reproducibility required for most molecular biology applications.[5] The thermostable dioxetanes developed by Schaap were stabilized by a protecting group such as phosphate or a sililoxy group. Upon chemical or enzymatic deprotection, an unstable intermediate is formed that produces light around 460 to 470 nm as it stabilizes.[7,8] The development of these compounds allowed dioxetanes to be used as probes for immobilized alkaline phosphatase. In addition to the stability of the parent compound, the light generated was intense and long-lived so that it could be easily captured on X-ray film or by charge-coupled device (CCD) camera.

To supplant radioactive detection as a routine method, alkaline phosphatase–based systems had to demonstrate comparable sensitivity. One advantage of enzymatic systems in general and alkaline phosphatase in particular is the turnover of the enzyme. Because alkaline phosphatase has a $k_{cat} = 4100\ s^{-1}$, a sufficient concentration and amount of substrate ensures an intense and long-lived production of light.[4,9] Radioactive methods with ^{32}P are able to detect 1 to 10×10^{-21} mol of immobilized DNA, representing a sensitivity that is substantially greater than the 1×10^{-18} mol needed for single gene detection.[4] By comparison, the phosphate protected 1,2-dioxetanes were able to detect at least 1×10^{-18} mol of alkaline phosphatase on nitrocellulose membranes.[4] Depending on the labeling method employed, this sensitivity could be augmented if more than one enzyme unit is attached to the DNA probe. Subsequent work was performed to develop enhancer systems for alkaline phosphatase systems that increase the sensitivity to approximately 2×10^{-21}, or roughly the same as radioactivity.[9]

Figure 12.1 shows a Southern blot where an immobilized sequence was hybridized to a probe labeled with alkaline phosphatase. The enzyme was detected with a commercially available dioxetane marketed by Applied Biosystems (Foster City, CA) as CDP-Star®. The sample was serially diluted so that the last lane visible contains 120 fg of DNA, translating to a sensitivity of 1.7×10^{-19} mol. Currently, there are a variety of commercially available 1,2-dioxetane substrates for alkaline phosphatase. These include CDP-Star, CSPD® (Applied Biosystems), and Lumigen® PPD, Lumi-Phos® 480, and Lumi-Phos® 530 manufactured by Lumigen, Inc. (Southfield, MI).

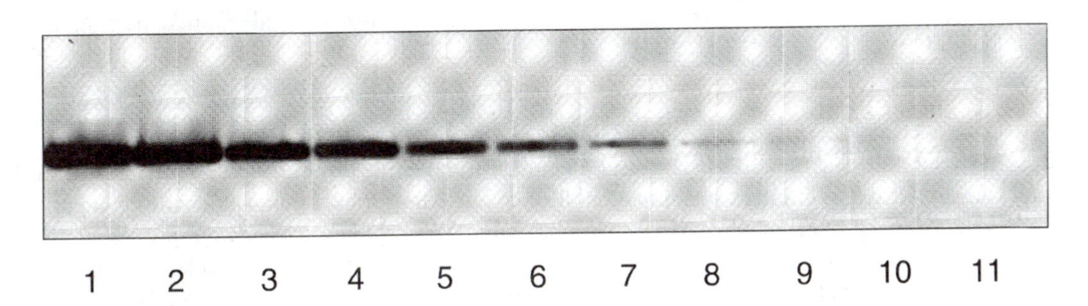

FIGURE 12.1 Chemiluminescent detection of Southern blot. A 1.1-kb fragment from pUC18::Vimentin was loaded into wells of a 1% agarose-TBE gel and electrophoresed at 120 V constant. Lane 1 represents the start of a 1:2 dilution series beginning with 62.5 ng of the 1.1-kb fragment and ending at Lane 11 with 120 fg. The DNA bands in the gel were transferred to Nytran SuperCharge nylon membrane (Schleicher & Schuell, Inc.) and hybridized with a psoralen–biotin-labeled 1.1-kb probe from pUC18::Vim. The blot was incubated with streptavidin–alkaline phosphatase, and detection was completed with the addition of CDP-Star.

12.2.2 Horseradish Peroxidase

HRP-induced luminescence derives from the generation of peroxide radical in the first step, and the oxidation of luminol in the second step.[3] The oxidative degradation of luminol produces light at 428 nm, but is not sustained. HRP-induced chemiluminescence was not viable until an enhancer system was developed so that the light output could be captured on X-ray film. Para-substituted phenolic compounds were found to be the most suitable enhancers for peroxidase-generated light.[3,6] The enhanced system described by Durrant et al.[19] increased the quantum output of the basic reaction by 1000-fold and altered the kinetics of the reaction so that the duration of light output was increased. A common commercial adaptation of this system is the Enhanced ChemiLuminescence or ECL™ kit (Amersham Int., Buckinghamshire, U.K.).

HRP systems utilizing enhanced chemiluminescence have been reported to have sensitivities comparable with alkaline phosphatase systems. This has allowed for their routine use in probing for membrane-bound DNA.[19–21] Depending on probe length, sequences have been detected at the 1 to 2 pg level, corresponding to 1×10^{-18} mol of 1-kb DNA; sensitivity fell by an order of magnitude if oligonucleotides were used as probes.[3] This observation is due in part to the direct labeling method routinely used for HRP. As described in the next section, the number of enzyme units per molecule is directly dependent on molecule length, with oligonucleotides containing one unit. This reduces the sensitivity of the assay system. Unlike the alkaline phosphatase dioxetane system, the light output with luminol is not sustained. Enhanced luminol systems reach a peak after approximately 10 min, and are stable for 1 to 2 h.[19] After this time, light output begins to fall. This is in comparison to dioxetane systems where light output is stable for several days.[4]

12.3 LABELING STRATEGIES

Regardless of which enzymatic system is used to detect immobilized DNA, the chosen enzyme must be attached to the probe. Sensitivity, signal-to-noise ratio, and linearity of signal are integral performance criteria that are, in part, dependent on the labeling method. Although total light emission is independent of enzyme concentration, the intensity of light produced during a given time period is dependent on enzyme concentration.[7] If the intensity is too low, then long exposures will be needed to visualize bound probe. This can result in high backgrounds, inability to look at multiple exposures, and the possibility that information will be missed. Alternatively, if the concentration of enzyme is too high, the signal that is generated may be beyond the linear range of the X-ray film. Thus, choosing a method for labeling probe DNA is not trivial and may be a highly empirical process.

12.3.1 Direct Labeling

Both alkaline phosphatase and HRP can be chemically linked to probe DNA.[22] This method of direct labeling involves covalent attachment of the enzyme to the probe through aldehyde chemistry. One system for the attachment of HRP to probes works by bringing a positively charged HRP complex in contact with a DNA backbone.[20,23] The complex interacts with the backbone approximately every 30 bases. Glutaraldehyde is then used to link the complexes covalently with the backbone. Once made, the probe is stable for at least 9 months in proper storage.[3]

Once the probe has been made, it can be added directly to the hybridization solution with no purification steps required. In addition, because the reactive enzyme unit is attached to the probe, no additional hybridization steps or washes are needed to visualize the product. Once hybridization has occurred, the appropriate light-producing substrate can be added for the detection of alkaline phosphatase or HRP.

Direct label systems offer the advantage of time savings relative to indirect systems. In one estimate, as much as 4 h can be saved with direct labeling compared with hapten-based systems.[20]

The time savings is primarily due to elimination of the second hybridization step where enzyme-linked antibody or streptavidin is introduced. Elimination of the second hybridization step also reduces the potential for nonspecific interactions, which increase background and reduce the signal-to-noise ratio.

The direct labeling systems currently available are recommended for probes that are greater than 300 bases long, although probes as short as 50 bases in length can be labeled.[19] Oligonucleotides can also be labeled if they are synthesized with a reactive thiol group.[20] Under these conditions, only one HRP complex is added per DNA molecule compared with multiple enzymes that can be added to longer DNA molecules. One problem associated with direct labeling is that the hybridization conditions must not destroy the enzymatic activity. For this reason, urea is recommended as a denaturant instead of the more traditional formamide, and temperature cannot be used to control stringency, since the enzyme has a limit of 42°C.[19]

12.3.2 Indirect Labeling

Indirect labeling methods take advantage of strong and specific interactions between two molecules, such as avidin and biotin, or antibodies and antigens. Probes are labeled, by any one of a variety of methods, with a hapten such as fluorescein, digoxygenin, or biotin. After they have been hybridized to the immobilized DNA, a second hybridization is performed with alkaline phosphatase or HRP that is coupled to streptavidin, to an antifluorescein antibody, or to an antidigoxygenin antibody.

Incorporation of haptens into DNA can be accomplished enzymatically or nonenzymatically. Enzymatic methods include end labeling with terminal transferase and random priming with DNA polymerase. Nonenzymatic methods include covalent attachment with *cis*-platinum derivatives and covalent cross-linking of psoralen derivatives. All these methods allow multiple haptens to be attached on each probe molecule, effectively allowing for signal amplification.

Indirect labeling methods require multiple steps for detection. Regardless of which hapten is used or the manner in which the hapten is coupled to the probe, once hybridization has occurred, a second hybridization must be performed. Although multiple hybridizations can increase the potential for background, they also effectively amplify the final signal. In addition, because the enzyme does not undergo the initial hybridization of probe to immobilized DNA, more stringent conditions can be used without fear of diminished enzymatic activity.

12.3.2.1 Enzymatic

Enzymatic labeling of DNA for chemiluminescent detection is similar to enzymatic labeling for radioactive detection. Modified nucleoside triphosphates containing a hapten must be incorporated into the probe by nick translation, by random priming, or by terminal tailing. Often, dUTP that has been coupled to a hapten is used as the modified nucleotide, although other modified nucleotides can be incorporated by these methods. Digoxygenin, fluorescein, and biotin are the most commonly used haptens incorporated into probes by these methods.

Nick translation is a common method used for incorporation of haptens into double-stranded DNA.[24] This method requires both *Escherichia coli* DNase I and DNA polymerase I. Low levels of DNase I are used to nick the DNA phosphodiester backbone randomly leaving free 3′-OH and 5′-PO_4 groups. The 5′ $\rightarrow$ 3′ exonuclease activity of DNA polymerase I removes bases from the nicks, and the polymerase activity resynthesizes the stretches of DNA that have been removed. If hapten-labeled nucleoside triphosphates are included in the reaction milieu, they will be incorporated into the nascent DNA strand. The efficiency of incorporation for any hapten-labeled nucleoside triphosphate will depend on how well it is recognized by DNA polymerase I. Most reactions involving haptens such as biotin or digoxygenin require longer incubation times than those for incorporation or radiolabeled nucleotides.[25,26]

Random-primer labeling is another popular method for synthesis of hapten-labeled probes.[24] In this technique, random hexa- or octanucleotides are used to prime the DNA that is to be labeled. First, the target DNA is heated to separate the strands. As the mix cools, the random primers anneal to complementary regions on the target strands. These new areas of double-stranded DNA serve as substrates for the Klenow fragment of DNA polymerase I. DNA is synthesized from the 3′ end of the random primer and is complementary to the original strands. Hapten incorporation by random priming generally yields better results than by nick translation, primarily because there is no need for exonuclease activity. Hapten-conjugated primers are ideally suited for use with random-priming methods. This ensures that at least one hapten will be present on each probe regardless of the incorporation efficiency during polymerization.

End labeling of probes is easily accomplished with deoxynucleotidyl transferase in the presence of dATP modified hapten.[24,27] This method results in the addition of several nucleotides to the end of the probe. Addition can be limited to one nucleotide by using ddATP (dideoxy-ATP)-modified hapten. Although the preferred substrate is single-stranded DNA, deoxynucleotidyl transferase will recognize double-stranded templates with protruding, blunt, or recessed ends. The efficiency of hapten incorporation will vary depending on the state of the DNA substrate.

12.3.2.2 Nonenzymatic

There are currently two available methods for nonenzymatic coupling of haptens to DNA. Both of these employ small molecules, which are highly active with virtually any DNA sequence, *cis*-platinum, and psoralen. The advantages of nonenzymatic coupling lie primarily in the efficiency of labeling and ease of use. Enzymatic methods reach a maximum level of incorporation, which does not increase even as more substrate DNA is added.[28] With nonenzymatic methods, however, incorporation of label is directly proportional to the amount of DNA that is present in the reaction milieu, resulting in no unlabeled species of probe.

cis-Platinum and psoralen are routinely used to label DNA with digoxygenin and biotin, respectively, although other haptens such as fluorescein can be used. The commercially available products based on *cis*-platinum and psoralen contain a linker arm between the active group and the hapten, decreasing the chance that the hapten will interfere with probe nucleotides. The *cis*-platinum method was originally developed by van Belkum et al.,[29] and was demonstrated to have similar sensitivity as enzymatic incorporation of digoxygenin.[29,30] Ross et al.[31] used the DIG Chem-Link labeling system (Boehringer Mannheim, Indianapolis, IN) to label cDNA from epidermal Langerhans cells with *cis*-platinum. The method was ideally suited for probing blots and arrays due to the high concentrations of probe that were achieved by this labeling method relative to enzymatic methods.

Psoralens are planar tricyclic compounds that intercalate into single- and double-stranded nucleic acids and, when exposed to long-wavelength ultraviolet light, form a covalent bond (Figure 12.2).[32,33] Hapten-coupled psoralens have been developed that have an improved affinity for nucleic

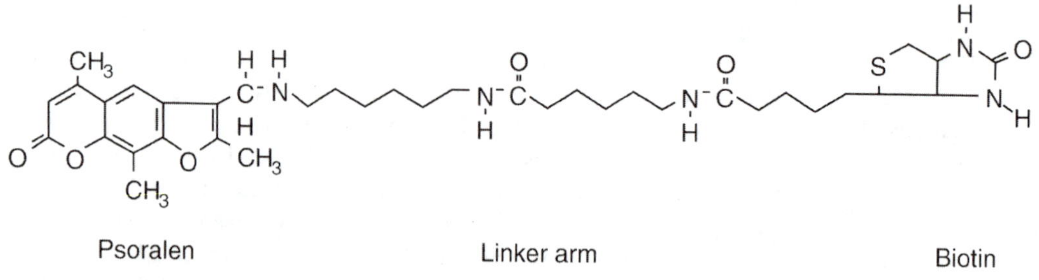

FIGURE 12.2 The structure of psoralen–biotin.

acids and can be used to create hybridization probes. Psoralen–hapten labeling of DNA can be accomplished over a wide range of template concentrations and is relatively insensitive to pH or salt conditions. One example of a commercially available psoralen-labeling system is the Rad-Free® (Schleicher & Schuell, Inc., Keene, NH) kit. In this system, psoralen is linked to a biotin molecule and can be used with streptavidin–alkaline phosphatase or with streptavidin–HRP. Haptenated-psoralens can also be incorporated into RNA probes. The BrightStar™ kit from Ambion, Inc. (Austin, TX) is a commercially available kit for labeling RNA with psoralen–biotin.

Because psoralens have no affinity for proteins or other macromolecules, contaminates have little impact on the labeling reaction. Unlike enzymatic labeling methods, no net DNA synthesis is required so the incoming template DNA does not act to dilute the specific activity of the final probe solution. This is a distinct advantage relative to enzymatic systems. Generally speaking, psoralen-labeled DNA will have one reporter group incorporated in every 20 to 40 nucleotides along the probe. Such probes have been used to detect picogram quantities of target sequence on membrane-based reactions.[34]

12.4 NOVEL TECHNIQUES AND USES OF CHEMILUMINESCENT DETECTION OF IMMOBILIZED DNA

Recently, the use of chemiluminescence as a technique to visualize immobilized DNA has progressed beyond traditional Southern blotting. Novel applications have been reported for both research and diagnostic purposes. In all these reports, the authors describe the advantages of chemiluminescence in terms of decreased waste disposal problems and the ability for virtually any laboratory personnel to conduct the experimentation. In addition, the current labeling and detection methodologies augmented the advantages of chemiluminescence by providing good sensitivity and specificity.

Two very interesting research applications with the alkaline phosphatase–CDP-Star system are cancer related, one involving the detection of minute amounts of apoptotic DNA and the other looking at point mutations in the p53 gene.[13,35] Identification of apoptotic DNA in mixed populations of cells is often difficult because of the small percentage of total DNA in the sample that may be fragmented. The traditional method to identify fragmented DNA is to stain an agarose gel with ethidium bromide.[36] Unfortunately, this is usually only successful with large amounts (≥400 ng) of fragmented DNA. One way to increase sensitivity is to use DNA polymerase to extend the fragments with ^{32}P nucleotides. López Blanco et al. adapted this method by extending the fragments with digoxygenin-dUTP, followed by agarose gel electrophoresis and blotting. The blots were probed with an alkaline phosphatase–linked antidigoxygenin antibody and visualized with CDP-Star. Fragmented DNA was observed with as little as 25 ng total DNA, a significant improvement over traditional ethidium staining.

Siefken et al.[35] used chemiluminescence for semiquantitative detection of ultraviolet-induced point mutations in *p53,* the most commonly mutated gene in cancer. The authors used a digoxygenin-labeled probe to examine blots that contained products from allele-specific polymerase chain reaction PCR (AS-PCR). The chemiluminescent products were visualized with instrumentation that allowed relative quantitation of the bands.

Molecular diagnostics has grown in utility as the techniques used in the research laboratory are adapted to the clinic. In clinical laboratories, the use of radioactivity is even more problematic than it is in research because of the number of personnel, safety, and disposal issues. One example of the use of chemiluminescent detection in molecular diagnostics that has been published is detection of maternal DNA in human cord blood by Poli et al.[11] Contamination of cord blood, used for bone marrow reconstitution, with maternal cells can lead to graft vs. host disease. In the report by Poli et al., probes were labeled during PCR by incorporation of biotinylated dCTP. Detection of immobilized sequences was accomplished with the CSPD substrate from Tropix, Inc., which gave sensitivities comparable to radioactive methods.

Detection of immobilized nucleic acids with chemiluminescence is routinely accomplished by visualizing the light that is produced on X-ray film. As with radioactivity, a black spot is produced on the film where silver grains have been excited by the light. Regardless of the system used to create the light, the signal appears the same. For a variety of reasons, including sample availability and quantitation, researchers routinely need to probe the same blot with multiple probes. The most common method of doing this is to hybridize and visualize one probe, then strip the blot and hybridize with a new probe. Although this method works, it is tedious and can lead to inconclusive results.

Recently, Reddy et al.[18] reported on a method for sequential visualizaton of the same blot that had been probed simultaneously with two different probes. The method takes advantage of the relative short duration of light produced by HRP and the robustness of alkaline phosphatase. The authors first hybridized a blot with two different probes, one labeled with biotin and one labeled with digoxygenin. Next, the blot was visualized with avidin–HRP and luminol. After the image had been generated, the blot was probed with antidigoxygenin-alkaline phosphatase. To ensure that the HRP signal was completely depleted at the time of alkaline phosphatase addition, the blot was washed with H_2O_2 to cause the HRP signal to cease. Using this method the authors simultaneously visualized three genotypes of the $CF\Delta F_{508}$ mutation.[18]

12.5 USE OF CHEMILUMINESCENT DETECTION WITH ARRAYS

Recently, advances in molecular biology and genomics have led to the increased use of arrays by researchers. Arrays are simply the ordered placement of bacterial colonies or DNA fragments on microporous or glass surfaces. The difference between arrays and traditional blots and libraries is the density at which the biomolecules can be placed on the surface, and the corresponding increase in the amount of information that can be garnered. For example, approximately 36,000 colonies representing 18,000 different genes or genomic inserts can be easily spotted on a 22×22 cm nylon filter; expression arrays are routinely generated with 1500 different genes represented on an 8×12 cm membrane or, more recently, on a microscope slide. These techniques have the potential to provide information to researchers at rates that are several orders of magnitude greater than with traditional methods.

Arrays can be probed with DNA in much the same way as Southern blots and traditional libraries. Genomic and cDNA arrays containing bacterial colonies are usually probed with radioactive probes. Recently, however, it has been demonstrated that chemiluminescence is a viable and perhaps preferred alternative. Guiliano et al.[17] used the NEBlot® Phototope® (New England Biolabs, Inc., Beverly, MA) random prime labeling kit to biotinylate probes that were used to screen cDNA arrays from *Brugia malayi*. Probes were detected using streptavidin-linked alkaline phosphatase and CDP-Star. The authors reported that as many as ten strippings and rehybridizations were performed on the filters without a loss in specificity.[17] Figure 12.3 shows a colony filter array from our laboratories that has been probed with DNA labeled with psoralen–biotin. A human BAC library was gridded onto Nytran® SuperCharge (Schleicher & Schuell) membrane and grown overnight. The next day, the array was probed with β-actin probe labeled with psoralen–biotin. The positive clones were visualized by adding streptavidin alkaline phosphatase and then generating light with CDP-Star. The array was exposed to X-ray film overnight.

Expression arrays are used to look at expression profiles of genes in tissue sets and to look at expression levels in response to different experimental conditions. Radiation is routinely used as the detection method of choice; however, on microarrays fluorescent detection is usually employed. Both of these methods are not without drawbacks. As discussed previously, radiation has the inherent problems of safety and disposal. Fluorescent detection is very powerful in terms of sensitivity and quantitation; however, it is expensive and requires image analysis capabilities

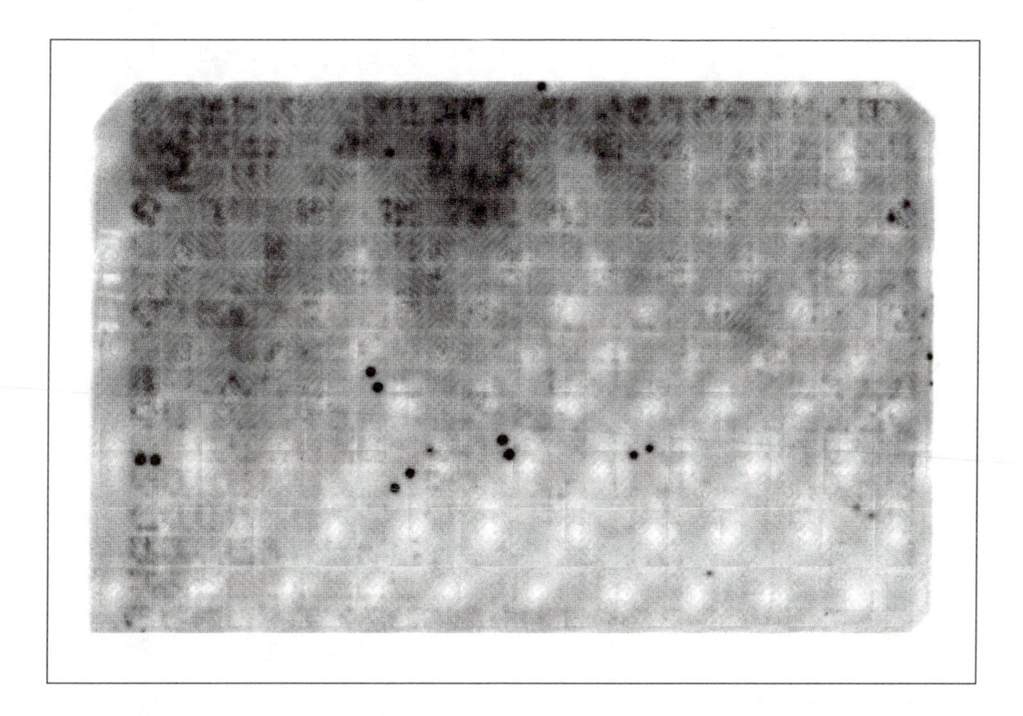

FIGURE 12.3 Chemiluminescent detection of β-actin clone on a human genomic BAC library filter array. Nytran SuperCharge 12 × 8 cm nylon membrane was arrayed in a standard 4 × 4, vector-defined double offset pattern using a 0.4-mm solid pin tool and BioRobotics Total Array System (Comberton, Cambridge, U.K.). The membrane contained 5568 colonies or 2784 clones from the CITB HSP C human BAC library (Research Genetics, Huntsville, AL). The filters were incubated overnight at 37°C, processed, and hybridized with 30 ng/ml psoralen–biotin β-actin probe. The positive clones were detected with alkaline phosphatase and CDP-Star and are present as distinct pairs of predefined vectors.

that are beyond the means of many laboratories. We and others have investigated the possibility of chemiluminescent detection of microarrays.[15,16,37] Some of the concerns with chemiluminescent detection of arrays include the perceived inability to multiplex and problems with resolution. Spots on arrays can be as small as 200 nm in diameter with only 400 nm from the center of one spot to the center of the next. A detection method that produced a signal without this resolution would not be useful.

Figure 12.4 shows a typical expression array that has been probed with cDNA directly labeled with HRP. The array was then visualized with a standard enhanced luminol system. By using this system, exceptional background and resolution was achieved and no false-positives were illuminated. Akhavan-Tafti et al.[15] reported the use of a new alkaline phosphatase substrate, Lumigen™ APS-5 (Lumigen, Inc., Southfield, MI), that is well suited for array detection. The substrate was coupled to an antidigoxygenin antibody and used to screen for hybridized probe linked to digoxygenin. Although resolution of spots was not a problem in this report, the spot size was 1 mm. This is substantially larger than spots that are placed robotically.

To address the issue of resolution, we looked at chemiluminescent detection of DNA arrayed on membrane-coated slides (Figures 12.5 and 12.6). Plasmid DNA was arrayed robotically in standard 5 × 5 arrays, where each spot was 200 μm in diameter with a pitch or center to center distance of 600 μm. In Figure 12.5, equivalent amounts of DNA were arrayed in each spot. The slide was hybridized with probe (40 ng/ml), direct labeled with HRP, and visualized with ECL. Excellent spot resolution and intensity was achieved.

In Figure 12.6, a titration series of plasmid DNA was arrayed on FAST™ (Schleicher & Schuell, Inc.) membrane-coated slides. Each 5 × 5 array contains two dilution series, one in the top ten

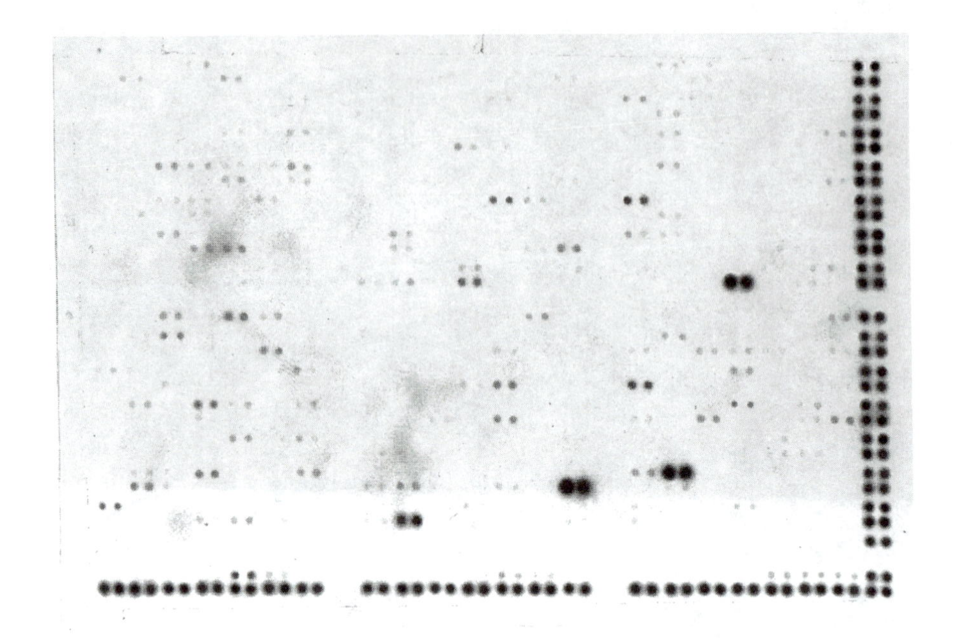

FIGURE 12.4 Expression analysis using chemiluminescent detection of a Human Atlas Array from ClonTech, Inc. (Palo Alto, CA). HRP direct-labeled cDNA was prepared from 2.5 mg poly A^+ RNA from the human leukemia cell line K-562 (ClonTech). The labeled cDNA was used to hybridize to the Atlas array. Detection was carried out with luminol substrate. The array contains 588 pairs of cDNAs representing important functional gene groups, and a series of negative and positive (housekeeping genes) controls. The spots on the far right and bottom of the array represent human genomic DNA spots for orientation purposes.

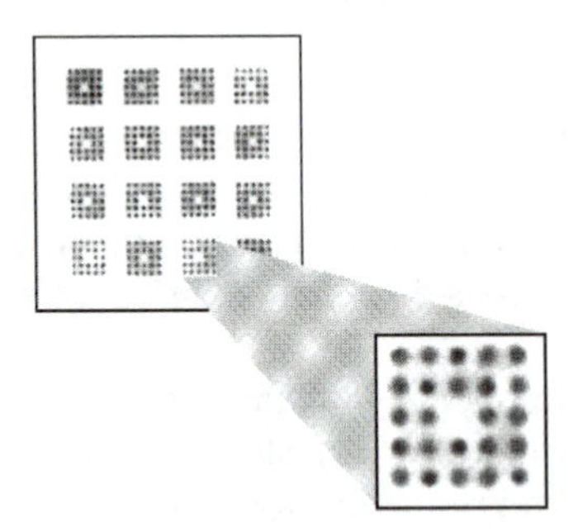

FIGURE 12.5 HRP detection of microarrays on a membrane slide from Schleicher & Schuell, Inc. The slide was microarrayed with pUC19 in a 5 × 5, 0.6-mm pitch, double-offset pattern using a 0.2-mm solid pin tool and BioRobotics Total Array System. The array was hybridized with 40 ng/ml HRP-labeled pUC19 probe and visualized with ECL. The blowout shows one 5 × 5 array; note the consistency of spot intensity.

spots and the other in the bottom ten spots. The slide was hybridized with plasmid (40 ng/ml) that was labeled with psoralen–biotin. The slides were then incubated with streptavidin–HRP and visualized with ECL. Under these conditions, extremely good background and spot resolution was achieved. In addition, a gradient of intensity can be seen on each array corresponding to the concentration gradient that was arrayed down. These results confirm the usefulness of chemiluminescent detection methodologies for DNA array technology.

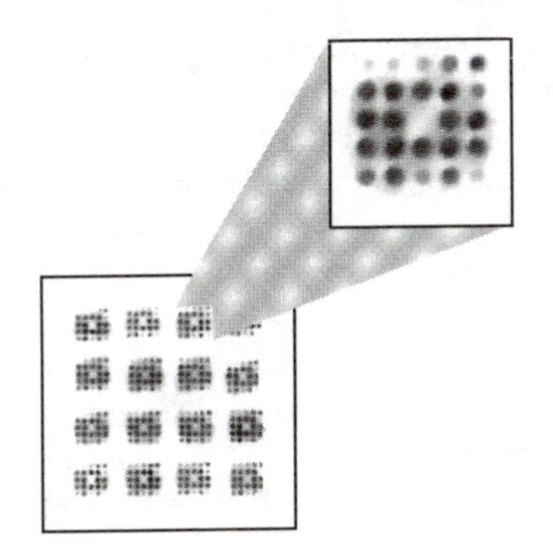

FIGURE 12.6 FAST™ membrane slide from Schleicher & Schuell, Inc. The slide was microarrayed with multiple concentrations of pUC19 in a 5 × 5, 0.6-mm pitch, double-offset pattern using 0.2-mm solid pin tool and BioRobotics Total Array System. The array was hybridized with 40 ng/ml pUC19 probe labeled with psoralen–biotin, followed by a second hybridization with streptavidin–HRP, and visualized with ECL. The blowout shows one 5 × 5 array, with a titration series from left to right in the top two rows and from right to left in the bottom two rows.

12.6 SUMMARY

During the last 10 years, chemiluminescent detection of immobilized nucleic acids has evolved from an interesting concept to a widely practiced technique. This evolution is due in part to the advantages of handling, safety, and disposal relative to radioactive methods. However, despite these advantages, widespread use of chemiluminescence would not have occurred without the demonstration that these methods provide the same sensitivity as radiation. Improvements in sensitivity were accomplished in part through improvements in light-producing reactions through the use of enhancers and longer-lived species, but also through improvements in labeling strategies. Direct labeling of probes with enzymes and nonenzymatic labeling systems have decreased the reliance on enzymatic labeling. Probes with high specific activity can be made efficiently and without the need for lengthy purification procedures that decrease yields.

Traditional Southern blots are a standard technique in molecular biology and will most assuredly remain so for the foreseeable future. However, if this were the only application for detection of immobilized DNA, the advantages of chemiluminescence are probably not sufficient to supplant radioactivity as the detection method of choice. The advent of array technology for screening bacterial and PCR libraries has presented a completely new field for use of chemiluminescence. The advantage of arrays is the tremendous wealth of information that can be obtained from a small sample. Chemiluminescence provides a viable alternative for screening arrays because of the efficiency of direct and nonenzymatic labeling, its sensitivity, and general ease of use. Undoubtedly, the use of chemiluminescent detection methodologies will increase as this developing field matures.

REFERENCES

1. Sambrook, J., Fritsch, E. F., and Maniatis, T., *Molecular Cloning: A Laboratory Manual,* 2nd ed., Cold Spring Harbor Laboratory Press, Cold Spring Harbor, NY, 1989, chaps. 9 and 10.
2. Schena, M., Shalon, D., Heller, R., Chai, A., Brown, P. O., and Davis, R. W., Parallel human genome analysis: microarray-based expression monitoring of 1000 genes, *Proc. Natl. Acad. Sci., U.S.A.,* 93, 10614, 1996.
3. Durrant, I., Light-based detection of biomolecules, *Nature,* 346, 297, 1990.
4. Pollard-Knight, D., Simmonds, A. C., Schaap, A. P., Akhavan, H., and Brady, M. A. W., Nonradioactive DNA detection on Southern blots by enzymatically triggered chemiluminescence, *Anal. Biochem.,* 185, 353, 1990.

5. Sheffield, J. S., Benjamin, W. H., and McDaniel, L. S., Detection of DNA in Southern blots by chemiluminescence is a sensitive and rapid technique, *BioTechniques,* 12, 836, 1992.
6. Whitehead, T. P., Thorpe, G. H. G., Carter, T. J. N., Groucutt, C., and Kricka, L. J., Enhanced luminescence procedure for sensitive determination of peroxidase labelled conjugates in immunoassay, *Nature,* 305, 158, 1983.
7. Schaap, A. P., Chemical and enzymatic triggering of 1,2-dioxetanes, *Photochem. Photobiol.,* 47, 50, 1988.
8. Schaap, A. P., Chemiluminescent substrates for alkaline phosphatase and β-galactosidase: application to ultrasensitive enzyme-linked immunoassays and DNA probes, *Biolumin. Chemilumin.,* 2, 253, 1988.
9. Schaap, A. P., Akhavan, H., and Romano, L. J., Chemiluminescent substrates for alkaline phosphatase: application to ultrasensitive enzyme-linked immunoassays and DNA probes, *Clin. Chem.,* 35, 1863, 1989.
10. Johnson, E. D. and Kotowski, T. M., Chemiluminescent detection of RFLP patterns in forensic DNA analysis, *J. Forensic Sci.,* 41, 569, 1996.
11. Poli, F., Sirchia, S. M., Scalamogna, M., Garagiola, I., Crespiatico, L., Pedranzini, L., Lecchi, L., and Sirchia, G., Detection of maternal DNA in human cord blood stored for allotransplantation by a highly sensitive chemiluminescent method, *J. Hematother.,* 6, 581, 1997.
12. Zhan, J., Fahimi, H. D., and Voelkl, A., Sensitive nonradioactive dot blot/ribonuclease protection assay for quantitative determination of mRNA, *BioTechniques,* 22, 500, 1997.
13. López Blanco, F., Gonzalez-Reyes, J., Fanjul, L. F., Ruiz de Galarreta, C. M., and Aguiar, J. Q., Chemiluminescence-based detection of minute amounts of apoptotic DNA, *BioTechniques,* 24, 354, 1998.
14. Lin, J. J., Ma, J., and Kuo, J., Chemiluminescent detection of AFLP markers, *BioTechniques,* 26, 344, 1999.
15. Akhavan-Tafti, H., Reddy, L. V., Siripurapu, S., Schoenfelner, B. A., Handley, R. S., and Schaap, A. P., Chemiluminescent detection of DNA in low- and medium-density arrays, *Clin. Chem.,* 44, 2065, 1998.
16. Rajeevan, M. S., Dimulescu, I. M., Unger, E. R., and Vernon, S. D., Chemiluminescent analysis of gene expression on high-density filter arrays, *J. Histochem. Cytochem.,* 47, 337, 1999.
17. Guiliano, D., Ganatra, M., Ware, J., Parrot, J., Daub, J., Moran, L., Brennecke, H., Foster, J. M., Supali, T., Blaxter, M., Scott, A. L., Williams, S. A., and Slatko, B. E., Chemiluminescent detection of sequential DNA hybridizations to high-density, filter-arrayed cDNA libraries: a subtraction method for novel gene discovery, *BioTechniques,* 27, 146, 1999.
18. Reddy, L. V., DeSilva, R., Handley, R. S., Schaap, A. P., and Akhavan-Tafti, H., Sequential chemiluminescent detection of target DNAs without stripping and reprobing, *BioTechniques,* 26, 710, 1999.
19. Durrant, I., Benge, L. C. A., Sturrock, C., Devenish, A. T., Howe, R., Roe, S., Moore, M., Scozzafava, G., Proudfoot, L. M. F., Richardson, T. C., and McFarthing, K. G., The application of enhanced chemiluminescence to membrane-based nucleic acid detection, *BioTechniques,* 8, 564, 1990.
20. Stone, T. and Durrant, I., Enhanced chemiluminescence for the detection of membrane-bound nucleic acid sequences: advantages of the Amersham system, *Genet. Anal. Tech. Appl.,* 8, 230, 1991.
21. Stone, T. and Durrant, I., Hybridization of horseradish peroxidase-labeled probes and detection by enhanced chemiluminescence, *Mol. Biotechnol.,* 6, 69, 1996.
22. Renz, M. and Kurz, C., A colorimetric method for DNA hybridization, *Nucl. Acids Res.,* 12, 3435, 1984.
23. Durrant, I. and Stone, T., Preparation of horseradish peroxidase-labeled probes, *Mol. Biotechnol.,* 6, 65, 1996.
24. Temsamani, J. and Agrawal, S., Enzymatic labeling of nucleic acids, *Mol. Biotechnol.,* 5, 223, 1996.
25. Langer, P. R., Waldrop, A. A., and Ward, D. C., Enzymatic synthesis of biotinylated polynucleotides: novel nucleic acid affinity probes, *Proc. Natl. Acad. Sci. U.S.A.,* 78, 6633, 1981.
26. Martin, R., Hoover, C., Grimme, S., Grogan, C., Holtke, J., and Kessler, C., A highly sensitive, nonradioactive DNA labeling and detection system, *BioTechniques,* 9, 762, 1990.
27. Deng, G. R. and Wu, R., Terminal transferase: use in the tailing of DNA and for *in vitro* mutagenesis, *Methods Enzymol.,* 100, 96, 1983.
28. Harvey, M. A., unpublished data, 1995.
29. Van Belkum, A., Linkels, E., Jelsma, T., Van den Berg, F. M., and Quint, W., Non-isotopic labeling of DNA by newly developed hapten-containing platinum compounds, *BioTechniques,* 16, 148, 1994.

30. Holz, H., Goldstein, C., Rein, R., and Holtke, H. J., DIG chem-link labeling and detection set, *Biochemica,* 3, 26, 1997.
31. Ross, R., Ross, X. L., Rueger, B., Laengin, T., and Reske-Kunz, A. B., Nonradioactive detection of differentially expressed genes using complex RNA or DNA hybridization probes, *BioTechniques,* 26, 150, 1999.
32. Levenson, C. H., Watson, R., and Sheldon, E. L., Biotinylated psoralen derivative for labeling nucleic acid hybridization probes, *Methods Enzymol.,* 184, 577, 1990.
33. Bush, C. N., King, T. H., Loy, R. S., Wenzel, A. Z., and Harvey, M. A., Non-enzymatic preparation of nucleic acid probes using a novel psoralen-biotin, *Protein Eng.,* 8(Suppl.), 91, 1995.
34. King, T., PCR amplification using psoralen biotinylated primers: PAGE analysis and direct chemiluminescent detection, *Am. Biotechnol. Lab.,* 13, 106, 1995.
35. Siefken, W., Hoppe, U., and Bergmann, J., Semiquantitative chemiluminescent detection of UV-B induced point mutations in the *p53* tumor-suppressor gene, *BioTechniques,* 26, 528, 1999.
36. Tonkinson, J. L., Marder, P. M., Andis, S. L., Schulz, R. M., Gossett, L. S., Shih, C., and Mendelsohn, L. G., Cell cycle effects of antifolate antimetabolites: implications for cytotoxicity and cytostasis, *Cancer Chemother. Pharmacol.,* 39, 521, 1997.
37. Parker, B. O. and Tonkinson, J. L., unpublished data, 1999.

13 Differential Display Polymerase Chain Reaction Using Chemiluminescent Detection

Gang An and Robert W. Veltri

CONTENTS

13.1 INTRODUCTION

Regulation of gene expression plays a fundamental role in many normal and abnormal cellular physiological processes, including normal cell differentiation and development, carcinogenesis, and host responses to many environmental insults and pathological changes. Identification of genes that are differentially expressed in different cell types or under various conditions is of great importance in modern biology. Traditionally, differential hybridization and subtractive hybridization have been used widely to identify genes that express differentially between different but closely related types of cells or tissues.[1,2] Although many genes have been successfully cloned by these methods, both approaches are rather labor-intensive and time-consuming, and allow simultaneous comparison of only two RNA populations. The recently described differential display polymerase chain reaction (DD-PCR) method[3,4] have gained popularity in the biomedical research community. The method allows for rapid and relatively complete survey of differential gene expression among various cell or tissue types. In the DD method, mRNAs from two or more cell types are divided into subpopulations and reverse-transcribed into cDNAs using 3′-anchored oligo(dT) primers. cDNAs from different cell types are then amplified by PCR using the same set of 3′-anchored primers in combination with a 5′ arbitrary primer. The radioactive cDNA products from PCR are resolved on a DNA sequencing gel in sequentially arranged lanes to allow rapid comparisons. Differentially expressed mRNA species are identified by comparing the autoradiographic intensities of cDNA bands on the gel. The identified cDNAs, which were significantly altered in their expression, are then reamplified and finally cloned into plasmid. Since the introduction of the DD-PCR method by

0-8493-0719-8/02/$0.00+$1.50

Liang and Pardee,[3] other similar methods utilizing different primer design strategies have been described, including RNA fingerprinting using arbitrarily primed PCR (RAP-PCR)[5] and palindromic primer-based DD-PCR.[6] Compared with DD-PCR, RAP-PCR uses arbitrary primers in both reverse transcription and PCR amplification, and palindromic primer-based DD-PCR uses only one arbitrary primer in PCR reactions.

The majority of the described methods use radioisotopes to detect band patterns on sequencing gels,[3–6] which has the technical disadvantages of handling hazardous radioactive material, potential contamination of the working area and thermal cycler, and the cost associated with Nuclear Regulatory Commission licensing as well as those costs related to waste disposal of the radioactive materials. For those who prefer non-isotopic approaches, there are some alternative methods from which to choose. For example, non-isotopic DD-PCR methods have been developed that use silver staining detection,[7] fluorescent detection,[8] or chemiluminescent detection.[9] Although both the silver staining and the fluorescent labeling DD-PCR methods are valuable alternatives to radioactive DD-PCR, the silver staining method is less sensitive and displays much fewer bands than radioisotopic-based method, and the fluorescent labeling method requires an expensive automatic sequencing machine and software.

We describe here a DD-PCR method employing the chemiluminescent detection method, which has all the benefits of a non-isotopic method without sacrificing the sensitivity of radioactive detection, and involves no specialized instrumentation. The method described uses four different 5′ biotinylated oligo(dT)-anchored primers to reverse-transcribe RNAs from different cell types into four pools of cDNAs. Each cDNA pool is then amplified by PCR using the same set of 5′ biotinylated oligo(dT)-anchored primer as in the reverse transcription, combining with an arbitrary primer. The PCR products are separated by standard denaturing gel electrophoresis. About 20% of the DNA fragments is then transferred to a nylon membrane by the capillary method. Following transfer, the DNA fragments are detected by SEQ-Light™ Chemiluminescent DNA Sequencing Detection System (Applied Biosystems, Foster City, CA). The DD-PCR band patterns are obtained by exposing the membrane to standard X-ray film. Exposure times are generally 30 to 60 min. After positive bands are identified, the X-ray film and the nylon membrane are used to locate the band position back to the gel. The remaining DNA (about 80%) on the gel in the identified position is recovered and used for reamplification and cloning. We have successfully used this method to clone genes differentially expressed in prostate cancer.

13.2 MATERIALS AND METHODS

13.2.1 Primer Design

The four biotinylated oligo(dT) anchored primers ($T_{12}VN$) are designed as follows where V represents mixed bases of A, G, and C:

5′-biotin-TTTTTTTTTTTTVG-3′
5′-biotin-TTTTTTTTTTTTVA-3′
5′-biotin-TTTTTTTTTTTTVT-3′
5′-biotin-TTTTTTTTTTTTVC-3′

The arbitrary decamers can be designed by randomly picking ten bases with about 60% GC content. Different arbitrary decamer kits are also available commercially (Operon Technologies, Alameda, CA).

13.2.2 RNA Isolation and Reverse Transcription

Total RNA is isolated from normal and prostate cancer specimens by the single-step, acid guanidinium thiocyanate–phenol–chloroform extraction method[10] or by TRIZOL reagent (GIBCO/BRL, Gaithersburg, MD) following manufacturer's directions. RNA (10 μg) from each tissue is treated with 5 units of RNase-free DNase I (GIBCO/BRL, Gaithersburg, MD) in the presence of 20 m*M* Tris-HCl, pH 8.4, 50 m*M* KCl, 2 m*M* $MgCl_2$, and 20 units of RNase inhibitor (Boehringer Mannheim, Indianapolis, IN). After extraction with phenol/chloroform and ethanol precipitation, the RNA is redissolved in DEPC-treated H_2O. Four pools of cDNA are generated using the four different biotinylated oligo(dT)-anchored primers. For each pool, five μg of the DNase-treated RNA from each tissue is reverse-transcribed into cDNA using one of the biotinylated oligo(dT)-anchored primer and M-MLV reverse transcriptase (GIBCO/BRL) in a total of 40 μl reaction. The reaction mixture contains 50 m*M* Tris-HCl, pH 8.3, 75 m*M* KCl, 3 m*M* $MgCl_2$, 10 m*M* DTT, 500 μM dNTP, 2 μM biotinylated oligo(dT)-anchored primer, and 400 U M-MLV reverse transcriptase. The reactions are incubated at room temperature (22°C) for 10 min, followed by a 50-min incubation at 42°C. A final 10-min incubation at 70°C is performed to inactivate the enzyme.

13.2.3 PCR

For each pool of cDNA from different cells or tissues (e.g., normal and cancer), PCR is performed using the same biotinylated oligo(dT)-anchored primer as in reverse transcription, combining with an arbitrary 10mer. Different primer combinations can be used for different PCR reactions to generate distinct band patterns. The PCR mixture contains 2 μl of cDNA, 10 m*M* (Tris)-HCl, pH 9.3, 50 m*M* KCl, 1.5 m*M* $MgCl_2$, 50 μM dNTPs, 1.25 U of Taq DNA polymerase, and 0.2 μM each of the biotinylated oligo(dT)-anchored primer and arbitrary 10mer in a total of 20 μl reaction. The amplification parameters include 40 cycles of reaction with 30 s denaturing at 94°C, 2 min annealing at 40°C, and 1 min extension at 72°C. A final extension at 72°C is performed for 15 min. After PCR, 12 μl of loading dye (95% formamide, 0.05% bromophenol blue, and 0.05% xylene cyanol) is added to each tube.

13.2.4 Electrophoresis and Chemiluminescent Detection

A 6% denaturing polyacrylamide DNA sequencing gel is prepared following established procedure.[1] The gel is prerun for about 1 h to warm up to about 50°C. The PCR products with loading dye are heated at 80°C for 3 min, and 5 μl of each is loaded immediately onto the gel. The gel is run at about 120 W constant power for about 2 h or until the xylene cyanol dye reaches about 20 cm from the bottom.

After electrophoresis, DNA fragments on the gel are transferred to the Tropilon-Plus™ nylon membrane (Tropix), or other positive-charged nylon membrane by capillary transfer. Fixing the gels or removing the urea is unnecessary. Approximately 20% of the DNA is transferred to the membrane, which can be easily detected using chemiluminescent detection procedure. The setup for capillary transfer is illustrated in Figure 13.1 and the procedure is as follows:

1. Disassemble the gel apparatus and separate the glass plates.
2. Lay a piece of Whatman® 3MM filter paper on top of the gel, making sure the paper is in good contact with the gel. Remove the gel from the glass plate with the filter paper. Place the paper with gel attached back on the plate with the gel side up.
3. Cut a piece of nylon membrane the same size as the region of the gel to be blotted and wet it thoroughly with TBE.
4. Carefully place the wet membrane on the gel. Remove any air bubbles by rolling a pipette over the membrane.
5. Place three pieces of dry Whatman filter paper on top of the membrane, another glass plate, and a 2-kg weight on top. Allow the transfer to proceed for 1 h.

Following the transfer step, orient the membrane and the gel by punching through using a needle with India ink in three locations. The membrane is then carefully separated from the gel. The transferred DNA is immobilized on the membrane by UV irradiation (total exposure: 120 mJoules), or baking at 80°C for 1 h.

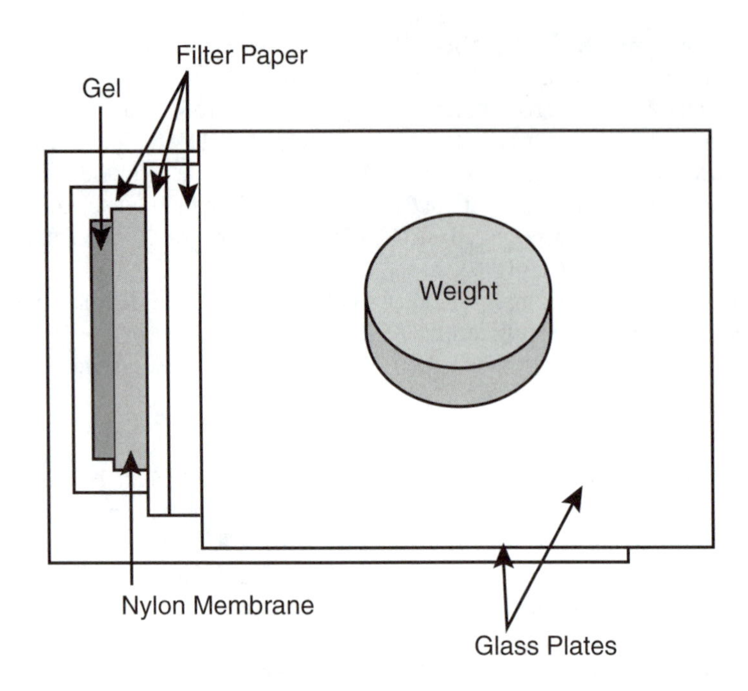

FIGURE 13.1 Illustration of the capillary transfer method.

The SEQ-Light™ chemiluminescent detection system (Applied Biosystems) is used to detect bands on the sequencing gel following manufacturer's instruction. After detection, DNA band patterns are captured on X-ray film, and differential expressed bands are identified.

13.2.5 Reamplification, Sequencing, and Confirmation of Expression

After positive bands are identified, the X-ray film and the nylon membrane are then used to locate the band position back to the gel. The remaining DNA (about 80%) on the gel in the identified position is recovered. Reamplification is performed following established procedure.[3,4] Briefly, each gel slice is soaked in 100 μl of H_2O in a 1.5-ml centrifuge tube and boiled for 10 min. After centrifuging for 2 min, the supernatant is transferred to a new tube. The DNA is precipitated by adding 10 μl of 3 *M* NaOAc (pH 5.2), 2 μl of glycogen (20 mg/ml), and 250 μl of ethanol, followed by incubating at dry ice or –80°C freezer for 20 min and then spinning for 10 min. The pellet is washed with 500 μl of 70% ethanol, dried, and resuspended in 10 μl of H_2O. Reamplification is performed using 4 μl of the DNA and the same set of primers and PCR conditions as in the display experiment. The reamplified DNA band can be used directly as a probe for Northern confirmation or a template for DNA sequencing.

Since all the antisense DNA molecule in the reamplified product has a biotin in its 5′ end, the reamplified DNA can be used directly as probe for non-isotopic Northern hybridization to confirm the differential expression of the gene. The reamplified band is purified by Qiaex kit (Qiagen, Chatsworth, CA) and eluted into 20 μl of H_2O. Northern hybridization is performed using established procedure.[1] Then, 10 μl of the purified DNA is boiled for 5 min, cooled on ice, and added to hybridization solution. Following hybridization and washing, the signal can be detected by chemiluminescent detection. The purified DNA band can also be used as a probe for non-isotopic cDNA library screening to clone full-length cDNA of the gene.

The purified band can be sequenced directly using the same biotinylated oligo(dT)-anchored primer and SeqLight® system. After DNA sequence data are obtained, primers can be designed and used in a relative reverse transcription-PCR (RT-PCR) experiment to confirm the differential expression of the gene, or used in a Rapid Amplification of cDNA End (RACE) experiment to clone the full-length cDNA of the gene.

13.3 RESULTS AND DISCUSSION

The described method is applied to identify genes differentially expressed between normal prostate and prostate cancer tissues. Figure 13.2 shows an example of the sequencing gel from the differential display experiments using chemiluminescent detection. In this particular experiment, biotinylated $T_{12}VG$ primer is used together with either OPB1 or OPB2 primer (Operon Technologies). The bands

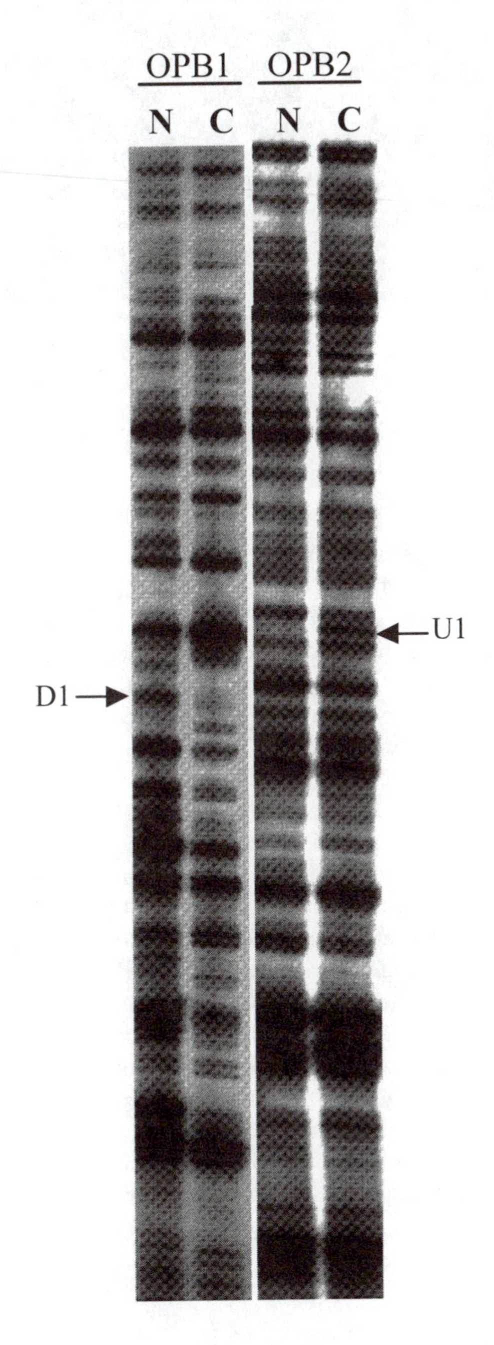

FIGURE 13.2 DD band patterns by chemiluminescent detection. Primers used: 5′-biotin-TTTTTTTTTTTT-VG-3′ and OPB1: 5′-GTTTCGCTCC-3′ or OPB2: 5′-TGATCCCTGG-3′. N, RNA from normal prostate tissue; C, RNA from prostate cancer tissue; arrows indicate differentially expressed bands.

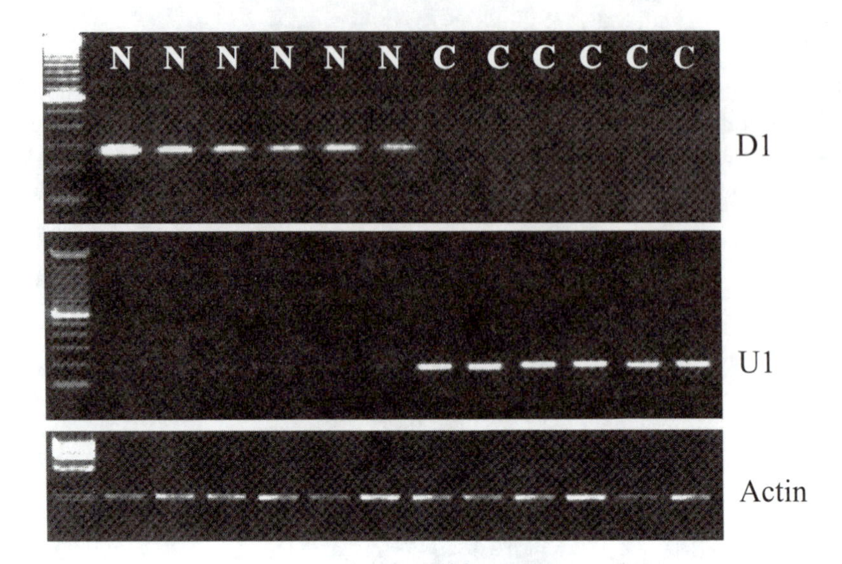

FIGURE 13.3 Confirmation of differential expression by RT-PCR. The results of RT-PCR for D1 and U1 transcripts and β-actin control are shown. N, normal prostate tissues; C, prostate cancer tissue. Upper panel shows the RT-PCR result of D1 transcript, middle panel shows the RT-PCR result of U1 transcript, and lower panel shows the β-actin control.

indicated by the arrows represent differentially expressed bands. Whereas the D1 band shows much higher intensity in normal prostate tissue, the U1 band shows much higher intensity in prostate cancer tissue. Overall, the quality, sensitivity, and the total number of bands displayed are very similar to that obtained by radioisotope detection method (data not shown). We routinely obtained 50 to 200 bands per reaction for each primer set. Although most of the bands show identical intensity among the tissues examined, one to three differentially expressed bands are seen from each experiment by a primer set.

Both the D1 and U1 bands were recovered and purified as described. Half the purified DNA was used as a template for chemiluminescent sequencing. Primers were then designed for both genes based on sequence information, and RT-PCR[11] was performed on six each of normal and prostate cancer tissues to confirm the differential expression of the genes. β-Actin RT-PCR was also performed as a control for normalization. As shown in Figure 13.3, D1 was confirmed to be downregulated in prostate cancer, and U1 was confirmed to be upregulated in prostate cancer. The confirmation rate of the differentially expressed cDNAs using our method is similar to that of DD using radioactive detection (data not shown).

In summary, the DD-PCR using a chemiluminescent detection method described here is a sensitive, reliable, cost-effective, and safe alternative to the radioisotopic differential display technique.[3,4] Also, it does not require any costly special equipment or software. Since no radioisotopes are involved, all the safety and waste disposal issues related to radioactive material handling requirements are eliminated. In our hands it has the same sensitivity, reproducibility, and confirmation rate as the original radioisotope-based DD-PCR method.

ACKNOWLEDGMENTS

We thank Dr. S. Mark O'Hara for helpful input during development of this method, and Ms. Guizhen Luo for providing technical assistance.

REFERENCES

1. Sambrook, J., Fritsch, E. F., and Maniatis, T., *Molecular Cloning, A Laboratory Manual,* Cold Spring Harbor Laboratory, Cold Spring Harbor, NY, 1989.
2. Lee, S., Tomasetto, C., and Sager, R., Positive selection of candidate tumor suppressor gene by subtractive hybridization, *Proc. Natl. Acad. Sci. U.S.A.,* 88, 2825, 1991.
3. Liang, P. and Pardee, A. B., Differential display of eukaryotic messenger RNA by means of the polymerase chain reaction, *Science,* 257, 967, 1992.
4. Liang, P., Averbouk, L., Keyomarsi, K., and Sager, R., Differential display and cloning of mRNAs from human breast cancer versus mammary epithelial cells, *Cancer Res.,* 52, 6966, 1992.
5. Welsh, J., Chada, K., Dalal, S. S., Cheng, R., Ralph, D., and McClelland, M., Arbitrarily primed PCR fingerprinting of RNA, *Nucl. Acids Res.,* 20, 4965, 1992.
6. Reddy, P. M. S., An, G., Di, Y. P., Zhao, Y. H., and Wu, R., A palindromic primer-based mRNA differential display method to isolate vitamin A-responsive genes in airway epithelium: characterization of nucleolin gene expression. *Am. J. Respir. Cell Mol. Biol.,* 15, 398, 1996.
7. Lohmann, J., Schickle, H., and Bosch, T. C. G., REN display, a rapid and efficient method for nonradioactive differential display and mRNA isolation, *BioTechniques,* 18, 200, 1995.
8. Smith, N. R., Aldersley, M., Li, A., High, A. S., Moynihan, T. P., Markham, A. F., and Robinson, P. A., Automated differential display using a fluorescent labeled universal primer, *BioTechniques,* 23, 274, 1997.
9. An, G., Luo, G. Z., Veltri, R. W., and O'Hara, S. M., A sensitive, non-radioactive differential display method using chemiluminescent detection, *BioTechniques,* 20, 342, 1996.
10. Chomczynski, P. and Sacchi, N., Single-step of RNA isolation by acid guanidinium thiocyanate-phenol-chloroform extraction, *Anal. Biochem.,* 162, 156, 1987.
11. An, G., Cazares, L., Luo, G., Miller, M. C., Wright, G. L., Jr., and Veltri, R. W., Differential expression of full length and a truncated Her-2/neu oncogene receptor in prostate cancer assessed using relative quantitative RT-PCR, *Mol. Urol.,* 2, 305, 1999.

14 Nonradioactive Labeled Amplified Fragment Length Polymorphisms for Application in Forest Trees

Birgit Ziegenhagen, Reinhold Brettschneider, Vivian Kuhlenkamp, and Matthias Fladung

CONTENTS

14.1 INTRODUCTION

The usage of radioactivity in basic molecular techniques is regarded hazardous when handling the relevant assays. In addition, radioactivity is burdening the environment for long time spans. Since Bronstein and McGrath[1] described forthcoming chemiluminescent approaches, these alternative assays have become widely used. The DIG-oxigenenine labeling and detection system is one of the systems that operates by chemiluminescent autoradiography.[2]

This study tests the applicability of DIG-AFLP technology for usage in forest trees as an alternative to radioactively labeled AFLPs. The universal multilocus fingerprinting AFLP technology was first described by Vos et al.[3] Since then the method has become a widely used DNA marker technology in plant genetics, and the efficiency of the technology has been appreciated in terms of high multiplex ratio (e.g., Powell et al.[4]). Recently, the technology has been used for creating linkage maps in a variety of plant species (e.g., Castiglione et al.,[5] Debener and Mattiesch[6]). Further applications in other fields of genetics, such as population genetics or in phylogeny (review in Robinson and Harris[7]), are emerging.

The aim of this study is to provide a universal DIG-AFLP protocol that can be used in laboratories that are neither equipped with isotopes nor willing to work with isotopes and that do not have automated sequencing facilities. The protocol has been established on tree species, which is often a challenge per se as the species are characterized by large genome sizes and many of them provoke difficulties in DNA isolation from the beginning.[8]

0-8493-0719-8/02/$0.00+$1.50

For testing the DIG-AFLP protocol, four tree species/genera were selected, two angiosperms and two gymnosperms. Two different primer-enzyme combinations (PECs) (*Eco* RI/*Mse* I and *Pst* I/*Mse* I, hereafter referred to as *Eco/Mse* and *Pst/Mse*) were applied that are supposed to generate principally different multibanding patterns.[9] In addition to the basic question of applicability, other questions were raised: Is there a sufficient degree of polymorphism already obtained among half-sibs? For a special application in *Populus* spp., is it possible to distinguish among clones and in our special case by a single PEC?

14.2 MATERIAL AND METHODS

14.2.1 MATERIAL

Leaf or needle material was analyzed from individuals/clones of the following tree species/genera:

Aspen: Three clones with four to eight ramets each were analyzed:

1. 'Brauna11,' a *P. tremula* clone, the ortet located in Saxon, Germany;
2. 'Esch5,' a hybrid clone of *P. tremula* × *P. tremuloides*; the hybrid mother is the above-mentioned *P. tremula* individual 'Brauna11,' and the hybrid father is the *P. tremuloides* individual 'Tur 141' (Woodstock, NH); and
3. 'Esch 5' *rol*C transgenes, the same hybrid clones as above transformed with *rol*C.[10]

As can be seen from this design, the clones are characterized by half-sibling family relationships.

Oak: The oak material originates from the Arboretum of the Institute for Forest Genetics and Forest Tree Breeding, Grosshansdorf. AFLP analysis was conducted on

1. A half-sibling family with a *Quercus robur* L. individual as the mother tree and individuals as father trees belonging to both species, *Q. robur* and *Q. petraea* Matt. Liebl.; and
2. Different adult individuals from different oak species.

Conifers: Individuals of silver fir, *Abies alba* Mill., and Norway spruce, *Picea abies* (L.). Karst., were analyzed for DIG-labeled AFLPs.

14.2.2 DNA EXTRACTION

Leaf or needle material, 50 to 100 mg (fresh weight), was homogenized in Eppendorf tubes under liquid nitrogen conditions using a mixer mill (Type MM2000, Retsch GmbH and CoKG, Haan, Germany; also see Ziegenhagen et al.[11]). For the different tree species analyzed, different protocols were chosen for total DNA extraction. These protocols were known to fulfill the different requirements of the relevant forest tree species regarding their more or less recalcitrant properties during the isolation procedure.

Broadleaf species: For aspen (*Populus* spp.), the classical CTAB-based protocol by Doyle and Doyle[12] was used with only slight modifications. Dumolin et al.[13] adapted this classical CTAB procedure to oaks (*Quercus* spp.). This protocol was also used in our studies on oaks, including an additional and final treatment of the dissolved DNA with 0.5 μg RNaseA (Boehringer Mannheim) at 37°C for 30 min.

Conifer species: For DNA extraction from silver fir (*A. alba* Mill.) a protocol was used that is based on the principle of acetic lysis by Dellaporta.[14] Our protocol is a minipreparation that was adapted to needles of *A. alba* by Ziegenhagen et al.[11] Total DNA from Norway spruce, *Picea abies* (L.) Karst., was isolated by means of the DNeasy Plant Mini Kit (QIAGEN GmbH, Hilden, Germany).

14.2.3 DIG-Oxigenine Labeling and Detection of AFLPs

DIG-labeled AFLP technology was adapted to tree species using a combination of different protocols. The basic protocol is the radioactive method by the patent holder for AFLP.[3] Our approach is a modification of a "radioactive" protocol by Meksem et al.[15] that was optimized for DIG-oxigenine application in maize by Brettschneider.[16]

Two different PECs were used. The characterization of restriction endonucleases, adapter sequences, and primer sequences are summarized in Table 14.1.

Step 1: Digestion/Ligation

The following mix was composed for digestion of total DNA: 500 ng total DNA, 5 U of each of the two double-digesting restriction endonucleases, 5 μl 10× "One-Phor-All" restriction buffer (Amersham, Pharmacia Biotech, Freiburg, Germany), and double-distilled water added to 50 μl final volume. The solution was incubated at 37°C for 1 h.

The 10-μl ligation volume was composed as follows: 1 μl of *Eco-* or *Pst*-adapters (5 p*M* each), 1 μl of *Mse*-adapters (50 p*M*), 1.2 μl ATP (10 m*M*, Boehringer Mannheim), 2.5 U T_4-DNA-ligase (MBI, St. Leon Roth, Germany), and 1 μl 10× "One-Phor-All" buffer. Double-distilled water was added to 10 μl. All 10 μl of the ligation was added to the digestion cocktail for further incubation at 37°C for 3 h.

TABLE 14.1
Enzymes, Adapters, and Primers Used for Digestion, Ligation, and PCR Amplification

PEC	*Eco/Mse*	*Pst/Mse*
Restriction endonuclease	*Eco* RI / *Mse* I	*Pst* I / *Mse* I
Adapter sequences	*Eco*-adapter[a]	*Pst*-adapter[a]
	5′ CTCGTAGACTGCGTACC 3′	5′ CTCGTAGACTGCGTACATGCA 3′
	3′ CTGACGCATGGTTAA 5′	3′ CATCTGACGCATGT 5′
	Mse-adapter[a]	*Mse*-adapter[a]
	5′ GACGATGAGTCCTGAG 3′	5′ GACGATGAGTCCTGAG 3′
	3′ TACTCAGGACTCAT 5′	3′ TACTCAGGACTCAT 5′
Primer sequences		
Preamplification	*Eco*-site[a]	*Pst*-site[a]
	5′ AGACTGCGTACCAATTC **A** 3′	5′ GACTGCGTACATGCAG **C** 3′
	Mse-site[a]	*Mse*-site[a]
	5′ GACGATGAGTCCTGAGTA **A** 3′	5′ GACGATGAGTCCTGAGTA **A** 3′
Final amplification	*Eco*-site[a,b]	*Pst*-site[a,b]
	5′ GACTGCGTACCAATTC **AAC** 3′	5′ GACTGCGTACATGCAG **CCA** 3′
	Mse-site[a]	*Mse*-site[a]
	5′ GATGAGTCCTGAGTAA **ACT** 3′	5′ GATGAGTCCTGAGTAA **ATA** 3′

[a] Sequence information by Keygene n.v. (Wageningen, the Netherlands; also see Vos et al.[3]). All primers except *Eco-* and *Pst*-site primers were synthesized by Gibco BRL (Life Technologies, Eggenstein, Germany).

[b] In the final amplification the *Eco*-site and *Pst*-site primers are DIG-labeled. The labeled primers were purchased from NAPS GmbH (Göttingen, Germany).

Note: Restriction endonucleases were purchased from either Gibco BRL or from New England Biolabs (Schwalbach/Taunus, Germany).

Step 2: PCR Pre-Amplification

A 50-μl PCR cocktail was prepared consisting of 5 μl solution obtained from step 1, 3 μl of each of the relevant primers for preamplification (concentration of the primers was 50 ng/μl), 5 μl of 2.5 m*M* dNTP, 1.5 μl $MgCl_2$ (50 m*M*), 1 U Taq polymerase, and 5 μl of 10× PCR buffer ($MgCl_2$, Taq polymerase, and 10× PCR buffer purchased from Eurogentec, Ougree, Belgium). Double-distilled water was added to 50 μl end volume.

PCR was run in a Thermal Cycler 1 (THC1, Perkin Elmer, Überlingen, Germany) with the following profile: 94°C for 5 min, followed by 20 cycles of 94°C for 30 s, 60°C for 30 s and 72°C for 1 min, last strand elongation (72°C) with an additional 10 min. Before final amplification, the preamplification products may be visualized and thus checked in agarose gels. The 20 μl of the PCR products are supposed to generate a faint smear in the relevant lanes of a 1.2% 0.5× TBE agarose gel.

Step 3: PCR Final Amplification

For final amplification, 20 μl of PCR cocktail was composed as follows: 5 μl of 1:5 diluted preamplification PCR product, 1 μl labeled *Eco*-site or *Pst*-site primer (10 ng/μl), 1.2 μl nonlabeled *Mse*-site primer (50 ng/μl), 2 μl dNTP (2.5 m*M*), 0.6 μl $MgCl_2$ (50 m*M*), 0.5 U Taq polymerase, 2 μl 10× PCR buffer. $MgCl_2$, Taq polymerase, and 10× PCR buffer were purchased from Eurogentec (Ougree, Belgium). Double-distilled water was added to 20 μl end volume.

PCR was run in the same PCR machine as mentioned above. The PCR profile followed a touchdown cycling with 94°C for 5 min, one cycle (94°C for 30 s, 65°C for 30 s, 72°C for 1 min), three cycles (94°C for 30 s, 64°C for 30 s, 72°C for 1 min), three cycles (94°C for 30 s, 62°C for 30 s, 72°C for 1 min), three cycles (94°C for 30 s, 60°C for 30 s, 72°C for 1 min), three cycles (94°C for 30 s, 58°C for 30 s, 72°C for 1 min), 24 cycles (94°C for 30 s, 56°C for 30 s, 72°C for 1 min), and a final elongation step at 72°C for 10 min.

Step 4: Gel Electrophoresis of Final PCR Products

Gel electrophoresis was performed in a 6% denaturating polyacrylamide gel (Rotiphor 40, 38:2 acrylamide:bisacrylamide, Roth, Karlsruhe, Germany), using a sequencing gel apparatus S2 by Gibco BRL. The two glass plates of the vertical chamber were pretreated, one with bind-silane, the other with repell-silane. The dimensions of the gels were 41.5 cm × 33.5 cm × 0.4 mm. In detail the gel solution was prepared as follows: 36 g urea (Sigma, Deisenhofen, Germany) was dissolved in 7.5 ml 10× TBE and 11.25 ml of the above-described polyacrylamide solution. Double-distilled water was added to a final volume of 75 ml. After complete solution of the ingredients, 375 μl APS (10%, Sigma) and 37.5 μl TEMED (Sigma) were added. Before adding APS and TEMED, gas bubbles must be removed from the gel.

Preparation of Samples

First, 20 μl loading buffer (98% formamide, 10 m*M* EDTA, 0.01% w/v bromophenol blue, 0.01% xylene cyanol) was added to the 20 μl PCR cocktail and everything well mixed. Then, 7 μl of this solution was subjected to denaturation at 95°C for 5 min, immediately cooled on ice, and 5 μl loaded onto the polymerized gel. As a size standard, 1 μl of a solution was used, which consisted of DIG-labeled size standard V (Boehringer), loading buffer and double-distilled water (1:10:10). The gels were prerun in 1× TBE buffer adjusted to pH 8.3 at 2000 V for 20 min. Then, 150 ml NaOAc (3 *M*) was poured into the lower buffer tank to start the main run, which lasts 2.5 to 3 h.

Step 5: Direct Blotting and Ultraviolet Cross-Linking

After electrophoresis, the repell-silane-treated glass plate was removed. From the gel still sticking to the bind-silane plate, the DNA was directly transferred to a neutral nylone membrane (Biodyne A, Gibco BRL). The membrane should be wetted by 1× TBE buffer. The transfer was allowed to stand about 20 h (overnight), with the membrane covered with five sheets of 3MM Whatman paper and a glass plate. UV cross-linking was performed in two steps in a Stratagene Crosslinker (UV Stratalinker 2400, Stratagene, Amsterdam, the Netherlands), first at 2500 μJ × 100; then the membrane was washed two times in 2× SSC (solution consisting of 3 *M* NaCl and 0.3 *M* sodium citrate at pH 7.0) and cross-linked again at 1200 μJ × 100. If a Stratagene Crosslinker is not available, a UV illuminator may be used as an alternative.

Step 6: Detection of DIG-Labeled DNA Banding Patterns

The DIG detection procedure followed the protocol according to the 1993 manual of the manufacturer (Boehringer).[2] To save chemicals and reduce solution volumes most of the procedure was carried out in plastic seals. The procedures, solutions, and duration of the single steps are given in detail in Table 14.2.

For chemiluminescence detection, the membrane was exposed to X-ray film (Hyperfilm MP by Amersham, Freiburg, Germany). For this purpose it was enclosed and kept in an X-ray cassette for 15 min. Development, fixation, and washing of the film was performed manually in tanks with chemicals used for X-ray developing and fixation. Developing was done for 5 min, fixation (5% acetic acid) for 1 min, and a final washing for 10 min (developing and fixation solutions were purchased by Adefo Chemie, Nürnberg, Germany).

Step 7: Documentation and Image Analysis

In addition to storing the data on X-ray films, the films were scanned with a laser densitometer and the images analyzed by the software ImagequaNT (both densitometer and software were purchased from Molecular Dynamics, Krefeld, Germany).

TABLE 14.2
Stepwise Procedure of DIG Detection

Procedure	Solutions	Volume, ml	Duration, min
Washing 1×	Washing buffer (Buffer I + 0.3% Tween)	200	5
Incubation	Buffer II (Buffer I + BS 1%)	180	20
Incubation	Buffer II + DIG antibody-conjugate (1:10000)	70	30
Washing 3×	Washing buffer, see above	200	3 × 15
Incubation	Buffer III	200	5
Chemiluminescent incubation	200 μl CSPD dissolved in 20 ml Buffer III	20	5
Incubation in the dark	Wet membrane in plastic seal at 37°C		15

BS = blocking reagent; Buffer I, II, III = solutions denoted according to the manual (for components, see Boehringer, 1993); CSPD = chemiluminescent substrate by Boehringer Mannheim.

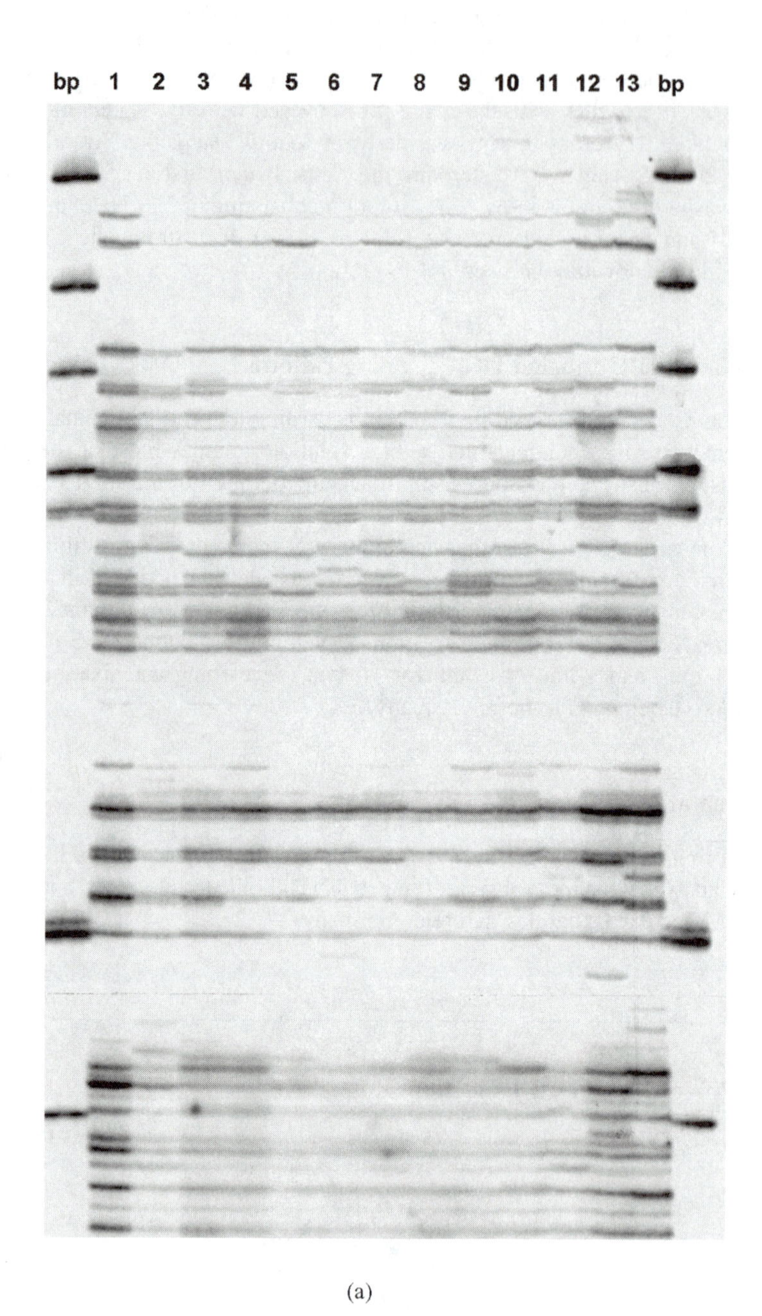

(a)

FIGURE 14.1 (a) DIG-AFLP analysis of an oak half-sibling family. Section of a gel demonstrating the banding patterns obtained by the PEC *Eco/Mse*. Lane 1: adult mother tree; lanes 2 to 13: half-siblings of the mother tree. bp = size standard V (Boehringer Mannheim); fragment sizes from top to bottom of this section: 267, 234, 213, 192, 184, 124, 123, 104 bp.

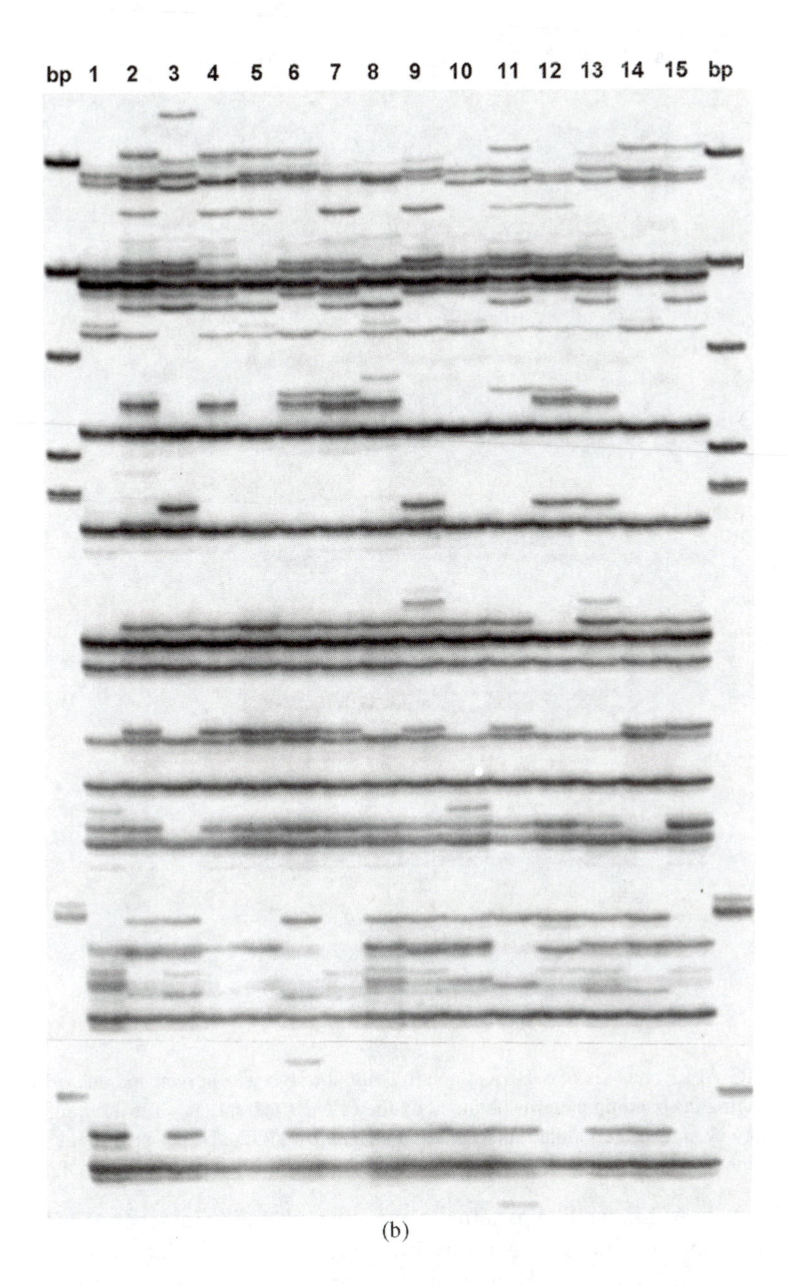

(b)

FIGURE 14.1 (*Continued*) (b) DIG-AFLP analysis of an oak half-sibling family. Section of a gel demonstrating the banding patterns obtained by the PEC *Pst/Mse*. Lane 1: anonymous oak adult tree; lane 2: mother tree; lanes 3 to 15: half-siblings of the mother tree. bp = size standard V (Boehringer Mannheim); fragment sizes from top to bottom of this section: 267, 234, 213, 192, 184, 124, 123, 104 bp.

14.3 RESULTS AND DISCUSSION

This study tests the applicability of nonradioactive DIG-labeled and -detected AFLP banding patterns in different tree species/genera, two angiosperms and two gymnosperms. Most AFLP protocols used previously operate on radioactively labeled primers, as suggested by the first describers.[3] The aim of this study is to provide a protocol that can be used universally in laboratories without isotopes or automated sequencing equipment, the latter being increasingly used for fragment

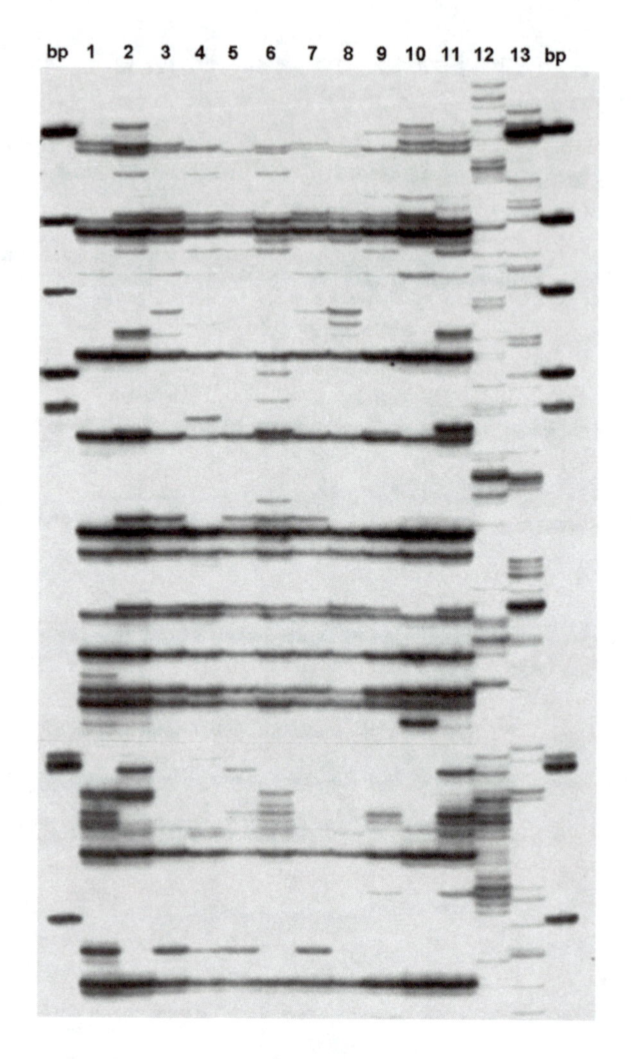

FIGURE 14.2 DIG-AFLP analysis of oaks, and one individual of Norway spruce and one of silver fir. Section of a gel demonstrating the banding patterns obtained by the PEC *Pst/Mse*. Lanes 1 to 11: adult oak individuals; lane 12: adult Norway spruce individual; lane 13: adult silver fir individual. bp = size standard V (Boehringer Mannheim); fragment sizes from top to bottom of this section: 267, 234, 213, 192, 184, 124, 123, 104 bp.

analysis as well as in AFLP banding patterns. A few other chemiluminescent AFLP protocols have already been established in different organisms, including herbaceous plant species (e.g., Lin et al.,[17] Vrieling et al.[18]). For forest trees a chemiluminescent approach has been reported by Greef et al.[19] They analyzed DIG-detected AFLPs citing a protocol that operates on the basis of the industrial kit AFLP Analysis System I (Gibco BRL). In contrast, our protocol is a combination of three basic experimental approaches.[3,15,16] It was designed as a flexible protocol, independent of the usage of kits.

In this study, tree species are tested whose genome sizes may be as large as 100 pg (2C-values) as, for example, is estimated for *Picea abies* (L.) Karst. (e.g., Murray[20]). This conifer species even exceed the so-called large-size genomes of *Alstroemeria* spp., for which a particular optimization of radioactively labeled AFLP fingerprinting has been elaborated.[21] As a first try, we decided to test two different PECs, *Eco/Mse* and *Pst/Mse* that, because of their different sensitivities to methylation, were suggested to sample different regions of the genome.[9] Furthermore, as an example of an application in the field of genetics, the potential of DIG-labeled AFLPs was evaluated for discrimination of individuals/clones in *Populus* spp.

Both PECs are characterized by a three-nucleotide overhang of the *Pst*-site and *Eco*-site primers used in the final PCR amplification. As demonstrated in Figures 14.1 through 14.3, they generated a scorable number of distinct bands. Reproducibility was tested by performing the procedures twice on the same DNA sample of each tree individual or ramet. The patterns were the same each time,

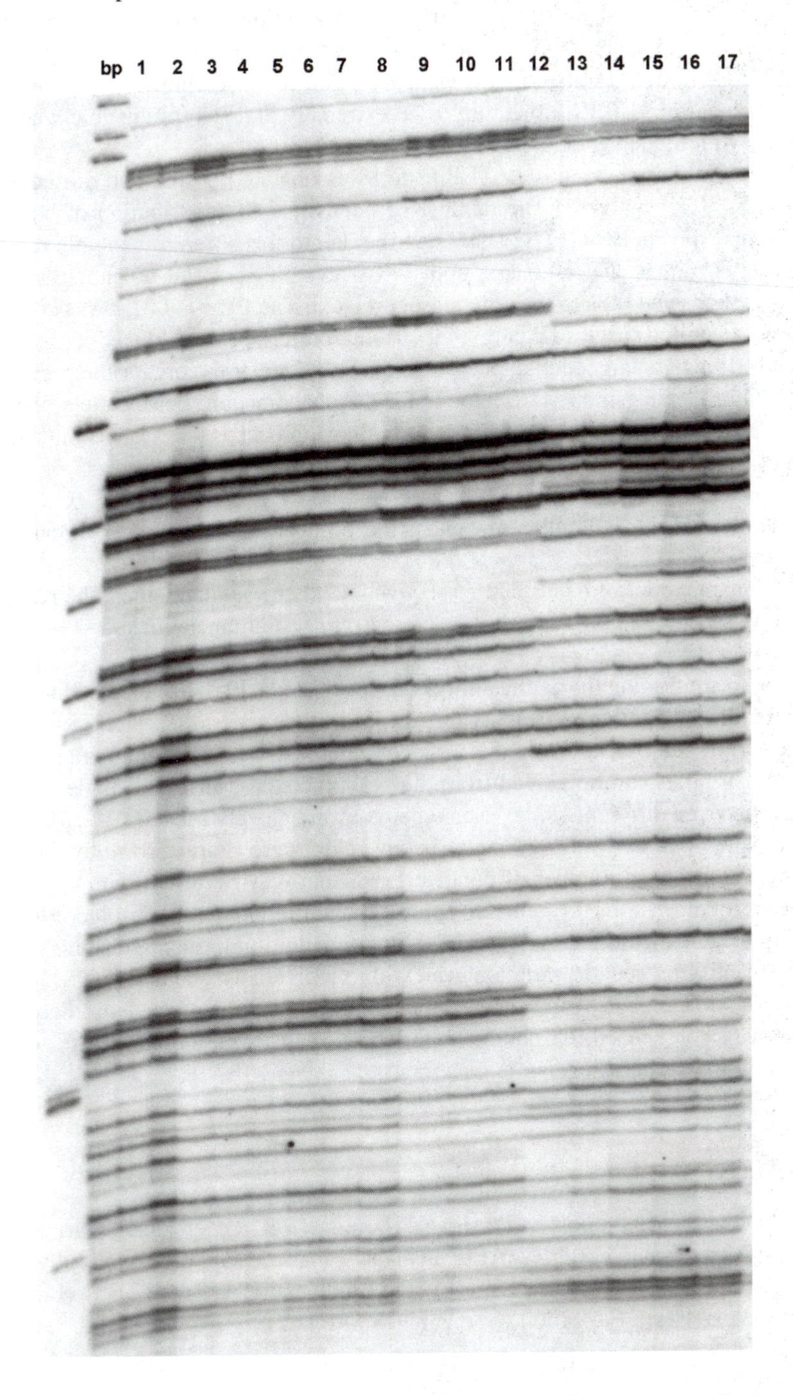

FIGURE 14.3 DIG-AFLP analysis of aspen clones. Section of a gel demonstrating the banding patterns obtained by the PEC *Eco*/*Mse*. Lanes 1 to 8: ramets of the clone 'Esch5 *rol*C transgenes'; lanes 9 to 12: ramets of the clone 'Esch5 wildtype'; lanes 13 to 17: ramets of clone 'Brauna11'. bp = size standard V (Boehringer Mannheim); fragment sizes from top to bottom of this section: 504, 458, 434, 267, 234, 213, 192, 184, 124, 123, 104 bp.

also suggesting that the respective isolation protocols provided DNA of sufficient quality. The latter is important to stress as forest tree species are often known to be recalcitrant species with regard to DNA extraction.[8] Reproducibility was also demonstrated when DNA extracts were used that were independently isolated from the same individual. In a case study with an oak individual, banding patterns were the same (data not shown). In addition, the identical banding patterns of ramets of the same clone may be regarded as indirect and further proof of reproducibility (see Figure 14.3). In sum, our results confirm what has been found earlier in a network of European laboratories. Thereafter, AFLPs turned out to be extremely highly reproducible when working on the same DNA with the same detection system.[22]

Figure 1a and b demonstrate the AFLP patterns as obtained by the two different PECs when the oak half-sibs were analyzed. The number of bands and the distribution of bands within the lanes are different for the two PECs. This might be due to the different genomic regions sampled by the different enzymes.[9] Nevertheless, both PECs reveal fingerprint potential as all individuals can be distinguished even though they are closely related. The PEC *Pst/Mse* was compared among oak and the two conifer species (Figure 14.2). Both conifer species showed a greater number of "denser" bands. It may be argued that their large genome sizes or the organism-specific sampling of the enzymes within their genomes are reasons for the phenomenon. For example, the genome size of oak is one third that of the two conifer species.[23] Two approaches may be recommended for large-sized conifer genomes:

1. For diminishing the number of bands, the number of nucleotide overhangs may be increased.[21]
2. If the same number of bands and respective loci are to be maintained, the resolution can be increased by longer gels (Markussen, personal communciation).

In addition to confirming the applicability of DIG-labeled AFLFs in selected forest tree species, this study provides an example for applying AFLPs in tree genetic research. Figure 14.3 reveals the potential of this multilocus method as a genetic marker for discriminating among individuals and clones of aspen. As can be judged from the gel, in this case just one single PEC is sufficient to distinguish between different clones even when they are closely related half-sibs. The reliability of the marker is high, as all ramets of the respective clones are characterized by the same banding patterns. Imagine the great number of putative PECs that may be tested with regard to sampling different regions of the genome as well as to approaching different numbers of loci. Here, a powerful, universally applicable methodology is described that in addition to distinguishing individuals may be useful for identifying genetic variation even within individuals, such as somaclonal, transgenic, or somatic variation. Our study is just one example of an application where extremely high variation of a genetic marker is needed. Of course, DIG-labeled AFLPs may be useful for other, common AFLP studies in trees, such as genome mapping.

ACKNOWLEDGMENTS

We greatly appreciate the support of Dr. Frank Hartung (IPK, Gatersleben, Germany), who gave helpful advice on experimental steps in AFLP analysis. Financial support is gratefully acknowledged from EU-Programme Biotechnology, project "Tree Biodiversity" CT 96 0703 (to B.Z.) and Federal Ministry for Research and Education, Germany (to M.F.).

ABBREVIATIONS

AFLP	amplified fragment length polymorphism
APS	ammonium persulfate

CTAB	hexadecyltrimethylammonium bromide
DIG	DIG-oxigenine chemiluminescent detection system according to Boehringer Mannheim
dNTP	2′-deoxynucleoside (A,T,C,G) 5′-triphosphates
EDTA	ethylenediaminetetraacetic acid
$MgCl_2$	magnesium chloride
NaCl	sodium chloride
NaOAc	sodium acetate solution
PEC	primer–enzyme combination
TBE	Tris-borate-EDTA
TEMED	*N,N,N′,N′*,-tetramethylethylene diamine

REFERENCES

1. Bronstein, J. and McGrath, P., Chemiluminescence lights up, *Nature,* 338, 599–600, 1989.
2. Boehringer Mannheim, The DIG System User's Guide for Filter Hybridization, three editions, 1989, 1993, 1995.
3. Vos, P., Hogers, R., Bleeker, M., Reijans, M., Van De Lee, T., Mornes, M., Frijters, A., Pot, J., Peleman, J., Kuiper, M., and Zabeau, M., AFLP: a new technique for DNA fingerprinting, *Nucl. Acids Res.,* 23, 4407–4414, 1995.
4. Powell, W., Morgante, M., Andre, C., Hanafey, M., Vogel, J., Tingey, S., and Rafalsky, A., The comparison of RFLP, RAPD, AFLP and SSR (microsatellite) markers for germplasm analysis, *Mol. Breeding,* 2, 225–238, 1996.
5. Castiglione, P., Ajmone-Marsan, P., Van Wijk, R., and Motto, M., AFLP markers in a molecular linkage map of maize: co-dominant scoring and linkage group distribution, *Theor. Appl. Genet.,* 99, 425–431, 1999.
6. Debener, T. and Mattiesch, L., Construction of a genetic linkage map for roses using RAPD and AFLP markers, *Theor. Appl. Genet.,* 99, 891–899, 1999.
7. Robinson, J. P. and Harris, S. A., Amplified fragment length polymorphisms and microsatellites: a phylogenetic perspective, in Which DNA Marker for Which Purpose? Final Compendium of the Research Project "Development, Optimisation and Validation of Molecular Tools for Assessment of Biodiversity in Forest Trees" in the European Union DGXII Biotechnology FW IV Research Programme "Molecular Tools for Biodiversity," available at http://webdoc.sub.qwdg.de, 1999.
8. Csaikl, U. M., Bastian, H., Brettschneider, R., Gauch, S., Meier, A., Schauerte, M., Scholz, F., Sperisen, C., Vornam, B., and Ziegenhagen, B., Comparative analysis of different DNA extraction protocols: a fast, universal maxi-preparation of high quality plant DNA for genetic evaluation and phylogenetic studies, *Plant Mol. Biol. Rep.,* 16, 69–86, 1998.
9. Young, W. P., Schupp, J. M., and Keim, P., DNA methylation and AFLP marker distribution in the soybean genome, *Theor. Appl. Genet.,* 99, 785–790, 1999.
10. Fladung, M., Kumar, S., and Ahuja, M. R., Genetic transformation of *Populus* genotypes with different chimeric constructs: transformation efficiency and molecular analysis, *Transgenic Res.,* 6, 111–121, 1997.
11. Ziegenhagen, B., Guillemaut, P., and Scholz, F., A procedure for mini-preparation of genomic DNA from needle tissue of silver fir (*Abies alba* Mill.), *Plant Mol. Biol. Rep.,* 24(2), 117–121, 1993.
12. Doyle, J. J. and Doyle, J. L., Isolation of plant DNA from fresh tissue, *Focus,* 12, 13–15, 1990.
13. Dumolin, S., Demesure, B., and Petit, R. J., Inheritance of chloroplast and mitochondrial genomes in pedunculate oak investigated with an efficient PCR method, *Theor. Appl. Genet.,* 91, 1253–1256, 1995.
14. Dellaporta, S. L., Wood, J., and Hicks, J. B., A plant DNA minipreparation: version II, *Plant Mol. Bio. Rep.,* 1, 19–21, 1983.
15. Meksem, K., Leister, D., Peleman, J., Zabeau, M., Salamini, F., and Gebhardt, C., A high resolution map on potato chromosome V based on RFLP and AFLP markers, *Mol. Gen. Genet.,* 249, 74–81, 1995.
16. Brettschneider, R., Non-radioactive AFLP method, based on Digoxigenine. *Mol. Screening News,* 9, 13–14, 1996.

17. Lin, J.-J., Ambrose, M., and Kuo, J., Chemiluminescent detection of AFLP™ fingerprints, *Focus,* 19, 36–38, 1997.
18. Vrieling, K., Peters, J., and Sandbrink, H., Amplified fragment length polymorphisms (AFLPs) detected with non-radioactive digoxigenine labelled primers in three plant species, *Plant Mol. Biol. Rep.,* 15, 255–262, 1997.
19. De Greef, B., Triest, L., De Cuiper, B., and Van Slyckens, J., Assessment of intraspecific variation in half-sibs of *Quercus petraea* (Matt.) Liebl. 'plus' trees, *Heredity,* 81, 284–290, 1998.
20. Murray, B. G., Nuclear DNA amounts in gymnosperms, *Ann. Bot.,* 82, 3–15, 1998.
21. Han, T. H., Van Eck, H. J., De Jeu, M. J., and Jacobsen, E., Optimization of AFLP fingerprinting of organisms with a large-sized genome: a study on *Alstroemeria* spp., *Theor. Appl. Genet.,* 98, 465–471, 1999.
22. Jones, C. J., Edwards, K. J., Castiglione, S., Winfield, M. O., Sala, F., Vander Wiel, C., Vosman, B. L., Matthes, M., Daly, A., Brettschneider, R., Maestri, E., Marmiroli, N., Rueda, J., Vazquez, A., and Karp, A., Reproducibility testing of AFLPs by a network of European Laboratories, in *Molecular Tools for Screening Biodiversity,* Karp, A., Isaac, P. G., and Ingram, D. S., Eds., Chapman & Hall, London, 1998, 191–192.
23. Bennett, M. D., Cox, A. V., and Leitel, I., Angiosperm DNA C-Values Database, available at www.rbgkew.org.uk/cval/database1.html, 1998.

15 Detection of Nucleic Acids Using Chemiluminescence: From Northerns to Southerns and Beyond

John C. Jackson, Beth A. Strachan, David von Schack, and Lee A. Sylvers

CONTENTS

15.1 INTRODUCTION

Many non-isotopic methods have been developed for nucleic acid detection, and recently some of these methods have been improved to the point where their sensitivity equals that of their isotopic counterparts. However, the robustness of these sensitive methods, usually based on alkaline phosphatase detection with 1,2-dioxetane substrates, tends to suffer from a high degree of signal-to-noise variability. Aside from this lack of robustness of the current methods, a further drawback arises from the inordinate amount of processing time involved in the procedure and the length of exposure time needed to obtain the desired sensitivity. These disadvantages undoubtedly contribute to the lack of wide acceptance of non-isotopic nucleic acid detection methods. We recently developed a complete system for the chemiluminescent detection of nucleic acids in Northern and Southern blot applications. This system combines a novel enhanced luminol substrate for horseradish peroxidase (HRP) with optimized hybridization and blocking steps that ensure consistent results with sensitivity equivalent to ^{32}P. This robust system also has the advantages of greatly reduced processing and film exposure time, and the high level of light output makes the system ideal for collecting data using cooled charge-coupled device (CCD) imaging systems. Posthybridization processing time has been reduced from the standard 2.5 to 1 h. Film exposure times range from 0.5 to 10 min with the substrate emitting light with relatively constant intensity over a 6-h period, thus allowing for multiple exposures.

0-8493-0719-8/02/$0.00+$1.50

15.2 MATERIALS AND METHODS

15.2.1 Equipment

UV Stratalinker from Stratagene (San Diego, CA), Nucleovision™ Chemiluminescent Workstation (NucleoTech Corp., Hayward, CA), and ChemiImager™ 4000 (Alpha Innotech, San Leandro, CA).

15.2.2 Reagents

Film-developing chemicals were purchased from Sigma Chemical (St. Louis, MO). Biotinylated and ^{32}P-labeled cRNA probes for the *Saccharomyces cerevisiae TCM1* and *INO2* genes were synthesized using either the North2South™ *in vitro* Transcription Kit from Pierce Chemical Company (Rockford, IL) or the MAXIscript™ system from Ambion (Austin, TX) and [α-^{32}P] UTP from Amersham Life Sciences (Arlington Heights, IL) from linearized plasmid pAB309Δ (*TCM1*) or pGEM-INO2 (*INO2*). DIG-labeled cRNA probes were synthesized using the DIG RNA Labeling System from Boehringer Mannheim (Indianapolis, IN). Biotinylated DNA probes were synthesized using the North2South™ Random Prime Labeling Kit from Pierce Chemical Company and a restriction fragment containing the gene encoding GFP protein. DIG-labeled probes were detected using the DIG High Prime Labeling and Detection Kit with CSPD® from Boehringer Mannheim. All other equipment and products were from Pierce Chemical Company.

15.2.3 Northern Blots

Total yeast RNA was isolated, fractionated on a formaldehyde/agarose gel, and transferred to Biodyne® B Nylon Membrane (Pall Corporation, East Hills, NY) as described previously.[1] DIG-labeled cRNA probes were hybridized and detected according to the manufacturer's specifications (Boehringer Mannheim). Radiolabeled cRNA probes (10^8 cpm/μg) were hybridized and detected as previously described.[1]

Biotinylated cRNA probes or DNA probes were hybridized and detected using the North2South Chemiluminescent Hybridization and Detection System as follows. Following capillary transfer, nucleic acids were UV-cross-linked to the membrane. The membranes were then prehybridized in hybridization buffer for at least 30 min at 65°C. Following prehybridization, cRNA probe was added to the hybridization buffer to a final concentration of 5 ng/ml and the membrane hybridized overnight at 65°C. The following day, the membranes were washed 3×20 min at 65°C in Stringency Wash Buffer. The membranes were then transferred to a plastic tray and blocked for 15 min in Blocking Buffer. Streptavidin–HRP conjugate was then added to the Blocking Buffer to a final concentration of 33 ng/ml and the membrane incubated for 15 min at room temperature with gentle agitation. The membrane was then washed 4×5 min each in Wash Buffer at room temperature with gentle agitation. Following washing, the membrane was incubated for 5 min in Substrate Equilibration buffer at room temperature. The working solution of North2South Chemiluminescent Substrate was prepared according to the manufacturer's recommendations and added to the membrane for 5 min. Membranes were exposed to film for various times as needed.

Multiple Tissue Northern blots (MTN®) and Multiple Tissue Expression Arrays (MTE™) were from CLONTECH Laboratories (Palo Alto, CA) and were hybridized and detected using the North2South Chemiluminescent Hybridization and Detection System (as described above). Biotinylated DNA probes for human β-actin and tyrosine aminotransferase (TAT) used in the MTN® and MTE™ experiment were synthesized using the Biotin Random Prime Labeling Kit (as described above).

15.2.4 Southern Blots

Genomic DNA from *S. cerevisiae* was isolated as described previously.[2] Genomic DNA was digested overnight with *Hind* III, *Eco*RI, or *Sal*I and Southern blots were carried out as previously described.[3] Biotinylated cRNA probes for the *INO2* gene were hybridized and detected using the North2South

Chemiluminescent Hybridization and Detection System (as described above except hybridizations and washes were carried out at 55°C and the probe was added to a final concentration of 30 ng/ml).

15.2.5 Plaque Lifts

Plaque lifts were carried out as described previously.[3] Biotinylated DNA probes for the gene encoding GFP were hybridized and detected using the North2South Chemiluminescent Hybridization and Detection System as described for Southern blots.

15.3 RESULTS AND DISCUSSION

The North2South Chemiluminescent Hybridization and Detection System was developed to address the current problems with chemiluminescent detection of nucleic acids. Although other commercially available systems now afford sensitivity equal to ^{32}P-labeled probes, they suffer from long processing time and typically display problematic background, most often manifested in experiment-to-experiment variability of signal-to-noise ratios. The North2South Chemiluminescent Hybridization and Detection System offers short process time, sensitivity equal to ^{32}P, and extremely low background with minimal to no variability from experiment to experiment.

To assess the sensitivity of various detection protocols, we compared Northern blots of yeast total RNA using the North2South Chemiluminescent Hybridization and Detection System, the DIG Labeling and Detection System, and radioactive labeling using ^{32}P. Blots of a dilution series of total RNA starting from 5 μg and serially diluting to 0.156 μg were hybridized with labeled *TCM1* probe, detected according to the various protocols, and exposed to film for various times. As depicted in Figure 15.1, the North2South Chemiluminescent Hybridization and Detection System offers greater sensitivity than the DIG-labeled probe and the ^{32}P-labeled probe in a fraction of the time.

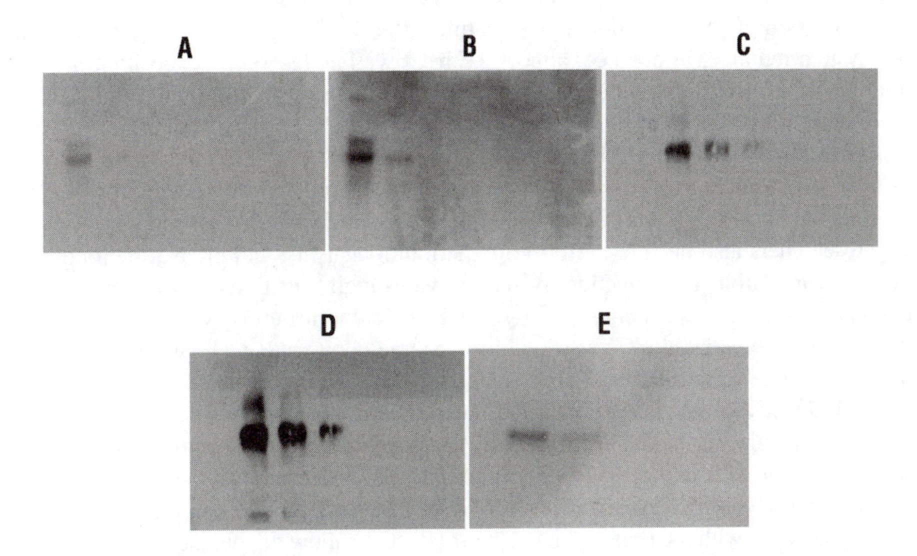

FIGURE 15.1 Northern blot comparison. A Northern blot of serial dilutions of total yeast RNA (5, 2.5, 1.25, 0.625, 0.313, and 0.156 μg) was performed as described to measure *TCM1* gene expression. In (A) and (B), a DIG-labeled *TCM1* cRNA probe was hybridized and detected using the DIG-labeling and detection system, in (C) and (D) a biotinylated *TCM1* cRNA probe was hybridized and detected using the North2South Chemiluminescent Hybridization and Detection System and in (E) a ^{32}P-labeled *TCM1* cRNA probe was hybridized and detected. Film was exposed for (A) 5 min, (B) 25 min, (C) 5 min, (D) 25 min, and (E) 72 h without intensifying screen. (From Jackson, J. C. et al., *Bioluminescence and Chemiluminescence Perspectives for the 21st Century,* John Wiley & Sons, 1999. With permission.)

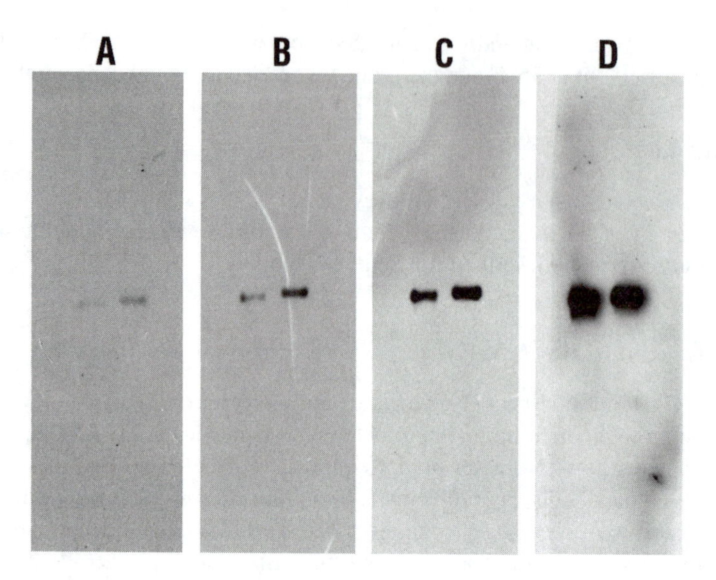

FIGURE 15.2 Northern blot demonstrating low background of the North2South Chemiluminescent Hybridization and Detection System. A Northern blot using 5 μg of yeast total RNA from different RNA preparations was performed as described. A biotinylated *TCM1* cRNA probe was then hybridized and detected as described previously. (A) 5-s film exposure, (B) 1-min film exposure, (C) 5-min film exposure, and (D) overnight film exposure. (From Jackson, J. C. et al., *Bioluminescence and Chemiluminescence Perspectives for the 21st Century,* John Wiley & Sons, 1999. With permission.)

In addition to a high level of sensitivity, the post-stringency wash process is relatively short using the North2South Chemiluminescent Hybridization and Detection System. The optimized protocol requires only 1 h of additional process time after the stringency washes, whereas the DIG detection system requires, in practice, almost 2.5 h, as well as far longer exposure times to film.

Although the North2South Chemiluminescent Substrate has an extremely high light output, the system also offers extremely low background. In Figure 15.2, a Northern blot of yeast total RNA was probed with a biotinylated *TCM1* cRNA probe, exposed to film initially for 5 s (A), 1 min (B), 5 min (C), and finally overnight (D). The strong signal and lack of background demonstrates an additional benefit of the North2South Chemiluminescent Hybridization and Detection System.

Many researchers also need to perform Southern blot analysis, and the North2South Chemiluminescent Hybridization and Detection System provides high sensitivity and extremely low background for this type of experiment. In Figure 15.3, a Southern blot of yeast genomic DNA was hybridized with a biotinylated cRNA probe for the single-copy yeast *INO2* gene. As demonstrated in Figure 15.3, the North2South Chemiluminescent Hybridization and Detection System also provides high sensitivity and low background for Southern blot applications.

Being able to carry out plaque lifts with low background often represents a daunting challenge for researchers even with using ^{32}P labeling and detection strategies. We tested the North2South Chemiluminescent Hybridization and Detection System for its usefulness in plaque lift procedures. Phage were plated out with bacteria and incubated at 37°C to allow the plaques to form. Phage DNA was then transferred to nylon membrane, probed with a biotinylated DNA probe, and detected using the North2South Chemiluminescent Hybridization and Detection System. As depicted in Figure 15.4, the low to nonexistent background afforded by the North2South Chemiluminescent Hybridization and Detection System allows for distinct plaques to be picked out even in areas of high signal density.

Recent advances in imaging technology have led to the widespread use of cooled CCD cameras for the collection of data from chemiluminescent blots. By reducing the camera operating temperature, these systems offer better image quality, increased sensitivity and signal-to-noise ratios, and a larger dynamic range than film. The enhanced luminol substrate in the North2South

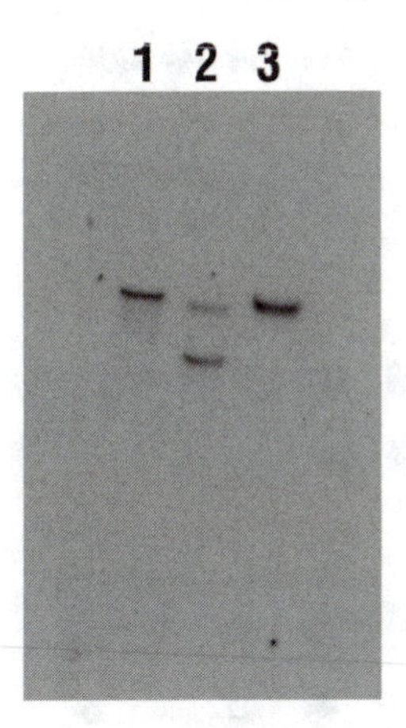

FIGURE 15.3 Southern blot detecting a single-copy gene (*INO2*). A Southern blot using 5 μg of yeast genomic DNA was performed as described. A biotinylated *INO2* cRNA probe was hybridized and detected as described previously. Lane 1: *Hin*d III digest; lane 2: *Eco*R I digest; and lane 3: *Sal* I digest. Film was exposed for 1 min. *Note*: The two bands shown in Lane 2 are due to the presence of an *Eco*R I site within the *INO2* gene. (From Jackson, J. C. et al., *Bioluminescence and Chemiluminescence Perspectives for the 21st Century*, John Wiley & Sons, 1999. With permission.)

FIGURE 15.4 Plaque lifts. Plaque lifts were performed as described previously. A biotinylated DNA probe for the gene encoding GFP was hybridized and detected using the North2South Chemiluminescent Hybridization and Detection System. Film was exposed for 1 min. (From Jackson, J. C. et al., *Bioluminescence and Chemiluminescence Perspectives for the 21st Century*, John Wiley & Sons, 1999. With permission.)

Chemiluminescent Hybridization and Detection System is ideally suited for cooled CCD camera systems because of its immediate and high-intensity light output. As demonstrated in Figure 15.5, the combination of low background and intense light output allows detection of extremely low levels of biotinylated oligonucleotide, down to 40 fg with a 30-min exposure using a cooled CCD camera system (assuming an average molecular weight of 320 for a deoxynucleotide, this corresponds

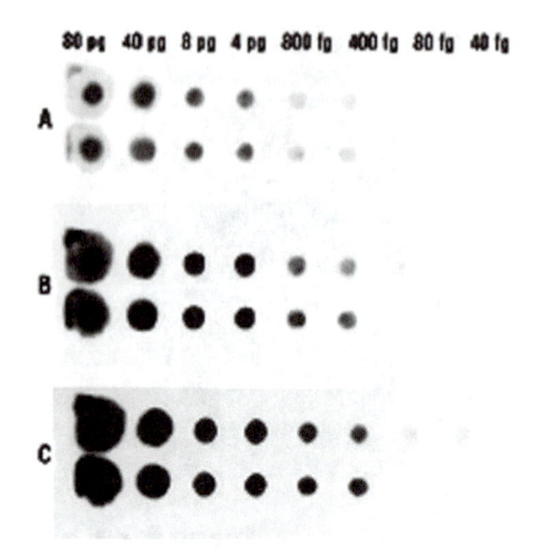

FIGURE 15.5 Detection of biotinylated 24-mer oligonucleotide using the North2South Chemiluminescent Hybridization and Detection System and a cooled CCD camera. Indicated dilutions of biotinylated oligo were spotted onto nylon membrane, detected using the North2South System and imaged for (A) 3 min, (B) 10 min, and (C) 20 min using a cooled CCD camera system provided by Nucleotech.

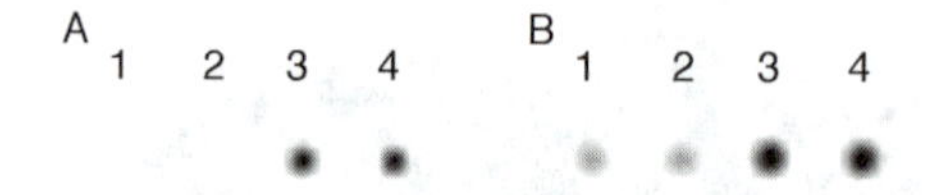

FIGURE 15.6 Detection of GFPuv mRNA under inducing and noninducing conditions using the North2South Chemiluminescent Hybridization and Detection System and a cooled CCD camera. Here, 2 μg of yeast total RNA was spotted onto a nylon membrane, hybridized with a biotinylated DNA probe for GFPuv, and detected using the North2South Chemiluminescent Hybridization and Detection System. In (A), exposure time was 5 min and in (B) 1 h using the ChemiImager 4000 System. Samples 1, 2: noninducing conditions. Samples 3, 4: inducing conditions.

to detection of 5.2 attomole of target). Under experimental conditions, this combination of intense light output and low background allows for the detection of very low levels of RNA using cooled CCD camera systems. To demonstrate this capability, a dot blot was performed using total RNA from the yeast *S. cerevisiae* expressing GFPuv under the control of the *CUP1* promoter. In the absence of copper, the yeast expressed very low, basal levels of GFPuv mRNA. The levels are nearly undetectable by standard Northern blot (data not shown), and no fluorescent GFPuv protein is detectable (data not shown). When the same yeast strain is induced with copper, GFPuv mRNA is expressed at a high level (Figure 15.6) and GFPuv protein fluorescence is readily detectable (data not shown). As demonstrated in Figure 15.6, the North2South Chemiluminescent Hybridization and Detection System has a very high signal-to-noise ratio, allowing for the detection of GFPuv mRNA under noninducing conditions after a 1 h exposure using a cooled CCD camera system.

Many new techniques have allowed for the isolation and determination of differentially expressed genes. Techniques such as SAGE (serial analysis of gene expression), differential display, and, more recently, the implementation of DNA chips have allowed for the rapid isolation of genes with novel expression patterns. However, such techniques suffer from limitations, including the inability to determine the size or tissue distribution of the novel transcript. To address these limitations, researchers are increasingly turning to the use of hybridization-ready Northern blots and multiple-tissue expression arrays. An increasing number of companies now offer hybridization-ready, multiple-tissue Northern blots. Use of these blots allows researchers both to verify differential expression of an mRNA species and to determine the size of the transcript. MTN blots are prepared using high-quality poly A$^+$ RNA from several different tissues and the amount of RNA on an MTN blot is adjusted so the β-actin hybridization signal is of comparable intensity in every lane. As demonstrated in Figure 15.7, the North2South Chemiluminescent Hybridization and Detection System functions very well with MTN blots, demonstrating both strong signal for a ubiquitously expressed mRNA (β-actin) and also high stringency for a tissue-specific mRNA (liver-specific gene for TAT). To define further tissue-specific expression, researchers can now purchase ready-to-hybridize multiple tissue expression arrays, such as the MTE arrays. These arrays enable researchers to study gene expression in as many as 76 different tissues, thus providing a resource that was previously unavailable and economically unfeasible. The high signal-to-noise ratio of the North2South Chemiluminescent Hybridization and Detection System is vitally important for its compatibility with arrays because background can make interpretation of the results difficult, thus leading to erroneous conclusions. As shown in Figure 15.8, the North2South Chemiluminescent Hybridization and Detection

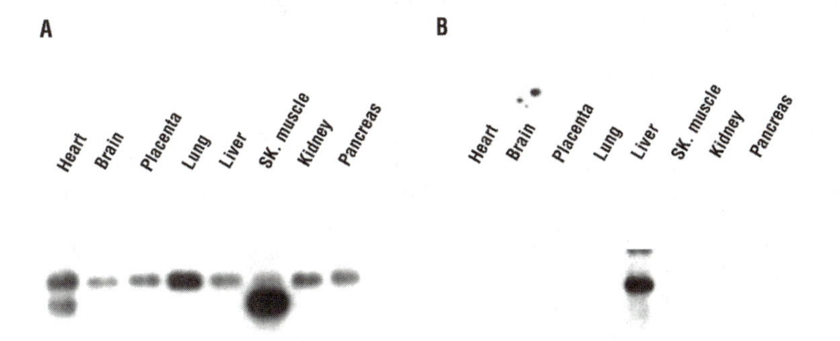

FIGURE 15.7 Compatibility of the North2South Chemiluminescent Hybridization and Detection System with MTN blots. MTN blots containing human poly A$^+$ RNA (from indicated tissues) were hybridized with random-prime-labeled DNA probe in (A) human β-actin and in (B) a liver-specific probe for TAT. Note in (A) the additional bands in the heart and skeletal muscle lanes, which correspond to a muscle-specific isozyme of β-actin (1.8 and 2.0 kb). Exposure times are as follows: (A) 2 min and (B) 15 min.

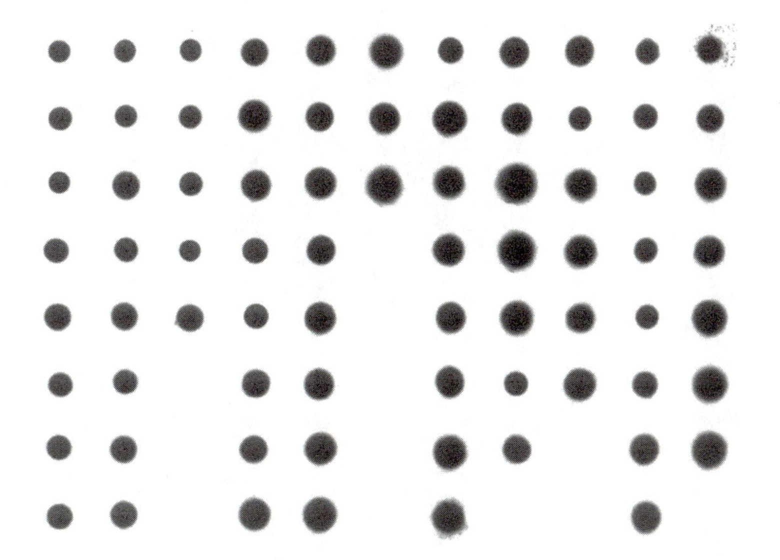

FIGURE 15.8 Using the North2South Chemiluminescent Hybridization and Detection System with an MTE. An MTE of poly A$^+$ RNA from 76 different tissues was hybridized with a biotinylated DNA probe for human β-actin.

System has an extremely low level of background and a strong signal, making it fully compatible with studies using MTE arrays.

The North2South Chemiluminescent Hybridization and Detection System provides a high level of sensitivity with extremely low background for Northern blots Southern blots, and plaque lifts, and is compatible with hybridization-ready multiple tissue Northern blots (such as MTN) and multiple tissue expression arrays (MTE). In addition, this system has a short process time, and the high level of light output allows for very short exposure times using cooled CCD camera systems or film.

ACKNOWLEDGMENTS

The authors thank Dr. John Lopes and Dave Eiznhamer (Wayne State University, Detroit, MI) for the gift of plasmids and yeast strains used in this study, Alison Martin of Nucleotech for supplying data generated using the Nucleovision Workstation, Holly Schubert for critical reading of the manuscript, and Pat Bremseth for assistance in generating figures.

REFERENCES

1. Hirsch, J. and Henry, S., Expression of the *Saccharomyces cerevisiae* inositol-1-phosphate synthase (*INO1*) gene is regulated by factors that affect phospholipid synthesis, *Mol. Cell. Biol.*, 6, 3320–3328, 1986.
2. Hoffman, C. and Winston, F., A ten-minute DNA preparation from yeast efficiently releases autonomous plasmids for transformation of *Escherichia coli, Gene,* 57, 267–272, 1987.
3. Sambrook, J., Fritsch, E., and Maniatis, T., Eds., *Molecular Cloning,* 2nd ed., Cold Spring Harbor Press, Plainview, NY, 1989.

16 Luciferase Assay: A Powerful Tool to Determine Toxic Metal–Induced NF-κB Activation

Fei Chen, Knox Van Dyke, Val Vallyathan, Vince Castranova, and Xianglin Shi

CONTENTS

16.1 INTRODUCTION

After several decades of intensive research, scientists have accumulated a rich and complex body of knowledge revealing that most toxic metals or metal-containing particles from either environmental sources or occupational sources act as human carcinogens.[1–4] Yet a detailed molecular mechanism(s) of metal-induced malignant transformation leading to cancer is still missing. Emerging evidence indicates that cellular transformation and tumorigenesis in humans is a multistep

0-8493-0719-8/02/$0.00+$1.50

process that requires both nongenetic and genetic alterations that promote the transformation of normal human cells into highly malignant ones.[5] However, currently it is still not clear which step or steps are effectively targeted by metals. Certainly, nuclear factor-κB (NF-κB) activation is involved in the process of carcinogenic transformation of cells, but how metals affect the signal transduction pathways leading to the activation of NF-κB is still poorly understood. It is unequivocal that oxidative stress resulting from metal-induced generation of reactive oxidative species (ROS) is an important mechanism for the activation of NF-κB by metal. However, a ROS-independent effect of metals on the cellular signaling pathway and on genomic stability may also account for this process.[6,7]

16.2 SIGNALING PATHWAYS FOR NF-κB ACTIVATION

16.2.1 General Pathway of NF-κB Activation

A wide range of signals, which typically include cytokines, mitogens, environmental and occupational particles or metals, intracellular stresses, viral and bacterial products, and ultraviolet (UV) radiation, induce expression of early response genes via the NF-κB family of transcription factors.[8,9] In resting cells, NF-κB is sequestered in the cytoplasm of most cells where it is bound to a family of inhibitory proteins, such as IκBα, IκBβ, IκBε, p105, and p100. Activation of the NF-κB signaling cascade results in complete degradation of IκB or the carboxy terminal partial degradation of the p105 and p100 precursors, allowing nuclear translocation of the NF-κB complexes (Figure 16.1). Activated NF-κB binds to specific DNA sequences in target genes, designated as κB-elements, and regulates transcription of genes mediating inflammation, carcinogenesis, and antiapoptotic reactions.

IκBα is the most abundant inhibitory protein for NF-κB.[9] The mechanisms of signal-induced IκBα degradation involve phosphorylation of two serine residues, S32 and S36. This phosphorylation leads to polyubiquitination of two specific lysines on IκBα (K21 and K22) by the SCF-β-TrCP complex and its degradation by the 26S proteasome.[10] The phosphorylation is accomplished by a specific IκB kinase (IKK) complex containing two catalytic subunits, IKKα and IKKβ, and a structural component named NEMO/IKKγ/IKKAP.[11,12] IKKα and IKKβ share a 50% sequence similarity. Both proteins contain an amino terminal kinase domain, a carboxy terminal region with a leucine zipper, and a helix-loop-helix domain. *In vitro* or *in vivo* studies indicate that both IKKα and IKKβ are capable of phosphorylating IκBα on ser32 and ser36, but IKKβ is more potent in IκBα phosphorylation induced by proinflammatory stimuli. Recent studies by several groups indicate the existence of an additional IKK-like kinase complex in T cells, named IKKi/ε,[13,14] which shares 27% homology with IKKα and IKKβ and possibly mediates NF-κB-activating kinase (NAK) signaling and PMA/PKCε-induced S36 phosphorylation of IκBα and NF-κB activation.

Diverse stimuli, with a possible exception of UV, activate NF-κB through activation of the kinase activity of IKK that phosphorylate IκBs, a critical step required for subsequent ubiquitination and degradation. UV-induced NF-κB activation may be via a mechanism that does not depend on IKK or N-terminal phosphorylation of IκB.[15,16] It is known that the activity of IKK can be stimulated by TNFα, IL-1, LPS, Tax protein, ionizing radiation, vanadate, and double-stranded RNA. However, how these diverse stimuli activate IKK remains elusive. A number of kinases, mostly based on transient overexpression in which wild-type forms activate IKK and catalytically inactive forms inhibit IKK, have been suggested to be able to phosphorylate and activate IKK. These kinases include MEKK1,[17] NIK,[11] TBK1/NAK,[14,18] Akt,[19,20] PKCθ,[21,22] PKCξ,[23] Cot kinase,[24] PKR,[25] and MLK3.[26] It is not clear whether each of these upstream kinases represents one of many routes to IKK activation by specific stimuli or represents a ubiquitous mediator leading the activation of IKK induced by diverse stimuli.

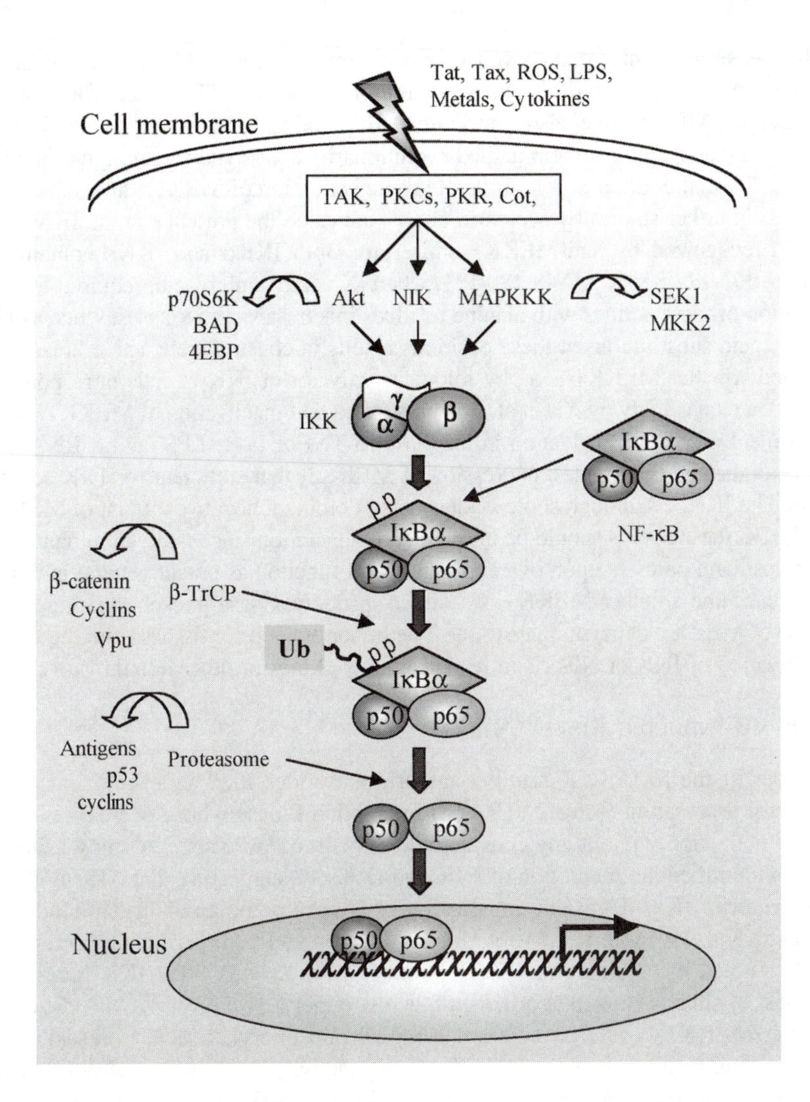

FIGURE 16.1 Signaling pathways of NF-κB activation. A number of extracellular stress inducers, including cytokines, ROS, the viral or bacterial products, stimulate IKK through upstream kinases directly or indirectly. Activated IKK phosphorylates N-terminal S32 and S36 residues of IκBα that is associated with the NF-κB p50 and p65 heterodimer. The SCF-β-TrCP complex recognizes phosphorylated IκBα and modifies IκBα with polyubiquitin chains. This is followed by proteasome-mediated degradation of IκBα. After degradation of IκBα, the activated NF-κB translocates into the nucleus, where it binds to the κB-sites of gene promoters or enhancers to upregulate target gene expression. Line arrows and filled arrows denote the NF-κB signaling pathways; open arrows denote the connections with bystanding signaling pathways.

16.2.2 MEKK1

MEKK1 is a mammalian serine/threonine kinase in the mitogen-activated protein kinase kinase kinase (MAPKKK) group. It was found that MEKK1 was a far more important activator for JNK signaling rather than ERK signaling as proposed originally.[27,28] The first evidence indicating the involvement of MEKK1 in signal-induced IKK activation was provided by Lee and co-workers.[17,29] In their studies, they reported that the addition of recombinant catalytic domain of MEKK1 (MEKK1Δ) to partially enriched fraction of nonstimulated HeLa cells stimulated an IKK-like kinase activity that phosphorylated IκBα at serines 32 and 36 and subsequent ubiquitination and degradation of IκBα. It was subsequently demonstrated that overexpression of MEKK1 stimulated NF-κB-dependent transcriptional reporter.[30,31] The activation of NF-κB by HTLV Tax protein was shown to involved MEKK1.[32] MEKK1 may also contribute to Toll- and IL-1 receptor-mediated IKK activation, as demonstrated by an adaptor protein named *e*volutionarily *c*onserved *s*ignaling *i*ntermediate in *T*oll pathways (ECSIT) that could promote the proteolytic activation of MEKK1 and subsequent activation of NF-κB.[33] Further studies by Mercurio et al.[12] indicated the presence in the IKK complex of a protein that is recognized by anti-MEKK1 antiserum. Both IKKα and IKKβ contain a MAPKK activation loop, 176/177 Ser-X-X-X-S 180/181, where X is any amino acid, required for their activation. Substitution of these serines with alanine residues inactivates both kinases, whereas phosphomimetic glutamic acid substitutions at these positions results in constitutively active kinases. It remains to be confirmed whether MEKK1 is a physiological activator of IKK in cells in response to various stimuli. Indeed, a recent study by Xia et al.[34] demonstrated that inactivation of MEKK1 does not result in any impairment of NF-κB activation in response to TNFα, IL-1, LPS, and dsRNA. In addition, several studies indicated that modest expression of MEKK1, that sufficient for JNK acctivation, had failed to stimulate IKK.[35] Although some studies using dominant negative mutant of MEKK1 showed inhibition of IKK, these results should be interpreted with caution since many dominant negatives can bind to their upstream partners upon overexpression and function as potent general inhibitors. A best approach to determine whether MEKK1 is a direct upstream kinase phosphorylating and activating IKK is the use of MEKK1-deficient mice to see whether or not cells or tissues from these mice display defects in activation of IKK or NF-κB in response to cytokines or other inflammatory stimuli.

16.2.3 NF-κB-Inducing Kinase (NIK)

NIK, a member of the MAPKKK family, was originally identified as a tumor necrosis factor-α (TNFα) receptor associating factor 2 (TRAF2)-interacting kinase whose overexpression results in potent NF-κB activation without any considerable effect on MAPKs.[36,37] A study using yeast two-hybrid screens identified an interaction of NIK and IKK,[11,38] suggesting that NIK might be a direct upstream activator of IKK. Transient transfection of NIK into human embryonic kidney 293 cells indicated that IKKα was more responsible for NIK, whereas IKKβ was slightly more responsible for MEKK1.[39] When the abilities of MEKK1 and NIK to activate total IKK kinase activity are compared, most of studies show that NIK is a much stronger activator of the NF-κB transcriptional reporter than is MEKK1.[37,40] NIK could preferentially phosphorylate IKKα on ser176 in the activation loop, leading to the activation of IKKα kinase activity. In contrast, MEKK1 was found to phosphorylate the corresponding serine residue, ser177, in the activation loop of IKKβ, preferentially. A dominant negative mutant of NIK blocked NF-κB activation by TNFα, interleukin-1 (IL-1), Fas,[36] Toll-like receptors 2 and 4,[41,42] LMP-1,[43] and CD3/CD28 stimulation.[24] Thus, NIK appears to be a general kinase mediating IKK activation induced by diverse stimuli. However, a recent analysis using the NIK-mutant mouse strain *alymphoplasia (aly)* contradicted this assumption. *Alymohoplasia* mouse strain failed to develop lymphoid organs, such as lymph nodes, Peyer's patches, as the result of a point mutation in the NIK locus.[44] The mutation of NIK locus results in a disruption of interactions between NIK and IKKα or TRAF proteins. Further analysis indicated that *aly* mutation did not affect TNFα-induced activation of NF-κB but only blocked lymphotoxin-mediated activation of NF-κB. Similarly, studies using cells derived form NIK-deficient mice have indicated that NIK appears to be dispensable in IKK activation induced by TNFα or IL-1. It raises the possibility that NIK may be specifically involved in IKK activation induced by lymphotoxin but not others.

16.2.4 NF-κB Activating Kinase (NAK)

Two groups independently identified a novel serine/threonine kinase possibly activating IKK through direct phosphorylation in cells stimulated with PMA.[14,18] This novel kinase was named NAK, TANK-binding kinase 1 (TBK1), and T2K, respectively. Pomerantz and Baltimore[18] cloned NAK by a yeast two-hybrid screen using the N-terminal stimulatory domain of TANK

1-190 fused to GAL4 as bait and a human B-cell library fused to the GAL4 activation domain. The same kinase was also identified by PCR using degenerate primers based on sequences common to IKKα and IKKβ.[14] The amino acid sequence analysis indicated that NAK protein contains a kinase domain at its N-terminus that exhibits about 30% identity to the corresponding kinase domains of IKKα and IKKβ and more than 60% identity to the corresponding kinase domain of IKKε/i. Whereas the report by Pomerantz and Baltimore[18] showed that NAK might form a ternary complex with TANK and TRAF2, suggesting that NAK functions distal upstream of the signal cascade leading to IKK activation, *in vitro* kinase activation assay by Tojima and co-workers[14] demonstrated that NAK was a direct upstream kinase phosphorylating IKKβ. Intriguingly, activation of endogenous NAK resulted in only ser36, but not ser32 phosphorylation of IκBα, a phenomenon similarly observed in recombinant IKKε/i-mediated IκBα phosphorylation.[45] Since both IKKα and IKKβ are able to phosphorylate both ser32 and ser36 of IκBα protein, it is unclear whether IKKε/i or a novel IKK isozyme functions as a downstream kinase of NAK to induce ser36 phosphorylation of IκBα. Transient transfection studies showed that dominant negative NAK inhibited NF-κB transcriptional reporter activity induced by PMA, Protein Kinase Cε (PKCε), and PDGF, but not by TNFα, IL-1β, LPS, and ionizing radiation. These results, therefore, suggest that NAK is likely a downstream kinase of PKCε or related isozymes and an upstream kinase of IKK in the signaling pathway through which growth factors, such as PDGF, stimulate NF-κB activity.

16.2.5 Akt

The pro-survival function of Akt has been well documented. The kinase activity of Akt is activated via the phosphoinositide-3-OH kinase (PI3K) and PI3K-dependent kinase 1/2 (PDK1/2) signaling pathway.[46] Overexpression or constitutive activation of Akt has been associated with tumorigenesis in a number of studies. As a serine/threonine kinase, Akt is able to phosphorylate pro-apoptotic protein Bad, anti-apoptotic protein Bcl-x, apoptotic protease caspase-9, Forkhead transcription factors, and eNOS.[46] However, considerable controversy remains regarding the involvement of Akt in signal-induced IKK activation. Studies by Ozes et al.[19] and Xie et al.[47] indicated that Akt was required for TNFα- or G protein activator–induced NF-κB activation by direct phosphorylating and activating IKKα in 293, HeLa, and ME-180 cells. A putative Atk phosphorylation site at amino acids 18 to 23 in both IKKα and IKKβ was identified. Akt induced Thr23 phosphorylation of IKKα both *in vitro* and *in vivo*. Mutation of Thr23 significantly decreased Akt-induced IKKα phosphorylation and TNFα-induced NF-κB activation in 293, HeLa, and ME-180 cells.[19] By contrast, Romashkova et al. [20] demonstrated that Akt was involved in PDGF-mediated but not in TNFα- or PMA-mediated NF-κB activation in human or rat fibroblasts. In this study, the authors suggested that upon PDGF stimulation, Akt could transiently associate with IKK and induce IKK activation, especially IKKβ activation. Several other studies, however, totally contradicting these reports, suggested that the effects of Akt on NF-κB did not occur on the level of IKK activation in several cell types. A study by Delhase and co-workers[48] indicated that Akt activation induced by IGF-1 failed to activate IKKα kinase activity, IκBα phosphorylation and degradation, and NF-κB DNA binding in HeLa cells, the same cell line used by Ozes et al.[19] Similarly, several recent studies showed that Akt was not involved in TNFα-induced NF-κB activation in human vascular smooth muscle cells, skin fibroblasts, and endothelial cells.[49,50] Rather, Akt might enhance the ability of the p65 (RelA) transactivation to induce transcription.[51,52] In Jurkat T-cells, Akt alone failed to activate NF-κB, but it was capable of potentiating NF-κB activation induced by PMA, partially by enhancing IκBβ degradation.[53] Thus, several questions remain to be answered: Does Akt phosphorylate IKK in a cell context- and stimulation-dependent manner? Do upstream kinases of Akt, such as PDK1 and PDK2, also activate IKK, as both IKKα and IKKβ contain a putative PDK1 phosphorylation site (S-F-X-G-T-X-X-Y-X-A-P-E) directly juxtaposed to the MAPKKK phosphorylation site?[54,55]

16.2.6 Mixed-Lineage Kinase 3 (MLK3)

MLK3, another member of the MAPKKK family, contains an N-terminal SH3 domain, followed by the catalytic domain and two tandem leucine/isoleucine zippers, a basic region, a Cdc43/Rac binding motif, and a proline-rich C terminus.[56] Based on these structurally characteristics, MLK3 may associate with a variety of protein modules. Studies by Hehner et al.[26] suggested that MLK3 could directly associate with IKK complex through its leucine zipper domain and phosphorylate Ser176 of IKKα and Ser177 and Ser181 of IKKβ. Transfection of Jurkat T cells with a kinase-mutated form of MLK3 blocked CD3-CD28 signal- and PMA-induced NF-κB transcriptional reporter. No significant influence of this mutated MLK3 was observed on either TNFα- or IL-1β-induced NF-κB activation. These results suggest that MLK3 may be important in mediating T-cell costimulation-induced activation of IKK and consequent NF-κB-dependent transcription. MLK3 has also been shown to form a complex with a JNK scaffold protein JIP and to stimulate JNK activation. Thus, MLK3 may function as an integral molecule between the signaling pathways leading to the activation of NF-κB and JNK, which would provide a molecular explanation why many stimuli induce NF-κB and JNK simultaneously under certain circumtances.

A variety of other kinases had been indicated to function upstream of IKK. Because of the lack of evidence of direct association of these kinases with IKK upon activation or specific phosphorylation site(s) of these kinases on IKK, whether these kinases are direct upstream kinases phosphorylating and activating of IKK or far more distal kinases indirectly activating IKK remains unsolved. These kinases include Cot,[24] PKCθ,[21,22] PKCα, PKCξ,[23] TAK1,[57,58] PKR,[25] etc. In light of the fact that a variety of kinases either related or unrelated can activate IKK, it seems likely that different cell types and stimuli may utilize distinct upstream kinases for the activation of IKK. An example to support this notion is provided by the observation that PKCθ and Cot kinase participate in CD3-CD28 costimulation signal-induced but not TNFα-induced activation of NF-κB.[22,24]

16.3 NF-κB Activation by Metals

16.3.1 V(V), As(III), and Cr(VI) Induce NF-κB

A number of reports during the last few years indicate that some metals are able to affect the activation or activity of NF-κB transcription factors.[59] Yet the results are not straightforward; both activation and inhibition by metals on NF-κB have been reported.[60–63] Several reports from different groups indicated that at a noncytotoxic dosage, chromium (VI) [Cr(VI)], arsenic trioxide [As(III)], vanadium (V) [V(V)], nickel, and cobalt are capable of activating NF-κB as monitored by either gel shift assay (reflecting the activation and nuclear translocation of NF-κB) or NF-κB-dependent reporter gene assay, an indicator of NF-κB activity. In contrast, several papers showed that Cr(VI), As(III), and other metals inhibited NF-κB activation by interfering with IKK kinase, DNA binding, or the interactions with nuclear cofactor cAMP-responsive element-binding protein-binding protein (CBP).[62,63] This controversy may largely result from the use of different doses of metals in each experimental system. All groups agree that a high concentration of metals is inhibitory to the activation and function of NF-κB.

16.3.2 Molecular Mechanism of Metal-Induced NF-κB

The mechanistic basis of metal-induced NF-κB activation or stimulatory activity is an unsolved issue, especially, for Cr(VI) and As(III). In the case of V(V)-induced NF-κB activation, early study by Imbert and co-workers[64] indicated that the activation of NF-κB by V(V) occurred independently of IκBα degradation. However, this observation could not be reproduced in several subsequent studies that suggest that V(V) did induce degradation of IκBα following the phosphorylation of serine or tyrosine.[65,66] In the mouse macrophage cell line RAW264.7 cells, we observed that V(V) induced IκBα degradation detected at 10 to 20 min with a peak degradation at 40 min.[65] In human myeloid U937 cells or epithelial HeLa cells, Mukhopadhyay et al.[66] noted that V(V)-induced IκBα degradation occurred at 30 min and reached maximum at 240 min. A similar result was achieved in Jurkat E6.1 cells and human B cell lymphoma line Ramos.[67,68] In addition, Imbert et al.[64] showed no resynthesis of IκBα after V(V) treatment. Our studies and those of others, however, clearly indicated that the resynthesis of IκBα occurred at 80 to 180 min after V(V) treatment.[65,66]

Phosphorylation on either tyrosine 42 or serine 32/36 sites of IκBα has been demonstrated in cells treated with V(V). The phosphorylation of IκBα on these sites may contribute to the subsequent degradation of this protein. Currently, there are no available data regarding the induction of IκBα phosphorylation and degradation in the cells in response to other metals, such as Cr(VI) and As(III). The phosphorylation of both tyrosine and serine residues on IκBα implied that certain tyrosine kinases and serine/theronine kinases are involved in metal-induced NF-κB activation. Knowing the fact that IKK is required for LPS- and inflammatory cytokine-induced IκBα phosphorylation and NF-κB activation, possibly the IKK may also contribute to certain metal-induced NF-κB activation. Indeed, in mouse macrophage cell line RAW264.7 cells, we observed that IKKβ kinase activation by V(V) is both dose- and time-dependent.[65] Compared with cytokine- or LPS-induced IKKβ kinase activation, V(V) likewise induced IKKβ activation in a relatively persistent manner.

16.3.3 ROS Involvement in NF-κB Activation Induced by Metal

It has been suggested that many metals can elicit an oxidative stress response of cells through the generation of ROS,[69–72] including superoxide anion, hydroxy radical, nitric oxide, and H_2O_2. Excessive amounts of ROS generated from chronic and acute inflammatory response or environmental stresses are cytotoxic. It is known that a limited production of ROS as a consequence of electron transfer reaction in cytosol, peroxisomes, and mitochondria are buffered or scavenged by both enzymatic and nonenzymatic antioxidant systems.[73] The mechanisms of ROS generation by metals under biologically relevant conditions may involve Fenton/Haber–Weiss chemistry and autoxidation.[72] In neutrophils and eosinophils or other phagocytes, another mechanism that is involved in metal-induced ROS generation is the production of hypochlorous acid, lipid peroxides, and nucleoside hydroperoxides.[74] Use of ROS scavengers or antioxidants suggests the involvement of cellular redox regulation of some kinases leading to the activation of NF-κB or others.[75] In an early model, the ROS-induced oxidative stress was proposed as a universal mechanism of NF-κB activation by diverse agents.[76] An unanswered and difficult question is which point(s) of the activation pathway of NF-κB is targeted by ROS. The signal transduction pathway, such as the upstream and proximal kinases, e.g., IKK, for NF-κB activation induced by TNF, LPS, and CD28, has recently been clearly identified.[8,13] However, no evidence has been presented to indicate that these kinase cascades would be oxidant responsive or redox regulated. Evidence provided by Li and Karin[77] demonstrated that ROI scavenger, *N*-acetyl-L-cysteine (NAC) could reduce TNFα-induced IκBα degradation and NF-κB DNA-binding activity, but failed to affect TNFα-induced IKK kinase activity in HeLa cells. These results raise the possibility that ROI may not target the activation pathway of IKK but, rather, interfere with the steps of ubiquination and degradation of IκB. The model proposed by Roederer and co-workers[78] suggested that NF-κB activation was controlled by intracellular thiol levels in which NF-κB inducers somehow potentiate oxidative stress by depleting glutathione levels. However, this model contradicts several later observations that showed that glutathione-oxidizing agent could inhibit NF-κB via interference with its DNA binding.[79–81] Several studies indicated that cysteine 62 of the p50 subunit of NF-κB is essential for NF-κB DNA binding and the oxidation of cysteine 62 inhibited NF-κB DNA binding activity. Thioredoxin, a ubiquitous dithiol-reducing enzyme, can reduce cysteine 62 and restore DNA-binding activity of NF-κB.[79] Therefore, the observed NF-κB activation by metal-induced ROI in our studies and others may depend on some alternative pathways, such as MAP kinases, Ras, and Rac1 with the potential to stimulate the IKKs. Indeed, JNK inhibition by a transient transfection of macrophages with a dominant negative SEK1, an upstream kinase of JNK, decreased vanadate-induced IκBα degradation.[65]

Many metals, including V(V) and As(III), and the oxidant H_2O_2 are potent inhibitors for protein tyrosine phosphatases (PTP).[82,83] All PTPs contain a signature active site characterized with the sequence of His-Cys-X-X-Gly-X-X-Arg-Ser/Thr, where X is any amino acid.[84] Oxidation of the cysteine residue in this signature motif, which is essential for the phosphatase activity, will inactivate PTPs. Protein serine/threonine phosphatases, such as PP1, PP2A, PP2B, and PP2C, may be also subject to redox regulation via the oxidative formation of disulfide bond between a conserved pair of cysteine residues.[85] Because the extent of protein phosphorylation, such as IκBα phosphorylation, reflects the balance between the opposing actions of protein kinases and phosphatases, changes in either side can consequently shift the equilibrium toward phosphorylation. It has been indicated that hypoxia and reoxygenation induce NF-κB activation through the induction of phosphorylation of tyrosine at position 42 on IκBα. Although this notion remains controversial, it will be important to determine whether tyrosine kinases are required for metal-induced NF-κB activation. Nevertheless, the question that remains unanswered is whether the effect(s) of metals on PTPs or tyrosine kinases is mediated directly by the metal itself or indirectly by the generation of ROS.

16.4 LUCIFERASE ANALYSIS USED IN NF-κB STUDIES

16.4.1 Methods Used to Determine NF-κB Activation or Activity

To delineate the mechanism of response to a variety of extracellular and intracellular stress signals, multiple assays are required for the determination of activation and function of NF-κB.[9] A number of methods have been developed and widely used experimentally to determine the activation and function of NF-κB. To date, three general methods are widely used to measure NF-κB activation, directly or indirectly:

1. Protein expression of NF-κB family members or cytoplasmic IκB degradation can be measured by immunoblot, using antibodies against NF-κB subunits, such as p50, p52, p65, and c-Rel, or IκB family proteins including IκBα, IκBβ, and IκBε. The advantage of this method is that it can determine the composition of activated NF-κB complexes and which inhibitory protein is involved in the activation of NF-κB by a given stimulating agent. This method, however, is time-consuming and does not allow for processing a large number of samples.
2. Since NF-κB will bind to target DNA elements after the activation and translocation into nuclei, the DNA-binding capacity of NF-κB can be assayed by gel shift assay, also called electrophoretic mobility shift assay (EMSA). General procedures for this assay include nuclear protein extraction, preparation of radioisotope ^{32}P-labeled double-stranded oligonucleotide probe containing the consensus sequence of κB element GGGAATTTCCC, incubation of the nuclear proteins with this labeled probe, electrophoresis of the reaction products of nuclear proteins and labeled probe, and, finally, autoradiography. This method may authentically reflect the activation status of NF-κB in addition to its relatively sensitivity. However, this assay suffers from several drawbacks; it is time-consuming and requires special precautions for handling radioactivity, which limit its usefulness as a screen for the functional status of NF-κB.
3. In comparison with other existing assays, it was believed that a reporter gene assay based on the expression of a κB-dependent reporter gene can provide one of the most specific and biologically relevant means to determine the functional aspect of NF-κB. This assay is based on reporter genes, typically firefly luciferase, chloramphenicol acetyltransferase (CAT), or β-galactosidase, placed under the control of a promoter containing NF-κB binding elements. This promoter can be artificial, made of two to five NF-κB binding elements and a TATA box, or natural, such as reconstructed HIV-1 long terminal repeat (LTR) sequence, which only contains two NF-κB binding elements and a TATA box.

This reporter gene construct is transfected transiently or stably into cells. Upon stimulation, the reporter gene activity in cells can be detected using ELISA, thin-layer chromatography, a luminometer, or a scintillation counter. The CAT and β-galactosidase-based reporter gene assay is being replaced by a much more sensitive and simple luciferase assay.

16.4.2 Procedures Required for NF-κB-Dependent Luciferase Reporter Gene Assay

Depending on the purposes of studies, many types of cells, such as T lymphocytes, B lymphocytes, macrophages, epithelial cells, and endothelial cells, can be used for the transfection of the NF-κB-dependent luciferase reporter construct. The transfection efficiency may vary among different type of cells. For transfection, cells are plated in six-well tissue culture plates at 1 to 5×10^5 cells/well and incubated until the cells are 50 to 80% confluent. This usually will take 18 to 48 h, but the time will vary among cell types. Transfection solution for each well is prepared by combination of NF-κB-dependent luciferase reporter construct (1 to 2 μg) and 2 to 25 μl of lipofectamine reagent in 200 μl of serum-free medium. The mixture is incubated at room temperature for 15 to 45 min to allow DNA–liposome complexes to form. At the end of this incubation, the mixture is overlaid onto the rinsed cells along with an additional 0.8 ml of serum-free medium. Following 2- to 24-h incubation, 2 ml of complete growth medium is added. Total cell extracts are prepared 24 to 72 h after the start of transfection, depending on cell type and stimulation. Total protein concentration of each extract is quantitated using a Bio-Rad protein assay reagent. Luciferase activity is determined with a liquid scintillation counter or luminometer using the luciferase assay kit.

Recently in our laboratory, stable cell lines have been established that express the NF-κB-dependent luciferase gene. A sensitive and rapid luciferase activity assay has been developed to determine the potency of individual stress inducers or antioxidants on NF-κB activation. A cell line derived from human bronchial epithelial cell line BEAS-2B was used for stable transfection. Cells were cotransfected with an expression vector carrying wild-type IKKβ or a kinase-mutated form of IKKβ (IKKβ-KM) and a modified pGL2-luciferase reporter gene construct in which two NF-κB-binding sites derived from 5′-LTR of HIV-1 gene had been inserted upstream of the luciferase gene. Since the expressing vector for IKKβ or IKKβ-KM also contains a neomycin resistance gene, stably transfected clones were first isolated in 700 μg/ml of G418 for 3 weeks and tested by both immunoblot for IKKβ or IKKβ-KM expression and luciferase activity assay. Cells stably expressing IKKβ/IKKβ-KM and luciferase were maintained in regular DMEM media supplemented with 200 μg/ml of G418.

16.4.3 Metals Induce NF-κB-Dependent Luciferase Activity

These cells stably express wild-type IKKβ and luciferase reporter genes, which produce a highly sensitive response to LPS and various metals, such as arsenite, Cr(VI), and vanadate. All of these inducers produced a dose-dependent induction of NF-κB-dependent luciferase activity (Figure 16.2). These cells are extremely responsive to arsenite, as luciferase induction in 24-well plates using 4 μg/ml of arsenite was increased to about 45-fold relative to solvent controls. Higher concentrations of arsenite or Cr(VI) exhibited inhibition on NF-κB-dependent luciferase activity due to its cytotoxicity (data not shown). Only marginal or no induction of NF-κB-dependent luciferase activity was observed in cells expressing the kinase-mutated form of IKKβ.

16.4.4 Inhibition of Doxycycline and CMT-3 on NF-κB

A number of chemical compounds that exhibited antioxidant or ROS scavenger activities have been shown to inhibit signal-induced NF-κB activation in a variety of experimental systems. However, none of these compounds is sufficient and specific. To explore possible novel NF-κB inhibitors,

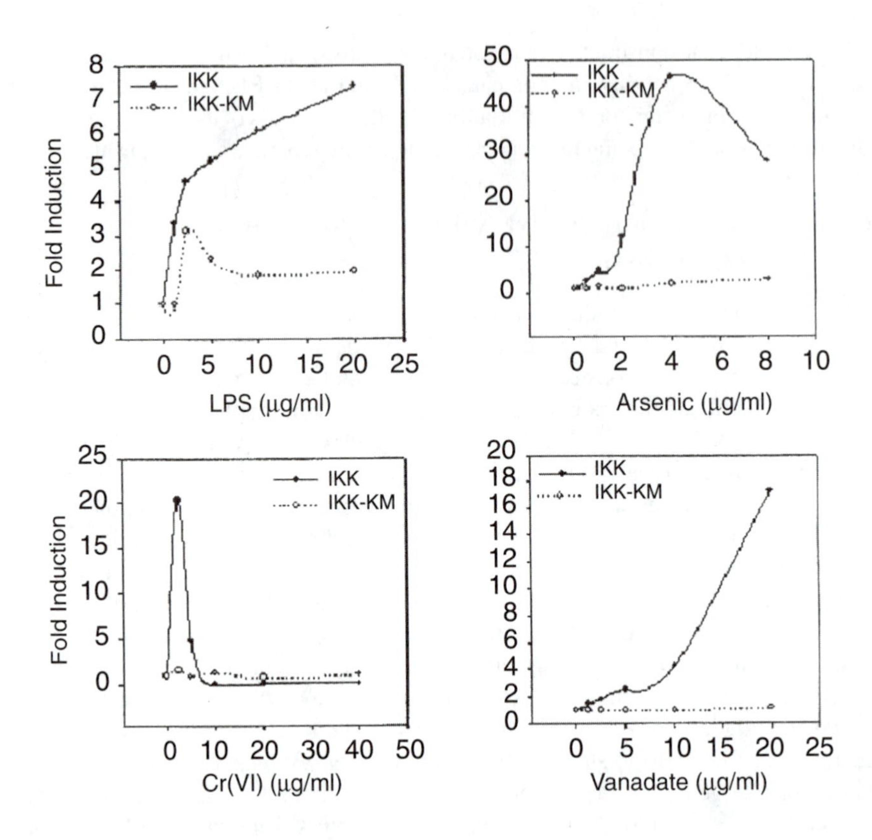

FIGURE 16.2 NF-κB-dependent luciferase activity assay in cellular response to LPS, arsenic, Cr(VI), and vanadate. Cells stably transfected with IKKβ or (IKK-KM) in combinations of NF-κB-luciferase reporter were treated with various concentrations of LPS, arsenic, Cr(VI), and vanadate as indicated for 12 h. Data are expressed in as the fold induction of relative luciferase activity ($n = 6$).

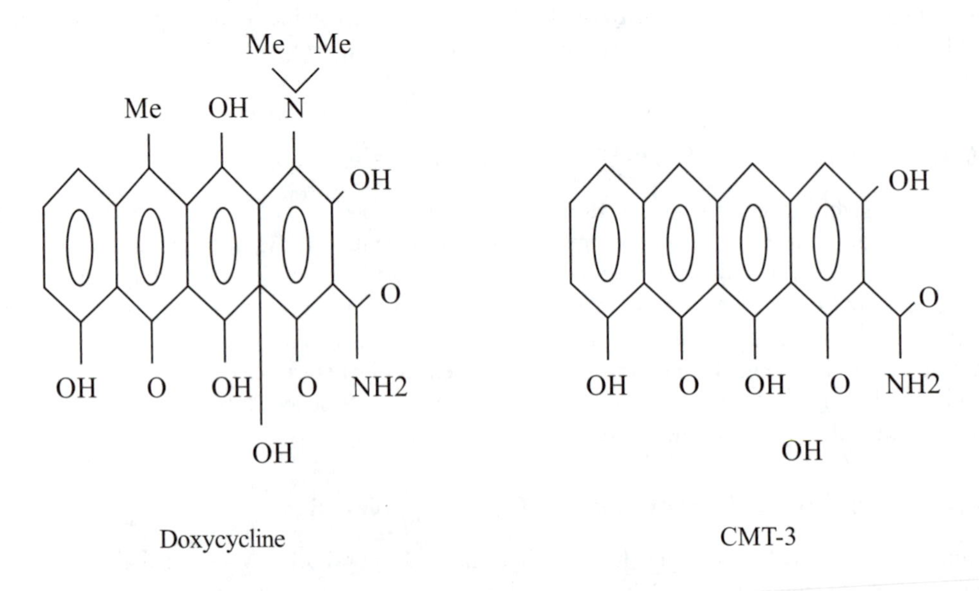

FIGURE 16.3 Structure of doxycycline and CMT-3.

we recently tested the effects of two tetracycline derivatives, doxycycline and CMT-3, on LPS- and vanadate-induced NF-κB-dependent luciferase activity. The chemical structures of both doxycycline and CMT-3 are depicted in Figure 16.3. Earlier study by Solomon and co-workers[86] showed that doxycycline was capable of inhibiting LPS-induced IL-1 mRNA expression, indicating doxycycline may suppress certain transcription factors, such as NF-κB, required for IL-1 gene transcription. Using cells stably expressing a NF-κB-dependent luciferase reporter gene, we observed that both doxycycline and CMT-3 inhibited basal and LPS- or vanadate-induced NF-κB activities in a dose-dependent manner (Figure 16.4). No cytotoxicity was observed at the highest dosage, 200 μg/ml, used in this experiment. We are currently determining mechanisms and specificity of these tetracycline derivatives on signal-induced NF-κB.

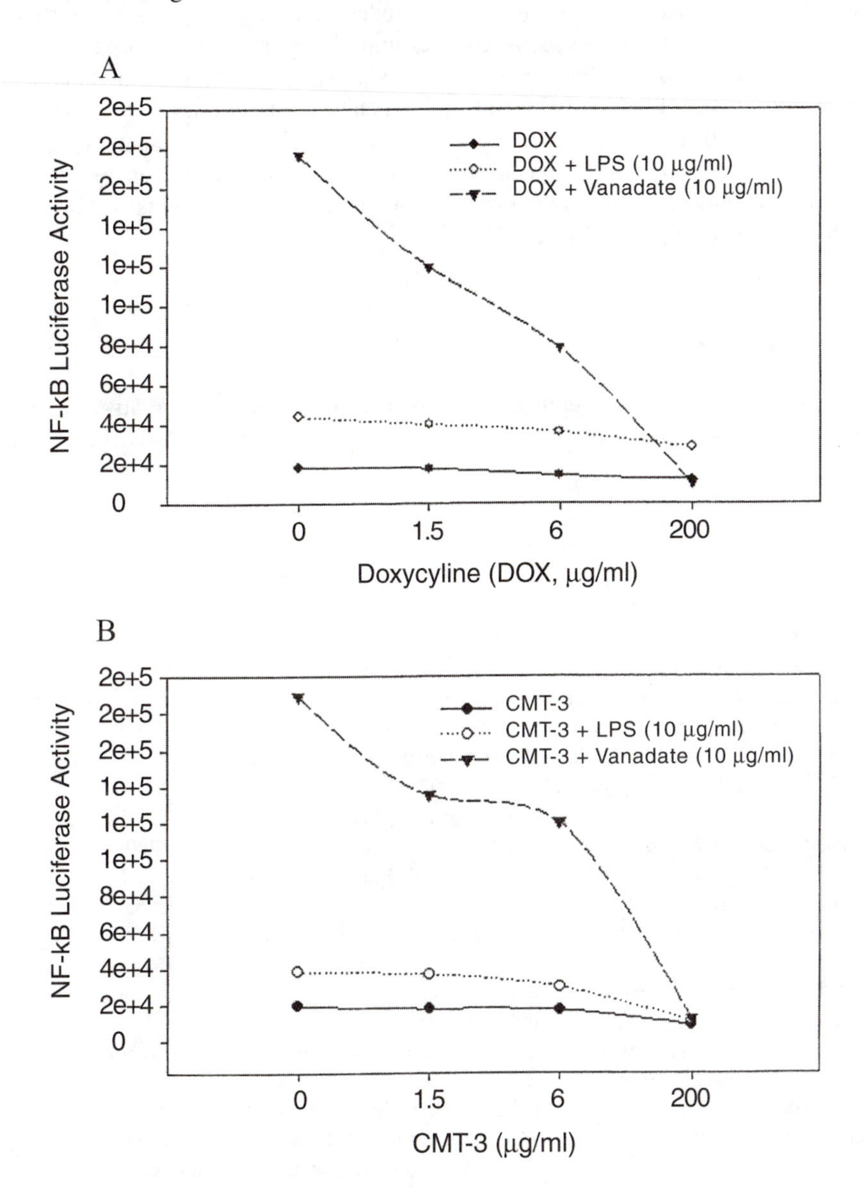

FIGURE 16.4 Doxycyline and CMT-1 inhibit NF-κB. Cells stably transfected with IKKβ or IKK-KM in combinations of NF-κB-luciferase reporter were pretreated with various concentrations of doxycline (A) or CMT-3 (B) as indicated for 2 h. Then the cells were incubated with LPS or vanadate for an additional 12 h. Relative luciferase activity was determined at the end of cell culture ($n = 6$).

16.5 SUMMARY

One of the major challenges in understanding the mechanisms of carcinogenic transformation of cells in response to toxic metals is to elucidate how the signal transduction pathway is activated, and how to determine the potency of individual metal. For example, why does the activation of the NF-κB, an anti-apoptotic transcription factor, coincide with obvious apoptotic features in cells treated with metals, whereas activation of NF-κB seems to be unfavorable for the induction of cell cycle arrest required for the maintenance of genomic stability. Since metals or their ROS derivatives are highly reactive but nonspecific molecules, targeting of only one signaling molecule is unlikely in cellular response to metals. Within a single signaling pathway, because of the highly reactive and nonspecific characteristics, metals or ROS can produce conflicting signals by interference with signaling molecules at different levels. A good example supporting this is that oxidative stress amplified or potentiated NF-κB activation, while oxidation of NF-κB proteins inhibited NF-κB function. No matter what effect has been elicited by metals at different levels of NF-κB signaling, determining the NF-κB-dependent luciferase activity will provide a convenient way to investigate the net effect of metals on NF-κB signaling. Translating the knowledge gained from this study may be a useful aid in identifying novel preventions or therapies for diseases caused by the exposure of environmental and occupational metals.

ACKNOWLEDGMENT

F.C. is supported by a Career Development Award on Genetics under a cooperative agreement from the Centers for Disease Control and Prevention through the Association of Teachers of Preventive Medicine.

REFERENCES

1. Sarkar, B., Metal replacement in DNA-binding zinc finger proteins and its relevance to mutagenicity and carcinogenicity through free radical generation, *Nutrition,* 11, 646, 1995.
2. Pollan, M. and Gustavsson, P., High-risk occupations for breast cancer in the Swedish female working population, *Am. J. Public Health,* 89, 875, 1999.
3. Maier, H. and Tisch, M., Epidemiology of laryngeal cancer: results of the Heidelberg case-control study, *Acta Otolaryngol. Suppl.,* 527, 160, 1997.
4. Snow, E. T., Metal carcinogenesis: mechanistic implications, *Pharmacol. Ther.,* 53, 31, 1992.
5. Hanahan, D. and Weinberg, R. A., The hallmarks of cancer, *Cell,* 100, 57, 2000.
6. Hamilton, J. W., Kaltreider, R. C., Bajenova, O. V., Ihnat, M. A., McCaffrey, J., Turpie, B. W., Rowell, E. E., Oh, J., Nemeth, M. J., Pesce, C. A., and Lariviere, J. P., Molecular basis for effects of carcinogenic heavy metals on inducible gene expression, *Environ. Health Perspect.,* 106(Suppl. 4), 1005, 1998.
7. Bartsch, H. and Nair, J., New DNA-based biomarkers for oxidative stress and cancer chemoprevention studies, *Eur. J. Cancer,* 36, 1229, 2000.
8. Karin, M. and Ben-Neriah, Y., Phosphorylation meets ubiquitination: the control of NF-κB activity, *Annu. Rev. Immunol.,* 18, 621, 2000.
9. Chen, F., Castranova, V., Shi, X., and Demers, L. M., New insights into the role of nuclear factor-κB, a ubiquitous transcription factor in the initiation of diseases, *Clin. Chem.,* 45, 7, 1999.
10. Palombella, V. J., Rando, O. J., Goldberg, A. L., and Maniatis, T., The ubiquitin-proteasome pathway is required for processing the NF-kappa B1 precursor protein and the activation of NF-κB, *Cell,* 78, 773, 1994.
11. Woronicz, J. D., Gao, X., Cao, Z., Rothe, M., and Goeddel, D. V., IκB kinase-β: NF-κB activation and complex formation with IκB kinase-α and NIK [see comments], *Science,* 278, 866, 1997.

12. Mercurio, F., Zhu, H., Murray, B. W., Shevchenko, A., Bennett, B. L., Li, J., Young, D. B., Barbosa, M., Mann, M., Manning, A., and Rao, A., IKK-1 and IKK-2: cytokine-activated IκB kinases essential for NF-κB activation [see comments], *Science,* 278, 860, 1997.
13. Israel, A., The IKK complex: an integrator of all signals that activate NF-κB? *Trends Cell Biol.,* 10, 129, 2000.
14. Tojima, Y., Fujimoto, A., Delhase, M., Chen, Y., Hatakeyama, S., Nakayama, K., Kaneko, Y., Nimura, Y., Motoyama, N., Ikeda, K., Karin, M., and Nakanishi, M., NAK is an IκB kinase-activating kinase, *Nature,* 404, 778, 2000.
15. Li, N. and Karin, M., Ionizing radiation and short wavelength UV activate NF-κB through two distinct mechanisms, *Proc. Natl. Acad. Sci. U.S.A.*, 95, 13,012, 1998.
16. Bender, K., Gottlicher, M., Whiteside, S., Rahmsdorf, H. J., and Herrlich, P., Sequential DNA damage-independent and -dependent activation of NF-κB by UV, *EMBO J.*, 17, 5170, 1998.
17. Lee, F. S., Hagler, J., Chen, Z. J., and Maniatis, T. Activation of the IκB α kinase complex by MEKK1, a kinase of the JNK pathway, *Cell,* 88, 213, 1997.
18. Pomerantz, J. L. and Baltimore, D., NF-κB activation by a signaling complex containing TRAF2, TANK and TBK1, a novel IKK-related kinase, *EMBO J.,* 18, 6694, 1999.
19. Ozes, O. N., Mayo, L. D., Gustin, J. A., Pfeffer, S. R., Pfeffer, L. M., and Donner, D. B., NF-κB activation by tumour necrosis factor requires the Akt serine-threonine kinase [see comments], *Nature,* 401, 82, 1999.
20. Romashkova, J. A. and Makarov, S. S., NF-κB is a target of AKT in anti-apoptotic PDGF signalling [see comments], *Nature,* 401, 86, 1999.
21. Sun, Z., Arendt, C. W., Ellmeier, W., Schaeffer, E. M., Sunshine, M. J., Gandhi, L., Annes, J., Petrzilka, D., Kupfer, A., Schwartzberg, P. L., and Littman, D. R., PKC-θ is required for TCR-induced NF-κB activation in mature but not immature T lymphocytes, *Nature,* 404, 402, 2000.
22. Lin, X., O'Mahony, A., Mu, Y., Geleziunas, R., and Greene, W. C., Protein kinase C-θ participates in NF-κB activation induced by CD3-CD28 costimulation through selective activation of IκB kinase β, *Mol. Cell. Biol.,* 20, 2933, 2000.
23. Lallena, M. J., Diaz-Meco, M. T., Bren, G., Paya, C. V., and Moscat, J., Activation of IκB kinase β by protein kinase C isoforms, *Mol. Cell. Biol.,* 19, 2180, 1999.
24. Lin, X., Cunningham, E. T., Jr., Mu, Y., Geleziunas, R., and Greene, W. C., The proto-oncogene Cot kinase participates in CD3/CD28 induction of NF-κB acting through the NF-κB-inducing kinase and IκB kinases, *Immunity,* 10, 271, 1999.
25. Zamanian-Daryoush, M., Mogensen, T. H., DiDonato, J. A., and Williams, B. R., NF-κB activation by double-stranded-RNA-activated protein kinase (PKR) is mediated through NF-κB-inducing kinase and IκB kinase, *Mol. Cell. Biol.,* 20, 1278, 2000.
26. Hehner, S. P., Hofmann, T. G., Ushmorov, A., Dienz, O., Wing-Lan Leung, I., Lassam, N., Scheidereit, C., Droge, W., and Schmitz, M. L., Mixed-lineage kinase 3 delivers CD3/CD28-derived signals into the IκB kinase complex, *Mol. Cell Biol.,* 20, 2556, 2000.
27. Minden, A., Lin, A., McMahon, M., Lange-Carter, C., Derijard, B., Davis, R. J., Johnson, G. L., and Karin, M., Differential activation of ERK and JNK mitogen-activated protein kinases by Raf-1 and MEKK, *Science,* 266, 1719, 1994.
28. Yan, M., Dai, T., Deak, J. C., Kyriakis, J. M., Zon, L. I., Woodgett, J. R., and Templeton, D. J., Activation of stress-activated protein kinase by MEKK1 phosphorylation of its activator SEK1, *Nature,* 372, 798, 1994.
29. Lee, F. S., Peters, R. T., Dang, L. C., and Maniatis, T., MEKK1 activates both IκB kinase α and IκB kinase β, *Proc. Natl. Acad. Sci. U.S.A.,* 95, 9319, 1998.
30. Meyer, C. F., Wang, X., Chang, C., Templeton, D., and Tan, T. H., Interaction between c-Rel and the mitogen-activated protein kinase kinase kinase 1 signaling cascade in mediating κB enhancer activation, *J. Biol. Chem.,* 271, 8971, 1996.
31. Hirano, M., Osada, S., Aoki, T., Hirai, S., Hosaka, M., Inoue, J., and Ohno, S., MEK kinase is involved in tumor necrosis factor α-induced NF-κB activation and degradation of IκB-α, *J. Biol. Chem.,* 271, 13,234, 1996.
32. Yin, M. J., Christerson, L. B., Yamamoto, Y., Kwak, Y. T., Xu, S., Mercurio, F., Barbosa, M., Cobb, M. H., and Gaynor, R. B., HTLV-I Tax protein binds to MEKK1 to stimulate IκB kinase activity and NF-κB activation, *Cell,* 93, 875, 1998.

33. Kopp, E., Medzhitov, R., Carothers, J., Xiao, C., Douglas, I., Janeway, C. A., and Ghosh, S., ECSIT is an evolutionarily conserved intermediate in the Toll/IL-1 signal transduction pathway, *Genes Dev.,* 13, 2059, 1999.
34. Xia, Y., Makris, C., Su, B., Li, E., Yang, J., Nemerow, G. R., and Karin, M., MEK kinase 1 is critically required for c-Jun N-terminal kinase activation by proinflammatory stimuli and growth factor-induced cell migration, *Proc. Natl. Acad. Sci. U.S.A.,* 97, 5243, 2000.
35. DiDonato, J. A., Hayakawa, M., Rothwarf, D. M., Zandi, E., and Karin, M., A cytokine-responsive IκB kinase that activates the transcription factor NF-κB [see comments], *Nature,* 388, 548, 1997.
36. Malinin, N. L., Boldin, M. P., Kovalenko, A. V., and Wallach, D., MAP3K-related kinase involved in NF-κB induction by TNF, CD95 and IL-1, *Nature,* 385, 540, 1997.
37. Song, H. Y., Regnier, C. H., Kirschning, C. J., Goeddel, D. V., and Rothe, M., Tumor necrosis factor (TNF)-mediated kinase cascades: bifurcation of nuclear factor-κB and c-jun N-terminal kinase (JNK/SAPK) pathways at TNF receptor-associated factor 2, *Proc. Natl. Acad. Sci. U.S.A.,* 94, 9792, 1997.
38. Regnier, C. H., Song, H. Y., Gao, X., Goeddel, D. V., Cao, Z., and Rothe, M., Identification and characterization of an IκB kinase, *Cell,* 90, 373, 1997.
39. Nakano, H., Shindo, M., Sakon, S., Nishinaka, S., Mihara, M., Yagita, H., and Okumura, K., Differential regulation of IκB kinase α and β by two upstream kinases, NF-κB-inducing kinase and mitogen-activated protein kinase/ERK kinase kinase-1, *Proc. Natl. Acad. Sci. U.S.A.,* 95, 3537, 1998.
40. Natoli, G., Costanzo, A., Moretti, F., Fulco, M., Balsano, C., and Levrero, M., Tumor necrosis factor (TNF) receptor 1 signaling downstream of TNF receptor-associated factor 2. Nuclear factor κB (NFκB)-inducing kinase requirement for activation of activating protein 1 and NFκB but not of c-Jun N-terminal kinase/stress-activated protein kinase, *J. Biol. Chem.,* 272, 26,079, 1997.
41. Kirschning, C. J., Wesche, H., Merrill Ayres, T., and Rothe, M., Human Toll-like receptor 2 confers responsiveness to bacterial lipopolysaccharide, *J. Exp. Med.,* 188, 2091, 1998.
42. Muzio, M., Natoli, G., Saccani, S., Levrero, M., and Mantovani, A., The human Toll signaling pathway: divergence of nuclear factor κB and JNK/SAPK activation upstream of tumor necrosis factor receptor–associated factor 6 (TRAF6), *J. Exp. Med.,* 187, 2097, 1998.
43. Sylla, B. S., Hung, S. C., Davidson, D. M., Hatzivassiliou, E., Malinin, N. L., Wallach, D., Gilmore, T. D., Kieff, E., and Mosialos, G., Epstein-Barr virus-transforming protein latent infection membrane protein 1 activates transcription factor NF-κB through a pathway that includes the NF-κB-inducing kinase and the IκB kinases IKKα and IKKβ, *Proc. Natl. Acad. Sci. U.S.A.,* 95, 10,106, 1998.
44. Shinkura, R., Kitada, K., Matsuda, F., Tashiro, K., Ikuta, K., Suzuki, M., Kogishi, K., Serikawa, T., and Honjo, T., Alymphoplasia is caused by a point mutation in the mouse gene encoding Nf-κ β-inducing kinase, *Nat. Genet.,* 22, 74, 1999.
45. Peters, R. T., Liao, S. M., and Maniatis, T., IKKε is part of a novel PMA-inducible IκB kinase complex, *Mol. Cell.,* 5, 513, 2000.
46. Khwaja, A., Akt is more than just a Bad kinase [news; comment], *Nature,* 401, 33, 1999.
47. Xie, P., Browning, D. D., Hay, N., Mackman, N., and Ye, R. D., Activation of NF-κ B by bradykinin through a Gα(q)- and Gβ γ-dependent pathway that involves phosphoinositide 3-kinase and Akt, *J. Biol. Chem.,* 275, 24,907, 2000.
48. Delhase, M., Li, N., and Karin, M., Kinase regulation in inflammatory response [comment], *Nature,* 406, 367, 2000.
49. Madge, L. A. and Pober, J. S., A phosphatidylinositol 3-kinase/Akt pathway, activated by tumor necrosis factor or interleukin-1, inhibits apoptosis but does not activate NFκB in human endothelial cells, *J. Biol. Chem.,* 275, 15,458, 2000.
50. Rauch, B. H., Weber, A., Braun, M., Zimmermann, N., and Schror, K., PDGF-induced Akt phosphorylation does not activate NF-κB in human vascular smooth muscle cells and fibroblasts, *FEBS Lett.,* 481, 3, 2000.
51. Sizemore, N., Leung, S., and Stark, G. R., Activation of phosphatidylinositol 3-kinase in response to interleukin-1 leads to phosphorylation and activation of the NF-κB p65/RelA subunit, *Mol. Cell. Biol.,* 19, 4798, 1999.
52. Madrid, L. V., Wang, C. Y., Guttridge, D. C., Schottelius, A. J., Baldwin, A. S., Jr., and Mayo, M. W., Akt suppresses apoptosis by stimulating the transactivation potential of the RelA/p65 subunit of NF-κB, *Mol. Cell. Biol.,* 20, 1626, 2000.

53. Kane, L. P., Shapiro, V. S., Stokoe, D., and Weiss, A., Induction of NF-κB by the Akt/PKB kinase, *Curr. Biol.,* 9, 601, 1999.
54. Pullen, N., Dennis, P. B., Andjelkovic, M., Dufner, A., Kozma, S. C., Hemmings, B. A., and Thomas, G., Phosphorylation and activation of p70s6k by PDK1 [see comments], *Science,* 279, 707, 1998.
55. Peterson, R. T. and Schreiber, S. L., Kinase phosphorylation: keeping it all in the family, *Curr. Biol.,* 9, R521, 1999.
56. Teramoto, H., Coso, O. A., Miyata, H., Igishi, T., Miki, T., and Gutkind, J. S., Signaling from the small GTP-binding proteins Rac1 and Cdc42 to the c-Jun N-terminal kinase/stress-activated protein kinase pathway. A role for mixed lineage kinase 3/protein-tyrosine kinase 1, a novel member of the mixed lineage kinase family, *J. Biol. Chem.,* 271, 27,225, 1996.
57. Sakurai, H., Miyoshi, H., Toriumi, W., and Sugita, T., Functional interactions of transforming growth factor β-activated kinase 1 with IκB kinases to stimulate NF-κB activation, *J. Biol. Chem.,* 274, 10,641, 1999.
58. Ninomiya-Tsuji, J., Kishimoto, K., Hiyama, A., Inoue, J., Cao, Z., and Matsumoto, K., The kinase TAK1 can activate the NIK-I κB as well as the MAP kinase cascade in the IL-1 signalling pathway, *Nature,* 398, 252, 1999.
59. Chen, F., Ding, M., Lu, Y., Leonard, S. S., Vallyathan, V., Castranova, V., and Shi, X., Participation of MAP kinase p38 and IκB kinase in chromium (VI)- induced NF-κB and AP-1 activation, *J. Environ. Pathol. Toxicol. Oncol.,* 19, 231, 2000.
60. Huang, C., Chen, N., Ma, W. Y., and Dong, Z., Vanadium induces AP-1- and NFκB-dependent transcription activity, *Int. J. Oncol.,* 13, 711, 1998.
61. Kaltreider, R. C., Pesce, C. A., Ihnat, M. A., Lariviere, J. P., and Hamilton, J. W., Differential effects of arsenic(III) and chromium(VI) on nuclear transcription factor binding, *Mol. Carcinog.,* 25, 219, 1999.
62. Roussel, R. R. and Barchowsky, A., Arsenic inhibits NF-κB-mediated gene transcription by blocking IκB kinase activity and IκBα phosphorylation and degradation, *Arch. Biochem. Biophys.,* 377, 204, 2000.
63. Shumilla, J. A., Broderick, R. J., Wang, Y., and Barchowsky, A., Chromium(VI) inhibits the transcriptional activity of nuclear factor-κB by decreasing the interaction of p65 with cAMP-responsive element-binding protein-binding protein, *J. Biol. Chem.,* 274, 36,207, 1999.
64. Imbert, V., Rupec, R. A., Livolsi, A., Pahl, H. L., Traenckner, E. B., Mueller-Dieckmann, C., Farahifar, D., Rossi, B., Auberger, P., Baeuerle, P. A., and Peyron, J. F., Tyrosine phosphorylation of IκB-α activates NF-κ, B without proteolytic degradation of I κ, B-α, *Cell,* 86, 787, 1996.
65. Chen, F., Demers, L. M., Vallyathan, V., Ding, M., Lu, Y., Castranova, V., and Shi, X., Vanadate induction of NF-κB involves IκB kinase β and SAPK/ERK kinase 1 in macrophages, *J. Biol. Chem.,* 274, 20,307, 1999.
66. Mukhopadhyay, A., Manna, S. K., and Aggarwal, B. B., Pervanadate-induced nuclear factor-κB activation requires tyrosine phosphorylation and degradation of IκBα, Comparison with tumor necrosis factor-α, *J. Biol. Chem.,* 275, 8549, 2000.
67. Krejsa, C. M., Nadler, S. G., Esselstyn, J. M., Kavanagh, T. J., Ledbetter, J. A., and Schieven, G. L., Role of oxidative stress in the action of vanadium phosphotyrosine phosphatase inhibitors. Redox independent activation of NF-κB, *J. Biol. Chem.,* 272, 11,541, 1997.
68. Barbeau, B., Bernier, R., Dumais, N., Briand, G., Olivier, M., Faure, R., Posner, B. I., and Tremblay, M., Activation of HIV-1 long terminal repeat transcription and virus replication via NF-κB-dependent and -independent pathways by potent phosphotyrosine phosphatase inhibitors, the peroxovanadium compounds, *J. Biol. Chem.,* 272, 12,968, 1997.
69. Shi, X., Chiu, A., Chen, C. T., Halliwell, B., Castranova, V., and Vallyathan, V., Reduction of chromium (VI) and its relationship to carcinogenesis, *J. Toxicol. Environ. Health B Crit. Rev.,* 2, 87, 1999.
70. Shi, X. L. and Dalal, N. S., Chromium (V) and hydroxyl radical formation during the glutathione reductase-catalyzed reduction of chromium (VI), *Biochem. Biophys. Res. Commun.,* 163, 627, 1989.
71. Shi, X. L. and Dalal, N. S., NADPH-dependent flavoenzymes catalyze one electron reduction of metal ions and molecular oxygen and generate hydroxyl radicals, *FEBS Lett.,* 276, 189, 1990.
72. Shi, X. L. and Dalal, N. S., The role of superoxide radical in chromium (VI)-generated hydroxyl radical: the Cr(VI) Haber-Weiss cycle, *Arch. Biochem. Biophys.,* 292, 323, 1992.
73. Davies, K. J., Oxidative stress: the paradox of aerobic life, *Biochem. Soc. Symp.,* 61, 1, 1995.

74. Buzard, G. S. and Kasprzak, K. S., Possible roles of nitric oxide and redox cell signaling in metal-induced toxicity and carcinogenesis: a review, *J. Environ. Pathol. Toxicol. Oncol.*, 19, 179, 2000.
75. Bowie, A. and O'Neill, L. A., Oxidative stress and nuclear factor-κB activation: a reassessment of the evidence in the light of recent discoveries, *Biochem. Pharmacol.*, 59, 13, 2000.
76. Schreck, R., Meier, B., Mannel, D. N., Droge, W., and Baeuerle, P. A., Dithiocarbamates as potent inhibitors of nuclear factor κB activation in intact cells, *J. Exp. Med.*, 175, 1181, 1992.
77. Li, N. and Karin, M., Is NF-κB the sensor of oxidative stress? *FASEB J.*, 13, 1137, 1999.
78. Roederer, M., Staal, F. J., Raju, P. A., Ela, S. W., and Herzenberg, L. A., Cytokine-stimulated human immunodeficiency virus replication is inhibited by *N*-acetyl-L-cysteine, *Proc. Natl. Acad. Sci. U.S.A.*, 87, 4884, 1990.
79. Matthews, J. R., Wakasugi, N., Virelizier, J. L., Yodoi, J., and Hay, R. T., Thioredoxin regulates the DNA binding activity of NF-κB by reduction of a disulphide bond involving cysteine 62, *Nucl. Acids Res.*, 20, 3821, 1992.
80. Matthews, J. R., Watson, E., Buckley, S., and Hay, R. T., Interaction of the C-terminal region of p105 with the nuclear localisation signal of p50 is required for inhibition of NF-κB DNA binding activity, *Nucl. Acids Res.*, 21, 4516, 1993.
81. Matthews, J. R., Kaszubska, W., Turcatti, G., Wells, T. N., and Hay, R. T., Role of cysteine62 in DNA recognition by the P50 subunit of NF-κB, *Nucl. Acids Res.*, 21, 1727, 1993.
82. Cavigelli, M., Li, W. W., Lin, A., Su, B., Yoshioka, K., and Karin, M., The tumor promoter arsenite stimulates AP-1 activity by inhibiting a JNK phosphatase, *EMBO J.*, 15, 6269, 1996.
83. Pugazhenthi, S., Tanha, F., Dahl, B., and Khandelwal, R. L., Inhibition of a Src homology 2 domain containing protein tyrosine phosphatase by vanadate in the primary culture of hepatocytes, *Arch. Biochem. Biophys.*, 335, 273, 1996.
84. Denu, J. M. and Dixon, J. E., Protein tyrosine phosphatases: mechanisms of catalysis and regulation, *Curr. Opin. Chem. Biol.*, 2, 633, 1998.
85. Fetrow, J. S., Siew, N., and Skolnick, J., Structure-based functional motif identifies a potential disulfide oxidoreductase active site in the serine/threonine protein phosphatase-1 subfamily, *FASEB J.*, 13, 1866, 1999.
86. Solomon, A., Rosenblatt, M., Li, D. Q., Liu, Z., Monroy, D., Ji, Z., Lokeshwar, B. L., and Pflugfelder, S. C., Doxycycline inhibition of interleukin-1 in the corneal epithelium, *Invest. Ophthalmol. Vis. Sci.*, 41, 2544, 2000.

17 Pyrosequencing: A Bioluminometric Method of DNA Sequencing*

Mostafa Ronaghi, Nader Pourmand, and Viktor Stolc

CONTENTS

17.1 INTRODUCTION

The development of DNA sequence determination techniques with enhanced speed, sensitivity, and throughput is of utmost importance for the study of biological systems. Conventional DNA sequencing relies on the elegant principle of the dideoxy chain termination technique first described more than two decades ago.[1] This multistep principle has gone through major improvements during the years to make it a robust technique that has been used for sequencing of several different bacterial, archeal, and eucaryotic genomes (www.ncbi.nlm.nih.gov, and www.tigr.org). However, this technique faces limitations in both throughput and cost for most future applications. Many research groups around the world have put efforts into the development of alternative principles for DNA sequencing. Three methods that hold great promise are sequencing by hybridization,[2–5] parallel signature sequencing based on ligation and cleavage,[6] and Pyrosequencing™ [7,8].

Pyrosequencing is based on the detection of released pyrophosphate during DNA synthesis. In a cascade of enzymatic reactions, visible light is generated that is proportional to the number of incorporated nucleotides. The cascade starts with a DNA polymerization reaction in which inorganic pyrophosphate (PPi) is released as a result of nucleotide incorporation by polymerase (Figure 17.1). The released PPi is subsequently converted to adenosine triphosphate (ATP) by ATP sulfurylase. The synthesized (ATP) provides the energy to luciferase to oxidize luciferin and generate light.

* Parts of this chapter are reproduced from M. Ronaghi, *Genome Res.*, 11(1), 2095–2103, 2001. With permission.

0-8493-0719-8/02/$0.00+$1.50

$$(\mathrm{DNA})_n + \mathrm{Nucleotide} \xrightarrow{\text{Polymerase}} (\mathrm{DNA})_{n+1} + \mathrm{PPi}$$

$$\mathrm{PPi} + \mathrm{APS} \xrightarrow{\text{ATP sulfurylase}} \mathrm{ATP} + \mathrm{SO_4^{2-}}$$

$$\mathrm{ATP} + \mathrm{Luciferin} + \mathrm{O_2} \xrightarrow{\text{Luciferase}} \mathrm{AMP} + \mathrm{PPi} + \mathrm{Oxyluciferin} + \mathrm{CO_2} + \mathrm{Light}$$

$$\mathrm{ATP} + \mathrm{dNTP} \xrightarrow{\text{Apyrase}} \mathrm{AMP} + \mathrm{dNMP} + 4\mathrm{Pi}$$

FIGURE 17.1 The general principle behind different Pyrosequencing reaction systems. A polymerase catalyzes incorporation of nucleotide(s) into a nucleic acid chain. As a result of the incorporation, a PPi molecule(s) is released and subsequently converted to ATP, by ATP sulfurylase. Light is produced in the luciferase reaction during which a luciferin molecule is oxidized. The left over nucleotides and the generated ATP are degraded by the enzyme apyrase. Nucleotides are added one by one within a minute interval and the target sequence is determined.

Unincorporated deoxy nucleotides and ATP are degraded by the enzyme apyrase. Because the identity of the added nucleotide is known at each step of the light signal generation, the sequence of the template can be determined. Standard Pyrosequencing uses the Klenow fragment of *Escherichia coli* DNA polymerase I, which is a relatively slow polymerase.[9] The ATP sulfurylase used in pyrosequencing is a recombinant version from the yeast *Saccharomyces cerevisiae.*[10] The luciferase is from the American firefly *Photinus pyralis*[11] and the nucleotide-degrading enzyme apyrase is from potato tubers, *Solanum tuberosum.*[12] The polymerization and detection reactions take place within 3 to 4 s at room temperature. Efficient enzymatic nucleotide removal takes approximately 20 to 30 s. Recent improvements in the pyrosequencing reaction include the use of dATPαS instead of dATP in polymerization,[7] inclusion of apyrase,[8] and inclusion of single-stranded DNA-binding protein.[13] These improvements have resulted in a read-length of more than 100 nucleotides. This chapter describes the use of this technique for accurate analysis of single nucleotide polymorphisms (SNPs), tag sequencing, and microbial typing.

17.2 MATERIALS AND METHODS

17.2.1 *In Vitro* Amplification and Template Preparation

Polymerase chain reaction (PCR) was performed according to the standard protocols using one 5′ biotinylated primer. The products were immobilized onto streptavidin-coated paramagnetic beads according to the suppliers recommendations (Dynal A.S., Oslo, Norway). Single-stranded DNA was obtained by removing the supernatant after incubating the immobilized PCR product in 0.10 *M* NaOH for 3 min. Sequencing primer was hybridized to the immobilized single-stranded DNA strand in 10 m*M* Tris-acetate pH 7.5, and 20 m*M* magnesium acetate. For multiplex Pyrosequencing, several sequencing primers were hybridized to a single template simultaneously.

17.2.2 Pyrosequencing

Pyrosequencing was performed at room temperature in a volume of 50 μl on a PSQ™ 96 system (www.pyrosequencing.com). The DNA template with hybridized primer was added to the Pyrosequencing reaction mixture containing: 10 U exonuclease-deficient Klenow DNA polymerase (Amersham Pharmacia Biotech, Uppsala, Sweden), 40 mU apyrase (Sigma Chemical Co., St. Louis, MO),

100 ng purified luciferase (BioThema, Dalarö, Sweden), 15 mU of recombinant produced ATP sulfurylase,[10] 0.1 *M* Tris-acetate (pH 7.75), 0.5 m*M* EDTA, 5 m*M* magnesium acetate, 0.1% bovine serum albumin, 1 m*M* dithiothreitol, 5 μM adenosine 5′-phosphosulfate (APS), 0.4 mg/ml polyvinylpyrrolidone (360 000), and 100 μg/ml D-luciferin (BioThema). For determination of long reads of DNA sequence (<100 nucleotides) and for Pyrosequencing of difficult templates with high GC content and potential for the formation of secondary structure, 0.5 μg SSB (Amersham Pharmacia Biotech) was added to the DNA template. For multiplex Pyrosequencing, three sequencing primers were hybridized on a single-stranded DNA template. The sequencing procedure was carried out by stepwise elongation of the primer strand upon sequential addition of the different deoxynucleoside triphosphates (Amersham Pharmacia Biotech) and simultaneous degradation of nucleotides by apyrase. The output of light resulting from nucleotide incorporation was detected by a charge-coupled device (CCD) camera in the PSQ 96 System.

17.3 RESULTS AND DISCUSSION

Pyrosequencing is emerging as a widely applicable technology for the detailed characterization of nucleic acids. The DNA template can be sequenced in real time enabling very fast analysis. Using 1 pmol of DNA, 6×10^{11} ATP molecules can be obtained, which, in turn, generate more than 6×10^{9} photons at a wavelength of 560 nm. This amount of light is easily detected by a photodiode, photomultiplier tube, or a CCD camera.

An automated version of a Pyrosequencing machine (PSQ 96 System) was recently developed (www.pyrosequencing.com). The automated version of Pyrosequencing uses a disposable ink-jet cartridge for precise delivery of small volume (200 nl) of six different reagents into a temperature-controlled microtiter plate (Figure 17.2). The microtiter plate is under continuous agitation to increase the rate of reagent mixing. A lens array is used to focus the generated luminescence efficiently from each individual well of the microtiter plate onto the chip of a CCD camera.

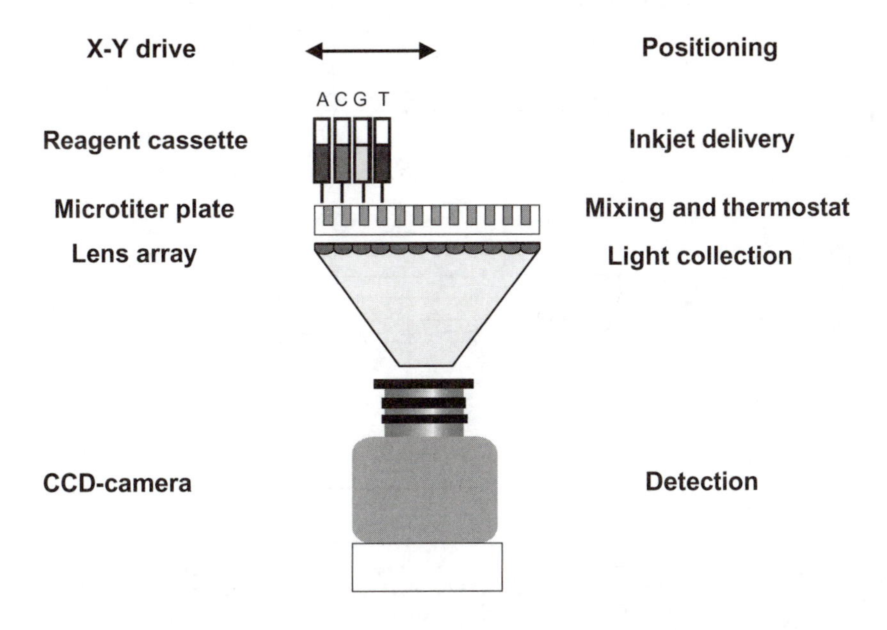

FIGURE 17.2 Schematic drawing of the automated system for Pyrosequencing. Four dispensers move on an *X-Y* robotics arm over the microtiter plate and add four different nucleotides, according to the prespecified order. The microtiter plate is agitated continuously to mix the added nucleotide. Generated light is directed to the CCD camera using a lens-array located exactly below the microtiter plate. (From Ronaghi, M., *Genome Res.*, 11(1), 2095–2103, 2001. With permission.)

Nucleotides are dispensed into alternating wells with a delay to minimize mixing of generated light between different wells. A cooled CCD camera images the plate every second to follow the exact process of the Pyrosequencing reaction. Data acquisition modules and an interface for PC connection are used in this instrument. Pyrosequencing software running in Microsoft Windows enables individual control of the dispensing order for each well in the PSQ 96 system.

Prior to initiating Pyrosequencing, the reagents and each of the four nucleotides are loaded into the ink-jet cartridge that is mounted in the instrument. A microtiter plate containing the DNA template with hybridized primer is placed into the Pyrosequencing machine, and after the enzymes and substrate have been delivered by the ink-jet, nucleotides are added to the solution according to the specified order. The Pyrosequencing signals in a pyrogram (Figures 17.3 through 17.5) show high-quality sequence data with high signal-to-noise ratio. The height of the peaks is proportional to the number of incorporated nucleotides. A 384-well microtiter plate-based high-throughput version of this machine is also under development, which will allow the analysis of up to 50,000 SNPs per 8 h and reduce the cost for genotyping below 20¢ per sample (www.pyrosequencing.com).

In addition to Pyrosequencing, the PSQ 96 System is a high-precision 96-well luminometer that can be used for any enzyme-screening assay that generates a light signal. For example, the PSQ 96 System can be used for any high-throughput ELISA-based immunoassay. It can also be used for screening of all enzyme activities found in yeast cells, by using genome-wide epitope-tagged yeast strains, each containing a different yeast open reading frame (ORF) fused to glutathione *S*-transferase (GST).[14,15] The PSQ 96 System has already been used for a number of applications.[16] Three main applications of this machine are presented and discussed below.

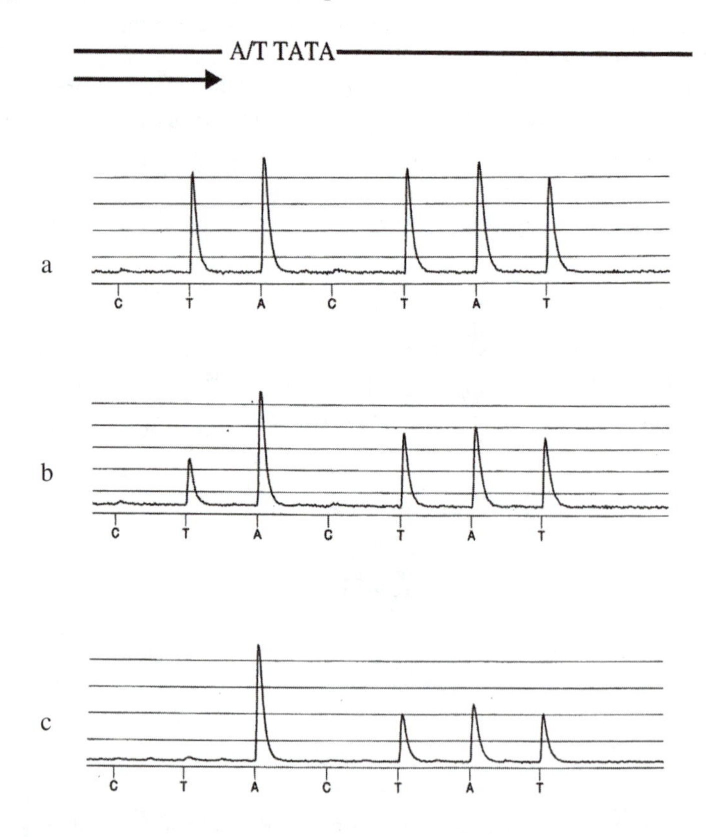

FIGURE 17.3 Pyrogram of the raw data obtained from Pyrosequencing of two different kinds of homozygous alleles (a and c) and a heterozygote DNA template (b). The order of nucleotide additions is indicated below the pyrogram, and the obtained sequence is indicated above the pyrogram. The height of the peaks determines the number of incorporated nucleotides. The time between addition of each nucleotide is 1 min.

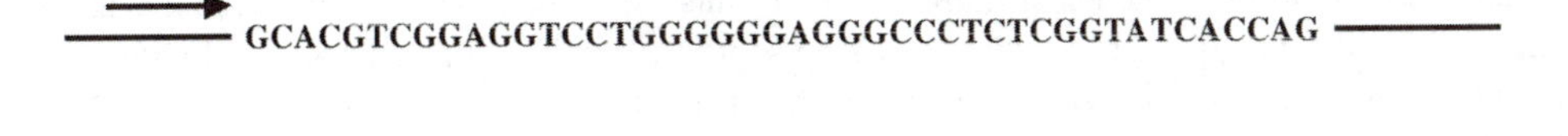

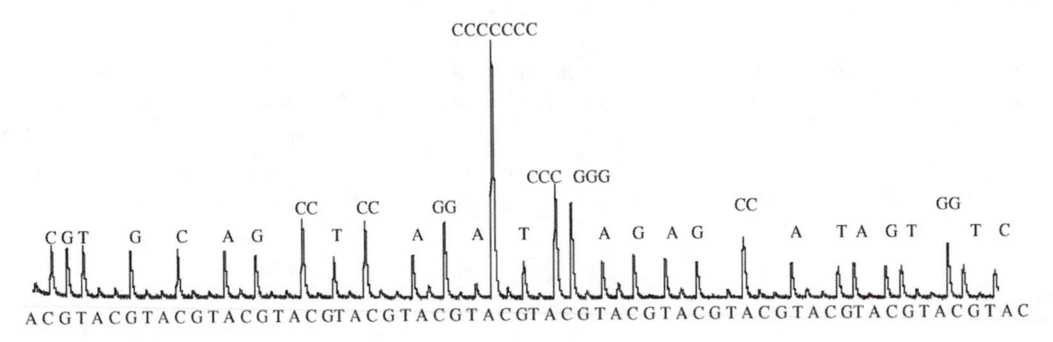

FIGURE 17.4 Pyrogram of the raw data obtained from *de novo* sequencing of a 230-nucleotide-long PCR product using the PSQ 96 System. The order of the nucleotide additions is indicated below the pyrogram and the obtained sequence is indicated above the pyrogram.

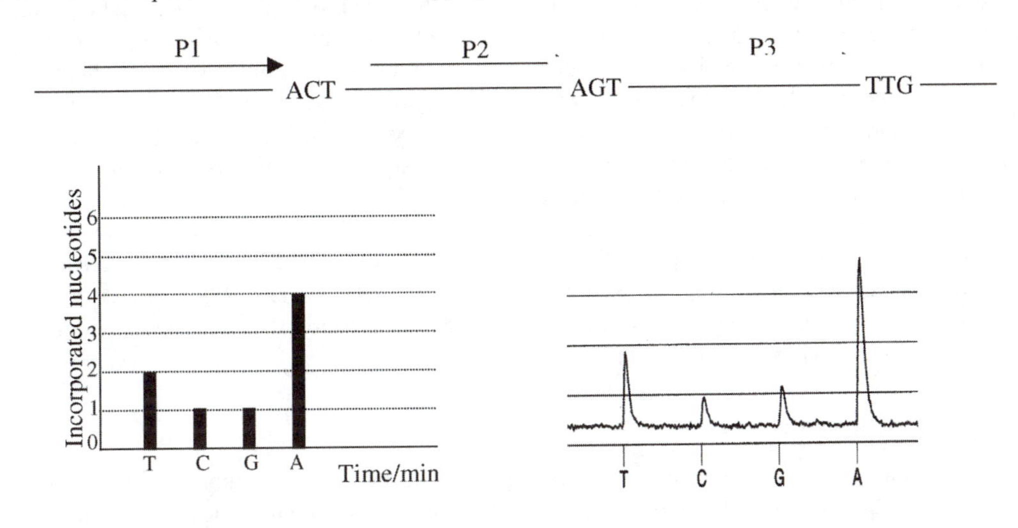

FIGURE 17.5 Pyrogram of the raw data obtained from multiplex Pyrosequencing of one of the subtypes of hepatitis virus C. Three sequencing primers (P1, P2, and P3) were hybridized to a single template (top panel), and Pyrosequencing resulted in a unique fingerprint for a specific genotype. The order of the nucleotide additions is indicated below the pyrogram and the obtained sequence is indicated above the pyrogram. The expected theoretical pattern (left) and the raw data (right) are shown. See Materials and Methods for more information.

17.3.1 Pyrosequencing for SNP Analysis

For analysis of SNPs by Pyrosequencing, the 3′-end of a primer is designed to hybridize one or a few bases 5′ adjacent to the polymorphic position. In a single tube, all the different variations can be determined as the region is sequenced. A striking feature of pyrogram readouts for SNP analysis is the clear distinction between the various genotypes of each allele combination. Thus, homozygous or heterozygous alleles result in a specific pattern.[17–20] This feature makes genotyping extremely accurate and easy. Relative standard deviation values for the ratio between key peaks of the respective SNPs and reference counterparts are 0.1 or lower.[18] Simple manual comparison of predicted SNP patterns and the raw data obtained from the PSQ 96 System can score an SNP, especially as no editing is needed (Figure 17.3). Because specific patterns can be readily achieved for the individual SNPs, it will also be possible to score the allelic status automatically with pattern recognition software. In a study based on results from three different laboratories, 26 different SNPs and more than 1600 DNA samples were analyzed. The algorithm classified the data from 94% of the samples as good or medium quality, and 99,4% of these were automatically assigned the expected genotypes. The primary source of any low data quality was insufficient signal-to-noise ratio, typically caused by low efficiency in PCR amplification. As Pyrosequencing signals are very quantitative, it is possible to use this strategy to examine allelic frequency in large population. Furthermore, Pyrosequencing enables determination of the phase of SNPs when they are in the vicinity of each other, allowing the detection of haplotypes.[21] This system allows more than 5000 samples to be analyzed in 8 h.

17.3.2 Pyrosequencing for Tag Sequencing

Tag sequencing can be used for *de novo* sequencing and resequencing. This includes applications such as partial cDNA sequencing,[22,23] microbial typing,[24] resequencing of disease gene,[25] and identification of oligonucleotide bar-code sequences (http://www-sequence.stanford.edu/group/yeast_deletion_project/deletions3.html).[15] For partial cDNA analysis, theoretically, eight or nine nucleotides in a row should define a unique sequence for every gene in the human genome. However, it has been found that to identify a gene uniquely from a complex organism such as a human, a longer sequence of DNA is needed.[22] In a pilot study, it was found that 98% of genes could be uniquely identified by sequencing a length of 30 nucleotides. Pyrosequencing was used to sequence this length for gene identification from a human cDNA library and the results were in complete agreement with longer sequence data obtained by Sanger DNA sequencing.

Pyrosequencing offers high-throughput analysis of cDNA libraries since 96 samples can be analyzed in less than 1 h. Like Sanger DNA sequencing, Pyrosequencing also has the advantage of library screening, because the original cDNA clone is directly available for further analysis (Figure 17.4). Pyrosequencing has also been used to sequence 10 to 50 nucleotides for semiconserved genes for viral, bacterial, and fungal typing.[24]

17.3.3 Multiplex Pyrosequencing

Pyrosequencing takes advantage of enzymatic reactions to determine the sequence of the DNA. Proportional signals obtained in a pyrogram represent the number of nucleotides incorporated by DNA polymerase. By using more than one sequencing primer in a Pyrosequencing reaction, a sequence-specific fingerprint can be obtained in which the height of each peak determines the number of incorporated nucleotides. The expected pattern for one subtype of hepatitis C virus and the raw data from Pyrosequencing presenting the subtype are demonstrated in Figure 17.5.

17.3.4 Challenges in Pyrosequencing

An inherent problem with the described method is *de novo* sequencing of polymorphic regions in heterozygous DNA material. In most cases, it will be possible to detect the polymorphism. If the polymorphism is a substitution, it will be possible to obtain a synchronized extension after the substituted nucleotide. If the polymorphism is a deletion or insertion of the same kind as the adjacent nucleotide on the DNA template, the sequence after the polymorphism will be synchronized. However, if the polymorphism is a deletion or insertion of another type, the sequencing reaction can become out of phase, making the interpretation of the subsequent sequence difficult. If the polymorphism is known, it is always possible to use programmed nucleotide delivery to keep the extension of different alleles synchronized after the polymorphic region. It is also possible to use a bidirectional approach,[26] whereby the complementary strand is sequenced to decipher the sequence flanking the polymorphism.

Another inherent problem is the difficulty in determining the number of incorporated nucleotides in homopolymeric regions, because of the nonlinear light response following incorporation of more than five to six identical nucleotides. The polymerization efficiency over homopolymeric regions has been investigated and the results indicate that it is possible to incorporate as many as ten identical adjacent nucleotides in the presence of apyrase.[13] However, to elucidate the correct number of incorporated nucleotides, it may be necessary to use specific software algorithms that integrate the signals. For resequencing, it is possible to add the same nucleotide twice for a homopolymeric region to ensure complete polymerization, and for the tag software to automatically sum the signals representing the correct number of incorporated nucleotides.[27]

17.4 CONCLUSION

Pyrosequencing technology is relatively new and there is much room for developments in both chemistry and in instrumentation. The technology is already time- and cost-competitive when compared with the existing sequencing methods. The current cost is 69¢ per sample using standard Pyrosequencing. With the 384-format machine, the cost will be reduced below 20¢. Work is under way to improve the chemistry further, to measure the sequencing efficiency at elevated temperatures, and to run the reaction in miniaturized formats. The advantage of Pyrosequencing in miniaturized formats may lie in the ease with which large numbers of high-density arrays can be manufactured and the future integration of sample preparation with these devices. Success in miniaturization of this technique into high-density microtiter plates, microarrays, or microfluidics[28] will reduce the cost and increase the throughput by one to two orders of magnitude, a crucial step for large-scale genetic testing.

ACKNOWLEDGMENTS

The authors are supported by National Institutes of Health grants and the Cancer Research Fund of the Damon Runyon–Walter Winchell Foundation. We thank Ronald Davis for valuable discussion. The authors also acknowledge *Genome Research* for permission to republish Figure 17.2 and some parts of the text of this chapter.

REFERENCES

1. Sanger, F., Nicklen, S., and Coulson, A. R., DNA sequencing with chain-terminating inhibitors, *Proc. Natl. Acad. Sci. U.S.A.,* 74, 5463–5467, 1977.
2. Bains, W. and Smith, G. C., A novel method for nucleic acid sequence determination, *J. Theoret. Biol.,* 135, 303–307, 1988.
3. Drmanac, R., Labat, I., Brukner, I., and Crkvenjakov, R., Sequencing of megabase plus DNA by hybridization: theory of the method, *Genomics,* 4, 114–128, 1989.
4. Khrapko, K. R., Lysov Yu, P., Khorlyn, A. A., Shick, V. V., Florentiev, V. L., and Mirzabekov, A. D., An oligonucleotide hybridization approach to DNA sequencing, *FEBS Lett.,* 256, 118–122, 1989.
5. Southern, E. M., Analysing polynucleotide sequences, Patent WO/10977, 1989.
6. Brenner, S., Williams, S. R., Vermaas, E. H., Storck, T., Moon, K., McCollum, C., Mao, J. I., Luo, S., Kirchner, J. J., Eletr, S., DuBridge, R. B., Burcham, T., and Albrecht, G., *In vitro* cloning of complex mixtures of DNA on microbeads: physical separation of differentially expressed cDNAs, *Proc. Natl. Acad. Sci. U.S.A.,* 97, 1665–1670, 2000.
7. Ronaghi, M., Karamohamed, S., Pettersson, B., Uhlen, M., and Nyren, P., Real-time DNA sequencing using detection of pyrophosphate release, *Anal. Biochem.,* 242, 84–89, 1996.

8. Ronaghi, M., Uhlen, M., and Nyren, P., A sequencing method based on real-time pyrophosphate, *Science,* 281, 363–365, 1998.
9. Benkovic, S. J. and Cameron, C. E., Kinetic analysis of nucleotide incorporation and misincorporation by Klenow fragment of *Escherichia coli* DNA polymerase I, *Methods Enzymol.,* 262, 257–269, 1995.
10. Karamohamed, S., Nilsson, J., Nourizad, K., Ronaghi, M., Pettersson, B., and Nyren, P., Production, purification, and luminometric analysis of recombinant *Saccharomyces cerevisiae* MET3 adenosine triphosphate sulfurylase expressed in *Escherichia coli, Prot. Exp. Purif.,* 15, 381–388, 1999.
11. DeLuca, M. and McElroy, W. D., Two kinetically distinguishable ATP sites in firefly luciferase, *Biochem. Biophys. Res. Commun.,* 123, 764–770, 1984.
12. Handa, M. and Guidotti, G., Purification and cloning of a soluble ATP-diphosphohydrolase (apyrase) from potato tubers (*Solanum tuberosum*), *Biochem. Biophys. Res. Commun.,* 218, 916–923, 1996.
13. Ronaghi, M., Improved performance of Pyrosequencing using single-stranded DNA-binding protein, *Anal. Biochem.,* 286, 282–288, 2000.
14. Martzen, M. R., McCraith, S. M., Spinelli, S. L., Torres, F. M., Fields, S., Grayhack, E. J., Phizicky, E. M. A biochemical genomics approach for identifying genes by the activity of their products, *Science,* 286, 1153–1155, 1999.
15. Eason, R., Corl, A., Sabnis, A., Webb, C., Karhanek, M., Davis, R., Stolc, V., Tag sequence-based characterization of the *Saccharomyces cerevisiae* yeast deletion strains, submitted.
16. Ronaghi, M., Pyrosequencing sheds light on DNA sequencing, *Genome Res.,* 11(1), 2095–2103, 2001.
17. Ahmadian, A., Gharizadeh, B., Gustafsson, A. C., Sterky, F., Nyren, P., Uhlen, M., and Lundeberg, J., Single-nucleotide polymorphism analysis by Pyrosequencing, *Anal. Biochem.,* 280, 103–110, 2000.
18. Alderborn, A., Kristofferson, A., and Hammerling, U., Determination of single nucleotide polymorphisms by real-time pyrophosphate DNA sequencing, *Genome Res.,* 10, 1249–1258, 2000.
19. Ekstrom, B., Alderborn, A., and Hammerling, U., Pyrosequencing for SNPs, *Prog. Biomed. Optics,* 1, 134–139, 2000.
20. Nordstrom, T., Nourizad, K., Ronaghi, M., and Nyren, P., Methods enabling Pyrosequencing on double-stranded DNA, *Anal. Biochem.,* 282, 186–193, 2000.
21. Ahmadian, A., Lundeberg, J., Nyren, P., Uhlen, M., and Ronaghi, M., Analysis of the *p53* tumor suppressor gene by pyrosequencing, *BioTechniques,* 28, 140–144, 2000.
22. Nordstrom, T., Gharizadeh, B., Pourmand, N., Nyren, P., and Ronaghi, M., Method enabling fast partial sequencing of cDNA clones, *Anal. Biochem.,* 292, 266–271, 2001.
23. Ronaghi, M., Pettersson, B., Uhlen, M., and Nyren, P., PCR-introduced loop structure as primer in DNA sequencing, *BioTechniques,* 25, 876–884, 1998.
24. Gharizadeh, B., Kalantari, M., Garcia, C. A., Johansson, B., Nyren, P., Typing of human papillomavirus by pyrosequencing, *Lab. Invest.,* 81, 673–679, 2001.
25. Garcia, A. C., Ahamdian, A., Gharizadeh, B., Lundeberg, J., Ronaghi, M., and Nyren, P., Mutation detection by Pyrosequencing: sequencing of exons 5 to 8 of the p53 tumour supressor gene, *Gene,* 253, 249–257, 2000.
26. Ronaghi, M., Nygren, M., Lundeberg, J., and Nyren, P., Analyses of secondary structures in DNA by pyrosequencing, *Anal. Biochem.,* 267, 65–71, 1999.
27. Ronaghi, M., Pyrosequencing: A Tool for Sequence-Based DNA Analysis, Doctoral thesis, Royal Institute of Technology, Stockholm, Sweden, 1998.
28. Eckersten, A., Örlefors, A., E., Ellström, C., Erickson, A., Löfman, E., Eriksson, A., Eriksson, S., Jorsback, A., Tooke, N., Derand, H. et al., High-throughput SNP scoring in a disposable microfabricated CD device, in *Proceeding of the Micro Total Analysis Systems,* Kluwer Academic, Amsterdam, 2000, 521–524.

18 Pathway-Specific cDNA Array Using Luminescent Detection

Knox Van Dyke

CONTENTS

18.1 INTRODUCTION

A variety of conventional methods exist to profile the expression of genes. Assays such as reverse transcription polymerase chain reaction (RT-PCR), Northern blot, and RNAse protection assay are generally concerned with one or several genes. At the other extreme, large-scale gene screens measure thousands of genes in an array on a single glass chip. The materials and instrumentation used in such systems are expensive. The high-density arrays, which are often constructed from whole plasmids or amplified inserts, can allow high levels of cross-hybridization. The use of oligo-dT_{18} or random primers can result in low sensitivity due to inefficient cDNA probe synthesis. Even when using gene-specific primers, sensitivity can be low because of the complexity of the primer mix required for reverse transcription of mRNA. Not surprisingly, the analysis and confirmation of these results can be time-consuming and disappointing. The GeArray™ represents a middle ground: an inexpensive method that offers high sensitivity in detecting multiple genes within a specific pathway.

18.2 GEARRAY TECHNOLOGY

GEArray (SuperArray, Inc., Bethesda, MD) is a nylon-based low-density array composed of duplicate spots of 23 different cDNAs specific to a particular biological pathway. Two housekeeping genes are included for data normalization (Figure 18.1). To perform an analysis, total cellular RNA is prepared and used to synthesize biotin-labeled cDNA probes. These probes are then hybridized with the cDNAs bound to the GEArray membrane. Alkaline phosphatase–streptavidin is added; then CDP-Star® is used to produce luminescence. Light intensity is measured using X-ray film or imaging. The relative expression level of each gene is determined by comparing its signal intensity to that of the two housekeeping genes (Figure 18.2).

The GEArray assay is quantitative and linear between 0.001 and 0.4% of total RNA (5 μg) with yields comparable to real-time quantitative PCR. This high sensitivity makes the GEArray particularly useful for detecting low abundance and rare mRNAs such as uninduced p53, $p21^{Waf1}$, and bax.

0-8493-0719-8/02/$0.00+$1.50

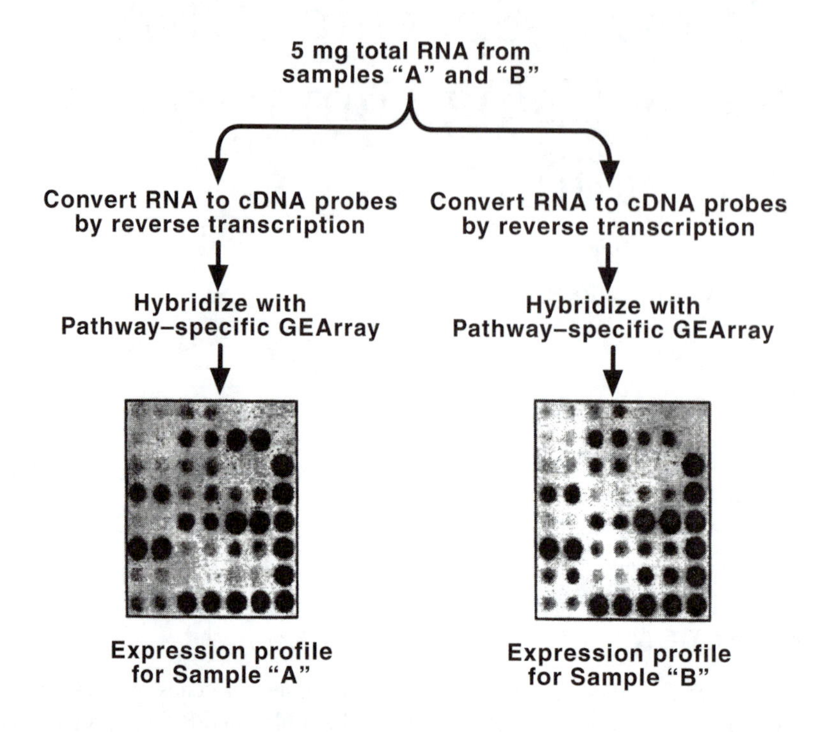

FIGURE 18.1 Hybridization expression array via luminescent assay.

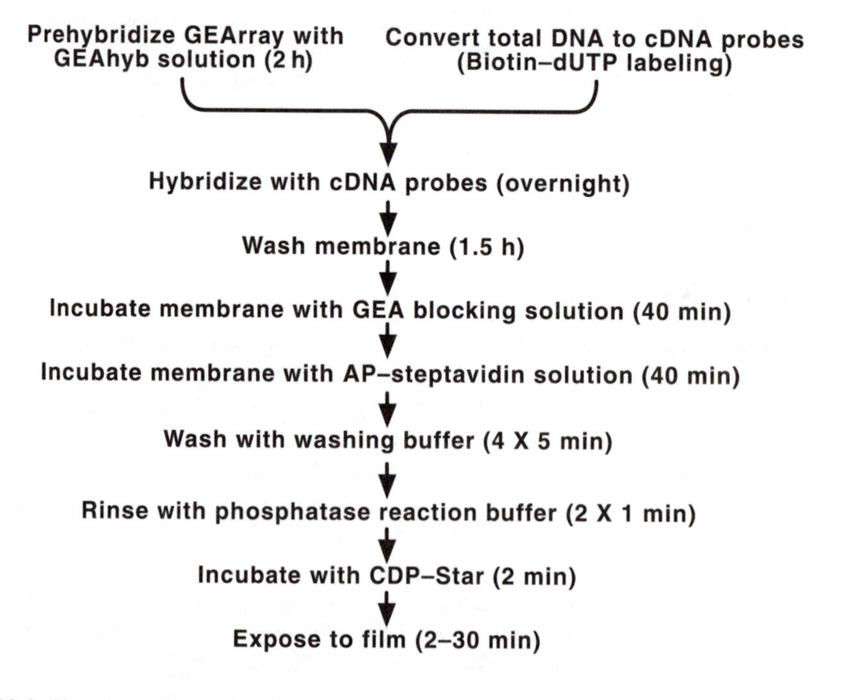

FIGURE 18.2 Flowchart of procedure for cDNA–RNA hybridization.

18.3 USES OF GEARRAY

More than 30 human and 15 mouse arrays are available. These arrays have been designed for studies in the following areas:

1. Apoptosis and stress/toxicity
2. Cell cycle
3. Oncogenes, tumor suppressor genes, angiogenesis, and metastasis
4. Signal transduction
5. Cytokines and the inflammatory response

The PathwayFinder array contains cDNAs from multiple signal transduction pathways and can be used to determine which pathways are affected by a particular process. Further analysis of the relevant pathways can then be performed by using more specific arrays (e.g., NF-κB, Jak-Stat, PI3 kinase/AKT). In addition, SuperArray will custom-manufacture a gene array provided by the investigator.

CONTACT INFORMATION

SuperArray, Inc.
P.O. Box 34494
Bethesda, MD 20827-0494, USA
Phone: 888-503-3187
Fax: 888-465-9859
Web: www.superarray.com
E-mail: info@superarray.com

Section IV

Cellular Luminescence

19 Application of Chemiluminescence in Phagocyte–Pathogen Interactions*

Susantha M. Gomis, Dale L. Godson, and Andrew A. Potter

CONTENTS

19.1 INTRODUCTION

19.1.1 PHAGOCYTES, PHAGOCYTOSIS, AND MICROBIAL KILLING

Phagocytosis is among the major effector mechanisms for the eradication of infectious agents and is a first line of defense. Using pattern recognition receptors that recognize characteristic microbial features, or receptors for other antimicrobial host factors bound to the microorganism, phagocytic

* Published with the permission of the Director of VIDO as journal series 275.

0-8493-0719-8/02/$0.00+$1.50

cells bind to microorganisms, internalize, and destroy them. Phagocytosis is performed by two different lineages of leukocytes, namely, polymorphonuclear granulocytes including neutrophils (PMN), basophils and eosinophils, and mononuclear phagocytes (MP) or macrophages. Because of their function, PMN and MP are often referred to as "professional phagocytes" to distinguish them from most other host cells that, although capable of limited uptake of extracellular material, are called "nonprofessional phagocytes."[1] To fulfill these functions, PMN and MP have specific receptors for microorganisms and specialized organelles and molecules for killing and degrading them. Stem cells derived from the bone marrow give rise to blood monocytes, and after migration into the various organs and tissues the monocytes become tissue macrophages. A major difference between PMN and MP is their life span; PMN live for few days and MP often survive for many weeks.[1]

Two forms of phagocytosis have been identified, conventional phagocytosis and coiling phagocytosis. In conventional phagocytosis, phagocyte pseudopodia move circumferentially and more or less symmetrically around the organism and fuse at the distal side. In coiling phagocytosis, phagocyte pseudopodia coil around the organism, as it is internalized.[2] Phagocytosis alone cannot eradicate invading microbes, and it must be followed by intracellular killing mechanisms. The binding and uptake of foreign material induces cellular processes to kill the microorganism.[3] The production of reactive oxygen species via activation of the nicotinamide-adenine dinucleotide phosphate (NADPH) oxidase of phagocytes is known as the respiratory burst.[3–5] NADPH oxidase is positioned in the cell membrane of phagocytes so that reactive oxygen species generated by these reactions is sequestered around foreign material during the inception of a phagocytic vacuole. Biochemically, the respiratory burst is characterized by rapid uptake of oxygen, enhanced metabolism of glucose, and release of a wide variety of reactive oxygen intermediates (ROI).[6,7] Glucose is metabolized principally through the hexose monophosphate shunt, with minor amounts oxidized via the oxidative phosphorylative pathways.[8–10] A large portion of the oxygen is used for energy metabolism, but substantial amounts are also reduced to superoxide anion.[10,11] Once generated, the anion spontaneously or catalytically dismutates to hydrogen peroxide. Superoxide and hydrogen peroxide then interact to generate an array of reactive derivatives of molecular oxygen, which includes superoxide anion (O_2^-), singlet oxygen ($^1O_2$), hydrogen peroxide (H_2O_2), and hydroxyl radicals (OH^-).[5,6] These ROI are important factors in the antimicrobial armamentarium of phagocytic cells.[12]

19.1.2 Assays to Measure Phagocytic Cell Function

The bactericidal or intracellular killing mechanism activity of phagocytes is preceded by adherence of bacteria to the phagocytic cell membrane followed by an ingestion or internalization phase. Measurement of phagocytosis is conducted by various techniques using light microscopy, fluorescent microscopy, and flow cytometry to assess ingestion of latex beads, or live or killed bacteria.

The conventional bactericidal activity of phagocytes is conducted by incubating phagocytes and bacteria together for a period of time, lysing the phagocytic cells and then determining the number of viable bacteria by counting bacterial colonies on agar plates. Because this procedure can be very time-consuming and laborious, several assays to measure the different enzymatic and nonenzymatic bactericidal mechanisms of phagocytes have been developed. The respiratory burst and nitric oxide–mediated killing mechanisms are the major nonenzymatic microbicidal mechanisms of phagocytes.

Nitric oxide (NO), a simple and unstable free radical, has recently been identified as involved in the killing of several intracellular parasites.[13–15] The toxicity of NO, which reacts with superoxide to produce peroxynitrite $(OONO)^-$, results from the inhibition of selected iron-dependent enzymatic pathways involved in cellular respiration, energy production, and DNA synthesis of target cells. Both macrophages and neutrophils produce NO;[16] however, activated macrophages produce much more $OONO^-$.

Various assays to measure the respiratory burst have been used, including reduction of nitroblue tetrazolium (NBT), superoxide and H_2O_2 generation, and chemiluminescence. The NBT assay can be used qualitatively with cells on a slide as a simple screening test for the "respiratory burst," as the clear yellow soluble dye is reduced to an insoluble dark blue formazan precipitate, or quantitatively if the formazan is solubilized and analyzed spectrophotometrically. The superoxide anion release assay measures the reduction of cytochrome *c*, with some reaction mixtures containing superoxide dismutase to confirm that reduction of cytochrome *c* was O_2^- dependent. The production of H_2O_2 is measured by the horseradish peroxidase (HRP)-dependent phenol red oxidation assay. Alternatively, flow cytometry has been used to detect the intracellular modification of the peroxide sensitive dye 2′,7′-dichlorofluoresceindiacetate (DCFH-DA). When used in conjunction with killed propidium iodide–labeled bacteria, it is possible to measure simultaneously both phagocytosis and hydrogen peroxide production by phagocytes.[17]

19.1.3 Chemiluminescence

Chemiluminescence was originally developed as a means of examining the intracellular respiratory burst. The respiratory burst in phagocytes is associated with the generation of light energy known as "native" chemiluminescence. A secondary chemiluminescence response in phagocytes can be magnified with secondary substrates, i.e., chemiluminigenic probes such as luminol (5-amino-2,3-dihydro-1,4-phthalazinedione) and lucigenin (10,10′-dimethyl-9,9′-biacridiniumdinitrate). These two probes can be used to measure different metabolic events within phagocytes. Luminol-enhanced chemiluminescence reflects primarily myeloperoxidase activity in neutrophils, whereas lucigenin-dependent chemiluminescence is myeloperoxidase independent and measures oxidase-associated oxygenation, essentially superoxide.[18]

Chemiluminescence has been used to measure the *ex vivo* function of leukocytes and to diagnose disorders such as chronic granulomatous disease, which is a metabolic disorder of granulocytes characterized by a heterogeneous group of deficiencies that prevent hexose-monophosphate shunt activity and the generation of superoxide and hydroxyl microbicidal radicals. This defect involves neutrophils, macrophages, and eosinophils.[19,20] All forms of this disease can be diagnosed by measuring the respiratory burst, and chemiluminescence provides a quantitative assessment of the activity.[19,20] Therefore, partial defects and a carrier state can be detected.

Increased activity of PMN by adding sera from patients with systemic lupus erythematosus (SLE) has been observed, leading investigators to conclude that circulating immune complexes cause neutrophil activation.[21] Synovial fluid from patients with rheumatoid arthritis has been shown to have an enhancing influence in assays of chemiluminescence.[22] Chemiluminescence of peripheral blood neutrophils is usually increased in patients with active rheumatic autoimmune disorders, including rheumatic arthritis,[23] SLE,[24] and progressive systemic sclerosis.[24,25]

Chemiluminescence can be utilized to measure phagocytosis indirectly. Animal studies have demonstrated enhanced chemiluminescence in host granulocytes during bacterial diseases.[26] Viral infections in animals and humans do not enhance and often depress the oxidative burst of PMN; however, this is dependent on the type virus.[27]

19.1.4 Phagocyte–Pathogen Interactions

19.1.4.1 Phagocytosis

Various receptors on phagocytes including complement, Fc, fibronectin, and mannose receptors mediate attachment to, and phagocytosis of pathogens. These receptors either induce or suppress certain metabolic events of phagocytes. Binding of IgG-coated particles with Fc receptors induces endocytosis and triggers multiple biological activities, including the production of ROI.[28] In contrast, the complement receptor pathway may provide intracellular parasites with safe passage into mononuclear phagocytes, since ligation of CR1 and CR3 receptors by particles coated with C3b or C3bi causes endocytosis but does not consistently stimulate the release of oxygen metabolites, such as hydrogen peroxide, or superoxide or the release of mediators of inflammation such as metabolites of arachidonic acid.[29–32] Consistent with this hypothesis, *Legionella pneumophila,*[33] *Mycobacterium leprae,*[34] and *Toxoplasma gondii*[35] have been found to elicit little or no metabolic burst upon entering mononuclear phagocytes. Because intracellular predators often enter their prey without inducing reactive oxygen metabolites, their growth in activated macrophages must ultimately be controlled by nonoxidative mechanisms.[36–39]

19.1.4.2 Pathogen Survival in Phagocytes

To survive within professional phagocytes, intracellular pathogens have a variety of evasion mechanisms. In principle, these include (1) interference with, or resistance to, oxidative killing, (2) inhibition of phagosome–lysosome fusion,[40–42] (3) interference with or resistance to lysosomal enzymes[43] or microbicidal peptides, and (4) escape into the cytoplasm.

Precise identification of the molecules responsible for intracellular survival is just beginning. Lipoarabinomannan in mycobacteria has been shown to scavenge cytotoxic oxygen free radicals.[44,45] Phenolic glycolipid-1 of *M. leprae* and lipophosphoglycan of *Leishmania donovani* scavenge reactive oxygen metabolites in a cell free system.[44,46] *Mycobacterium avium* survives intracellularly by inhibiting superoxide anion production by macrophages,[47,48] as well as phagolysosomal fusion following phagocytosis.[49,50]

19.1.4.3 Activation of Phagocyte Function

Phagocytic function can be regulated by cytokines. Tumor necrosis factor (TNF) can stimulate bacteriostatic and bactericidal activities in macrophages,[51] which may be important in the immune response against *M. avium,*[52,53] *M. tuberculosis,*[1,54] *Chlamydia trachomatis,*[55] *Listeria monocytogenes,*[56,57] *Candida albicans,*[58] *Legionella pneumophila.*[59] It has been proposed that interferon-γ (IFN-γ) produced by T cells activates macrophages to secrete TNF which, in turn, sensitizes the same or other macrophages to secrete ROI that rapidly destroy parasites through a process of lipid peroxidation.[60]

IFN-γ also directly enhances macrophage antimicrobial[61,62] and oxidative[63,64] activities in such a way that they kill or inhibit the multiplication of *L. pneumophila,*[65,66] *Toxoplasma gondii,*[67–69] *Chlamydia psittaci,*[70] *Leishmania donovani,*[71,72] *L. major,*[73] and *Trypanosoma cruzi.*[74]

Macrophages and monocytes exposed to rIFN-γ develop the ability to secrete ROI, as well as to kill nonspecifically obligate or facultative intracellular microorganisms.[68,75,76] Growth inhibition of *M. bovis* by IFN-γ-activated macrophages is an oxygen-independent process and may involve phagosome–lysosome fusion.[76,77] For some facultative intracellular pathogens, such as *Listeria monocytogenes, M. avium*, and *Salmonella typhimurium*, IFN-γ is not capable of sufficiently activating the antibacterial effector function.[78,79]

19.1.4.4 Illustration of Bacterial–Phagocyte Interaction with *Haemophilus somnus*

Haemophilus somnus is a Gram-negative, small, fastidious bacterium that has been recognized as a facultative intracellular pathogen of cattle. It causes a wide variety of clinical syndromes in cattle including myocarditis, pneumonia, encephalitis, and arthritis, which are collectively known as bovine hemophilosis.[80–85] *Haemophilus somnus* is able to interact with a number of cells types in the host, including phagocytes. *In vitro* studies have established that bovine PMN are unable to kill *H. somnus*, and that *H. somnus* can replicate within bovine monocytes.[83] Moreover, *H. somnus* is able to suppress the chemiluminescence activity of both bovine PMN and monocytes.[85,86] These findings suggest that the ability of the bacterium to persist and proliferate within these cells could contribute to the pathogenesis of hemophilosis. The objective of this study was to use chemiluminescence and phagocytosis assays to investigate the phagocytic function and ROI-mediated killing mechanisms of bovine mononuclear phagocytes following interaction with *H. somnus*.

19.2 MATERIALS AND METHODS

19.2.1 BACTERIA AND ISOLATION OF BOVINE MONONUCLEAR PHAGOCYTES

Logarithmically growing *H. somnus* (HS25) and *Staphylococcus aureus* were prepared as previously described.[85] Bovine peripheral blood mononuclear cells (PBMC) were obtained from clinically normal 6- to 12-month-old beef calves as previously described.[85] Lung lavages were performed for bovine alveolar macrophages (BAM) on anesthetized 2- to 3-month-old calves. Isolated PBMC or BAM were suspended in macrophage SFM medium with glutamine (Gibco BRL, Life Technologies, Inc., Grand Island, NY) supplemented with 10% fetal bovine serum (Gibco BRL, Life Technologies, Inc.).

In some cases, PBMC and BAM were obtained from 6- to 8-month-old beef calves that had been challenged intravenously with approximately 5×10^8 colony-forming units (CFU) of a field isolate of *H. somnus.*[85,87] Blood samples were taken from experimentally infected animals on day 1, 6, 7, 9, and 12 postinfection. Lung lavages were performed after euthanasia of moribund animals to obtain BAM.

19.2.2 CHEMILUMINESCENCE ASSAY

This assay was conducted in 3.5-ml, 55 × 12 mm tubes (Sarstedt, Nümbrecht, Germany). Each tube contained 4×10^6 BBM or BAM, 700 μl of Hank's balanced salt solution (HBSS) containing opsonized *S. aureus*, 200 μl of HBSS containing *H. somnus*, and 15 μl of 5-amino-2,3-dihydro-1,4-phthalazinedione (luminol) (Sigma Chemical Co., St. Louis, MO). HBSS was substituted for *H. somnus* in the controls. Chemiluminescence readings were measured on the Picolite Model 6500 luminometer (United Technologies Packard, Downers Grove, IL) every 3 min for 10 s each over a period of 90 min. The luminol-dependent chemiluminescence response (LDCL) represented the total amount of light emitted over 90 min from excited oxygen species during the bovine mononuclear phagocyte (BMP) respiratory burst.

$$\%\ \text{Activity Remaining} = \frac{\text{results of BMP with } H.\ somnus}{\text{results of BMP without } H.\ somnus} \times 100$$

19.2.3 FLOW CYTOMETRIC PHAGOCYTOSIS ASSAY

This assay was conducted in 96-well microtiter plates (Corning Glass Works, Corning, NY) as previously described.[86,88] Briefly, to each microtiter well, 70 μl of opsonized *S. aureus* (red fluorescence), 20 μl of PKH2-labeled unopsonized *H. somnus* (green fluorescence), and 100 μl of isolated PBMC or BAM were added. Plates were then incubated for 1 h at 37°C. After incubation, 20 μl of 300-m*M* ethylenediamine-tetraacetic acid (EDTA) (J.T. Baker Chemical Co., Phillipsburg, NJ) was added. The supernatant was discarded and cell pellets were washed twice with 200 μl of phosphate buffered saline with 0.2% gelatin (PBSAg). After washing, 200 μl of lysostaphin (Sigma) (44 U/ml) were added. As the final step, plates were placed on ice for 1 min before resuspending the cells in PBSAg containing 2% formalin. Fluorescent microscopic examination of BBM and BAM demonstrated phagocytosis of *S. aureus* and *H. somnus* (Figure 19.1); 5000 monocytes or alveolar macrophages were routinely examined by flow cytometry (FC) for each sample. Red fluorescence from propidium iodide (PI) and green fluorescence from PKH2-was displayed on a quadrant structure such that quadrant 1 represented cells that had actively phagocytized PI-labeled *S. aureus*; quadrant 2 represented cells that had both actively phagocytized *S. aureus* and *H. somnus*; quadrant 3 represented nonphagocytic BBM or BAM; and quadrant 4 represented cells that had phagocytized PKH2-labeled *H. somnus* but not PI-labeled *S. aureus*. FC studies were performed using the Becton Dickinson FACScan flow cytometer (Becton Dickinson, Canada, Inc., Mississauga, Ontario) with a 15-MW argon laser light source. To examine the effect of soluble products of *H. somnus* vs. contact with *H. somnus*, phagocytes were separated from *H. somnus* by a 0.4-μm membrane in transwell plates (Costar, Cambridge, MA.)[85]

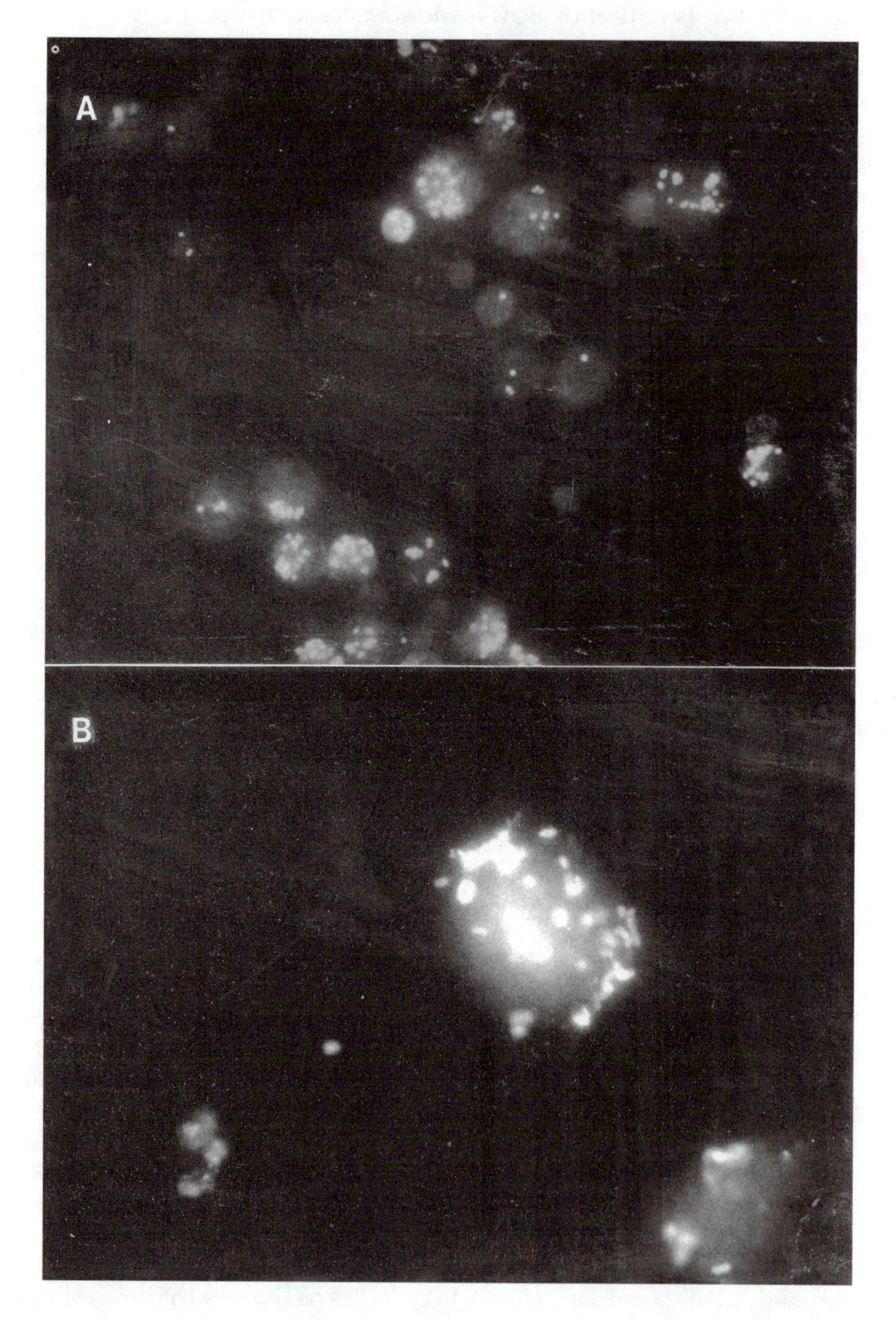

FIGURE 19.1 (Color figure follows p. 266.) Photomicrographs of alveolar macrophages containing fluorescent-labeled bacteria. (A) PI-labeled *S. aureus* (red fluorescence) in alveolar macrophages. (B) PKH2-labeled *H. somnus* (green fluorescence) and PI-labeled *S. aureus* (yellow fluorescence) in alveolar macrophages. (The red fluorescence changed to yellow fluorescence when color filters were adjusted for green fluorescence in the fluorescent microscope.)

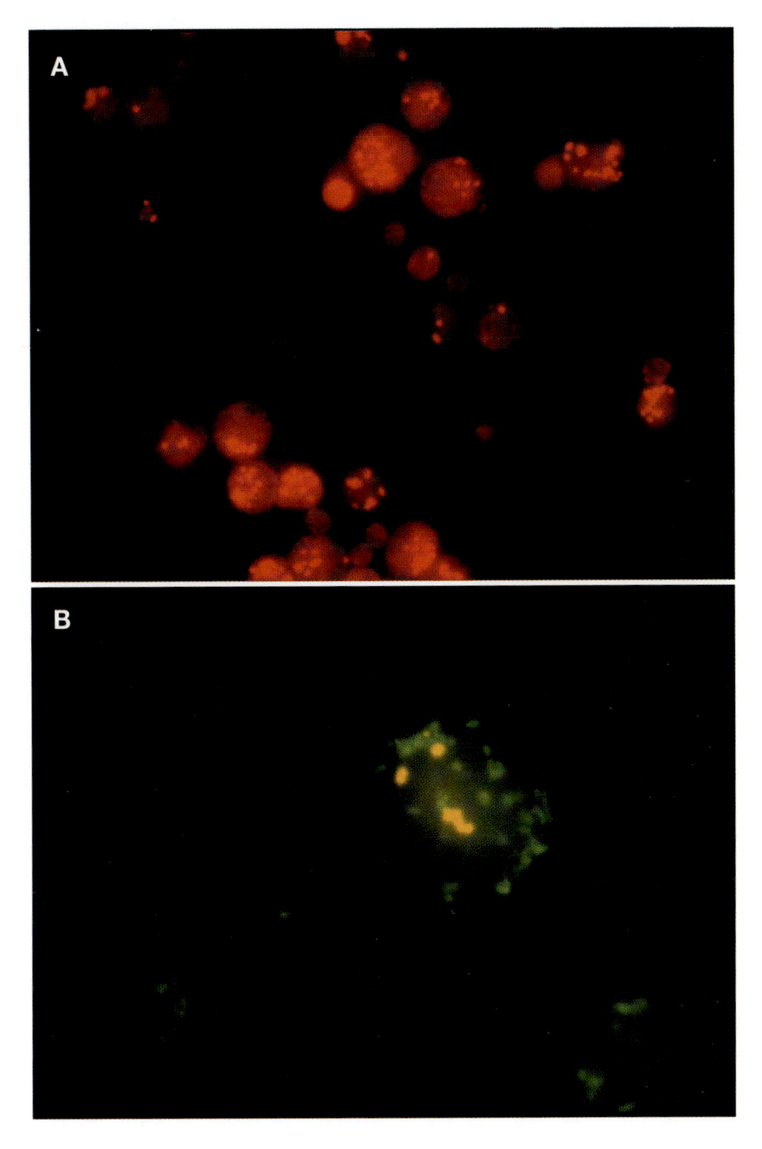

COLOR FIGURE 19.1 Photomicrographs of alveolar macrophages containing fluorescent-labeled bacteria. (A) PI-labeled *S. aureus* (red fluorescence) in alveolar macrophages. (B) PKH2-labeled *H. somnus* (green fluorescence) and PI-labeled *S. aureus* (yellow fluorescence) in alveolar macrophages. (The red fluorescence changed to yellow fluorescence when color filters were adjusted for green fluorescence in the fluorescent microscope.)

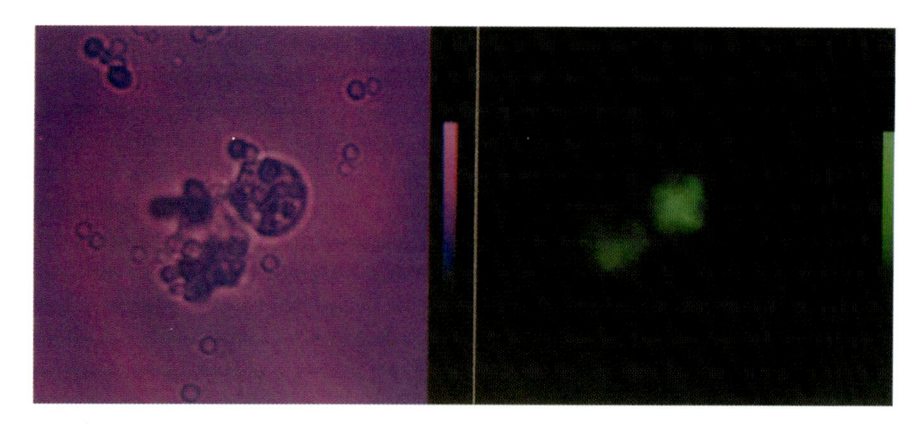

COLOR FIGURE 22.6 Image analysis of phagocytosis of HPPA-bound microspheres by PMN. The image processed by IMRAS in real time was recorded on videotape. Left: Microscope image on screen from the videotape. Right: Fluorescent microscope image on screen from the videotape. (From Suzuki, K. et al., *Bioimages*, 1, 13, 1993. With permission.)

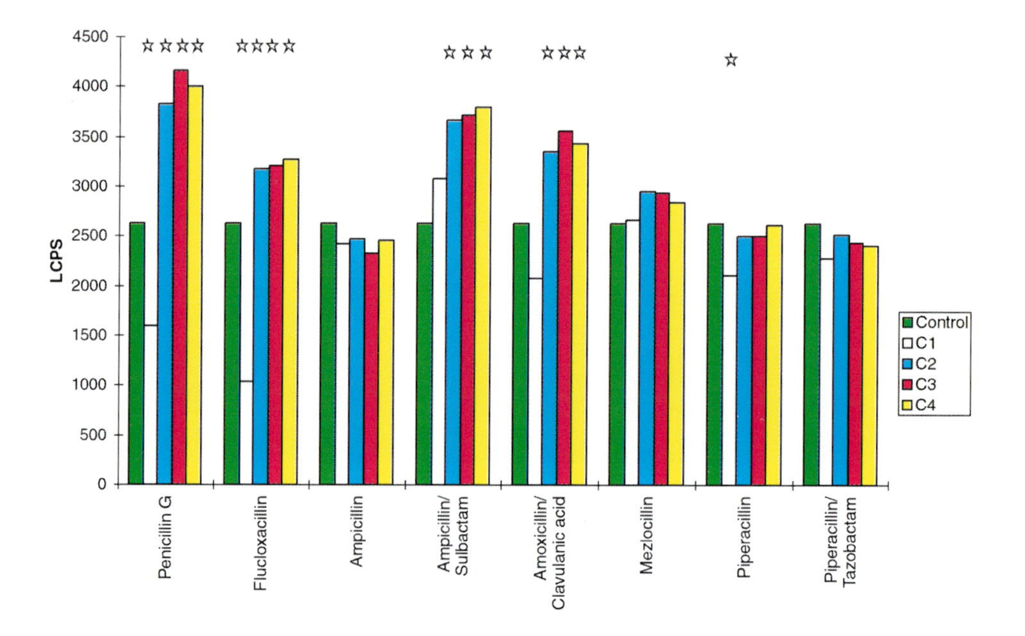

COLOR FIGURE 23.1 Penicillins. C1, normal concentration × 10; C2, normal concentration; C3, normal concentration × 0.1; C4, normal concentration × 0.01; Control, without antibiotics. Significant changes against control marked with a star.

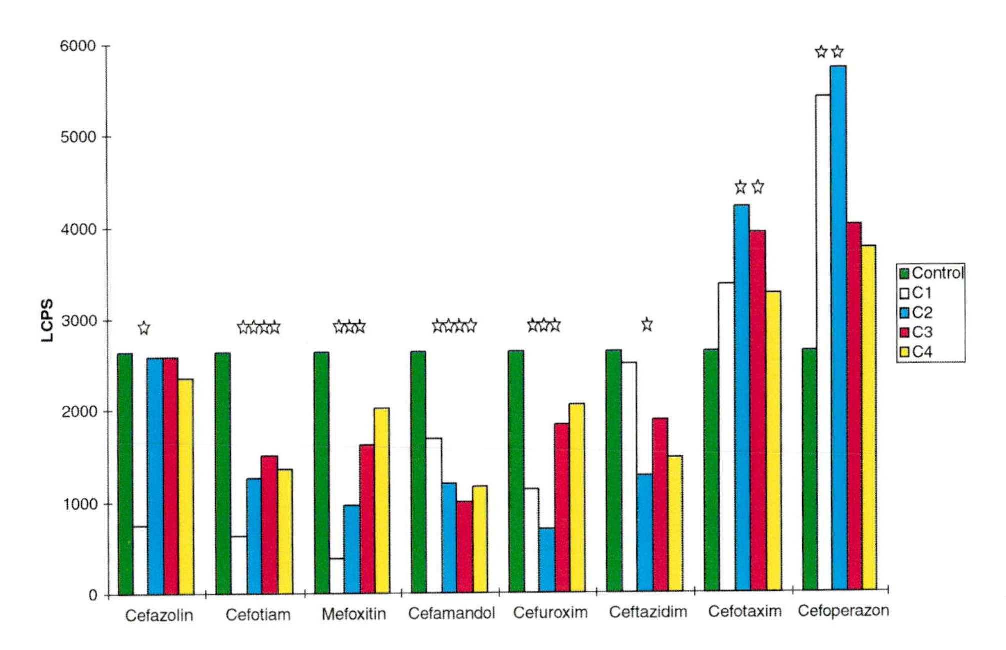

COLOR FIGURE 23.2 Cephalosporins. C1, normal concentration × 10; C2, normal concentration; C3, normal concentration × 0.1; C4, normal concentration × 0.01; Control, without antibiotics. Significant changes against control marked with a star.

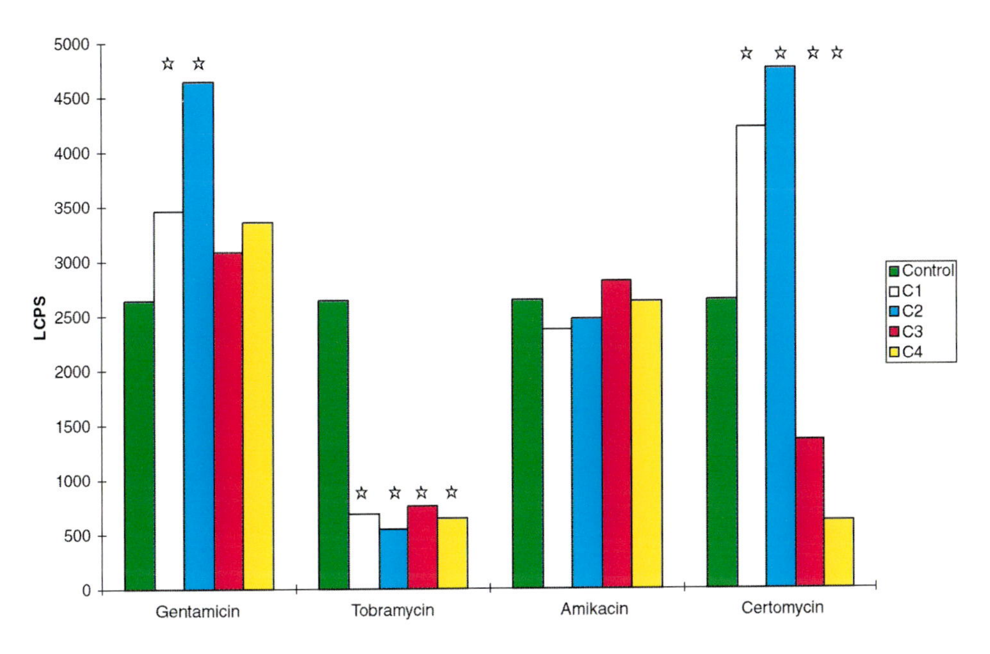

COLOR FIGURE 23.3 Aminoglycosides. C1, normal concentration × 10; C2, normal concentration; C3, normal concentration × 0.1; C4, normal concentration × 0.01; Control, without antibiotics. Significant changes against control marked with a star.

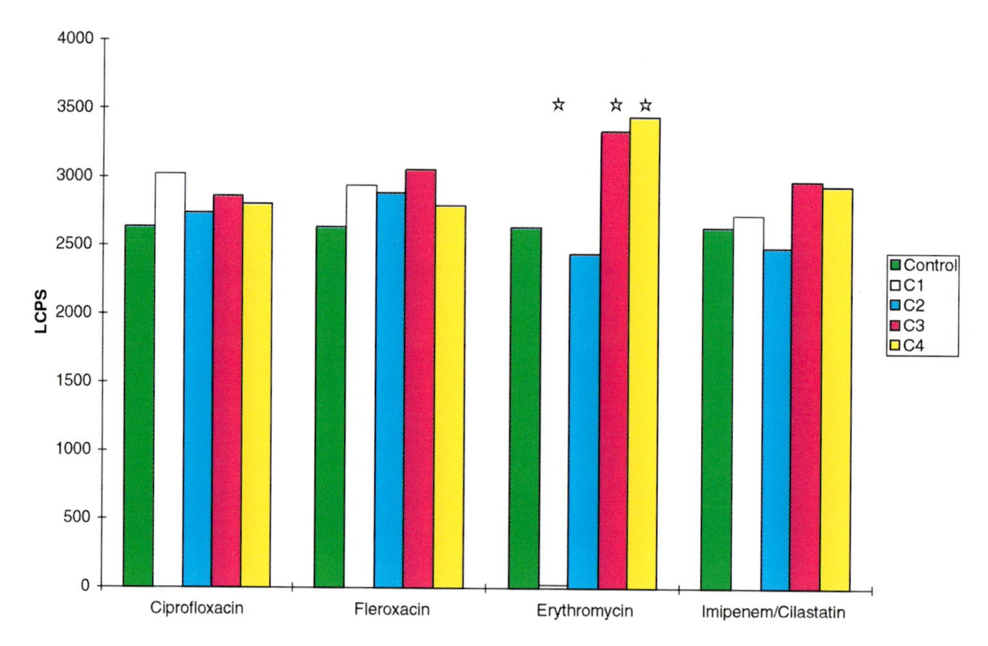

COLOR FIGURE 23.4 Gyrase inhibitor, macrolides, carbapenems. C1, normal concentration × 10; C2, normal concentration; C3, normal concentration × 0.1; C4, normal concentration × 0.01; Control, without antibiotics. Significant changes against control marked with a star.

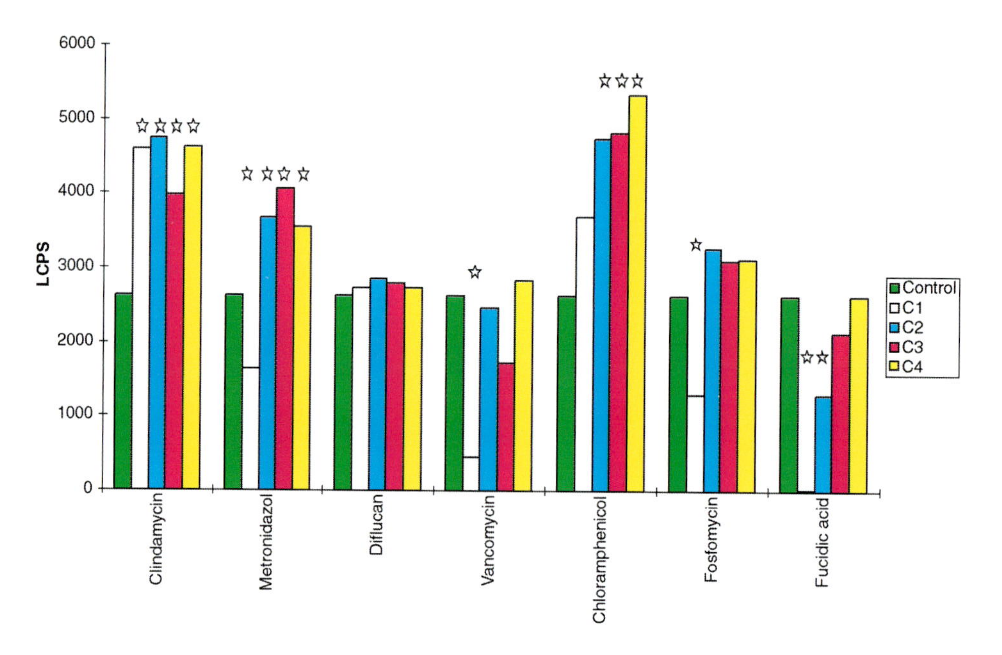

COLOR FIGURE 23.5 Others. C1, normal concentration × 10; C2, normal concentration; C3, normal concentration × 0.1; C4, normal concentration × 0.01; Control, without antibiotics. Significant changes against control marked with a star.

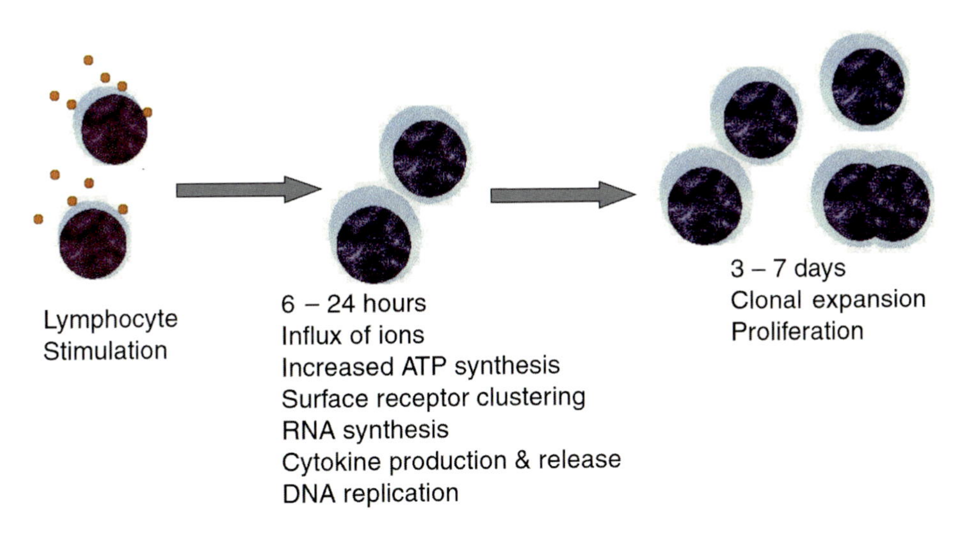

COLOR FIGURE 24.1 Lymphocyte response.

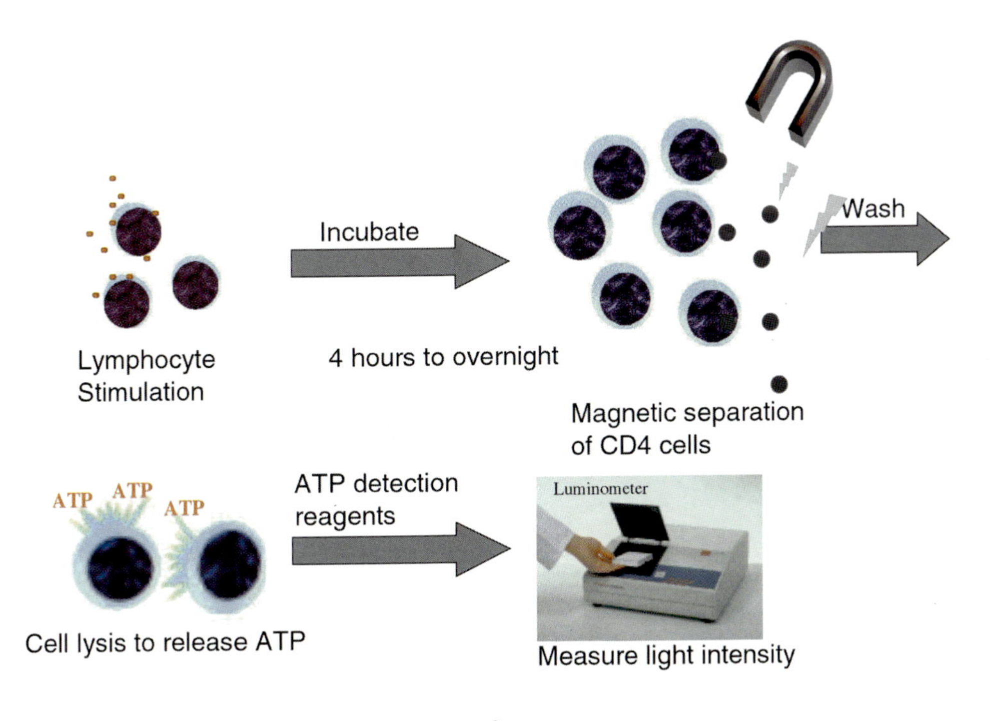

COLOR FIGURE 24.2 *In vitro* CMI assay format.

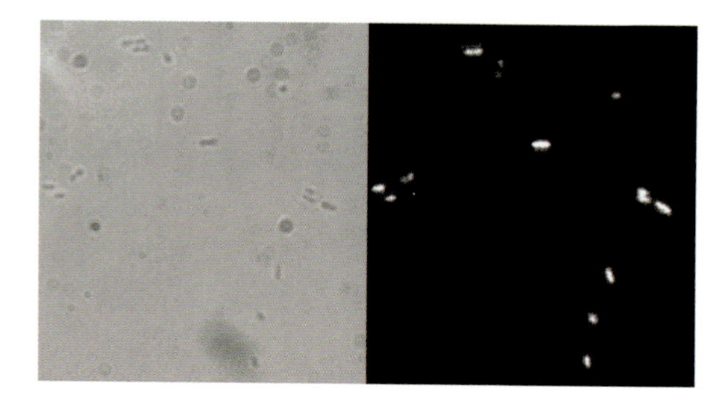

COLOR FIGURE 36.5 Bacterial chemiluminescence imaged through the microscope. Natural bacterial chemiluminescence imaged through the microscope (100× lens, 1.3 NA) using a VersArray 1300B LN-cooled CCD camera system. The image on the left is a brightfield reference image and the image on the right is a bioluminescence image of *Vibrio harveyi* (2.5-min exposure). (Data were taken by Irina Mihalcescu in the laboratory of Stan Leibler, Princeton University.)

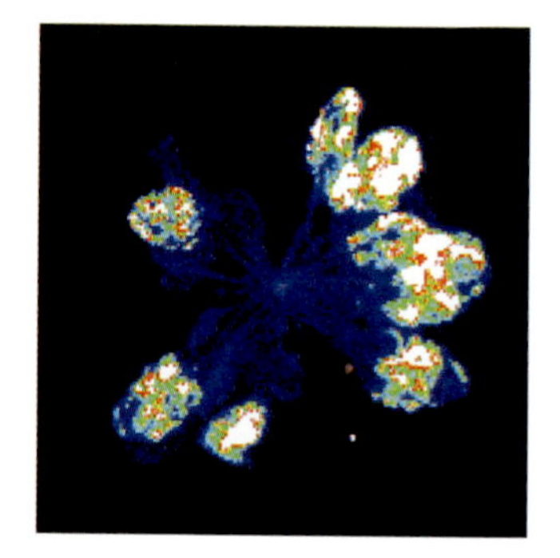

COLOR FIGURE 36.6 Plant chemiluminescence image. Bacteria expressing luciferase were used to infect *Arabidopsis* plants and then whole plants were examined. The blue represents a very low signal, whereas the white represents a high signal. A brightfield reference image is shown on the left and the chemiluminescence image is shown on the right. (Data are from Dr. Jian-Min Zhou, Kansas State University.)

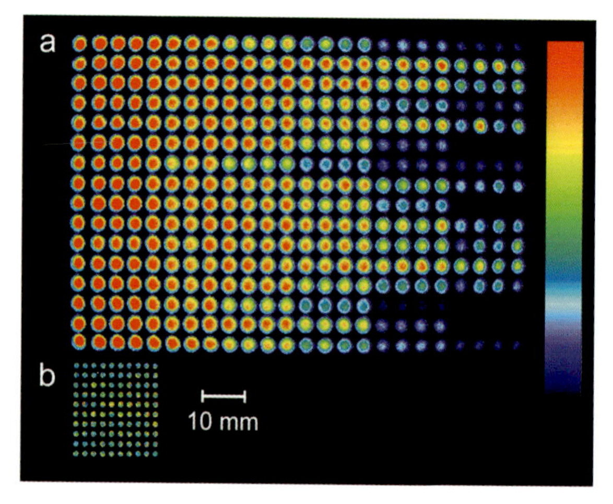

COLOR FIGURE 37.3 Pseudocolored CL images of a 384-well microtiter plate (a) and a bidimensional array of HRP spots (spot diameter 1.0 mm) deposited on cellulose paper (b).

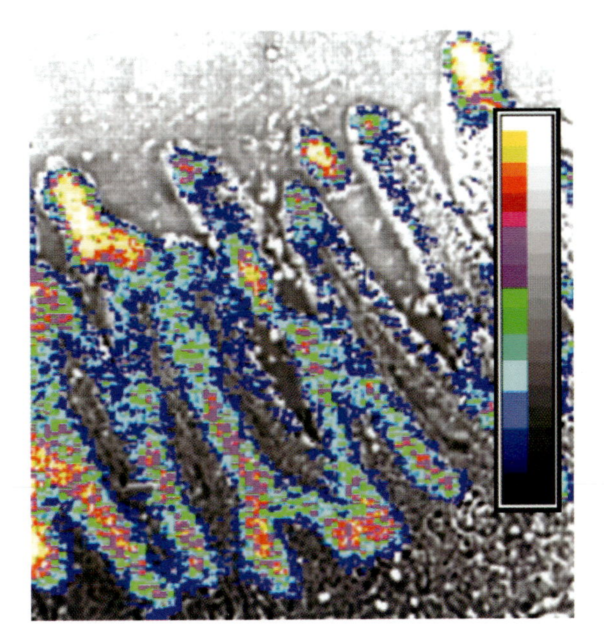

COLOR FIGURE 37.4 CL localization of endogenous alkaline phosphatase in rabbit intestinal mucosa cryosection. Pseudocolor ruler on the right shows the relative light intensity.

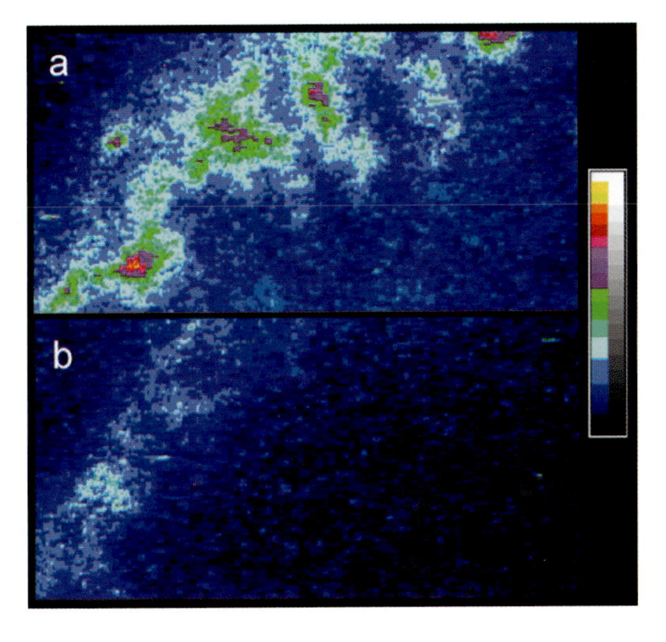

COLOR FIGURE 37.5 Immunochemiluminescent localization of IL-8 in gastric mucosa cryosections of a patient infected with *H. pylori* (a) and an *H. pylori* negative control subject (b). The pseudocolor-processed CL signals are reported, with the pseudocolor ruler on the right showing the relative light intensity. (Courtesy of Dr. J. E. Crabtree, St. James's University Hospital, Leeds, U.K.)

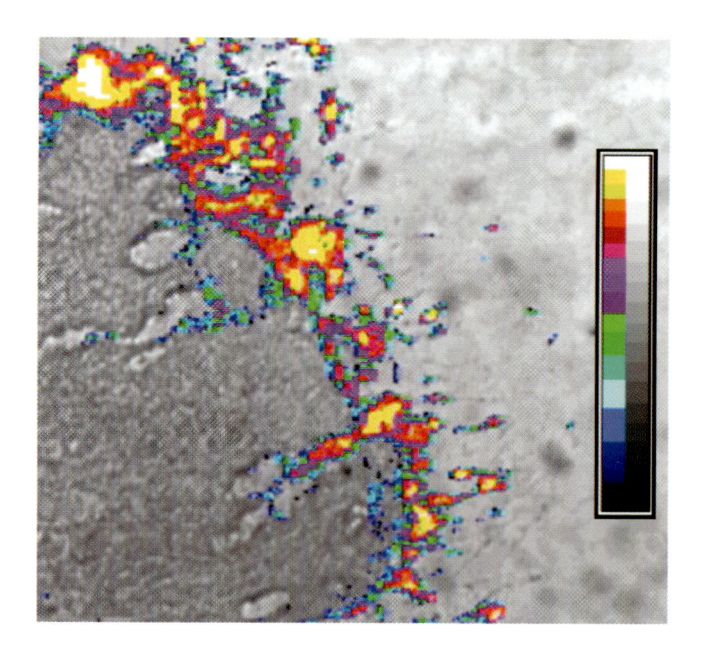

COLOR FIGURE 37.6 Immunochemiluminescent localization of NGAL in *H. pylori*-infected gastric mucosa cryosection. Pseudocolor ruler on the right shows the relative light intensity. (Courtesy of Dr. J. E. Crabtree, St. James's University Hospital, Leeds, U.K.)

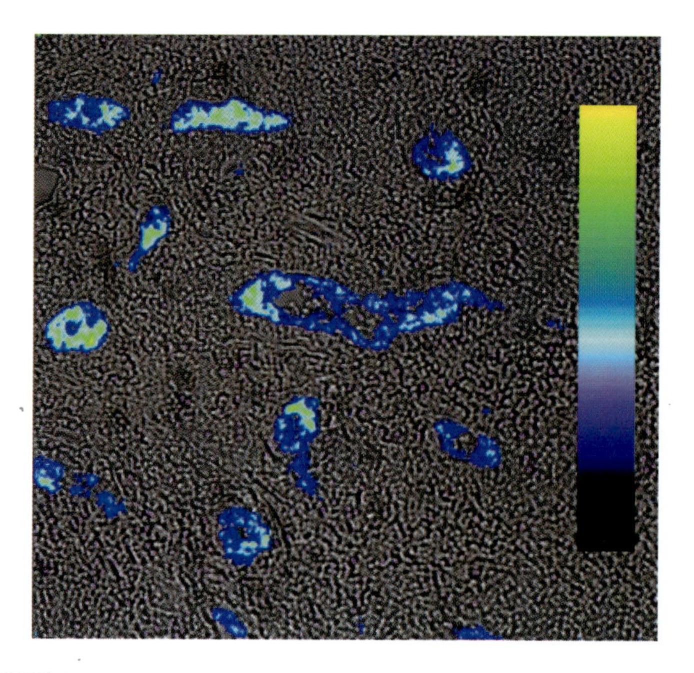

COLOR FIGURE 37.7 Immunochemiluminescent localization of von Willebrand factor in endothelial cells of a paraffin-embedded section of human tonsil tissue. Pseudocolor ruler on the right shows the relative light intensity. (Courtesy of Dr. P. Chieco, Institute of Oncology "F. Addarii," Bologna, Italy.)

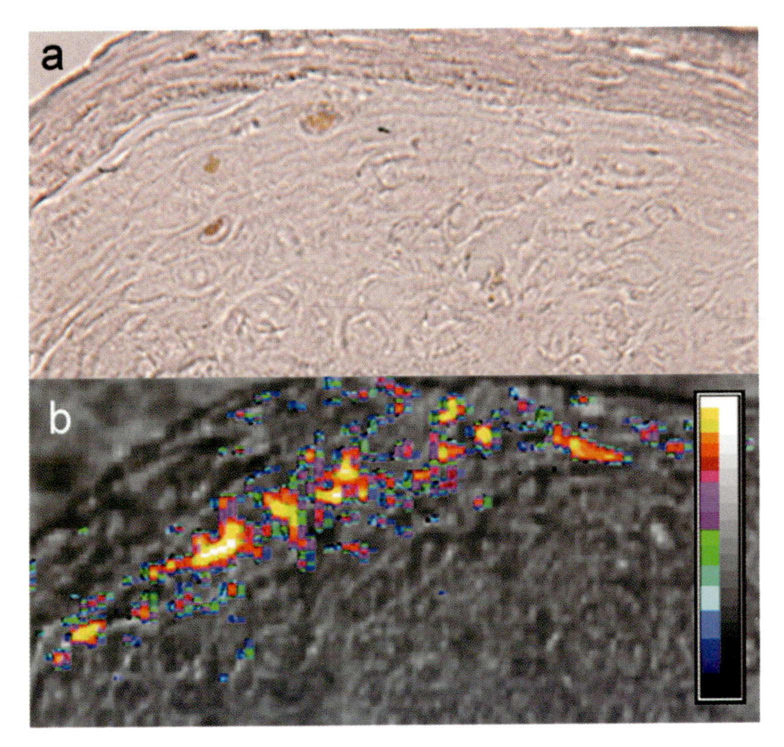

COLOR FIGURE 37.8 ISH for the detection of human papillomavirus DNA in skin tissue cryosections with colorimetric (a) and CL (b) detection. Pseudocolor ruler on the right of b shows the relative light intensity.

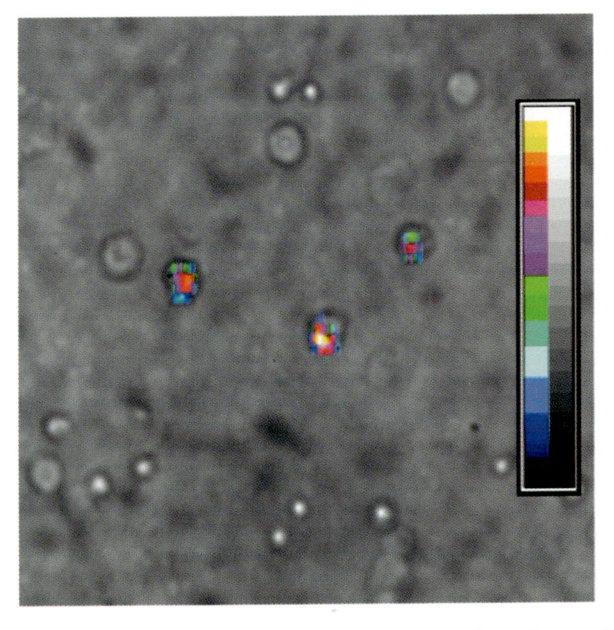

COLOR FIGURE 37.9 Chemiluminescent ISH for the detection of parvovirus B19 DNA in bone marrow cells. Pseudocolor ruler on the right shows the relative light intensity.

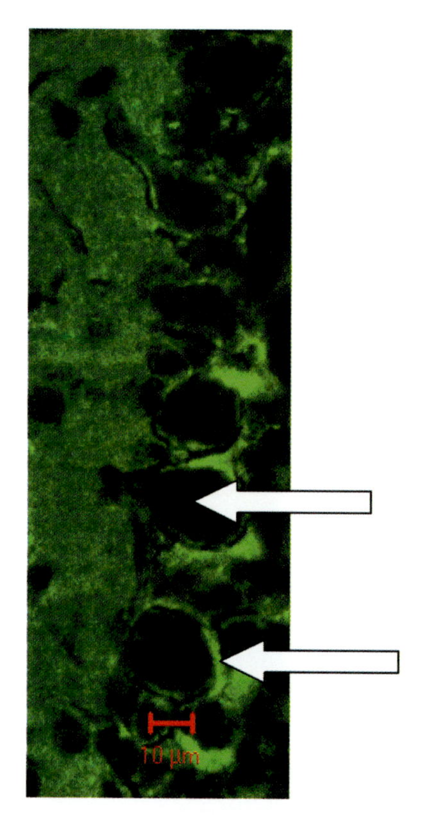

COLOR FIGURE 38.2 400× confocal micrograph (2048 × 2048 pixels) of cerebellar Purkinje cells from a 19-day-old animal stained with anti-α_3 subunit antibody XVIF9-G10. The upper arrow points to the nuclear region of a Purkinje cell. The lower arrow points to the Purkinje cell soma.

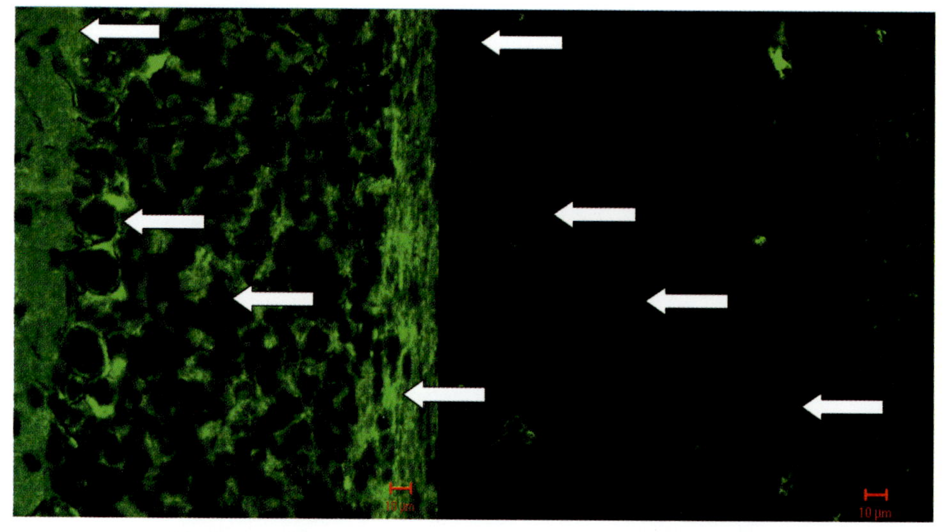

19 days old **13 days old**

COLOR FIGURE 38.3 400× confocal micrographs (2048 × 2048 pixels each) of sagittal sections of the cerebella from 13- and 19-day-old animals stained with anti-α_3 subunit antibody XVIF9-G10. The arrows point to corresponding structures in both 13- and 19-day-old sections. From top to bottom the arrows point to: the molecular layer, the Purkinje cell layer, the granular cell layer, and the white matter.

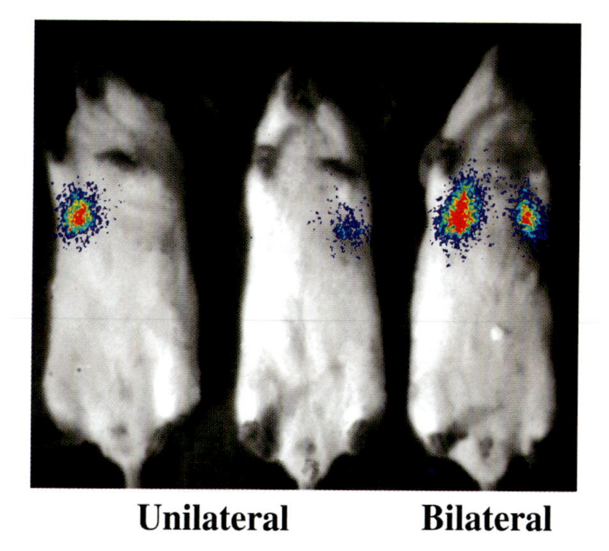

COLOR FIGURE 40.1 Murine models of bacterial pneumonia. Mice were inoculated by intranasal inoculation using a labeled strain of *P. aeruginosa*. Immediately after infection, the mice were imaged using an intensified CCD camera (model C2400-32, Hamamatsu, Japan).

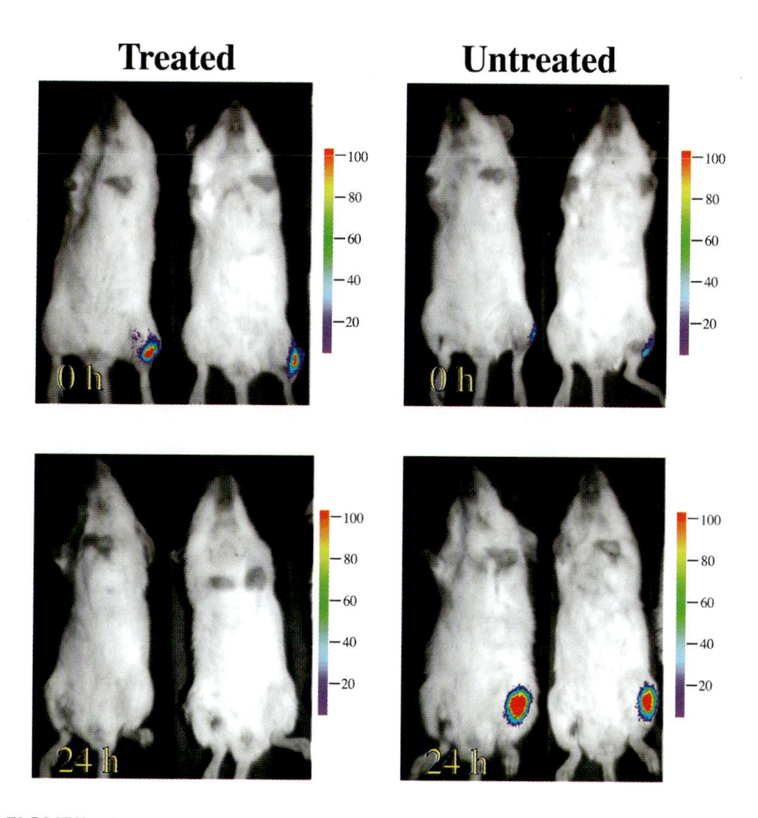

COLOR FIGURE 40.2 Monitoring the effects of amoxicillin on bioluminescent *S. aureus* in mice. Shown are two treated and two untreated *S. aureus* 8325-4 pMK4 luxABCDE P1-infected mice, imaged ventrally for 5 min at 0 and 24 h postinfection using the intensified CCD camera.

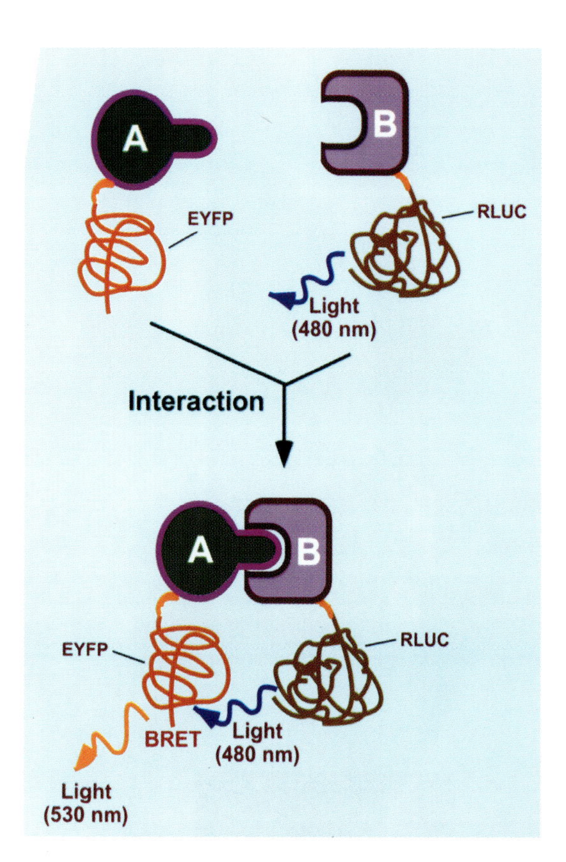

FIGURE 41.1 A diagram of BRET for a protein–protein interaction assay. One protein of interest (B) is genetically fused to the donor luciferase RLUC, and the other candidate protein (A) is fused to the acceptor fluorophore EYFP. Interaction between the two fusion proteins can bring FLUC and EYFP close enough for BRET to occur, with an emission of longer-wavelength light.

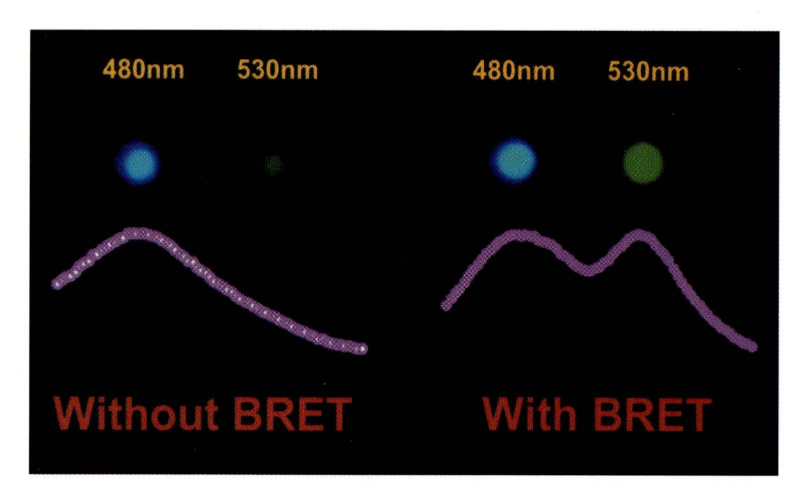

FIGURE 41.2 Comparison of complete BRET spectra using a fluorescence spectrophotometer with camera images of *E. coli* cells. Top: Cultures imaged with a CCD camera through filters transmitting light of 480 or 530 nm from the transformed *E. coli* strains coexpressing fusion proteins exhibiting BRET on the right side (RLUC::DaiB and EYFP::KaiB) or fusion proteins that are not exhibiting BRET on the left side (RLUC::KaiB and EYFP::KaiA). Bottom: Luminescence emission spectra measured continuously from 440 to 580 nm for the same strains.

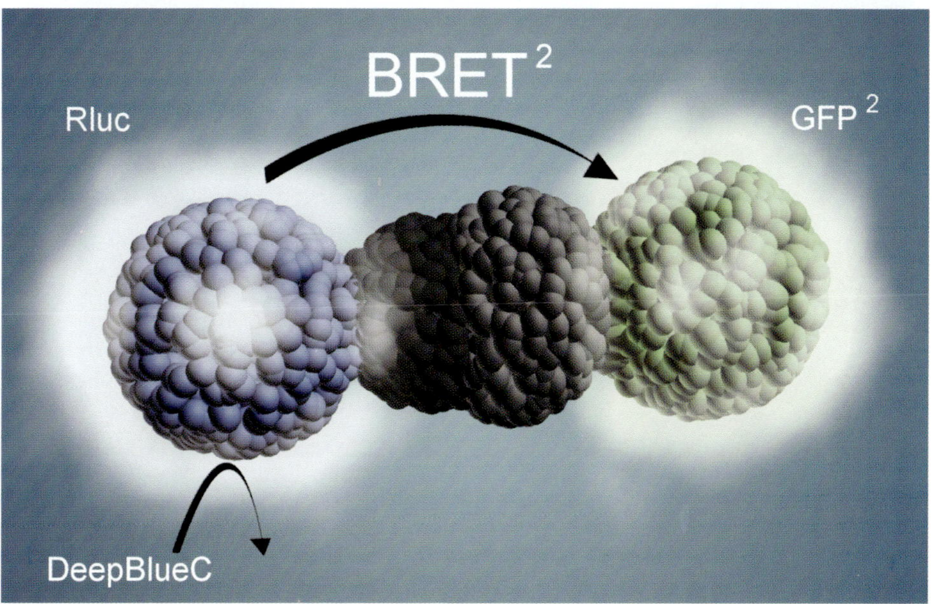

COLOR FIGURE 42.1 Schematic representation of BRET principle. In the presence of its substrate DeepBlueC (a coelenterazine derivative), Rluc emits blue light (400 nm). When Fluc and GFP^2 are brought into close proximity by means of a biological interaction, e.g., interaction of proteins genetically fused to Rluc (Protein A) and GFP^2 (Protein B), blue light energy is transferred to GFP^2, which re-emits green light (505 nm). The BRET signal is calculated as a ratio of the signal at 505 nm vs. that at 400 nm.

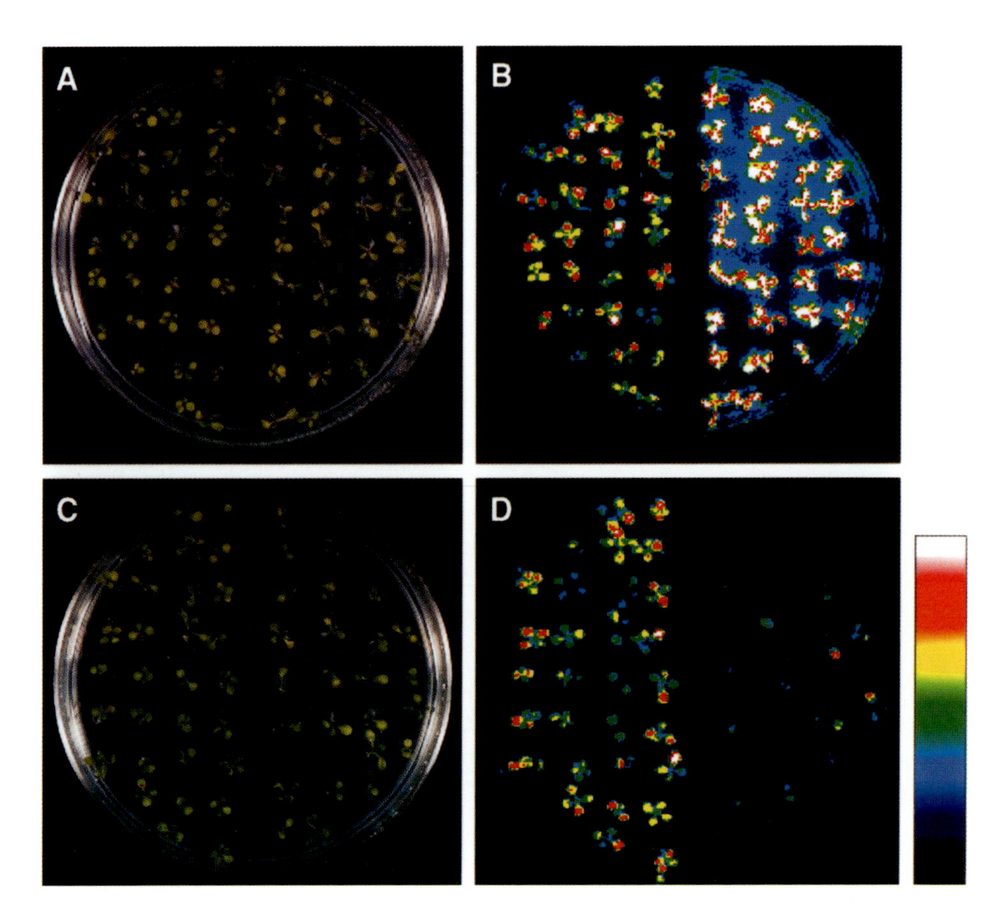

COLOR FIGURE 43.3 Stress mutants showing altered RD29A::LUC expression. (A) The plate corresponding to luminescence image in B; (B) Mutants (right half) showing higher RD29A::LUC expression than wild-type (left half) after ABA treatment; (C) The plate corresponding to luminescence image in D; (D) Mutants (right half) showing lower RD29A::LUC expression than wild-type (left half) after BA treatment. Right spectrum bar shows color changes depending on luminescence intensity; as intensity goes higher, color changes from black to white.

19.2.4 Labeling of Bacteria

PI (Sigma) labeling of *S. aureus* was carried out according to the method of Pfeifer et al.[88] PKH2 labeling of *H. somnus* was carried out according to a previously described method.[89] The fluorescent intensity of log and stationary phase *H. somnus* was measured by FC before each experiment to monitor the consistency of the labeling procedure. To determine the survival and purity of *H. somnus* in BAM or BBM, at time 0 and 72 h, samples from microtiter plates were cultured on 5% sheep blood agar plates in duplicate.

19.3 Results

19.3.1 Chemiluminescent Response of BBM and BAM Interacting with *H. somnus*

The LDCL response of BBM was significantly reduced, by live *H. somnus*, either logarithmically growing (30.5 ± 7.78%, $P < 0.0001$) or stationary phase (54.5 ± 2.12%, $P < 0.0001$) (Figure 19.2) when the phagocyte-to-*H. somnus* ratio was 1:100. The inhibitory effect seen with live *H. somnus* was completely absent when killed or opsonized (with hyperimmune serum) *H. somnus* was used. If fact, the LDCL response of BBM tended to increase when *H. somnus* was either killed or opsonized, reaching a peak level of 150 to 200% with opsonized *H. somnus* (Figure 19.2).

For BAM, the LDCL response also was significantly reduced by either live, logarithmically growing (14.25 ± 3.30%, $P < 0.001$) or stationary phase *H. somnus* (49 ± 7.87%, $P < 0.006$) (Figure 19.3)

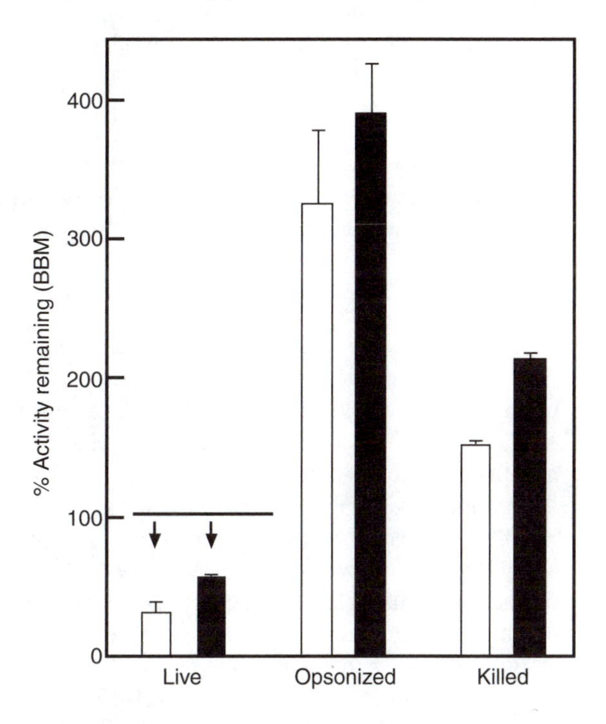

FIGURE 19.2 LDCL response of BBM following interaction with *H. somnus*. Results are expressed as the percentage of the response to *S. aureus* alone (100% = 3.2×10^7 total counts). All samples contained *S. aureus* in addition to live, killed, or opsonized *H. somnus*. Live log or stationary phase *H. somnus* inhibited the LDCL response of BBM, whereas *H. somnus* opsonized with hyperimmune serum or heat-killed *H. somnus* enhanced the LDCL response. Each bar represents the mean and standard deviation of four animals. The ratio of phagocytes to *H. somnus* was 1:100. Solid bar (■) indicates stationary phase *H. somnus*, and open bars (□) indicate log phase *H. somnus*. (From Gomis, S. M. et al., *Microb. Pathog.*, 23, 327, 1997. With permission.)

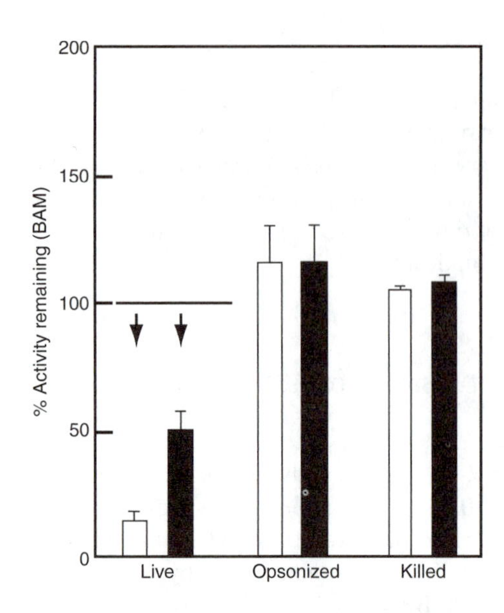

FIGURE 19.3 LDCL response of BAM following interaction with *H. somnus*. Results are expressed as the percentage of the response to *S. aureus* alone (100% = 9.3×10^7 total counts). All samples contained *S. aureus* in addition to live, killed, or opsonized *H. somnus*. Live log or stationary phase *H. somnus* inhibited the LDCL response of BAM, whereas *H. somnus* opsonized with hyperimmune serum or heat-killed *H. somnus* enhanced the LDCL response. Each bar represents the mean and standard deviation of four animals. The ratio of phagocytes to *H. somnus* was 1:100. Solid bar (■) indicates stationary phase *H. somnus*, and open bars (□) indicate log phase *H. somnus*. (From Gomis, S. M. et al., *Microb. Pathog.*, 23, 327, 1997. With permission.)

when the phagocyte-to-*H. somnus* ratio was 1:100. As with BBM, the inhibitory effect of *H. somnus* on BAM was abrogated either by killing or opsonizing *H. somnus*. The control samples containing BBM or BAM with *H. somnus* alone had poor chemiluminescence response (data not shown).

19.3.2 Phagocytosis by BAM and BBM Interacting with *H. somnus*

Bovine mononuclear cells were gated according to cell size and granularity to differentiate alveolar macrophages or monocytes from lymphocytes. A quadrant system was used for analysis of red fluorescence for actively phagocytizing cells vs. "nonresponders." Quadrant 1 represented cells that were actively phagocytizing *S. aureus*. Quadrant 3 represented cells that were not active and remained fluourescently negative (Figure 19.4a and b).

The presence of live, logarithmically growing *H. somnus* decreased phagocytosis of opsonized *S. aureus* by BAM (21 ± 1.58%) compared with their normal phagocytosis (52.4 ± 3.36%) ($P <$ 0.0001) when the phagocyte-to-bacteria ratio was 1:100 (Figure 19.4). Phagocytosis by BAM was not decreased when either heat or formalin killed log phase *H. somnus* or *in vitro* passaged *H. somnus* were used (data not shown). The inhibitory effect was not seen when *H. somnus* and BAM were partitioned by a 0.4 μm pore size membrane. Stationary phase *H. somnus* increased the phagocytosis of opsonized *S. aureus* compared with normal levels (71.2 ± 15.42%, 52.4 ± 3.36%, respectively; $P < 0.031$).

In contrast to BAM, phagocytosis of *S. aureus* by BBM was increased both by logarithmically growing and stationary phase *H. somnus*. There was no decrease in phagocytosis of *S. aureus* by BBM even with a high ratio of bacteria to phagocytes (Figure 19.5).

Since *in vitro* addition of *H. somnus* modulated the function of bovine mononuclear phagocytic cells from healthy calves, the effect of *H. somnus* in animals with the clinical disease was studied. For this study, BAM and BBM were obtained from experimentally infected cattle. Since this study was conducted to monitor *ex vivo* activity, no *in vitro* addition of *H. somnus* was made. BBM of

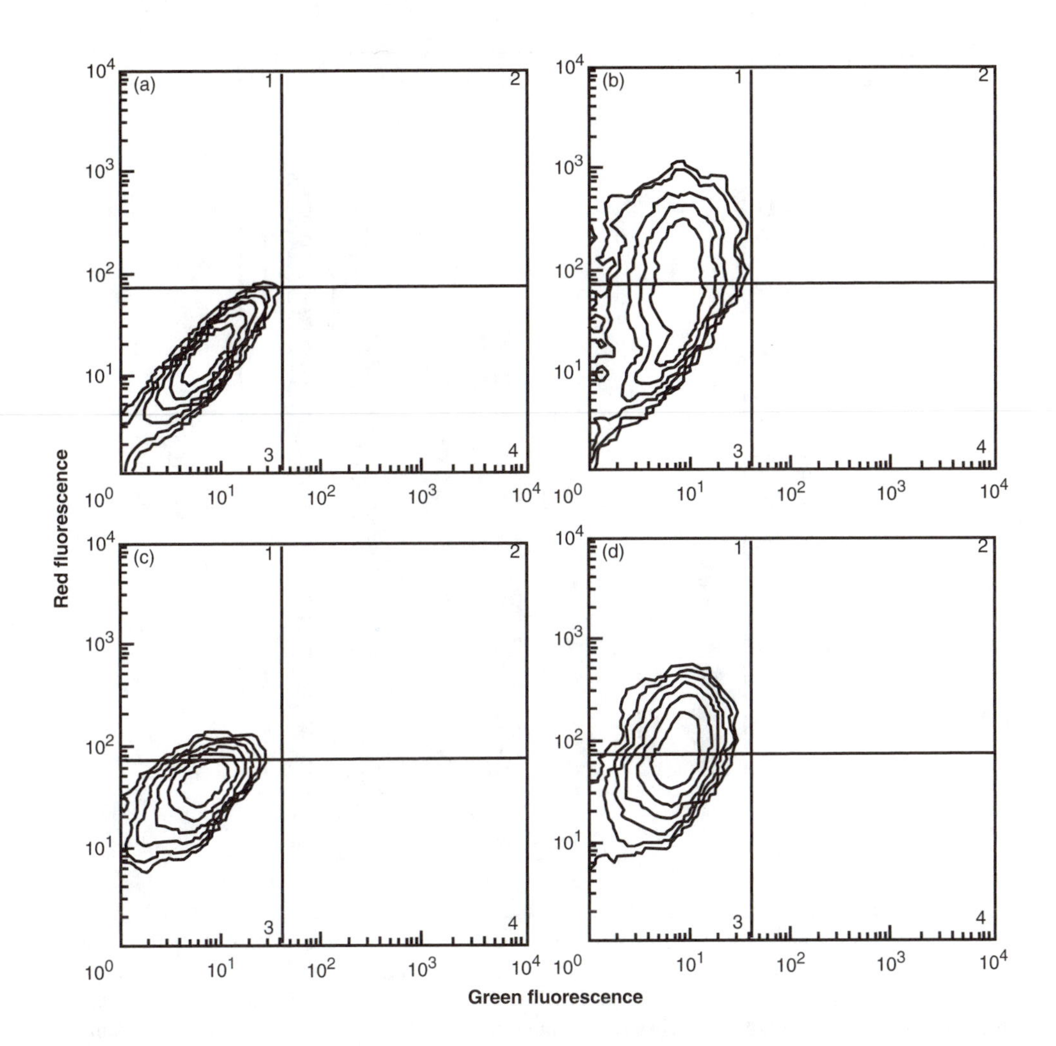

FIGURE 19.4 The effect of log and stationary phase *H. somnus* on BAM phagocytic function of a single animal. BAM were incubated with PI-labeled *S. aureus* and unlabeled log or stationary phase *H. somnus* at a BAM-to-*H. somnus* ratio of 1:100. The *y* axis represents the red fluorescence. (a) Control BAM without addition of labeled bacteria were used to set thresholds for quadrants. (b) BAM incubated with PI-labeled *S. aureus*; actively phagocytizing alveolar macrophages have increased red fluorescence and are found in quadrant 1 (50% of the total BAM). (c) BAM incubated with PI-labeled *S. aureus* and log phase *H. somnus*. There was a significant decrease in phagocytic function with quadrant 1 containing only 13% of the BAM. (d) BAM incubated with PI-labeled *S. aureus* and stationary phase *H. somnus*. There was a slight increase in phagocytic function of BAM by stationary phase *H. somnus*, with quadrant 1 containing 56% of the BAM. (From Gomis, S. M. et al., *Microb. Pathog.*, 22, 13, 1997. With permission.)

both experimentally infected animals and clinically normal animals responded similarly. The phagocytosis by BAM of experimentally infected animals was not decreased as with the *in vitro* studies with logarithmically growing *H. somnus*.

19.3.3 Uptake of *H. somnus* by Mononuclear Phagocytes

Since log phase *H. somnus* inhibited phagocytosis of *S. aureus* by BAM, the FC assay was modified by using PKH2-labeled *H. somnus* to determine the actual uptake of *H. somnus*. The PKH2 labeling did not have any effect on viability of either logarithmically growing or stationary phase *H. somnus*. Fluorescent intensity was similar between logarithmically growing and stationary phase *H. somnus*.

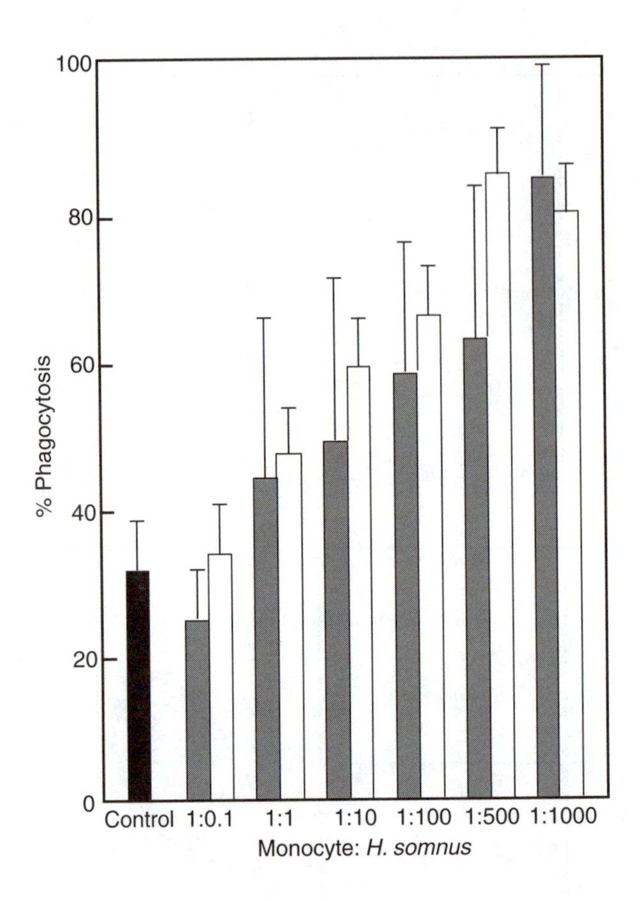

FIGURE 19.5 Effect of log phase and stationary phase *H. somnus* on phagocytosis of killed opsonized *S. aureus* by BBM. In the presence of log or stationary phase *H. somnus*, phagocytosis of *S. aureus* by BBM was significantly increased compared to normal levels when the phagocyte-to-*H. somnus* ratio was 1:100 or above ($P < 0.001$). Each bar shown is the mean of BBM that were positive for ingestion of *S. aureus* and SD of three animals. ■ = control; ▦ = log phase of *H. somnus*; □ = stationary phase of *H. somnus*. (From Gomis, S. M. et al., *Microb. Pathog.*, 22, 13, 1997. With permission.)

Logarithmically growing *H. somnus* was able to enter BAM while selectively inhibiting *S. aureus* uptake (Figure 19.6a). However (Figure 19.6b), stationary phase *H. somnus* entered BAM without any effect on *S. aureus* uptake. In contrast, both logarithmically growing and stationary phase *H. somnus* were able to enter BBM without any inhibitory effect on *S. aureus* uptake (data not shown). In fact, *S. aureus* uptake by BBM was slightly increased, as illustrated in Figure 19.5. Opsonization with hyperimmune serum did not enhance phagocytosis of *H. somnus*.

19.4 DISCUSSION

Qualitative and quantitative non-isotopic analytical procedures and immunoassays that rely on the phenomenon of chemiluminescence are now widely used in human and veterinary medicine, forensic medicine, agriculture, and the food industry. Chemiluminescence is the process by which certain reactions produce energy in the form of photons rather than heat, and it has many advantages as an analytical technique.[90] The extreme sensitivity of chemiluminescent analysis, resulting from the low background, is the most important advantage. Chemiluminescence has a large linear response. Since it is an emission process, the response is proportional to the concentration, from the minimum detectable concentration to the point where it is no longer possible to maintain an excess of reactants relative to the analyte. Most chemiluminescent assays can be performed faster

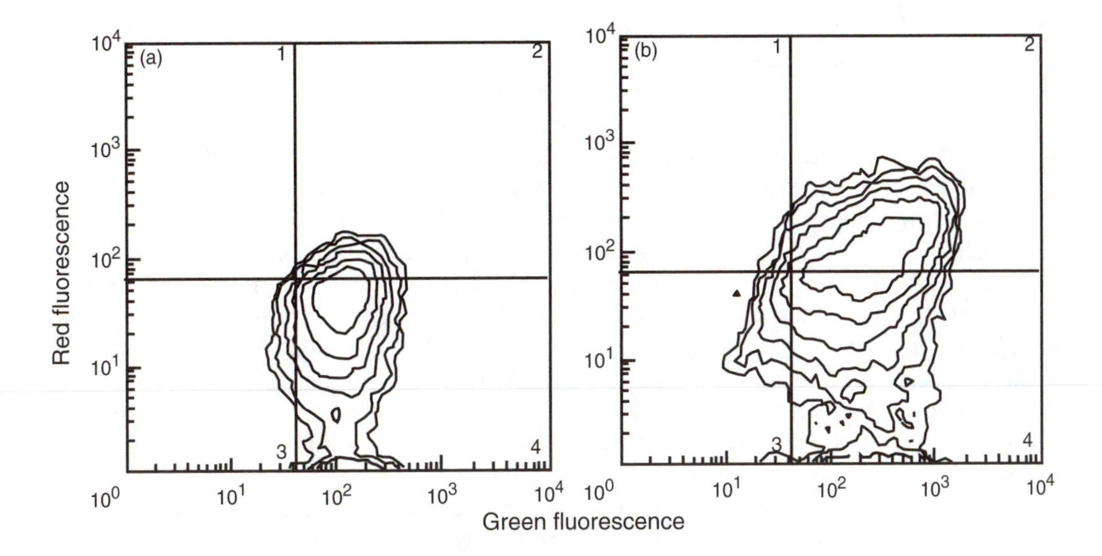

FIGURE 19.6 Effect of log and stationary phase *H. somnus* on BAM phagocytosis of *S. aureus* and *H. somnus* of a single animal. The FC assay was identical to Figure 19.4 except *H. somnus* was labeled with PKH2. The *y* axis represents the red fluorescence (uptake of *S. aureus*), and the *x* axis represents green fluorescence (uptake of *H. somnus*). (a) BAM incubated with PI-labeled *S. aureus* and PKH2-labeled log phase *H. somnus* (quadrant 1 + 2 = 12%, quadrant 2 + 4 = 97%). (b) BAM treated with PI-labeled *S. aureus* and stationary phase *H. somnus* (quadrant 1 + 2 = 51%, quadrant 2 +4 =96%). (From Gomis, M. S. et al., *Microb. Pathog.*, 22, 13, 1997. With permission.)

than nonchemiluminescent assays. Most chemiluminescent reagents and conjugates are stable. Bioluminescent reactions are highly specific because of the involvement of enzymes.

To illustrate the chemiluminescence activity of phagocytes following interaction with a pathogenic bacterium, we used a bovine pathogen, *H. somnus*. The interactions between bovine mononuclear phagocytes and *H. somnus* are known to be complex and poorly understood; therefore, the killing mechanisms of bovine mononuclear phagocytes following interaction with *H. somnus* were studied. Although *H. somnus* is a catalase-negative organism, Sample and Czuprynski[91] demonstrated that *H. somnus* is capable of removing H_2O_2 from an aqueous solution in an energy-dependent manner that is strictly cell associated. Reduction of LDCL activity of BBM and BAM infected with *H. somnus* could be explained by H_2O_2 removal. Moreover, the absence of reduced LDCL activity of BBM or BAM with heat killed *H. somnus* correlates with the observation of Sample and Czuprynski[91] that viable *H. somnus* were required to remove H_2O_2. It is also clear that *H. somnus* was not a poor stimulus for bovine mononuclear phagocytes, but that the organism actively interfered with the oxidative burst, as heat killed or opsonized *H. somnus* increased the LDCL activity of BBM or BAM. It has also been demonstrated that *H. somnus* is able to inhibit LDCL activity of bovine PMN.[88,92–94] In addition, PMN from calves experimentally infected with *H. somnus* show reduced LDCL activity when compared with PMN from healthy calves.[86]

A FC assay using fluorescent-labeled bacteria was developed to monitor *H. somnus* phagocyte interactions, which are important in the development of immunity and survival of the bacteria.[96–99] In this assay, uptake of PI-labeled, killed, opsonized *S. aureus* was used as an indicator of phagocytic function. *Haemophilus somnus* was able to modulate the phagocytic function of BAM as shown by the ability of log phase *H. somnus* to cause inhibition of *S. aureus* uptake. This inhibition was not due to competition by *H. somnus* for receptors on the BAM since formalin- or heat-killed, *in vitro* passaged, and stationary phase *H. somnus* did not cause any decrease in phagocytosis and actually tended to stimulate *S. aureus* uptake. Thus, there are two possibilities to explain the inhibitory effect on *S. aureus* uptake by BAM: first, there may be a product secreted by log phase

H. somnus that mediates this action; or, second, phagocytized *H. somnus* may directly inhibit BAM. By using a membrane to separate *H. somnus* from BAM, it was possible to rule out a secretory product-mediated inhibition. Therefore, it appeared that contact of live, log phase *H. somnus* with BAM was required for the suppression of *S. aureus* uptake. The ability of *H. somnus* to interfere with the phagocytic function of BAM may contribute to its ability to survive and multiply within these cells and is likely important in the pathogenesis of hemophilosis.

Bovine hemophilosis is typically a subacute to chronic disease syndrome, indicating that innate host defense mechanisms, such as phagocytosis, are unable to eliminate the organism during the early stages of infection.[83] Moreover, *H. somnus* can invade a variety of organ systems in the body, including the lungs, heart, and brain, indicating an ability of the organism to escape from the killing mechanism of phagocytes, and disseminate in the body. Although logarithmically growing *H. somnus* suppressed phagocytic function of BAM, both live logarithmically growing and stationary phase *H. somnus* were able to suppress LDCL activity of both BBM and BAM. The ability of *H. somnus* to suppress the respiratory burst and to escape from NO killing[95] is likely important in sustaining and disseminating the infection in the body.

Suppression of chemiluminescence activity of phagocytes by pathogenic bacteria has been demonstrated as an important virulent mechanism in many bacteria–phagocyte models. For example, virulent strains of *Legionella pneumophila* fail to induce the same degree of peak chemiluminescence and superoxide anion production, as do their avirulent counterparts.[100] This phenomenon may be due either to active suppression of PMN responses by virulent strains or possibly to varying degrees of opsonization. With regard to the former, Lochner et al.[101] described a small-molecular-weight cytotoxin from *L. pneumophila*, which alters PMN oxidative metabolism, and later Saha et al.[102] described the blockage of superoxide anion production in human PMN by the activity of an acid phosphatase isolated from *L. micdadei*. Similarly, virulent *Salmonella typhi* exhibit significantly smaller oxidative metabolism by LDCL compared with avirulent *S. typhi*.[103] The clinical spectrum of meningococcal infection is wide, ranging from asymptomatic carriers of the bacteria, to mild, local infections, to meningitis, and to fulminated, fatal septicemia.[104–107] When opsonized with serum from healthy adults without a history of meningococcal disease, the pathogenic meningococcal strains of serogroups A, B, C, Y (Orebro variant), and W-135 give significantly lower chemiluminescence indexes than the apathogenic ones of serogroups Y, Z, and 29E and of non-groupable strains.[108]

Helicobacter pylorus is recognized as the principal cause of antral gastritis and is essential in the development of peptic ulcers. Ammonia, found to be elevated in the gastric aspirates of patients infected with *H. pylori*,[109] not only neutralizes acidity in the stomach to allow bacterial colonization but also enhances the cytotoxicity of neutrophil-derived oxidants.[110] Moreover, ammonia is able to reduce the chemiluminescence of PMN.[111]

In some cases, chemiluminescence measurement has shown that bacteria, such as *S. typhimurium*, induce a dose-dependent oxidative burst in PMN, but are still able to survive and multiply in monocytes.[112] Similarly, mycobacteria are capable of inducing ROI production by both mononuclear phagocytes[113,114] and PMN.[115] As expected, *Mycobacterium bovis* (BCG) stimulates macrophages to release large amounts of superoxide anions. Thus, chemiluminescent studies have shown that survival of these bacteria is dependent on mechanisms other than interference with the respiratory burst.

PMN obtained from patients with periodontal disease challenged with the periodontopathic organism *Fusobacterium nucleatum* have an elevated chemiluminescence response compared with PMN from controls without disease. The elevated chemiluminescence response of subjects in the diseased groups suggests that the activated PMN may be contributing to tissue damage by lysosomal enzymes.[116]

Chemiluminescence assays are a very sensitive way to monitor the oxygen metabolism of phagocytes following interaction with bacteria. Because of the short lifetime of excited states, light emission almost instantaneously follows the activation reaction. Chemiluminescence is the easiest way to assess the metabolic burst of granulocytes and mononuclear phagocytes over a long period of time without the need to add artificial substrates. Thus, the application of chemiluminescence in immunoassays furnishes techniques useful in functional and metabolic studies of phagocytes.

REFERENCES

1. Kaufmann, S. H. E. and Reddehase, M. J., Infection of phagocytic cells, *Curr. Opin. Immunol.,* 2, 43, 1989.
2. Horwitz, M. A., Phagocytosis of Legionnaires' disease bacterium (*Legionella pneumophila*) occur by a novel mechanism; engulfment within a pseudopod coil, *Cell,* 36, 27, 1984.
3. Klebanoff, S. J., Oxygen metabolism and toxic properties of phagocytes, *Ann. Intern. Med.,* 93, 480, 1980.
4. Babior, B. M., Oxygen-dependant microbial killing by phagocytes, *N. Engl. J. Med.,* 298, 659, 1978.
5. Berton, G., Dusi, S., and Bellavite, P., The respiratory burst of phagocytes, in *The Respiratory Burst and Its Physiological Significance,* A. J. Sbarra and R. R. Strauss, Eds., Plenum Press, New York, 1988, 33.
6. Badwey, J. A. and Karnovesky, K. L., Active oxygen species and the function of phagocytic leucocytes, *Annu. Rev. Biochem.,* 49, 695, 1980.
7. Romeo, D., Zabucchi, G., and Marzi, T., Kinetic and enzymatic features of metabolic stimulation of alveolar and peritoneal macrophages challenged with bacteria, *Exp. Cell. Res.,* 78, 423, 1973.
8. Montarroso, A. M. and Myrvik, Q. N., Oxidative metabolism of BCG activated alveolar macrophages, *J. Reticuloendothel. Soc.,* 25, 559, 1979.
9. Papermaster, B. G., Whitcomb, M. E., and Sagone, A. L., Characterization of metabolic responses of the human alveolar macrophage, *J. Reticuloendothel. Soc.,* 28, 129, 1980.
10. Roos, D. and Van der Stijl-Neijenhuis, J. S., The oxygen metabolism of human blood monocytes and neutrophils, in *Mononuclear Phagocytes: Functional Aspects,* Part II, Van Furth, R., Ed., Martinus Nijhoff, The Hague, the Netherlands, 1980, 1219.
11. Rossi, F., Bellavite, P., and Dobrina, A., Oxidative metabolism of mononuclear phagocytes, in *Mononuclear Phagocytes: Functional Aspects,* Part II, Van Furth, R., Ed., Martinus Nijhoff, The Hague, the Netherlands, 1980, 1187.
12. Nathan, C. F., Murray, W. H., and Cohn, Z. A., The macrophage as an effector cell, *N. Engl. J. Med.,* 303, 622, 1980.
13. Moncada, S., Palmer, R. M. J., and Higgs, E. A., Biosynthesis of nitric oxide from L-arginine. A pathway for the regulation of cell function and communication, *Biochem. Pharmacol.,* 138, 1709, 1989.
14. Collier, J. and Vallance, P., Second messenger role for NO widens to nervous and immune system, *Trends Pharmacol. Sci.,* 123, 427, 1989.
15. Nathan, C. F. and Hibbs, J. B., Jr., Role of nitric oxide synthesis in macrophage antimicrobial activity, *Curr. Opin. Immunol.,* 3, 65, 1991.
16. Hibbs, J. B., Taintr, R. R., Vavrin, Z., and Rachlin, E. M., Nitric oxide: a cytotoxic activated macrophage effector molecule, *Biochem. Biophys. Res. Commun.,* 157, 87, 1988.
17. Hasui, M., Hirabayashi, Y., and Kobayashi, Y., Simultaneous measurement by flow cytometry of phagocytosis and hydrogen peroxide production of neutrophils in whole blood, *J. Immunol. Meth.,* 117, 53, 1989.
18. Steele, R. W., Clinical applications of chemiluminescence of granulocytes, *Rev. Infect. Dis.,* 13, 918, 1991.
19. Allen, R. C., Stjernholm, R. L., Reed, M. A., Harper, T. B., Gupta, S., Steele, R. H., and Waring, W. W., Correlation of metabolic and chemiluminescent responses of granulocytes from three female siblings with chronic granulomatous disease, *J. Infect. Dis.,* 136, 510, 1977.
20. Capsoni, F., Minonzio, F., Venegoni, E., Lazzarin, A., Galli, M., Silvani, C., Ongari, A. M., and Zanussi, C., Chronic granulomatous disease in a 23-year-old female, *J. Clin. Lab. Immunol.,* 19, 149, 1986.
21. Via, C. S., Allen, R. C., and Welton, R. C., Direct stimulation of neutrophil oxygenation activity by serum from patients with systemic lupus erythematosus: a relationship to disease activity, *J. Rheumatol.,* 11, 745, 1984.

22. Bender, J. G., Van Epps, D. E., Searles, R., and Williams, R. C., Jr., Altered function of synovial fluid granulocytes in patients with acute inflammatory arthritis; evidence for activation of neutrophils and mediation by a factor present in synovial fluid, *Inflammation,* 10, 443, 1986.
23. Magaro, M., Altomonte, L., Zoli, A., Mirone, L., DeSole, P., DiMario, G., Lippa, S., and Oradei, A., Influence of diet with different lipid composition on neutrophil chemiluminescence and disease activity in patients with rheumatoid arthritis, *Ann. Rheum. Dis.,* 47, 793, 1988.
24. Czirjak, L., Danko, K., Sipka, S., Zeher, M., and Szegedi, G., Polymorphonuclear neutrophil function in systemic sclerosis, *Ann. Rheum. Dis.,* 46, 302, 1987.
25. Kovacs, I. B., Thomas, R. H. M., Macvkay, A. R., Rustin, M. H. A., and Kirby, J. D. T., Increased chemiluminescence of polymorphonuclear leucocytes from patients with progressive systemic sclerosis, *Clin. Sci.,* 70, 257, 1986.
26. McCarthy, J. P., Bodroghy, R. S., Jahrling, P. B., and Sobocinnski, P. Z., Differential alterations in host peripheral polymorphonuclear leukocyte chemiluminescence during the course of bacterial and viral infections, *Infect. Immun.,* 30, 824, 1980.
27. Bellanti, J. A., Krasner, R. I., Bartelloni, P. J., Yang, M. C., and Beisel, W. R., Sandfly fever: sequential changes in neutrophil biochemical and bactericidal functions, *J. Immunol.,* 108, 142, 1972.
28. Mellman, I., Relationships between structure and functions in the Fc receptor family, *Curr. Opin. Immunol.,* 1, 16, 1988.
29. Wright, S. D. and Silverstein, S. C., Receptors for C3b and C3bi promote phagocytosis but not the release of toxic oxygen from human phagocytes, *J. Exp. Med.,* 158, 2016, 1983.
30. Yamamoto, K. and Johnston, R. B., Jr., Dissociation of phagocytosis from stimulation of the oxidative metabolic burst in macrophages, *J. Exp. Med.,* 159, 405, 1984.
31. Aderem, W., Wright, S. D., Silverstein, S. C., and Cohn, Z. A., Ligated complement receptors do not activate the arachidonic acid cascade in resident peritoneal macrophages, *J. Exp. Med.,* 161, 617, 1985.
32. Detmers, P. A. and Wright, S. D., Adhesion promoting receptors on leucocytes, *Curr. Opin. Immunol.,* 1, 10, 1988.
33. Jacobs, R. F., Locksley, R. M., Wilson, C. B., Hass, J. E., and Klebanoff, S. J., Interaction of primate alveolar macrophages and *Legionella pneumophila, J. Clin. Invest.,* 73, 1515, 1984.
34. Holzer, T. J., Nelson, K. E., Schauf, V., Crispen, R. G., and Anderson, B. R., *Mycobacterium leprae* fails to stimulate phagocytic cell superoxide generation, *Infect. Immun.,* 51, 514, 1986.
35. Wilson, C. B., Tsai, V., and Remington, J. S., Failure to trigger the oxidative burst by normal macrophages. Possible mechanism for survival of intracellular pathogens, *J. Exp. Med.,* 151, 328, 1980.
36. Flesch, I. A. and Kaufmann, S. H. E., Attempts to characterize the mechanisms involved in mycobacterial growth inhibition by gamma-interferon-activated bone marrow macrophages, *Infect. Immun.,* 56, 1464, 1988.
37. Kagaya, K., Watanabe, K., and Fukazawa, Y., Capacity of recombinant gamma interferon to activate macrophage for *Salmonella* killing activity, *Infect. Immun.,* 57, 609, 1989.
38. Wolf, J. E., Abegg, A. L., Travis, S. J., Kobayashi, G. S., and Little, J. R., Effect of *Histoplasma capsulatum* on murine macrophage functions: inhibition of macrophage priming, oxidative burst, and antifungal activities, *Infect. Immun.,* 57, 513, 1989.
39. Mor, N., Gorden, M. B., and Pabst, M. J., *Mycobacterium lepraemurium* activates macrophages but fails to trigger release of superoxide anion, *J. Immunol.,* 140, 3956, 1989.
40. Armstrong, J. A. and Hart, P. D., Response of cultured macrophages to *Mycobacterium tuberculosis* with observations on fusion of lysosomes with phagosomes, *J. Exp. Med.,* 134, 713, 1971.
41. Horwitz, M. A., Formation of a novel phagosome by the Legionnaires' disease bacterium (*Legionella pneumophila*) in human monocytes, *J. Exp. Med.,* 158, 1319, 1983.
42. Horwitz, M. A., The legionnaires' disease bacterium (*Legionella pneumophila*) inhibits phagosome-lysosome fusion in human monocytes, *J. Exp. Med.,* 158, 2108, 1983.
43. Corrollo, M. E. W., Jackett, P. S., Aber, V. R., and Lowrie, D. B., Phagolysosomal formation, cyclic adenosine 3′,5′-monophosphate and the fate of *Salmonella typhimurium* within mouse peritoneal macrophages, *J. Gen. Microbiol.,* 110, 421, 1979.
44. Chan, J., Fujiwara, T., Brennan, P., McNeil, M., Turco, S. J., Sibille, J. C., Snapper, Aisen, P., and Bloom, B. R., Microbial glycolipids: possible virulence factors that scavenge oxygen radicals, *Proc. Natl. Acad. Sci. U.S.A.,* 86, 2453, 1989.

45. Chan, J., Fan, X., Hunter, S. W., Brennan, P. J., and Bloom, B. R., Lipoarabinomannan, a possible virulence factor involved in persistence of *Mycobacterium tuberculosis* within macrophages, *Infect. Immun.,* 59, 1755, 1991.
46. Neil, M. A. and Klebanoff, S. J., The effect of phenolic glycolipid-1 from *Mycobacterium leprae* on the antimicrobial activity of human macrophages, *J. Exp. Med.,* 167, 30, 1988.
47. Mayer, B. K. and Falkinham, J. O., Superoxide dismutase activity of *Mycobacterium avium, M. intracellulare*, and *M. scrofulaceum, Infect. Immun.,* 53, 631, 1986.
48. Bermudez, L. E. and Young, L. S., Oxidative and non-oxidative intracellular killing of *Mycobacterium avium* complex, *Microb. Pathog.,* 7, 289, 1989.
49. Black, C., Bermudez, L. E., Young, L. S., and Remington, J., Co-infection of macrophages modulates IFN-γ and TNF-induced activation against intracellular pathogens, *J. Exp. Med.,* 172, 977, 1990.
50. Crowle, A. J., Dahl, R., Ross, E., and May, M. H., Evidence that vesicles containing living, virulent *Mycobacterium tuberculosis* or *Mycobacterium avium* in cultured human macrophages are not acidic, *Infect. Immun.,* 59, 1823, 1991.
51. Bermudez, L. E., Petrofsky, M., Wu, M., and Young, L. S., Interleukin-6 antagonizes tumor necrosis factor-mediated mycobacteriostatic and mycobactericidal activities in macrophages, *Infect. Immun.,* 60, 4245, 1992.
52. Bermudez, L. E. M. and Young, L. S., Tumor necrosis factor, alone or in combination with IL-2, but not IFN-γ, is associated with macrophage killing of *Mycobacterium avium* complex, *J. Immunol.,* 140, 3006, 1988.
53. Bermudez, L. E. and Young, L. S., Oxidative and non-oxidative intracellular killing of *Mycobacterium avium* complex, *Microb. Pathog.,* 7, 289, 1989.
54. Rook, G. A. W., Steele, M., Ainsworth, M., and Champion, B. R., Activation of macrophages to inhibit proliferation of *Mycobacterium tuberculosis:* comparison of the effects of recombinant γ-interferon on human monocytes and murine peritoneal macrophages, *Immunology,* 59, 333, 1986.
55. Williams, D. M., Magee, D. M., Bonewald, L. F., Smith, J. G., Bleicker, C. A., Byrne, G. I., and Schacter, J., A role *in vivo* for tumor necrosis factor alpha in host defence against *Chlamydia trachomatis, Infect. Immun.,* 58, 1572, 1990.
56. Nakane, A., Minagawa, T., and Kato, K., Endogenous tumor necrosis factor (cachetin) is essential to host resistance against *Listeria monocytogenes* infection, *Infect. Immun.,* 56, 2563, 1988.
57. Desiderio, J. V., Kiener, P. A., Lin, P. F., and Warr, G. A., Protection of mice against *Listeria monocytogenes* infection by recombinant tumor necrosis factor alpha, *Infect. Immun.,* 57, 1615, 1989.
58. Djeu, J. Y., Blanchard, D. K., Halkias, D., and Friedman, H., Growth inhibition of *Candida albicans* by human polymorphonuclear neutrophils: activation by IFN-γ and tumor necrosis factor, *J. Immunol.,* 137, 2980, 1986.
59. Blanchard, D. K., Djeu, J. Y., Klein, T. W., Friedman, H., and Steward, W. E., Protective effect of TNF in experimental *Legionella pneumophila* infection of mice via activation of PMN function, *J. Leucocyte Biol.,* 43, 429, 1988.
60. Clark, I. A., Hunt, N. H., and Cowden, W. B., Oxygen derived free radicals in the pathogenesis of parasitic diseases, *Adv. Parasitol.,* 25, 1, 1986.
61. Murray, H. W. and Cohn, Z. A., Macrophage oxygen-dependant antimicrobial activity. Enhanced oxidative metabolism as an expression of macrophage activation, *J. Exp. Med.,* 152, 1596, 1980.
62. Nathan, C. F., Murray, W. H., and Cohn, Z. A., The macrophage as an effector cell, *N. Engl. J. Med.,* 303, 622, 1980.
63. Klebanoff, S. J., Oxygen dependent cytotoxic mechanisms of phagocytes, in *Advances in Host Defence Mechanisms,* J. I. Gallin and A. S. Fauci, Eds., Raven Press, New York, 1982, 111.
64. Ito, M., Karmali, R., and Krim, M., Effect of interferon on chemiluminescence and hydroxyl radical production in murine macrophages stimulated by PMA, *Immunology,* 56, 533, 1985.
65. Bhardwaj, N., Nash, T. W., and Horwitz, M. A., Interferon-gamma activated human monocytes inhibit the intracellular multiplication of *Legionella pneumophila, J. Immunol.,* 137, 2662, 1986.
66. Nash, T. W., Libby, D. M., and Horwitz, M. A., IFN-γ activated human alveolar macrophages inhibit the intracellular multiplication of *Legionella pneumophila, J. Immunol.,* 140, 3978, 1988.
67. Nathan, C. F., Murray, H. W., Wiebe, M. E., and Rubin, B. Y., Identification of interferon-γ as the lymphokine that activates human macrophage oxidative metabolism and antimicrobial activity, *J. Exp. Med.,* 158, 670, 1983.

68. Nathan, C. F. and Tsunawaki, S., Enzymatic basis of macrophage activation, *J. Biol. Chem.*, 69, 4305, 1984.
69. Wilson, C. B. and Westfall, J., Activation of neonatal and adult human macrophages by alpha, beta, and gamma interferons, *Infect. Immun.*, 49, 351, 1985.
70. Rothermel, C. D., Rubin, B. Y., and Murray, H. W., γ-Interferon is the factor in lymphokine that activates human macrophages to inhibit intracellular *Chlamydia psittaci* replication, *J. Immunol.*, 131, 2542, 1983.
71. Murray, H. W., Rubin, B.Y., and Rothermel, C. D., Killing of intracellular *Leishmania donovani* by lymphokine-stimulated human phagocytes. Evidence that interferon-gamma is the activating lymphokine, *J. Clin. Invest.*, 72, 1506, 1983.
72. Hoover, D. L., Nacy, C. A., and Meltzer, M. S., Human monocyte activation for cytotoxicity against intracellular *Leishmania donovani* amastigotes: induction of microbicidal activity by interferon gamma, *Cell. Immunol.*, 94, 500, 1985.
73. Passwell, J. H., Shor, R., and Shoham, J., The enhancing effect of interferon-beta and gamma on the killing of *Leishmania tropica major* in human mononuclear phagocytes *in vitro*, *J. Immunol.*, 136, 3062, 1986.
74. Reed, S. G., Nathan, C. F., Pipl, D. L., Rodricks, P., Shanebeck, K., Conlon, P. J., and Grabstein, K. H., Recombinant granulocyte/macrophage colony-stimulating factor activates macrophages to inhibit *Trypanasoma cruzi* and release hydrogen peroxide, comparison with interferon γ, *J. Exp. Med.*, 166, 1734, 1987.
75. Murray, H. W., Rubin, B. Y., Carriero, S. M., Harris, A. M., and Jaffee, F. A., Human mononuclear phagocyte antiprotozoan mechanisms: oxygen dependent vs. oxygen independent activity against intracellular *Toxoplasma gondii*, *J. Immunol.*, 134, 1982, 1985.
76. Flesch, I. and Kaufmann, S. H. E., Mycobacterial growth inhibition by inteferon-γ activated bone marrow macrophages and differential susceptibility among strains of *Mycobacterium tuberculosis*, *J. Immunol.*, 138, 4408, 1987.
77. Flesch, I. A. and Kaufmann, S. H. E., Attempts to characterize the mechanisms involved in mycobacterial growth inhibition by gamma-interferon-activated bone marrow macrophages, *Infect. Immun.*, 56, 1464, 1988.
78. van Dissel, J. T., Stikkelbroeck, J. J. M., Michel, B. C., van den Barselaar, M. T., Leijh, P. C. J., and van Furth, R., Inability of recombinant interferon-γ to activate the antibacterial activity of mouse peritoneal macrophages against *Listeria monocytogenes* and *Salmonella typhimurium*, *J. Immunol.*, 139, 1673, 1987.
79. Tsuyuguchi, I., Kawasumi, H., Takashima, T., Tsuyuguchi, T., and Kishimoto, S., *Mycobacterium avium–Mycobacterium intracellulare* complex-induced suppression of T-cell proliferation *in vitro* by regulation of monocyte accessory cell activity, *Infect. Immun.*, 58, 1369, 1990.
80. Corstvet, R. E., Panciera, R. J., Rinker, H. B., Starks, B. L., and Howard, C., Survey of tracheas of feedlot cattle for *Haemophilus somnus* and other selected bacteria, *J. Am. Vet. Med. Assoc.*, 163, 870, 1973.
81. Crandell, R. A., Smith, A. R., and Kissil, M., Colonization and transmission of *Haemophilus somnus* in cattle, *Am. J. Vet. Res.*, 38, 1749, 1977.
82. Humphrey, J. D., Little, P. B., Barnum, D. A., Doig, P. A., Stephens, L. R., and Thorsen, J., Occurrence of *Haemophilus somnus* in bovine semen and in the prepuce bulls and steers, *Can. J. Comp. Med.*, 46, 215, 1982.
83. Lederer, J. A., Brown, J. F., and Czuprynski, C. J., "*Haemophilus somnus*," a facultative intracellular pathogen of bovine mononuclear phagocytes, *Infect. Immun.*, 55, 381, 1986.
84. Harris, F. W. and Janzen, E. D., The *Haemophilus somnus* disease complex (haemophilosis): a review, *Can. Vet. J.*, 30, 816, 1989.
85. Gomis, S. M., Godson, D. L., Beskorwayne, T., Wobeser, G. A., and Potter, A. A., Modulation of phagocytic function of bovine mononuclear phagocytes by *Haemophilus somnus*, *Microb. Pathog.*, 22, 13, 1997.
86. Pfeifer, C. G., Campos, M., Beskorwayne, T., Babiuk, L. A., and Potter, A. A., Effect of *Haemophilus somnus* on phagocytosis and hydrogen peroxide production by bovine polymorphonuclear leukocytes, *Microb. Pathog.*, 13, 191, 1992.

87. Harland, R. J., Potter, A. A., and Schuh, J. C. L., Development of an intravenous challenge model for *Haemophilus somnus* disease in beef calves in *71st Conf. Res. Work. Anim. Dis.,* 1990, 29.
88. Hasui, M., Hirabayashi, Y., and Kobayashi, Y., Simultaneous measurement by flow cytometry of phagocytosis and hydrogen peroxide production of neutrophils in whole blood, *J. Immunol. Methods,* 117, 53, 1989.
89. Raybourne, R. B. and Bunning, V. K., Bacterium–host cell interactions at the cellular level: fluorescent labelling of bacteria and analysis of short-term bacterium–phagocyte interaction by flow cytometry, *Infect. Immun.,* 62, 665, 1994.
90. Van Dyke, K., McCapra, F., and Behesti, I., in *Bioluminescence and Chemiluminescence, Instruments and Applications,* Van Dyke, K., Ed., CRC Press, Boca Raton, FL, 1985, 1.
91. Sample, A. K. and Czuprynski, C. J., Elimination of hydrogen peroxide by *Haemophilus somnus,* a catalase-negative pathogen of cattle, *Infect. Immun.,* 59, 2239, 1991.
92. Czuprynski, C. J. and Hamilton, H. L., Bovine neutrophils inject but do not kill *Haemophilus somnus in vitro, Infect. Immun.,* 50(2), 431, 1985.
93. Hubbard, R. D., Kaeberle, M. L., Roth, J. A., and Chiang, Y. W., *Haemophilus somnus*-induced interference with bovine neutrophil functions, *Vet. Microbiol.,* 12, 77, 1986.
94. Chiang, Y. W., Kaeberle, M. L., and Roth, J. A., Identification of suppressive components in "*Haemophilus somnus*" fractions which inhibit bovine polymorphonuclear leukocyte function, *Infect. Immun.,* 52, 792, 1986.
95. Fields, P. I., Swanson, R. V., Haidaris, C. G., and Heffron, F., Mutants of *Salmonella typhimurium* that cannot survive within the macrophage are virulent, *Proc. Natl. Acad. Sci. U.S.A.,* 83, 5189, 1986.
96. Berche, P., Gaillard, J. L., and Sansonetti, P. J., Intracellular growth of *Listeria monocytogenes* as a prerequisite for *in vivo* induction of T cell-mediated immunity, *J. Immunol.,* 138, 2266, 1987.
97. Brunt, L. M., Portnoy, D. A., and Unanue, E. R., Presentation of *Listeria monocytogenes* to CD8 T cells requires secretion of haemolysin and intracellular bacterial growth, *J. Immunol.,* 145, 3540, 1990.
98. Bancroft, G. J., Schreiber, R. D., and Unanue, E. R., Natural immunity: a T-cell-dependent pathway of macrophage activation defined in the scid mouse, *Immunol. Rev.,* 124, 5, 1991.
99. Gomis, S. M., Godson, D. L., Wobeser, G. A., and Potter, A. A., Effect of *Haemophilus somnus* on nitric oxide production and chemiluminescence response of bovine blood monocytes and alveolar macrophages, *Microb. Pathog.,* 23, 327, 1997.
100. Summersgill, J. T., Raff, M. J., and Miller, R. D., Interactions of virulent and avirulent *Legionella pneumophila* with human polymorphonuclear leukocytes, *Microb. Pathog.,* 5, 41, 1988.
101. Lochner, J. E., Bigley, R. H., and Lglewski, B. H., Defective triggering of polymorphonuclear leukocyte oxidative metabolism by *Legionella pneumophila* toxin, *J. Infect. Dis.,* 151, 42, 1985.
102. Saha, A. K., Dowling, J. N., and LaMarco, K. L., Properties of an acid phosphatase from *Legionella micdadei* which block superoxide anion production by human neutrophils, *Arch. Biochem. Biophys.,* 43, 150, 1985.
103. Kossack, R. E., Luerrant, R. L., Densen, P., Schadelin, J., and Mandell, G. L., Diminished neutrophil oxidative metabolism after phagocytosis of virulent *Salmonella typhi, Infect. Immun.,* 31(2), 674, 1981.
104. De Voe, I. W., The meningococcus and mechanisms of pathogenicity, *Microb. Rev.,* 46, 162, 1982.
105. Peltola, H., Meningococcal disease; still with us, *Rev. Infect. Dis.,* 5, 71, 1983.
106. Hagman, M. and Danielsson, D., Increased adherence to vaginal epithelial cells and phagocytic killing of gonococci and urogenital meningococci with heat modifiable proteins, *APMIS,* 97, 839, 1989.
107. Berg, S., Trolfors, B., Alestig, K., and Jodal, U., Incidence, serogroups and case-fatality rate of invasive meningococcal infections in the Swedish region 1975–1989, *Scand. J. Infect. Dis.,* 24, 333, 1992.
108. Fredlund, H., Serum factors and polymorphonuclear leukocytes in human host defense against *Neisseria meningitidis.* Studies of interactions with special reference to a chemiluminescence technique, *Scand. J. Infect. Dis.,* 87, 1, 1993.
109. Marshall, B. J. and Langton, S. R., Urea hydrolysis in patients with *Campylobacter pyloridis* infection, *Lancet,* 8487I, 965, 1986.
110. Grisham, M. B., Jefferson, M. M., and Melton, D. F., Chlorination of endogenous amines by isolated neutrophils, *J. Biol. Chem.,* 259, 10404, 1984.
111. Mayo, K., Held, M., Wadstrom, T., and Megraud, F., *Helicobacter pylori*-human polymorphonuclear leukocyte interaction in the presence of ammonia, *Eur. J. Gastroenterol. Hepatol.,* 9, 457, 1997.

112. Riber, U. and Lind, P., Interactions between *Salmonella typhimurium* and phagocytic cells in pigs; phagocytosis, oxidative burst and killing in polymorphonuclear leukocytes and monocytes, *Vet. Immunol. Immunopathol.,* 67, 259, 1999.
113. Grisham, M. B., Jefferson, M. M., and Melton, D. F., Chlorination of endogenous amines by isolated neutrophils, *J. Biol. Chem.,* 259, 10404, 1984.
114. Jackett, P. S., Andrew, P. W., and Lowrie, D. B., Release of superoxide and hydrogen peroxide from guinea-pig alveolar macrophages during phagocytosis of *Mycobacterium bovis* BCG, *Adv. Exp. Med. Biol.,* 155, 687, 1982.
115. May, M. E. and Spagnuolo, P. J., Evidence for activation of a respiratory burst in the interaction of human neutrophils with *Mycobacterium tuberculosis, Infect. Immun.,* 55, 2304, 1987.
116. Seymour, G. J., Whyte, G. J., and Powell, R. N., Chemiluminescence in the assessment of polymorphonuclear leukocyte function in chronic inflammatory periodontal disease, *J. Oral. Pathol.,* 15, 125, 1986.

20 L-012: An Ultrasensitive Reagent for Detecting Cellular Generation of Reactive Oxygen Species

Isuke Imada, Eisuke F. Sato, Yuzo Ichimori, Yoshio Aramaki, and Masayasu Inoue

CONTENTS

20.1 INTRODUCTION

Reactive oxygen species play important roles in the defense mechanism against infection and in the pathogenesis of various diseases. As shown in Figure 20.1, reactive oxygen species are produced by various biological systems, such as NADH dehydrogenase and succinate-cytochrome *c* reductase in mitochondria,[1,2] NADPH oxidase in neutrophils,[3] and xanthine oxidase in vascular endothelial cells.[4] Superoxide is generated by one electron reduction of molecular oxygen and is dismutated into hydrogen peroxide, which can be converted to hydroxyl radical either by Haber–Weis reaction or Fenton reaction.[5] Hydrogen peroxide is also metabolized to hypochlorite by myeloperoxidase in leukocytes.[6] The reaction of hydrogen peroxide with hypochlorite generates singlet oxygen.[7,8]

Reactive oxygen species have been determined by various methods including electron-spin resonance spectroscopy,[9] spectrophotometry of cytochrome *c* and nitroblue tetrazolium,[10,11] and chemiluminescence of luminol,[12,13] lucigenin,[13] and luciferin analogue, 2-methyl-6-[*p*-methoxyphenyl]-3, 7-dihydroimidazo[1,2-*α*]pyrazin-3-one (MCLA).[14] Chemiluminescence is one of the most convenient and highly sensitive methods for the analysis of reactive oxygen species generated by biologically complex systems, such as cells and blood samples. Although the chemiluminescence probes described above have been used for the analysis of reactive oxygen species in blood samples, sensitivities of these probes are fairly low, particularly under physiological conditions.[13,15,16] We synthesized 8-amino-5-chloro-7-phenylpyrido[3,4-*d*] pyridazine-1,4(2H,3H)dione (L-012, Figure 20.2) as a novel chemiluminescence probe[17] and used it for measuring the production of reactive oxygen species by activated EoL-1 cells.[18] Under physiological conditions, opsonized zymosan-dependent chemiluminescence intensities of L-012 in human blood and rat peritoneal neutrophils were much higher than that of

0-8493-0719-8/02/$0.00+$1.50

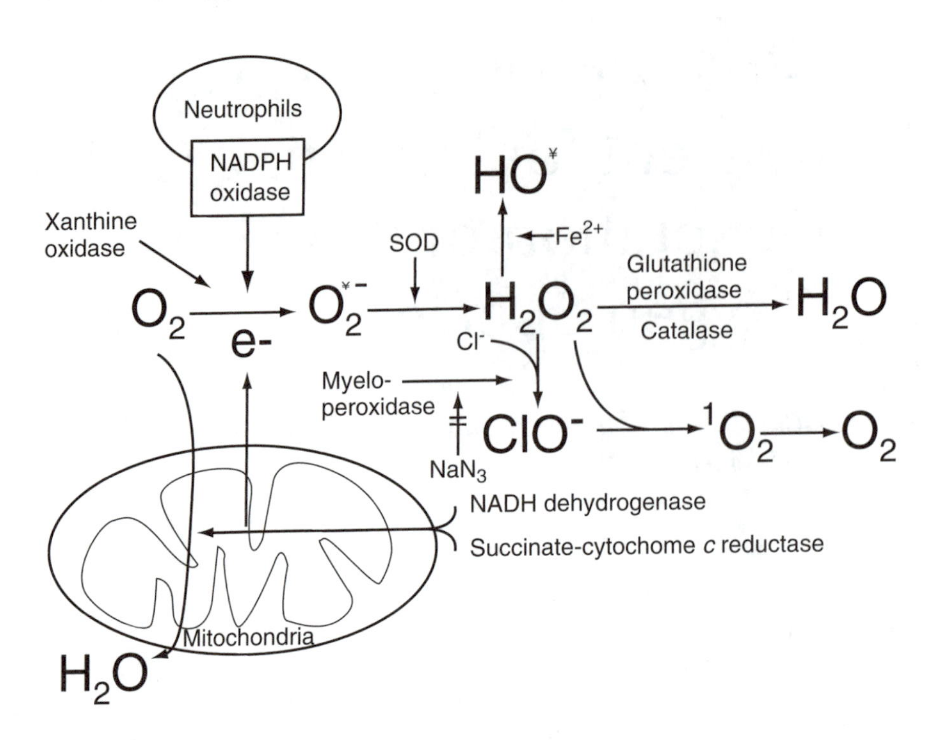

FIGURE 20.1 Generation and metabolism of reactive oxygen.

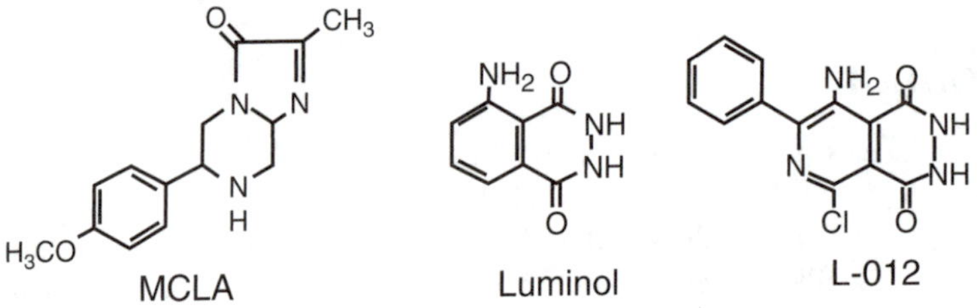

MCLA Luminol L-012

FIGURE 20.2 Structures of L-012 and other chemiluminescence probes.

luminol and MCLA.[19] Phorbol ester-dependent chemiluminescence of L-012 in oral neutrophils was also higher than that of luminol and MCLA. Kinetic analysis revealed that L-012 developed chemiluminescence predominantly by reacting with hydroxyl radical and hypochlorite, and revealed that this probe is useful for the study of the metabolism and role of reactive oxygen species in complex biological systems.

20.2 PROPERTIES OF CHEMILUMINESCENCE PROBE L-012

Similar to luminol, L-012 is a cyclic hydrazide derivative with 2,3-dihydro-1,4-pyridopyridazine dione in its structure. Therefore, L-012 has been expected to develop chemiluminescence by stabilizing its oxidatively formed dianion. In fact, L-012 developed a strong chemiluminescence in the presence of hydrogen peroxide and horseradish peroxidase.[20,21] The presence of either superoxide dismutase (SOD) or catalase decreased the chemiluminescence intensity of L-012 (Table 20.1). Thus, L-012 might also react with superoxide, hydrogen peroxide, or their metabolite(s).

TABLE 20.1
Effects of Various Agents on L-012 Chemiluminescence

Addition	Xanthine Oxidase		Fe^{2+}/H_2O_2	
	cpm × 10^3	(%)	cpm × 10^4	(%)
None	11.6 ± 2.8	(100)	781 ± 136	(100)
SOD	2.0 ± 1.0	(17)	n.d.	
Deferoxamine	1.2 ± 0.1	(10)	8.0 ± 1.5	(1.0)
DMSO	n.d.		21 ± 2.7	(2.7)
Mannitol	n.d.		200 ± 3.6	(25.6)

Note: Reactive oxygen species were generated in cell-free systems at 37°C either by xanthine oxidase (6.4 mU/ml) or $FeSO_4/H_2O_2$ (50 *μM*/1 m*M*). The concentrations of SOD, deferoxamine, DMSO, and mannitol were 32 mU/ml, 0.1 m*M*, 8 m*M*, and 8 m*M*, respectively.

n.d. = not determined.

Source: Imada, I. et al., *Anal. Biochem.*, 271, 53, 1999. With permission.

20.3 ASSAY FOR THE GENERATION OF REACTIVE OXYGEN SPECIES BY CELL FREE SYSTEMS

To clarify the chemical properties of reactive oxygen species responsible for L-012-dependent chemiluminescence, reactive oxygen species were generated in cell-free systems as shown in Table 20.1, and the resulting chemiluminescence was determined in the presence or absence of various agents. The superoxide radical was generated by a hypoxanthine–xanthine oxidase system at 37°C. A reaction mixtures contains 1 m*M* hypoxanthine and xanthine oxidase (6.4 mU) as described previously,[22] in 0.9% NaCl solution containing 6 m*M* KCl, 6 m*M* $MgCl_2$, and 10 m*M* phosphate buffer, pH 7.4 (KRP). Concentrations of L-012, luminol, and MCLA used for the assay were 4, 100, and 4 *μM*, respectively. Xanthine oxidase–generated reactive oxygen species increased L-012-dependent chemiluminescence by some mechanism that was inhibited by either SOD or deferoxamine. The hydroxyl radical was generated by incubating a solution in KRP containing 100 *μM* diethylenetriamine pentaacetic acid, 1 m*M* H_2O_2, and 50 *μM* $FeSO_4$ at 37°C.[23] The presence of Fe^{2+} and H_2O_2 also increased the L-012-dependent chemiluminescence by a mechanism that was inhibited by deferoxamine, DMSO, or mannitol.

20.4 GENERATION OF REACTIVE OXYGEN SPECIES BY HUMAN NEUTROPHILS

Reactive oxygen species that are generated by leukocytes have been studied extensively, but methods available for their analysis are not sufficiently sensitive. Although luminol shows relatively high intensities of chemiluminescence around its optimal pH of 9.5,[12] the chemiluminescence intensity of this compound is extremely low under physiological conditions at pH 7.5.[20] The sensitivity of MCLA is relatively high and, hence, this compound also has been used for analyzing reactive oxygen species generated by isolated cells and cell-free systems.[16] However, because of high blank values of MCLA, it cannot be used for the determination of reactive oxygen species in whole blood. The present work shows that L-012 is a highly sensitive chemiluminescence probe for analyzing reactive oxygen species generated not only by simple, cell-free system and isolated neutrophils, but also by complex biological systems such as whole blood.

Human blood samples and oral neutrophils were obtained from healthy volunteers as described previously.[24] For example, venous blood was obtained from normal healthy humans with one tenth volume of 3.8% sodium citrate. The blood samples (10 to 50 *μl*) were incubated in 0.5 ml of KRP

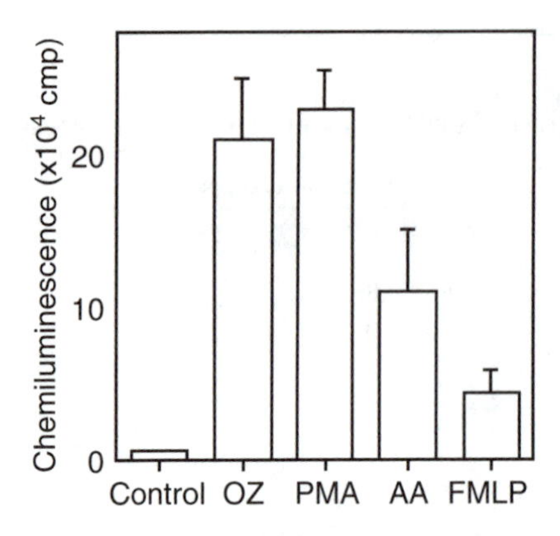

FIGURE 20.3 Effects of various stimuli on L-012 chemiluminescence in human blood. Reaction mixtures contained in a final volume of 500 μl KRP, 10 μl of blood samples, and 400 μM L-012. Reactive oxygen species were generated by adding 2 μg/ml opsonized zymosan (OZ), or 3 nM PMA, or 7.5 μM arachidonic acid (AA), or 125 μM FMLP. Control group received 10 μl of saline. During the incubation at 37°C, chemiluminescence intensity was recorded continuously for 20 to 40 min using Luminescence Reader BLR-201. Data show maximum intensity of chemiluminescence (mean ± SD, n = 3). (From Imada, I. et al., *Anal. Biochem.,* 271, 53, 1999. With permission.)

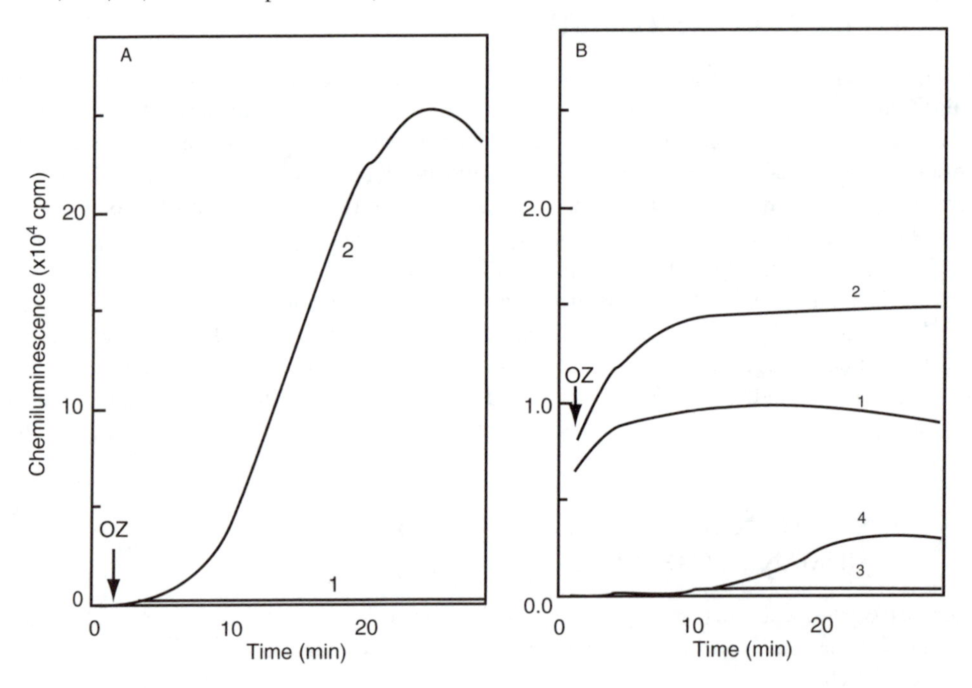

FIGURE 20.4 Effect of opsonized zymosan (OZ) on the chemiluminescence of human blood. Reaction mixtures contained in a final volume of 500 μl KRP, 10 μl of blood, and 400 μM L-012 (A, 1 and 2), or 4 μM MCLA (B, 1 and 2), or 1 mM luminol (B, 3 and 4). At the indicated times (arrows), 2 μg/ml of OZ (2 and 4) or saline (1 and 3) was added to the mixture. Chemiluminescence intensities were recorded as described in Figure 20.3. (From Imada, I. et al., *Anal. Biochem.,* 271, 53, 1999. With permission.)

in the presence of 400 μM L-012, or 1 mM luminol, or 4 μM MCLA. After incubation of the mixture for 3 min at 37°C, the reaction was started by adding 2 μg/ml of opsonized zymosan, or 1.25 μM *N*-formylmethionylleucylphenylalanine (FMLP), or 12 pM phorbol-12-myristate-13-acetate (PMA), or 15 nM arachidonic acid. During the incubation, chemiluminescence intensity was recorded continuously for 20 to 40 min using the Luminescence Reader BLR-201 (Aloka, Tokyo, Japan). Among various stimulants used, opsonized zymosan, PMA, and arachidonic acid strongly increased L-012-dependent chemiluminescence in human blood (Figure 20.3). FMLP also increased chemiluminescence intensity, although its potency was significantly lower than that of other stimulants.

When stimulated by opsonized zymosan, chemiluminescence intensity of L-012 strongly increased in a time-dependent manner (Figure 20.4A). In contrast, chemiluminescence intensities of luminol and MCLA were about 1 and 5% of that of L-012, respectively (Figure 20.4B). Even under physiological conditions, chemiluminescence intensity of L-012 was significantly larger than that of luminol and MCLA. Opsonized zymosan activates receptor-mediated generation of reactive oxygen species, phagocytosis, and degranulation of neutrophils, whereas arachidonic acid and PMA have been used as membrane perturbers and activators for Ca^{2+}- and phospholipid-dependent protein kinase C, respectively.[25] Thus, the type of stimuli is also an important factor for active oxygen metabolism in neutrophils. Among these probes used, L-012 exhibited the strongest chemiluminescence irrespective of the ligands used for stimulating blood samples and isolated neutrophils.

Because neutrophils in the oral cavity are already primed during and/or after infiltrations to the mucosal surface, they spontaneously generate reactive oxygen species without added stimuli.[26] Under unstimulated conditions, chemiluminescence intensity of L-012 was about 3 and 50 times greater than that of MCLA and luminol, respectively (Figure 20.5). When stimulated by PMA, reactive oxygen species generation by neutrophils from the oral cavity increased in a time-dependent manner. PMA enhanced L-012-dependent chemiluminescence by about 12-fold. The enhanced chemiluminescence of L-012 was about 10 and 100 times stronger than that of MCLA and luminol, respectively. Thus, highly sensitive L-012 permits studies of the mechanism of reactive oxygen species

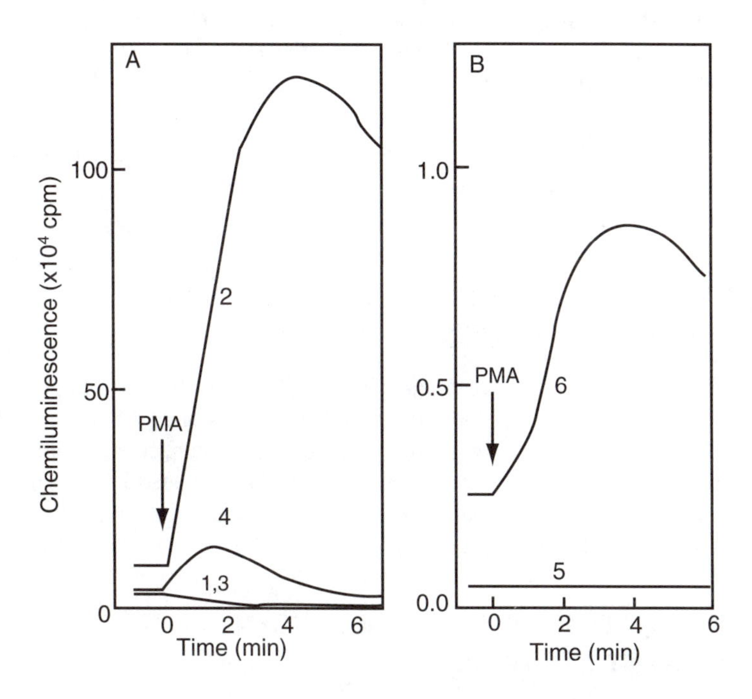

FIGURE 20.5 Chemiluminescence of PMA-stimulated neutrophils from the oral cavity. Reaction mixtures contained in a final volume of 500 μl KRP, 1×10^5 OPMN, and 400 μM L-012 (A, 1 and 2), or 4 μM MCLA (A, 3 and 4), or 1 m*M* luminol (B, 5 and 6). At the indicated times (arrows), either 10 n*M* PMA (2, 4, 6) or saline (1, 3, 5) was added to the mixture at 37°C. Chemiluminescence intensities were recorded as described in Figure 20.3. (From Imada, I. et al., *Anal. Biochem.*, 271, 53, 1999. With permission.)

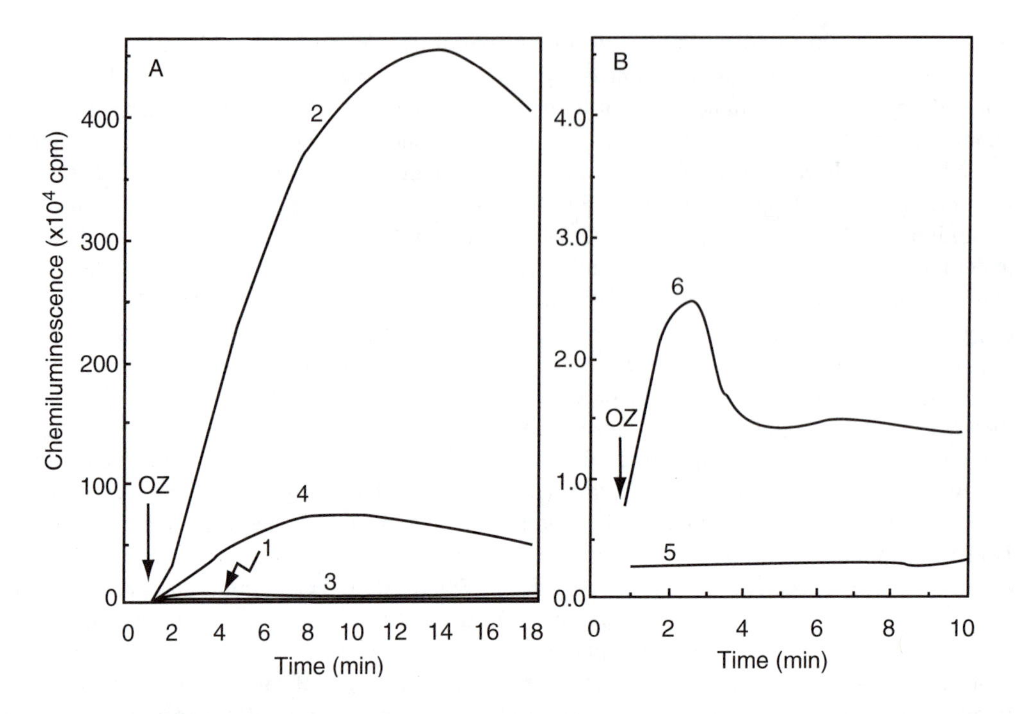

FIGURE 20.6 Chemiluminescence of rat peritoneal neutrophils. Reaction mixtures contained in a final volume of 500 μl KRP, 1×10^5 rat peritoneal neutrophils, and 400 μM L-012 (A, 1 and 2), or 4 μM MCLA (A, 3 and 4), or 1 m*M* luminol (B, 5 and 6). At the indicated times (arrows), either 2 μg/ml opsonized zymosan (OZ) (2, 4, 6) or saline (1, 3, 5) was added to the mixture. During the incubation at 37°C, chemiluminescence was monitored as shown in Figure 20.3. (From Imada, I. et al., *Anal. Biochem.*, 271, 53, 1999. With permission.)

generation by leukocytes and of the role of reactive oxygen species in the pathogenesis of various diseases. The present work describes that, under physiological conditions, L-012 reveals strong chemiluminescence in the presence of various types of activated neutrophils.

20.5 GENERATION OF REACTIVE OXYGEN SPECIES BY RAT PERITONEAL NEUTROPHILS

Neutrophils were obtained from the peritoneal cavity of the rat 16 h after intraperitoneal injection of 2% casein as described previously.[27] When stimulated by opsonized zymosan, chemiluminescence intensity of L-012 also increased with time (Figure 20.6). Again, chemiluminescence intensity of L-012 was stronger than that of MCLA and luminol. When stimulated by opsonized zymosan, L-012- dependent chemiluminescence of isolated rat peritoneal neutrophils was 18- and 5-fold greater than that of stimulated blood samples, respectively.

20.6 CHEMICAL NATURE OF REACTIVE OXYGEN SPECIES GENERATED BY NEUTROPHILS

To elucidate the chemical nature of the reactive oxygen species responsible for L-012-dependent chemiluminescence, the effects of various scavengers and inhibitors on chemiluminescence intensity were studied. As shown in Figure 20.7A, L-012-dependent chemiluminescence of rat peritoneal neutrophils was decreased significantly by the presence of SOD, catalase, uric acid, deferoxamine, or azide; the inhibitory effect was strongest with azide. Chemiluminescence intensities of neutrophils in the oral cavity and human blood samples (Figures 20.7B and 20.7C, respectively) were also

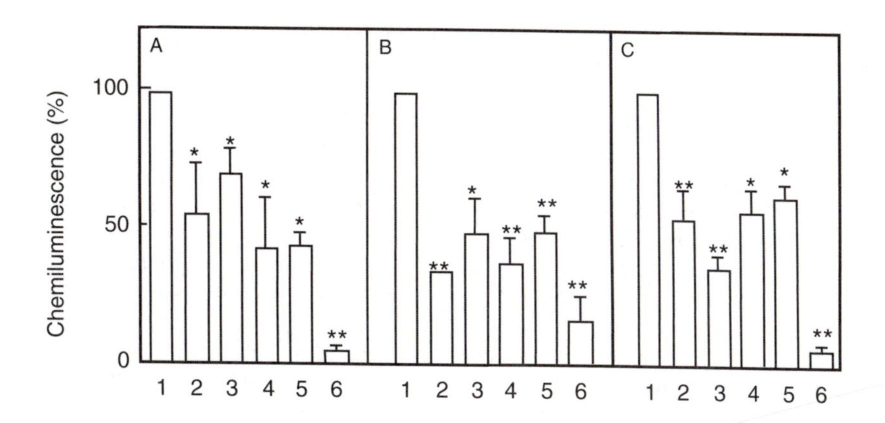

FIGURE 20.7 Effects of various agents on L-012 chemiluminescence of neutrophils. Reaction mixtures contained in a final volume of 500 μl KRP, 400 μM L-012, and 1×10^5 rat peritoneal neutrophils (A), or neutrophils from the oral cavity (B), or 10 μl human blood (C). Chemiluminescence intensity was determined in the absence (1) or presence of 300 U/ml SOD (2), 0.22 U catalase (3), 0.5 μM uric acid (4), or 0.1 mM deferoxamine (5), or 20 μM azide (6). The reaction was started by adding 1 μg of opsonized zymosan (A and C) or 3 nM PMA (B). Other conditions were the same as in Figure 20.3. The data show mean ± SD ($n = 3$ or 4). * $p < 0.05$, ** $p < 0.01$ compared with 1. (From Imada, I. et al., *Anal. Biochem.*, 271, 53, 1999. With permission.)

suppressed by these scavengers and inhibitors. Because neutrophil-dependent chemiluminescence was strongly inhibited by azide, a specific inhibitor of myeloperoxidase, hypochlorite might also underlie the mechanism of L-012-dependent chemiluminescence. The chemiluminescence intensity of L-012 was also suppressed by deferoxamine and uric acid, an iron chelator and a scavenger for hydroxyl radical, respectively.[25,28] Thus, hydroxyl radical might also be involved in L-012-dependent chemiluminescence. This possibility was also confirmed by experiments using cell-free systems for generating reactive oxygen species, such as hypoxanthine–xanthine oxidase and Fe^{2+}–H_2O_2 systems (see Table 20.1). Similar results were also confirmed by an Electron Spin Resonance (ESR) analysis (data not shown). Because azide and uric acid also scavenge singlet oxygen,[28,29] the possible involvement of this species in L-012-dependent chemiluminescence should be studied further.

REFERENCES

1. Boveris, A., Cadenas, E., and Stoppani, A. O. M., Role of ubiquinone in the mitochondrial generation of hydrogen peroxide, *Biochem. J.*, 156, 435, 1976.
2. Takeshige, K. and Minakami, S., NADH- and NADPH-dependent formation of superoxide anions by bovine heart submitochondrial particles and NADH-ubiquinone reductase preparation, *Biochem. J.*, 180, 129, 1979.
3. Babior, B. M., Oxygen-dependent microbial killing by phagocytes, *N. Engl. J. Med.*, 298, 659, 1978.
4. Parks, D. A. and Granger, D. N., Ischemia-induced vascular changes: role of xanthine oxidase and hydroxyl radicals, *Am. J. Physiol.*, 245, G285, 1983.
5. Czapsky, G., Reaction of •OH, *Methods Enzymol.*, 105, 209, 1984.
6. Harrison, J. E. and Schultz, J., Studies on the chemiluminescence chlorinating activity of myeloperoxidase, *J. Biol. Chem.*, 251, 1371, 1976.
7. Foote, C. S. and Wexler, S., Oxidations with excited singlet molecular oxygen, *J. Am. Chem. Soc.*, 86, 3879, 1964.
8. Mascio, P. D., Bechara, E. J. H., Medeiros, M. H. G., Briviba, K., and Sies, H., Singlet molecular oxygen production in the reaction of peroxynitrite with hydrogen peroxide, *FEBS Lett.*, 355, 287, 1994.
9. Rosen, G. M. and Rauckman, E. J., Spin trapping of superoxide and hydroxyl radicals, *Methods Enzymol.*, 105, 198, 1984.

10. Nakagawara, A., Shibata, Y., Takeshige, K., and Minakami, S., Action of cytochalasin E on polymorphonuclear leucocytes of guinea pig peritoneal exudates, *Exp. Cell Res.,* 101, 225, 1976.
11. Beyer, W. F., Jr. and Fridovich, I., Assaying for superoxide dismutase activity: some large consequences of minor changes in conditions, *Anal. Biochem.,* 161, 559, 1987.
12. Misra, H. P. and Squatrito, P. M., The role of superoxide anion in peroxidase-catalyzed chemiluminescence of luminol, *Arch. Biochem. Biophys.,* 215, 59, 1982.
13. Allen, C., Phagocytic leukocyte oxygenation activities and chemiluminescence: a kinetic approach to analysis, *Methods Enzymol.,* 133, 449, 1986.
14. Nishida, A., Kimura, H., Nakano, M., and Goto, T., A sensitive and specific chemiluminescence method for estimating the ability of human granulocytes and monocytes to generate O_2^-, *Clin. Chim. Acta,* 179, 177, 1989.
15. DeChatelet, L. R. and Shirley, P. S., Evaluation of chronic granulomatous disease by a chemiluminescence assay of microtiter quantities of whole blood, *Clin. Chem.,* 27, 1739, 1981.
16. Nakano, M., Determination of superoxide radical and singlet oxygen based on chemiluminescence of luciferin analogs, *Methods Enzymol.,* 186, 585, 1990.
17. Masuya, H., Kondo, K., Aramaki, Y., and Ichimori, Y., Pyridopyridazine compounds and their use, Eur. patent appl., 491,477, 1992.
18. Nishinaka, Y., Aramaki, Y., Yoshida, H., Masuya, H., Sugawara, T., and Ichimori, Y., A new sensitive chemiluminescence probe, L-012, for measuring the production of superoxide anion by cells, *Biochem. Biophys. Res. Commun.,* 193, 554, 1993.
19. Imada, I., Sato, E. F., Miyamoto, M., Ichimori, Y., Minamiyama, Y., Konaka, R., and Inoue, M., Analysis of reactive oxygen species generated by neutrophils using a chemiluminescence probe L-012, *Anal. Biochem.,* 271, 53, 1999.
20. Ii, M., Yoshida, H., Aramaki, Y., Masuya, H., Hada, T., Terada, M., Hatanaka, M., and Ichimori, Y., Improved enzyme immunoassay for human basic fibroblast growth factor using a new enhanced chemiluminescence system, *Biochem. Biophys. Res. Commun.,* 193, 540, 1993.
21. Ji, X., Kondo, K., Aramaki, Y., and Kricka, L. J., Effect of enhancers on the pyridopyridazine-peroxide-HRP reaction, *J. Biolumin. Chemilumin.,* 11, 1, 1996.
22. Nakano, M., Assay for superoxide dismutase based on chemiluminescence of luciferin analog, *Methods Enzymol.,* 186, 227, 1990.
23. Floyd, R. A., Observations on nitroxyl free radicals in arylamine carcinogenesis and on spin-trapping hydroxyl free radicals, *Can. J. Chem.,* 60, 1577, 1982.
24. Sato, E. F., Utsumi, K., and Inoue, M., Human oral neutrophils: isolation and characterization, *Methods Enzymol.,* 268, 503, 1996.
25. Takahashi, R., Edashige, K., Sato, E. F., Inoue, M., Matsuno, T., and Utsumi, K., Luminol chemiluminescence and active oxygen generation by activated neutrophils, *Arch. Biochem. Biophys.,* 285, 325, 1991.
26. Yamamoto, M., Saeki, K., and Utsumi, K., Isolation of human salivary polymorphonuclear leukocytes and their stimulation-coupled responses, *Arch. Biochem. Biophys.,* 289, 76, 1991.
27. Morimoto, Y. M., Sato, E. F., Nobori, K., Takahashi, R., and Utsumi, K., Effect of calcium ion on fatty acid-induced generation of superoxide in guinea pig neutrophils, *Cell Struct. Funct.,* 11, 143, 1986.
28. Britigan, B. E., Pou, S., Rosen, G. M., Lilleg, D. M., and Buettner, G. R., Hydroxy radical is not a product of the reaction of xanthine oxidase and xanthine, *J. Biol. Chem.,* 265, 17,533, 1990.
29. Uehara, K., Hori, K., Nakano M., and Koga, S., Highly sensitive chemiluminescence method for determining myeloperoxidase in human polymorphonuclear leukocytes, *Anal. Biochem.,* 199, 191, 1991.

21 Avoiding the Jaws of Using Lucigenin as a Chemiluminigenic Probe for Biological Superoxide: Determining When It's Safe to Go Back into the Water

Michael A. Trush and Yunbo Li

CONTENTS

21.1 INTRODUCTION

One of the products of cellular metabolism is the generation of the reactive oxygen species (ROS) superoxide and hydrogen peroxide.[1–3] For some cells, the generation of ROS is a major facet of their biochemistry and is essential to their overall function. This is particularly true of phagocytic cells of the host defense system like polymorphonuclear leukocytes (PMNs), monocytes, and macrophages. However, for the majority of other cells found in the human body, ROS generation is not a major component of their overall biochemistry and function, but can be a product of specialized enzymes within various organelles. For example, cytochrome P-450 in the endoplasmic reticulum utilizes ROS in the biotransformation of xenobiotics. Similarly, ROS is generated following

0-8493-0719-8/02/$0.00+$1.50

the release of electrons from various complexes of the mitochondrial electron transport chain.[2,4] In fact, the mitochondria appear to be the predominant source of ROS generation in cells. A number of other enzymes have the ability to generate ROS in isolated situations,[1] but the extent they contribute to ROS generation within the cellular milieu is not clear at this time.

Because of the concern over the last several years that ROS contribute to a plethora of human disease processes,[1,5] there is much interest in detecting ROS generation by cells from various organs. To this end, a useful and sensitive approach has been the monitoring of excited state generation or chemiluminescence (CL). This approach was initiated by Allen to assess the biochemistry and function of both normal and chronic granulomatous disease PMNs.[6,7] Subsequently, this approach was applied to a number of different cell types and even intact organs.[8–10] Because of limitations with either the cell number available or instrumentation sensitivity, amplification techniques using the chemiluminigenic probes lucigenin and luminol came into play: luminol-derived CL to assess hydrogen peroxide and/or peroxidase activity and lucigenin-derived CL to monitor the generation and presence of superoxide.[11] Initially, these probes were utilized with phagocytic cells, but eventually their use extended to other cell types.[12–15] Recently, the use of lucigenin as a chemiluminigenic probe for detecting superoxide in biological systems has come into question as a result of the demonstration that lucigenin can redox-cycle and, as such, become a secondary and possibly even a primary source of superoxide.[16,17] The purpose of this chapter is not to review the extensive literature on the use of lucigenin in biological systems or to assess whether it was used correctly in the past. Rather, it is to examine the current aspects of its validity as a superoxide-detecting probe and to describe its application as a chemiluminigenic probe for assessing intramitochondrial superoxide.

21.2 DETERMINING WHETHER OR NOT LUCIGENIN IS REDOX-CYCLING IN A BIOLOGICAL SYSTEM

The demonstration that through its redox cycling lucigenin becomes a potential source of superoxide within biological systems certainly brings the following question to the forefront: Does lucigenin-derived CL reflect superoxide originating from a biological source(s), from its own redox cycling, or from both? From our experience, the answer to this question is that it depends on the biological system and the concentration of lucigenin being employed.[18] Although the initial paper by Liochev and Fridovich[16] was comprehensive in the scope of enzymes examined, one of the aspects that is evident about the data is that the concentration of lucigenin varied from one enzymatic system to the other, and the concentration of lucigenin used was always a redox-cycling concentration. Consequently, they concluded that lucigenin should not be used as a superoxide probe in biological systems. By using several of the same enzyme systems used by Liochev and Fridovich,[16] as well as isolated organelles and intact cells, varying the concentration of lucigenin, and monitoring oxygen consumption, a hallmark for demonstrating chemical redox cycling, the patterns illustrated in Figure 21.1 emerged. In these studies the concentration of lucigenin varied from 0 to 100 μM and, in addition to oxygen consumption, DEPMPO spin trapping was used as another assessment of superoxide generation. As illustrated in Figure 21.1A, as the concentration of lucigenin was increased, lucigenin-derived CL increased but without any concomitant increase in oxygen consumption. This was the pattern seen with a xanthine oxidase–xanthine system (see figure 2 in Reference 18) and TPA-stimulated macrophages (see figure 10 in Reference 18). In addition, with both systems the presence of 50 μM lucigenin did not stimulate but rather inhibited the DEPMPO superoxide-dependent signal. This is noteworthy in that the concentration of DEPMPO was in the millimolar range, whereas the concentration of lucigenin was in the micromolar range. Overall, these results demonstrate that lucigenin does not redox cycle with these systems. It was, however, subsequently shown with the xanthine oxidase system that there was increased cytochrome *c* reduction in the presence of lucigenin.[19] The interpretation of this observation was then reevaluated by Afanas'ev et al.,[20] who concluded that this increased reduction in cytochrome *c* was not due to lucigenin redox cycling. Subsequently, this conclusion was challenged by Spasojevic et al.[21]

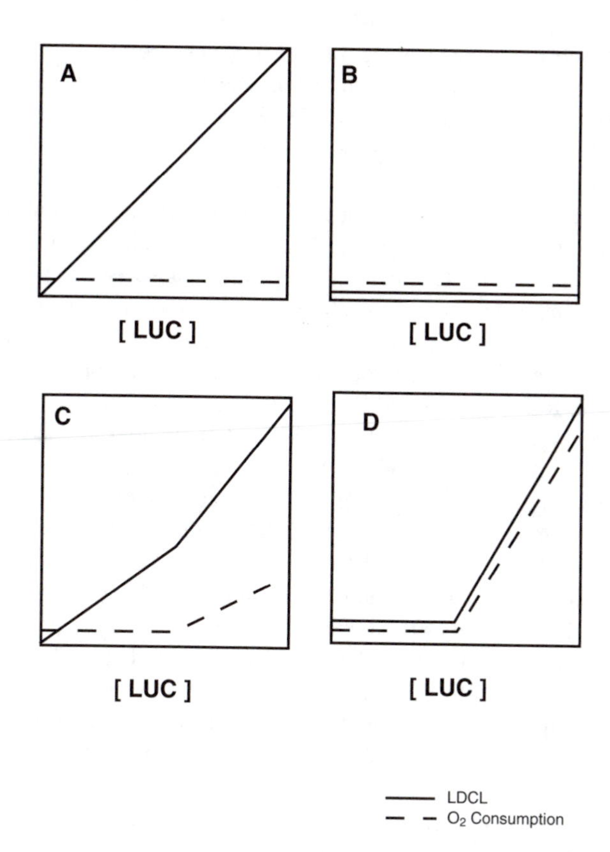

FIGURE 21.1 The diagrams illustrated in panels A to D depict the lucigenin-derived chemiluminescence (solid line) and oxygen consumption (dashed line) observed as a function of varying the lucigenin concentration [LUC]. The response in A demonstrates that as the [LUC] increases there is an increase in lucigenin-derived CL without a concomitant increase in oxygen consumption. Hence, in this situation lucigenin is not redox-cycling and lucigenin-derived CL is reflective of biological superoxide. In B, there is neither an increase in lucigenin-derived CL or an increase in oxygen consumption as the [LUC] increases. In this situation the biological system is not generating superoxide or reducing lucigenin to its cation radical. In C a biphasic response is illustrated whereby lucigenin begins to redox-cycle as indicated by increase in both lucigenin-derived CL and oxygen consumption at some [LUC]. Finally, the response depicted in D is one in which both lucigenin-derived CL and oxygen consumption are entirely attributable to lucigenin redox cycling. No lucigenin-derived CL is observed at non-redox-cycling concentrations of lucigenin. Biological systems that elicit each of the patterns illustrated in panels A to D are discussed in Section 21.2.

Nonetheless, using oxygen consumption and DEPMPO spin trapping, we could not demonstrate redox cycling by lucigenin with the xanthine oxidase–xanthine system and more importantly with an intact cellular superoxide-generating system, NADPH oxidase.

Figure 21.1B illustrates a pattern where over a wide range of lucigenin concentrations neither CL nor a stimulation of oxygen consumption was observed. This was the pattern we observed with glucose–glucose oxidase, a system that primarily generates hydrogen peroxide. What this system indicated was that glucose oxidase cannot reduce lucigenin to its radical form; thus, lucigenin redox cycling did not occur. Similarly, there was no increase in the DEPMPO superoxide-dependent signal in either the absence or presence of 50 μM lucigenin (see figure 5 in Reference 18).

The patterns illustrated by Figures 21.1C and D both indicate that, as the concentration of lucigenin was increased, at some point there was a concomitant stimulation in oxygen consumption, indicative of lucigenin redox cycling. The difference between the patterns observed in panels C and D was that panel C reflects a biological system that generates superoxide and panel D, a system that does not generate substantial superoxide (table 2 in Reference 18). Both systems can, however, apparently reduce lucigenin to its radical form. Experimentally, the pattern depicted in panel C is illustrated by a lipoamide dehydrogenase (LADH)–NADH system (see figure 3 in Reference 18) and isolated mitochondria driven by either pyruvate–maleate or succinate (see figure 7 in Reference 18, figure 3 in Reference 22, and figure 3 in Reference 23). With the LADH–NADH system lucigenin redox cycling, as indicated by an increase in both lucigenin-derived CL and oxygen consumption, was observed at a lucigenin concentration of 50 μM or greater. Similarly, with 50 μM lucigenin, there was now an increase in the DEPMPO superoxide-dependent signal (see figure 3C in Reference 18), and, with isolated mitochondria, there was a simulation of oxygen consumption by lucigenin at concentrations of 50 μM or greater. However, the stimulation in oxygen consumption by 100 μM lucigenin was not equivalent to that observed with 5 μM of the redox-cycling benzo[*a*]pyrene-1,6-quinone (see figure 7 in Reference 18 and figure 3 in Reference 22).

In the absence of lucigenin, the LADH–NADH system does reduce cytochome *c* that is inhibitable by superoxide dismutase (SOD) (table 1 in Reference 18). On the other hand, there is minimal SOD-inhibitable reduction of cytochrome *c* by xanthine oxidase–NADH, which probably accounts for the pattern observed in Figure 21.1D. Thus, the steep concomitant increase in both oxygen consumption and lucigenin-derived CL at 20 μM lucigenin was due entirely to the redox cycling of lucigenin. Similarly, with the xanthine oxidase–NADH system a DEPMPO superoxide-dependent signal was observed only at redox cycling concentrations of lucigenin.

Clearly, there is no doubt that lucigenin has the potential to redox-cycle and, as such, to become a source of superoxide. Chemically, this makes sense since lucigenin does undergo a one-electron reduction to form the lucigenin cation radical.[16,21] Typically, chemical redox cycling is not considered to be destructive to the chemical. As long as there is a source of electrons and a supply of molecular oxygen, most chemicals that redox-cycle undergo a reduction to a radical form and then an oxidation back to the parent chemical.[3] However, remember it is the lucigenin radical that reacts with superoxide, ultimately resulting in the breakdown of lucigenin into two *N*-methylacridone molecules. Because of structural similarities, lucigenin is often compared with the herbicide paraquat, a chemical used experimentally as a prototypic redox cycler. An interesting observation that has been lost in the literature is the demonstration that the positively charged paraquat radical can also react with superoxide resulting in the breakdown of paraquat.[24] However, to our knowledge this reaction has been shown to occur only in a pure chemical system. Whether paraquat reacts with superoxide in a biological system does not appear to have been either considered or comprehensively examined and, as such, remains a possibility.

The question then arises: When does the lucigenin radical stop reacting with biological superoxide and begin donating an electron to molecular oxygen thereby becoming a source of chemically generated superoxide? The patterns illustrated in Figures 21.1C and D in fact provide an answer to this question: when the concentration of biologically derived superoxide becomes limiting. This is exactly what was seen with the xanthine oxidase–NADH system (see figure 6 in Reference 18) and illustrated by Figure 21.1D. This system generates very little superoxide but apparently can reduce lucigenin to its radical form. Once the concentration of the lucigenin radical has built up sufficiently, there was both an increase in oxygen consumption and the occurrence of lucigenin-derived CL, with both events occurring at the same lucigenin concentration. However, with the LADH–NADH system and with isolated mitochondria, significant lucigenin-derived CL is observed without observing a concurrent stimulation of oxygen consumption. However, when biological superoxide becomes limiting, then a concomitant increase in both oxygen consumption and lucigenin-derived CL should be observed, which is exactly what is experimentally observed and illustrated in Figure 21.1C.

Figure 21.2 will now be used to illustrate further the concept that lucigenin contributes to superoxide generation in biological systems when biological superoxide becomes limiting. The key relationship to consider is the ratio of the concentration of the lucigenin radical to the concentration of biological superoxide. When this ratio is less than 1, lucigenin should not be a source of

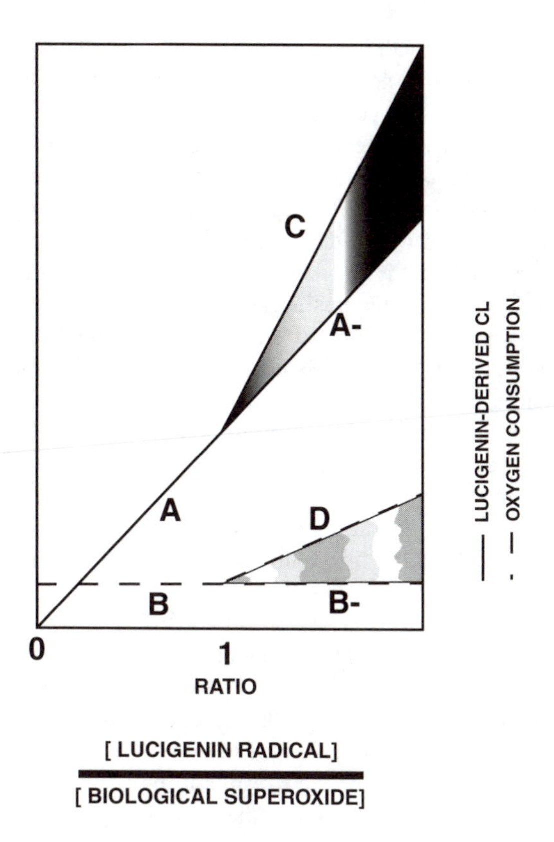

FIGURE 21.2 This figure depicts the situation whereby lucigenin-derived CL is initially reflective of biological superoxide (segment A) but then becomes reflective of both biological superoxide and lucigenin-derived superoxide (segment C). As shown, as the ratio of the [lucigenin radical] to [biological superoxide] becomes greater than 1, lucigenin redox cycling begins to occur resulting in an increase in both lucigenin-derived CL (segment C) and oxygen consumption (segment D). Line A, A− represents lucigenin-derived CL resulting from the interaction between the lucigenin radical and biological superoxide. Similarly, line B, B− represents oxygen consumption arising from the biological system. The shaded areas between C and A− and D and B− represent the increased responses attributable to lucigenin redox cycling.

superoxide and lucigenin-derived CL should reflect the reaction of the lucigenin radical with biologically derived superoxide. This is illustrated in Figure 21.2 by the segment of the solid line designated as A. However, once this ratio becomes greater than 1, i.e., the concentration of the lucigenin radical becomes greater than the concentration of biologically derived superoxide, the lucigenin radical autoxidizes, leading to both an increase in oxygen consumption and superoxide generation. These responses are represented in Figure 21.2 by lines D and C, respectively. The shaded area between A− and C represents the increase in lucigenin-derived CL as a result of lucigenin now becoming a second source of superoxide, and the shaded area between B− and D represents the increase in oxygen consumption as a result of lucigenin radical autoxidation. Line A, A− represents lucigenin-derived CL resulting from the reaction of the lucigenin radical with only biological superoxide, whereas line B, B− represents oxygen consumption resulting from the biological system. Lines A, A− and B, B− are identical in pattern to the response depicted in Figure 21.1A and shown experimentally with xanthine oxidase–xanthine (see figure 2 in Reference 18) and macrophage NADPH oxidase (see figure 10 in Reference 18).

A methodological difficulty arises in trying to quantify the concentration of the lucigenin radical within biological systems. Unlike the paraquat radical, which is stable and can be determined under anaerobic conditions by electron-spin resonance spectroscopy, the lucigenin radical undergoes a second reduction under anaerobic conditions.[16] Along these lines, what we have been able to show with isolated mitochondria is that at non-redox-cycling concentrations of lucigenin, 20 *μM* and below, a linear relationship exists between the concentration of lucigenin added and the concentration of mitochondrial *N*-methylacridone measured and between the mitochondrial concentrations of either lucigenin or *N*-methylacridone and lucigenin-derived CL (see figures 8 and 9 in Reference 23). Such linear relationships would not occur if lucigenin were redox-cycling at these concentrations to a significant extent. If lucigenin was in fact redox-cycling, one would expect a biphasic relationship between lucigenin and *N*-methylacridone similar to the relationship between lucigenin and oxygen consumption, as illustrated in Figures 21.1C and 21.2.

Another biological system that appears to follow the pattern depicted by Figure 21.1C is the assessment of superoxide by vascular tissue. Skatchkov et al.[25] demonstrated that 250 *μM*, but not 5 *μM* of lucigenin increased vascular superoxide as detected by spin trapping studies. Similarly, the addition of 50 or 250 *μM* lucigenin to bovine aortic endothelial cells (BAEC) increased hydrogen peroxide extracellularly, a response not observed with 10 *μM* lucigenin.[26] A more recent study showed that 5 *μM* lucigenin detected basal superoxide production by BAEC when assessed in a Packard liquid scintillation counter operated in the out-of-coincidence mode.[27] In the same study, they concluded that NADH or NADPH lucigenin-enhanced CL is artifactual based on the observation that the addition of lucigenin and NAD(P)H decreased acetycholine-induced relaxation in aortic vascular rings, a result interpreted to be due to toxic effect of lucigenin-derived superoxide. Another concern of these investigators is that the majority of lucigenin-derived CL in their system may arise intracellularly, when they were trying to assess superoxide extracellularly. It should be noted that oxygen uptake studies were not done in the presence of lucigenin, nor was the intracellular presence of lucigenin assessed. An overall criticism of most of these studies that have questioned the use of lucigenin is the extrapolation of their results to biological systems not investigated in their studies. A prime example has been questioning the use of lucigenin to assess intramitochondrial superoxide generation.[21,26] This is further discussed in Section 21.4.

21.3 LUCIGENIN-DERIVED CL: COMPARISON WITH OTHER PROBES FOR ASSESSING BIOLOGICALLY DERIVED SUPEROXIDE

Presented in the previous section and in our recent series of studies[18,22,23] is the finding that by assessing lucigenin CL over a range of lucigenin concentrations, one can identify a concentration of lucigenin that does not redox-cycle. Hence, the CL observed reflects biological superoxide reacting with the lucigenin radical. The question then arises regarding how this lucigenin-derived CL compares with other probes for assessing biological superoxide. Our experimental data demonstrate that lucigenin-derived CL and SOD-inhibitable cytochrome *c* reduction correlate quite well (see figures 4 and 11 in Reference 18) in two different biological systems, xanthine oxidase–xanthine and TPA-stimulated macrophages. Similarly, with a submitochondrial particle preparation supported by NADH, there was very good correlation between SOD-inhibitable epinephrine oxidation, a common method used for determining mitochondrial superoxide,[4] and lucigenin-derived CL (see figure 3 in Reference 23). In a study by Tarpey et al.,[26] 50 or 250 *μM* lucigenin resulted in increases in extracellular hydrogen peroxide from BAEC. Unfortunately, at the 10 *μM* concentration of lucigenin, which did not increase hydrogen peroxide release, they did not show if they could observe luminescence from a non-redox-cycling concentration of lucigenin. It would have been interesting to correlate hydrogen peroxide release and lucigenin-derived CL over the same concentration range, something they were technically capable of doing. Similarly, it would have been interesting to compare the CL responses over the same wide range of concentrations between the luminophore coelenterazine and lucigenin. In this regard, using electron-spin resonance studies, Skatchkov et al.[28] compared 250 *μM* lucigenin to 1 *μM* coelenterazine in terms of redox cycling capability, and concluded that coelenterazine does not redox-cycle whereas lucigenin does. Subsequently, Skatchkov et al.[25] compared 5 to 250 μM lucigenin with aortic vascular rings and concluded that the 5 μM concentration seems to be a sensitive and valid probe for assessing superoxide in vascular tissue. Probably the most comprehensive study examining different methods for assessing extracellular superoxide anion release by vascular endothelial cells was done by Barbacanne et al.[27] In this study, they used cytochrome *c* reduction, DMPO spin trapping, hydroethidine fluorescence, and lucigenin-enhanced CL. They observed that lucigenin-enhanced CL appeared to be more sensitive than the other three techniques, but they were concerned because of the unusual results observed with lucigenin in the presence of NAD(P)H. In all the above studies, the scientific community would have benefited from experiments comparing lucigenin-derived CL with the other superoxide probes under identical experimental conditions. What many of these studies basically concluded is a condemnation of the use of lucigenin that is not warranted.[16,19,21,27–30]

21.4 LUCIGENIN-DERIVED CL AS A PROBE FOR DETECTING INTRAMITOCHONDRIAL SUPEROXIDE

One of the initial applications of lucigenin-derived CL with biological systems was for the assessment of NADPH oxidase activity in phagocytic cells,[11,31,32] an application proven to be quite useful. There was, however, a feeling that because lucigenin has two positive charges, it would not enter cells and thus its CL would be reflective of extracellular superoxide. In 1989, we reported that with unstimulated rat alveolar macrophages the observance of lucigenin-derived CL was mitochondrial in origin.[33] This has subsequently been verified by our studies with both isolated mitochondria and intact cells,[18,22,23,33–38] as well as studies by other investigators.[39] This section will examine the experiments that indicate when non-redox-cycling concentrations (20 μM or less) are used lucigenin-derived CL is indeed reflective of intramitochondrial superoxide.

Mitochondria have long been considered the major organelle for superoxide and hydrogen peroxide production within cells.[1,4] Experimentally, submitochondrial particles have been primarily employed for determining superoxide generation by mitochondria,[4] whereas intact mitochondria have been more likely to be used to assess hydrogen peroxide release from mitochondria. Staniek and Nohl[40] have recently questioned whether mitochondrial respiration will supply electrons for superoxide generation. Moreover, they stated, "since mitochondrial superoxide formation under homeostatic conditions could not be demonstrated *in situ* so far, conclusions drawn from isolated mitochondria must be considered with precaution." To the contrary, our data demonstrate that lucigenin-derived CL can be used to assess intramitochondrial superoxide generation under homeostatic conditions, that is, when the cell is supplying reducing equivalents in the form of NADH or succinate to the mitochondrial electron transport chain (METC).

As illustrated in Figure 21.3, there are four critical processes required for the detection of intramitochondrial superoxide by lucigenin. The first step ① is the ability of lucigenin to enter cells, and the second ② is its accumulation in the mitochondria. The first process is really not well investigated except for a study by Braakman et al.,[41] who described a vesicular cation uptake of lucigenin in rat hepatocytes. We have, however, observed with both rat alveolar macrophages and ML-1 cell-derived macrophages that addition of lucigenin results in its accumulation in the mitochondria. With the former cells this was shown by confocal microscopy[34] and with the latter cells by an ion-pair HPLC method that we developed.[23] The reason the positively charged lucigenin accumulates in the mitochondria is that the mitochondria generate a negative membrane potential. As was discussed previously, over a range of non-redox-cycling concentrations the accumulation of lucigenin is linear (see figure 6 in Reference 23). Moreover, the addition of FCCP, an agent that uncouples mitochondria and results in a reduced membrane potential, results in a significantly reduced accumulation of lucigenin into this organelle (see figure 6 in Reference 23).

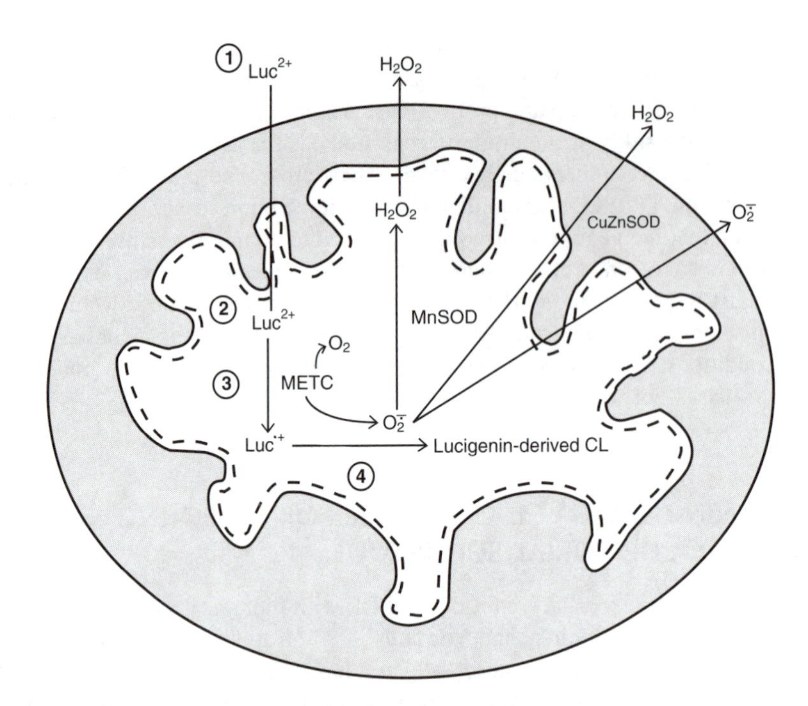

FIGURE 21.3 This diagram illustrates the steps required for lucigenin to be able to detect intramitochondrial superoxide. Step ① depicts the entry of lucigenin into a cell, and step ② depicts the accumulation of lucigenin into the mitochondria. The accumulation of the positively charged lucigenin is due to the generation of a negative membrane potential (–––) by the mitochondria. Once in the mitochondria, lucigenin is reduced to its radical cation, step ③, by the METC located in the inner mitochondrial membrane. The lucigenin radical then can interact with METC-derived $O_2^{\bar{\cdot}}$ resulting in lucigenin-derived CL, step ④. Also illustrated in the diagram are the superoxide-detoxifying enzymes MnSOD, located in the mitochondrial matrix, and CuZnSOD, located in the intermembrane space (V. Culotta, personal communication). Also shown is the presence of $O_2^{\bar{\cdot}}$ and H_2O extramitochondrially. Extramitochondrial $O_2^{\bar{\cdot}}$ or H_2O_2 can be detected by SOD-inhibitable DEPMPO spin trapping or horseradish peroxidase–mediated H_2O_2 dependent luminol-derived CL, respectively. The experiments supporting each of these steps are discussed in Section 21.4.

The third ③ required step is the reduction of lucigenin by the METC to produce the lucigenin cation radical. It is this radical rather than lucigenin itself that reacts with biological superoxide, process ④, to generate lucigenin-derived CL. Process three ③ was demonstrated by using submitochondrial particles and observing that both complex I and complex III could reduce lucigenin to its radical as assessed by METC-mediated lucigenin redox cycling (see figure 7 in Reference 23). It should be noted that the concentrations of lucigenin used in this experiment, 50 and 150 μM, far exceed the concentrations of lucigenin (1 to 5 μM) we typically used with our intact cell or isolated mitochondria experiments to assess intramitochondrial superoxide.[18,22,23,33–39] As previously pointed out, we can detect lucigenin redox cycling in isolated mitochondria at concentrations of 50 μM or greater.[18,22,23] With ML-1 cell-derived differentiated macrophages even at 50 μM lucigenin we did not see a stimulation of KCN-resistant oxygen consumption, but did with 5 μM benzo[*a*]pyrene-1,6-quinone.[18,22,23] Remember with isolated mitochondria reducing equivalents are being added. Therefore, it is quite likely with such substrate input to the METC, as the lucigenin concentration is increased, that the concentration of the lucigenin radical will eventually exceed the concentration of superoxide made by the METC (see Figure 21.2). However, with intact cells, substrate input to the METC is likely to be controlled by cellular demands, and thus it is less likely that a ratio of 1 will be exceeded as described in the Section 21.2. Evidence that concentrations of lucigenin less than 50 μM are not redox-cycling in mitochondria is provided by the linear relationships that exist between the following measurements: the amounts of lucigenin and *N*-methylacridone; the amount of lucigenin in mitochondria and lucigenin-derived CL; and the amount of *N*-methylacridone found in mitochondria and lucigenin-derived CL (see figures 8 and 9 in Reference 23).

There are two key observations that indicate that lucigenin-derived CL in mitochondria is a superoxide-dependent process ④. The first was the demonstration that SOD could inhibit lucigenin-derived CL with a submitochondrial preparation and that CuDIPs, a lipophilic SOD-mimetic inhibited lucigenin-derived CL with intact rat alveolar macrophages.[32] CuZnSOD, which is not likely to enter mitochondria, *does not* inhibit lucigenin-derived CL with isolated mitochondria (see table 2 in Reference 22 and figure 4 in Reference 23). On the other hand, the membrane-permeable SOD-mimetics TEMPO or Mn(III) TMPyP *do* inhibit lucigenin-derived CL with isolated mitochondria. The second indication was actually demonstrating the presence of *N*-methylacridone in mitochondria,[23] since *N*-methylacridone is the product of the reaction of the lucigenin radical with superoxide. Moreover, at non-redox-cycling concenterations of lucigenin there should be a linear relationship between the amount of lucigenin found in mitochondria and the amount of *N*-methylacridone measured. Indeed, this is the relationship that exists.[23]

Demonstrating that the above four processes do occur is important from two perspectives. The first is that we now have, with interpretation nuances that will be discussed below, a procedure to investigate the homeostatic control of mitochondiral ROS production in cells. The second is to address the criticisms that have been put forth with regard to using this technique for assessing mitochondrial-generated superoxide.[21,26] For example, Spasojevic et al.[21] stated:

> [T]he artifactual interpretation of Luc^{2+} luminescence continues. Thus, Li et al. [their references 15 and 16] recently reported that Luc^{2+} luminescence is caused by mitochondria and interpret the intensity of this luminescence to be a measure of intramitochondrial $O_2^{\cdot-}$ concentration. What they may have actually demonstrated is that Luc^{2+} can be reduced to $Luc^{\cdot+}$ by mitochondrial enzymes. How much of the $O_2^{\cdot-}$ was endogenously produced and how much as a consequence of the autooxidation of $Luc^{\cdot+}$ remains unknown.

To the contrary, what our published data *clearly* show is that with isolated mitochondria lucigenin *does not* redox-cycle (autoxidize) at concentrations of 20 μM or less[18,22,23] and with intact cells at concentrations even up to 50 μM. Hence, we have identified non-redox-cycling concentrations of lucigenin that show endogenous production of superoxide. In a similar vein to the comments of Spasojevic et al.,[21] Tarpey et al.[26] wrote the following:

> It remains possible that lucigenin chemiluminescence may serve as a nonspecific redox-sensitive marker for the enhanced electron transfer activities that are present in various pathologies, with the caveat that (from the present and previous observations) the estimation of $O_2^{\cdot-}$ formation from such electron transfer processes may be artifactually enhanced. This is especially the case for mitochondria, organelles rich in flavoproteins that can directly reduce lucigenin to the mono-cation radical. This will enhance nonrespiratory mitochondrial oxygen consumption [their reference 32] because of lucigenin-mediated redirection of reducing equivalents to $O_2^{\cdot-}$ and H_2O_2 formation [their reference 30]. Thus, it would be expected that a significant fraction of mitochondrial lucigenin-dependent chemiluminescence is due to direct respiratory chain reduction of lucigenin, extramitochondrial diffusion of the mono-cation radical, inhibitable by various electron transport and flavoprotein inhibitors, as well as by exogenously added SOD [their reference 63]. Added SOD would not be expected to inhibit lucigenin chemiluminescence induced by "true" respiratory chain-derived $O_2^{\cdot-}$, since it is a mitochondria-impermeable macromolecule and would not be anticipated to outcompete the already significant levels of respiratory chain-associated endogenous mitochondrial MnSOD.

The above statements fail to acknowledge the following: (1) the enhancement of nonrespiratory mitochondrial oxygen consumption only occurs at lucigenin concentrations of 50 μM or greater; and (2) their reference 63 (present Reference 35) does not present any data at all showing exogenously added SOD inhibits lucigenin-derived CL with mitochondria. We do agree with their comment that various electron transport and flavoproteins inhibitors should inhibit METC-mediated lucigenin-derived CL. In fact complex I and complex III inhibitors are very effective at modulating both lucigenin reduction and lucigenin-derived CL (see figure 7 in Reference 23) and superoxide-derived hydrogen peroxide production (see figures 7 and 8 in Reference 22).

With regard to the interpretation of lucigenin-derived CL results with mitochondria, it is important to remember that this CL response is critically dependent on two processes (see Figure 21.3): the accumulation of lucigenin in the mitochondria and the reduction of lucigenin by components of the METC. Anything that interferes with either of these two processes could result in decreased lucigenin-derived CL. As such, it could then be interpreted that mitochondria are producing less superoxide.[38] This may or may not be the case. To verify that the mitochondria are making less superoxide, horseradish peroxidase–mediated hydrogen peroxide–dependent luminol-derived CL or SOD-inhibitable DEPMPO spin trapping can be performed. If these responses is also decreased, then the mitochondria are indeed producing less superoxide and as a result less superoxide and hydrogen peroxide is being released from the mitochondria. This, in fact, was observed with mitochondria isolated from the livers of genetically obese ob/ob mice.[42] These mitochondria demonstrate increased expression of uncoupling protein-2 (UCP-2), and upon addition of succinate a decrease in lucigenin-derived CL was observed. Conversely, greater luminol-derived CL occurred as compared with mitochondria prepared from normal mice.[42]

A surprising decrease in mitochondrial-dependent lucigenin-derived CL was also observed upon addition of TPA to rat alveolar macrophages, but not with rat peritoneal macrophages or rat PMNs.[33] Peritoneal macrophages and PMNs exhibit very little or no mitochondrial respiration, respectively. Using mitochondria isolated from differentiated ML-1 cells, we now know that TPA addition, in fact, results in dramatically increased superoxide and hydrogen peroxide release as demonstrated by luminol-derived CL and SOD-inhibitable DEPMPO spin trapping.[43,44] Like the activation of NADPH oxidase activity by protein kinase C (PKC), this modulatory effect of TPA on mitochondrial-derived ROS also appears to be mediated by PKC.[43,44]

It is clear from studies with alveolar macrophages and ML-1 cell-derived macrophages that with resting or unstimulated cells lucigenin-derived CL originates primarily from the mitochondrial compartment.[18,44–46] With the ML-1 cell model, differentiation of the cells in the presence of either chloramphenicol or ethidium bromide results in an inhibition of the development of the METC but not the expression and activity of NADPH oxidase. Such cells are in fact devoid of mitochondrial oxygen consumption because of the absence of METC components necessary for mitochondrial respiration and thus lack the ability to elicit lucigenin-derived CL from this organelle. These ML-1 cell-derived macrophages lacking mitochondrial respiration can, however, elicit lucigenin-derived CL via activation of NADPH oxidase following TPA stimulation.

Because, with either resting rat alveolar macrophages or ML-1 cell-derived macrophages, the mitochondrial-dependent lucigenin-derived CL is so predominant a process, it is difficult to assess TPA-stimulated NADPH oxidase activity by lucigenin-derived CL. However, it is possible to assess this enzyme activity if mitochondrial respiration is blocked (see figure 9 in Reference 18) or by using macrophages devoid of a functional METC.[44,45]

21.5 APPLICATIONS OF MITOCHONDRIAL-DEPENDENT LUCIGENIN-DERIVED CL IN CELL BIOLOGY AND TOXICOLOGY

Based on the discussion in the previous section that lucigenin-derived CL can be used to assess intramitochondrial superoxide generation in cells under normal homeostatic conditions, we would now like to describe some of the applications for which we have utilized this technique.

21.5.1 Assessment of Mitochondrial Maturation during Cell Differentiation

One of the distinguishing phenotype characteristics of macrophages as compared with monocytes is the presence of mitochondrial respiration.[47] Mitochondrial respiration occurs as a result of maturation of the METC during the differentiation from monocytes to macrophages. This process can be easily assessed using lucigenin-derived CL, which is inhibited by doxycycline, a tetracycline derivative that inhibits mitochondrial protein synthesis.[37] In this regard, we have used the ML-1 cell differentiation model to examine further factors that modulate mitochondrial maturation during cell differentiation.[44–46] Such factors include inhibitors of signal transduction systems,[46] as well as metabolites of environmental toxins.[44–46,48]

21.5.2 Assessment of Altered Mitochondrial Gene Expression

Recently, Chen et al.[49] observed that treatment of hepatocytes with ethinyl estradiol resulted in increased expression of components of the METC. Accordingly, using lucigenin-derived CL we have verified that with both intact cells and isolated mitochondria with an altered METC result in increased mitochondrial superoxide generation.[50,51] In both genetically obese ob/ob mice and regenerating liver, there is increased expression of mitochondrial UCP-2,[52,53] which results in altered ROS generation by liver mitochondria.[42] The expression of UCP-2 itself also appears to be regulated by ROS. This was demonstrated by showing that the addition of benzo[*a*]pyrene-1,6-quinone (BPQ), a redox cycling chemical, to hepatocytes increased expression of this protein.[42]

21.5.3 Assessment of METC-Mediated Chemical Redox Cycling

One of the mechanisms of chemical-induced toxicity to cells is via redox cycling generating ROS.[3,9] Recently, we observed that benzo[*a*]pyrene-derived quinone derivatives elicit toxicity to bone marrow stromal cells through altered mitochondrial function.[54] Data presented in Figures 21.4 and 21.5 verify that indeed BPQ undergoes redox cycling and ROS production with isolated mitochondria. There are two aspects of the data in Figure 21.5 that are of particular note: (1) the redox cycling is observed at very low concentrations of BPQ and (2) different response patterns are observed with lucigenin-derived CL as compared to luminol-derived CL over the range of BPQ

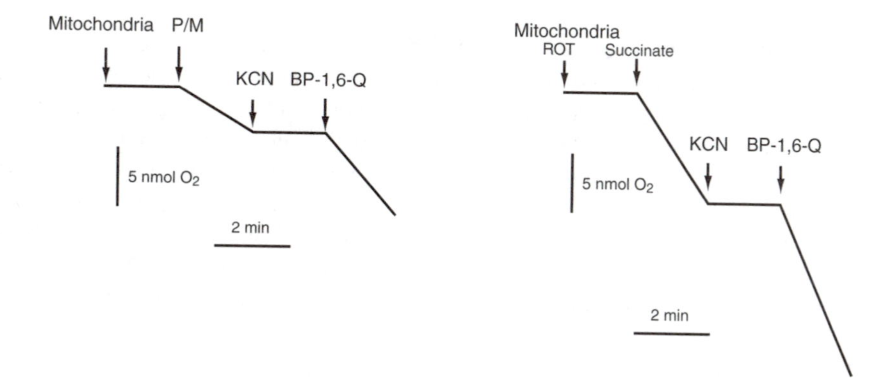

FIGURE 21.4 Stimulation of KCN-resistant O_2 consumption by BP-1,6-quinone (BP-1,6-Q) with mitochondria isolated from ML-1 cell-derived macrophages. The KCN-resistant O_2 consumption was detected following incubation of the pyruvate/malate or succinate supported mitochondria with BP-1,6-Q at 10 μM. The sequence of addition of the various separate agents to the mitochondria is indicated. Data represent the mean from three experiments.

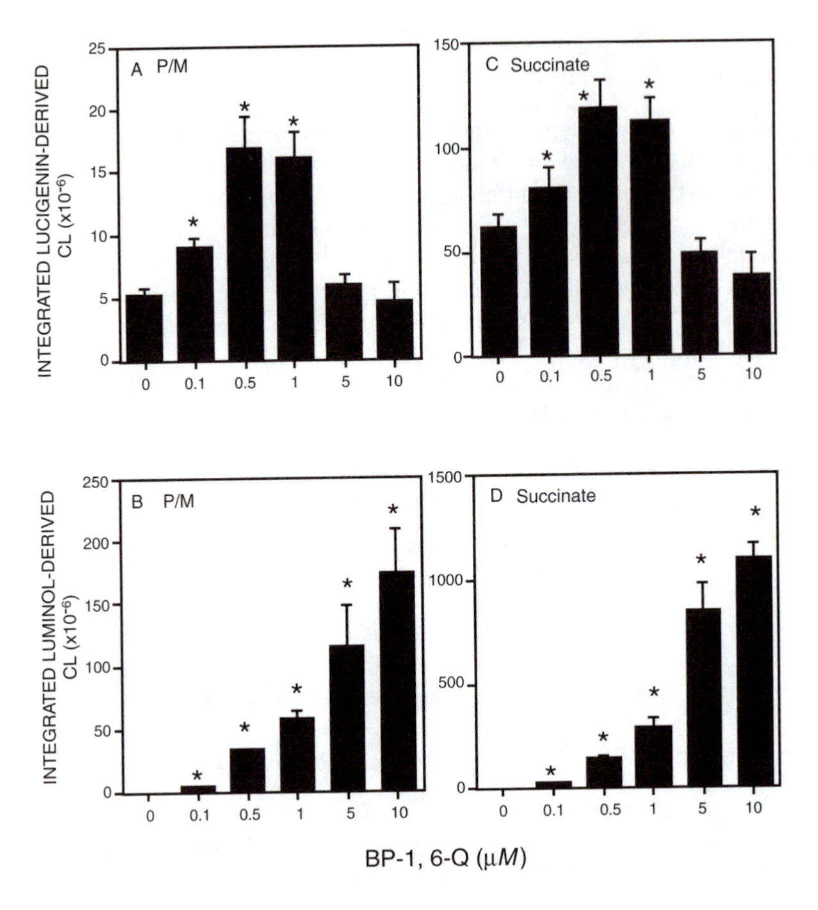

FIGURE 21.5 Concentration-dependent stimulation of mitochondrial ROS production by BP-1,6-Q with isolated mitochondria. The mitochondria were incubated with the indicated concentrations of BP-1,6-Q in the presence of pyruvate–malate (A and B) or succinate (C and D). Production of mitochondrial ROS was detected by lucigenin-derived (A and C) and luminol-derived (B and D) CL. Data represent mean ± S.E. from at least three experiments. *Indicates significantly different from 0 μM BP-1,6-Q at $p < 0.05$.

concentrations used. As shown, at 5 or 10 μM BPQ the lucigenin-derived CL is decreased as compared with the 0.1, 0.5, and 1 μM concentrations. On the other hand, horseradish peroxidase–dependent luminol-derived CL, which assesses extramitochondrial hydrogen peroxide, increases in a concentration-dependent fashion. Thus, we have a situation like that previously described: a decrease in lucigenin-derived CL with an increase in luminol-derived CL. If we were assessing only the lucigenin response, we could have mistakenly concluded that ROS generation is inhibited at 5 or 10 μM BPQ. However, the data in Figure 21.4 clearly show that redox cycling is occurring with 10 μM BPQ. Thus, the 5 and 10 μM concentrations of BPQ have apparently altered the mitochondria in some fashion that now limits the ability of lucigenin to detect intramitochondrial superoxide.

21.5.4 Assessment of Mitochondrial Function during Aging

There is a body of literature that METC activity changes with aging.[55–57] Sohal's laboratory in particular has monitored extramitochondrial hydrogen peroxide to demonstrate that mitochondrial ROS is also changed with aging.[58,59] Recently, we have shown that Leydig cells isolated from the testis of aged male rats exhibit increased lucigenin-derived CL.[60] This is of interest because the function of Leydig cells is to synthesize testosterone, an activity that decreases with aging in men.

21.5.5 Assessment of Mitochondrial Function in Disease States

There have been several studies to date that indicate that lucigenin-derived CL can be used to assess mitochondrial function in cells isolated from patients with disease.[61,62] Recently, we observed that there is an increased lucigenin-derived CL in monocyte-derived macrophages obtained from children with Down syndrome.[63] We used macrophages instead of the monocytes because of the increased mitochondrial respiration that accompanies this differentiation process as discussed in Section 21.5.1. It is of interest to note that neurons from a mouse model of Down syndrome exhibited increased mitochondrial ROS.[64] Similarly, there is increased ROS detected during differentiation of human neurons obtained from a fetus with Down syndrome as compared with neurons obtained from a normal fetus.[65]

Of all the applications discussed here for lucigenin-derived CL, the assessment of mitochondrial function in disease states may have the most appeal and the greatest potential, but it is the one that would require the greatest caution in interpreting results in terms of altered ROS. This is because of the nuances that were discussed previously in terms of factors that can modify the ability of lucigenin superoxide to detect mitochondrial superoxide but do not affect ROS production. However, as discussed above, there are ways to deal with this problem.

21.6 SUMMARY: THE MARGIN OF SAFETY AND LUCIGENIN

Because lucigenin clearly has the *potential* to redox-cycle, it has been suggested that lucigenin should not be used to assess superoxide generation in biological systems.[16,17,19,21,26,29,30] If the same rationale were applied to pharmaceutical agents, for the most part, drugs to treat many human maladies would be lacking. All drugs have the potential to elicit toxicity. The classic phrase used in toxicology is "the dose makes the poison." For drugs, a therapeutic index, which is the ratio of the dose required to produce a toxic effect as compared to the dose needed to elicit the desired therapeutic effect, is determined, hence, a margin of safety.[66] Defining a margin of safety is accomplished through careful analysis of dose–response relationships for both efficacy and toxicity. These same broad concepts can be applied to the use of lucigenin as a probe for detecting biological superoxide. As illustrated in Figures 21.1 and 21.2 and described in Section 21.2, different patterns of lucigenin-derived CL can be observed with different biological systems. As pointed out in this chapter, through doing comprehensive concentration–response relationship analysis of lucigenin-derived CL and oxygen consumption, it is experimentally possible to define the concentration where lucigenin becomes toxic, i.e., where it begins to redox-cycle and can no longer be used as an effective probe for assessing biological superoxide (Figure 21.2). Just as there can be variation for the dose of a drug that elicits toxicity between different individuals, the concentration where lucigenin redox cycles is likely to vary between different cell types and organelles from different tissues. If lucigenin is to be used, the onus then falls on the users to define the "safe" or non-redox-cycling concentration of lucigenin for their particular use and system.

Another consideration to be taken into account is that when redox-cycling concentrations of lucigenin are used, adverse cellular effects can be observed.[26–28,30,67] For example, we have observed that at 50 μM lucigenin with isolated mitochondria we can detect a decrease in mitochondrial aconitase activity.[23] This does not occur at concentrations of 20 μM or less. Similarly, redox-cycling concentrations of lucigenin have been shown to alter the ability of vascular tissue to relax.[25–27] Again, this does not occur at non-redox-cycling concentrations of lucigenin. It is generally assumed that these effects are due to the increase in superoxide derived from lucigenin redox cycling. However, simultaneous with an increase in lucigenin-derived superoxide there is a concomitant increase in lucigenin-derived CL. CL is a photoemissive process, and there is an extensive body of literature implicating photoemissive species as contributing to alterations in biomolecules.[68–72] In this regard, it is important to remember that lucigenin is being primarily used as a short-term probe for directly detecting the presence of biologically derived superoxide and indirectly the biological process that generated the superoxide.

TABLE 21.1
Recommendations to Overcoming Some Concerns and/or Limitations in the Use of Lucigenin-Derived CL for Assessing Biologically Derived Superoxide

Potential Concern/Limitation	Recommendation
Is the lucigenin-derived CL reflective of biologically derived superoxide?	Perform a concentration–response relationship with lucigenin to determine if the pattern is like that in panel A, C, or D of Figure 21.1.
Is a decrease in lucigenin-derived CL due to a decrease in superoxide derived from the biological system?	Use another method to assess ROS such as luminol-derived CL or electron spin resonance.
Does the lack of a lucigenin-derived CL response indicate that the biological system does not generate superoxide?	Determine if the luminometer is sensitive enough to detect lucigenin-derived CL. Use the xanthine oxidase–xanthine system with lucigenin as a positive control. With cellular systems, use another method to assess ROS generation by the biological system that is not limited by increasing the concentration of the probe.
Is the lucigenin-derived CL mitochondrial in origin?	Determine if the lucigenin-derived CL can be modulated/inhibited by inhibitors of the mitochondrial electron transport chain. Assess presence of lucigenin in mitochondria by confocal microscopy or HPLC.

The above issues aside, the other issue that needs to be appreciated about lucigenin is that it can enter cells and subsequently be accumulated in mitochondria. It is quite likely that many of the lucigenin-derived CL responses observed by investigators using various cells is mitochondrial in origin but has not been recognized as such. This can be viewed either positively or negatively, depending on the cellular source of superoxide one is trying to assess through CL. Similarly, as was pointed out, mitochondrial-dependent lucigenin-derived CL can be affected by factors that absolutely have nothing to do with altering the generation of ROS by the mitochondria, but affects either the reduction or accumulation of lucigenin in mitochondria. As was discussed in Section 21.5, other probes can be used to evaluate further the status of mitochondria ROS generation and overcome these limitations with the use of lucigenin. A concern that may limit the use of lucigenin, or any chemiluminigenic probe for that matter, is the availability of instrumentation sensitive enough to detect sufficient lucigenin-derived CL at non-redox-cycling concentrations, particularly in biological systems that generate low levels of ROS.

Finally, there remain many potential applications for the use of lucigenin-derived CL in biomedical research. In Section 21.5 we described a few of these applications we have had the opportunity to examine. If readers believe that lucigenin-derived CL may be useful for their purposes, we trust the information provided in this chapter will be a useful road map to its proper use (Table 21.1).

ACKNOWLEDGMENTS

This chapter is dedicated to the memory of Dr. Mark E. Wilson, friend and colleague of Michael A. Trush, who passed away in 2000. Mark utilized chemiluminigenic probing to assess PMN function. Dr. Trush would also like to acknowledge the following students and postdoctoral fellows who contributed to the authors' understanding of utilizing lucigenin-derived CL to assess intramitochondrial superoxide: Drs. Russell Esterline, Steven J. Rembish, Hong Zhu, Hongchin He, Kevin Stansbury, and James Kim. In addition, the following collaborators are acknowledged for their studies applying lucigenin-derived CL to assess mitochondrial function: Drs. Jinqiang Chen, James D. Yager, Anna Mae Diehl, Barry Zirkin, Haolin Chen, and George Capone. The following sources of support are also acknowledged: CAAT, American Cancer Society SIG-3, a pilot grant from the Johns Hopkins Oncology Center, ES 03760, ES 03819, ES 05131, ES 07141, and ES 08078.

REFERENCES

1. Freeman, B. A. and Crapo, J. D., Free radicals and tissue injury, *Lab. Invest.,* 47, 412, 1982.
2. Chance, B., Sies, H., and Boveris, A., Hydroperoxide metabolism in mammalian organs, *Physiol. Rev.,* 59, 527, 1979.
3. Trush, M. A., Mimnaugh, E. G., and Gram, T. E., Activation of pharmacologic agents to radical intermediates: implications for the role of free radicals in drug action and toxicity, *Biochem. Pharmacol.,* 31, 3335, 1982.
4. Boveris, A., Determination of the production of superoxide radicals and hydrogen peroxide in mitochondria, *Methods Enzymol.,* 105, 429, 1984.
5. Thomas, C. E. and Kalyanaraman, B., *Oxygen Radicals and the Disease Process,* Hargood Academic Publishers, Amsterdam, the Netherlands, 1997.
6. Allen, R. C., Stjernholm, R. L., and Steele, R. H., Evidence for the generation of an electronic excitation state(s) in human polymorphonuclear leukocytes and its participation in bactericidal activity, *Biochem. Biophys. Res. Commun.,* 47, 679, 1972.
7. Stjernholm, R. L., Allen, R. C., Steele, R. H., Waring, W. W., and Harris, J. A., Impaired chemiluminescence during phagocytosis of opsonized bacteria, *Infect. Immun.,* 7, 313, 1973.
8. Cadenas, E. and Sies, H., Low-level chemiluminescence as an indicator of singlet molecular oxygen in biological systems, *Methods Enzymol.,* 105, 221, 1984.
9. Trush, M. A., Chemiluminescence as a probe to investigate chemical-cell interactions: a toxicological perspective, in *Cellular Chemiluminescence,* Vol. III, Van Dyke, K. and Castranova, V., Eds., CRC Press, Boca Raton, FL, 1987, 185.
10. Cadenas, E., Low-level chemiluminescence of biological systems, in *Bioluminescence and Chemiluminescence: New Perspectives,* Scholmerich, J., Andreesen, R., Kapp, A., Ernst, M., and Woods, W.G., John Wiley & Sons, New York, 1987, 33.
11. Trush, M. A., Wilson, M. E., and Van Dyke, K., The generation of chemiluminescence (CL) by phagocytic cells, *Methods Enzymol.,* 75, 462, 1978.
12. Kahl, R., Weimann, A., and Hildebrant, A. G., Detection of active oxygen in rat hepatocyte suspensions with chemilumigenic probe lucigenin, *Biochem. Biophys. Res. Commun.,* 140, 468, 1986.
13. Maly, F. E., Cross, A. R., Jones, D. T., Wolf-Vorbeck, G., Walker, C., Dahinden, C. A., and DeWeck, A. L., The superoxide generating system of B cell lines. Structural homology with the phagocytic oxidase and trigerring via surface Ig, *J. Immunol.,* 140, 2334, 1988.
14. McKinney, K. A., Lewis, S. E., and Thompson, W., Reactive oxygen species in human sperm: luminol and lucigenin chemiluminescence probes, *Arch. Androl.,* 36, 119, 1996.
15. Ohoi, I., Sone, K., Tobari, H., Kawano, E., and Nakamura, K., A simple chemiluminescence method for measuring oxygen-derived free radicals generated in oxygenated rat myocardium, *Jpn. J. Pharmacol.,* 61, 101, 1993.
16. Liochev, S. I. and Fridovich, I., Lucigenin (bis-*N*-methylacridinium) as a mediator of superoxide anion production, *Arch. Biochem. Biophys.,* 337, 115, 1997.
17. Vasquez-Vivar, J., Hogg, N., Pritchard, K. A., Jr., Martasek, P., and Kalyanaraman, B., Superoxide from lucigenin: an electron spin reasonance spin-trapping study, *FEBS Lett.,* 403, 127, 1997.
18. Li, Y., Zhu, H., Kuppusamy, P., Rouband, V., Zweier, J. L., and Trush, M. A., Validation of lucigenin (bis-*N*-methylacridinium) as a chemilumigenic probe for detecting superoxide anion radical production by enzymatic and cellular systems, *J. Biol. Chem.,* 273, 2015, 1998.
19. Liochev, S. I. and Fridovich, I., Lucigenin as mediator of superoxide production: revisited, *Free Radical Biol. Med.,* 25, 926, 1998.
20. Afanas'ev, I. B., Ostrachovitch, E. A., and Korking, L. G., Lucigenin is a mediator of cytochrome *c* reduction but not of superoxide production, *Arch. Biochem. Biophys.,* 366, 267, 1999.

21. Spasojevic, I., Liochev, S. I., and Fridovich, I., Lucigenin: redox potential in aqueous media and redox cycling with $O_2^{\bar{\cdot}}$ production, *Arch. Biochem. Biophys.*, 373, 447, 2000.
22. Li, Y., Zhu, H., and Trush, M. A., Detection of mitochondria-derived reactive oxygen species production by the chemilumigenic probes lucigenin and luminol, *Biochem. Biophys. Acta*, 1428, 1, 1999.
23. Li, Y., Stansbury, K. H., Zhu, H., and Trush, M. A., Biochemical characterization of lucigenin (bis-*N*-methylacridinium) as a chemiluminescent probe for detecting intramitochondrial superoxide anion production, *Biochem. Biophys. Res. Commun.*, 262, 80, 1999.
24. Nanni, E. J., Jr., Anelis, C. T., Dickson, J., and Sawyer, D. T., Oxygen activation by radical coupling between superoxide ion and reduced methyl viologen, *J. Am. Chem. Soc.*, 103, 4268, 1981.
25. Skatchkov, M. P., Sperling, D., Hink, U., Mulsch, A., Harrison, D. G., Sinderman, I., Meinertz, T., and Munzel, T., Validation of lucigenin as a chemiluminescent probe to monitor vascular superoxide as well as basal vascular nitric oxide production, *Biochem. Biophys. Res. Commun.*, 254, 319, 1999.
26. Tarpey, M. M., White, C. R., Suarez, E., Richardson, G., Radi, R., and Freeman, B. A., Chemiluminescent detection of oxidants in vascular tissue: lucigenin but not coelenterazine enhances superoxide formation, *Circ. Res.*, 84, 1203, 1999.
27. Barbacanne, M., Souchard, J., Darblade, B., Fliou, J., Nepveu, F., Pipy, B., Bayard, F., and Arnal, J., Detection of superoxide anion released extracellularly by endothelial cells using cytochrome C reduction, ESR, fluorescence and lucigenin-enhanced chemiluminescence techniques, *Free Radical Biol. Med.*, 29, 388, 2000.
28. Skatchkov, M. P., Sperling, D., Hink, U., Anggard, E., and Munzel, T., Quantification of superoxide radical formation in intact vascular tissue using a Cypridina luciferin analog as an alternative to lucigenin, *Biochem. Biophys. Res. Commun.*, 248, 382, 1998.
29. Heiser, I., Muhr, A., and Elstner, F. F., Production of OH-radical-type oxidant by lucigenin, *Z. Naturforsch. [C]*, 53, 9, 1998.
30. Sohn, H. Y., Keller, M., Gloe, T., Crause, P., and Pohl, U., Pitfalls of using lucigenin in endothelial cells: implications for NAD(P)H dependent superoxide formation, *Free Radical Res.*, 32, 265, 2000.
31. Allen, R. C., Phagocytic oxygenation activities: quantitative analysis based on luminescence, in *Bioluminescence and Chemiluminescence: New Perspectives,* Scholmerich, J., Andreesen, R., Kapp, A., Ernst, M., and Woods, W. G., Eds., John Wiley & Sons, New York, 1987, 13.
32. Jacobshagen, U. and Andreesen, R., Respiratory burst formation by human macrophages at different stages of maturation: dissociation of the generation of particular oxygen radicals, in *Bioluminescence and Chemiluminescence: New Perspectives,* Scholmerich, J., Andreesen, R., Kapp, A., Ernst, M., and Woods, W. G., John Wiley & Sons, New York, 1987, 77.
33. Esterline, R. and Trush, M. A., Lucigenin chemiluminescence and its relationship to mitochondrial respiration in phagocytic cells, *Biochem. Biophys. Res. Commun.*, 159, 584, 1989.
34. Rembish, S. J. and Trush, M. A., Further evidence that lucigenin-derived chemiluminescence monitors mitochondrial superoxide generation in rat alveolar macrophages, *Free Radical Biol. Med.*, 17, 117, 1994.
35. Li, Y. and Trush, M. A., Diphenyleneiodonium, an NAD(P)H oxidase inhibitor, also potently inhibits mitochondrial reactive oxygen species production, *Biochem. Biophys. Res. Commun.*, 253, 295, 1998.
36. He, H., Wang, X., Gorospe, M., Holbrook, N., and Trush, M. A., Phorbol ester-induced mononuclear cell differentiation is blocked by the mitogen-activated protein kinase kinase (MEK) inhibitor PD98059, *Cell Growth Diff.*, 10, 307, 1999.
37. Rembish, S. J., Yang, Y., Esterline, R. L., Seacat, A., and Trush, M. A., Lucigenin-derived chemiluminescence as a monitor of mitochondrial maturation and modulation in mononuclear cells, in *In Vitro Toxicology: Mechanisms and New Technology,* Goldberg, A. M., Ed., Mary Ann Liebert, New York, 1991, 463.
38. Rembish, S. J., Yang, Y., and Trush, M. A., Inhibition of mitochondrial superoxide generation in rat alveolar macrophages by 12-tetradecanoylphorbol-13-acetate: potential role of protein kinase C, *Res. Commun. Mol. Pathol. Pharmacol.*, 85, 115, 1994.
39. Hennet, T., Richter, C., and Peterhan, E., Tumor necrosis factor-alpha induces superoxide anion generation in mitochondria of L929 cells, *Biochem. J.*, 15, 587, 1993.
40. Staniek, K. and Nohl, H., Are mitochondria a permanent source of reactive oxygen species? *Biochem. Biophys. Acta,* 1460, 268, 2000.
41. Braakman, I., Pijning, T., Verest, D., Weert, B., Meiger, D. K., and Groothuis, G. M., Vesicular uptake system for the cation lucigenin in the rat hepatocyte, *Mol. Pharmacol.*, 36, 537, 1989.

42. Yang, S. Q., Zhu, H., Li, Y., Lin, H. Z., Gabrielson, K., Trush, M. A., and Diehl, A. M., Mitochondrial adaptations to obesity-related oxidant stress, *Arch. Biochem. Biophys.,* 378, 259, 2000.
43. Li, Y., Zhu, H., Kuppusamy, P., Zweier, J. L., and Trush, M. A., A role for mitochondria-derived reactive oxygen species in protein kinase C–mediated signal transduction, *Proc. Am. Assoc. Cancer Res.,* 41, 233, 2000.
44. Li, Y., Mitochondria-Derived Reactive Oxygen Species: Detection, Involvement in Mononuclear Cell Signal Transduction and Dysfunction, dissertation, Johns Hopkins University, Baltimore, MD, 1999.
45. Rembish, S. J., An *In Vitro* Mononuclear Cell Differentiation Model and Its Application to Toxicology, dissertation, Johns Hopkins University, Baltimore, MD, 1994.
46. He, H., Molecular Mechanisms of Chemical Modulation of Mononuclear Cell Differentiation, dissertation, Johns Hopkins University, Baltimore, MD, 1998.
47. Cohn, Z. A., The structure and function of monocytes and macrophages, *Adv. Immunol.,* 9, 163, 1968.
48. Trush, M. A., Twerdok, L. E., Rembish, S. J., Zhu, H., and Li, Y., Analysis of target cell susceptibility as a basis for the development of a chemoprotective strategy against benzene-induced hematotoxicities, *Environ. Health Perspect.,* 104(Suppl. 6), 1227, 1996.
49. Chen, J.-Q., Schwartz, D. A., Young, T. A., Norris, J. S., and Yager, J. D., Identification of genes whose expression is altered during mitosuppression in livers of ethinyl estradiol-treated female rats, *Carcinogenesis,* 17, 2783, 1996.
50. Chen, J.-Q., Gokhale, M., Li, Y., Trush, M. A., and Yager, J. D., Enhanced levels of several mitochondrial mRNA transcripts and mitochondrial superoxide production during ethinyl estradiol-induced heepatocarcinogenesis and after estrogen treatment of HepG2 cells, *Carcinogenesis,* 19, 2182, 1998.
51. Chen, J. Q., Li, Y., Lavigne, J. A., Trush, M. A., and Yager, J. D., Increased mitochondrial superoxide production in rat lives mitochondria, rat hepatocytes, and HepG2 cells following ethinyl estradiol treatment, *Toxicol. Sci.,* 51, 224, 1999.
52. Fleury, C., Neverova, M., Collins, S., Raimbault, S., Champigny, D., Levi-Meyrueis, C., Bouillard, F., Seldin, M. F., Surwit, R. S., Ricquier, D., and Warden, C. H., Uncoupling protein-2: a novel gene linked to obesity and hyperinsulinema, *Nat. Genet.,* 15, 269, 1997.
53. Lee, J. F.-Y., Li, Y., Zhu, H., Yang, S. Q., Lin, H. Z., Trush, M. A., and Diehl, A. M., Tumor necrosis factor induces expression of uncoupling protein-2 in the regenerating liver, *Hepatology,* 29, 677, 1999.
54. Zhu, H., Li, Y., and Trush, M. A., Characterization of benzo[*a*]pyrene quinone-induced toxicity to primary cultured bone marrow stromal cells from DBA/2 mice: potential role of mitochondrial dysfunction, *Toxicol. Appl. Pharmacol.,* 130, 108, 1995.
55. Cutlar, R. G., Packer, L., Bertram, J., and Mori, A., Eds., *Oxidative Stress and Aging,* Birkhauser Verlag, Basel, Switzerland, 1995.
56. Beckman, K. B. and Ames, B. N., The free radical theory of aging matures, *Physiol. Res.,* 78, 547, 1998.
57. Sohal, R. S. and Dubey, A., Mitochondrial oxidative damage, hydrogen peroxide release, and aging, *Free Radical Biol. Med.,* 16, 621, 1994.
58. Kwong, L. K. and Sohal, R. S., Substrate and site specificity of hydrogen peroxide generation in mouse mitochondria, *Arch. Biochem. Biophys.,* 350, 118, 1998.
59. Kwong, L. K. and Sohal, R. S., Age related changes in activities of mitochondrial electron transport complexes in various tissues of the mouse. *Arch. Biochem. Biophys.,* 373, 16, 2000.
60. Chen, H., Cangello, D., Benson, S., Folmer, J., Zhu, H., Trush, M. A., and Zirkin, B. R., Age-related increase in mitochondrial superoxide generation in the testosterone-producing cells of brown Norway rat testes: relationship to reduced steriodogenic function? *Exp. Gerontol.,* 36, 1361, 2001.
61. Miesel, R., Murphy, M. P., and Kroger, H., Enhanced mitochondrial radical production in patients which rheumatoid arthritis correlates with elevated levels of tumor necrosis factor alpha in plasma, *Free Radical Res.,* 25, 151, 1996.
62. Trulson, A., Nilsson, S., Brekkan, E., and Venge, P., Patients with renal cancer have a larger proportion of high-density blood monocytes with increased lucigenin-enhanced chemiluminescence, *Inflammation,* 18, 99, 1994.
63. Capone, G., Kim, P., Jovanovich, S., Payne, L., Freund, L., Welch, K., Miller, E., and Trush, M., Evidence for increased mitochondrial superoxide production in Down syndrome, *Life Sci.,* in press.
64. Schuchmann, S. and Heinemann, U., Increased mitochondrial superoxide generation in neurons from trisomy 16 mice: a model of Down's syndrome, *Free Radical Biol. Med.,* 28, 235, 2000.

65. Busciglio, J. and Yankner, B., Apoptosis and increased generation of reactive oxygen species in Down's syndrome neurons *in vitro, Nature,* 378, 776, 1998.
66. Eaton, D. L. and Klaassen, C. D., Principles of toxicology, in *Casarett and Doull's Toxicology: The Basic Science of Poisons,* Klaassen, C. D., Ed., McGraw-Hill, New York, 1996, chap. 2.
67. Liochev, S. I. and Fridovich, I., Lucigenin luminescence as a measure of intracellular superoxide dismutase activity in Escherichia coli, *Proc. Natl. Acad. Sci. U.S.A.,* 94, 2891, 1997.
68. Cilento, G., Dioxetanes as intermediate in biological processes, *J. Theor. Biol.,* 55, 471, 1975.
69. Faljoni, A., Haun, M., Hoffman, M. E., Meneghini, R., Duran, N., and Cilento, G., Photochemical-like effects in DNA caused by enzymically energized triplet carbonyl compounds, *Biochem. Biophys. Res. Commun.,* 80, 490, 1978.
70. Cilento, G. and Adam, W., Photochemistry and photobiology without light, *Photochem. Photobiol.,* 48, 361, 1988.
71. Cilento, G., Generation of triplet carbonyl compounds during peroxidase catalysed reactions, *J. Biolumin. Chemilumin.,* 4, 193, 1989.
72. Cilento, G. and Adam, W., From free radicals to electronically excited species, *Free Radical Biol. Med.,* 19, 103, 1995.

22 Chemiluminescent Polymer Microspheres for Measuring Reactive Oxygen Species

Shuntaro Hosaka and Takafumi Uchida

CONTENTS

22.1 INTRODUCTION

In 1972, Allen et al.[1] reported that weak chemiluminescence (CL) was detected in the phagocytosis of opsonized bacteria by human polymorphonuclear leukocytes (PMN) and proposed that the CL reflected the generation of singlet oxygen. In a subsequent paper, they discussed the relationship of superoxide anion production, singlet oxygen generation, and CL.[2] This finding was followed by other researchers' observations of CL in the phagocytosis of other phagocytic cells, namely, activated monocytes,[3] eosinophils,[4] and alveolar macrophages.[5] The CL spectra were analyzed to identify the species of reactive oxygen.[6–14] However, from the standpoint of clinical applications, the extreme weakness of natural CL in phagocytosis makes the application of CL measurement in the examination of cell functions difficult.

Allen and Loose[15] found that high-intensity CL could be measured in the phagocytosis of *Escherichia coli* by rabbit alveolar or peritoneal macrophages in the presence of luminol but not in the absence of this compound. Since then, most measurements of reactive oxygen species (ROS) in phagocytosis have been made in the presence of a CL sensitizer, which reduces the number of phagocytes and the instrument sensitivity necessary.

In addition to luminol, lucigenin[16,17] and analogues of *Cypridina* luciferin CLA[18,19] were found to be excellent CL sensitizers in the measurement of ROS. In this chapter, these CL sensitizers are called CL probes. The types of ROS measured with these CL probes will be discussed later.

In the examination of the function of phagocytic cells, a stimulus for ROS production is needed. The stimulants that have been used for this purpose have been either particles, such as microbial cells, or soluble substances, such as oligopeptides. Supposedly, the type of stimulant may affect

0-8493-0719-8/02/$0.00+$1.50

the kind and location of the released ROS. The authors observed the location of the released ROS because phagocytizing cells release ROS not only into phagosomes but also outside the cells.

ROS released into phagosomes kill microorganisms, acting as a self-defense mechanism of living bodies. In contrast, ROS released extracellularly injure tissues and genes, and are thus harmful to living bodies. It is therefore important to measure extracellular ROS and phagosomal ROS separately. As conventional CL probes were used in a solution, supposedly only a small amount of a CL probe could enter the cells. Accordingly, the authors thought that the probability of the reaction of ROS with a CL probe in a solution is far greater outside the cells than within the cells. This means CL generated by use of a CL probe solution reflects the degree of extracellular release of ROS. For this reason, the authors attempted to prepare CL microspheres as a stimulant and to measure phagosomal ROS attacking phagocytized microspheres as working models of microorganisms.

22.2 LUMINOL-BOUND MICROSPHERES

Uchida et al.[20,21] bound luminol to polymer microspheres by covalent bonding and measured CL in the phagocytosis of prepared luminol-bound microspheres by mouse macrophages. The carrier for the luminol was a microspheric copolymer of glycidyl methacrylate. Briefly, the method of preparation was as follows. Epoxy groups in the copolymer were converted into formyl groups through hydrolysis with dilute sulfuric acid, followed by periodate oxidation. Luminol was bound to the copolymer, forming Schiff base that was reduced by sodium borohydride. Macrophages were taken from the peritoneal cavity of mice that had been stimulated by intraperitoneal injection of Freund's complete adjuvant (FCA) 4 days prior. Luminol-bound microspheres were used without opsonization.

Figure 22.1a shows CL over time when luminol-bound microspheres were incubated with FCA-activated mouse macrophages. Cytochalasin B strongly inhibited CL. Figure 22.1b shows CL over

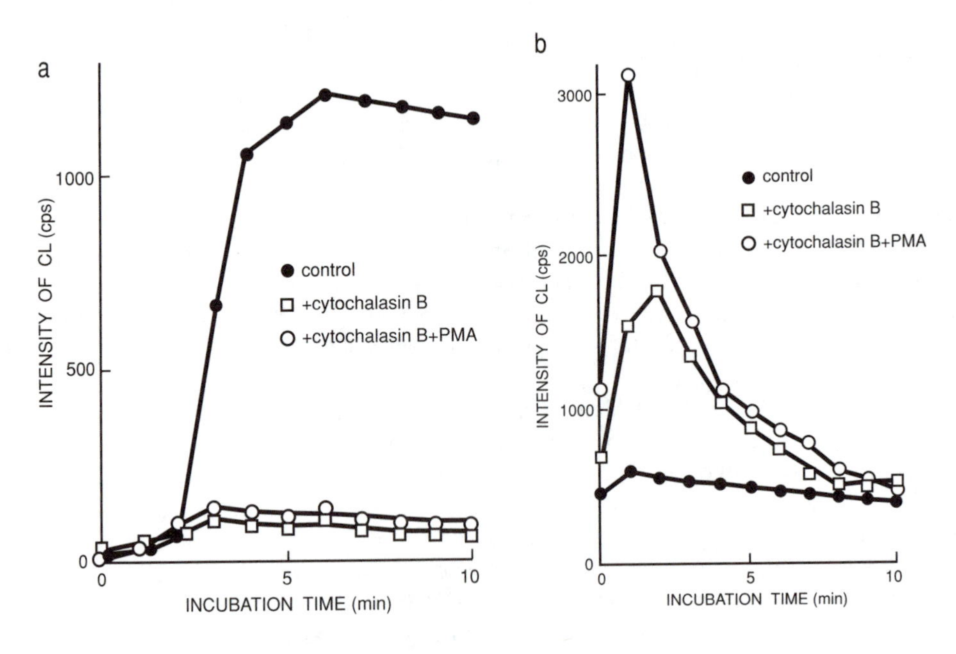

FIGURE 22.1 (a) Measurement of CL over time when luminol-bound microspheres were incubated with FCA-activated mouse macrophages. The macrophages were either treated or not treated with cytochalasin B and PMA. (b) Measurement of CL over time when plain microspheres were incubated with FCA-activated macrophages in the presence of free luminol. The macrophages were either treated or not treated with cytochalasin B and PMA. (From Uchida, T. et al., in *Macrophage Biology*, Richard, S. and Kojima, M., Eds., Alan R. Liss, New York, 1985, 545. With permission.)

time when plain microspheres (not bound to luminol) were incubated with the counterpart cells in the presence of free luminol (luminol solution). In the latter case, CL was very weak unless the macrophages were treated with cytochalasin B. Additionally, an application of phorbol myristate acetate (PMA) to the cells further enhanced CL. The treatment of the cells with cytochalasin B conversely affected the CL intensity shown in Figure 22.1a and b.

These results can be interpreted as follows. The treatment of cells with cytochalasin B inhibited phagocytosis, resulting in the almost complete depression of CL from luminol-bound microspheres, while the same treatment of cells enhanced the extracellular release of ROS, resulting in the intensification of CL caused by free luminol outside the cells. Observation by microscope supports this interpretation. Macrophages without the cytochalasin B treatment were found to contain phagocytized microspheres within 2 min after the start of incubation, at which time the CL intensity rose steeply. By contrast, few microspheres were phagocytized when the macrophages were treated with cytochalasin B. In addition, it was already known that cytochalasin B affects the membranes of phagocytic cells, causing inhibition of phagocytosis and an increase in the amount of ROS released extracellularly.[22–24] Therefore, the authors concluded that CL generated during the incubation of luminol-bound microspheres with macrophages reflected the amount of ROS released into the phagosomes, that is, the microbicidal activity of the cells.[25]

Table 22.1 shows the relationship between CL generation and the microbicidal activity of macrophages. The higher the CL intensity, the lower the *Candida* survival, although the relationship was not proportional. It should be noted that phagocytic activity does not always reflect microbicidal activity. Table 22.2 compares CL intensity and the average number of phagocytized microspheres per cell. As the table shows, thioglycollate-activated mouse macrophages exhibited about the same

TABLE 22.1
Relationship between Candidacidal Activity and the Intensity of CL Elicited by Luminol-Bound Microspheres in Murine Peritoneal Macrophages

Macrophage	Integrated Relative CL Intensity	Viable *Candida* (% survival)
FCA-activated	100 ± 11	12 ± 5
Thioglycollate-activated	9 ± 1	27 ± 9
Resident	3 ± 0	45 ± 18

Source: Uchida, T. et al, *J. Bioactive Compatible Polym.*, 1, 172, 1986. With permission from Technomic Publishing Co., Inc., © 1986.

TABLE 22.2
Relationship between Phagocytosis and Phagosomal ROS-Dependent CL in Murine Peritoneal Macrophages

Macrophage	Integrated Relative CL Intensity	Phagocytosis (microsphere/cell)
FCA-activated	100 ± 4	6.8 ± 0.5
Thioglycollate-activated	11 ± 2	7.5 ± 1.1
Resident	3 ± 1	0.3 ± 0.1

Source: Uchida, T. and Hosaka, S., *J. Immunol. Methods*, 77, 55, 1985. With permission from Elsevier Science.

phagocytic activity as the FCA-activated macrophages, but the intensity of their CL was only about 11% as great.

Uchida et al.[21] studied the effect of lipopolysaccharide (LPS) on macrophages using luminol-bound microspheres. Mouse macrophages activated *in vivo* with FCA were further treated overnight with LPS *in vitro*, then incubated with luminol-bound microspheres or a luminol solution containing plain microspheres. The CL intensity decreased in the former case but increased in the latter case, depending on the concentration of LPS, as shown in Table 22.3. Although phagocytosis was reduced with the LPS concentration, the degree of reduction was more remarkable in the CL from luminol-bound microspheres. These results show that LPS inhibited the generation of ROS in the phagosomes, whereas it enhanced the extracellular oxidative burst.

To study the effect of alcohol on the activity of macrophages, Yamamoto et al.[26] used luminol-bound microspheres to measure the ROS within macrophages from alcohol-stimulated mice. The mean values of CL caused by ROS generated within the macrophages from alcohol-injected mice and alcohol-fed mice were higher than those of the controls, as shown in Table 22.4. The phagocytic activity of the macrophages from alcohol-injected mice was also higher than that of controls.

TABLE 22.3
Effect of Preincubation with LPS on Phagocytosis and Phagocytosis-Associated CL by FCA-Activated Murine Peritoneal Macrophages

	Relative CL Intensity		
LPS conc. (μg/ml)	Plain Microspheres with Free Luminol	Luminol-Bound Microspheres	Phagocytosis[a]
0	100 ± 13	100 ± 17	100 ± 4
0.001	100 ± 21	97 ± 21	Not done
0.01	102 ± 11	87 ± 12	99 ± 10
0.1	102 ± 7	85 ± 18	89 ± 3
1	105 ± 13	84 ± 15	78 ± 3
10	160 ± 11	57 ± 11	71 ± 5
100	237 ± 30	21 ± 12	68 ± 4

[a] Relative value of phagocytized luminol-bound microspheres in a cell.

Source: Uchida, T. et al., in *Macrophage Biology*, Richard, S. and Kojima, M., Eds., Alan R. Liss, New York, 1985, 545. With permission.

TABLE 22.4
Effect of Alcohol Administration *in Vivo* on Phagocytosis and Phagocytosis-Associated CL by Murine Peritoneal Macrophages

Source of Macrophages	Phagocytosis (microspheres/cell)	Integrated Relative CL Intensity
Alcohol-injected mice	3.76 ± 0.22	5110 ± 391
Control (saline injected)	0.98 ± 0.12	617 ± 133
Alcohol-fed mice	Not done	1884 ± 326
Control	Not done	259 ± 104

Source: Yamamoto, H. et al., *Jpn. J. Alcohol Stud. Drugs Dependence,* 21, 183, 1986. With permission.

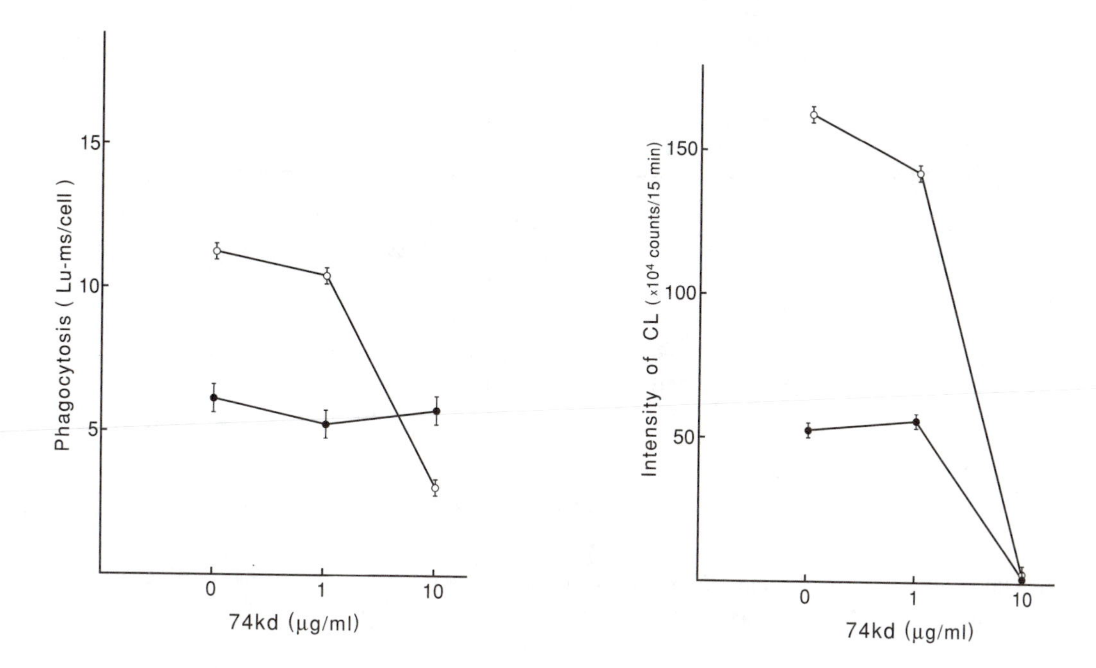

FIGURE 22.2 Effects of the 74-kDa glycoprotein in monocyte (●) and neutrophil (○) functions. The values are expressed as mean ± SD (n = 3). (From Uchida, T. et al., *Acta Haematol. Jpn.*, 71, 759, 1988. With permission.)

Uchida et al.[27] applied luminol-bound microspheres to the investigation of a substance suppressing the function of phagocytes in the sera of leukemia patients. The substance was a 74-kDa glycoprotein separated by Jacquemin[28] from a patient with chronic myelocytic leukemia. Figure 22.2 shows the effect of the 74-kDa glycoprotein on the functions of monocytes and neutrophils from normal donors. Although the CL was drastically inhibited by the 74-kDa glycoprotein, phagocytosis was not blocked to a similar extent. This result suggests that the 74-kDa glycoprotein acted directly on the CL generation system within the phagocytes. The researchers reported the relationship between the concentration of the 74-kDa glycoprotein in 23 patients with leukemia and the intensity of CL generated when luminol-bound microspheres treated with the patients' sera were incubated with normal phagocytic cells. The correlation coefficient of the CL intensity and the 74-kDa glycoprotein concentration was –0.55 for the monocytes and –0.51 for the neutrophils. Although these values were not high, they did suggest that the 74-kDa glycoprotein might reduce the generation of ROS in phagocytes.

Using luminol-bound microspheres and luminol solution, Masuko et al.[29] studied the influence of contact with biomedical materials on neutrophils, as schematically shown in Figure 22.3. Phagocytes were stimulated by a membranous material and released ROS extracellularly, which were measured by the CL from the free luminol in the solution. In contrast, phagosomal ROS released in phagocytes stimulated by a material were estimated by the CL from the phagocytized luminol-bound microspheres. The researchers compared the influence on human neutrophils of poly(methyl-methacrylate) (PMMA) and cellulose, treated and not treated with blood plasma proteins.

Figure 22.4 compares intracellular and extracellular ROS release from neutrophils stimulated with membrane materials. PMMA caused a greater degree of extracellular release of ROS than did cellulose without any treatment. In contrast, neutrophils in contact with PMMA exhibited weaker CL than those in contact with cellulose from luminol-bound microspheres. From these results, it might be presumed that PMMA is not suitable for biomedical use. It seems that contact of the material with blood may enhance the extracellular release of ROS, possibly injuring living body tissues and also reducing the release of ROS into phagosomes, which reflects the microbicidal functions of phagocytes. This situation is reversed, however, by the treatment of the material with

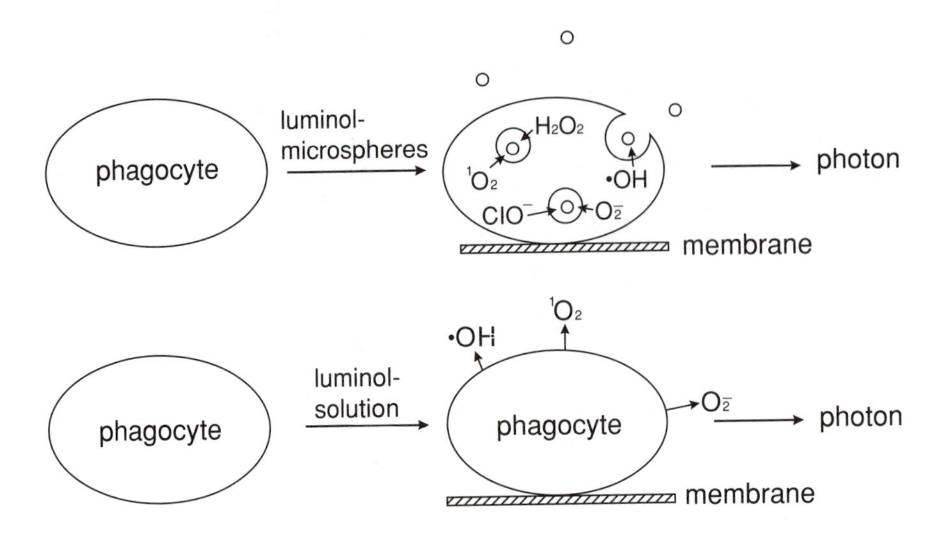

FIGURE 22.3 Schematic diagram of the measurement of ROS generation by phagocytes in contact with membrane materials. (From Masuko, S. et al., *Artif. Organs Today,* 1, 245, 1991. With permission.)

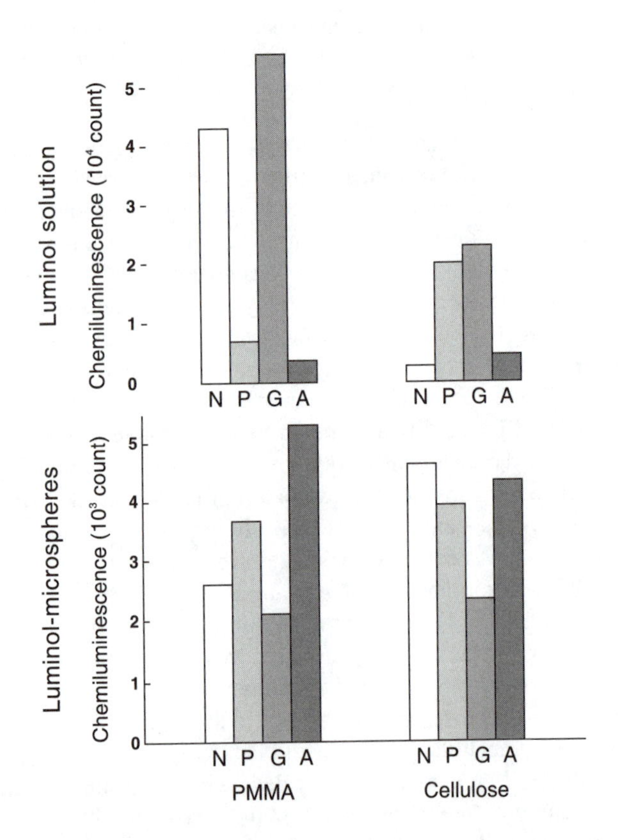

FIGURE 22.4 Comparison of intracellular and extracellular ROS release from neutrophils stimulated with membrane materials either treated or not treated with blood plasma proteins. N = not treated, G = gamma globulin, P = plasma, A = serum albumin. (From Masuko, S. et al., *Artif. Organs Today,* 1, 245, 1991. With permission.)

blood plasma or serum albumin, as shown in the figure. The intensity of CL from the luminol solution was significantly lower for the PMMA than for the cellulose. A gamma-globulin treatment of the materials enhanced CL from the luminol solution and reduced CL from the luminol-bound microspheres in both materials. The effect of plasma treatment or serum albumin is explained by the greater extent of the adsorption of serum albumin to PMMA than to cellulose.

Other examples of the application of luminol-bound microspheres were the study of the effect of interferon on ROS production in human monocytes and the study of the effect of neutrophil-activating substances on intracellular generation of ROS. Using luminol-bound microspheres, Matsumoto and Imanishi[30] examined the effects of recombinant human interferon-γ, having Met-Gln at the N terminal, on the *in vitro* production of phagosomal and extracellular ROS by human monocytes after phagocytosis. They compared the results with those for natural human interferon-α and -β.[30] Takeuchi et al.[31] investigated the effects of neutrophil-activating substances including *N*-formyl-L-leucyl-L-phenylalanine (FMLP), concanavalin A (Con-A), and PMA on the intracellular generation of CL using luminol-bound microspheres, and discussed the relationship between the generation of ROS and metabolic changes inside the cells.

22.3 FLUOROGEN-BOUND MICROSPHERES

Suzuki et al.[32,33] bound fluorogenic substrates of enzymes to polymer microspheres and measured enzyme activity within phagocytizing cells by the fluorescence of the enzymatic reaction product. In their first work, 4-methylumbelliferyl-β-D-glucuronide (4MUGL), a substrate of β-D-glucuronidase, was covalently bound to polymer microspheres.[32,33] 4MUGL-bound microspheres liberated 4-methylumbelliferone (4MU) into extracellular fluid from human PMN during phagocytosis. In this case, the measurement target was β-D-glucuronidase in phagolysosomes, but 4MU liberated by enzymatic hydrolysis of the substrate was released extracellularly. The distribution of 4MU in the extracellular fluid was estimated to be 75% of the total. This result prompted the suggestion that extracellular 4MU might be produced from 4MUGL-bound microspheres outside cells by the action of extracellularly secreted β-D-glucuronidase. However, Suzuki and his colleagues inferred that 4MU produced in the phagosomes instantly permeated through the cell membrane into the extracellular fluid. They based this conclusion on the observation that the leakage of β-D-glucuronidase into the extracellular fluid was slight under their experimental conditions.

Uchida et al.[34,35] prepared decanoylfluorescein-bound microspheres that were then used to measure esterase activity within the phagosomes. In this case, the fluorescent substance produced by the enzymatic action was theoretically retained on the microspheres, as illustrated in Figure 22.5. In other words, nonfluorescent microspheres could be converted into fluorescent ones. The incubation mixture of decanoylfluorescein-bound microspheres and human neutrophils was analyzed by flow cytometry. This confirmed that decanoylfluorescein bound to the microspheres was hydrolyzed within the cells and not outside them. In addition, it was found that there were two classes in the neutrophils: one class with high phagocytic activity and one with low activity.

Suzuki et al.[36] succeeded in imaging myeloperoxidase (MPO) release into phagolysosomes of human PMN using 3-(*p*-hydroxyphenyl) propionic acid (HPPA)-bound microspheres. They confirmed that HPPA, which had been known as a substrate for sensitive assay of horseradish peroxidase, could also be used as a substrate of MPO, and bound this compound to the microspheres of an aminated derivative of a glycidyl methacrylate copolymer by making an amide bond between them. When HPPA-bound microspheres were phagocytized by PMN, fluorescence of the ingested microspheres was observed. The production of fluorescence in the microspheres was promoted with added free HPPA. The researchers observed fluorescence from the microspheres within PMN using a newly developed image analyzer, IMRAS. The image of phagocytosis of the microspheres by PMN on a glass slide was simultaneously recorded by image processing in real time through IMRAS onto videotape. Figure 22.6 shows the TV microscope screen image of PMN-phagocytizing

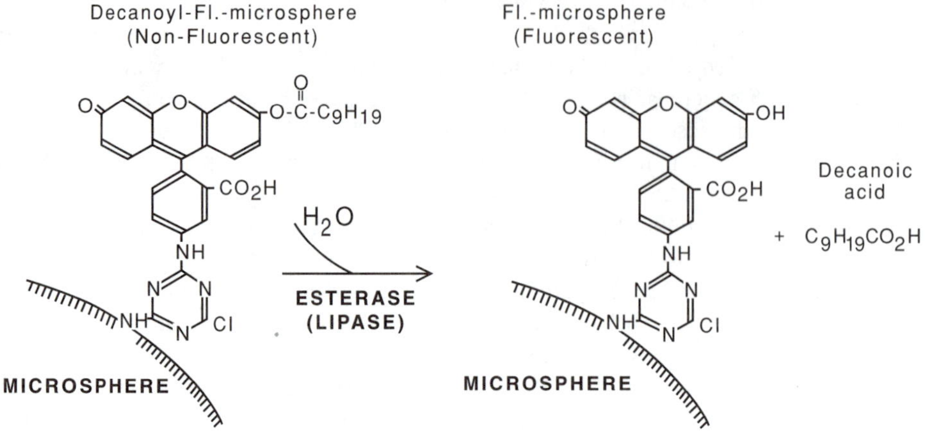

Decanoyl-Fl.-microsphere (Non-Fluorescent)

Fl.-microsphere (Fluorescent)

H_2O ESTERASE (LIPASE)

Decanoic acid + $C_9H_{19}CO_2H$

MICROSPHERE

FIGURE 22.5 Reaction scheme of the conversion of nonfluorescent microspheres (substrate) into fluorescent ones. (From Uchida, T. et al., *Biochem. Biophys. Res. Commun.,* 127, 584, 1985. With permission.)

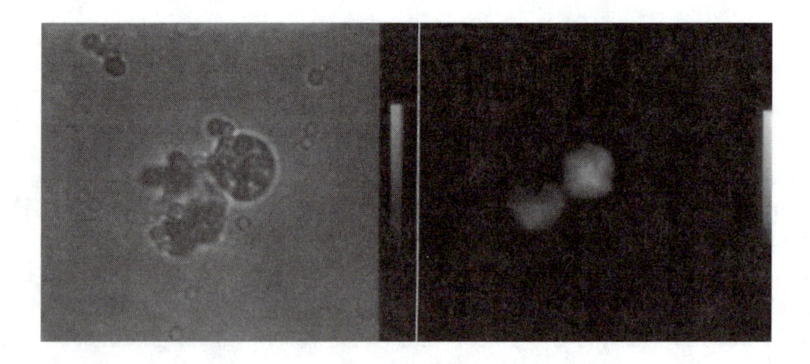

FIGURE 22.6 (Color figure follows p. 266.) Image analysis of phagocytosis of HPPA-bound microspheres by PMN. The image processed by IMRAS in real time was recorded on videotape. Left: Microscope image on screen from the videotape. Right: Fluorescence microscope image on screen from the videotape. (From Suzuki, K. et al., *Bioimages,* 1, 13, 1993. With permission.)

microspheres and the phagocytized fluorescent microspheres. It should be noted that the microspheres that were not phagocytized and that were located outside the cells did not fluoresce.

22.4 ABEI-BOUND MICROSPHERES

ABEI-bound microspheres have been commercially available in Japan under the brand name Lumisphere® (Toray Research Center, Inc., Tokyo, Japan) for about 10 years. Some users misinterpret these to be luminol-bound microspheres, but the differences between them do not seem important. Yokoo et al.[37] reported that the opsonization of Lumisphere microspheres was required to elicit CL from human neutrophils. However, other researchers have used Lumisphere microspheres without opsonization. Using Lumisphere microspheres without opsonization, Kohashi et al.[38] investigated phagocytic and CL-inducing activities of human PMN of normal adults and of patients with several respiratory diseases. Nishijima et al.[39] investigated the effects of protease inhibitors on the intracellular production of ROS by measuring CL elicited from phagocytized Lumisphere microspheres in human neutrophils without opsonization. According to their report, gabexate mesylate and urinastatin, urinary trypsin inhibitors, increased intracellular CL from Lumisphere microspheres in a dose-dependent manner. In contrast, protease inhibitors in a dose-dependent fashion reduced extracellularly released ROS from stimulated neutrophils, assayed by conventional measurement of free luminol-dependent CL.

Actually, opsonization undoubtedly enhances the phagocytosability of Lumisphere microspheres, leading to an increase of CL intensity. Furthermore, since opsonization means the binding of complement fragments such as C3b, and the adsorption of immunoglobulins to the surface of the microspheres, some investigators would prefer the use of opsonization if they consider the binding of the stimulant particles to the C3b or Fc receptor on cells to be important.

Lumisphere microspheres were used in the investigation of the effect of drugs on the function of phagocytes, for example, the effect of cystein ethyl ester hydrochloride (Cystanin) on mouse peritoneal macrophages and leukocytes,[40] or the effect of new quinolones on the intracellular production of ROS in human PMN.[41] Lumisphere microspheres were also used in clinical studies, for example, in the measurement of ROS production by phagocytes in patients on hemodialysis[42] or the analysis of the function of PMN in patients with periodontitis.[43]

22.5 SELECTIVITY OF LUMINOL- OR ABEI-BOUND MICROSPHERES TO ROS

Although CL microspheres such as luminol- or ABEI-bound microspheres have an advantage in that CL observation is mainly of microspheres undergoing phagocytosis, it has not yet been clarified what ROS are measured by their use. Allen and Loose[15] assumed that ROS eliciting CL from luminol on stimulating phagocytes was among the superoxide anions, hydrogen peroxide, hydroxyl radicals, and singlet oxygen. Nakano and Ushijima[44] presumed CL was not generated by the oxidation of luminol with superoxide anions because the luminol generated negligibly weak CL with a xanthine oxidase–acetaldehyde system although it generated strong CL in the phagocytosis of opsonized zymosan by granulocytes. They inferred that the ROS-oxidizing luminol was hypochlorite or hypochloric acid, which was the reaction product of hydrogen peroxide and chloride catalyzed by MPO. Their inference was based on the observation that CL from luminol in the phagocytosis of opsonized zymosan by granulocytes was inhibited by sodium azide and that granulocytes from MPO-deficient patients did not elicit CL from luminol even if stimulated with opsonized zymosan or PMA.

Recently, Hosaka et al.[45] compared the CL intensities in the reactions of several kinds of CL probes with ROS. The results are summarized in Tables 22.5 and 22.6. At pH 7.2, clearly recognizable CL was elicited from luminol and luminol Na with hydrogen peroxide/hypochlorite, a claimed generator of singlet oxygen. The CL must be caused by singlet oxygen because it was inhibited by azide, which is said to be a quencher of singlet oxygen.[46,47] Further, as luminol did not generate CL with any other ROS generator, it was concluded that luminol is a selective CL probe for singlet oxygen at pH 7.2. It should be noted that hypochlorite did not elicit CL from luminol at this pH although it does do so strongly at basic conditions. It should also be added that no other CL probe was selective for singlet oxygen. At pH 5.3, the CL intensity of luminol was too small to reach any conclusion.

Because ABEI could not be dissolved in the neutral medium of physiological pH, ABEI-bound microspheres (ABEI-ms) were examined in a state of suspension. Figure 22.7 shows the CL elicited from ABEI-ms over time with hydrogen peroxide/MPO, which was measured at pH 5.6 near the optimum of MPO. The CL elicited from ABEI-ms with hydrogen peroxide/MPO was thought to be due to only singlet oxygen under this condition for the following reasons. Because hydrogen peroxide did not cause CL without MPO and the combination of MPO and hypochlorite did not cause CL (Figure 22.7), the possibility was that the CL was caused by singlet oxygen, which was the product of the MPO-catalyzed reaction of hydrogen peroxide and chloride. This conclusion was supported by the fact that the addition of azide to the reaction mixture completely inhibited CL.

The authors thus concluded that ABEI-ms constitute a selective probe for singlet oxygen. Although the sensitivity of ABEI-ms to singlet oxygen was not high, 2×10^5 cells of neutrophils or macrophages were sufficient for measurement. However, Van Dyke (personal communication) told

TABLE 22.5
CL Intensity in the Reaction of Dissolved CL Probes with ROS under Neutral Conditions[a]

Claimed ROS	Generator	CL Intensity/10^3 Count Lucigenin (1 mmol/l)	MCLA (0.1 μmol/l)	PMAC (0.1 μmol/l)	Luminol Na (0.1 μmol/l)	Luminol (0.1 μmol/l)
Blank	PBS	0.003	11.430	0.079	–0.001	–0.005
H_2O_2	H_2O_2	0.652	78.436	1189.433	0.177	0.749
	$H_2O_2 + NaN_3$	0.654	74.303	754.801	0.231	1.643
OCl^-	NaOCl	0.010	0.044	0.052	0.045	–0.004
$^1O_2$	$NaOCl/H_2O_2$	1.209	90.168	12.287	6.413	7.297
	$NaOCl/H_2O_2 + NaN_3$	0.369	9.272	6.488	0.214	0.413
$^1O_2$	H_2O_2/MPO	0.186	48.257	485.504	1.032	—
	H_2O_2/MPO + NaN_3	0.155	96.443	316.454	0.647	—
$^1O_2$[b]	H_2O_2/MPO	1.582	344.257	6320.979	6.010	—
	H_2O_2/MPO + NaN_3	1.319	584.650	4394.369	3.705	—
O_2^-	KO_2	28.153	22.545	98.395	0.054	0.026
	KO_2 + SOD	21.405	14.423	34.590	0.002	0.122
O_2^-	HPX/XOD	17.038	486.120	55.560	0.161	0.218
	HPX/XOD + SOD	0.036	10.614	1.952	–0.001	–0.004
O_2^-[b]	HPX/XOD	51.875	540.165	343.027	0.299	0.500
	HPX/XOD + SOD	0.558	25.057	48.295	–0.004	–0.003

[a] pH 7.2 ± 0.2.
[b] Measured for 600 s.

Source: Hosaka, S. et al., *Luminescence,* 14, 349, 1999. With permission.

the authors in private letters that he was almost totally sure that singlet oxygen has nothing to do with luminol-dependent luminescence at or near neutral pH. He wrote the results of the application of particular reactions that change specifically with reactions involving singlet oxygen to $MPO/H_2O_2/Cl^-$ reactions indicated that singlet oxygen was not involved these reactions. Further, he added that electrically generated singlet oxygen actually quenches luminol. He cited Brestel's work[48] and assumed that chloroperoxy anion ($OOCl^-$) might be formed from the reaction of hydrogen peroxide and hypochlorite and react with luminol, although this intermediate has not yet been isolated.

Oosthuizen et al.[49] also attended to the Brestel's claim[50] that the luminol-dependent CL was not due to singlet oxygen generation by $NaOCl/H_2O_2$. They tried to determine which component could elicit CL in this system but could not reach a conclusion. Therefore, the authors cannot say, at present, more than that luminol is a selective probe for hydrogen peroxide/MPO/Cl^- under acidic conditions.

22.6 ACRIDINIUM ESTER-BOUND MICROSPHERES

As described above, ABEI-ms are not sensitive to superoxide anions. As a detector of superoxide anions, lucigenin first attracted attention. Greenlee et al.[51] reported the production of excited species by the reaction of lucigenin radicals with superoxide anions in a xanthine–xanthine oxidase system. CL was observed when macrophages or neutrophils were incubated with opsonized zymosan in the presence of lucigenin.[16,17] Because it was enhanced by the addition of azide in this case, the CL was not generated by MPO-dependent ROS. It was known that lucigenin also generates CL by hydrogen peroxide.[52] However, it was presumed that CL from the mixture of granulocytes and

TABLE 22.6
CL Intensity in the Reaction of Dissolved CL Probes with ROS under Acidic Conditions[a]

Claimed ROS	Generator	CL Intensity/10^3 Count Lucigenin (1 mmol/l)	MCLA (0.1 μmol/l)	PMAC (0.1 μmol/l)	Luminol Na (0.1 μmol/l)	Luminol (0.1 μmol/l)
Blank	PBS	0.012	2.042	0.021	0.011	0.017
H_2O_2	H_2O_2	0.047	28.456	9.797	0.052	0.018
	$H_2O_2 + NaN_3$	0.044	30.496	6.013	0.037	0.032
OCl^-	NaOCl	0.020	0.012	0.016	0.016	0.014
$^1O_2$	$NaOCl/H_2O_2$	0.169	1.018	4.702	0.242	0.047
	$NaOCl/H_2O_2 + NaN_3$	0.147	7.975	3.729	0.084	0.053
$^1O_2$	H_2O_2/MPO	0.013	54.402	3.647	0.160	—
	H_2O_2/MPO + NaN_3	0.011	53.483	1.942	0.000	—
$^1O_2$[b]	H_2O_2/MPO	0.048	281.297	48.568	1.980	—
	H_2O_2/MPO + NaN_3	0.039	281.101	31.845	0.021	—
O_2^-	KO_2	0.131	19.607	0.135	0.058	0.026
O_2^-	HPX/XOD	0.107	268.625	0.164	0.018	0.014
	HPX/XOD + SOD	0.022	3.105	0.018	0.006	−0.003
O_2^-[b]	HPX/XOD	1545	592.485	4.280	0.166	0.171
	HPX/XOD + SOD	0.084	37.631	0.362	0.041	0.003

[a] pH 5.3 ± 0.1.
[b] Measured for 600 s.

Source: Hosaka, S. et al., *Luminescence,* 14, 349, 1999. With permission.

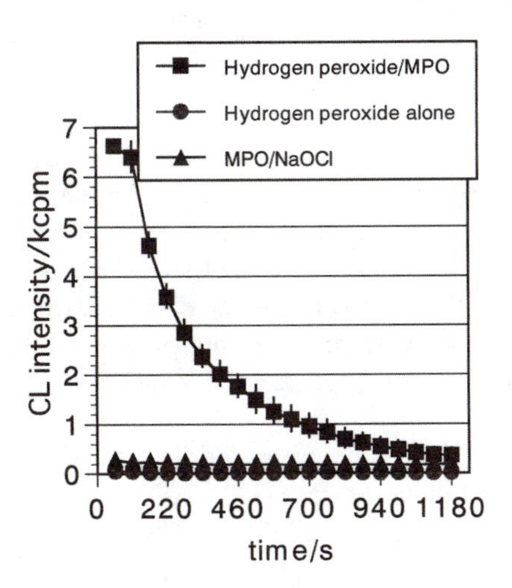

FIGURE 22.7 Measurement of CL over time elicited from ABEI-ms with hydrogen peroxide/MPO at pH 5.6. (From Hosaka, S. et al., *Luminescence,* 14, 349, 1999. With permission.)

lucigenin was principally caused by the superoxide anions because it was remarkably suppressed by superoxide dismutase (SOD).[16]

Goto and Takagi[53] found that strong CL was elicited from a *Cypridina* luciferin analogue (CLA) by a xanthine–xanthine oxidase system, and the CL was suppressed by SOD. Nakano et al.[18,19]

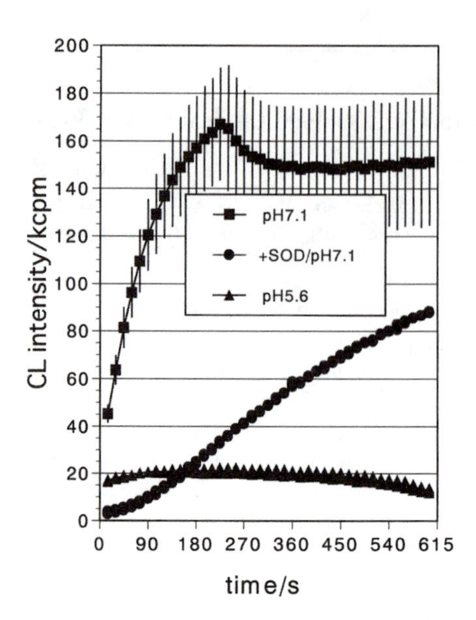

FIGURE 22.8 Measurement of CL over time from acridinium ester-bound microspheres (AE-ms) incubated with hypoxanthine–xanthine oxidase (HPX/XOD). (From Hosaka, S. et al., *Luminescence,* 14, 349, 1999. With permission.)

observed CL in a mixture of the CLA and macrophages[18] or neutrophils[19] stimulated with opsonized zymosan. In addition, a derivative of the CLA, methyl-Cypridina luciferin analogue (MCLA), was also found to produce CL by reacting with singlet oxygen.[54,55] They studied CL generation by the CLA and the MCLA in detail, and concluded that these *Cypridina* luciferin analogues are excellent CL-probes for the measurement of superoxide anions. Now that these compounds are commercially available, many researchers have used them.

Recently, Oosthuizen et al.[49] reported that CL response to superoxide anions or singlet oxygen was very low in either case when luminol was used as a probe because the CL optima for this probe was above pH 9.0, and they preferred the MCLA for physiological assessments. In the authors' experiment, as well, the MCLA showed a very high sensitivity to superoxide anions, as seen in Tables 22.5 and 22.6. However, the MCLA cannot be chemically bound to polymer microspheres. Therefore, the authors have attempted to produce another type of chemiluminescent microsphere, which is highly sensitive to ROS at physiological pH values, by covalently binding an acridinium ester to polymer microspheres.

Figure 22.8 shows the CL over time from acridinium ester-bound microspheres (AE-ms) incubated with hypoxanthine–xanthine oxidase (HPX/XOD). CL was measured at pH 5.6 as well as pH 7.2 taking into consideration the possibility that the pH within phagocytizing cells drops below 6.[36,56] Although the intensity of CL was lower at pH 5.6 than at pH 7.2 by a factor of eight, it still was strong enough to be measured. When SOD was added at pH 7.2, CL was initially inhibited, but the intensity gradually increased and exceeded half the CL intensity without SOD after 10 min. This time-dependent increase of CL in the presence of SOD was thought to be due to hydrogen peroxide produced by the dismutation of superoxide catalyzed by SOD. It was then confirmed that AE-ms generated strong CL with hydrogen peroxide (Figure 22.9). Accordingly, it can be concluded that the CL occurring at the initial stage was due to the superoxide anions. Then, CL caused by the accumulated hydrogen peroxide increased over time. These results indicate that AE-ms can be an excellent probe for superoxide anions and hydrogen peroxide under neutral conditions.

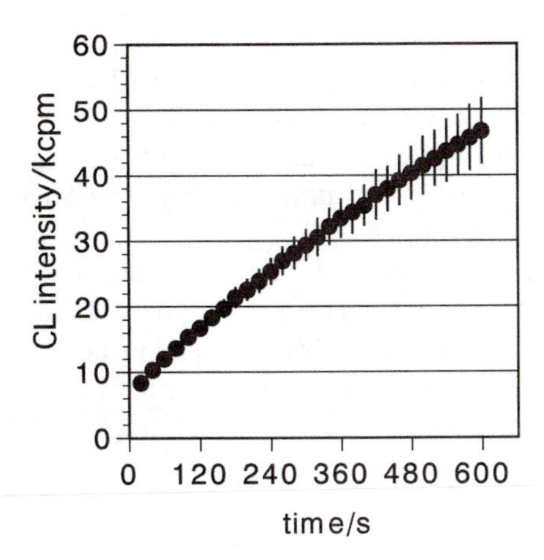

FIGURE 22.9 Measurement of CL over time from acridinium ester-bound microspheres (AE-ms) incubated with hydrogen peroxide. (From Hosaka, S. et al., *Luminescence,* 14, 349, 1999. With permission.)

22.7 LUMINOL- OR LUCIGENIN-COATED POLYSTYRENE MICROBEADS

Van Dyke et al.[57] developed two types of polystyrene round microbeads linked to either luminol or lucigenin with a diameter of 1 μm. These CL polystyrene microbeads were found to be useful for studying phagocytosis without opsonization by measuring cellular CL. They compared luminol beads and lucigenin beads in the measurement of cellular CL with human neutrophils, human monocytes, and differentiated promyelocytic leukemic cells (HL60). When MPO was present, luminol beads evoked higher CL from neutrophils than lucigenin beads. But when little or no MPO was present, lucigenin beads produced higher amounts of light, as seen in the case of monocytes. These results are understandable because it is well known that luminol reacts with peroxidases such as MPO while lucigenin reacts with superoxide anion.

They found an interesting difference between luminol and lucigenin. Luminol beads gave essentially the same reaction as a luminol solution accompanied by plain beads if the luminol concentration was sufficiently high in the measurement of cellular CL. In both cases, the CL intensity gradually increased and reached the maximum after about 15 min. On the other hand, a lucigenin solution added with plain beads yielded different reactions than lucigenin beads. The lucigenin solution added with plain beads gave an early peak at about 1 min, which was not seen with the lucigenin beads, and then the light continued to decrease. However, the light from the lucigenin beads continued to increase until the 15-min point or longer. They explained this difference in the following manner. Lucigenin is a di-cation that normally would not penetrate the neutrophilic cell membrane. If lucigenin is bound to the beads, the lucigenin enters the cell. On the contrary, luminol is hydrophobic and is not charged at physiological pH. Therefore, little difference exists between luminol bound to beads and free luminol added with plain beads regarding its penetration through the cell membrane.

They measured the effect of indomethacin addition to CL from luminol beads in human neutrophils. From the results they stated luminol beads can be effective as a reagent to activate the oxidative burst of neutrophils, and in turn the activated neutrophils can be used to detect the activity of nonsteroidal anti-inflammatory drugs.

22.8 CONCLUSION

The authors suggested that luminol-bound microspheres and ABEI-ms are suitable CL probes for measuring MPO activity in phagocytizing cells and AE-ms can be an excellent probe for superoxide anions and hydrogen peroxide.

It should be noted, however, that nitric monoxide (NO) is reported to react with hydrogen peroxide to form peroxynitrite which oxidizes luminol, producing CL.[58] Kikuchi et al.[59] applied this luminol-dependent CL to the real-time measurement of NO produced *ex vivo*. Van Dyke et al.[60] measured the antioxidant effect of phenolic substances in the oxidation system including luminol and peroxynitrite. Since nitrogen oxides are formed in phagocytes, reactions of the CL microspheres with them should be investigated.

Further, it should be added that Fäldt et al.[61] recently reported that luminol, but not isoluminol, can act as an inhibitor of neutrophil NADPH-oxidase activity. As mentioned above, the author has assumed that there is no significant difference between luminol and isoluminol except for sensitivity. Since ABEI is a derivative of isoluminol, any differences between them should be investigated.

Another topic for research is the preparation of CLA-bound microspheres.

REFERENCES

1. Allen, R. C., Stijernholm, R. J., and Steele, R. H., Evidence for the generation of electronic excitation state(s) in human polymorphonuclear leukocytes and its participation in bactericidal activity, *Biochem. Biophys. Res. Commun.,* 47, 679, 1972.
2. Allen, R. C., Yevich, S. J., Orth, R. W., and Steele, R. H., The superoxide anion and singlet molecular oxygen: their role in microbicidal activity of the polymorphonuclear leukocytes, *Biochem. Biophys. Res. Commun.,* 60, 909, 1974.
3. Sagone, A. R., Jr., King, G. W., and Metz, E. N., A composition of metabolic response to phagocytosis in human granulocytes and monocytes, *J. Clin. Invest.,* 57, 1352, 1975.
4. Clebanoff, S. K., Durack, D. T., Rose, H., and Clark, R. A., Functional studies on human peritoneal eosinophils, *Infect. Immun.,* 17, 167, 1977.
5. Miles, P. R., Lee, P., Trush, M. A., and Van Dyke, K., Chemiluminescence associated with phagocytosis of foreign particles in rabbit alveolar macrophages, *Life Sci.,* 20, 165, 1977.
6. Cheson, B. D., Vhristensen, R. L., Sperling, R., Kohler, B. E., and Babior, B. M., The origin of the chemiluminescence of phagocytosing granulocytes, *J. Clin. Invest.,* 58, 789, 1976.
7. Ushijima, Y. and Nakano, M., Kinetic aspect of luminescence in activated leukocyte systems, *J. Appl. Biochem.,* 2, 138, 1980.
8. Khan, A. U. and Kasha, M., Chemiluminescence arising from simultaneous transitions in pairs of singlet oxygen molecules, *J. Am. Chem. Soc.,* 92, 3293, 1970.
9. Cadenas, E., Daniele, R. P., and Chance, B., Low level chemiluminescence of alveolar macrophages, *FEBS Lett.,* 123, 225, 1981.
10. Kanofsky, J. R., Wright, J., Miles-Richardson, G. E., and Tauber, A. I., Biochemical requirement for singlet oxygen production by purified human myeloperoxidase, *J. Clin. Invest.,* 74, 1489, 1984.
11. Vladimirov, Yu. A., Roshchupkin, D. I., and Fesenko, E. E., Photochemical reactions in aminoacid residues and inactivation of enzymes during U.V.-irradiation. A review, *Photochem. Photobiol.,* 11, 227, 1970.
12. Takahashi, A., Totsune-Nakano, H., Nakano, M., Mashiko, S., Suzuki, N., Ohma, C., and Inaba, H., Generation of O_2^- and tyrosine cation-mediated chemiluminescence during the fertilization of sea urchin eggs, *FEBS Lett.,* 246, 117, 1989.
13. Ushijima, Y., Nakano, M., Takyu, C., and Inaba, H., Chemiluminescence in L-tyrosine-H_2O_2-horseradish peroxidase system: possible formation of tyrosine cation radical, *Biochem. Biophys. Res. Commun.,* 128, 936, 1985.
14. Kobayashi, S., Sugioka, K., Nakano, M., Takyu, C., Yamagishi, A., and Inaba, H., Excitation of indole acetate in myeloperoxidase-hydrogen peroxide H_2O_2 system: possible formation of indole acetate cation radical, *Biochim. Biophys. Res. Commun.,* 93, 967, 1980.
15. Allen, R. C. and Loose, D., Phagocytic activation of a luminol-dependent chemiluminescence in rabbit alveolar and peritoneal macrophages, *Biochem. Biophys. Res. Commun.,* 69, 245, 1976.
16. Williams, A. J. and Cole, P. J., The onset of polymorphonuclear leukocyte membrane-stimulated metabolic activity, *Immunology,* 43, 733, 1981.
17. Seim, S., Role of myeloperoxidase in the luminol-dependent chemiluminescence response of phagocytosing human monocytes, *Acta Pathol. Microbiol. Immunol. Scand. Sect. C,* 91, 123, 1983.

18. Sugioka, K., Nakano, M., Kurashige, S., Akuzawa, Y., and Goto, T., A chemiluminescent probe with a Cypridina luciferin analogue, 2-methyl-6-phenyl-3,7-dihydroimidazo[1,2-a]pyrazin-3-one, specific and sensitive for O_2^- production in phagocytizing macrophages, *FEBS Lett.,* 197, 27, 1986.
19. Nakano, M., Sugioka, K., Ushijima, Y., and Goto, T., Chemiluminescence probe with Cypridina luciferin analogue, 2-methyl-6-phenyl-3,7-dihydroimidazo[1,2-a]pyrazin-3-one for estimating the ability of human granulocytes to generate O_2^-, *Anal. Biochem.,* 159, 363, 1986.
20. Uchida, T., Kanno, T., and Hosaka, S., Direct measurement of phagosomal reactive oxygen by luminol-binding microspheres, *J. Immunol. Methods,* 77, 55, 1985.
21. Uchida, T. and Hosaka, S., Direct measurement of reactive oxygen by microsphere-bound luminol, in *Macrophage Biology,* Richard, S. and Kojima, M., Eds., Alan R. Liss, New York, 1985, 545.
22. Malawista, S. E., Gee, J. B. L., and Bensch, K. G., Cytochalasin B reversibly inhibits phagocytosis: functional, metabolic, and ultrastructural effects in human blood leukocytes and rabbit alveolar macrophages, *Yale J. Biol. Med.,* 44, 286, 1971.
23. Goldstein, I. M., Roos, D., Kaplon, A. B., and Weisman, G., Complement and immunoglobulins stimulate superoxide production by human leukocytes independently of phagocytosis, *J. Clin. Invest.,* 56, 1155, 1975.
24. Root, R. K. and Metcalf, J. A., H_2O_2 release from human granulocytes during phagocytosis, *J. Clin. Invest.,* 60, 1266, 1977.
25. Uchida, T., Masuko, S., Hosaka, S., and Tanzawa, H., The application of luminol-bound microspheres for quantitative analysis of toxic oxygen within phagosomes, *J. Bioactive Compatible Polym.,* 1, 172, 1986.
26. Yamamoto, H., Mori, H., Okano, H., Sassa, R., and Uchida, T., The effect of alcohol on the activity of macrophage: assessment of macrophage activity by determination of intracellular reactive oxygen, *Jpn. J. Alcohol Stud. Drugs Dependence,* 21, 183, 1986.
27. Uchida, T., Sasada, M., Hosaka, S., and Kubo, A., Suppressive substance for the function of phagocytes in the serum of leukemic patients, *Acta Haematol. Jpn.,* 51, 759, 1988.
28. Jacquemin, P. C. and Strijickmans, P., Detection of a retrovirus–related glycoprotein in immune complexes from patients with hematopoietic disorders, *Int. J. Cancer,* 36, 535, 1985.
29. Masuko, S., Hosaka, S., Uchida, T., and Tanzawa, H., Measurement of reactive oxygen produced through the contact of biomedical materials and human neutrophils, *Artif. Organs Today,* 1, 245, 1991
30. Matsumoto, S. and Imanishi, H., Effect of interferon on activated oxygen production in human monocytes *in vitro*: direct measurement of phagosomal activated oxygens by a new method using luminol-binding microspheres, *Chem. Pharm. Bull.,* 34, 4775, 1986.
31. Takeuchi, A., Shimizu, A., Hashimoto, T., Uchida, T., Masuko, S., and Hosaka, S., Effect of neutrophil activating substances on cellular generation of phagocytic chemiluminescence by means of luminol-bound microspheres., *Int. J. Tissue React.,* 10, 169, 1988.
32. Suzuki, K., Uchida, T., Sakatani, T., Sasagawa, S., Hosaka, S., and Fujikura, T., A simple method for phagocytosis and lysosomal enzyme release of human polymorphonuclear leukocytes, using enzyme substrate-conjugated microspheres [in Japanese], *Igaku no Ayumi,* 132, 741, 1985.
33. Suzuki, K., Uchida, T., Sakatani, T., Sasagawa, S., Hosaka, S., and Fujikura, T., Measurement of active phagocytosis by polymorphonuclear leukocytes by fluorescence liberation from phagocytized microspheres, *J. Leukocyte Biol.,* 39, 475, 1986.
34. Uchida, T., Hosaka, S., and Miura, K., Direct measurement of phagolysosomal esterase activity, *Biochem. Biophys. Res. Commun.,* 127, 584, 1985,
35. Uchida, T., Suzuki, K., Hosaka, S., and Fujikura, T., Direct assay of phagolysosomal hydrolase by fluorogenic substrate-binding microspheres, in *Polymer in Medicine II,* Chiellini, E., Giusti, P., Migliaresi, C., and Nicolais, L., Eds., Plenum Press, New York, 1986, 149.
36. Suzuki, K., Hosaka, S., Morikawa, K., Suzuki, M., Suzuki, S., Miyoshi, M., Fujita, M., and Mizuno, M., Bioimaging of myeloperoxidase release into phagolysosome of viable polymorphonuclear leukocytes and reaction of the enzyme with microspheres using a highly sensitive analyzer system, IMRAS, *Bioimages,* 1, 13, 1993.
37. Yokoo, T., Hayashi, K., Yanai, M., and Tsuji, Y., Chemiluminescence assay of polymorphonuclear leukocytes using the luminol-binding microspheres [in Japanese], *Igaku no Ayumi,* 142, 348, 1987.
38. Kohashi, O., Kohashi, Y., Shobata, M., Uchida, T., Hosaka, S., Hashimoto, S., Mitsuyama, T., and Shigematsu, N., Lumisphere-induced phagocytosis and chemiluminescence of human polymorphonuclear leukocytes of normal adults and the patients with various respiratory diseases [in Japanese], *Jpn. J. Inflammation,* 8, 41, 1986.

39. Nishijima, J., Hiraoka, N., Murata, A., Oka, Y., Kitagawa, K., Tanaka, N., Toda, H., and Mori, T., Protease inhibitors (gabexate mesylate and urinastatin) stimulate intracellular chemiluminescence in human neutrophils, *J. Leukocytes Biol.,* 52, 262, 1992.
40. Hisadome, M., Fukuda, T., and Terasawa, M., Effect of cystein ethyl ester hydrochloride on host defense mechanisms, V. Potentiation of nitroblue tetrazolium reduction and chemiluminescence in macrophages or leukocytes of mice or rats, *Jpn. J. Pharmacol.,* 53, 57, 1990.
41. Aoki, M., Ono, Y., Kunii, O., and Goldstein, E., Effect of newer quinolones on the extra- and intracellular chemiluminescence response of human polymorphonuclear leukocytes, *J. Antimicrob. Chemother.,* 34, 383, 1994.
42. Ueda, U., Production of oxygen free radicals by phagocytes in patients on hemodialysis [in Japanese], *Kansenshogakuzasshi [J. Jpn. Assoc. Infect. Dis.],* 63, 997, 1989.
43. Katsuragi, H., Suzuki, A., Tomii, N., Hasegawa, A., and Saito, K., The lymphocyte subsets, specific serum antibodies and the function of polymorphonuclear leukocytes in 30–35 year old periodontal patients (in Japanese), *Nippon Shishubyo Gakkai Kaishi [J. Jpn. Assoc. Periodontol.],* 35, 661, 1993.
44. Nakano, M. and Ushijima, Y., Bactericidal activity and active oxygen species in neutrophils [in Japanese], *Jpn. J. Inflammation,* 4, 191, 1984.
45. Hosaka, S., Kuramitsu, Y., and Itagaki, T., Selectivity and sensitivity in the measurement of reactive oxygen species by use of chemiluminescent microspheres prepared by the binding of acridinium ester or ABEI to polymer microspheres, *Luminescence,* 14, 349, 1999
46. Nakano, M., Scavenging or quenching of reactive oxygen [in Japanese], in *Reactive Oxygen,* Niki, T. and Shimazaki, H., Eds., Ishiyaku Shuppan, Tokyo, 1987, 73.
47. Saito, I. and Matsuura, T., Physicochemical aspects of molecular oxygen and active oxygen species [in Japanese], in *Chemistry of Active Oxygen Species,* Chemical Society of Japan, Gakkai Shuppan Center, Tokyo, 1990, 7
48. Brestel, E. P., Mechanism of cellular chemiluminescence, in *Cellular Chemiluminescence,* Van Dyke, K. and Castranova, V., Eds., CRC Press, Boca Raton, FL, 1987, 93.
49. Oosthuizen, M. M. J., Engelbrecht, M. E., Lambrechts, H., Greyling, D., and Levy, R. D., The effect of pH on chemiluminescence of different probes exposed to superoxide and singlet oxygen generators, *J. Biolumin. Chemilumin.,* 12, 277, 1997.
50. Brestel, E. P., Co-oxidation of luminol by hypochlorite and hydrogen peroxide: implications for neutrophil chemiluminescence, *Biochem. Biophys. Res. Commun.,* 126, 482, 1985.
51. Greenlee, L., Fridovich, I., and Handler, P., Chemiluminescence induced by operation of iron-flavoproteins, *Biochemistry,* 1, 779, 1962.
52. Kamiya, I., in *Chemiluminescence* [in Japanese], Kodansha, Tokyo, 1972, 77.
53. Goto, T. and Takagi, T., Chemiluminescence of a Cypridina luciferin analogue, 2-methyl-6-phenyl-3,7-dihydroimidazo[1,2-a]pyrazin-3-one, in the presence of the xanthine-xanthine oxidase system, *Bull. Chem. Soc. Jpn.,* 53, 833, 1980.
54. Sugioka, K., Sawada, H., and Nakano, M., Generation of singlet molecular oxygen in peroxidase-H_2O_2 system with KBr and linoleic acid hydroperoxide-Fe^{2+} system, in *Medical and Biochemical and Chemical Aspects of Free Radicals, Proc. 4th Biennial General Meeting of Soc. Free Radical Res.,* 1998, 899.
55. Nakano, M., Singlet oxygen and luminescence [in Japanese], in *Active Oxygen & Chemiluminescence,* Nakano, M. and Yoshikawa, T., Eds., Nihon Igakukan, Tokyo, 1990, 47.
56. Cech, P. and Leher, R. I., Phagolysosomal pH of human neutrophils, *Blood,* 63, 88, 1984.
57. Van Dyke, K., Allender, P., Wu, L., Gutierrez, J., Garcia, J., Ardekani, A., and Karo, W., Luminol- or lucigenin-coated micropolystyrene beads, a single reagent to study opsonin-independent phagocytosis by cellular chemiluminescence with human neutrophils, monocytes, and differentiated HL60 cells, *Microchem. J.,* 41, 196, 1990.
58. Kikuchi, K., Nagano, T., Hayakawa, H., Hirata, Y., and Hirobe, M., Detection of nitric oxide production from a perfused organ by a luminol-H_2O_2 system, *Anal. Chem.,* 65, 1794, 1993.
59. Kikuchi, K., Nagano, T., Hayakawa, H., Hirata, Y., and Hirobe, M., Real time measurement of nitric oxide produced *ex vivo* by luminol-H_2O_2 chemiluminescence method, *J. Biol. Chem.,* 268, 23,106, 1993.
60. Van Dyke, K., Sacks, M., and Qazi, N., A new screening method to detect water-soluble antioxidants: acetaminophen (tylenol) and other phenols react as antioxidants and destroy peroxynitrite-based luminol-dependent chemiluminescence, *J. Biolumin. Chemilumin.,* 13, 339, 1998.
61. Fäldt, J., Ridell, M., Karlsson, A., and Dahlgren, C., The phagocyte chemiluminescence paradox: luminol can act as an inhibitor of neutrophil NADPH-oxidase activity, *Luminescence,* 14, 153, 1999.

23 The Effect of Antibiotics on Phagocytic Function of Granulocytes

Paul Hengster, Thomas Eberl, Franz Allerberger, Marialuise Kunc, Manfred P. Dierich, and Raimund Margreiter

CONTENTS

23.1 INTRODUCTION

Polymorphonuclear neutrophil granulocytes form the first line of defense against bacterial invaders. In the clinical setting these cells are usually supported by antibiotics, which are known to exert a powerful inhibitory or even lytic effect on bacteria. Some antibiotics do not enter phagocytes, whereas others are able to enter phagocytes and even accumulate, thereby supporting intracellular killing[1] When the concentration of the antibiotic in the extracellular space is below intracellular levels, the antibiotic may leave the phagocyte again. The presence of antibiotics may therefore alter phagocyte function.[2–10] It has been shown that they may influence the expression of membrane receptors.[11] Furthermore, one of the cephalosporins, cefodizime, is able to stimulate granulocyte movement efficiently.[12,13]

Macrophages and granulocytes release oxygen free radicals following phagocytosis, a process known as oxidative burst.[14,15] Under abnormal conditions, such as in chronic granulomatous disease, granulocytes may be able to phagocytize but are not effective in killing the phagocytized bacteria because of a decreased ability or lack of ability to produce reactive oxygen species.[16] The release of reactive oxygen species is of particular interest because this mechanism is one of the main host defense mech-anisms against bacteria, but may also cause tissue injury.[17] Phagocytosis is dependent on the degree of opsonization as well as the hydrophobia and surface charge of the bacteria.[12,18] The generation of a chemiluminescence (CL) signal during the process of phagocytosis is closely linked

0-8493-0719-8/02/$0.00+$1.50

to the bactericidal capacity of these cells.[19] Detection of a CL signal is a simple and easy method for the evaluation of granulocyte function. Use of microplates allows us to study the impact of various groups of antibiotics on granulocytes after stimulation with bacteria within a short time.[20]

23.2 MATERIAL AND METHODS

23.2.1 Sample Preparation

Freshly prepared buffy coats from volunteer donors were used as leukocyte source. In the procedure, 35 ml of buffy coat was layered on 15 ml Lymphoprep^{R} (Nycomed Pharma, Oslo, Norway) in a conical 50-ml tube (Falcon, Becton Dickinson Labware, Franklin Lakes, NJ) at room temperature and centrifuged at 1000 *g* for 20 min (Hettich Rotanda/TRC, Tuttlingen, Germany). The supernatant containing platelet and mononuclear cells was discarded and the pellet containing erythrocytes and granulocytes resuspended in 2% gelatin-PBS (phosphate-buffered saline) 0.2 g KCL, 8 g NaCl, 0.2 g KH_2PO_4, 1.42 g Na_2HPO_4 in 1 l of distilled water, at a pH of 7.4. Erythrocytes were seeded at 37°C for 45 min and the granulocyte-rich supernatant was removed. Cells were pelleted at 600 *g* and remaining erythrocytes lysed with 0.84% ammonium chloride at 37°C for 20 min. After washing twice in PBS, cells were counted (Neubauer counting chambers). Cell count was adjusted to 1×10^6/ml and 100 μl of cell suspension was seeded to each well of a white 96-well microplate (Nunc 437842, Roskilde, Denmark). This gives a final amount of 10^5 cells/well. Granulocytes were approximately 95% pure and more than 97% of granulocytes were viable as checked by trypan blue exclusion. Granulocytes were then incubated with various antibiotics in a humidified incubator at 5% CO_2 at 37°C for 3 h, 1.5 h, and 45 min in four different concentrations (Table 23.1): peak serum concentration, ten times peak serum concentration, and 0.1 and 0.01% of peak serum concentration.

23.2.2 Chemiluminescence Measurement

After adding of 100 μl luminol at a final concentration of 0.57 m*M*, the first measurement was performed to determine the baseline of cell stimulation. Part of the cells were stimulated with 0.86 m*M* PMA to establish the maximal CL signal by a chemically defined substance independent of opsonization of bacteria as an activation control.

For CL counting, white 96-well microplates (Nunc 437842), were used. Standard measurements were performed on triplicates at 20°C and pH 7.6 in a MicroBeta Plus™ (Wallac, Turku, Finland). Counting time per well was 2 s, the energy range between 0 and 2000 keV. Temperature was maintained constant by a temperature control unit (Wallac 1220 Temperature Control Unit, Turku, Finland). Results are expressed in luminescence counts per second (LCPS). For the experiments, cells were activated with opsonized *Staphylococcus aureus* and nonopsonized bacteria as a control. Stimulated cells and unstimulated controls were measured at intervals of 10 to 70 min.

23.2.3 Chemiluminescence Reagents

The luminescence buffer (LB) contained NaCl 0.14 *M*, KCl 2.7 m*M*, Na_2HPO_4 12 m*M*, KH_2PO_4 1.5 m*M*, Ca_2Cl_2 0.9 m*M*, $MgCl_2$ 0.49 m*M* at pH 7.6 (20°C). In the procedure, 1.77 mg of Luminol (Sigma) was dissolved in 1 ml dimethylsulfoxide (DMSO). This stock solution was diluted 1:100 in LB to give a final luminol concentration of 0.1 m*M*; 400 μl of this solution was added to each well containing 100 μl cell suspension. A 3-m*M* PMA stock solution in DMSO was diluted in LB 1:200, of which 50 μl was added to controls.

23.2.4 Bacteria

The experiments were performed with *S. aureus* ATCC 25923. For the experiments with antibiotics they were opsonized with homologous serum. Bacteria were used at a concentration of 10^5/ml. The bacteria solution was adjusted to an optical density of 1.0 at 595 nm (Beckmann DU 640,

TABLE 23.1
Concentrations of Substances Used

Name (Generic)	Conc. 1, mg/ml	Conc. 2, mg/ml	Conc. 3, mg/ml	Conc. 4, μg/ml
Penicillin G	12	1.2	0.12	12
Flucloxacillin	4	0.4	0.04	4
Ampicillin	4	0.4	0.04	4
Amoxicillin + clavulanic acid	4.4	0.44	0.044	4.4
Ampicillin + sulbactam	6	0.6	0.06	6
Mezlocillin	10	1	0.1	10
Piperacillin	8	0.8	0.08	8
Piperacillin + tazobactam	8	0.8	0.08	8
Cefotiam	4	0.4	0.04	4
Cefoperazon	4	0.4	0.04	4
Cefoxitin	4	0.4	0.04	4
Cefamandol	4	0.4	0.04	4
Cefotaxim	4	0.4	0.04	4
Cefuroxim	3	0.3	0.03	3
Ceftriaxon	4	0.4	0.04	4
Ceftazidim	4	0.4	0.04	4
Cefazolin	4	0.4	0.04	4
Gentamycin	0.72	0.072	0.0072	0.72
Tobramycin	0.72	0.072	0.0072	0.72
Amikacin	1	0.1	0.01	1
Netilmicin	0.72	0.072	0.0072	0.72
Erythromycin	4	0.4	0.04	4
Imipenem	1	0.1	0.01	1
Ciprofloxacin	0.4	0.04	0.004	0.4
Fleroxacin	0.8	0.08	0.008	0.8
Fluconazol	0.4	0.04	0.004	0.4
Clindamycin	1.2	0.12	0.012	1.2
Metronidazol	1	0.1	0.01	1
Fusidic acid	1	0.1	0.01	1
Chloramphenicol	4	0.4	0.04	4
Vancomycin	4	0.4	0.04	4
Fosfomycin	8	0.8	0.08	8

Fullerton, CA) in luminescence buffer. The optical density corresponds with 10^9 bacteria/ml when plated on agar plates. Bacteria were then opsonized with human serum for 30 min at 37°C. 50 μl of opsonized bacteria or nonopsonized bacteria (controls) were added to each well.

23.2.5 Antibiotics Tested

Table 23.1 contains the information about the antibiotics tested.

23.2.6 Statistical Methods

For analysis and graphics, the Excel program was used. All values were obtained in triplicate. Mean values are listed in all results. Statistical differences ($p = 0.05$) were determined by univariate variant analysis.

TABLE 23.2
Names and Origin of Substances Used

Name (Generic)	Product	Company	Location
Penicillin G	Penicillin G	Biochemie	Vienna, Austria
Flucloxacillin	Floxapen^{R}	Smith Kline Beecham	Heppignies, Belgium
Ampicillin	Standacillin^{R}	Tyrol Pharma	Kundl, Austria
Ampicillin + sulbactam	Unasyn^{R}	Pfizer	Latina, Italy
Amoxicillin + clavulanic acid	Augmentin^{R}	Smith Kline Beecham	Heppignies, Belgium
Mezlocillin	Baypen^{R}	Bayer	Leverkusen, Germany
Piperacillin	Pipril^{R}	Cyanamid	Wolfratshausen, Germany
Piperacillin + tazobactam	Tazonam^{R}	Cyanamid	Wolfratshausen, Germany
Cefazolin (1st gen.)	Gramaxin^{R}	Cyanamid	Wolfratshausen, Germany
Cefotiam (2nd gen.)	Spizef^{R}	Tyrol Pharma	Kundl, Austria
Cefoxitin (2nd gen.)	Mefoxitin^{R}	Merck Sharp and Dohme	Haarlen, the Netherlands
Cefamandol (2nd gen.)	Mandokef^{R}	Lilly	Giessen, Germany
Cefuroxim (2nd gen.)	Curocef^{R}	Glaxo Wellcome	Vienna, Austria
Ceftriaxon (3rd gen.)	Rocephin^{R}	Biochemie	Vienna, Austria
Ceftazidim (3rd gen.)	Fortum^{R}	Glaxo Wellcome	Vienna, Austria
Cefoperazon (3rd gen.)	Cefobid^{R}	Pfizer	Latina, Italy
Cefotaxim (3rd gen.)	Claforan^{R}	Usiphar Roussel Uclaf	Combiege, France
Gentamicin	Refobacin^{R}	Tyrol Pharma	Kundl, Austria
Netilmicin	Certomycin^{R}	Aesca	Traiskirchen, Austria
Tobramycin	Tobrasix^{R}	Lilly	Giessen, Germany
Amikacin	Biklin^{R}	Bristol Meyers Squibb	Latina, Italy
Erythromycin	Erythrocin^{R}	Abbott	St. Rémy-sur-Avre, France
Imipenem/cilastatin	Zienam^{R}	Merck Sharp and Dohme	Haarlen, the Netherlands
Ciprofloxacin	Ciproxin^{R}	Bayer	Leverkusen, Germany
Fleroxacin	Quinodis^{R}	Hoffmann La Roche	Basel, Switzerland
Fosfomycin	Fosfocin^{R}	Biochemie	Vienna, Austria
Clindamycin	Dalacin C^{R}	Pharmacia UpJohn	Crawley, England
Metronidazol	Metronidazol^{R}	Nycomed	Gedon Richter Chem. Fabrik, Budapest, Hungary
Fucidic acid	Fucidin^{R}	Leo Pharmac. Products	Ballerup, Denmark
Chloramphenicol	Paraxin^{R}	Biochemie	Vienna, Austria
Vancomycin	Vancomycin^{R}	Lilly	Giessen, Germany
Doxyclyclin	Vibravenös^{R}	Pfizer	Amboise, France
Fluconazol	Diflucan^{R}	Pfizer	Amboise, France

23.3 RESULTS

Prior to the testing of various antibiotics it was confirmed that the granulocytes did not change their ability to produce reactive oxygen species during incubation. No relevant difference in CL was found after stimulation of granulocytes with PMA even when tested 24 h later. The lowest number of granulocytes required to obtain reproducible results was 10^6/ml. The use of 100 μl of this cell suspension resulted in 10^5 cells/well. *Staphylococcus aureus* used for granulocyte stimulation released reactive oxygen species only after opsonization. For the test system used, 10^5 turned out to be the optimal number of bacteria.

No CL signal above background of approximately 50 LCPS was generated by unstimulated granulocytes. Nonopsonized bacteria or luminol alone did not cause any reactive oxygen release from granulocytes. As expected, chemical stimulation with PMA caused an immediate rise of CL within seconds, whereas biological stimulation with opsonized bacteria increased CL counts after a delay of 3 min to reach a maximum at 20 min and to return to baseline after up to 4 h. Interestingly, in the presence of most antibiotics the effect of PMA on reactive oxygen release was enhanced up to 200 times. The addition of antibiotics alone increases the number of CL signals of resting granulocytes, which, however, did not exceed 200 LCPS.

Results obtained of antibiotic-treated cells stimulated by the addition of opsonized *S. aureus* are different from the observed inhibition of antibiotic substances on resting cells. In general, there was an increase of reactive oxygen species. However, there was no big difference in results in the antibiotic dilutions 2 to 4 resembling the normally reached or an even lower serum concentration. The influence on granulocytes at the highest concentration of some substances differed widely from the other concentrations.

Penicillins at most dilutions did not alter the release of reactive oxygen species from granulocytes when compared with untreated cells (Figure 23.1). Only at high concentrations was there as light decrease in CL signals detected. Most cephalosporins (Figure 23.2) had little effect on CL signals. Most cephalosporins showed an inhibition that was even more pronounced at high concentrations. Cefotaxim and cefoperazon, on the contrary, had a stimulatory effect at all concentrations. Each amino glycoside (Figure 23.3) showed a somewhat different behavior. Gentamicin was slightly stimulating while amikacin had no effect. With tobramycin, a strong inhibition at all concentrations was found. Certomycin at high concentration exerted a higher oxidative burst, whereas concentrations below the normally reached levels were inhibitory. Gyrase inhibitors, erythromycin, and imipenem (Figure 23.4) did not alter the reactive CL signal, except erythromycin at the highest concentration.

The strongest stimulatory effect was seen with clindamycin and chloramphenicol (Figure 23.5). Vancomycin and fosfomycin had no influence; only at the highest concentration was an inhibition found (Figure 23.5). Metronidazole was stimulating except at the highest concentration where a suppression was seen. Fucidic acid exerted a dose-dependent inhibition at all concentrations. No influence was observed with the antifungal substance fluconazol (Figure 23.5).

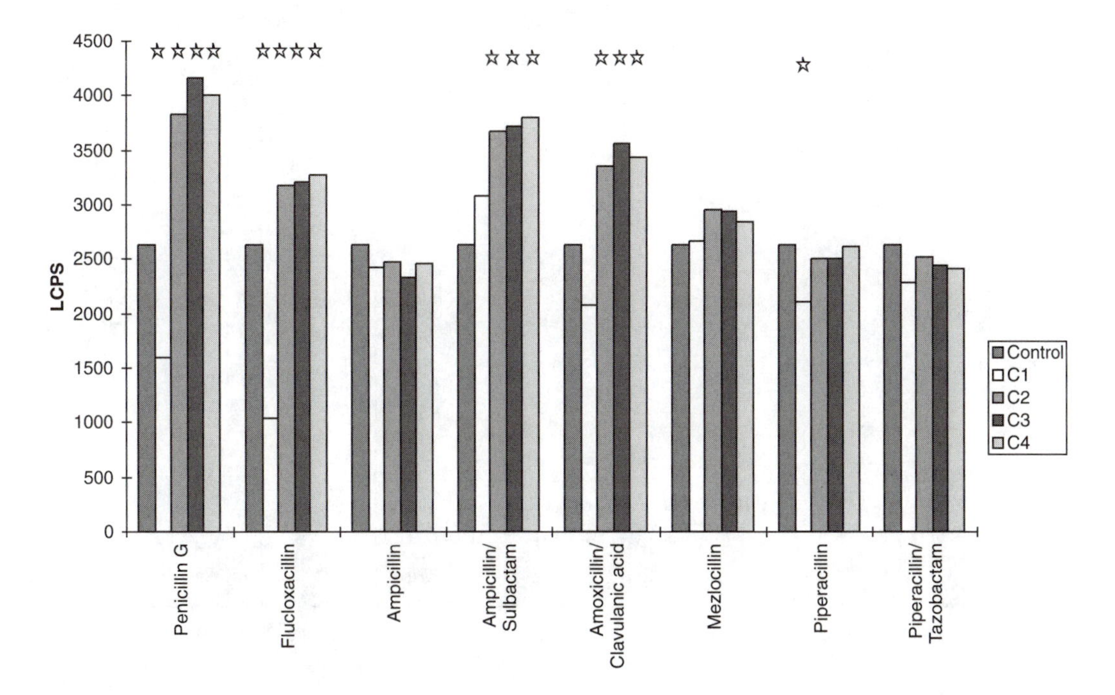

FIGURE 23.1 (Color figure follows p. 266.) Penicillins. C1, normal concentration × 10; C2, normal concentration; C3, normal concentration × 0.1; C4, normal concentration × 0.01; Control, without antibiotics. Significant changes against control marked with a star.

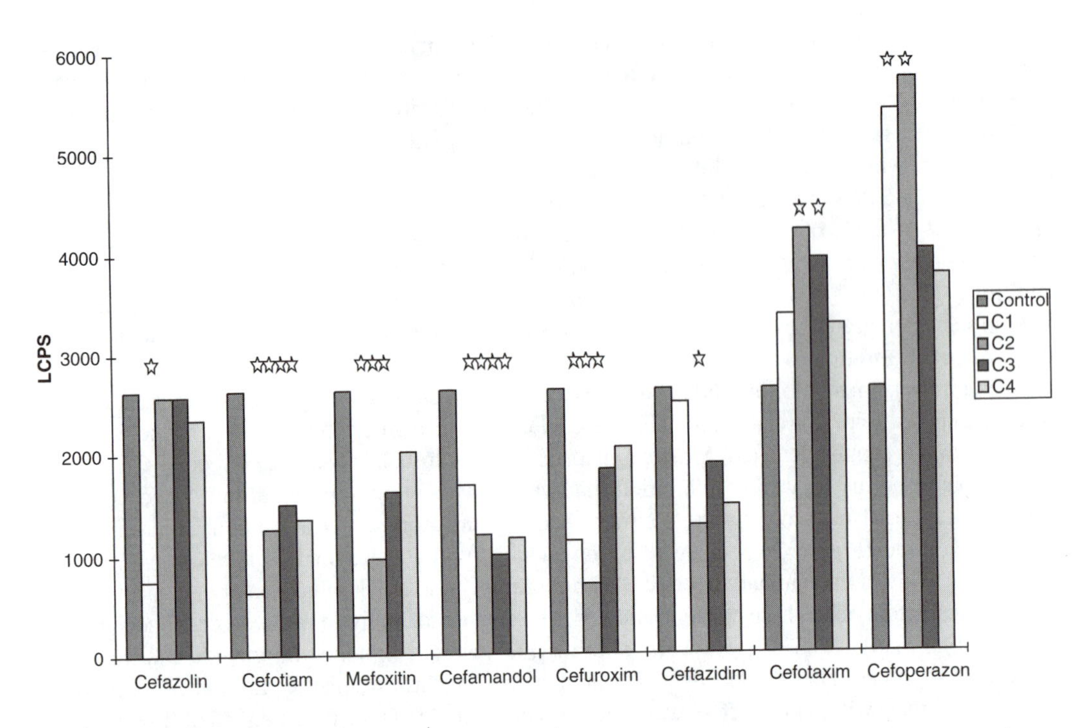

FIGURE 23.2 (Color figure follows p. 266.) Cephalosporins. C1, Normal concentration × 10; C2, normal concentration; C3, normal concentration × 0.1; C4, normal concentration × 0.01; Control, without antibiotics. Significant changes against control marked with a star.

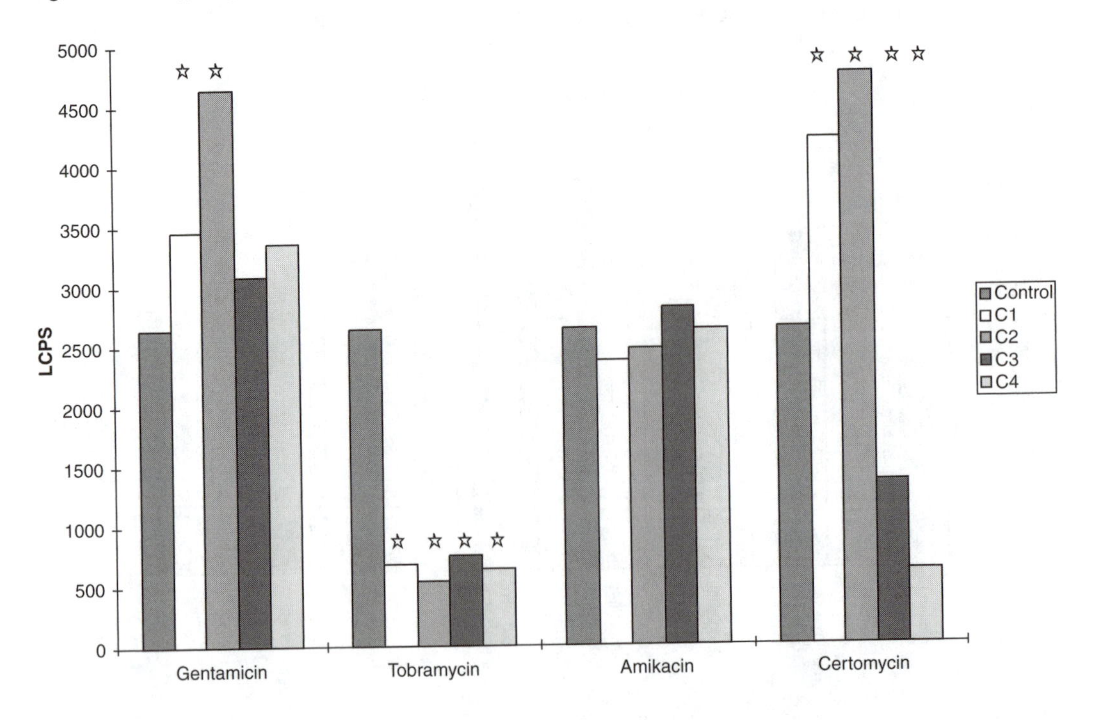

FIGURE 23.3 (Color figure follows p. 266.) Aminoglycosides. C1, normal concentration × 10; C2, normal concentration; C3, normal concentration × 0.1; C4, normal concentration × 0.01; Control, without antibiotics. Significant changes against control marked with a star.

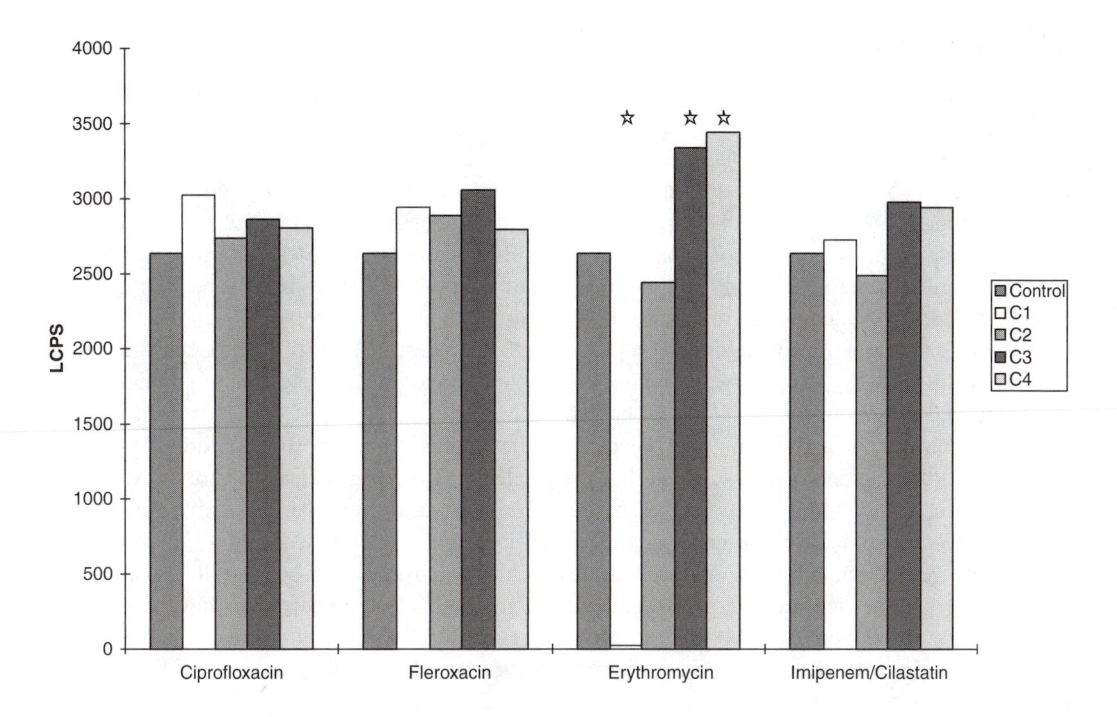

FIGURE 23.4 (Color figure follows p. 266.) Gyrase inhibitor, macrolides, carbapenems. C1, normal concentration × 10; C2, normal concentration; C3, normal concentration × 0.1; C4, normal concentration × 0.01; Control, without antibiotics. Significant changes against control marked with a star.

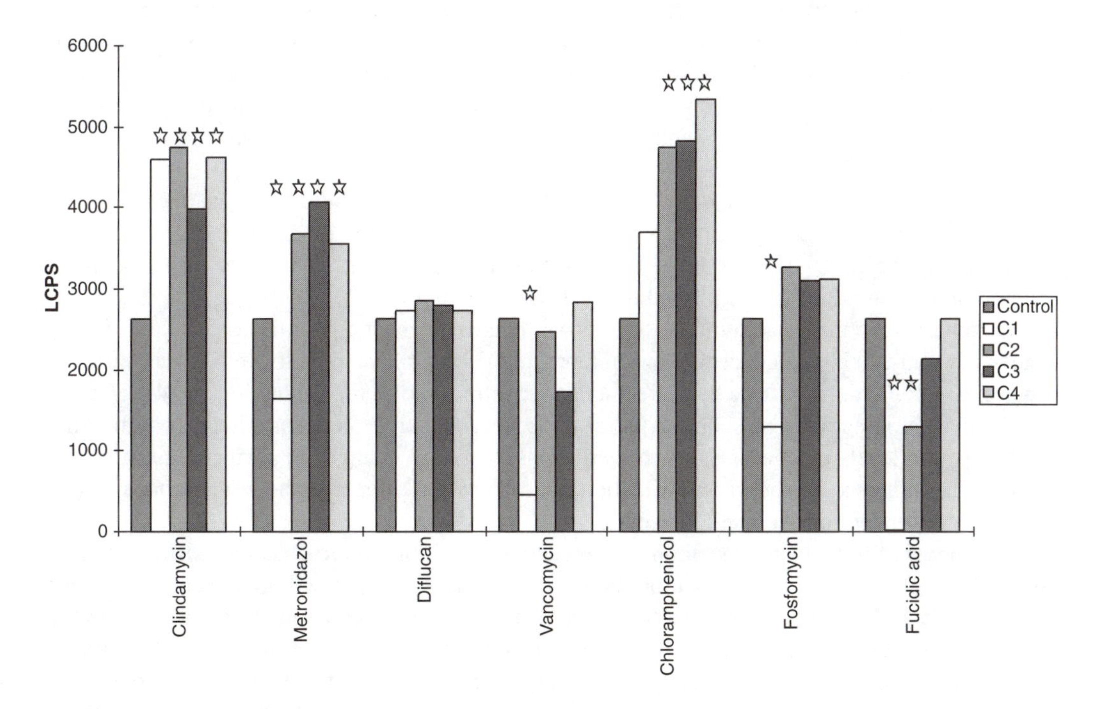

FIGURE 23.5 (Color figure follows p. 266.) Others. C1, normal concentration × 10; C2, normal concentration; C3, normal concentration × 0.1; C4, normal concentration × 0.01; Control, without antibiotics. Significant changes against control marked with a star.

23.4 DISCUSSION

Phagocytosis with subsequent intracellular digestion and the release of free oxygen radicals from these cells is a main defense mechanism against bacterial invaders. Mainly granulocytes are responsible for controlling bacterial infections, for example, with *S. aureus*, which led to our model. Phagocytosis of *S. aureus* is opsonin dependent and may be impaired by the presence of protein A on its surface.[21] A special mechanism for altered phagocytosis may be the neutralization of the bacterial surface charge.[18] Luminol-enhanced CL is a efficient method for detecting reactive oxygen release. Luminol enters the cell, and therefore results show the total intra- and extracellular respiratory burst. To exclude the impact of the varying radical production by cells at different time points all our investigations were done comparatively. The lowest number of granulocytes to obtain comparable results was 10^5 cells/well. With a lower number, the limit of detection would have been surpassed and usage of more cells would have been impracticable because of the amount of cells within one buffy coat and the number needed for the tests. In our experiments, only opsonized bacteria stimulated granulocytes sufficiently for the release of oxygen radicals. There are some CL data available for polymorphonuclear cells, but our investigation compared a large number of antibiotic substances within one test system. In addition, because we first added all the antibiotic substances, we maintained the same concentration of these substances throughout the experiment. Opsonized bacteria were then added and alterations of the bacteria occurred within the reaction vessel. This must be mentioned because all previous reports used antibiotic-pretreated bacteria. For example, washing of cells may liberate the antibiotics in a rapid fashion and preincubation may leave fragments of the prokaryotic cells.[22]

Antibiotics used for treatment of these infections may reach the interior of phagocytes in varying amounts and may influence the function of phagocytes.[23] In fact, little is known about the influence of these substances on the unspecific immune system. Antibiotics themselves may alter the release of radicals from polymorphonuclear granulocytes. But we found that antibiotics themselves without the presence of bacteria alter CL measurement only moderately.[12] For rifampicin, tetracycline, and trimethoprim-sulfamethoxazole an inhibitory action was previously noted.[24] But, conversely, stimulatory effects for cefotaxim and clindamycin and also trimethoprim-sulfamethoxazole were also reported.[25] One report showed an inhibition by ampicillin, cephalothin, cephalexin, tetracycline, doxycycline, gentamicin, and trimethoprime.[5]

In the presence of *S. aureus* most β-lactams have little influence on CL; nevertheless, at the highest concentration used and in the presence of bacteria inhibition on CL is seen.[26] One report showed a significant O_2 release in cefotaxime but not cefodizime in another test system, an observation that could not be confirmed in our test system.[27]

Macrolides at normal concentration were shown to have no effect on CL release, and this was the same in our study. At higher concentrations of more than 0.1 mg/ml an inhibitory effect was reported; however, we could demonstrate this effect at a concentration of 4 mg/ml but not at 0.4 mg/ml.[23,28]

The most striking influences on the CL of phagocytes occurs at concentrations far higher than the therapeutic levels reached *in vivo*. Nevertheless, there are changes at normal concentrations *in vitro*. The influence on the *in vivo* situation is not known. Differences between various studies are mainly dependent on the method used.

The release of free oxygen metabolites is essential for killing bacteria but also destroys healthy tissue. Although in chronic inflammation the release should be suppressed, in an acute situation it is more likely to have a beneficial effect. In our setting, it must be assumed that a reduction of release by an antibiotic substance will help the bacteria to survive while an increased liberation helps to eliminate the invader more rapidly. Because granulocytes are short-lived cells present in a very high number, an exhaustive stimulation must not be assumed to be of interest for the macroorganism. The dosage of antibiotics can be determined either by the minimal inhibitory concentration or by the stimulatory or inhibitory concentration for phagocytic cells. There are no studies so far that clearly show an influence of the antibiotic substance used and the outcome of severe infections.

REFERENCES

1. Prokesch, R. C. and Hand, W. L, Antibiotic entry into human polymorphonuclear leukocytes, *Antimicrob. Agents Chemother.,* 21, 373–380, 1982.
2. Oxford, J., Immunomodulating effects of antimicrobial agents, *J. Antimicrob. Chemother.,* 6, 691–699, 1980.
3. Forsgren, A. and Schmeling, D., Effects of antibiotics on chemotaxis of human leucocytes, *Antimicrob. Agents Chemother.,* 11, 580–584, 1977.
4. Milatovic, D., Antibiotics and phagocytosis, *Eur. J. Clin. Microbiol.,* 2, 414–425, 1983.
5. Welch, W. D., Davis, D., and Thrupp, L. D., Effect of antimicrobial agents on human polymorphonuclear leukocyte microbicidal function, *Antimicrob. Agents Chemother.,* 20, 15–20, 1981.
6. Horwitz, M. A., Phagocytosis of microorganisms, *Rev. Infect. Dis.,* 4, 104–123, 1982.
7. Briheim, G. and Dahlgren, C., Influence of antibiotics on formylmethioninyl-leucyl-phenylalanine-induced leucocyte chemiluminescence, *Antimicrob. Agents Chemother.,* 31, 763–767, 1987.
8. Mandell, L. A., Effect of antimicrobial and antineoplastic drugs on the phagocytic and microbicidal function of the polymorphonuclear leucocyte, *Rev. Infect. Dis.,* 4, 683–697, 1983.
9. Milatovic, D., Antibiotics and phagocytosis, *Eur. J. Clin. Microbiol.,* 2, 414–425, 1997.
10. Daschner, F. D., Antibiotics and host defense with special reference to phagocytosis by human polymorphonuclear leukocytes. *J. Antimicrob. Chemother.,* 16, 135–141, 1985.
11. Hauser, W. E. and Remington, J. S., Effect of antibiotics on the immune response, *Am. J. Med.,* 72, 711–716, 1992.
12. Muratsugu, M., Tomonaga, M., Miyake, Y., Terayama, K., and Ishida, N., Electrophoretic mobility of cefodizime-treated *Staphylococcus aureus* and chemiluminescence of human polymorphonuclear leucocytes, *J. Antimicrob. Chemother.,* 28, 887–896, 1991.
13. Labro, M. T., Cefodizime as a biological response modifier: a review of its *in vivo, ex vivo* and *in vitro* immunomodulatory properties, *J. Antimicrob. Chemother.,* 26, 37–47, 1990.
14. Tengler, R. S., Furukawa, K., de Weck, A. L., and Maly, F. E., Chemiluminescence of mononuclear cells is enhanced during antigen recognition, *J. Biolumin. Chemilumin.,* 8, 159–167, 1993.
15. Trush, M. A., Wilson, M. E., and Van Dyke, K., The generation of chemiluminescence (CL) by phagocytic cells, *Methods Enzymol.,* 57, 462–494, 1978.
16. Roesler, J., Hockertz, S., Vogt, B., and Lohmann Matthes, M. L., Staphylococci surviving intracellularly in phagocytes from patients suffering from chronic granulomatous disease are killed *in vitro* by antibiotics encapsulated in liposomes, *J. Clin. Invest.,* 88, 1224–1229, 1991.
17. Babior, B. M., Oxidants from phagocytes: agents of defense and destruction, *Blood,* 64, 959–966, 1984.
18. Nomura, S., Kuroiwa, A., and Nagayama, A., Changes of surface hydrophobicity and charge of *Staphylococcus aureus* treated with sub-MIC of antibiotics and their effects on the chemiluminescence response of phagocytic cells, *Chemotherapy,* 41, 77–81, 1995.
19. Schroeder, H. R., Boguslaski, R. C., Carrico, R. J., and Buckler, R. T., Monitoring specific protein-binding reactions with chemiluminescence, *Methods Enzymol.,* 57, 424–445, 1996.
20. Hengster, P., Kunc, M., Linke, R., Eberl, T., Steurer, W., Öfner, D., Berthold, F., and Margreiter, R., Optimization of phagocyte chemiluminescence measurements using microplates and vials, *Luminescence,* 14, 91–98, 1999.
21. Peterson, P. K., Verhoef, J., Sabath, L. D., and Quie, P. G., Effect of protein A on staphylococcal opsonization, *Infect. Immn.,* 15, 760–764, 1977.
22. Dette, G. A. and Knothe, H., Kinetics of erythromycin uptake and release by human lymphocytes and polymorphonuclear leucocytes, *J. Antimicrob. Chemother.,* 18, 73–82, 1986.
23. Labro, M. T., el Benna, J., and Babin Chevaye, C., Comparison of the *in vitro* effect of several macrolides on the oxidative burst of human neutrophils, *J. Antimicrob. Chemother.,* 24, 561–572, 1989.

24. Siegel, J. P. and Remington, J. S., Effect of antimicrobial agents on chemiluminescence of human polymorphonuclear leukocytes in response to phagocytosis, *J. Antimicrob. Chemother.,* 10, 505–515, 1982.
25. Oleske, J. M., de la Cruz, A., Ahdieh, H., Sorvino, D., La Braico, J., Cooper, R., Singh, R., Lin, R., and Minnefor, A., Effects of antibiotics on polymorphonuclear leukocyte chemiluminescence and chemotaxis, *J. Antimicrob. Chemother.,* 12(Suppl. C), 35–38, 1983.
26. Labro, M. T., Immunomodulation by antibacterial agents. Is it clinically relevant? *Drugs,* 45, 319–328, 1993.
27. Labro, M. T., Babin-Chevaye, C., and Hakim, J., Effects of cefotaxime and cefodizime on human granulocyte functions *in vitro*, *J. Antimicrob. Chemother.,* 18, 233–237, 1986.
28. Dumas, R., Brouland, J. P., Tedone, R., and Descotes, J., Influence of macrolide antibiotics on the chemiluminescence of zymosan-activated human neutrophils, *Chemotherapy,* 36, 381–384, 1990.

24 *In Vitro* CMI: Rapid Assay for Measuring Cell-Mediated Immunity

Judith Britz, Peter Sottong, and Richard Kowalski

CONTENTS

24.1 INTRODUCTION

The cellular immune response to an infectious agent occurs rapidly following exposure to the organism. This activation triggers a cascade of important immunological events including the division of lymphocytes and other lymphoid cells. With many infectious organisms, particularly intracellular pathogens, the cellular immune response is critical to protective immunity. Measurement of cell-mediated immunity to a specific antigen can be useful in diagnosing infectious diseases, measuring hypersensitivity to certain agents, assessing exposure to immunologically reactive drugs, or testing vaccine efficacy. Measuring lymphoid cell responses to certain stimulants can assess the overall immunological competence of lymphocytes, the fundamental cellular defects or effects of chronic disease, and general or specific immunodeficiencies. Despite its importance, the standardized measurement of cell-mediated immunity using either *in vivo* or *in vitro* methods has been difficult. Cell-mediated immunity was originally defined *in vivo* by the ability to adoptively transfer protective immunity to a disease with lymphocytes rather than serum. Once successfully transferred to the host, skin testing has been used to test for the presence of cell-mediated immunity. However, interpretation of skin test results as edematous, erythematous reactions vs. indurated are highly subjective and also require the patient to return 24 to 48 h after testing.

0-8493-0719-8/02/$0.00+$1.50

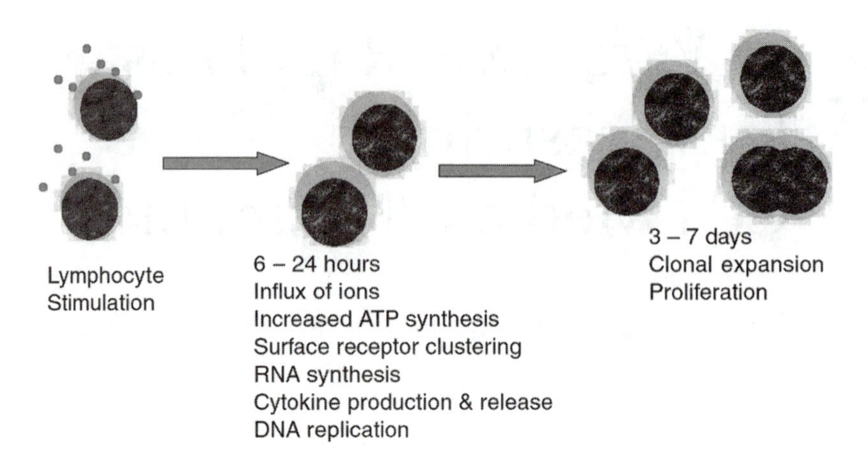

FIGURE 24.1 (Color figure follows p. 266.) Lymphocyte response.

Lymphocytes stimulated by mitogens or antigens divide in response to a series of activation events. These include clustering of cell surface receptors, increased uptake of metabolites, calcium ion fluxes, increased turnover of phospholipids, and increases in intracellular ATP levels. These activation events precede cytokine synthesis and proliferation but are also expected to be earlier indicators of foreign recognition (Figure 24.1).

The most prevalent *in vitro* method in use for the last 25 years has been to measure cellular proliferation of cultured lymphocytes. Lymphocytes in peripheral blood express receptors on the cell surface that bind with mitogens or specific antigens presented in conjunction with major histocompatibility antigens. Exposure to the antigen results in activation and expansion of the population of lymphocytes reactive to that antigen. A lymphoproliferation assay (LPA) first requires purification of peripheral blood mononuclear cells (PBMC) from whole blood and then sterile incubation for 5 to 6 days in the presence of exogenous serum and mitogens or antigens to stimulate division. On day 6, cells are pulsed with ^{3}H thymidine. Radioactive thymidine incorporation into DNA is then determined by harvesting the cells on filters and counting each filter in scintillant. Radioactive waste disposal is then required. Betensky et al.[1] recently reported that overnight transport of samples often lowered the reactivity of patient samples to recall antigens. The conclusion from this work is that samples could not be transported without compromise; therefore, laboratories need to perform the tests near the site of sample acquisition. This necessitates the implementation of methods that can be performed in 24 h or less, preferably without radioactivity.

Other *in vitro* methods currently used for measurement of immune function include enumeration of T cells and T-cell subsets, measurement of cytotoxic activity by radioactive chromium (^{51}Cr) release, and detection of intracellular cytokines by flow cytometry or their secretion by enzyme-linked immunosorbent assays (EIA/ELISPOT). In general, these methods require extensive manipulation of the cells, including separation of PBMCs from whole blood, long incubation times, and use of radioactive substances in some systems. For these reasons, many of these methods are not suitable for clinical applications.

Given the value of anticipating a change in a patient's clinical course based on immune function information, there is a need for a more rapid, standardized *in vitro* procedure that is adapted to the clinical laboratory. The *in vitro* CMI assay is described, which uses bioluminescent detection of ATP as an early indicator of lymphocyte activation. This assay is an alternative to classical lymphoproliferation techniques, and provides the first standardized, rapid, non-isotopic, and clinically accessible test for cell-mediated immunity.[2] The application of bioluminescence, a nonradioactive method of measuring the nucleotide adenosine triphosphate (ATP), for analysis of blood cells has been reviewed elsewhere.[3]

Bioluminescent techniques for measuring ATP in cells have also been similarly described to estimate biomass.[4,5] These methods showed improved sensitivity and reproducibility to other non-isotopic cellular techniques, such as using tetrazolium salts.[6] The validity of measuring ATP to calculate activation of proliferating and nonproliferating cells has also been demonstrated.[7] Measured luminescence also correlates with cell number,[8,9] and the degree of activation of PBMCs,[10] and thymocytes.[11] There is a linear relationship between cell concentration and ATP level in a sample. Estimates of cell concentration may be calculated if it is assumed that the ATP content per viable cell remains fairly constant in resting cells. In general, viable somatic cells contain about 1 pg (10^{-12} g) or approximately 2 fmol (2×10^{-15} mol) of ATP per cell.[5,12] ATP released by fewer than 10, or as many as 2×10^5 viable somatic cells (or a sample containing from 400 to 8×10^6 cells/ml) is measurable.[13] When certain cells are stimulated, the concentration of ATP per cell increases and provides an indicator of cellular activation.

ATP is synthesized within the respiratory cycle and is used as a basic energy source within cells. Changes in the ATP energy status of cells can be measured using bioluminescence systems consisting of the reaction of the substrate luciferin and the enzyme luciferase. For example, cell injury or oxygen substrate depletion results in a rapid decrease in cytoplasmic ATP, thereby providing an index of cell viability similar to trypan blue exclusion tests. Early changes in intracellular ATP levels of lymphocytes by mitogen stimulation are correlated to cell activation and subsequent cellular proliferation.[14] A decrease in ATP level soon after exposure to the plant mitogen phytohemagglutinin (PHA) can be transiently demonstrated, presumably because cell metabolism becomes more anaerobic initially.[15] Thereafter, large increases in the intracellular ATP level occur.

Increases in the concentration of intracellular ATP can measure immune activation as a function of stimulation by a variety of mitogens and specific antigens.[16] For most antigens or mitogens less than 24-h exposure is sufficient to measure activation in the *in vitro* CMI assay and significant activation can be measured within 4 to 6 h of exposure to strong mitogens such as PHA. The assay utilizes either whole blood or PBMCs, is non-isotopic, achieves results in less than 24 h, and has the additional advantage of identifying the specific T-cell subset involved in an immune response. The method involves incubating a dilution of whole blood or PBMCs with a mitogen or antigen, separating the desired subset of cells by means of monoclonal antibody–coated paramagnetic particles, washing the cells to remove any unbound cells or other interfering substances, lysis of the separated cells, and measurement of ATP (Figure 24.2). The amount of intracellular ATP is increased if the cells have responded to the stimulus (see Figure 24.1).

The *in vitro* CMI assay utilizes monoclonal antibodies attached to paramagnetic particles to facilitate separation of specific lymphocyte subsets. The utility of this approach has been previously demonstrated.[17,18] Separation is based on selective binding of the monoclonal antibody with the subset of interest and forming aggregates or complexes, which are removed from the blood sample by exposure to a strong magnet. Because the magnet is located at the side of a microtiter plate well, it is possible to separate and wash away nonspecific materials, leaving an enriched population of a specific cell subset.

The separated cell population is then lysed using a hypotonic detergent solution and the level of ATP in the solution is measured by the addition of firefly luciferase and luciferin in the presence of magnesium ions. The assay is based on the requirement of luciferase for ATP in producing light (emission maximum ~560 nm at pH 7.8) from the reaction:[19]

$$\text{Luciferin} + \text{ATP} + \text{O}_2 \xrightarrow[\text{Luciferase}]{\text{Mg}^{+2}} \text{oxyluciferin} + \text{AMP} + \text{pyrophosphate} + \text{CO}_2 + \text{light}$$

The amount of light produced by the reaction is read using a luminometer. Unlike older flash kinetics, the *in vitro* CMI assay reagent produces a rapidly rising glow of light that is sustained for about 15 min. An ATP calibrator is used to generate a standard curve to which all samples can

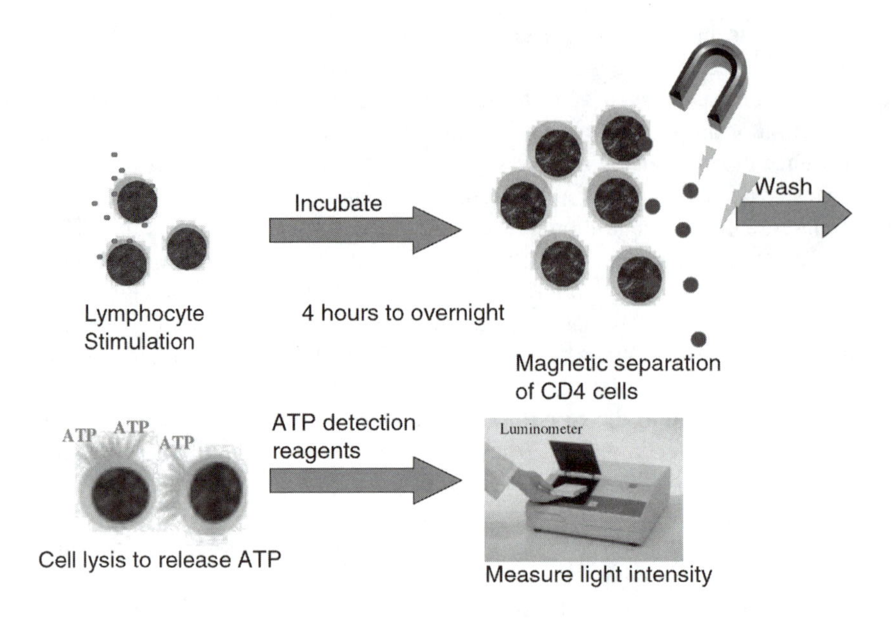

FIGURE 24.2 (Color figure follows p. 266.) *In vitro* CMI assay format.

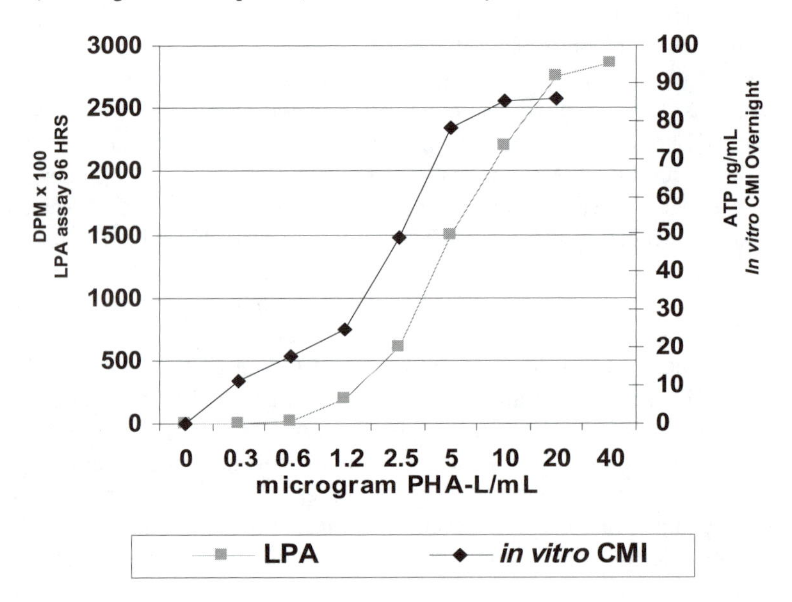

FIGURE 24.3 Comparison with LPA.

be referred. The direct luminometer output (expressed in relative light units, RLUs, or millivolts, mV) is converted to ATP concentration (nanograms per milliliter) to determine the amount of cellular stimulation that has occurred, allowing comparisons between tests and from day to day.

The assay combining all these principles was first described for measuring responses of T lymphocytes to PHA, interleukin-2 (IL-2), and recall antigens to assess T-cell responses to Q/Fever antigens in immunized mice at a relatively early time (7 days postimmunization).[20] *In vitro* CMI human T-cell responses to mitogens and recall antigens (tetanus toxoid) have been correlated with cell proliferation using whole blood.[16,21]

Figure 24.3 illustrates the correlation of dose responses of T lymphocytes to PHA using both the *in vitro* CMI assay and lymphocyte proliferation. Unlike the 96-h incubation required for assessing proliferation, the immune activation of CD4 T-helper cells was apparent after only 18 h. Since no radioactive materials are employed in the *in vitro* CMI assay, handling and waste disposal are simplified. Although the CD4 subset of T cells was exclusively measured in this study, other subsets could be as easily selected for measurement simultaneously, by using appropriate antibody-coated paramagnetic particles.

As immune reconstitution becomes more important in the assessment of disease management and as cytokine therapies become readily available, there will be an increasing need to measure cellular immune function more rapidly. The *in vitro* CMI assay system provides a faster, easier-to-use method of measuring cell activation in response to a variety of stimuli. It has clear applications in monitoring of infectious diseases like HIV and hepatitis, transplant acceptance, response to cancer therapy, and management of autoimmune disorders.

24.2 APPLICATIONS OF *IN VITRO* CMI

The range of potential applications for *in vitro* CMI is extremely broad given the fundamental importance of cellular immunity, the wide array of diseases and conditions in which it plays a critical role, and the absence of any alternative, broadly applicable analytical method suitable for routine clinical use in this area.

Infectious diseases develop frequently as a result of immune system failure. Antibiotic and antiviral therapies have been developed as "magic bullets" intended to kill the target organism without damage to the host cells. In the course of infection, however, perturbations of the immune response occur. Total recovery often requires immune reconstitution. HIV is the prototypical example.

24.2.1 HIV Disease Management: Monitoring Pathogenesis

HIV-infected patients are routinely monitored for viral load and CD4 lymphocyte counts. Without therapy, viral load levels can be expected to increase over time, concurrent with a decline in CD4 levels. Clerici et al.[22] found that up to 1 year before the decline in CD4 cells, the loss of various functional markers of immunity preceded the loss in T-cell number. These markers included *in vitro* reactivities to mitogens (e.g., PHA), recall antigens, and alloantigens. In collaboration with Farzadegan and colleagues,[23] responses to mitogens and a variety of recall antigens were tested in HIV-infected drug users and uninfected controls using the *in vitro* CMI assay. Figure 24.4 demonstrates that the

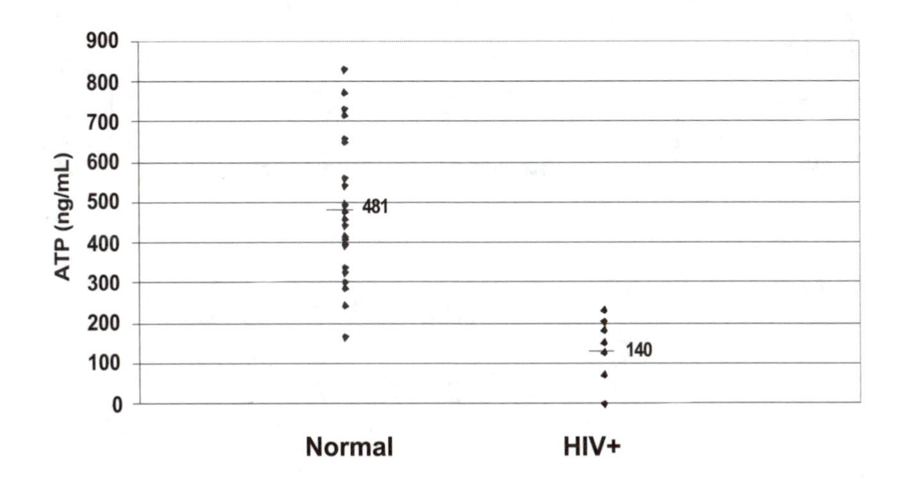

FIGURE 24.4 PHA distribution of normal and HIV-positive patients.

response to a strong nonspecific stimulator like PHA is dramatically suppressed in these HIV-infected untreated patients. Dolan et al.[24] further showed that low PHA reactivity in the lymphoproliferation assay was associated with poor clinical outcomes.

24.2.2 Monitoring Efficacy of Therapy

Patients receiving highly active antiretroviral therapy (HAART) frequently demonstrate increases in $CD4^+$ cells with declining viral load. Perrin and Telenti[25] showed in their cohort of HAART-treated adult patients that 45% showed an increase in $CD4^+$ cell levels concomitant with a decline in viral load. In a significant number of patients, however, there is discordance between the viral load and CD4 trends relative to the clinical status of the patient. In fact, at the 1999 Conference on Retroviruses and Opportunistic Infection,[26] it was reported that despite successful viral suppression, 9 to 15% of patients fail to demonstrate sufficient functional immunity to reduce their risk of opportunistic infections. Loechelt et al.[27,28] conducted longitudinal studies of pediatric patients with AIDS receiving HAART therapy at Children's Hospital in Washington, D.C. Figures 24.5 and 24.6 show the results of monitoring two patients from their larger study for $CD4^+$ cell levels, viral load, and the *in vitro* CMI response to PHA as a functional marker of immunity. The patient in Figure 24.5 was not progressing on a positive clinical course, thus resulting in a therapy change. The patient responded positively to the change in the course of treatment, which was concomitant with her improvement in functional immune response to PHA (see arrow). The patient in Figure 24.6 was healthy at the onset of the study but was noncompliant with his therapy. Within 3 months, his functional immune response to PHA was significantly declining and paralleled a clinical crisis. Neither patient showed any significant change in either their CD4 levels or viral load throughout the same 6-month period. Figure 24.7 illustrates the contribution of measuring cell-mediated immunity to the current parameters for managing patients with AIDS.

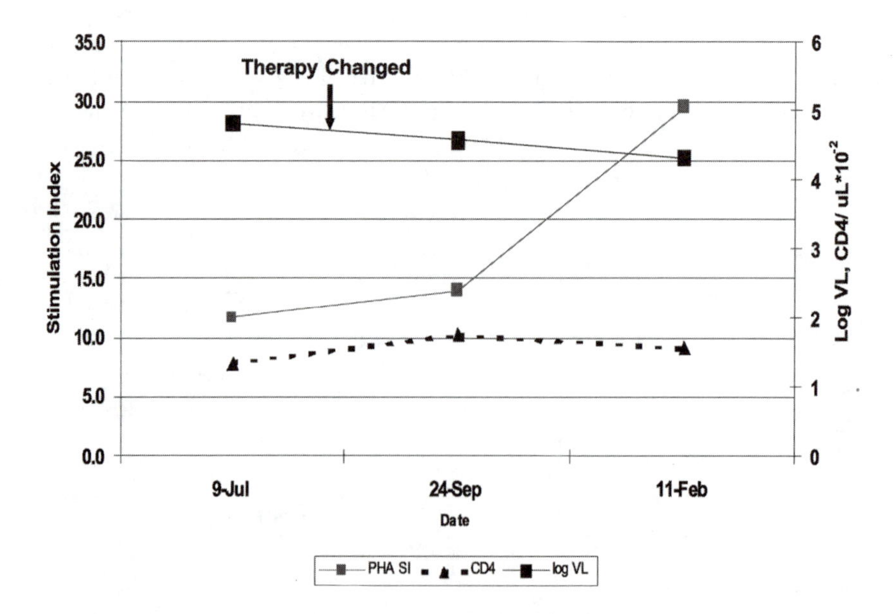

FIGURE 24.5 Monitoring of pediatric patients with HIV therapy change.

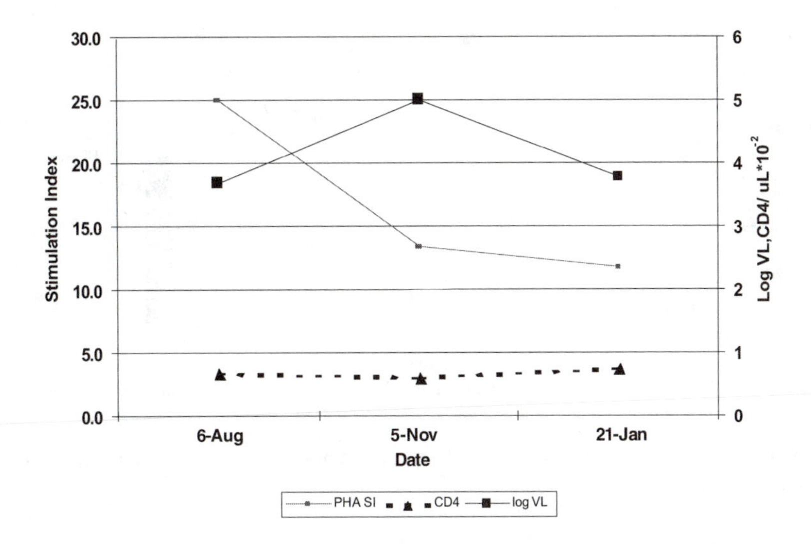

FIGURE 24.6 Monitoring of pediatric patients with HIV noncompliant.

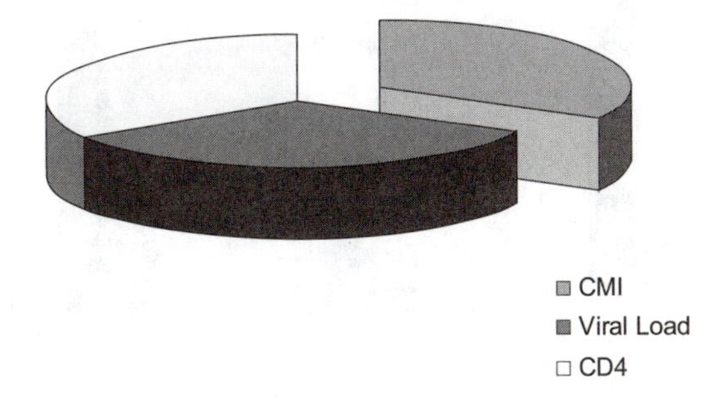

FIGURE 24.7 Management of patients with AIDS.

24.2.3 Immune Reconstitution

Rosenberg et al.[29] reported that patients with AIDS designated as long-term nonprogressors (LTNP) have measurable lymphoproliferative responses to HIV p24 antigen, whereas lymphocytes of the typical patient with AIDS are unresponsive to stimulation with p24. Cunningham-Rundles et al.[30] conducted a study of pediatric patients with AIDS receiving HAART and monitored their ability to respond in the *in vitro* CMI assay to recombinant p24 antigen. In Figure 24.8, children classified as responsive to therapy were able to mount a p24 response *in vitro* as measured by intracellular ATP accumulation. Children who were unresponsive to therapy did not respond to p24. More recently, Rosenberg et al.[31] demonstrated that patients acutely infected with HIV will retain their ability to respond in the lymphoproliferation assay to p24 antigen, if therapy is initiated early after infection. These same patients show an ability to mount even stronger p24 lymphoproliferative responses following multiple regimens of structured treatment interruption.

24.2.4 Response to Vaccination

Since most patients with AIDS lose their ability to respond to HIV-specific antigens throughout the course of disease, potential therapeutic approaches include active immunization of these patients with HIV antigens. Leandersson et al.[32] recently used the *in vitro* CMI assay together with lymphoproliferation, blast cell formation, and intracellular cytokine measurements to demonstrate that

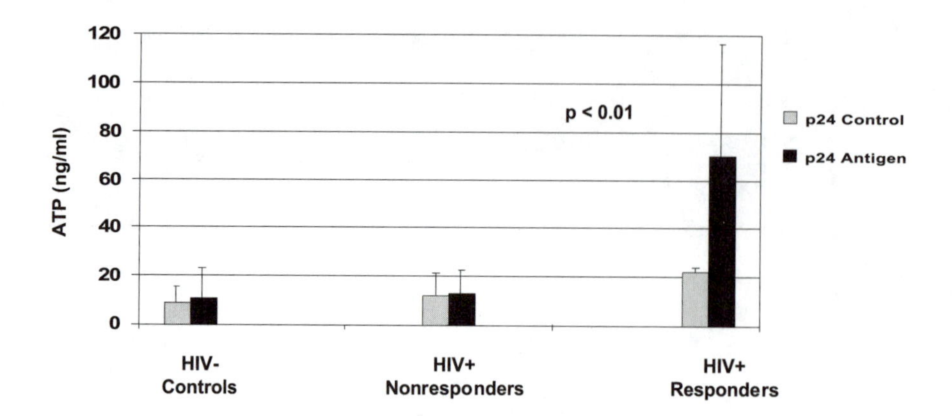

FIGURE 24.8 Comparison of treatment responders and nonresponders to p24 antigen in HIV infection compared with controls.

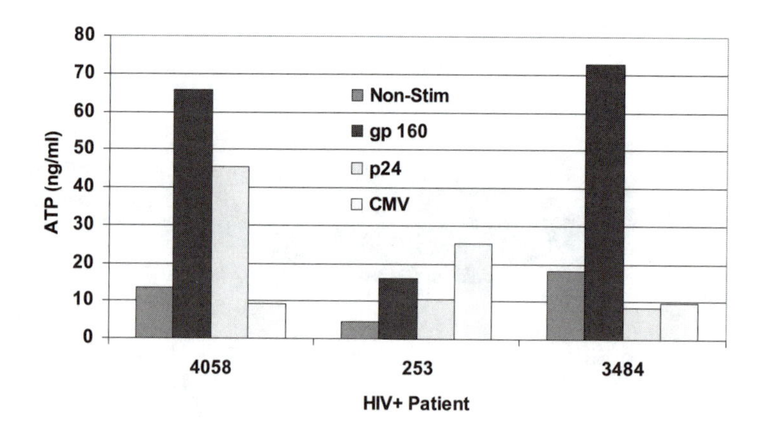

FIGURE 24.9 ATP response in gp160 vaccinated HIV-positive patients.

HIV-infected patients immunized with an HIV gp160 vaccine were able to respond following immunization. In Figure 24.9, three vaccinated patients produced immune reactivity of various levels to gp160, p24, or CMV antigens in the *in vitro* CMI assay. Similar results were obtained with each of the other assays mentioned above. Whether these immunizations generate protective immunity in patients is still under study. The assay has been shown to be useful in quantifying the immune response to vaccination with an influenza vaccine.[33]

24.2.5 Immune-Based Therapy: Screening Assays

In HIV and other infectious diseases, such as hepatitis C, antiviral therapies are increasingly being combined with immune response modifying therapeutics. IL-2 is currently in clinical trials (NIAID AIDS Clinical Trial Group) for the treatment of AIDS. Another use of the *in vitro* CMI assay could be for assessing the impact of immune response modifiers on the *in vitro* lymphocyte response to mitogens and antigens. IL-2 was coincubated with mitogens or antigens and the blood from HIV-infected and uninfected individuals overnight and tested for its impact on the ATP response. Since patients with AIDS are generally anergic to stimulation with antigens, the question was whether IL-2 (or other co-stimulators like IL-15, anti-CD28, anti-CD49d) could restore this reactivity. After testing eight healthy volunteers and ten HIV-infected patients the general observation was that IL-2 nonspecifically stimulated resting lymphocytes in both populations, but did not generally restore responses. There were two exceptions among the HIV-positive group: (1) one patient demonstrated

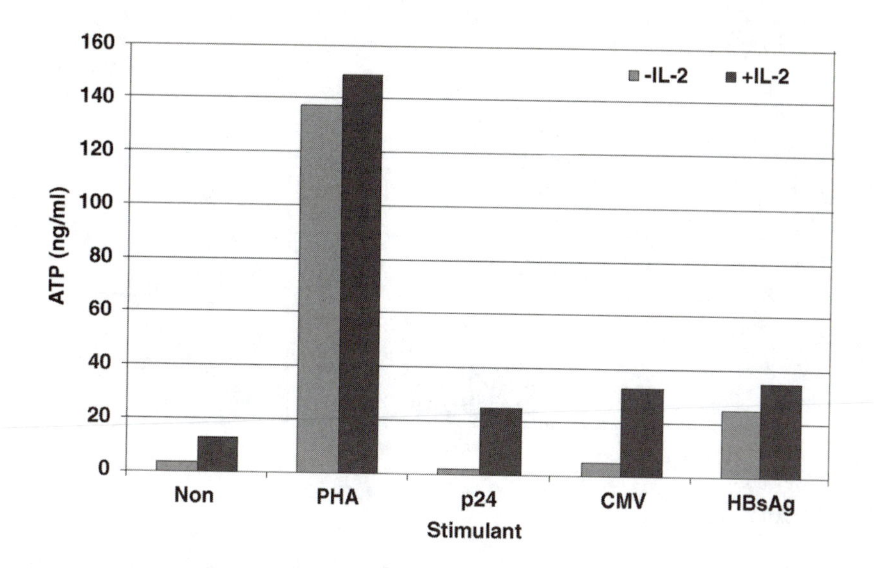

FIGURE 24.10 *In vitro* costimulation of immune response in a patient with HIV by addition of IL-2.

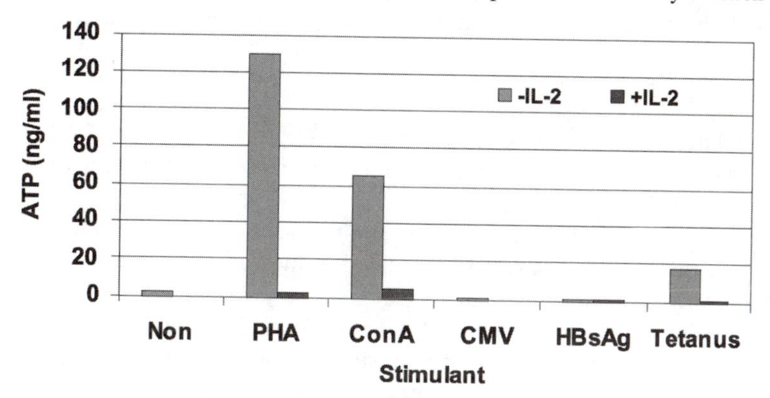

FIGURE 24.11 *In vitro* suppression of immune response in a patient with HIV by addition of IL-2.

enhanced recall responses to p24 and CMV in the presence of IL-2 (Figure 24.10); (2) more dramatically, one patient's response to mitogens as well as recall antigens was completely obliterated in the presence of IL-2 (Figure 24.11). These results suggest that individual differences in functional immune reactivity may be critical to determining the efficacy of any chosen therapy and emphasize the importance of monitoring an individual's response to therapy. These data also suggest that the *in vitro* CMI assay may be used in the development of new drugs as a screening method for assessing the impact of a proposed drug on immune system activity.

Although applications of the *in vitro* CMI assay have been described for disease management of HIV-infected patients, it provides merely a model for other infectious disease where immunodeficiency or immunostimulation may contribute to the pathogenesis of the disease.

24.2.6 Hepatitis C Virus

Over 200 million people are infected worldwide with hepatitis C virus (HCV). In the United States, there are 4 million chronic carriers of hepatitis C with 70,000 new cases each year. This disease, which is more prevalent than AIDS, develops chronic symptoms in 85% of infected individuals. Cirrhosis develops in 10 to 15% of infected carriers and hepatocellular carcinoma is the sequela in 1 to 5% of patients. Interferon alpha, a biopharmaceutical, is the preferred treatment because of

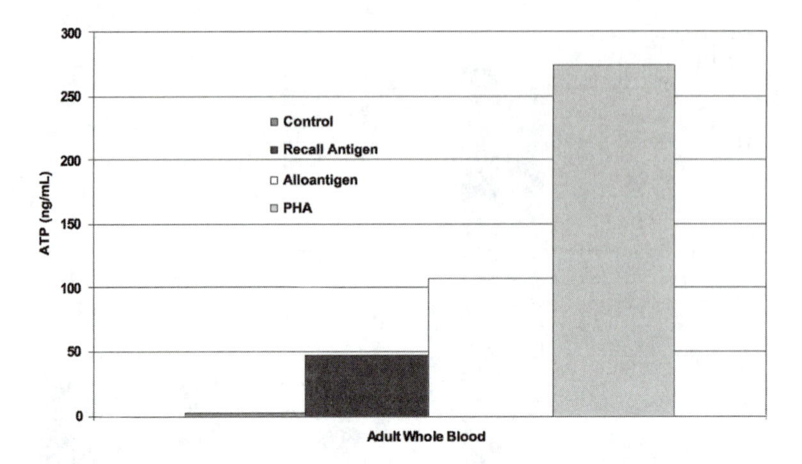

FIGURE 24.12 Alloantigen response.

its immunomodulatory activities. However, this drug is extremely expensive and only effective in 20% of patients. The ability to predict patients who would be responsive to therapy by the *in vitro* CMI assay would have a tremendous impact on managing the cost of the disease.

24.2.7 Organ Transplantation

Each year, 55,000 transplants are performed worldwide. Approximately half of these are performed in the United States at one of the 250 transplant clinics. The most common type of organ transplant is kidney, followed by liver, heart, and lung.

Prior to transplantation, matching of donors and recipients is required to assure the longevity of the graft. The *in vitro* CMI assay can be adapted to measure T cells activated in response to alloantigen, using an assay analogous to the mixed lymphocyte reaction (MLR). Figure 24.12 shows alloantigen stimulation of diluted whole blood using irradiated HLA mismatched splenocytes. After transplant, the goal is to calibrate immunosuppressive therapy so that the transplant is not rejected but does not leave the patient susceptible to opportunistic infection.

Figure 24.13 illustrates the immunosuppressive effect of cyclosporine A in an apparently healthy adult whole-blood sample as measured by the *in vitro* CMI assay. For this figure, diluted whole-blood samples were preincubated with various concentrations of cyclosporine then stimulated with PHA. The figure clearly shows suppression of PHA in a dose-dependent manner. A wider array of drugs beyond cyclosporine is now becoming available to manage the acute and chronic phases of rejection.

Once immunosuppressed, patients remain on therapy for the rest of their lives. Today monitoring is driven by assessing drug toxicity and organ failure. Functional immune response testing holds the promise of identifying rejection early enough to intervene with additional treatments. Immune response monitoring also has the potential to anticipate opportunistic infections and to assist the physician regarding the use and timing of prophylactic drugs.

The Holy Grail of transplantation is to induce tolerance. Measuring the patient's alloreactivity post-transplant provides the opportunity to remove the patient from life-long immunosuppressive therapy once reactivity to donor alloantigens is lost.

24.2.8 Cancer

Cancer probably originates when the immune system fails to recognize newly emerging transformed cells. Once a tumor becomes established, the cancer cells frequently produce high levels of immunosuppressive cytokines. Of course, many of the therapies designed to kill rapidly dividing

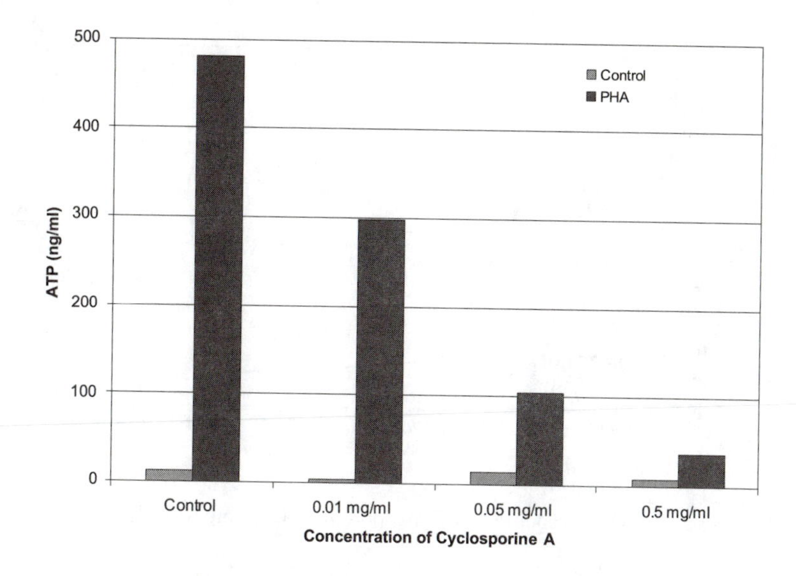

FIGURE 24.13 *In vitro* immunosuppression of PHA response by cyclosporine A.

cancer cells also eliminate the precursors of immunity, further compromising the patient's ability to control the cancer immunologically. For these reasons, cancer is a fertile area for clinical applications of immune response monitoring. Cancer vaccines show promise in stimulating the patient's own specific immunity. The *in vitro* CMI assay can be used as a tool to accelerate the development of these vaccines and to provide a mechanism for monitoring the patient's progress postvaccination.

24.2.9 Autoimmunity

The symptoms associated with autoimmunity are frequently nonspecific and can take years to diagnose. The current methodologies are based on detecting antibodies in the serum to very crude extracts from cells, for which there is no standardization. Although the humoral side of autoimmunity has been studied fairly extensively, there is a cellular component that is much less understood.

Schulz has recently described a model for Crohn's disease in which he used the patient's own intestinal flora as the immune stimulant in the *in vitro* CMI assay.[34] Compared to healthy patients, who are not reactive to self-bacterial flora, Crohn's patients are reactive.

24.2.10 Vaccine Development

Although there are a large number of potentially infectious agents, the number of vaccines in use is relatively small. One of the explanations for this is that traditional vaccine development has been directed toward boosting a serological response. When immunity is effective at a humoral level, this strategy is successful. The majority of diseases, however, require a significant cellular response to be protective. Vaccine development has been limited by the lack of methods as simple as serology to assess cellular immunity.

Feasibility of the *in vitro* CMI methodology was initially demonstrated in 1996 in a prototype assay developed for the Department of Defense to evaluate T-cell response to Q fever, a potential biological warfare agent.[20] Since then, it has been possible to monitor T-cell activation in the *in vitro* CMI assay in response to immunization with influenza, OspA antigen (Lyme disease), HBsAg recombinant vaccine, and HIV antigens. Responses can be detected as early as a week postimmunization and are usually sustained for several weeks before declining to the level observed with recall antigens (Figure 24.14). The value of the technology in this case is the acceleration of vaccine

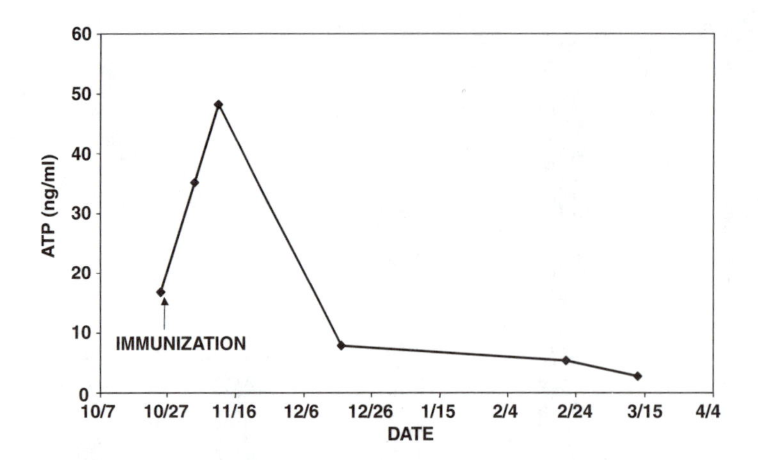

FIGURE 24.14 CD4 *in vitro* CMI response to HBsAg pre- and postimmunization.

development, rather than routine patient monitoring. It is especially useful to assess the potency of vaccines during the manufacturing process. This methodology provides an alternative to costly animal models and can be applied to the development of vaccines against biological warfare agents, as demonstrated by Q fever.

24.2.11 Biopharmaceutical Development

For companies developing new therapeutic drugs, the *in vitro* CMI assay can be adapted for use in high-throughput screening. Every new drug that is developed may be classified as neutral, stimulatory, or suppressive to the immune system. The *in vitro* CMI assay can provide an efficient primary screen for potential effects on immune function *in vitro* without the time-consuming and costly adaptation to animal models. In some cases, if immune restoration is the desired effect of the drug therapy, patient monitoring can also be performed. The increased adoption of herbal therapies, available without prescription, also warrants study for their impact on immune system function. One of the major challenges facing therapies derived from the human genome involves how to insert new genes into the patient or patient cells without initiating immune-mediated attack. *In vitro* CMI can provide an early method of assessing the compatibility of the vector in the host. The individualized nature of genomic therapy is also consistent with the advantages offered by understanding the specific patient's immune repertoire and how it changes over time in response to therapy.

24.3 SUMMARY

The ability of cells to regulate the synthesis of ATP in response to environmental factors allows the use of bioluminescence as a marker of cellular metabolic status. Cells of the immune system respond to foreign invaders by initially increasing synthesis of ATP followed by producing various cytokines and ultimately proliferating. By detecting this initial burst of ATP production with bioluminescence, the *in vitro* CMI platform rapidly measures cellular responses. This platform is designed to assess responses directly in specific cellular subsets such as T-helper cells ($CD4^+$), cytotoxic T cells ($CD8^+$), natural killer cells ($CD56^+$), B-lymphocytes ($CD19^+$), and a multitude of others. The *in vitro* CMI assay is the first technology for immune function testing with clinical accessibility. Its advantages over lymphoproliferation are clear in terms of time (1 vs. 7 days) and no radioactivity. Other research methods involving cytokines frequently require costly instrumentation (flow cytometry or imaging for ELISPOT) and highly skilled operators to perform the testing.

Given the ease of use and tremendous potential for clinical utility, the *in vitro* CMI platform has the essential elements for adoption as a standard of care in the management of a wide range of diseases through the functional assessment of immunity. *In vitro* CMI assay is for "research use only."

REFERENCES

1. Betensky, R. A., Connick, E., Devers, J., Landay, A. L., Nokta, M., Plaeger, S., Rosenblatt, H., Schmitz, S. L., Valentine, F., Wara, D., Weinberg, A., and Lederman, H., Shipment impairs lymphocyte proliferative responses to microbial antigens, *Clin. Diagn. Lab. Immunol.,* 7(5), 763, 2000.
2. Halsey, J. F., Evaluation of a T-cell proliferation test, presented at American College of Asthma, Allergy, and Immunology, Seattle, WA, 2000.
3. Harker, M., Application of luminescence in medical microbiology and hematology, in *Clinical and Biochemical Luminescence,* Vol. 12, L. J. Kricha and T. J. N. Carter, Eds., Marcel Dekker, New York, 1982, 189.
4. Kangas, L., Gronroos, M., and Neiminen, A. L., Bioluminescence of cellular ATP: a new method for evaluating cytotoxic agents *in vitro, Med. Biol.,* 62, 343, 1984.
5. Lundin, A., Hasenson, M., Persson, J., and Pousette, A., Estimation of biomass in growing cell lines by ATP assay, *Methods Enzymol.,* 133, 42, 1986.
6. Petty, R. D., Sutherland, L., Hunter, E., and Cree, I., Comparison of MTT and ATP-based assays for the measurement of viable cell number, *J. Biolumin. Chemilumin.,* 10, 29, 1995.
7. Andreotti, P. E., Linder, D., Hartmann, D. M., Johnson, L. J., Harel, E. A., and Thaker, P. H., ATP lymphocyte activation assay application for lymphokines and cytokines, *Biolumin. Chemilumin.,* 261, 1993.
8. Crouch, S. P. M., Kozlowski, R., Slater, K. J., and Fletcher, J., The use of ATP bioluminescence as a measure of cell proliferation and cytotoxicity, *J. Immunol. Methods,* 160, 81, 1993.
9. Ishizaka, A., Tono-oka, T., and Matsumoto, S., Evaluation of the proliferative response of lymphocytes by measurement of intracellular ATP, *J. Immunol. Methods,* 127, 1984.
10. Campbell, A. K., *Chemiluminescence: Principles and Applications in Biology and Medicine,* Ellis Horwood, New York, 1988.
11. Wrogemann, K., Weidemann, M. J., Ketelson, U.-P., Wekerle, H., and Fischer, H., Chemiluminescence and immune cell activation II. Enhancement of concanavalin A-induced chemiluminescence following *in vitro* preincubation of rat thymocytes; depending on macrophase–lymphocyte interactions, *Eur. J. Immunol.,* 10, 36, 1980.
12. Stanley, P. E., Extraction of adenosine triphosphate from microbial and somatic cells, *Methods Enzymol.,* 133, 14, 1986.
13. Sigma Technical Bulletin BSCA-1.
14. White, A. G., Raju, K. T., Keddie, S., and Abouna, G. M., Lymphocyte activation: changes in intracellular adenosine triphosphate and deoxyribonucleic acid synthesis, *Immunol. Lett.,* 22, 47, 1989.
15. Cunningham-Rundles, S., Hansen, J. A., and Dupont, B., Lymphocyte transformation *in vitro* in response to mitogens and antigens, in *Clinical Immunology,* Vol. 3, F. H. Bach and R. A. Good, Eds., Academic Press, New York, 1976, 151.
16. Sottong, P. R., Rosebrock, J. A., Britz, J. A., and Kramer, T. R., Measurement of T-lymphocyte responses in whole blood cultures using newly synthesized DNA and ATP, *Clin. Diagn. Lab. Immunol.,* 7, 307, 2000.
17. Brinchmann, J. E., Leivestad, T., and Vartdal, F., Quantification of lymphocyte subsets based on positive immunogenetic selection of cells directly from blood, *J. Immunogenet.,* 16, 177, 1989.
18. DelGratta, C., Penna, S. D., Battista, B., DiDonato, L., Virtullo, P., Romani, G. L., and DiLuzio, S., Detection and counting of specific cell populations by means of magnetic markers linked to monoclonal antibodies, *Phys. Med. Biol.,* 40, 671, 1995.
19. DeLuca, M., Firefly luciferase, *Adv. Enzymol.,* 44, 37, 1976.
20. Wier, M., Rapid assay of cellular immunity in Q fever, presented at 96th General Meeting of the American Society for Microbiology, New Orleans, LA, 1996.
21. Kramer, T. and Rosebrock, J. R., Proliferative responsiveness of lymphocytes in whole-blood cultures to phytohemagglutinin (PHA-L), concanavalin A (ConA), and tetanus toxoid (TT), presented at General Meeting of Experimental Biology Federation, San Francisco, CA, 1998.

22. Clerici, M., Stocks, N., Zajac, R. A., Boswell, R. N., Lucey, D. R., Via, C. S., and Shearer, G. M., Detection of three different patterns of T helper cell dysfunction in asymptomatic human HIV+ patients, *J. Clin. Invest.,* 84, 1892, 1989.
23. Britz, J. A., Kowalski, R., Schneider, M. C., Sottong, P., and Farzadagan, H., A rapid assay for the assessment of immunity in AIDS patients using cellular ATP: an alternative to lymphoproliferation, presented at 2000 International Meeting of the Institute of Human Virology, Baltimore, MD, 2000.
24. Dolan, M. J., Clerici, M., Blatt, S. P., Hendrix, C. W., Melcher, G. P., Bosell, R. N., Freeman, T. M., Ward, W., Hensley, R., and Shearer, G. M., *In vitro* T cell function, delayed type hypersensitivity skin testing and CD4+ T cell phenotyping independently predict survival time in patients infected with human immunodeficiency virus, *J. Infect. Dis.,* 172, 79, 1995.
25. Perrin, L. and Telenti, A., HIV treatment failure: testing for HIV resistance in clinical practice, *Science,* 280, 1872, 1998.
26. Conference on Retroviruses and Opportunistic Infection, 1999 Abstract 330.
27. Loechelt, B. J., Chan, M., Rosebrock, J. A., Kramer, T. R., Sottong, P. R., and Chen, W., Evaluation of specific immune reconstitution in HIV infected patients, presented at 1999 International Meeting of the Institute of Human Virology, Baltimore, MD, 1999.
28. Loechelt, B. J., Chan, M., Rosebrock, J. A., Sottong, P. R., and Britz, J. A., Immune reconstitution in pediatric HIV infected patients as measured by the Luminetics™ technology—an alternative to radioactive lymphoproliferation, presented at 16th Annual Clinical Virology Symposium and Annual Meeting Pan American Society for Clinical Immunology, Clearwater Beach, FL, 2000.
29. Rosenberg, E. S., Billingsley, J. M., Caliendo, A. M., Boswell, S. L., Sax, P. E., Kalams, S. A., and Walker, B. D., Vigorous HIV-1 specific CD4+ T cell responses associated with control of viremia, *Science,* 278, 1447, 1997.
30. Cunningham-Rundles, S., Califano, S. C., Chen, L., Marshall, F., Dunn, A. M., and Cervia, S. S., Detection of early activation of CD4+ T cells in response to p24 antigen in pediatric HIV disease, presented at the 1999 International Meeting of the Institute for Human Virology, Baltimore, MD, 1999.
31. Rosenberg, E. S., Altfeld, M., Poon, S. H., Phillips, M. N., Wilkes, B. M., Edridge, R. L., Robbins, G. K., D'Aquila, R. T., Goulder, P. J. R., and Walker, B. D., Immune control of HIV-1 after early treatment of acute infection, *Nature,* 408, 523, 2000.
32. Leandersson, A.-C., Hejderman, B., Britz, J. A., Gaines, H., Fredriksson, M., Bratt, G., Sandstrom, E., and Wahren, B., Comparison of different methods to measure specific T helper cell responses in HIV infected individuals, presented at 2000 International Meeting of the Institute of Human Virology, Baltimore, MD, 2000.
33. Britz, J. A., Chen, W., Schneider, S., and Sottong, P., Use of the Luminetics™ Assay for T cell activation for monitoring cellular immune response to vaccination: a novel alternative assay to lymphoproliferation, presented at the American Association of Immunologists and Clinical Immunology Society Joint Annual Meeting, Immunology, Seattle, WA, 2000.
34. Schultz, M., Oral therapy with lactobacillus GG alters the activation response of PBMC and CD4 T-lymphocytes toward Bacteroides SP, presented at 39th ASCB Annual Meeting, Washington, D.C., 1999.

25 Protective Effects of Native But Not Oxidized High-Density Lipoprotein against Proinflammatory Respiratory Burst Activities of Polymorphonuclear Leukocytes Induced by Hypochlorite-Oxidized Low-Density Lipoprotein

Steffi Kopprasch, Jens Pietzsch, and Jürgen Gräßler

CONTENTS

25.1 INTRODUCTION

Atherosclerosis, the major cause of cardiovascular morbidity and mortality, is a chronic inflammatory disease characterized by complex cellular and molecular processes not only in the intima of arterial walls but also in the circulation and the endothelium.[1] Elevated plasma cholesterol levels and increased accumulation of cholesterol within the arterial wall are key elements in the pathogenesis

0-8493-0719-8/02/$0.00+$1.50

of atherosclerosis. As such, intensive research is directed toward alterations in lipoprotein metabolism that promote atherosclerotic modification of the vascular intima.

Low-density lipoprotein (LDL) and high-density lipoprotein (HDL) are responsible for the transport of cholesterol in the organism. Both lipoproteins are extremely prone to oxidative damage. There is experimental and clinical evidence that oxidatively altered LDL plays a significant role in atherogenesis.[2,3] Oxidized LDL (oxLDL) but not native LDL (natLDL) serves as a ligand for delivering cholesterol to developing foam cells (monocytes/macrophages, smooth muscle cells) via scavenger receptors and other oxLDL receptors. Moreover, oxLDL exerts considerable biological activities that could force proinflammatory reactions in atherogenesis. oxLDL is chemotactic for monocytes,[4–6] smooth muscle cells,[7] and polymorphonuclear leukocytes,[8] and it stimulates the expression of adhesion molecules in vascular endothelial cells and leukocytes including P-selectin,[9] MCP-1,[10] VCAM-1,[11] and L-selectin.[12] Furthermore, oxLDL has been shown to modulate the expression of cytokines and chemokines in both positive and negative fashion.[13–16] oxLDL has also been reported to induce apoptosis,[17] the expression of heat shock proteins in human endothelial cells,[18] and the synthesis of collagen in smooth muscle cells.[19]

The *in vivo* presence of oxLDL in atherosclerotic lesions has been demonstrated using antibodies against oxLDL.[20,21] For a long time it was proposed that LDL could be present in its oxidized form only within the arterial wall. This suggestion was based preferentially on two findings: (1) oxLDL injected intravenously disappeared from the plasma compartment with a very short half-life[22] and (2) plasma is provided with a substantial antioxidative potential that prevents LDL oxidation *in vitro*.[23–25] Recent studies, however, revealed that mildly oxidized LDL can circulate in plasma for a prolonged time and that oxLDL in addition to native lipoprotein enters and accumulates in the arterial intima.[26]

The pathogenetic impact of oxidatively modified LDL is underlined by clinical trials demonstrating an association between elevated plasma levels of oxLDL or antibodies against oxLDL and the severity of cardiovascular diseases.[27–29] Despite the great progress in the research about the atherogenic significance of oxLDL substantial questions are still unresolved. In particular, the location and the mechanism(s) by which natLDL is oxidized *in vivo* as well as the metabolic fate of circulating oxLDL are uncertain. Generally, oxLDL can develop its damaging proinflammatory activities also in the circulation. Therefore, both an effective protection against oxidative modification and an efficient elimination from the circulation of oxLDL are considered to be essential parts of the host defense against oxidative stress.

One candidate to counteract substantially the oxidation of LDL and the subsequent proinflammatory activities of oxLDL are HDLs. Epidemiological studies have established an inverse relationship between plasma levels of HDL and the risk of coronary heart disease.[30,31] The exact mechanisms behind the protective effects of HDL are not fully elucidated yet. It seems, however, that the ability of HDL to stimulate the efflux of cholesterol from foam cells,[32] the suppression of LDL oxidation by HDL-associated proteins,[33] and the removal of oxidized phospholipids from oxLDL[34,35] could significantly contribute to the antiatherogenic properties of HDL.

Polymorphonuclear leukocytes (PMNL) are equipped with a broad-spectrum arsenal of defense mechanisms that easily could oxidize native LDL in the circulation. Indeed, *in vitro* investigations revealed that incubation of stimulated PMNL with natLDL leads to oxLDL, which is recognized by scavenger receptors of macrophages.[36] Alternatively, PMNL as a classical effector cell could also serve as a target of oxLDL. Neutrophils, normally circulating in a resting state, can be activated by a wide variety of soluble and particulate stimuli.[37] It has been shown previously that oxidized lipoproteins stimulate the formation of reactive oxygen species (ROS) in different cell types including monocytes[38] and juxtaglomerular cells,[39] suggesting that oxLDL-induced stimulation of cellular ROS generation may be a general phenomenon.

Therefore, it was of interest to investigate the direct potential stimulatory effects of oxidatively modified LDL on human PMNL respiratory burst activity and the ability of oxLDL to modulate neutrophil ROS generation evoked by other stimuli. The intracellular and extracellular ROS generation was estimated by luminol-amplified chemiluminescence (CL). This method offers the possibility of measuring changes of neutrophil oxidative metabolism with great sensitivity. Furthermore, the study presented on this chapter examines whether natHDL is able to prevent oxLDL-induced changes. Because of the high susceptibility of HDL to oxidative modification[40] and the fact that HDL may lose its antiatherogenic properties following oxidation,[41,42] the study also tested if oxidation of HDL would diminish its protective role against the proinflammatory effects of oxLDL.

25.2 MATERIALS AND METHODS

25.2.1 Preparation of LDL, HDL, and Apolipoprotein AI

Blood samples from healthy normolipidemic middle-aged volunteers were analyzed. Exclusion criteria were the intake of antioxidants such as probucol or a supplementation with vitamins A or E. The subjects had given their written consent, and the study protocol was approved by the Ethics Committee of the Medical Faculty Dresden.

Blood was collected after overnight fasting for at least 12 h into polypropylene vacuum tubes containing sodium ethylene diamine tetraacetate (EDTA) at a final concentration of 0.1%. Plasma was recovered after centrifugation at 4°C (10 min at 3000 × *g*). Sodium azide (0.01%) and aprotinin (200 kIU/ml) were added immediately. Plasma aliquots were stored at 4°C and used to determine lipids and concentrations of apolipoprotein B-100 and apolipoprotein AI (apoAI).

LDL ($1.006 < d < 1.063$ kg/l) and HDL ($1.063 < d < 1.210$ kg/l) were isolated by sequential very fast ultracentrifugation.[43,44] We used the tabletop ultracentrifuge OptimaTM TLX with fixed angle rotor TLA-120.2 and thick-wall polycarbonate tubes without cap (Beckman Instruments, Palo Alto, CA). The run conditions were full speed (625,000 *g*) and 18°C. The periods chosen for flotation were 1.6 h for LDL and 4.3 h for HDL. Density media were made oxygen free by degassing and purging with argon. The recoveries of cholesterol, triglycerides, and protein in the lipoprotein particles averaged 96, 97, and 98%, respectively. Notably, LDL and HDL prepared by this technique were obtained free of albumin.

Lipid and protein constituents in plasma and lipoprotein fractions were measured as previously described.[43] The EDTA content, sodium azide, and salt from the density gradient were removed using a size exclusion column (Econo-Pac 10DG, Bio-Rad Laboratories, Hercules, CA). The cholesterol and protein recovery of this procedure averaged 93 and 91%, respectively.

apoAI was isolated by delipidation of freshly prepared human HDL as described by Scanu and Edelstein.[45] In brief, HDL was dialyzed against 0.15 mol/l NaCl, 1 mmol/l EDTA, pH 7.4, and delipidated three times with ethanol:diethyl ether (3:1, v/v) at −20°C. The delipidated samples were dissolved in 30 mmol/l Tris-HCl, pH 8.0, 6 mol/l urea, and applied to an anion exchange chromatography column (Mono Q HR-Sepharose 10/10, Amersham Pharmacia Biotech, Rainham, U.K.). For elution served a linear salt gradient (0 to 0.15 mol/l NaCl, 30 mmol/l Tris-HCl, pH 8.0, 6 mol/l urea) at a flow rate of 4 ml/min to obtain pure apoAI. Eluted fractions were stored in lyophilized form at −70°C.[46] Prior to use, the purified apolipoprotein fractions were resolubilized in 6 mol/l guanidine–HCl, dialyzed extensively against oxygen-free phosphate-buffered saline (PBS), and stored at −20°C. The purity of apolipoprotein fractions was checked by sodium dodecylsulfate polyacrylamide gel electrophoresis (SDS-PAGE) as previously described.[43]

25.2.2 Oxidation of Lipoproteins

LDL and HDL were oxidized by sodium hypochlorite (NaOCl). Hypochlorous acid (HOCl) is a potent non-metal-dependent oxidant that is formed by myeloperoxidase (MPO, EC 1.11.1.7) from H_2O_2 and chloride ions. Upon stimulation, phagocytic white cells release MPO from azurophil granules either into the phagosome or the extracellular space. Recent studies have demonstrated that the exposure of isolated human LDL to hypochlorite (^{-}OCl) oxidizes the lipoprotein efficiently, producing an aggregated high-uptake form for macrophages.[47–50] Moreover, myeloperoxidase,[51] the myeloperoxidase-specific oxidation product, 3-chlorotyrosine,[52] and ^{-}OCl-oxidized proteins including apolipoprotein B-100[21] have been shown to be present in human atherosclerotic lesions, raising the possibility that myeloperoxidase may promote lipoprotein oxidation *in vivo*. In accordance, it has been shown recently that HDL is also extremely susceptible to oxidation by hypochlorite that induces physicochemical changes similar to those observed for oxLDL.[53]

Prior to oxidation, lipoprotein solutions were diluted to 0.2 g/l protein. A 20 mmol/l NaOCl (Aldrich) solution was prepared freshly each day from a stock solution. The concentration of hypochlorite was determined photometrically at pH = 12.0 using $\varepsilon_{290} = 350\ M^{-1}\ cm^{-1}$.[54] Oxidation of lipoproteins was carried out at 37°C for 40 min with a final NaOCl concentration of 1 mmol/l unless otherwise stated.

To measure the kinetics of hypochlorite consumption during lipoprotein oxidation a CL assay system was used. It has been shown previously that an intense CL flash emerges from the oxidation of luminol by hypochlorous acid lasting about 2 s.[55] At different oxidation times 50 μl of lipoprotein suspension was added to 850 μl PBS prewarmed at 37°C. Immediately thereafter, 100 μl of luminol (5×10^{-5} mol/l final concentration, pH = 7.4) was injected into the reaction vial and light emission was recorded for 10 s. The integral counts were proportional to the NaOCl concentration in the reaction vial and calculated from a calibration curve. CL measurements were performed with an Autolumat LB 953 (Berthold Co., Wildbad, Germany).

In addition to oxidative modification, natHDL and apoAI suspensions (0.2 g/l protein) were treated with trypsin–EDTA (Sigma; 278 μg trypsin per mg HDL or apoAI protein) for 1 h at 37°C. The degradation of apoAI after proteolysis was assessed by SDS-PAGE using a Tris–glycine buffer system as previously published.[43]

25.2.3 Physicochemical Properties of Native and Oxidized Lipoproteins

For determination of relative electrophoretic mobility (REM), samples of lipoproteins were loaded onto "ready for use" agarose gels (Laboratoires Sebia, Issy-les-Moulineaux, France) and electrophoresed at 25 V/gel for 90 min, fixed, dried, and stained with the lipid stain Sudan Black. The distance of lipoprotein migration was then measured to the center of the band. Migration was initially expressed relative to native lipoprotein from the same donor and is presented as a percentage of the control.

Additionally, modifications of apolipoprotein components were assessed by determination of fluorescent products with an emission maximum at 430 nm when excited at 360 nm.[56] In a typical assay, native and oxidized LDL and HDL were diluted with PBS to a final concentration of 0.04 g/l protein and the fluorescence was measured with a spectralfluorometer (SFM 25, Kontron Instruments, Zurich, Switzerland). The fluorescence intensity was expressed as the ratio relative to that of quinine sulfate (0.1 μmol/l in 0.1 N H_2SO_4) measured at a fluorescence maximum of 448 nm and an excitation maximum of 352 nm.

The level of lipid oxidation was monitored as the accumulation of thiobarbituric acid-reactive substances (TBARS) as described by Yagi.[57] Concentrations of TBARS were related to apolipoprotein content.

Additionally, lipid oxidation was monitored continuously by the formation of conjugated dienes during an observation period of 180 min. For this purpose, cuvettes with lipoprotein solutions (0.2 g/l protein) were placed in a spectrophotometer (Beckman DU 7400) and oxidation was started by the addition of NaOCl ranging from 0.2 to 1.0 mmol/l final concentration. Absorbance changes were recorded at 234 nm.

25.2.4 Determination of Paraoxonase Activity

Paraoxonase (PON) activity was measured spectrophotometrically in an arylesterase assay using phenylacetate as substrate.[58] Briefly, 200 μl HDL or apoAI (native and trypsinized) with a protein concentration of 0.2 g/l were added to 800 μl Tris-HCl buffer (9 mmol/l, pH = 8.0) containing 0.9 mmol/l $CaCl_2$ and 1.25 mmol/l phenylacetate. The generation of phenol was followed at 25°C for 60 s at 270 nm (ε_{270} = 1306 $M^{-1} \times cm^{-1}$). PON activity is expressed as μmol × $min^{-1} \times mg^{-1}$ HDL or apoAI protein.

25.2.5 Isolation of Polymorphonuclear Leukocytes and Chemiluminescence Assays

Parallel with the LDL and HDL oxidation, PMNL were isolated from heparinized blood of healthy volunteers by density gradient centrifugation with equal volumes of Histopaque-1077 and Histopaque-1119 (Sigma). Cells were washed twice in Krebs-Ringer phosphate buffer (KRG; 120 mmol/l NaCl, 4.9 mmol/l KCl, 1.2 mmol/l $MgSO_4$, 1.7 mmol/l KH_2PO_4, 8.3 mmol/l Na_2HPO_4, 10 mmol/l glucose, pH = 7.3). In an additional step erythrocytes were lysed with ice-cold water. Finally, the cells were resuspended in KRG plus 1 mmol/l $CaCl_2$. The suspension was adjusted to 1×10^7 cells/ml for CL measurements. The cell suspension was stored on ice until use (usually within 120 min).

CL was measured using disposable polypropylene tubes with a 1.0 ml reaction mixture. For determination of respiratory burst activity 1×10^6 PMNL were preincubated with KRG plus calcium and luminol (100 μmol/l final concentration) for 5 min at 37°C. Then, the cells were activated by 70 nmol/l natLDL or oxLDL and light emission was recorded for 30 min.

To investigate the influence of native or modified (oxidized or trypsinized) HDL or apoAI on the oxLDL-induced ROS-generating activity, either PMNL or oxLDL was preincubated with the corresponding lipoprotein or protein suspensions at a final protein concentration of 0.04 g/l and after completion of the reaction mixture light emission was recorded.

To assess the effects of native and modified lipoproteins on neutrophils activated by other inflammatory stimuli cells were preincubated for 5 min with native and/or oxidized lipoprotein suspensions followed by activation with the chemotactic peptide formyl-methionyl-leucyl-phenylalanine (FMLP, 10^{-7} mol/l final concentration), the particulate stimulus opsonized zymosan (0.5 mg/ml final concentration) or phorbol myristate acetate (PMA, 5×10^{-8} mol/l final concentration).

Neutrophil CL activity was expressed as maximum counts/min per 10^6 cells.

25.3 RESULTS

Hypochlorite-induced changes of physicochemical properties of LDL and HDL are summarized in Table 25.1. Exposure of natLDL to NaOCl with a NaOCl/LDL particle ratio of 2900 resulted in considerable protein and lipid oxidation. Modifications of apolipoprotein B-100 were detected by a 4.8-fold enhanced electrophoretic mobility relative to natLDL and an increased generation of autofluorescence products. Monitoring the fluorescence intensity of LDL at 360/430 nm has been

TABLE 25.1
Physicochemical Characterization of Native and Oxidized LDL and HDL

	Relative Electrophoretic Mobility	Fluorescence Intensity (360/430 nm, RF in %)	TBARS (μmol/g protein)
natLDL	—	9.3 ± 1.1 (n = 10)	5.4 ± 2.9 (n = 10)
oxLDL	4.8 ± 0.3 (n = 7)	13.3 ± 1.3 (n = 10)	28.7 ± 3.8 (n = 10)
natHDL	—	6.9 ± 2.9 (n = 10)	1.3 ± 0.8 (n = 10)
oxHDL	1.8 ± 0.1 (n = 7)	25.1 ± 5.2 (n = 10)	2.4 ± 0.8 (n = 5)

RF = relative fluorescence; TBARS = thiobarbituric acid-reactive substances; results as means ± SD of n determinations.

shown to be an easy and sensitive method for evaluating the degree of oxidation of LDL.[59] It has been reported that the fluorescence increase during LDL oxidation is mainly due to a derivatization reaction of the free amino groups of apolipoprotein B.[60] It is proposed that the fluorophore results preferentially from a reaction of the apolipoprotein lysine residues with aldehydes or hydroperoxides thereby forming a Schiff base.[56]

Lipid oxidation was accompanied by accumulation of TBARS. Compared with native LDL, the TBARS levels were increased fivefold in oxLDL. natHDL was oxidized by the same protein/NaOCl ratio (to 0.2 g/l protein 1.0 mmol/l NaOCl were added) as natLDL. Following oxidation, HDL showed an 1.8-fold increased electrophoretic mobility and a 3.6-fold enhanced fluorescence intensity as signs of oxidation of the apolipoprotein component. As compared with LDL, HDL had a lower baseline TBARS content (1.3 vs. 5.4 μmol/g protein) with only moderate 1.8-fold increase during the 40-min oxidation period.

In addition to changes in TBARS levels, NaOCl-induced lipid peroxidation in LDL and HDL was followed continuously by formation of conjugated dienes. The corresponding time traces show an immediate and sharp increase in the absorbance at 234 nm without any lag phase following the addition of the oxidant. After about 10 min (HDL) and 60 min (LDL), respectively, the curves reached a plateau with only minor further changes during the next 2 h. As demonstrated in Figure 25.1, the cumulative diene formation during a 3-h period increased with increasing concentrations of hypochlorite used for lipoprotein oxidation. Addition of 0.2 to 1.0 mmol/l NaOCl to the HDL and LDL lipoprotein suspensions (0.2 g/l protein) was associated with a nearly linear increase in diene accumulation. The oxidant concentration–dependent diene formation did not differ substantially in both HDL and LDL preparations (see Figure 25.1).

Figure 25.2 shows the kinetics of hypochlorite consumption during oxidation of natLDL and HDL. Immediately after addition to the lipoprotein suspension the initial 1 mmol/l NaOCl content declined exponentially and the oxidant was no longer detectable after 10 min. From these results, we concluded that the amount of $^{-}$OCl used in our experiments for LDL and HDL oxidation was totally consumed after the entire oxidation period of 40 min and that remaining hypochlorite could not distort the following CL experiments with polymorphonuclear leukocytes.

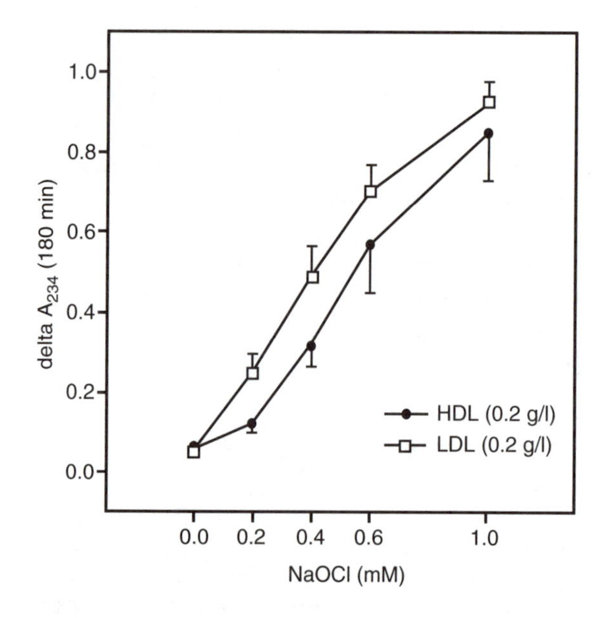

FIGURE 25.1 NaOCl concentration-dependent diene accumulation after a lipoprotein oxidation period of 180 min. Formation of dienes was measured spectrophotometrically at 234 nm. Data as means ± SEM of 5 to 10 determinations obtained with different LDL and HDL preparations.

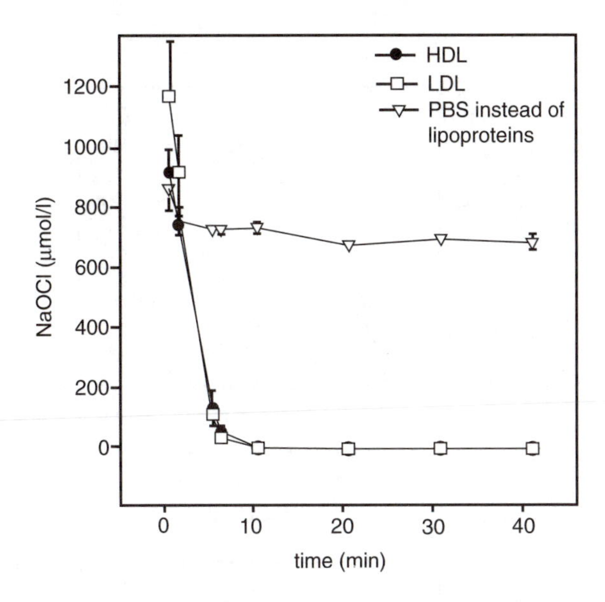

FIGURE 25.2 Kinetics of NaOCl decline during oxidation of native lipoproteins. 1 mmol/l NaOCl was added to native LDL or HDL (0.2 g/l protein) and the NaOCl consumption was followed for 40 min at 37°C by means of CL. The data are means ± SD of three different oxidations. Error bars are partially within symbol size.

Oxidized LDL, but not native LDL and HDL or oxidized HDL at final concentrations of 0.04 g/l protein evoked a considerable CL response in PMNL with one peak at around 10 to 12 min indicating a stimulation of the neutrophil respiratory burst by oxidatively modified LDL. Both the oxidation time of LDL by NaOCl and the concentration of the oxidant significantly influenced the subsequent PMNL light generation. To examine the impact of oxidation time, aliquots of natLDL were oxidized at specific times prior to CL measurements using 1 mmol/l NaOCl. The subsequent measurement of PMNL CL activity revealed a clearly time-dependent increase of CL response (Figure 25.3a). Prolongation of oxidation time of LDL up to 2 h produced only minor further increases. Increasing the concentration of NaOCl from 0.1 to 1.0 m*M* for LDL oxidation was found to be similarly effective in elevating the PMNL light emission (Figure 25.3b).

To assess the relation between extracellularly and intracellularly generated ROS following oxLDL-induced stimulation of PMNL activity, a CL assay system originally developed by Dahlgren[61] was used. According to this method, the intracellularly localized CL activity was measured in a reaction mixture containing in additional superoxide dismutase (SOD, 200 U/ml) and/or catalase (CAT, 2000 U/ml). These two large-molecular scavengers of superoxide anions and hydrogen peroxide, respectively, do not have access to intracellular sites. In the presence of SOD and CAT only extracellular $O_2^{\bar{\cdot}}$ and H_2O_2 will be removed. Although it is not proved that ROS scavenged in the extracellular space have been also exclusively generated there, it is unlikely that $O_2^{\bar{\cdot}}$ or H_2O_2 generated intracellularly could move through the cytoplasm and pass the plasma membrane to get outside the cell. First, it has been shown that the cytosol has a large capacity to consume hydrogen peroxide that is probably mediated by catalase and glutathione peroxidase.[62] Second, superoxide anion is dismutated in the cytoplasm and, moreover, its passage through membranes is restricted and requires specific anion channels.[63] Thus, the assay system containing SOD and CAT provides a measurement of intracellularly generated oxygen metabolites.

The extracellular CL response can be calculated by subtracting the intracellular response from the total CL activity using the peak height for calculation. The results obtained for oxLDL were compared with the CL response induced by zymosan, another particulate stimulus evoking similar time traces of light emission after incubation with PMNL.

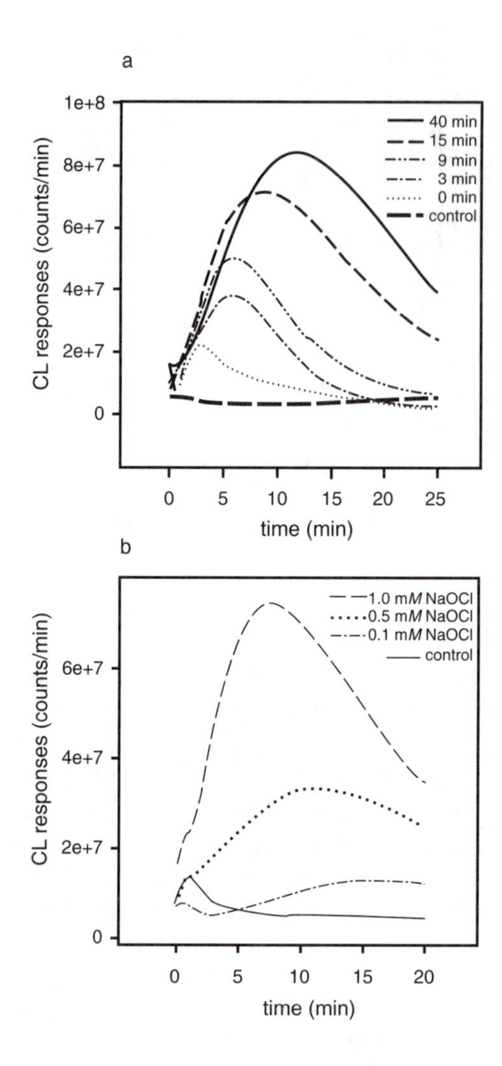

FIGURE 25.3 Impact of oxidation time (a) and NaOCl concentration (b) on LDL-induced CL response of 10^6 PMNL. (a) Aliquots of natLDL were oxidized at specific times prior to CL measurements by the addition of 1 mmol/l NaOCl (final concentration). (b) natLDL were oxidized for 15 min with different amounts of NaOCl and immediately thereafter assayed for CL activity. The figure shows typical time traces that were obtained in at least three independent experiments. (From Kopprasch, S. et al., *Atherosclerosis,* 136, 321, 1998. With permission from Elsevier Science.)

As shown in Figure 25.4, SOD and to a lesser extent CAT diminished the oxLDL-induced CL response. The simultaneous addition of both enzymes to the reaction mixture reduced the PMNL light generation by about 70%, suggesting that in contrast to zymosan-stimulated cells a large amount of generated oxygen metabolites was released into the extracellular environment. Cytochalasin B, an inhibitor of phagocytosis, reduced the CL activity of both zymosan- and oxLDL-stimulated PMNL to nearly 40% of control values without additive.

As already mentioned above, natHDL alone did not stimulate the respiratory burst of polymorphonuclear leukocytes. It has been shown previously that oxidation of LDL is suppressed in the presence of natHDL.[33] It was of interest to investigate if HDL would also attenuate the subsequent biological activities of oxLDL on phagocytes. This was tested by incubating 1×10^6 PMNL with natHDL (0.04 g/l protein final concentration) for 5 min at 37°C. Following stimulation with oxLDL at the same protein concentration, there was a marked reduction in the CL response to about 30% of the levels achieved with oxLDL alone. These observations indicate that natHDL are provided

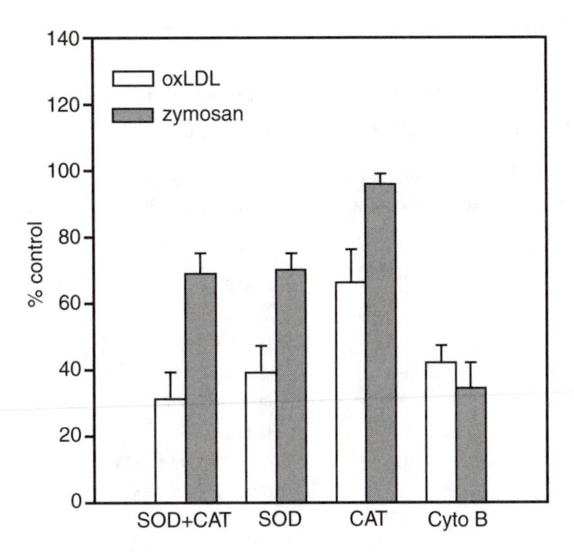

FIGURE 25.4 Effect of SOD, CAT, and cytochalasin B (Cyto B) on oxLDL- and zymosan-stimulated PMNL CL. PMNLs (10^6/ml) were preincubated with SOD (200 U/ml), CAT (2000 U/ml), or Cyto B (10 μmol/l) for 5 min at 37°C and then stimulated with opsonized zymosan (0.5 mg/ml) or oxLDL. Controls without additives were run in parallel. Data as means ± SEM from three LDL preparations used in experiments with nine different PMNL preparations. (From Kopprasch, S. et al., *Atherosclerosis,* 136, 321, 1998. With permission from Elsevier Science.)

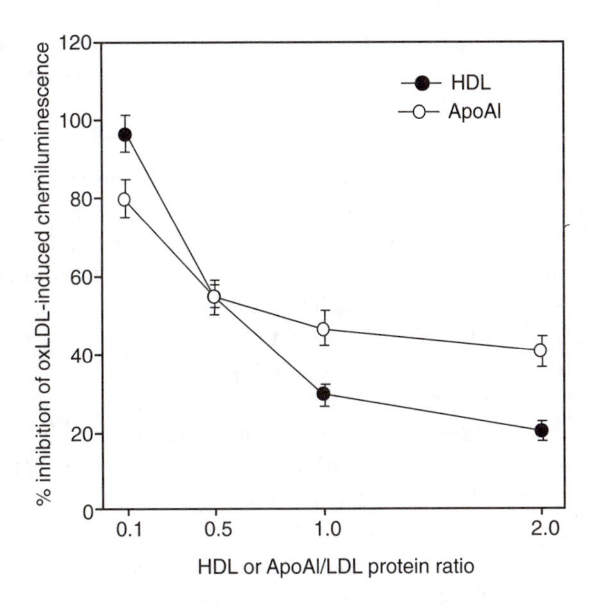

FIGURE 25.5 Dose-dependent inhibition of oxLDL-induced CL by natHDL and apoAI. PMNL (1×10^6/ml) were preincubated for 5 min with HDL or apoAI and then stimulated with oxLDL. CL activity was recorded over a time period of 30 min. Results are expressed as means ± SEM of six determinations obtained with two distinct HDL and apoAI and six PMNL preparations.

with fast-acting mechanism(s), efficiently preventing the stimulatory effects of oxLDL. The protective effect of natHDL was dose dependent. Maximal CL suppression to nearly 20% of the oxLDL-induced light generation was achieved at a HDL-to-LDL ratio of 2.0 (Figure 25.5). In contrast, neither oxHDL or natLDL were able to blunt the activation of PMNL induced by oxLDL.

TABLE 25.2
Effect of Preincubation of PMNL with Native or Oxidized Lipoproteins on Respiratory Burst Activation Induced by OxLDL

Preincubation Mixture	Time of Contact (min) during Preincubation	Substance Used to Stimulate PMNL	PMNL Activity (max. counts/min × 10^6)
—	—	natLDL	1.3 ± 0.2
—	—	natHDL	1.9 ± 0.3
—	—	oxHDL	2.6 ± 0.6
—	—	oxLDL	24.1 ± 1.9
PMML + natHDL	5	oxLDL	7.1 ± 1.3
PMML + oxHDL	5	oxLDL	23.0 ± 3.8
PMML + natLDL	5	oxLDL	24.9 ± 5.5
natHDL + oxLDL	20	Preincubation mixture	5.2 ± 1.2
oxHDL + oxLDL	20	Preincubation mixture	22.9 ± 4.3
natLDL + oxLDL	20	Preincubation mixture	34.6 ± 8.9
—	—	natHDL + oxLDL (added simultaneously)	8.8 ± 1.8
—	—	oxHDL + oxLDL (added simultaneously)	21.8 ± 4.4
—	—	natLDL + oxLDL (added simultaneously)	35.9 ± 8.6

PMNL (10^6 cells/ml) were preincubated with lipoproteins (0.04 g/l protein final concentration) at 37°C. In another vial natHDL and oxLDL (0.2 g/l protein each) were preincubated together and afterward added to PMNL (0.04 g/l protein final concentration). PMNL activity was measured by means of luminol-enhanced CL. Maximum activity usually was achieved between 5 and 20 min after PMNL stimulation. Data expressed as means ± SEM of 5 to 21 experiments.

Theoretically, the observed effect of HDL could result either from interaction with PMNL or from interaction with oxidized LDL, or from both. To test these hypotheses, additional experiments have been performed in which the incubation conditions prior to activation of cells were modified. In one experimental setting, oxLDL was preincubated with natHDL, oxHDL, or natLDL for 20 min at 37°C. Then, PMNL were stimulated with the preincubation mixture. In a second assay, HDL and oxLDL were prewarmed to 37°C separately and than added to PMNL simultaneously. The effects of different preincubation conditions on PMNL respiratory burst activity are summarized in Table 25.2. The results clearly demonstrate that neither the time of contact between the individual components of the cell activation system nor the composition of the preincubation mixture significantly modulated the protective effect of natHDL against activation of PMNL. Additionally, the CL experiments with varying incubation conditions confirm the fact that both natLDL and hypochlorite-oxidized HDL are not able to prevent enhanced ROS generation by oxLDL.

The inhibitory effects of HDL on PMNL activity were restricted to its native form. To relate the suppressive effects with the subfractions of the HDL particle, natHDL was delipidated, and the inhibitory action of apoAI was compared with that of natHDL. Moreover, both HDL and apoAI were trypsinized for 1 h at 37°C and thereafter tested for their protective effects on oxLDL-induced PMNL activation. As shown in Figure 25.5, delipidated apoAI was able to block oxLDL-induced CL response dose-dependently although it was less effective than intact HDL. Trypsinization of the HDL protein moieties weakened the suppression of both HDL and apoAI (Figure 25.6). Whereas trypsinized apoAI had only minimal residual inhibitory activity, trypsinized HDL was still able to considerably suppress oxLDL-induced CL by about 60%.

To relate the inhibitory power of HDL and its subfractions to paraoxonase, an antioxidative enzyme tightly associated with the apoAI component, PON activities were measured in native and

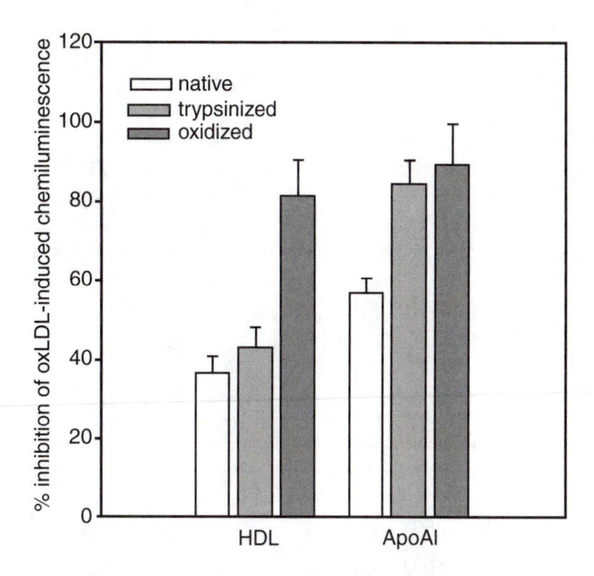

FIGURE 25.6 Protective effects of native, trypsinized, and oxidized HDL and apoAI against stimulatory action of oxLDL on PMNL. Data expressed as means ± SEM of 6 to 12 determinations.

trypsinized HDL and apoAI. With phenylacetate as substrate, natHDL had a PON activity of 1.51 ± 0.14 μmol × min^{-1} × mg^{-1} HDL protein (mean ± SEM, $n = 10$). In trypsinized HDL, native apoAI, and trypsinized apoAI, no significant paraoxonase activity could be detected.

After detecting the ability of oxLDL to activate the respiratory burst of resting PMNL, we were interested whether it would also modulate ROS generation of activated cells. For this purpose, PMNL were incubated with native or oxidized LDL for 5 min at 37°C and thereafter stimulated with the soluble chemoattractant FMLP, the particulate matter zymosan, or PMA. Additionally, the inhibitory activity of HDL in these assay systems was tested.

The results in Figure 25.7 show that oxLDL but not natLDL very distinctly affected PMNL respiratory burst activity in dependence on the stimulus used. The most-pronounced effects were observed with FMLP-treated cells. In this case, the presence of oxLDL in the reaction mixture was followed by a dramatic 14-fold increase of the PMNL CL response. Conversely, oxLDL inhibited zymosan-stimulated CL by nearly 50% and had no effect on PMA-induced ROS generation.

In all three assay systems the presence of natHDL (0.04 g/l protein) alone with the cells and their stimulus in the reaction vial did not interfere with PMNL light generation. natHDL did, however, modulate the effect of oxLDL in that they markedly blunted the oxLDL-induced amplification of FMLP-CL and slightly counteracted the oxLDL-mediated decrease of zymosan CL.

25.4 DISCUSSION

A growing body of evidence suggests that biomolecules, physicochemically modified as a result of oxidative stress, are involved in the pathogenesis of various inflammatory diseases including atherosclerosis. In this respect, the implications of oxidized lipoproteins for the initiation and development of atherosclerotic lesions have raised special interest.[2,3] It has been shown that oxidized LDL behaves not only as an inert "substrate" for foam cell formation but also develops properties of a signal molecule with the potential to support inflammatory processes in the circulation and the vascular subendothel.

The results of the present study confirm these findings. We could show that hypochlorite-oxidized LDL can serve as a potent stimulator of PMNL respiratory burst activity, and, moreover, it modulates ROS generation of activated PMNL in a direction toward increased proinflammatory action.

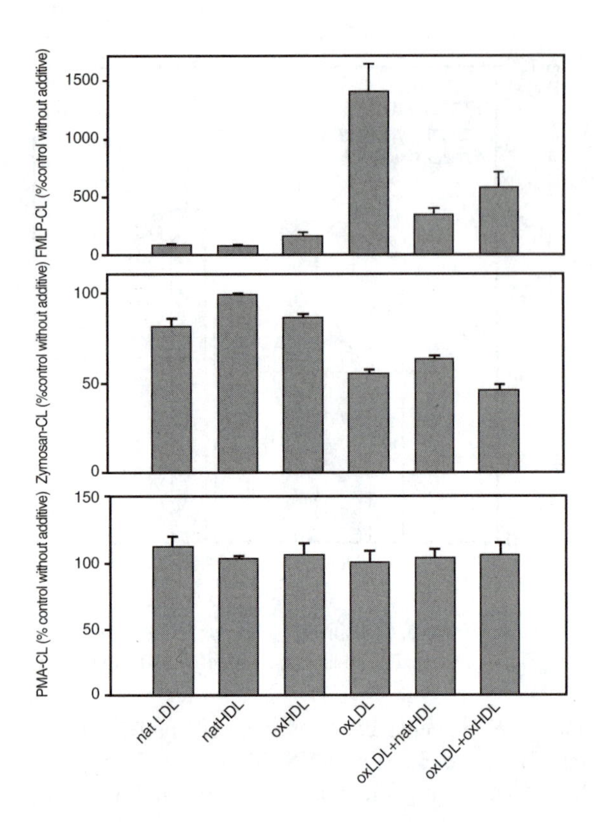

FIGURE 25.7 Effect of native and oxidized lipoproteins on stimulus-induced neutrophil respiratory burst activity. PMNLs were preincubated for 5 min at 37°C with the lipoprotein suspensions (0.04 g/l protein final concentration each) and thereafter stimulated with FMLP, zymosan, or PMA. Data as means ± SEM of ten (FMLP), seven (zymosan), and five (PMA) determinations.

PMNL are crucial for host defense. They are armed with various biochemical mechanisms including a large ROS-generating capacity that has the potential to eliminate infectious agents but also to damage the host itself. As such, PMNL may initiate LDL oxidation, e.g., by myeloperoxidase-derived hypochlorous acid, but they can also, as classical effector cells, serve as a target for biologically active oxLDL.

To prevent oxidation of biomolecules aerobic organisms are equipped with a broad spectrum of lipophilic and hydrophilic antioxidants.[64,65] These antioxidants significantly delay or inhibit oxidation of various substrates. Therefore, they constitute the first line of defense against ROS. HDLs can be placed into the second line of defense against oxidative damage. They remove and inactivate oxidized lipid species from LDL[34] and reverse or inhibit oxLDL-mediated proinflammatory and cytotoxic effects on macrophages,[35] endothelial cells,[66,67] and juxtaglomerular cells.[39] Therefore, it was of interest to investigate the potential protective effects of HDL onto oxLDL-induced PMNL activation.

The oxidant hypochlorous acid has been shown to be a candidate for lipoprotein oxidation in physiological conditions. Hypochlorite-oxidized proteins,[68] including apolipoprotein B-100,[21] have been shown to be present in human atherosclerotic lesions. Treatment of isolated HDL with reagent or enzymatically generated HOCl resulted in enhanced turnover rates of oxHDL by macrophages and decreased cholesterol efflux capacity.[53]

In the present study native human LDL and HDL were oxidized by an identical oxidation protocol after adjusting the protein concentrations to 0.2 g/l. Exposure of lipoproteins to 1 m*M* NaOCl for 40 min at 37°C resulted in marked physicochemical changes of both the lipid and protein moieties.

Modifications of the protein components were detected by enhanced electrophoretic mobility on agarose gels and increased generation of autofluorescence products at 360/430 nm. It is well established that cysteine and methionine are primary targets for HOCl.[69] In addition, reaction of HOCl with proteins gives nitrogen-centered radicals, which arise from lysine side-chain amino groups.[70] It has been shown recently that during LDL oxidation by hypochlorite these radicals initiate antioxidant consumption within the lipoprotein followed by lipid oxidation via tocopherol-mediated peroxidation.[71]

The results from our study document a considerable oxidation of the lipid components in LDL and HDL. Although there was no significant difference in the accumulation of conjugated dienes in both lipoproteins TBARS levels were tenfold higher in oxLDL when compared with oxHDL. This could be explained by their different lipid composition.

To follow the biological activities of modified lipoproteins on PMNL functional activity the respiratory burst was measured by luminol-amplified CL. The CL technique represents an ideal tool to monitor changes in PMNL ROS generation.[72–74] It allows measurement of the kinetics of respiratory burst activation and differentiation of intracellular from extracellular ROS generation.[61,75] CL assays with luminol and PMNL are characterized by high sensitivity and reproducibility. In the present study, incubation of PMNL with oxLDL but not natLDL, natHDL, or oxHDL produced a significant enhancement of the CL response. The data obtained in the present study using the CL method for ROS detection clearly indicate that a substantial part of oxygen metabolites produced during the activation of the respiratory burst by oxLDL is of extracellular origin. The alterations in PMNL ROS-generating activity were dependent upon the dose, oxidation time, and extent of LDL oxidation. Furthermore, cytochalasin B, an inhibitor of actin polymerization and phagocytosis, substantially suppressed the oxLDL-evoked CL response, raising the possibility that the oxLDL were phagocytosed upon incubation with neutrophils.

natHDL, but not oxHDL, dose-dependently diminished the effect of oxLDL on PMNL respiratory burst activities. A direct ROS scavenging activity of natHDL could be excluded from experiments with activated neutrophils. Theoretically, the observed effect of HDL could result from (1) interaction with PMNL leading to increased cellular resistance against activation of NADPH oxidase, (2) interaction with oxLDL, thereby removing or neutralizing signaling compounds for cell activation or inducing conformational changes that restrict contact to PMNL. Our preincubation studies revealed that HDL exerted its inhibitory effects independently of whether it was in previous contact with the cells or with oxLDL suggesting a fast-acting protective mechanism. By contrast, Suc et al.[67] observed protecting action of HDL against the cytotoxic effects of oxLDL on endothelial cells only after a preincubation of HDL with cells for 10 h and longer. From their experimental results they concluded that HDL acts at the cellular level by enhancing the resistance against the toxicity of oxLDL, which involves the synthesis of yet-unknown protein(s). The exact mechanisms behind the protective actions of HDL against oxLDL-induced proinflammatory actions are not known. Considering the results of Suc et al.[67] and our data, it seems, however, that both fast-acting and prolonged mechanisms could play a role in the defense against oxidative damage.

There is a growing body of evidence for the activation of signaling pathways following the interaction between HDL and cells.[76] In this respect, apoAI, the major protein constituent of HDL, could be of central importance. Competitive binding studies with PMNL indicate that these cells are provided with an apoAI receptor.[77] The data reported here support a protective role of the apoAI component. ApoAI per se was able to inhibit oxLDL-induced PMNL CL. The effect of apoAI was dependent on the integrity of the molecule since trypsinized and oxidized apoAI had no substantial inhibitory effect. However, apoAI was considerably less effective when compared with natHDL. The diminished inhibitory efficacy is supposed to be due to structural modifications during the delipidation procedure or to the presence of additional protective factors in the whole HDL particle. The latter would be consistent with our finding that not only native, but also trypsinized HDL, inhibited oxLDL-evoked respiratory burst activation. Further support for an important role of the lipid moiety in the protective effects of the HDL particle comes from studies demonstrating an antioxidant[78] and a cytoprotective[79] effect of HDL that is associated with the HDL phospholipid fraction. In experiments using [^{14}C]-lysophosphatidylcholine (LPC), Nilsson et al.[79] showed that HDL-like phospholipid particles but not apoAI alone effectively bind LPC. As a result, the cytotoxic effects of LPC on smooth muscle cells were completely abolished. LPC, a degradation product of phosphatidylcholine, is enriched in oxLDL.[80] LPC alters the physical properties of plasma membranes, thereby modifying the kinetics of transmembrane ion transport[81] and the activity of membrane-bound enzymes.[82,83] Therefore, it is conceivable that oxLDL exerts its stimulatory effects on resting PMNL with subsequent activation of the membrane-bound NADPH oxidase by incorporation of oxLDL-derived LPC into PMNL. natHDL or trypsinized HDL (with remaining phospholipid fraction) then could prevent this stimulation by removing LPC from oxLDL or from PMNL.

The elimination of LPC from oxLDL or from cells is a rapid process. Matsuda et al.[66] demonstrated that HDL reversed the inhibiting effects of oxLDL on endothelium-dependent arterial relaxation by removing LPC from oxLDL already after a period of 30 min. Our studies revealed that HDL affected oxLDL-induced PMNL respiratory burst activity in a similar short time period suggesting that the removal of LPC might be a possible mechanism contributing to the fast-acting protective actions of HDL. An alternative protective mechanism involves the action of HDL-associated enzymes including paraoxonase,[84] platelet activating factor (PAF) acetylhydrolase,[85] and lecithin cholesterol acyltransferase.[86] PON has been suggested as a factor largely responsible for the antioxidant role of HDL.[87] Active PON is able to degrade those multioxygenated phospholipids in LDL that evoke characteristic inflammatory responses in cells induced by oxLDL.[34,88] In the present study, significant PON activity was present only in natHDL. Trypsin-mediated proteolysis of HDL and of apoAI was followed by complete loss of PON activity. Moreover, in our delipidated apoAI batches no PON activity was found. Together with the findings that not only natHDL but also, although to a lesser degree, trypsinized HDL and apoAI inhibited oxLDL-induced PMNL respiratory burst activation, these results indicate a lack of a close association between PON activity and protective effects of HDL. However, the demonstration of a PON-independent protective mechanism for HDL in our experiments does not exclude any role of PON, since natHDL with intact PON activity exhibited a greater inhibitory effect against oxLDL action than trypsinized HDL or apoAI without PON activity. Rather, our results indicate that more than one mechanism contributes to the protective action of HDL against oxLDL-induced PMNL activity modulation.

Besides activating resting PMNL, oxLDL modulated the respiratory burst of activated cells depending on the stimulus used. CL response of FMLP-stimulated cells when first treated with oxLDL increased 14-fold. By contrast, zymosan-induced CL was decreased by 50% in the presence of oxLDL. Finally, oxLDL had no effect on PMA-induced neutrophil CL. The mechanisms by which oxLDL either directly activates PMNL oxidant generation or indirectly modulates respiratory burst activities are not known. A specific receptor for oxLDL on neutrophils has not yet been identified.

Since FMLP and PMA activate the PMNL NADPH oxidase at least partially through different signal transduction pathways and oxLDL had distinct effects on FMLP- and PMA-induced CL it could be supposed that the modified lipoprotein or one of its biologically active constituents like LPC affect neutrophil signal transduction.

FMLP stimulates NADPH oxidase via binding to a G protein–coupled receptor followed by the signal cascade phosphatidylinositol 3-kinase → protein kinase C (and perhaps other protein kinases) → p47phox phosphorylation → superoxide generation[89,90] and concomitant increase in cytosolic Ca^{2+} levels.[91] It is suggested that mitogen-activated protein kinases are not involved in the FMLP-induced activation of neutrophil respiratory burst.[92] PMA bypasses the pathways upstream of protein kinase C (PKC) and directly activates the enzyme. oxLDL has been shown to modify signal transduction in several cell types including macrophages,[93] monocytes,[94] and endothelial cells.[95] Recently, Martens et al.[94] demonstrated a twofold increase in phosphatidylinositol-3-kinase activity in a human monocyte-like cell line incubated with extensively oxidized LDL.

Because oxLDL stimulated only the FMLP-induced CL but did not influence PMA-mediated CL response, it may be suggested that the oxLDL-amplified signal transduction steps are located upstream of PKC.

Zymosan opsonized with serum opsonins (IgG, C3b) interacts with various receptors on neutrophils: IgG receptors, FcγRII, FcγRIII, and the receptor for C3b, C3bR. The signal transduction triggered by opsonized zymosan involves tyrosine kinase activity, extracellular-signal-regulated kinases, and cytosolic phospholipase A_2.[96] The role of PKC in zymosan-induced signaling pathway is discussed controversially. Whereas Sergeant et al.[97] suggested that PKC regulates the assembly and activation of NADPH oxidase, Hazan et al.[96] reported a PKC-independent pathway of respiratory burst activation.

The causes for the interference of oxLDL with zymosan-induced oxidant generation are unknown. Principal possible mechanisms of oxLDL effects include inhibition of the zymosan-specific intracellular signal transduction, interaction with the zymosan particles that prevents receptor coupling, or competition between oxLDL and zymosan for the Fc receptors on neutrophils.

Similar to oxLDL-mediated activation of resting PMNL natHDL significantly attenuated the effects of oxLDL on activated cells, especially when PMNL were stimulated with FMLP.

Activation of PMNL respiratory burst leads to both intracellular and extracellular generation of oxygen-derived reactive metabolites. Whereas intracellular oxidant production results mainly in the destruction of phagocytosed pathogens, the extracellular generation of ROS has the potential to initiate and force various inflammatory processes. The ratio between intracellular and extracellular oxidant production generated by PMNL strongly depends on the stimulus used. It has been shown by us previously[98] that FMLP causes cells to generate proportionally more extracellular reactive oxygen metabolites while stimulation with zymosan is associated with preferential intracellular ROS production. The results obtained in our study show that hypochlorite-modified LDL acts on PMNL by amplifying extracellular ROS generation and suppressing intracellular oxidant production. Thus, hypochlorite-oxidized LDL is able to modify PMNL activity toward increased destructive capacity of surrounding tissue by oxygen metabolites and decreased phagocytotic capability. By modulating the extracellular and intracellular ROS generation hypochlorite-modified LDL could amplify the inflammatory process present in the early stages of atherogenesis. natHDL, but not oxHDL, rapidly and efficiently inhibited the deleterious effects of oxLDL on neutrophil oxidant generation.

REFERENCES

1. Ross, R., Atherosclerosis—an inflammatory disease, *N. Engl. J. Med.*, 340, 115, 1999.
2. Steinberg, D., Low density lipoprotein oxidation and its pathobiological significance, *J. Biol. Chem.*, 272, 20,963, 1997.
3. Ylä-Herttuala, S., Is oxidized low-density lipoprotein present *in vivo*? *Curr. Opin. Lipidol.*, 9, 337, 1998.
4. Frostegard, J., Nilsson, J., Haegerstrand, A., Hamsten, A., Wigzell, H., and Gidlund, M., Oxidized low density lipoprotein induces differentiation and adhesion of human monocytes and the monocytic cell line U937, *Proc. Natl. Acad. Sci. U.S.A.*, 87, 904, 1990.
5. Quinn, M. T., Parthasarathy, S., Fong, L. G., and Steinberg, D., Oxidatively modified low density lipoproteins: a potential role in recruitment and retention of monocyte/macrophages during atherogenesis, *Proc. Natl. Acad. Sci. U.S.A.*, 84, 2995, 1987.
6. Kume, N., Cybulsky, M. I., and Gimbrone, M. A., Lysophosphatidylcholine, a component of atherogenic lipoproteins, induces mononuclear leukocyte adhesion molecules in cultured human and rabbit arterial endothelial cells, *J. Clin. Invest.*, 90, 1138, 1992.
7. Autio, I., Jaakkola, O., Solakivi, T., and Nikkari, T., Oxidized low-density lipoprotein is chemotactic for arterial smooth muscle cells in culture, *FEBS Lett.*, 277, 247, 1990.

8. Woenckhaus, C., Kaufmann, A., Bußfeld, D., Gemsa, D., Sprenger, H., and Gröne, H. J., Hypochlorite-modified LDL: chemotactic potential and chemokine induction in human monocytes, *Clin. Immunol. Immunopathol.,* 86, 27, 1998.
9. Vora, D. K., Fang, Z. T., Liva, S. M., Tyner, T. R., Parhami, F., Watson, A. D., Drake, T. A., Territo, M. C., and Berliner, J. A., Induction of P-selectin by oxidized lipoproteins, *Circ. Res.,* 80, 810, 1997.
10. Cushing, S. D., Berliner, J. A., Valente, A. J., Territo, M. C., Navab, M., Parhami, F., Gerrity, R., Schwartz, C. J., and Fogelman, A. M., Minimally modified low density lipoprotein induces monocyte chemotactic protein 1 in human endothelial cells and smooth muscle cells, *Proc. Natl. Acad. Sci. U.S.A.,* 87, 5134, 1990.
11. O'Brien, K. D., Allen, M. D., McDonald, T. O., Chait, A., Harlan, J. M., Fishbein, D., McCarty, J., Ferguson, M., Hudkins, K., Benjamin, C. D., Lobb, R., and Alpers, C. E., Vascular cell adhesion molecule-1 is expressed in human coronary atherosclerotic plaques, *J. Clin. Invest.,* 92, 945, 1993.
12. Liao, L., Starzyk, R. M., and Granger, D. N., Molecular determinants of oxidized low-density lipoprotein-induced leukocyte adhesion and microvascular dysfunction, *Arterioscler. Thromb. Vasc. Biol.,* 17, 437, 1997.
13. Hamilton, T. A., Ma, G. P., and Chisolm, G. M., Oxidized low density lipoprotein suppresses the expression of tumor necrosis factor-α mRNA in stimulated murine peritoneal macrophages, *J. Immunol.,* 144, 2343, 1990.
14. Fong, L. G., Fong, T. A., and Cooper, A. D., Inhibition of lipopolysaccharide-induced interleukin-1 beta mRNA expression in mouse macrophages by oxidized low density lipoprotein, *J. Lipid. Res.,* 32, 1899, 1991.
15. Terkeltaub, R., Banka, C. L., Solan, J., Santoro, D., Brand, K., and Curtiss, L. K., Oxidized LDL induces monocytic cell expression of interleukin-8, a chemokine with T-lymphocyte chemotactic activity, *Arterioscler. Thromb.,* 14, 47, 1994.
16. Rajavashisth, T. B., Andalibi, A., Territo, M. C., Berliner, J. A., Navab, M., Fogelman, A. M., and Lusis, A. J., Induction of endothelial cell expression of granulocyte and macrophage colony-stimulating factors by modified low-density lipoproteins, *Nature,* 344, 254, 1990.
17. Dimmeler, S., Haendeler, J., Galle, J., and Zeiher, A. M., Oxidized low-density lipoprotein induces apoptosis of human endothelial cells by activation of CPP32-like proteases, *Circulation,* 95, 1760, 1997.
18. Zhu, W., Roma, P., Pirillo, A., Pellegatta, F., and Catapano, A. L., Human endothelial cells exposed to oxidized LDL express hsp70 only when proliferating, *Arterioscler. Thromb. Vasc. Biol.,* 16, 1104, 1996.
19. Jimi, S., Saku, K., Uesugi, N., Sakata, N., and Takebayashi, S., Oxidized low density lipoprotein stimulates collagen production in cultured arterial smooth muscle cells, *Atherosclerosis,* 116, 15, 1995.
20. Ylä-Herttuala, S., Palinski, W., Rosenfeld, M. E., Parthasarathy, S., Carew, T. E., Butler, S., Witztum, J. L., and Steinberg, D., Evidence for the presence of oxidatively modified low density lipoprotein in atherosclerotic lesions of rabbit and man, *J. Clin. Invest.,* 84, 1086, 1989.
21. Hazell, L. J., Arnold, L., Flowers, D., Waeg, G., Malle, E., and Stocker, R., Presence of hypochlorite-modified proteins in human atherosclerotic lesions, *J. Clin. Invest.,* 97, 1535, 1996.
22. Nagelkerke, J. F., Havekes, L., van Hinsbergh, V. W., and van Berkel, T. J., *In vivo* catabolism of biologically modified LDL, *Arteriosclerosis,* 4, 256, 1984.
23. Patterson, R. A. and Leake, D. S., Human serum, cysteine and histidine inhibit the oxidation of low density lipoprotein less at acidic pH, *FEBS Lett.,* 434, 317, 1998.
24. Dabbagh, A. J. and Frei, B., Human suction blister interstitial fluid prevents metal ion-dependent oxidation of low density lipoprotein by macrophages and in cell-free systems, *J. Clin. Invest.,* 96, 1958, 1995.
25. Leake, D. S. and Rankin, S. M., The oxidative modification of low-density lipoproteins by macrophages, *Biochem. J.,* 270, 741, 1990.
26. Juul, K., Nielsen, L. B., Munkholm, K., Stender, S., and Nordestgaard, B. G., Oxidation of plasma low-density lipoprotein accelerates its accumulation and degradation in the arterial wall *in vivo*, *Circulation,* 94, 1698, 1996.
27. Wu, R., Nityanand, S., Berglund, L., Lithell, H., Holm, G., and Lefvert, A. K., Antibodies against cardiolipin and oxidatively modified LDL in 50-year-old men predict myocardial infarction, *Arterioscler. Thromb. Vasc. Biol.,* 17, 3159, 1997.

28. Holvoet, P., Stassen, J. M., Van Cleemput, J., Collen, D., and Vanhaecke, J., Oxidized low density lipoproteins in patients with transplant-associated coronary artery disease, *Arterioscler. Thromb. Vasc. Biol.,* 18, 100, 1998.
29. Toikka, J. O., Niemi, P., Ahotupa, M., Niinikoski, H., Viikari, J. S. A., Rönnemaa, T., Hartiala, J. J., and Ratakari, O. T., Large-artery elastic properties in young men: relationships to serum lipoproteins and oxidized low-density lipoproteins, *Arterioscler. Thromb. Vasc. Biol.,* 19, 436, 1999.
30. Miller, N. E., Associations of high-density lipoprotein subclasses and apolipoproteins with ischemic heart disease and coronary atherosclerosis, *Am. Heart J.,* 113, 589, 1987.
31. Johansson, J., Carlson, L. A., Landou, C., and Hamsten, A., High density lipoproteins and coronary atherosclerosis. A strong inverse relation with the largest particles is confined to normotriglyceridemic patients, *Arterioscler. Thromb.,* 11, 174, 1991.
32. Brown, M. S., Ho, Y. K., and Goldstein, J. L., The cholesteryl ester cycle in macrophage foam cells. Continual hydrolysis and re-esterification of cytoplasmic cholesteryl esters, *J. Biol. Chem.,* 255, 9344, 1980.
33. Parthasarathy, S., Barnett, J., and Fong, L. G., High-density lipoprotein inhibits the oxidative modification of low-density lipoprotein, *Biochim. Biophys. Acta,* 1044, 275, 1990.
34. Watson, A. D., Berliner, J. A., Hama, S. Y., La Du, B. N., Faull, K. F., Fogelman, A. M., and Navab, M., Protective effect of high density lipoprotein associated paraoxonase. Inhibition of the biological activity of minimally oxidized low density lipoprotein, *J. Clin. Invest.,* 96, 2882, 1995.
35. Sakai, M., Miyazaki, A., Hakamata, H., Sasaki, T., Yui, S., Yamazaki, M., Shichiri, M., and Horiuchi, S., Lysophosphatidylcholine plays an essential role in the mitogenic effect of oxidized low density lipoprotein on murine macrophages, *J. Biol. Chem.,* 269, 31,430, 1994.
36. Wieland, E., Brandes, A., Armstrong, V. W., and Oellerich, M., Oxidative modification of low density lipoproteins by human polymorphonuclear leukocytes, *Eur. J. Clin. Chem. Clin. Biochem.,* 31, 725, 1993.
37. Smith, J. A., Neutrophils, host defense, and inflammation: a double-edged sword, *J. Leukocyte Biol.,* 56, 672, 1994.
38. Riis Hansen, P., Kharazmi, A., Jauhiainen, M., and Ehnholm, C., Induction of oxygen free radical generation in human monocytes by lipoprotein(a), *Eur. J. Clin. Invest.,* 24, 497, 1994.
39. Galle, J., Heinloth, A., Schwedler, S., and Wanner, C., Effect of HDL and atherogenic lipoproteins on formation of $O_2^{\bar{\cdot}}$ and renin release in juxtaglomerular cells, *Kidney Int.,* 51, 253, 1997.
40. Hurtado, I., Fiol, C., Gracia, V., and Caldu, P., *In vitro* oxidised HDL exerts a cytotoxic effect on macrophages, *Atherosclerosis,* 125, 39, 1996.
41. Nagano, Y., Arai, H., and Kita, T., High density lipoprotein loses its effect to stimulate efflux of cholesterol from foam cells after oxidative modification, *Proc. Natl. Acad. Sci. U.S.A.,* 88, 6457, 1991.
42. Therond, P., Bonnefont-Rousselot, D., Laureaux, C., Vasson, M. P., Motta, C., Legrand, A., and Delattre, J., Copper oxidation of *in vitro* dioleolylphosphatidylcholine-enriched high-density lipoproteins: physicochemical features and cholesterol effluxing capacity, *Arch. Biochem. Biophys.,* 362, 139, 1999.
43. Pietzsch, J., Subat, S., Nitzsche, S., Leonhardt, W., Schentke, K. U., and Hanefeld, M., Very fast ultracentrifugation of serum lipoproteins: influence on lipoprotein separation and composition, *Biochim. Biophys. Acta,* 1254, 77, 1995.
44. Leonhardt, W., Pietzsch, J., Julius, U., and Hanefeld, M., Recovery of cholesterol and triacylglycerol in very-fast ultracentrifugation of human lipoproteins in a large range of concentrations, *Eur. J. Clin. Chem. Clin. Biochem.,* 32, 929, 1994.
45. Scanu, A. M. and Edelstein, C., Solubility in aqueous solutions of ethanol of the small molecular weight peptides of the serum very low density and high density lipoproteins: relevance to the recovery problem during delipidation of serum lipoproteins, *Anal. Biochem.,* 44, 576, 1971.
46. Weisweiler, P., Friedl, C., and Ungar, M., Isolation and quantitation of apolipoproteins A-I and A-II from human high-density lipoproteins by fast-protein liquid chromatography, *Clin. Chim. Acta,* 169, 249, 1987.
47. Arnhold, J., Wiegel, D., Richter, O., Hammerschmidt, S., Arnold, K., and Krumbiegel, M., Modification of low density lipoproteins by sodium hypochlorite, *Biomed. Biochim. Acta,* 50, 967, 1991.
48. Hazell, L. J. and Stocker, R., Oxidation of low-density lipoprotein with hypochlorite causes transformation of the lipoprotein into a high-uptake form for macrophages, *Biochem. J.,* 290, 165, 1993.

49. Hazell, L. J., van den Berg, J. J. M., and Stocker, R., Oxidation of low-density lipoprotein by hypochlorite causes aggregation that is mediated by modification of lysine residues rather than lipid oxidation, *Biochem. J.,* 302, 297, 1994.
50. Stelmaszynska, T., Kukovetz, E., Egger, G., and Schaur, R. J., Possible involvement of myeloperoxidase in lipid peroxidation, *Int. J. Biochem.,* 24, 121, 1992.
51. Daugherty, A., Dunn, J. L., Rateri, D. L., and Heinecke, J. W., Myeloperoxidase, a catalyst for lipoprotein oxidation, is expressed in human atherosclerotic lesions, *J. Clin. Invest.*, 94, 437, 1994.
52. Hazen, S. L. and Heinecke, J. W., 3-Chlorotyrosine, a specific marker of myeloperoxidase-catalyzed oxidation, is markedly elevated in low density lipoprotein isolated from human atherosclerotic intima, *J. Clin. Invest.*, 99, 2075, 1997.
53. Panzenboeck, U., Raitmayer, S., Reicher, H., Lindner, H., Glatter, O., Malle, E., and Sattler, W., Effects of reagent and enzymatically generated hypochlorite on physicochemical and metabolic properties of high density lipoproteins, *J. Biol. Chem.,* 272, 29,711, 1997.
54. Morris, C. J., The acid ionization constant of HOCl from 5 to 35°, *J. Phys. Chem.,* 70, 3798, 1966.
55. Arnhold, J., Panasenko, O. M., Schiller, J., Vladimirov, Y. A., and Arnold, K., The action of hypochlorous acid on phosphatidylcholine liposomes in dependence on the content of double bonds. Stoichiometry and NMR analysis, *Chem. Phys. Lipids,* 78, 55, 1995.
56. Maeba, R., Shimasaki, H., and Ueta, N., Conformational changes in oxidized LDL recognized by mouse peritoneal macrophages, *Biochim. Biophys. Acta,* 1215, 79, 1994.
57. Yagi, K., A simple fluorometric assay for lipoperoxide in blood plasma, *Biochem. Med.,* 15, 212, 1976.
58. Dantoine, T. F., Debord, J., Charmes, J. P., Merle, L., Marquet, P., Lachatre, G., and Leroux-Robert, C., Decrease of serum paraoxonase activity in chronic renal failure, *J. Am. Soc. Nephrol.,* 9, 2082, 1998.
59. Cominacini, L., Garbin, U., Davoli, A., Micciolo, R., Bosello, O., Gaviraghi, G., Scuro, L. A., and Pastorino, A. M., A simple test for predisposition to LDL oxidation based on the fluorescence development during copper-catalyzed oxidative modification, *J. Lipid. Res.,* 32, 349, 1991.
60. Esterbauer, H., Jürgens, G., Quehenberger, O., and Koller, E., Autoxidation of human low density lipoprotein: loss of polyunsaturated fatty acids and vitamin E and generation of aldehydes, *J. Lipid. Res.,* 28, 495, 1987.
61. Dahlgren, C., Effects on extra- and intracellularly localized, chemoattractant-induced, oxygen radical production in neutrophils following modulation of conditions for ligand-receptor interaction, *Inflammation,* 12, 335, 1988.
62. Lundqvist, H., Follin, P., Khalfan, L., and Dahlgren, C., Phorbol myristate acetate-induced NADPH oxidase activity in human neutrophils: only half the story has been told, *J. Leukocyte Biol.,* 59, 270, 1996.
63. Gennaro, R. and Romeo, D., The release of superoxide anion from granulocytes: effect of inhibitors of anion permeability, *Biochem. Biophys. Res. Commun.,* 88, 44, 1979.
64. Frei, B., Stocker, R., and Ames, B. N., Antioxidant defenses and lipid peroxidation in human blood plasma, *Proc. Natl. Acad. Sci. U.S.A.,* 85, 9748, 1988.
65. Halliwell, B. and Gutteridge, J. M. C., The antioxidants of human extracellular fluids, *Arch. Biochem. Biophys.,* 280, 1, 1990.
66. Matsuda, Y. K., Hirata, K., Inoue, N., Suematsu, M., Kawashima, S., Akita, H., and Yokoyama, M., High density lipoprotein reverses inhibitory effect of oxidized low density lipoprotein on endothelium-dependent arterial relaxation, *Circ. Res.,* 72, 1103, 1993.
67. Suc, I., Escargueil-Blanc, I., Troly, M., Salvayre, R., and Negre-Salvayre, A., HDL and ApoA prevent cell death of endothelial cells induced by oxidized LDL, *Arterioscler. Thromb. Vasc. Biol.,* 17, 2158, 1997.
68. Fu, S., Davies, M. J., Stocker, R., and Dean, R. T., Evidence for roles of radicals in protein oxidation in advanced human atherosclerotic plaque, *Biochem. J.,* 333, 519, 1998.
69. Winterbourn, C. C., Comparative reactivities of various biological compounds with myeloperoxidase-hydrogen peroxide-chloride, and similarity of the oxidant to hypochlorite, *Biochim. Biophys. Acta,* 840, 204, 1985.
70. Hawkins, C. L. and Davies, M. J., Hypochlorite-induced damage to proteins: formation of nitrogen-centred radicals from lysine residues and their role in protein fragmentation, *Biochem. J.,* 332, 617, 1998.
71. Hazell, L. J., Davies, M. J., and Stocker, R., Secondary radicals derived from chloramines of apolipoprotein B-100 contribute to HOCl-induced lipid peroxidation of low-density lipoproteins, *Biochem. J.,* 339, 489, 1999.

72. Faden, H., Luminol-dependent whole blood chemiluminescence assay, in *Cellular Chemiluminescence,* Van Dyke, K. and Castranova, V., Eds., Vol. 2, CRC Press, Boca Raton, FL, 1987, 184.
73. Stevens, D. L, Bryant, A. E., Huffman, J., Thompson, K., and Allen, R. C., Analysis of circulating phagocyte activity measured by whole blood luminescence: correlations with clinical status, *J. Infect. Dis.,* 170, 1463, 1994.
74. Kopprasch, S., Graessler, J., Kohl, M., Bergmann, S., and Schröder, H. E., Comparison of circulating phagocyte oxidative activity measured by chemiluminescence in whole blood and isolated polymorphonuclear leukocytes, *Clin. Chim. Acta,* 253, 145, 1996.
75. Kopprasch, S., Müller, S., Köhler, T., and Schröder, H. E., Characterization of the chemiluminescence response of polymorphonuclear leukocytes stimulated with urate crystals, *J. Biolumin. Chemilumin.,* 11, 213, 1996.
76. Fidge, N. H., High density lipoprotein receptors, binding proteins, and ligands, *J. Lipid. Res.,* 40, 187, 1999.
77. Blackburn, W. D., Dohlman, J. G., Venkatachalapathi, Y. V., Pillion, D. J., Koopman, W. J., Segrest, J. P., and Anantharamaiah, G. M., Apolipoprotein A-I decreases neutrophil degranulation and superoxide production, *J. Lipid. Res.,* 32, 1911, 1991.
78. Graham, A., Hassall, D. G., Rafique, S., and Owen, J. S., Evidence for a paraoxonase-independent inhibition of low-density lipoprotein oxidation by high-density lipoprotein, *Atherosclerosis,* 135, 193, 1997.
79. Nilsson, J., Dahlgren, B., Ares, M., Westman, J., Hultgardh Nilsson, A., Cercek, B., and Shah, P. K., Lipoprotein-like phospholipid particles inhibit the smooth muscle cell cytotoxicity of lysophosphatidylcholine and platelet-activating factor, *Arterioscler. Thromb. Vasc. Biol.,* 18, 13, 1998.
80. Steinbrecher, U. P., Parthasarathy, S., Leake, D. S., Witztum, J. L., and Steinberg, D., Modification of low density lipoprotein by endothelial cells involves lipid peroxidation and degradation of low density lipoprotein phospholipids, *Proc. Natl. Acad. Sci. U.S.A.,* 81, 3883, 1984.
81. Sedlis, S. P., Corr, P. B., Sobel, B. E., and Ahumada, G. G., Lysophosphatidylcholine potentiates Ca^{2+} accumulation in rat cardiac myocytes, *Am. J. Physiol.,* 244, H32, 1983.
82. Owens, K., Kennett, F. F., and Weglicki, W. B., Effects of fatty acid intermediates on Na^+-K^+-ATPase activity of cardiac sarcolemma, *Am. J. Physiol.,* 242, H456, 1982.
83. Shier, W. T., Baldwin, J. H., Nilsen-Hamilton, M., Hamilton, R. T., and Thanassi, N. M., Regulation of guanylate and adenylate cyclase activities by lysolecithin, *Proc. Natl. Acad. Sci. U.S.A.,* 73, 1586, 1976.
84. Mackness, M. I., Arrol, S., and Durrington, P. N., Paraoxonase prevents accumulation of lipoperoxides in low-density lipoprotein, *FEBS Lett.,* 286, 152, 1991.
85. Watson, A. D., Navab, M., Hama, S. Y., Sevanian, A., Prescott, S. M., Stafforini, D. M., McIntyre, T. M., La Du, B. N., Fogelman, A. M., and Berliner, J. A., Effect of platelet activating factor-acetylhydrolase on the formation and action of minimally oxidized low density lipoprotein, *J. Clin. Invest.,* 95, 774, 1995.
86. Liu, M. and Subbaiah, P. V., Hydrolysis and transesterification of platelet-activating factor by lecithin-cholesterol acyltransferase, *Proc. Natl. Acad. Sci. U.S.A.,* 91, 6035, 1994.
87. Shih, D. M., Gu, L., Hama, S., Xia, Y. R., Navab, M., Fogelman, A. M., and Lusis, A. J., Genetic-dietary regulation of serum paraoxonase expression and its role in atherogenesis in a mouse model, *J. Clin. Invest.,* 97, 1630, 1996.
88. Aviram, M., Rosenblat, M., Bisgaier, C. L., Newton, R. S., Primo-Parmo, S. L., and La Du, B. N., Paraoxonase inhibits high-density lipoprotein oxidation and preserves its functions. A possible peroxidative role for paraoxonase, *J. Clin. Invest.,* 101, 1581, 1998.
89. Vlahos, C. J., Matter, W. F., Brown, R. F., Traynor-Kaplan, A. E., Heyworth, P. G., Prossnitz, E. R., Ye, R. D., Marder, P., Schelm, J. A., Rothfuss, K. J., Serlin, B. S., and Simpson, P. J., Investigation of neutrophil signal transduction using a specific inhibitor of phosphatidylinositol 3-kinase, *J. Immunol.,* 154, 2413, 1995.
90. Kodama, T., Hazeki, K., Hazeki, O., Okada, T., and Ui, M., Enhancement of chemotactic peptide-induced activation of phosphoinositide 3-kinase by granulocyte-macrophage colony-stimulating factor and its relation to the cytokine-mediated priming of neutrophil superoxide-anion production, *Biochem. J.,* 337, 201, 1999.

91. Foyouzi-Youssefi, R., Petersson, F., Lew, D. P., Krause, K. H., and Nüsse, O., Chemoattractant-induced respiratory burst: increase in cytosolic Ca^{2+} concentrations are essential and synergize with a kinetically distinct second signal, *Biochem. J.,* 322, 709, 1997.
92. Coffer, P. J., Geijsen, N., M'Rabet, L., Schweizer, R. C., Maikoe, T., Raaijmakers, J. A. M., Lammers, J. W. J., and Koenderman, L., Comparison of the roles of mitogen-activated protein kinase kinase and phosphatidylinositol 3-kinase signal transduction in neutrophil effector function, *Biochem. J.,* 329, 121, 1998.
93. Matsumura, T., Sakai, M., Kobori, S., Biwa, T., Takemura, T., Matsuda, H., Hakamata, H., Horiuchi, S., and Shichiri, M., Two intracellular signaling pathways for activation of protein kinase C are involved in oxidized low-density lipoprotein-induced macrophage growth, *Arterioscler. Thromb. Vasc. Biol.,* 17, 3013, 1997.
94. Martens, J. S., Reiner, N. E., Herrera-Velit, P., and Steinbrecher, U. P., Phosphatidylinositol 3-kinase is involved in the induction of macrophage growth by oxidized low density lipoprotein, *J. Biol. Chem.,* 273, 4915, 1998.
95. Li, D., Yang, B., and Mehta, J. L., Ox-LDL induces apoptosis in human coronary artery endothelial cells: role of PKC, PTK, bcl-2, and Fas, *Am. J. Physiol.,* 275, H568, 1998.
96. Hazan, I., Dana, R., Granot, Y., and Levy, R., Cytosolic phospholipase A_2 and its mode of activation in human neutrophils by opsonized zymosan. Correlation between 42/44 kDa mitogen-activated protein kinase, cytosolic phospholipase A_2 and NADPH oxidase, *Biochem. J.,* 326, 867, 1997.
97. Sergeant, S. and McPhail, L. C., Opsonized zymosan stimulates the redistribution of protein kinase C isoforms in human neutrophils, *J. Immunol.,* 159, 2877, 1997.
98. Kopprasch, S., Gatzweiler, A., Graessler, J., and Schröder, H. E., Beta-adrenergic modulation of FMLP- and zymosan-induced intracellular and extracellular oxidant production by polymorphonuclear leukocytes, *Mol. Cell. Biochem.,* 168, 133, 1997.

26 Peroxynitrite-Based Luminol Luminescence of Macrophages and Enchancement of the Signal

Knox Van Dyke, Paul McConnell, Michael Taylor, and Laura Frost

CONTENTS

26.1 HISTORY OF MACROPHAGE CHEMILUMINESCENCE

Initially, it was thought that macrophages could not produce light, but using luminol we and others[1,2] showed that macrophages can produce luminol-dependent chemiluminescence (LDCL). Macrophages produce much less light in the unstimulated state than do neutrophils (~1/100). In fact, macrophages produce light without the addition of luminol as do neutrophils, but without luminol it takes many more cells to observe CL.[3] Luminol when oxidized at about neutral pH emits light in the blue region about 425 nm. This corresponds to the highest sensitivity of light detection for bialkali photomultiplier tubes used in a standard scintillation counters—the instrument originally used to measure light in out-of-coincidence mode.[1] The original published work used luminol in in-coincidence mode with saturating amounts of luminol.[2] These conditions lead to nonreproducible quantitative conditions regarding phagocytic cells and light. In our first experiment, we actually ruined a photomultiplier tube because the correct conditions for the use of luminol and cells were not known. We solved the problem by using out-of-coincidence mode and smaller concentrations

0-8493-0719-8/02/$0.00+$1.50

of luminol. High concentrations of luminol with high cell numbers (neutrophils) result in an unusual metabolic state that can activate cells without a stimulant of the oxidative burst. The correct conditions of pH and the use of bystander molecules or enhancers must be maintained to yield interpretable results. There are a variety of times when the isolated cells are few in number or the cells are not available from humans or animals in sufficient quantities to perform characterization studies to ascertain the origin of a particular toxic chemistry relevant to a study. There is a distinct need to develop the most sensitive assay available.

26.2 IMPORTANCE OF MACROPHAGE MEASUREMENT IN DISEASES

Most of the studies of the immune system have been directed toward T and B lymphocytes. It was generally thought that these cells formed the basis for most of our immunity to foreign invaders, e.g., bacteria, fungi, and viruses. Certainly these cells are important in protection, but they are far from the majority of the cells acting in this capacity. Phagocytic cells—monocytes/macrophages and granulocytes, neutrophils, basophils, eosinophils—produce killing mechanisms that control the amount of invading organisms and that often trigger inappropriate chemistry that can induce many disease states that are related to inflammation or immunity. An example of this is the study of human immunodeficiency disease. It was thought that because certain T lymphocytes decreased or disappeared after HIV infection, T cell loss must be the cause not the effect of the disease. However, T cells are not involved in the majority of the killing of invaders because, when their numbers decrease to zero, many people are still healthy. But if the monocyte/macrophage cells become decreased or their function has been impaired, the host will certainly die from overwhelming infections even of trivial origin.[4] This is clearly the case in immunodeficiency of genetic origin not caused by viral infection or from chronic granulomatous disease caused by a series of genetic deficiences in the mechanism that produces superoxide, i.e., NADPH oxidase. This defect causes a deficiency in production of peroxynitrite superoxide and, in turn, hydrogen peroxide, which allows certain bacteria to flourish. Macrophages are a mainstay of the host immune system and they produce the necessary interleukin, which allows the T lymphocytes to maintain their number.

Macrophages are an initial processing cell for antigens. The killing and control of killing belong more to the macrophage. The T cell appears to be mostly a bystander in the disease. Phagocytic cells have taken a back seat in the minds of many immunologists. Assays that measure the killing mechanisms are the key to understanding how the immune system protects us. The key to the chemistry of the killing reactions is the reactions peroxynitrite/peroxidase and hypochlorite/myeloperoxidase. The former occurs in macrophages and both occur in neutrophils or polymorphonuclear leukocytes. NO synthase II may not be as inducible in neutrophils.

26.3 CHEMISTRY OF LIGHT IS MEASUREMENT OF KILLING REACTIONS USING LUMINOL AS A BYSTANDER LIGHT EMITTER

The chemistry of killing mechanisms is only beginning to be understood. This is illustrated in Figure 26.1. We know it takes an extremely toxic chemical to kill foreign organisms. Further, the killing would have to be controlled and/or compartmentalized to kill the invader selectively and to leave the host cells intact. It should be a selective toxicity directed toward invading microorganisms and not the host tissue. It should not be surprising that the same chemistry that could kill an organism would be a strong oxidizer and that this same mechanism(s) could oxidize luminol to produce blue light! The measurement of the oxidation of luminol at neutral pH is tantamount to measuring the ability of the cell to kill foreign invaders (Figure 26.2). Instead of killing an invading organism, the peroxynitrite or its carbon dioxide derivative (peroxynitrite reacts with available

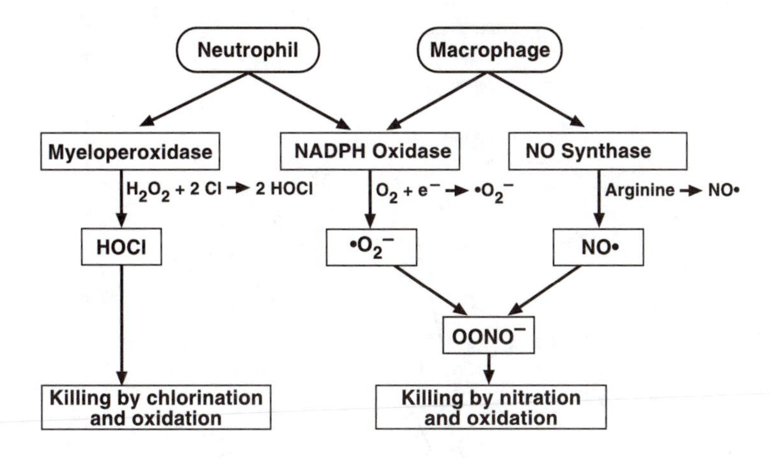

FIGURE 26.1 Host defense killing mechanisms.

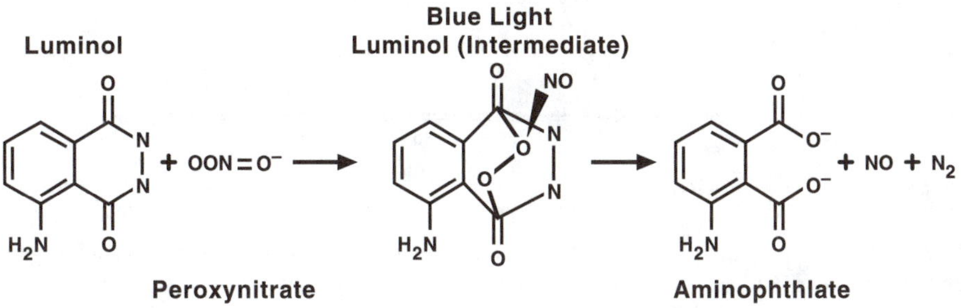

Luminol + $OON{=}O^-$ → Blue Light Luminol (Intermediate) → Aminophthlate + NO + N_2

Peroxynitrate

FIGURE 26.2 Possible peroxynitrite reactions with luminol to produce blue light.

carbon dioxide) reacts with luminol, which is raised to an excited state and produces light.[5,6] Therefore, the luminol acts as a bystander molecule producing light that is easily measured using a sensitive light-gathering instrument known as a luminometer. For best results, the photomultiplier tube of the instrument should match the light detection from a set color or wavelength of light produced from the chemical reaction (luminol produces blue light at approximately 425 nm) to the highest sensitivity of the photomultiplier tube. Therefore, a best signal-to-noise (background) ratio with correct electronic gain produces optimal sensitivity from a given instrument. If a chemical reaction could be enhanced even further, e.g., by boosting the excited state of luminol, we could maximize the efficiency of detection of emitted light. (See Chapter 2 for details.)

26.4 ENHANCED LIGHT MEASURES MACROPHAGE KILLING AND HEALTH OF MACROPHAGES

Enhanced cellular luminescence allows the measurement of macrophage cellular killing reactions at 10- to 50-fold or higher sensitivity (Figures 26.3 through 26.5). In addition, the light that is emitted is a sensitive measurement of the health of the macrophage. One can measure a decrease in light emitted by phagocytic cells as the cell ages over time. The luminol light will decrease an order of magnitude earlier than the measurement of cellular toxicity using trypan blue exclusion.[7] This would apply to luminol-enhanced light, but with enhancement, fewer cells are necessary.

Enhancement of the cellular reaction depends on the genetic activation of the chemistry of oxidation. At first, we measured luminol-dependent light emitted from macrophages that we isolated

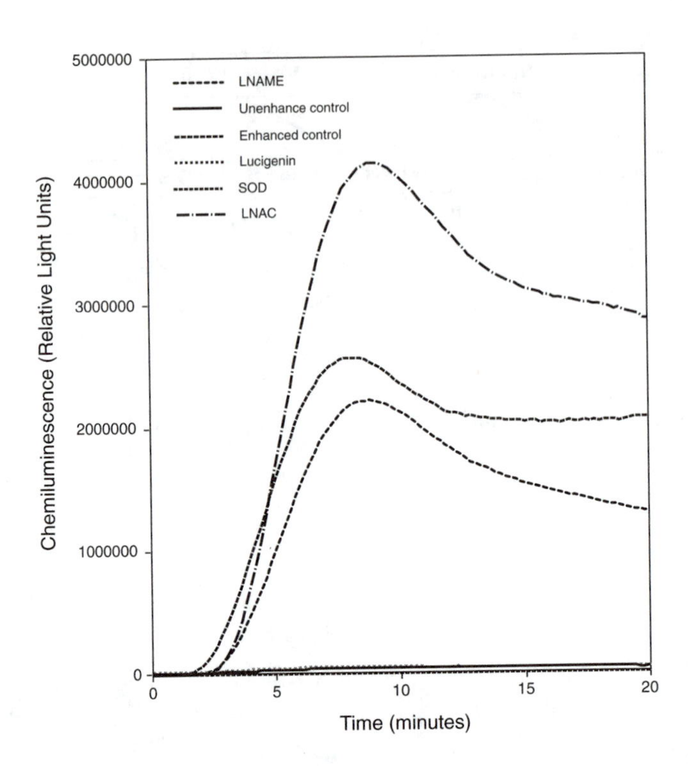

FIGURE 26.3 Enhanced luminol dependent luminescence of U937 monocytes 5 days after γ-interferon stimulated phorbol myristate acetate with and without inhibitors. One curve has lucigenin replacing luminol. The inhibitors are !mg/ml L-NAME (NO synthase II), superoxide dismutase-I mg/ml (superoxide), and *N*-acetylcysteine (antioxidant). Assay done immediately.

from lung that had not undergone activation via inflammation. It gave only a small fraction of the light per cell compared with stimulated neutrophils.[3] Since it was known that macrophages played an important role in killing and inflammation, why was so little light produced? Later, when we were studying macrophage-hybridoma cells activated by γ-interferon and lipopolysaccharide a, 30- to 40-fold activation of luminol light occurred.[8] It was as if the activation of macrophages greatly stimulated this macrophage and therefore light or killing mechanisms could be turned on whenever necessary, almost akin to a light switch. What produced all this extra light?

When we studied the effect of silica instillation into the lungs of animals and measured the luminol-dependent light from macrophages 24 h later, we found that the light per cell had been activated 10- to 20-fold. Clearly, the genetic mechanisms inside the macrophages, and possibly the neutrophils as well, had been induced to handle the foreign silica.[9,10] What were these mechanisms? Since the luminol-dependent light was inhibited by superoxide dismutase and the NO synthase inhibitor L-NAME, peroxynitrite (OON=O–) was a likely candidate. Indeed, it was shown that NO synthase II was induced in these cells.[11] The steroidal anti-inflammatory drug dexamethasone given prior to silica instillation could block the induction of NO synthase and probably NADPH oxidase as well. Therefore, the steroid prevented the increased production of light via a peroxynitrite-dependent luminol light via a NF-κb mechanism.[12,13] This nuclear genetic promoter mechanism is known to be involved with the induction of NO synthase II (Figure 26.6a). The NADPH oxidase system, a membrane-bound enzyme that reduces molecular oxygen into superoxide, must also be increased since the free radical superoxide O_2^- reacts with the free radical nitric oxide that is formed from inducible NO synthase II to form peroxynitrite anion. Figure 26.6b is a diagram of the nonactivated NADPH oxidase. The major protein pieces of the enzyme are

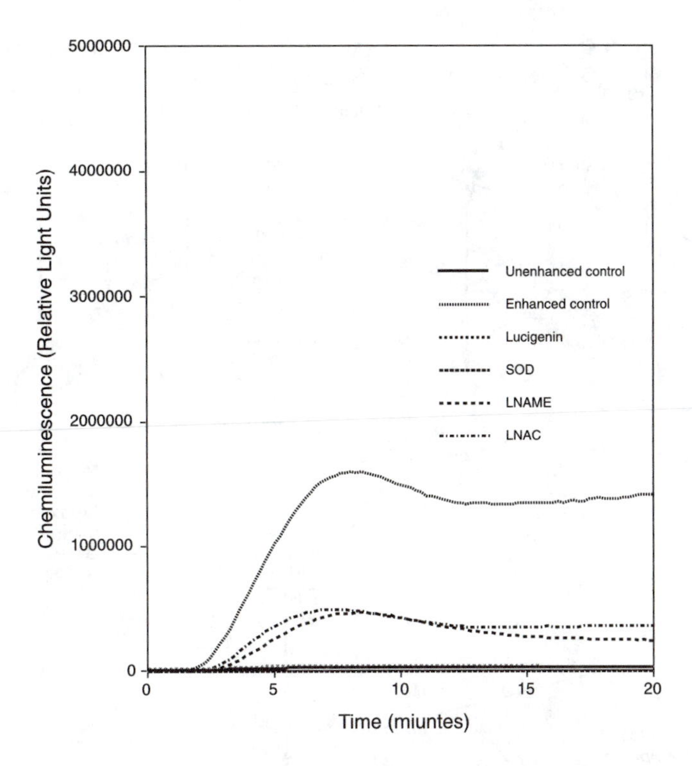

FIGURE 26.4 Same conditions as Figure 26.3 except assay done 1 h later than Figure 26.3.

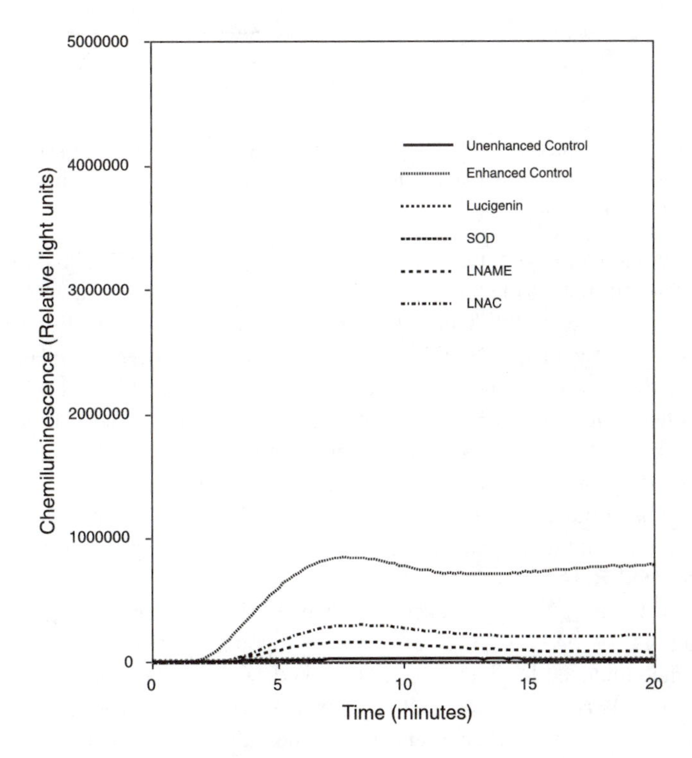

FIGURE 26.5 Same conditions as Figures 26.3 and 26.4 except assay done 1 h after Figure 26.4.

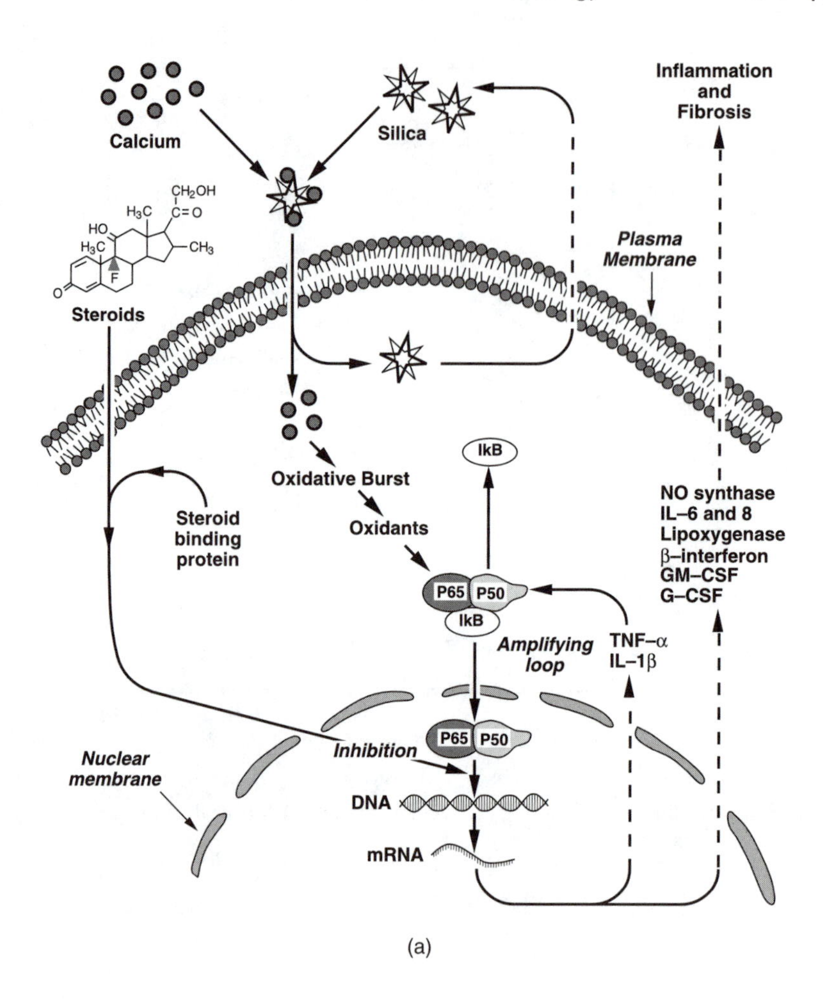

FIGURE 26.6 (a) Diagram depicts some of the possible pathways of nuclear factor kappa B, which controls NF-κ-B binding sites on promoters belonging to several key inflammatory cytokines and growth factors involved in the inflammatory process.

bound in the membrane of phagocytic cells and are gp 91phox and p22-phox. When the enzyme is activated it is believed that various protien such as p40, p67, p47 Rap 1A, and p2/rac1/2 migrate to the membrane and attach in the manner shown in the diagram. An electron from the cofactor NADPH is trasferred to oxygen and $NADP^+$ and H^+ are produced. When this electron is transferred to molecular oxygen, it produces the free radical superoxide anion ($O_2^{\bar{\cdot}}$). This avidly reacts with the free radical nitric oxide and produces peroxynitrite. If luminol is added the peroxynitrite or its carbon dioxide derivative reacts to produce blue light at 425 nm. It is known that the free radicals (NO and $O_2^{\bar{\cdot}}$) have a high attraction for each other and, if found in sufficient quantity, they will produce substantial amounts of $OONO^-$ (Figure 26.7). This strong oxidant, peroxynitrite, attacks all the major biochemical entities, namely, carbohydrates, lipids, proteins, enzymes, and nucleic acids DNA and RNA.

Some antioxidants are quickly attacked by peroxynitrite such as sulfhydryl-containing compounds and mono-, di-, tri-, and polyphenols.[14] These compounds protect against peroxynitrite attack by destroying it, and enhanced luminescence is inhibited, which allows one to measure the effects of these antioxidants easily with fewer cells and in real time. Recently, it has been shown that peroxynitrite reacts with carbon dioxide and the compound that results from this interaction can react with luminol. Whether the luminol reacts directly with peroxynitrite or the carbon dioxide derivative, CL or light results.

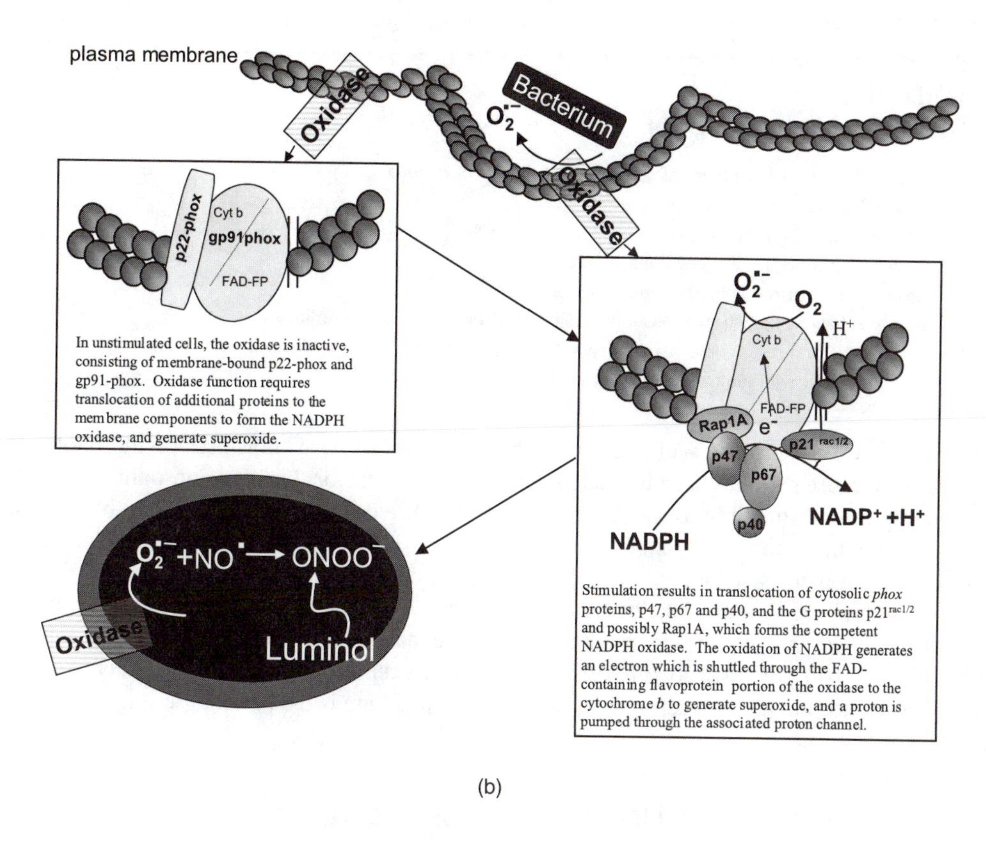

FIGURE 26.6 (*Continued*) (b) Diagram of NADPH oxidase (in active and inactive states).

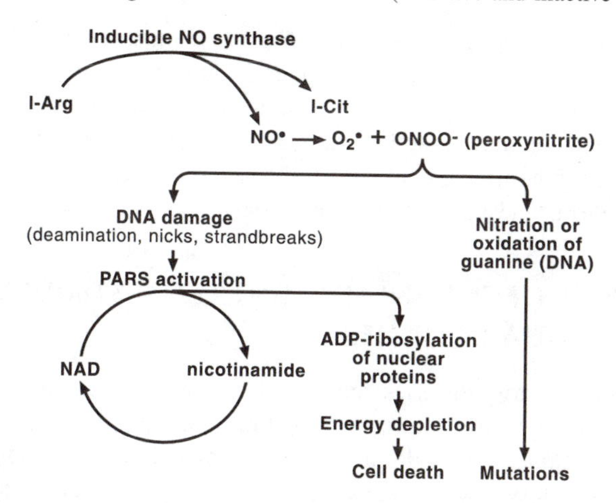

FIGURE 26.7 Important metabolic pathways involved in the production and metabolic attack of DNA causing strand breaks resulting in energy depletion and cell death. This produces apoptosis or necrosis depending on conditions.

26.5 THE CHEMISTRY OF PEROXYNITRITE AND ENHANCEMENT WITH LUMINOL IS SELECTIVE WITH MACROPHAGES

In addition to measuring the luminescence of peroxynitrite from macrophages, luminol produces light from neutrophils. The main source of light in these cells comes from the myeloperoxidase (MPO) reaction that occurs from the reaction of hydrogen peroxide with chloride ion to produce hypochlorite. In the prescence of MPO and the reaction product, luminol is oxidized at pH = 7.4.

TABLE 26.1
Uses of Stratagene Enhancer

To measure the metabolic activity of cells, i.e., cell illness or death assay
To measure Kupffer cell luminescence
To measure peritoneal macrophages
To measure Kupffer cell–enhanced luminescence
To measure bone marrow–enhanced luminescence
To measure enhanced luminescence using a cooled CCD camera in *x–y* coordinates
In neonatal cells to measure defects in phagocytosis of bacteria or sepsis problems
In a flow cell sorter with luminescence from an on–off laser light

Myeloperoxidase comprises 5% of the dry weight of neutrophils, also known as polymorphonuclear leukocytes. Because particles must be coated with recognition factors known as opsonins (antibodies and/or complement), the light assay can be used to measure defects in cells or serum by mix-and-match experiments with normal control cells and serum. However, inflammatory stimulation does increase with an increase of the number of neutrophils, but the light from the neutrophil does not appear to be enhanced with the Stratagene® enhancer. It appears that enhancement of light is selective for macrophages. Because we know that the enhancer does not produce a large enhancement with SIN-1 (a precursor to peroxynitrite), the enhancement of light must be occurring with a secondary factor unique to macrophages. A possible candidate is the peroxidase originating from the macrophage.

26.6 ENHANCEMENT OF LIGHT FROM TISSUE-BOUND MACROPHAGES

Recently, we developed an assay for tissue-bound alveolar macrophages that are trapped in the lungs.[9] These macrophages are activated and trapped in lung tissue. When a stimulant like silica is instilled or inhaled into the lung, it produces induced luminol CL. If the lung tissue is finely chopped into small cubes, the luminol-activated light can be easily measured. Again the Stratagene enhancer (Table 26.1) may be used to enhance these macrophages that are bound *in situ* to lung tissue. This allows one to use less tissue in the reaction.

26.7 MEASUREMENT OF ENHANCED LIGHT FROM STIMULATED BRAIN MICROGLIA IN DISHES

Injury, infection, or inflammatory diseases can stimulate peroxynitrite production in the brain and produce death of surrounding neurons. Such an effect may be the basis for Alzheimer's disease,[15–17] Parkinson's disease,[18] and many other diseases of neurological origin. In addition, many diseases such as diabetes, inflammatory bowel disease, kidney disease, and many others thought to be of immunological origin may actually be caused by activated macrophages (Figures 26.8 through 26.10). We have shown clearly that microglia stimulated by human γ-interferon can produce luminol CL and this light can be enhanced by the Stratagene enhancer. In these studies, we use a luminometer to measure the luminescence emanating from macrophages attached to a dish.[19] This was the first time that microglia have been shown to produce peroxynitrite. It is also the first time that interferon-stimulated microglia have been shown to produce enhanced luminescence using the Stratagene enhancer in a dish rather than a tube. This procedure allows us to measure luminescence without attacking the cells with proteolytic enzymes and damaging or killing the cells while attempting to detach them from the dish; dishes can cause cells to respond differently from free-floating cells (Table 26.2).[20]

TABLE 26.2
Assay

Concentration	Volume	Assay Microglia
1×10^{-6} *M*	1.0 ml luminol	1. 10 ml luminol (1:10) dilution from stock. Note: stock luminol = 10 mg/100 ml PBS
1×10^{-6} *M*	0.5 ml PMA[a]	2. 0.5 ml PMA (1/50 m*M* in PBS)
Varies with drug	0.5 ml drug or PBS	3. 0.5 ml saline or SOD 5 mg/ml in PBS or L-name 5 mg/ml in PBS
20 μM	10 μl enhancer	4. 10 μl enhancer (4 m*M*) Stratagene

[a] Indicates that this compound is added last.

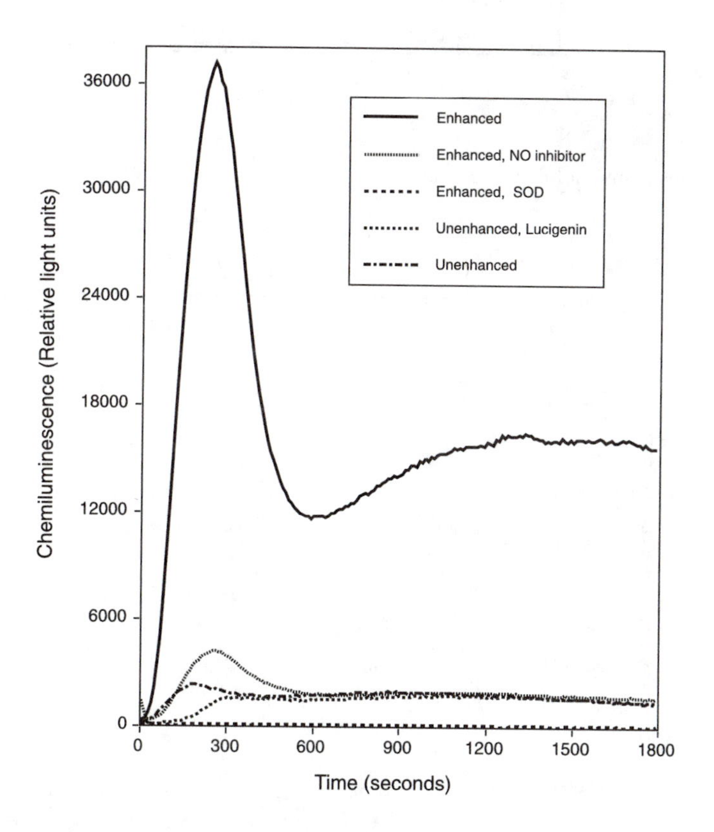

FIGURE 26.8 Enhanced microglia luminol-dependent luminescence from rat brain stimulated by γ-interferon and activated by phorbol myristate acetate in the presence and absence of inhibitors. In one curve luminol is replaced by lucigenin, the reaction was measured on a luminometer (Zylux FB-12) that can measure light from cells attached to 35-mm dishes.

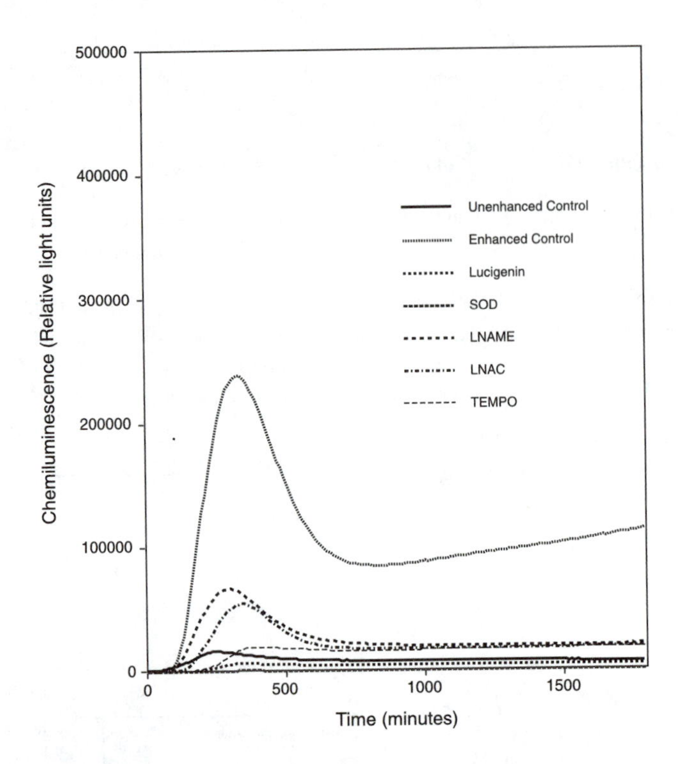

FIGURE 26.9 Enhanced luminol-dependent luminescence from microglia (rat) induced by γ-interferon with inhibitors.

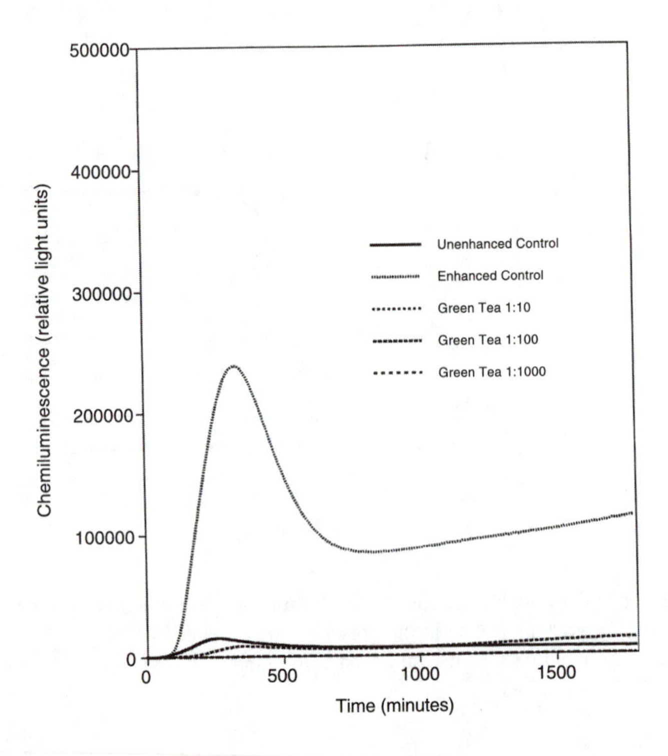

FIGURE 26.10 Microglia under similar conditions as Figure 26.9, inhibited by extracts of green tea.

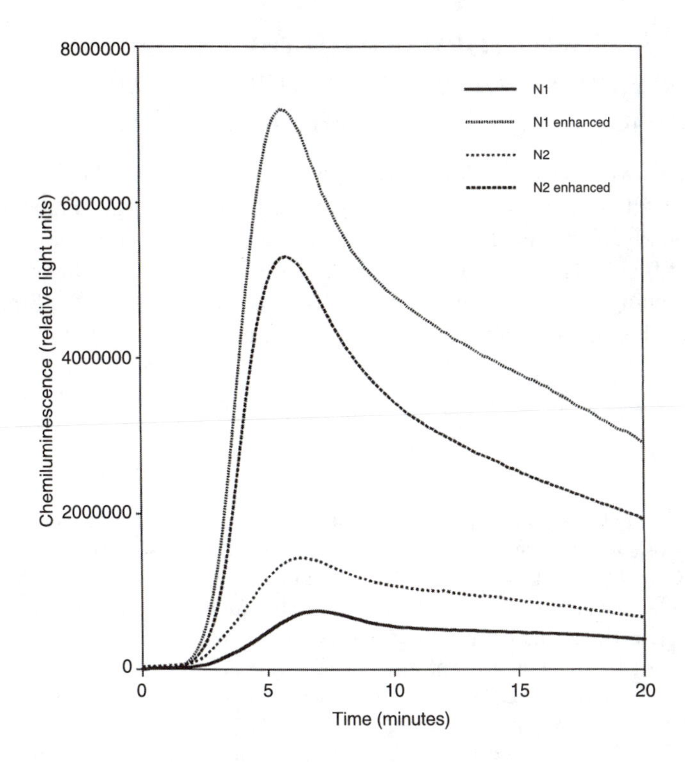

FIGURE 26.11 Enhanced luminescence of alveolar macrophages activated by phorbol myristate acetate.

26.8 USE OF ENHANCERS TO MEASURE ALVEOLAR MACROPHAGE-ENHANCED LUMINESCENCE

We have measured the enhancement of cells washed from the lungs of rats in an unstimulated state (almost 100% macrophages) and measured an almost 40-fold enhancement (Figure 26.11). If macrophage cells are stimulated by disease, the degree of enhancement is even greater. However, if neutrophils numbers increase greatly, enhancement of light is not seen. If a significant proportion of the population is macrophages, then enhancement appears in proportion to the number of macrophages. Furthermore, enhanced peroxynitrite-based light associated with nonstimulated macrophages produces a basal high degree of enhancement (20- to 30-fold), but if induction of the genetic mechanisms occurs due to stimulation, e.g., with γ-interferon, then the set point for luminescence is raised.

26.9 MEASUREMENT OF PEROXYNITRITE-ENHANCED LUMINOL LUMINESCENCE FROM CANCER CELLS STIMULATED BY γ-INTERFERON U937 MACROPHAGES

We have demonstrated that U937 macrophages stimulated with γ-interferon can produce peroxynitrite-activated luminol light, which can be enhanced using the Stratagene enhancer. This light is inhibited using superoxide dismutase and L-NAME, an inhibitor of nitric oxide synthase II. This indicates that superoxide anion free radical and the nitric oxide free radical are producing peroxynitrite, which can then react with the luminol, thereby stimulating the production of light. In addition, since the enhancer stimulates this light 20- to 30-fold or more, we are able to conserve the number of cells that we use for each sample.

26.10 USE OF A 96-WELL FORMAT WITH THE STRATAGENE ENHANCER/LUMINOL AS A HIGH-THROUGHPUT ASSAY FOR ANTIOXIDANTS AND GENETIC INHIBITORS

It is always helpful to measure cellular chemistry in a format that allows maximal efficiency. Certainly, the 96-well system is an excellent way of performing a large number of measurements quickly and efficiently. However, the small wells make efficient measurement of light from a small amount of cells difficult. This is a situation where one would need an efficient luminometer and a method of enhancing the light output as much as possible. Certainly, the Stratagene enhancer could be used for this purpose where there are a small number of cells with a large number of small wells.

REFERENCES

1. Van Dyke, K. et al., Luminol dependent chemiluminescence analysis of cellular and humoral defects of phagocytosis using a chem-gio photometer, *Microchem. J.,* 2, 463–474, 1977.
2. Alien, R. C. and Loose, L. D., Phagocytic activation of a luminol-dependent chemiluminescence in rabbit alveolar and peritoneal macrophages, *Biochem. Biophys. Res. Commun.,* 69, 245–252, 1976.
3. Miles, P. P., Lee, P., Trush, M. A., and Van Dyke, K., Chemiluminescence associated with phagocytosis of foreign particles in rabbit alveolar macrophages, *Life Sci.,* 20, 165, 1977.
4. Orenstein, J. M., Fox, C., and Wahl, S. M., Macrophages as a source of HIV during opportunistic infectious, *Science,* 276, 1857–1861, 1997.
5. Radi, R., Cosgrove, T. P., Beckman, J. S. et al., Peroxynitrite induced luminol chemiluminescence, *Biochem.,* 290, 51–57, 1993.
6. Wang, J. F., Komarov, P., Sies, H., and DeGroot, H., Contribution of nitric oxide synthase to luminol dependent chemiluminescence by phorbol ester activated Kupffer cells, *Biochem. J.,* 279, 311–314, 1991.
7. Lee, P., Walker, E. R., Miles, P. P., and Castranova, V., Differential generation of chemiluminescence from various cellular fractions obtained by dog lung lavage, in *Cellular Chemiluminescence,* Van Dyke, K. and Castranova, V., Eds., CRC Press, Boca Raton, FL, 1989, 53–60.
8. Ardekani, A. M., DeBaetselier, P., and Van Dyke, K., Chemiluminescence response in murine peritoneal macrophage hybridization cell line (2C11-12) primed with LPS or interferon B, *Microchem. J.,* 40, 139–151, 1989.
9. Antonini, J. M., Van Dyke, K., Ye, Z., DiMatteo, M., and Reasor, M. J., Introduction of luminol dependent chemiluminescence as a method to study silica inflammation in the tissue and phagocytic cells of rat lung, *Environ. Health Perspect.,* 102(Suppl. 10), 37–42, 1994.
10. Van Dyke, K., Antonini, J., Wu, L., Ye, Z., and Reasor, M. J., The inhibition of silica-induced lung inflammation by dexamethasone as measured by bronchoalveolar lavage fluid parameters and peroxynitrite-dependent chemiluminescence, *Agents Actions,* 41, 44–49, 1994.
11. Xie, Q.-W., Kashiwabara, Y., and Nathan, C., Role of transcription factor NF-κB Rel in induction of NO synthase, *Biochemistry,* 269, 4705–4708, 1994.
12. Sacks, M., Antioxidation and NF-κB: A Practical Consideration, dissertation, Department of Forestry and Agriculture, West Virginia University, Morgantown, WV, 1998.
13. Blackwell, T. S., Blackwell, T. R. et al., *In vivo* antioxidant treatment suppresses nuclear factor Kb activation and neutrophilic lung inflammation, *J. Immunol.,* 157, 1630–1637, 1996.
14. Van Dyke, K., McConnell, P., and Marquardt, L., Green tea extract and its polyphenols markedly inhibit luminol dependent chemiluminescence activated by peroxynitrite or SIN-1, *Luminescence,* 15, 37–43, 2000.
15. Van Dyke, K., The possible role of peroxynitrite in Aizheimer's disease—a simple hypothesis that could be tested more thoroughly, *Med. Hypothesis,* 48, 375–380, 1997.
16. Van Dyke, K., Birkle, D., Reasor, M. J. et al., Production of peroxynitrite from activated microglia; could this be linked to Aizheimer's and other inflammatory brain damage? *Toxicol. Sci.,* 42, 148, 1998.

17. Szabo, C., Physiological and pathophysiological roles of nitric oxide in the central nervous system, *Brain Res. Bull.,* 41, 131–141, 1996.
18. Good, P. P., Hsu, A., Werner, P. et al., Protein nitration in Parkinson's disease, *J. Neuropathol. Exp. Neurol.,* 57(4), 338–342, 1998.
19. Van Dyke, K., Birkle, D., Konat, G., Reasor, M., McConnell, P., Marquardt, L., Taylor, M., Pingerelli, P., and Kotturi, G., Presented at the meeting of the American Society of Cell Biology, 1998.
20. Nathan, C., Mechanisms and modulation of macrophage activation, *Mol. Biol. Cell* (Suppl.), 27, 25, 1998.

Section V

Oxidative Stress

27 Overview of Oxidative Stress

Knox Van Dyke

CONTENTS

27.1 INTRODUCTION TO OXIDATIVE STRESS

The concept of "oxidative stress" (OS) is attributed to the German scientist Helmut Seis.[1] OS occurs when there is a higher concentration of oxidants than of the opposing antioxidants, i.e., there is a shift in the ratio of oxidants to antioxidants to greater than 1. Basically, our body uses oxygen for the maintenance of life. Oxygen is used in the oxidation of carbohydrates to create energy to live and in the creation of *oxidized* lipids and nucleic acids to produce the four bases that form the code of life. Substances that protect us from infection are derived from oxygen molecules in one form or another. However, utilization of oxygen produces a long-term risk. This risk is related to attack on our cellular macromolecules, e.g., protein, DNA, RNA, and lipid, from the myriad of different oxygen-linked substances. Some of these are quite toxic depending on the form of oxygen and the atoms linked to it, which create new and possibly more toxic compounds. For example, oxygen can exist in a ground state or, if it absorbs energy, in two different excited states, epsilon and delta. When an electron is added to oxygen, superoxide ($O_2^{\cdot -}$) is produced. When it becomes further reduced, it can form hydrogen peroxide (H_2O_2). It can exist in a highly toxic and energetic state called hydroxyl free radical ($OH^{\cdot}$), which is formed in the presence of reduced iron ion (Fe^{2+}) and hydrogen peroxide in a Fenton reaction. Additionally, if superoxide reacts with nitric oxide it can form ($OONO^-$), or peroxynitrite. Another strong oxidizing entity is formed by the reaction of hypochlorite and hydrogen peroxide to produce an oxidizing peroxide, possibly peroxyhypochlorite ($OOClO^-$). See Tables 27.1 and 27.2 for half-lives of oxygen species and serum concentrations of antioxidants.

27.2 OXIDATION/ANTIOXIDATION

Oxidation can be defined as a loss of electrons, a gain of oxygen, or a loss of hydrogen. *Antioxidation* (reduction) is the reverse of oxidation—a gain of electrons, a gain of hydrogen, or a loss of oxygen. Most of the body's defense reactions (if not all) that are protective (defensive) are in fact oxidations. However, there must be an opposing force that counterbalances the oxidation and this is called antioxidation. The body must be maintained in an equilibrium or status quo, called *homeostasis*, where the two opposing forces must be roughly balanced if not weighted toward antioxidants for protection against the ravages of oxidation. An analogous situation might be missile attack where incoming missiles enter an area (oxidation) and the area is protected by antimissiles (antioxidation). The maximum oxidized state (at least when compared to life) is probably death itself. But as the

0-8493-0719-8/02/$0.00+$1.50

TABLE 27.1
Estimated Half-Life of Reactive Oxygen Species

Oxygen Species	Estimated Half-Life (μs)
Hydroxyl radical	10^{-3}
Alkoxyl radical	1
Peroxyl radical	7000
Hydrogen peroxide	Enzymic
Superoxide radical	Enzymic
Singlet oxygen	1
Semiquinone radical	Days
Nitric oxide radical	10^6–10^7
Peroxynitrite	0.3×10^6–10^6
Hypochlorite	0.3×10^6–10^6

TABLE 27.2
Nonenzymic Antioxidants from Human Plasma

Plasma Content (μmol%)	
Water-Soluble	
Ascorbic acid	3–15
Uric acid	16–45
Glutathione	0.1–0.2
Lipid-Soluble	
α-Tocopherol	1.5–4
γ-Tocopherol	0.3–0.5
α-Carotene	0.05–0.1
β-Carotene	0.03–0.06
Ubiquinol-10	0.04–0.01

body disintegrates even more, continuous oxidation must take place (perhaps because of a continued depletion of antioxidants). The Chinese recognized this long ago, calling these opposing forces *yin* and *yang*. One would think that there might be a downside to having excess antioxidants in the body. Although there are examples of a few individuals who have had problems with excessive antioxidants, they appear to be few and far between. Even a loaf of bread tends to spoil quickly without the addition of antioxidants, often called preservatives.

27.3 OXYGEN AND DISEASES

Oxygen is essential to life as we know it. It is used by the body in normal metabolic oxidation mechanisms where various reactive oxygen species are produced. For every 10^{12} oxygen molecules entering a cell each day, it is estimated that 1/100 damages protein and 1/200 damages DNA or RNA. In addition, membrane lipids are oxidatively attacked. Some of the damage is probably repaired but some is not, and over time the aging process is most likely associated with the damaged molecular species. This damage arises from excessive oxidant species and is almost certainly linked to the basis of a variety of degenerative diseases that occur in old age. These are diseases such as cancer, stroke, heart attacks, and neurodegenerative diseases associated with advancing age, e.g., Alzheimer's, Huntington's, amyotrophic lateral sclerosis (Lou Gehrig's disease), and Parkinson's disease, and various dementias. Atherosclerosis and diabetes also have been shown to have major oxidative-stress components. Antioxidants can play a significant role in preventing the oxidant damage to cells and tissue. By increasing the entire spectrum of antioxidants we can raise our level of protection against the damaging consequences of oxidants over time. Perhaps if we were to increase the entire group of antioxidants, we could slow the ravages of time and live healthier lives for a longer time. However, it would take continuous treatment over a lifetime to find out, and no such experiment has been reported in humans, simply because of the constraints of time itself. We are not good at making overview observations over long periods of time, but it seems intuitively obvious that we need to protect ourselves from the ravages of oxidative stress continuously. Because we tend to think and act moment to moment, day to day, or year to year does not alter the fact that every second of our lives we breathe in large quantities of oxygen, which we know can have long-term consequences. Perhaps it is because without oxygen there are no consequences and no life either.

27.4 ANTIOXIDANTS

What are the key antioxidants in the body? See Table 27.2. Apparently, the key natural anti-oxidant is the end product of purine metabolism, uric acid. Uric acid for most people is probably the main antioxidant representing from 75 to 95% of antioxidant protection in the blood. Vitamin C or ascorbic acid is an important water-soluble antioxidant and the various tocopherols (vitamin E) are excellent fat-soluble antioxidants, particularly for membranes of cells. There are certain proteins that can act like antioxidants, and the tripeptide glutathiones (reduced form) are the main intercellular endogenous/exogenous antioxidants (or, at least, are the ones usually mentioned). There are a variety of enzymes that function as antioxidants, e.g., the superoxide dismutases, glutathione peroxidase, and catalase, and various peroxidases metabolize different oxygen-based oxidants.

Luminescence assays can be used to find new antioxidants that have the potential to increase the antioxidant load above and beyond that which exists from the antioxidants previously mentioned. This is because the light produced from the oxidation of luminol occurs as the result of attack by known, physiologically relevant oxidants such as peroxynitrite and hypochlorite/hydrogen peroxide. By using the light-producing reaction between the known oxidants, e.g., peroxynitrite or hypochlorite, and luminol to produce oxidation-based blue light, we can identify antioxidants; antioxidants inhibit this light. Using this simple chemical system, we have found many polyphenolic compounds that are strong antioxidants. These antioxidants originate in plants, e.g., green or black tea, grapes, blueberries, cranberries, etc. A variety of research has shown that these polyphenolic substances are protective, possibly as anticancer substances. A major reason that this may occur can be seen in the attack of peroxynitrite or hypochlorite/hydrogen peroxide on the amino acid tyrosine alone or as a polymerized amino acid in proteins. Peroxynitrite nitrates or oxidizes tyrosine; hypochlorite oxidizes or chlorinates tyrosine. Therefore, the polyphenols act as bystander molecules that can be readily attacked by oxidants but that destroy the ability of the oxidants to attack anything else because the ability of chemical oxidation has been spent on attacking phenols. We present this work in Chapters 28 through 35.

REFERENCE

1. Seis, H., in *Oxidative Stress,* H. Seis, Ed., Academic Press, London, 1985, 1–8.

28 Luminescence Analysis of Oxidative Stress Induced by Organ Transplantation

Antonín Lojek and Milan Číž

CONTENTS

28.1 INTRODUCTION

Phagocytes (neutrophils, monocytes, macrophages, eosinophils), which constitute a front-line of the body's defense mechanisms against invading microbial pathogens, generate reactive oxygen species (ROS) as one of their instruments for killing foreign organisms. Neutrophils are the first cells to invade a site of inflammation following infection. Upon stimulation with both soluble and particulate matter, the neutrophil oxidative metabolism is activated resulting in a respiratory burst accompanied by a reduction of molecular oxygen to superoxide via a special electron transport system (NADPH-oxidase). Superoxide radicals form hydrogen peroxide in a dismutation reaction catalyzed by the enzyme superoxide dismutase. Hydrogen peroxide serves as a substrate for the myeloperoxidase reaction, in which a variety of highly toxic metabolites, including hypochlorite, are generated. These ROS have strong microbicidal effects and they are of great importance for the proper functioning of the immune system.[1–3]

Beside the beneficial role of neutrophils, accumulating evidence suggests that a high-level production of neutrophil-derived ROS participates in a number of different diseases, including inflammatory conditions that can ultimately lead to tissue injury.[1,4] One of these diseases is ischemia/reperfusion

0-8493-0719-8/02/$0.00+$1.50

of organs and tissues.[5–7] A typical clinical situation, where a certain degree of ischemia cannot be avoided, is organ transplantation. The organ graft is subjected to both cold and warm ischemia periods during organ storage and implantation.[8] Despite the fact that a reperfusion of the organ with oxygenated donor blood is necessary to maintain cell functions and viability, mediators released during ischemia/reperfusion can promote polymorphonuclear cell adhesion, diapedesis, and the production of ROS, proteolytic enzymes, and other bioactive substances. Proinflammatory cytokines, such as interleukin 6 (IL-6), IL-8, and tumor necrosis factor-α (TNF-α), are among the inflammatory mediators known to be produced during human organ transplantation. These cytokines, which are released during transplantation, may prime neutrophils to produce increased amounts of ROS, and their effects on neutrophils could exacerbate both local and remote tissue damage.[6,9–11] The extent of the transplanted organ damage resulting from such an injury may be significant for the outcome of the transplantation procedure.

To protect itself against deleterious effects of the ROS produced, the human body possesses complex protective antioxidant systems. Tissues are equipped with both intracellular and extracellular antioxidant defense systems. Among the intracellular antioxidants, enzymes such as superoxide dismutase, catalase, glutathione peroxidase, etc. are of major importance. On the other hand, nonenzymatic compounds such as uric acid, thiols, ascorbic acid, α-tocopherol, albumin, ceruloplasmin, bilirubin, transferrin, etc. represent major antioxidants in extracellular fluids.[12,13] Detailed knowledge of the activity of antioxidants in biological fluids is rather important when investigating a relationship between oxidative stress and disease.

It is quite evident from the arguments mentioned above, that functional analysis of phagocytes and the monitoring of the antioxidant capacity of blood plasma are of key importance for understanding ischemia/reperfusion-induced oxidative stress. The most convenient methods of evaluating both parameters mentioned above are based on measuring chemiluminescence (CL) activity.

28.1.1 Chemiluminescence of Phagocytes

When phagocyte-derived ROS are released, electronically excited states are produced. They emit photons on relaxation to the ground state. This emission is referred to as phagocyte CL. Since the native CL of phagocytes is very weak, the light production is usually amplified by luminophors like luminol or lucigenin. Luminophors interact with ROS to produce larger, more measurable amounts of light. Lucigenin is thought to be highly specific for the superoxide anion and thus reflects the activity of NADPH-oxidase, whereas luminol-derived CL is considered to be dependent on myeloperoxidase activity. Neutrophils, eosinophils, and monocytes have both NADPH-oxidase and myeloperoxidase activities, thus generating both luminol- and lucigenin-enhanced CL. Matured macrophages have decreased contents of myeloperoxidase and thus produce diminished luminol-enhanced CL.[1,2] However, when macrophages are stimulated luminol-dependent peroxynitrite CL becomes prominent. Differential cell counts in healthy volunteers showed that 50 to 70% of the leukocytes are neutrophils and only 2 to 6% are monocytes. Since monocytes are not as active as neutrophils, it is possible to conclude that whole-blood CL is nearly totally due to the activity of neutrophils.

A majority of phagocyte CL studies are performed using separated cells. However, cell separation is laborious and time-consuming. Furthermore, the expression of immunoglobulin and complement receptors on the cell membranes and also the functional properties of the cells are changed during the isolation processes.[14–17] All these problems can be avoided by measuring the CL response directly in diluted fresh whole blood.[18,19]

28.1.2 Chemiluminescence Analysis of Total Antioxidant Capacity

Several methods are available for the measurement of total antioxidant activity of biological fluids. The principle of these methods is based on the production of free radicals at a known rate and an evaluation of the capability of a sample to inhibit this radical production.[12] One of the methods used most widely is CL detection developed by Metsä-Ketelä.[13] The method is based on the discovery of Wayner and co-workers[20] that the thermal decomposition of 2,2-azo-bis-2-amidino-propane hydrochloride (ABAP) yields peroxyl radical at a known and constant rate. Its reaction with the CL substrate luminol leads to the formation of luminol radicals that emit light detectable by a luminometer. Antioxidants in a sample inhibit this CL for a time that is directly proportional to the total antioxidant potential of the sample.[12]

This chapter describes the results of CL analyses of neutrophil oxidative burst and plasma total antioxidative capacity obtained for samples collected from patients during liver, heart, or kidney transplantation and in the course of their first postoperation week.

28.2 MATERIALS AND METHODS

28.2.1 PATIENTS

In the study, 58 patients who underwent orthotopic liver transplantation (8 males and 8 females, mean age 36 years, range 8 to 57 years), kidney transplantation (14 males and 4 females, mean age 44 years, range 25 to 61 years), and heart transplantation (21 males and 3 females, mean age 46 years, range 10 to 61 years) were monitored during operation and followed up to 7 days after a successful transplantation. Control groups of healthy subjects were also included in the study for whole-blood CL activity ($n = 23$) and plasma total antioxidative capacity ($n = 14$) measurements.

28.2.2 SAMPLING

Heparinized peripheral blood samples from transplant recipients were taken before surgery, at the beginning of reperfusion, and 30 min and 4 h after the reperfusion. Peripheral blood was also taken 1, 3, 5, and 7 days after the surgery. The obtained plasma samples were frozen immediately and stored at –20°C.

28.2.3 CHEMILUMINESCENCE ASSAY OF THE OXIDATIVE BURST OF PHAGOCYTES

Luminol (Sigma Chemical Co., St. Louis, MO) was dissolved in dimethyl sulfoxide (Koch-Light Laboratories Ltd, Colnbrook, U.K.), stored at –20°C, and diluted with MEM (Eagle's Minimal Essential Medium) to a final concentration of 5×10^{-4} M immediately prior to use. The suppliers for other chemicals were SEVAC (Czech Republic) for MEM and Lachema (Czech Republic) for starch.

The total number of leukocytes in blood and the percentage of polymorphonuclear leukocytes (PMNLs) (both band cells and segmented cells) were counted using hemocytometer and stained blood smears. Luminol-dependent enhanced CL was used to follow ROS production. CL was measured on Luminometer 1251 (Bio-Orbit, Turku, Finland), at 37°C. The samples contained 100 μl blood, 100 μl luminol, 100 μl starch suspension (1%), and MEM in a total volume of 700 μl. Reaction was started by adding starch. Spontaneous CL was also measured with MEM added to the reaction mixture instead of starch. Total 1-h production of ROS was evaluated (integral of the CL curve). Both MEM and dimethyl sulfoxide were verified not to affect the level of CL.

28.2.4 CL ASSAY OF THE TOTAL ANTIOXIDATIVE CAPACITY OF THE PLASMA

The reaction was initiated by mixing 475 μl phosphate-buffered saline (PBS), 50 μl 10^{-4} M luminol, and 50 μl 400 mM ABAP in PBS. This mixture was incubated in the temperature-controlled sample carousel of the BioOrbit 1251 luminometer (37°C) for 15 min to produce peroxyl radicals while light emission was measured in intervals of 71 s. During this period, a steady state of the CL signal was reached. Then, 20 μl of plasma was added directly into the cuvettes and the samples were measured for another 30 min. The total antioxidative capacity (TRAP, total peroxyl radical-trapping potential) was defined as the time necessary for a 50% recovery of the original steady-state CL signal. This 50% level was chosen according to previous experiments in our laboratory (unpublished data). All other solutes used had no radical-scavenging properties. Trolox (Aldrich Chemical Co., Milwaukee, WI), a water-soluble analogue of tocopherol, was used as a reference inhibitor.

28.2.5 Statistical Analysis

The detailed information on statistical methods used to evaluate the results of the present study can be found in the legends to figures.

28.3 RESULTS AND DISCUSSION

28.3.1 Chemiluminescence Activity in Whole Blood of Liver, Heart, or Kidney Recipients

Basic CL activity measured in the whole blood of healthy volunteers is shown in Table 28.1. When samples of peripheral blood collected before the transplantation were analyzed, the whole-blood CL activity of patients indicated for organ transplantation was much higher than in healthy subjects especially in the liver and heart recipients (Figure 28.1). Since there were no substantial differences among the healthy population and organ recipients in leukocyte counts, the obtained results signify augmented metabolic activity of neutrophils and an elevated ROS production. This oxidative stress was reflected in the rise of lipid peroxidation (data not shown).

Other samples were collected at the beginning of reperfusion, and 30 min, 4 h, and 1, 3, 5, and 7 days after the operation. The activated CL in liver recipients was significantly lower during the early reperfusion period (up to 4 h), and after that the CL activity did not differ from the preoperational values. The activated CL in kidney recipients was not significantly changed during the first 30 min after reperfusion, and it significantly increased in all later time intervals. The activated CL started to rise even at the time of reperfusion in heart recipients. This increase became significant at 4 h after reperfusion and this trend lasted during the following 5-day period (see Figure 28.1).

On the other hand, when the CL response was corrected for neutrophil numbers (it rose in all three types of transplantation; data not shown), there was an obvious and significant decrease in the whole-blood CL of liver recipients. The decrease in corrected CL activity was also observed in heart and kidney recipients. The shift was less obvious than in the case of liver recipients chiefly as a result of lower corrected CL activity in heart and kidney recipients (see Figure 28.1). The reduction in CL activity with a parallel increase in neutrophil numbers means that the metabolic activity of neutrophils expressed as ROS generation returned to physiological values.

TABLE 28.1
Total and Corrected CL Activity of Whole Blood and Total Peroxyl Radical-Trapping Potential of Plasma Obtained from Healthy Volunteers

	CL [$mV \cdot s \cdot 10^3$]	CL/10^5 Neutrophils [$mV \cdot s \cdot 10^3$]	TRAP [$\mu mol \cdot l^{-1}$]
Mean ± SEM	63 ± 5	14 ± 1	1121 ± 44
n	23	23	14
(min; max)	(30; 102)	(7; 23)	(857; 1348)

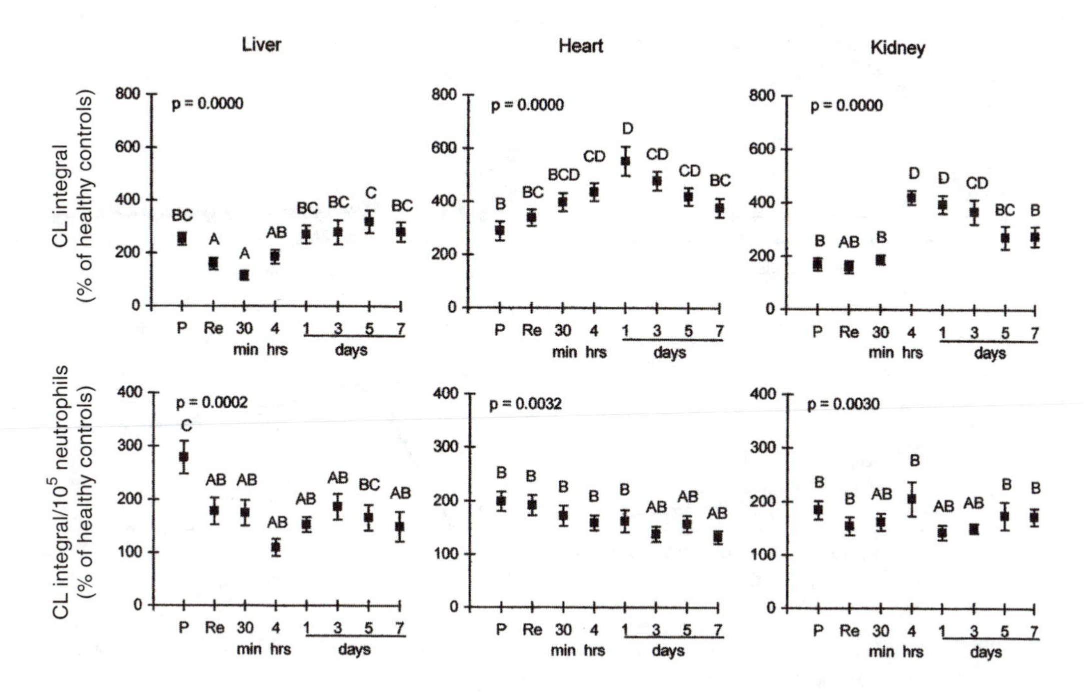

FIGURE 28.1 Total whole-blood CL and CL corrected per 10^5 neutrophils in liver, heart, or kidney transplant recipients expressed as percentages of healthy controls. Data are expressed as means ± standard errors of the mean. Symbols P and Re represent samples collected before the operation (P) and at the time of reperfusion (Re). A one-way analysis of variance (ANOVA) and the Newman–Keuls test were used to evaluate differences of the mean among groups. Groups marked with the same capital letter did not differ significantly.

Organ transplantation is associated with systemic inflammatory response, which is an extremely complex phenomenon combined with the release of various cytokines.[9,10] As a matter of fact, we observed a positive correlation between the levels of proinflammatory cytokines IL-6, IL-8, and TNF-α and the CL activity of neutrophils (data not shown). Cytokine release can be triggered by many factors, such as surgical procedure, release of endotoxin, duration of ischemic period and reperfusion of the donor ischemic organ, and exposure of the blood to the extracorporeal circuit.[8] Although both absolute and corrected CL activities of blood samples collected immediately prior to operation were elevated in comparison with blood samples from healthy volunteers, the actual transplantation continued raising the absolute levels of ROS. However, after correcting CL for neutrophil numbers, it became clear that the oxidative bursts of individual neutrophils stopped increasing and the actual elevation of ROS production was mainly due to increased neutrophil numbers.

Somewhat different results were observed in our previous study where the metabolic activity of phagocytes collected from the coronary sinus during open heart surgery was analyzed in patients with ischemic heart disease. The preoperational level of neutrophils did not significantly differ from the level of healthy volunteers. After releasing the aortal clamp and subsequent reperfusion with oxygenated blood, there was a fast activation of phagocytes that reached a threefold level compared with the control value before the surgery. The relationship between the leukocyte numbers (which increased as well) and their activity during operation implies that apparently both factors participate in the increased ROS production; i.e., increased phagocyte accumulation occurs within the myocardium consequent to an activation of chemotactic mechanisms, and, in addition, these phagocytes are strongly stimulated to produce more ROS. Similar effects were also expressed in the peripheral blood until 24 h after the surgery.[21]

To minimize the blood volume needed for CL, it was necessary to design a further miniaturization of the procedure. MEM was replaced with Hank's balanced salt solution and luminol was

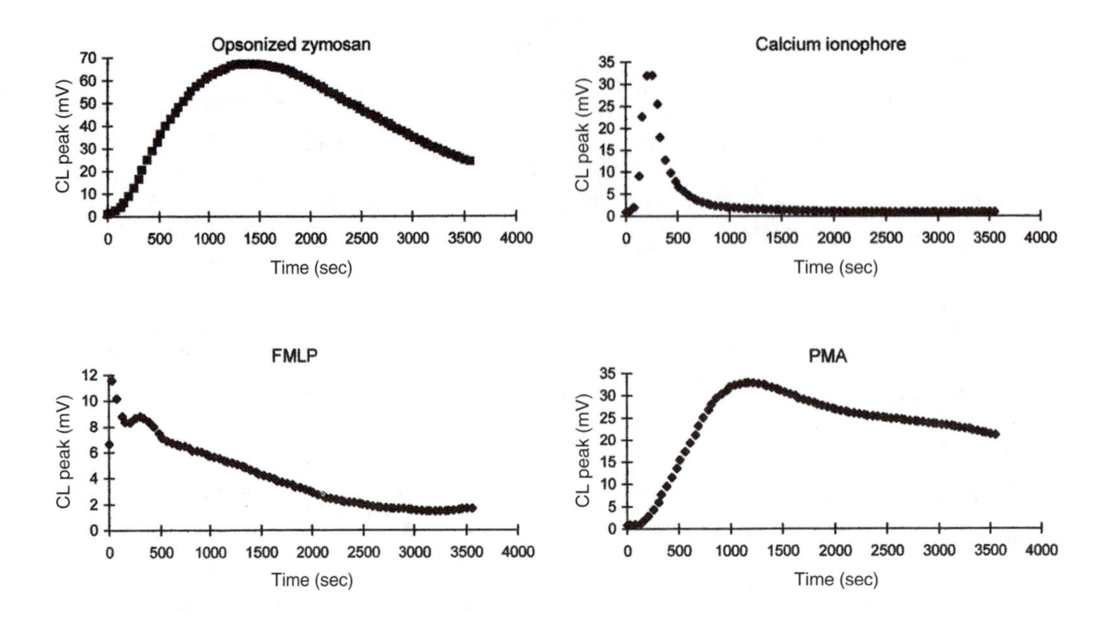

FIGURE 28.2 Kinetics of CL response activated with OZ, PMA, Ca-I, and FMLP in the whole blood of healthy volunteers. See text for details.

dissolved in borate buffer (pH = 9.0). The optimal final concentration of such dissolved luminol was determined to be 1 m*M*, whereas higher concentrations were found to be toxic. To induce the oxidative burst of phagocytes, the following substances were used: opsonized zymosan (OZ, binding to the complement and immunoglobulin receptors), phorbol myristate acetate (PMA, direct activation of protein kinase C, PKC), calcium ionophore A23187 (Ca-I, increasing calcium ions penetration into the cell) and *N*-formyl-methionyl-leucyl-phenylalanine (FMLP, a bacterial polypeptide with chemotactic effects). The optimum concentrations for the activators were found to be 0.3 μg/ml for PMA, 10 μg/ml for FMLP, and 30 μg/ml for Ca-I. Concerning OZ, a sufficient concentration inducing a satisfactory CL response appears to be that of 50 μg/ml although an optimum was not found, since the response increased along the whole concentration range studied (up to 500 μg/ml). It was found that 2 μl of whole blood was enough to obtain sufficient CL activity. The curves of CL response of whole blood neutrophils for the activators mentioned above are illustrated in Figure 28.2. The data obtained with this miniaturized technique have not yet been finalized, and their evaluation will receive special attention.

28.3.2 TRAP in the Plasma of Liver, Heart, or Kidney Recipients

TRAP was measured in patients indicated for liver, heart, and kidney transplantation. Whereas the preoperational total antioxidant capacity of the plasma in heart and liver recipients was approximately the same as in healthy controls and it did not change during the time intervals studied, a decreased preoperational level of plasma antioxidant capacity was observed in patients with regular hemodialysis treatment indicated for kidney transplantation (Figure 28.3). In kidney recipients, the TRAP parameter remained at around the preoperational level during the early postischemic period and it increased during the late postischemic period, reaching the level of healthy volunteers on the day 5 after transplantation.

The CL method used in our experiments detects the activity of chain-breaking antioxidants in serum, which means that the increased total antioxidant activity measured in our study is related to the protection from lipid peroxidation. Although the TRAP values in the preoperational samples

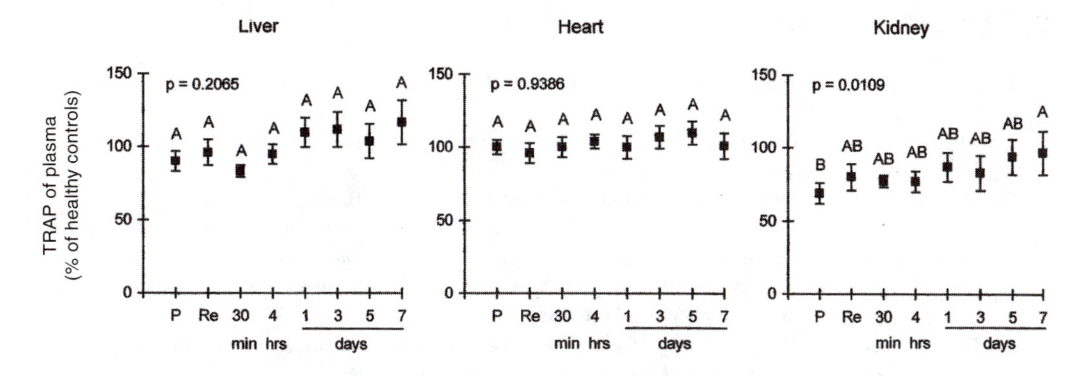

FIGURE 28.3 TRAP measured in plasma samples obtained from liver, heart, or kidney recipients. Data are expressed as means ± standard errors of the mean. Symbols and explanations are identical to Figure 28.1.

from liver and heart recipients were the same as those from healthy controls, oxidative stress induced by the increased oxidative burst of phagocytes was expressed by an elevated level of lipid peroxidation. Despite the fact that the TRAP values improved gradually in all three groups of patients during the first week after transplantation, plasma antioxidant capacity was not sufficient to eliminate the transplantation-induced oxidative stress, and the lipid peroxidation further increased.

Detailed insight into any changes in plasma antioxidant capacity will also require determination of the representation of major antioxidants in the analyzed samples. This type of analysis has only been performed using a model of experimentally occluded and reperfused rat superior mesenteric artery.[22] The most important findings are that the natural antioxidative mechanisms of the body are mobilized as early as at the end of 45-min ischemia due to the oxidative challenge and that the mobilization lasts for at least 4 h after reperfusion. On days 1 to 4 of the postischemic period, the antioxidant activity of serum oscillated around the preoperational level. When TRAP values were correlated against the serum concentrations of individual antioxidants (urate, ascorbate, α-tocopherol, albumin), the most promising antioxidant appeared to be urate (correlation coefficient r = 0.74). Nevertheless, when the ideal sum of TRAP values for these antioxidants was calculated from their concentrations and stoichiometric factors, it was much lower than the TRAP measured experimentally.[22] This suggests that some other antioxidant(s) contributed to the increase in the total antioxidant capacity of serum. Their identification remains to be elucidated.

28.4 CONCLUSION

It can be concluded that CL methods were confirmed to be a very useful tool for the assessment of phagocyte-generated ROS and of the total radical-trapping antioxidant parameter of plasma (serum) as markers of oxidative stress induced by ischemia/reperfusion of organs and tissues. Information on the intensity of ROS production and on the total peroxyl radical-trapping potential is of great importance in other pathophysiological conditions as well, because it provides basic data needed when indicating proper anti-inflammatory and antioxidative therapy, thus preventing oxidative damage to organs and tissues with serious clinical consequences.

ACKNOWLEDGMENTS

This study was supported by Grants 308/97/1141 and 524/98/0190 of the Grant Agency of the Czech Republic and by Grant 4796-3 of the Internal Grant Agency of the Ministry of Health, Czech Republic.

REFERENCES

1. Lilius, E.-M. and Marnila, P., Photon emission of phagocytes in relation to stress and disease, *Experientia,* 48, 1082, 1992.
2. Allen, R. C. and Stevens, D. L., The circulating phagocyte reflects the *in vivo* state of immune defense, *Curr. Opin. Infect. Dis.,* 5, 389, 1992.
3. McCord, J. M., Oxygen-derived free radicals, *New Horizons,* 1, 70, 1993.
4. Halliwell, B., Current status review: free radicals, reactive oxygen species and human disease: a critical evaluation with special reference to atherosclerosis, *Br. J. Exp. Pathol.,* 70, 737, 1989.
5. Schoenberg, M. H. and Beger, H. G., Oxygen radicals and postischemic organ damage—pathophysiology, clinical relevance and therapy, *Zentralbl. Chir.,* 120, 174, 1995.
6. Toledo-Pereyra, L. H. and Suzuki, S., Neutrophils, cytokines, and adhesion molecules in hepatic ischemia and reperfusion injury, *J. Am. Coll. Surg.,* 179, 758, 1994.
7. Dinerman, J. L. and Mehta, J. L., Endothelial, platelet and leukocyte interactions in ischemic heart disease: insights into potential mechanisms and their clinical relevance, *J. Am. Coll. Cardiol.,* 16, 207, 1990.
8. Robson, S. C., Candinas, D., Hancock, W. W., Wrighton, C., Winkler, H., and Bach, F. H., Role of endothelial cells in transplantation, *Int. Arch. Allergy Immunol.,* 106, 305, 1995.
9. Mueller, A. R., Platz, K. P., Haak, M., Undi, H., Müller, C., Köttgen, E., Weidemann, H., and Neuhaus, P., The release of cytokines, adhesion molecules, and extracellular matrix parameters during and after reperfusion in human liver transplantation, *Transplantation,* 62, 1118, 1996.
10. Pirenne, J., Pirenne-Noizat, F., deGroote, D., Vrindts, Y., Lopez, M., Gathy, R., Jacquet, N., Meurisse, M., Honore, P., and Franchimont, P., Cytokines and organ transplantation. A review, *Nucl. Med. Biol.,* 21, 545, 1994.
11. Lojek, A., Černý, J., Němec, P., Ničovský, J., Soška, V., Číž, M., Slavíková, H., and Kubala L., Phagocyte-induced oxidative stress in patients with haemodialysis treatment and organ transplantation, *Biofactors,* 8, 165, 1998.
12. Alho, H. and Leinonen, J., Total antioxidant activity measured by chemiluminescence method, *Methods Enzymol.,* 299, 1, 1999.
13. Uotila, J. T., Kirkkola A.-L., Rorarius, M., Tuimala, R. J., and Metsä-Ketelä, T., The total peroxyl radical-trapping ability of plasma and cerebrospinal fluid in normal and preclamptic parturients, *Free Radical Biol. Med.,* 16, 581, 1994.
14. Glasser, L. and Fiederlein, R. L., The effect of various cell separation procedures on assays of neutrophil function. A critical appraisal, *Am. J. Clin. Pathol.,* 93, 662, 1990.
15. Fearon, D. T. and Collins L. A., Increased expression of C3b receptors on polymorphonuclear leukocytes induced by chemotactic factors and purification procedures, *J. Immunol.,* 130, 370, 1983.
16. Saniabadi, A. R. and Nakano, M., Improved preparation of leukocytes for chemiluminescent study of human phagocytic leukocyte-generated reactive oxygen species, *J. Biolumin. Chemilumin.,* 8, 207, 1993.
17. Tennenberg, S. D., Zemlan, F. P., and Solomkin J. S., Characterization of *N*-formyl-methionyl-leucyl-phenylalanine receptors on human neutrophils. Effects of isolation and temperature on receptor expression and functional activity, *J. Immunol.,* 141, 3937, 1988.
18. Stevens, D. L., Bryant, A. E., Huffman, J., Thompson, K., and Allen, R. C., Analysis of circulating phagocyte activity measured by whole blood luminescence: correlations with clinical status, *J. Infect. Dis.,* 170, 1463, 1994.
19. Lojek, A., Číž, M., Marnila, P., Dušková, M., and Lilius, E.-M., Measurement of whole blood phagocyte chemiluminescence in the Wistar rat, *J. Biolumin. Cheminlumin.,* 12, 225, 1997.
20. Wayner, D. D., Burton, G. W., Ingold, K. U., and Locke, S., Quantitative measurement of the total, peroxyl radical-trapping antioxidant capability of human blood plasma by controlled peroxidation. The important contribution made by plasma proteins, *FEBS Lett.,* 187, 33, 1985.
21. Lojek, A., Černý, J., Pillich, J., Číž, M., Kubíčková, D., Pavlíček, V., Němec, P., and Wagner, R., Human neutrophil mobilization during open heart surgery, *Physiol. Res.,* 41, 431, 1992.
22. Slavíková, H., Lojek, A., Hamar, J., Dušková, M., Kubala, L., Vondráček, J., and Číž, M., Total antioxidant capacity of serum increased in early but not late period after intestinal ischemia in rats, *Free Radical Biol. Med.,* 25, 9, 1998.

29 Acceleration of Tissue Aging in Chickens Caused by Oxidative Stress Using Allopurinol and Detected by Cellular Humoral Chemiluminesence

Hillar Klandorf, Dinesh Rathore, Muhammad Iqbal, Xianglin Shi, Melvin Simoyi, and Knox Van Dyke

CONTENTS

29.1 INTRODUCTION

Most avian species live significantly longer than mammals of comparable body size.[1] The maximum longevity in birds is known to range from 4 years for blue jays (*Cyanocitta cristata*) to 64 years for macaws (*Ara macao*). The longevity of birds is somewhat surprising, since they exhibit many traits that should render them more susceptible to the degenerative processes of aging. These traits have been reviewed[2] and include (1) metabolic rates as much as 2 to 2.5 times greater than those

0-8493-0719-8/02/$0.00+$1.50

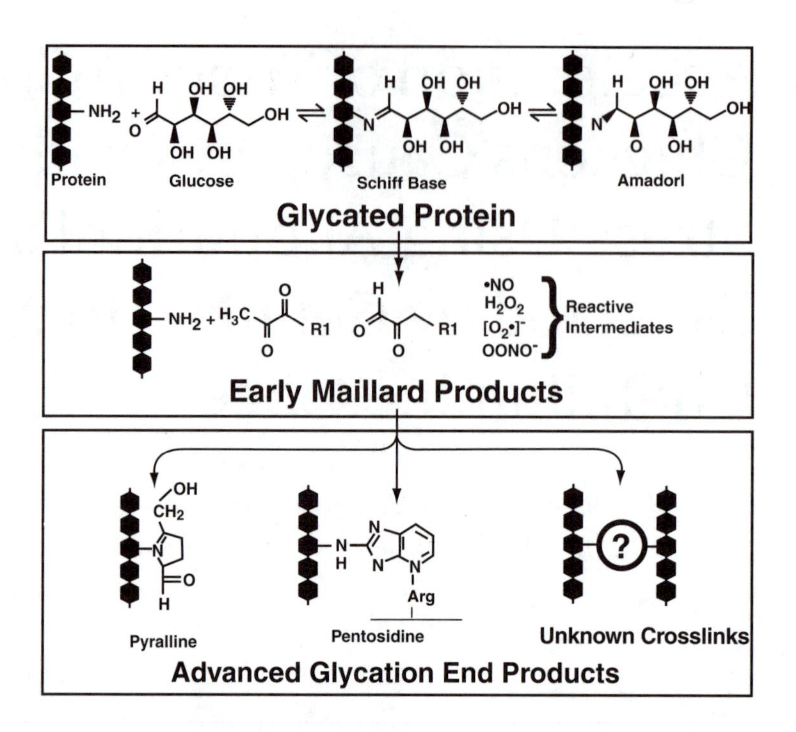

FIGURE 29.1 Formation of glucose-derived cross-links begins when glucose attaches to an amino ($-NH_3$) group of a plasma or tissue protein (top) such as collagen. The initial unstable product, known as a Schiff base, is soon transformed into a more stable Amadori product, which further rearranges itself into early and advanced glycation end products in the presence of free radicals. These accumulated end products form covalent cross-links within a particular protein or with other matrix proteins.

of mammals of similar body size, (2) blood sugar levels typically two to six times higher than those of mammals, and (3) body temperatures about 3°C higher than mammals. Each of these factors should expose them to a higher rate of free oxygen radical production and accelerate the formation of Maillard products (Figure 29.1). Without special protective mechanisms against the potential for oxidative damage, birds should be comparatively short lived and age more rapidly than mammals. However, it has been reported that avian species have higher levels of circulating antioxidants (α-tocopherols, carotenoids, and uric acids) compared with comparably sized mammals.[3,4] Apparently birds have evolved mechanisms to limit the damage caused by the production of increased amounts of oxygen free radicals or oxidants emanating from radicals, e.g., peroxynitrite.

Uric acid is one of the circulating antioxidants that is ubiquitous and demonstrates a positive correlation with maximum life span across species[5–7] and a potent scavenger of free radicals in human and many animal tissues.[8,9] Humans, the longest-living primate, have comparatively high levels of uric acid because they lack uricase, the terminal degradative enzyme present in monkeys and other mammals.[4,5] In support of this concept, the lower levels of uric acid in macaques (1/3 that of humans) correlates with a shorter life span compared to primates with high levels of uric acid.[10] It has also been demonstrated[9,11] that urate, *in vitro,* has the ability to scavenge peroxides, hydroxyl radical species, and hypochlorous acid. Following the reperfusion injury that occurs after a myocardial infarction, uric acid concentrations increase in plasma. Whether this represents a compensatory response remains unclear.[8] For this reason, it has been proposed that lower tissue concentrations of the glycoxidation product pentosidine, in birds as compared to mammals, are due to a more efficacious avian antioxidant system,[12,13] which includes concentrations of uric acid approximately twofold greater than that measured in humans.[14,15] Skin pentosidine (Ps) is a tissue cross-link that forms within the matrix of a protein that has been linked with an increased rate of aging in birds.[12,13]

In the current study, the role of uric acid as an antioxidant was evaluated in chickens. The specific objectives were to determine the effects of allopurinol and hemin on uric acid levels in broiler chickens and on the accumulation of Ps and shear values (SV) of pectoralis major muscle. Allopurinol is a structural analogue of the natural purine base hypoxanthine, and it and its metabolite oxypurinol are potent inhibitors of xanthine oxidase, the enzyme involved in the conversion of xanthine to uric acid.[16] Hemin, on the other hand, increases the concentration of uric acid.[17] The present studies sought to determine the capacity of uric acid to scavenge oxygen free radicals and enhance measures of oxidative injury in chickens with reduced uric acid levels.

29.2 MATERIALS AND METHODS

29.2.1 Birds and Management

The purpose of the first study was to establish the dose of allopurinol and hemin required to manipulate concentrations of plasma uric acid in broiler chicks. Broiler chicks ($n = 90$; Ross × Ross; mixed sex) approximately 6 weeks of age were obtained from Ross Breeders and maintained under standard husbandry practices. These included recommended brooders and temperatures, bell drinkers, and pan feeders. Specifications for space, temperature, light, and husbandry was adhered to.[18] At 8 weeks of age, 50 broilers were divided into five groups: control, allopurinol fed (5 mg/kg and 10 mg/kg BW), and hemin fed (5 mg/kg and 10 mg/kg BW). Blood was sampled for plasma uric acid measurements at day 3 and 10. Another group comprising 30 broilers was divided into three treatments: control and allopurinol fed (40 and 50 mg/kg BW). These birds were fed high allopurinol diets for 3 weeks and blood samples were taken weekly.

The principal experiment involved birds of 12 weeks of age. After the determination of appropriate dose, the main trial started at week 12. The birds were divided into two groups, diet restricted (DR) and *ad libitum* (AL), with three treatments within each group (control, allopurinol fed, and hemin fed). The diet-restricted birds were fed with a limited allowance diet (Table 29.1). Allopurinol- and hemin-supplemented feeds were prepared every week based on the weight of the birds the previous week. Birds were killed at 4 and 10 weeks after the onset of the trial.

TABLE 29.1
The Amount of Feed (g/day) Provided on a Daily Individual Basis for the Diet-Restricted and *ad Libitum*-fed Groups

Age, weeks	*Ad Libitum*	Diet Restricted
12	203	130
13	232	150
14	220	160
15	208	165
16	211	165
17	208	165
18	201	165
19	208	165
20	202	165
21	201	165
22	206	165

29.2.2 Electron Spin Resonance Measurement

Electron spin resonance (ESR) spin trapping was used to determine *in vitro* the possible contribution of allopurinol and oxypurinol to antioxidant activity in the treated animals by measuring short-lived free radical intermediates.[19] This technique involves the interaction of a short-lived oxygen free radical with a diamagnetic compound (spin trap) to form a relatively long-lived free radical product, which can be detected by ESR. The intensity of the adduct signal corresponds to the amount of short-lived oxygen free radicals that have formed an adduct within the spin trap. Reactants are mixed in test tubes in total final volume of 0.50 ml and transferred to a flat cell for ESR measurements. All ESR measurements were made using a Varian E4 spectrometer and a flat cell assembly. Hyperfine splitting was measured to 0.1 G directly from magnetic field separations using potassium tetraperoxochromate (K_3CrO_8) and 1,1-diphenyl-2-picrylhydrazyl (DPPH) as standards. The relative radical concentration was estimated by multiplying half of the peak height by multiplying by (ΔH_{pp}).[19] ΔH_{pp} represents peak-to-peak width.

29.2.3 Pentosidine Determination

Birds ($n = 5$) from the second study were randomly selected from each dietary group at 16 and 22 weeks of age and killed by electrical stunning. Approximately 1 g of skin was removed from the abdominal area, washed with normal saline, and stored at –80°C for analysis. A collagen digest was prepared as described by Monnier et al.[19,20] and Sell et al.[21] Briefly, this technique involves skin preparation (removal of the epidermis and adipose layers and very fine mincing), delipidation in a chloroform-methanol mixture (2:1), and rehydration in 50% methanol followed by hydrolysis in 6 *N* HCl at 110°C for 18 h under nitrogen. The samples were placed into a Speed Vac Centrifuge and vacuum desiccated samples were reconstituted with 250 μl dd H_2O and filtered using a Costar Spin-X® centrifuge tube filter. Collagen content was estimated by the modified Stagmen and Stalder method using a hydroxyproline standard. This method assumes that hydroxyproline makes up 14% of the total collagen.[22] The measurement of Ps was accomplished by reverse-phase high-performance liquid chromatography (HPLC);[12] 1 mg of skin collagen digest in 100 μl water/0.01 *M* heptaflurobutyric acid (HFBA) was injected into a 0.46 × 25 cm Vydac 218TP104 (10 μm) C-8 column connected to a Waters HPLC. The apparatus consisted of two pumps (Waters™ 600 Controller), an auto sampler (Waters™ 717), and a scanning fluorescence detector (Waters™ 474). Separations were achieved by a linear gradient of 12 to 42% acetonitrile from 0 to 25 min in water and 0.01 *M* HFBA at a flow rate of 1 ml/min. The Ps peak is detected by an online scanning fluorescence detector at an excitation wavelength of 325 nm/emission wavelength 370 nm. Quantification of Ps was made by comparison of peak areas with a Ps standard injected under identical conditions. A software package (Millennium 2.1) was used for peak integration.

29.2.4 Breast Weight, Cooking Time, and Shear Value Evaluation

Birds were electrically stunned and bled using a modified Kosher technique.[23] The pectoralis major muscle was isolated, refrigerated at 4°C for 4 h, vacuum-packed, and stored at –20°C until further processed. The breast muscle was cooked to an internal temperature of 70°C on a Farberware Smokeless Indoor Grill. The end point internal temperature was monitored with an industrial data logger, equipped with a copper-constant thermocouple. Cooked muscle was cooled to room temperature and refrigerated overnight at 4°C. Slices of approximately 1.2 to 1.3 cm were cut perpendicular to fiber orientation of the muscle. From each sample, four to five cores were removed from the thickest portion of the cooked muscle. Shear values (SV) were determined by using an Instron Universal Mechanical Machine. A Warner-Bratzler Apparatus was attached to a 50-kg load cell and tests were performed at a cross head speed of 127 mm/min. Output from a Linear Variable Displacement Transformer (LVDT) conditioner was acquired by a computer equipped with a DT 2805 data acquisition board. Signals were processed with the HP-VEE software package.

29.2.5 Uric Acid and Glucose Determination

Plasma uric acid ($n = 5$ per treatment group) was determined using a commercially available Uric Acid Reagent. Plasma glucose was measured using a YSI 2700 Select Biochemistry Analyzer.

29.2.6 Luminol-Based Chemiluminescence as a Measurement of Oxidative Stress

Chemiluminescence techniques are functional assays to quantify the release of oxidants from cells or tissues.[24,25] Luminol-based chemiluminescence (LBCL) oxidative stress was used to define the amount of oxidation stress as described by Iqbal et al.[13] Blood (1 ml) from 16- and 22-week-old birds ($n = 5$) was carefully layered on 5 ml in mono-polyresolving medium in a 13 × 100 mm #10 Falcon tube (2027) (ICN 16-980-49) and leukocytes were isolated by centrifugation at 300 *g* for 5 min. The total number of leukocytes was counted using a routine hemocytometric technique. To a 3-ml luminometer tube was then added 100 μl of leukocytes (10^6/ml), 100 μl luminol (10^{-5} *M*) solution, 200 μl phosphate-buffered saline (PBS) (pH 7.4), and 100 μl phorbol myristate acetate (PMA). Luminol reacts with hypochlorite with the production of photons. The tube was placed into a luminometer with the temperature control set at 37°C. Oxidative stress was determined by measuring the integrated luminescence generated over 20 min using KINB software and a Berthold luminometer. Results were reported as counts per minute (CPM). Measured luminescence was corrected based on the number of leukocytes present in each reaction vessel.

29.2.7 Statistical Analysis

Data were analyzed by the general linear models procedure.[26] The Student-Newman-Keuls Multiple Range test was used to estimate the significance of difference between means.

29.3 RESULTS

29.3.1 Study 1

At day 10, compared to controls, plasma uric acid was decreased 57% ($P < 0.05$) when allopurinol was fed at a concentration of 10 mg/kg BW (data not shown). In contrast, there was a 20% increase at day 10 in plasma uric acid when hemin was fed, at a concentration of 10 mg/kg BW. There was no significant effect of the treatments on uric acid at day 3. By week 2, plasma uric acid concentration had decreased 44% ($P < 0.05$) in the group fed 40 mg allopurinol/kg BW and 65% ($P < 0.01$) in the 50-mg group (Figure 29.2) when compared with the control. Chemiluminescence-dependent oxidative stress increased twofold ($P < 0.05$) in the 40-mg group whereas in the 50-mg allopurinol/kg BW group there was a fourfold increase ($P < 0.001$) after 2 weeks of feeding as compared with the control (Figure 29.3). Figure 29.4 shows the effect of H_2O_2 on the $^*OH$ generation from a reaction mixture recorded 3 min after reaction initiation from a pH of 7.4 phosphate buffer containing 1 m*M* DMPO, 0.1 m*M* H_2O_2, and 0.02 m*M* Fe(II). The intensity of the $^*OH$ generation decreased ninefold when a saturated solution of allopurinol (Figure 29.4b) and oxypurinol (Figure 29.4c) was added. There was approximately a 40 and 30% reduction in intensity of the $^*O$ radical production with allopurinol and oxypurinol addition, respectively (Figure 29.5).

29.3.2 Study 2

29.3.2.1 Plasma Uric Acid and Glucose

Allopurinol ($P < 0.05$) reduced plasma concentrations of uric acid, both in AL and DR birds (Figure 29.6). The reduction ranged from 26 to 74%, being more pronounced by week 22. Uric acid concentration was marginally higher in the AL birds compared with the DR controls at week 16,

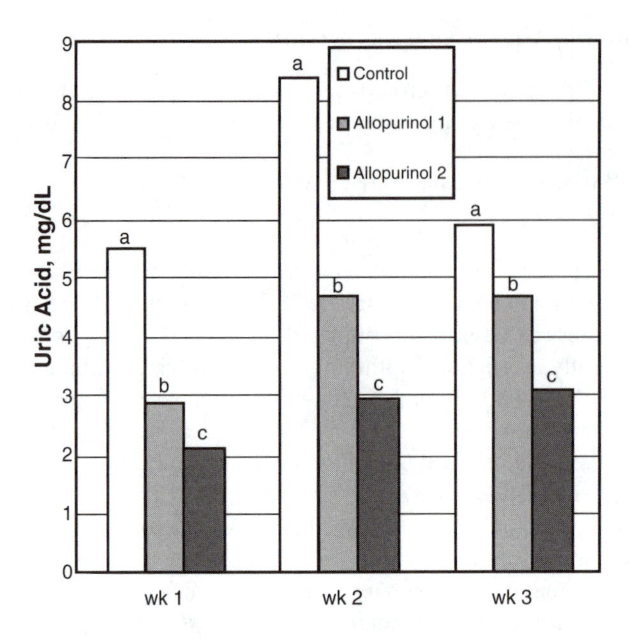

FIGURE 29.2 The effect of two levels of allopurinol on plasma uric acid concentrations on broiler chicks from 8 to 11 weeks old. Allopurinol 1–40 mg/kg BW, Allopurinol 2–50 mg/kg BW. Means with no common letters differ significantly ($P < 0.05$).

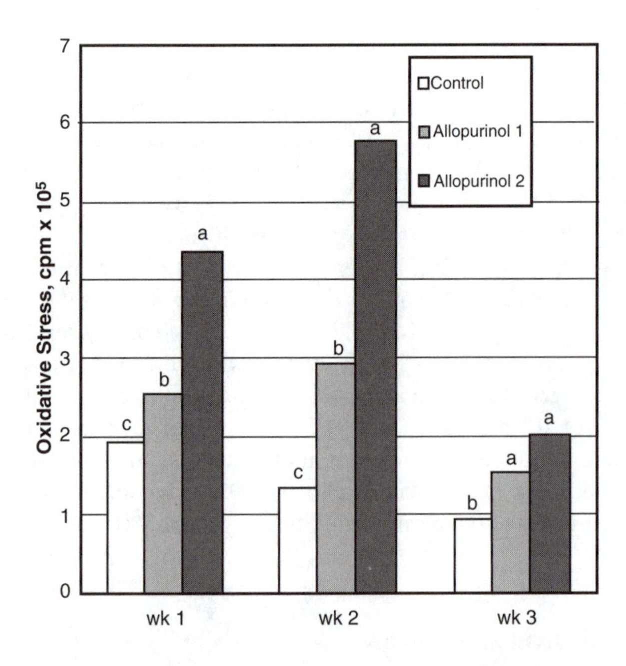

FIGURE 29.3 Effect of two levels of allopurinol on luminol-induced chemiluminescence detected oxidative stress on broiler chicks from 8 to 11 weeks old. Allopurinol 1–40 mg/kg BW, allopurinol 2–50 mg/kg BW. Means with no common letter differ significantly ($P < 0.05$).

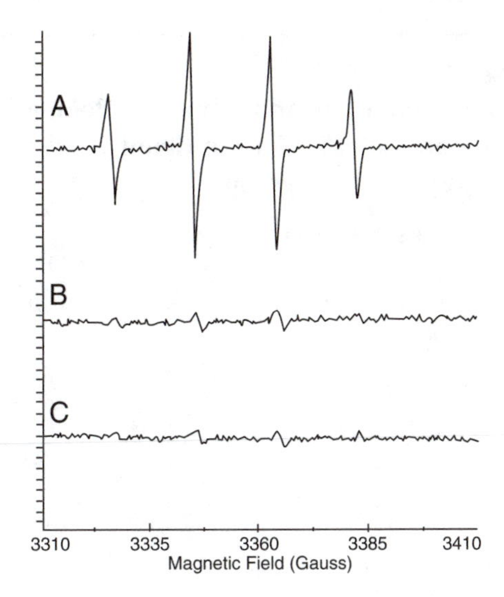

FIGURE 29.4 The effect of H_2O_2 on the hydroxyl radical generation. (A) ESR spectrum recorded 3 min after reaction initiation from a pH of 7.4 phosphate buffer containing 1 m*M* DMPO, 0.1 m*M* H_2O_2, and 0.02 m*M* Fe(II). (B) Same as A, but with saturated allopurinol. (C) Same as A, but with saturated oxypurinol.

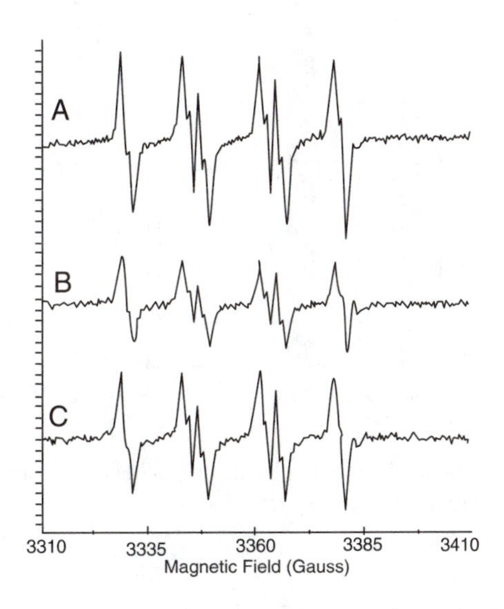

FIGURE 29.5. The intensity of superoxide radical generation with xanthine oxidase. (A) ESR spectrum recorded 3 min after reaction initiation from a pH of 7.4 phosphate buffer containing 50 m*M* DMPO, 0.35 m*M* xanthine, 0.05 unit xanthine oxidase, and 0.3 m*M* Datapac. (B) Same as A, but with saturated allopurinol. (C) Same as A, but with saturated oxypurinol.

although the increase was not significant. Hemin increased uric acid concentrations at week 16, with an increase of 11 to 16%. At week 22, there was an increase (14%) only in the DR group. Glucose concentrations were not ($P < 0.05$) different nor did they show any consistent trend throughout the study (Table 29.2).

TABLE 29.2
Effect of Allopurinol and Hemin on Plasma Glucose and Total Leukocyte Count (n = 5 per treatment group)

Group	Plasma Glucose (mg/dl)		TLC × 10^5	
	wk 16	wk 22	wk 16	wk 22
DR-C	231 ± 4	234 ± 5	15.1 ± 0.0*	24.6 ± 7.4
DR-A	229 ± 7	239 ± 8	32.2 ± 2.4*	66.3 ± 14.2*
DR-H	224 ± 3	231 ± 7	42.3 ± 2.3*	97.8 ± 35*
AL-C	233 ± 3	239 ± 8	22.8 ± 2.0	25.3 ± 5.0
AL-A	224 ± 10	226 ± 7	37.08 ± 2.2*	73.6 ± 21.0*
AL-H	235 ± 4	233 ± 5	43.0 ± 1.7*	130.7 ± 29*

*$P < 0.05$ compared within a column.

DR = diet restriction, AL = *ad libitum*, C = *control*, A = allopurinol, H = hemin. Means ± SE.

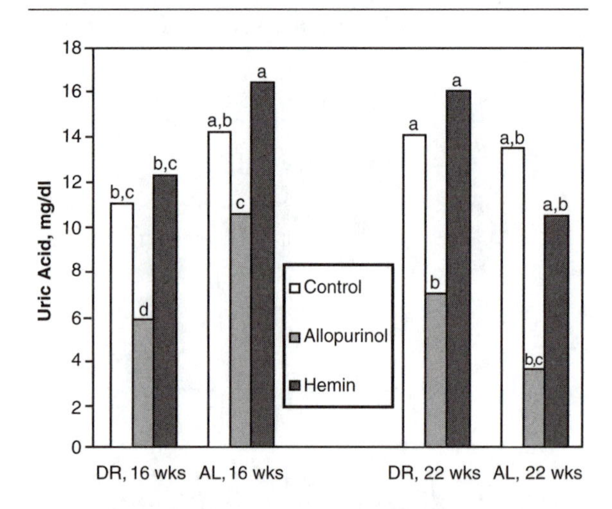

FIGURE 29.6 Effect of allopurinol and hemin on plasma uric acid concentrations at 16 and 22 weeks of age. Means with no common letters differ significantly ($P < 0.05$).

29.3.2.2 Skin Pentosidine

Skin concentrations of Ps were increased sevenfold ($P < 0.05$) in AL birds fed allopurinol at 22 weeks of age, whereas DR limited this increase (Figure 29.7). There was no effect of allopurinol on concentrations of Ps at week 16. Similarly, there was no effect of hemin on either DR or AL fed birds at week 16 whereas by week 22 it was detectable in the DR birds and markedly increased ($P < 0.05$) in the AL birds.

29.3.2.3 Chemiluminescence Measurement of Oxidative Stress

Hemin increased the chemiluminescence-dependent oxidative stress ($P < 0.05$) at 22 weeks in both the DR and AL groups and in the AL birds at 16 weeks. Allopurinol elevated ($P < 0.05$) the oxidative stress in the AL group at week 22 group (Figure 29.8). Both hemin and allopurinol increased the total leukocyte counts both at 16 and 22 weeks. This effect was much more pronounced ($P < 0.05$)

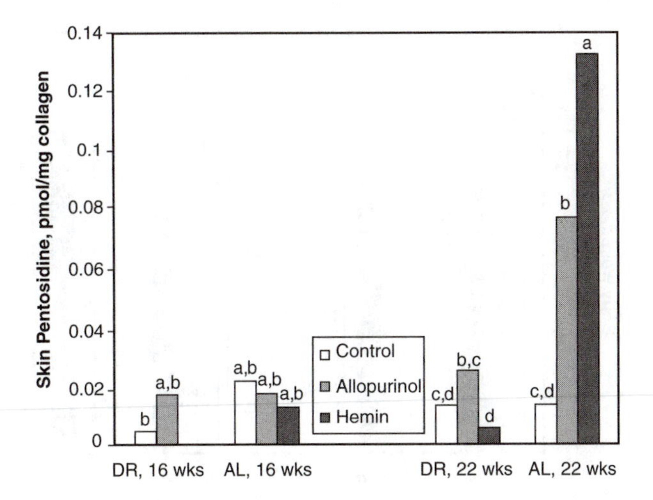

FIGURE 29.7 Effect of allopurinol and hemin on skin pentosidine, Ps, at 16 and 22 weeks of age. Means with no common letters differ significantly ($P < 0.05$).

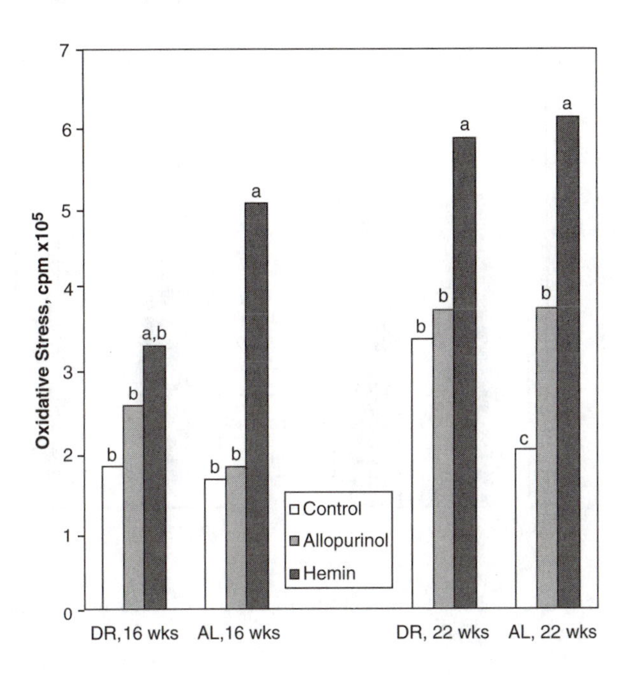

FIGURE 29.8 Effect of allopurinol and hemin luminol-induced chemiluminescence-detected oxidative stress at 16 and 22 weeks of age. Means with no common letters differ significantly ($P < 0.05$).

in hemin-fed birds. There was 2.7- and 4.0-fold increase in the total leukocyte count at week 16 and 22, respectively, in the hemin-fed birds (see Table 29.2).

29.3.2.4 Body Weight and Shear Force

Allopurinol reduced both the body weight, in both groups at both 16 and 22 weeks (Figure 29.9). The reduction in body weight was more pronounced at week 22 (Figure 22.9). Mortality was 40% in the allopurinol-fed birds by the end of the trial, compared to 7 and 13% in the control and hemin-fed birds. Allopurinol-fed birds had increased ($P < 0.05$) shear force values compared with the

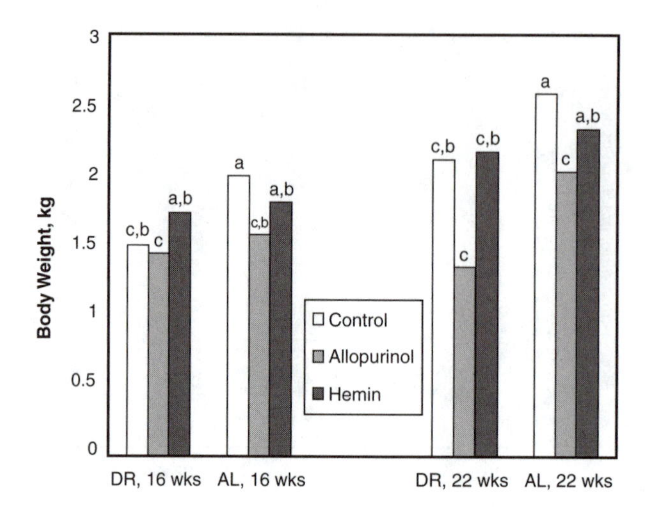

FIGURE 29.9 Effect of allopurinol and hemin on body weight at 16 and 22 weeks of age. Means with no common letters differ significantly ($P < 0.05$).

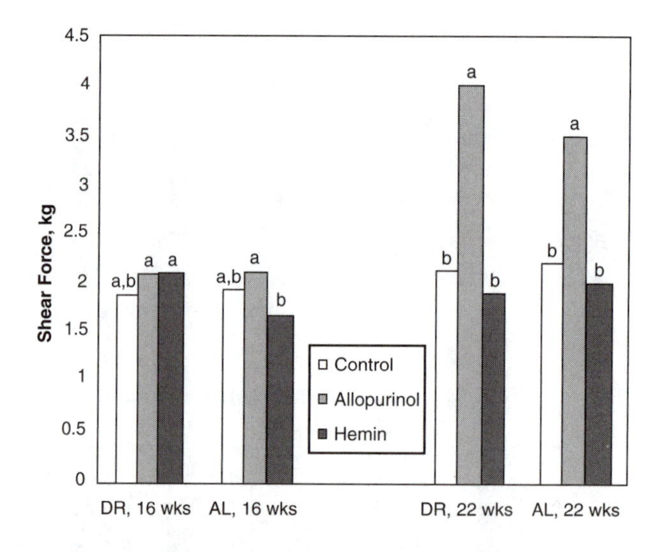

FIGURE 29.10 Effect of allopurinol and hemin on shear values (SV) at 16 and 22 weeks of age. Means with no common letters differ significantly ($P < 0.05$).

control birds. There was an 88 and 58% increase in the DR and AL group at week 22 (Figure 29.10). However, there was no significant effect of allopurinol at 16 weeks. Hemin marginally reduced the shear values at week 22 without significance. No diet effects were seen for hemin, at either week 16 or 22.

29.4 DISCUSSION

Previous studies have established that birds have higher concentrations of antioxidants in their bodies and appear more efficient in dealing with the oxidative stress compared to mammals.[3,27] Iqbal et al.[12,13] hypothesized that uric acid plays an important role in limiting oxidative stress and subsequent accumulation of advanced glycosylation end products (AGEs), such as Ps in birds. In previous studies, dose rates of allopurinol from 2 to 50 mg/kg were used in laboratory rats,[28]

6.5 mg/kg in dogs, and varying dose rates, from 30 to 50 mg/kg, in ethanol-fed turkey pullets.[29] In the present study, the administration of allopurinol to broiler chickens at 10 mg/kg was shown to decrease plasma uric acid concentrations. In Study 2, the reduction ranged from 26 to 74% and was more pronounced with the longer duration of treatment, but was unexpectedly associated with an increase in mortality. Results from Study 1 did not suggest that the dose selected was toxic although the duration of treatment was shorter. It is also evident that the selected dose was not toxic as birds were able to withstand 40 and 50 mg allopurinol/kg BW for a period of 3 weeks. The lowering of the uric acid is explained on the basis of the primary action of allopurinol, inhibition of xanthine oxidase, an enzyme that is involved in the conversion of hypoxanthine to xanthine and xanthine to uric acid. This indicates that broiler chickens are responsive to allopurinol treatment and that the major metabolite of allopurinol, oxypurinol, is probably responsible for the inhibition of xanthine oxidase. A decrease in the uric acid concentration was accompanied by a decrease in the body weight and an increase in tissue aging as evidenced by an increase in shear force values (88 and 58% increase) in the DR and AL group, respectively. In humans, many conditions result in hyperuricemia (gout, alcoholism, diuretics, cardiac myopathy, etc.) and in the search for appropriate animal models, poultry can also be a valuable tool since the uric acid levels can be manipulated.

Allopurinol and oxypurinol are known to have free radical scavenging activity. However the concentration of allopurinol in such experiments exceeds physiological concentrations by several orders of magnitude. For example, the hydroxyl radical scavenging activity of allopurinol[30] required concentrations in excess of 0.5 m*M* and daily administration of allopurinol to cats (50 mg/kg BW) produced serum concentrations of only 12.0 f*M*. Faure et al.[31] reported that allopurinol inhibited radiation-induced lipid peroxidation in rat erythrocytes, which suggested that it or its metabolic oxypurinol has a direct antioxidant action. In their experiments, the allopurinol concentration was reported to be 20 m*M*, nearly 200 times the usual level achieved in serum with pharmacological doses of allopurinol. In the current study, the antioxidant effects of allopurinol are unlikely to be present as the concentrations achieved were much too low. Allopurinol significantly ($P < 0.001$) depressed body weight, particularly at week 22, both DR and AL, and at week 16, in the AL birds. There was also increased (40%) mortality by week 22. There are several reports in the literature documenting toxicity of allopurinol in humans. In support of these findings was our observation that allopurinol-treated birds showed a reduction in feed consumption. This may have been due to development of hypertonicity. Similar results have been reported in turkey pullets[32] and humans.[33] The reason seems to be that plasma oxypurinol concentrations become several times greater than that required to prevent uric acid formation. Other investigators surmise that decreased uric acid excretion during long-term therapy with allopurinol altered the tubular reabsorption of oxypurinol if uric acid and oxypurinol share the same transport mechanism.[34] Allopurinol is readily absorbed and is rapidly converted to oxypurinol that is excreted in urine unchanged.[35] In humans, a decrease in the intake of protein alters tubular transport of uric acid.[36] Although it minimally affected renal clearances of allopurinol, it has a marked effect on renal clearance of oxypurinol. Other reported side effects of allopurinol include gastrointestinal intolerance and vasculitis.[36,37] Additional studies are required to determine whether any of these side effects may be responsible for the allopurinol toxicity observed in our studies and in the turkey pullets.

Hemin (10 mg/kg BW) marginally increased the concentrations of uric acid compared with controls, in both the DR and AL birds. This increase ranging from 11 to 16% was associated with only a slightly reduced shear values at week 22. Interestingly, the hemin-associated increase in uric acid did not reduce the oxidative stress but rather enhanced it. Hemin is a blood product and a source of iron. A growing body of evidence indicates that transition metals such as iron catalyze the formation of reactive oxygen species and stimulate lipid peroxidation.[38] The relationship among metal ions, oxygen radicals, and tissue damage has been reviewed.[39,40] In our studies, hemin administration resulted in oxidative stress ($P < 0.001$) and increased total leukocyte count (TLC) in the AL-fed 16- and 22-week-old and DR 16-week-old birds. These results are thus in agreement with previous studies[39–41] that ascribe an increased oxidative stress capacity to hemin.

Changes in plasma uric acid concentrations also were associated ($P < 0.001$) with changes in the accumulation of Ps concentrations, particularly with duration of treatment. In agreement with previous studies in birds, concentrations of Ps were lower in DR birds compared to AL-fed birds.[13] Further, concentrations of Ps, and possibly other AGE, were not associated with elevated concentrations of plasma glucose even though conditions of high body temperature and metabolic rate are conducive to their formation.[13] The present studies show no significant effects of either allopurinol or hemin on plasma glucose concentrations. This is in agreement with previous findings in birds[13] but not with results in mammals where variable effects of DR have been reported. For example, a 10% decrease in blood glucose concentrations in DR rats was associated with a decrease in the glycosylation of proteins, which might be caused by the decreased oxidative stress in DR animals.[42,43] Although several studies have firmly established that collagen glycosylation is increased with age, others have failed to demonstrate anything but a cursory relationship between protein cross-linking and glycosylation, either *in vivo* or *in vitro*.[44–47] As an example, the amount of glycosylated Hb in hummingbirds, which have plasma glucose concentrations in excess of 650 mg/dl, is only 2 to 5%, much lower than that measured in mammals, which have levels ranging from 6 to 8%.[48]

The reduction in uric acid concentrations, at week 22 in both the DR and AL birds, as a result of allopurinol feeding was associated with an increase in SF value of the pectoralis major muscle. This observation is consistent with our hypothesis that uric acid is an important antioxidant in birds. The reason for the increase in SF values has not been established although the increase in the concentration of the intramolecular cross-link Ps may be associated with an increased level of oxidative stress and glycation. A reduction in the concentration of antioxidants accelerated the formation of glycoxidation products. This finding is supported by the work of others.[49–51,52]

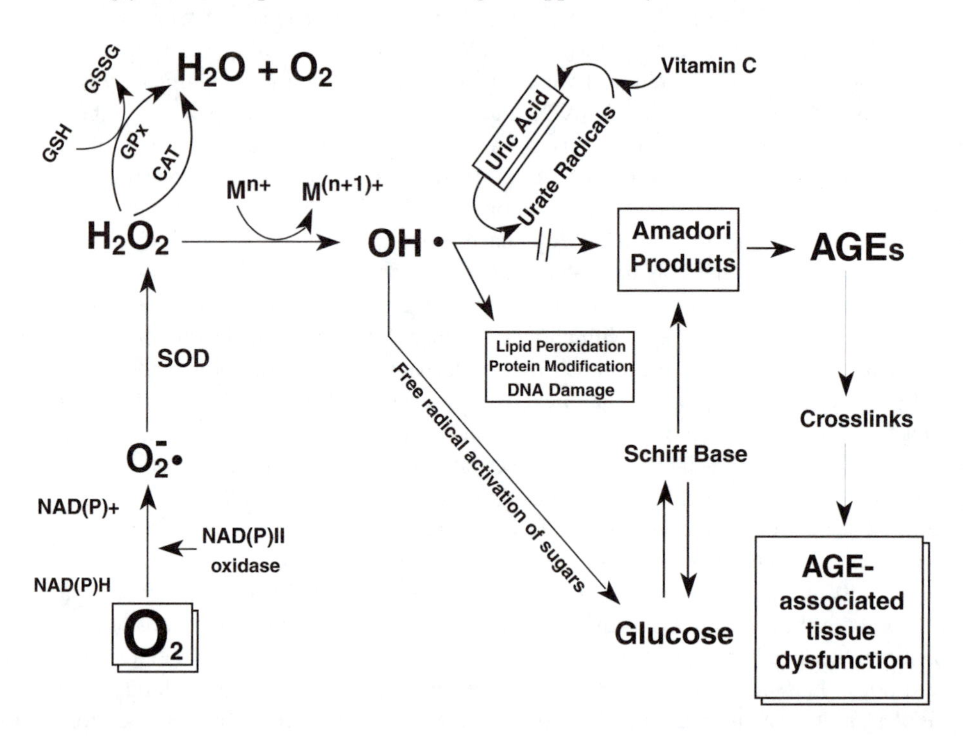

FIGURE 29.11 Generation of reactive oxygen species (ROS) and associated tissue dysfunction in aves, and the mechanisms utilized against damage by reactive oxygens. O_2, molecular oxygen; O_2^-, superoxide anion; $OH^•$, hydroxyl radical; H_2O_2, hydrogen peroxide; GSH/GSSG, reduced/oxidized glutathione; GP_x, glutathione peroxidase; NADPH/NADP$^+$, reduced/oxidized nicotinamide adenine dinucleotide phosphate; SOD, superoxide dismutase; CAT, catalase; metal (Fe, Ti, Cu, Cr); AGEs, advanced glycosylated end products.

In conclusion, our investigation has documented that a reduction in uric acid concentration in chickens is associated with increase in oxidative stress, the accumulation of the glycoxidation product Ps, and a decline in the SF value of the pectoralis major muscle. In view of these findings we propose that it is the availability of uric acid in part that regulates the generation of glycoxidation products in the tissues of birds (Figure 29.11). Consequently, a reduction in uric acid concentration accelerates the formation of cross-links and tissue dysfunction. Diet restriction was effective in limiting both the allopurinol-induced increase in oxidative stress as well as the accumulation of Ps. In view of our findings about the toxicity of allopurinol and oxypurinol both the dosage and duration (short vs. extended time scale) and route of administration require additional evaluation. Hemin, while effective at increasing uric acid concentrations, was demonstrated to be an inducer of oxidative stress. The response of the birds to the increase in uric acid was likely masked by hemin-induced oxidative stress and discrimination between the two responses is complicated.

ACKNOWLEDGMENTS

The authors thank Vincent M. Monnier, Institute of Pathology, Case Western Reserve University for supplying the pentosidine standard and other technical guidance. We are also grateful to Nabil Al-Humadi, NIOSH for his valuable assistance in the HPLC analyses, Hakan Kocamis for technical assistance, Edwin C. Townsend for his help with the statistical analysis, and Gertrude Elion for the gift of allopurinol.

REFERENCES

1. Lindstedt, S. and Calde, W., Body size and longevity in birds, *Condor,* 78, 91–94, 1976.
2. Holmes, J. D. and Austad, S. N., Birds as animal models for the comparative biology: prospectus, *J. Gerontol. Biol. Sci.,* 50A, B59–B66, 1995.
3. Ku, H. and Sohal, R. S., Comparison of mitochondrial prooxidant generation and antioxidant defenses between rat and pigeon: possible basis of variation in longevity and metabolic potential, *Mech. Aging Dev.,* 72, 67–76, 1993.
4. Schreiber, G. W., Tiemeyer, W., Flurer, C. I., and Zucker, H., Purine metabolites in serum of higher primates, including man, *Int. J. Primatol.,* 7, 521–531, 1986.
5. Ames, B. N., Cathcart, R., Schwiers, E., and Hochstein, P., Uric acid provides an antioxidant defense in humans against oxidant- and radical-caused aging and cancer: a hypothesis, *Proc. Natl. Acad. Sci. U.S.A.,* 78, 6858–6862, 1981.
6. Cutler, R. G., Antioxidants, aging, and longevity, in *Free Radicals in Biology,* Vol. VI, Pryor, W. A., Ed., Academic Press, Orlando, 1984, 371–428.
7. Cutler, R. G., Urate and ascorbate: their possible roles as antioxidants in determining longevity of mammalian species, *Arch. Gerontol. Geriatr.,* 3, 321–348, 1984.
8. Parmley, L., Mufti, A. G., and Downey, J. M., Allopurinol therapy of ischemic heart disease with infarct extension, *Can. J. Cardiol.,* 8, 280–286, 1992.
9. Hellsten, Y., Tullson, P. C., Richter, E. A., and Bangsbo, J., Oxidation of urate in human skeletal muscle during exercise, *Free Radical Biol. Med.,* 22, 169–174, 1997.
10. Short, R., Williams, D. D., and Bowden, D. M., Circulating antioxidants as determinants of the rate of biological aging in Pigtailed Macaques (*Macaca nemestrina*), *J Gerontol. Biol. Sci.,* 52A, B26–B30, 1997.
11. Becker, F. B., Towards the physiological function of uric acid, *Free Radical Biol. Med.,* 14, 615–631, 1993.
12. Iqbal, M., Probert, L. L., and Klandorf, H., Effect of dietary aminoguanidine on tissue pentosidine and reproductive performance in broiler breeder hens, *Poult. Sci.,* 76, 1574–1579, 1997.
13. Iqbal, M., Probert, L. L., Al-humadi, N. H., and Klandorf, H., Protein glycosylation and advanced glycosylation endproducts (AGEs): an avian solution, *J. Gerontol. Biol. Sci.,* 54, B171–B176, 1999.
14. Bishop, M. L., Duben-Engelkink, J. L., and Fody, E. P., *Clinical Chemistry, Principles, Procedures, Correlations,* 2nd ed., Lippincott, Philadelphia, PA, 1992.

15. Shapiro, F., Mahagna, M., and Nir, I., Stunting syndrome in broilers: effect of glucose or maltose supplementation on digestive organs, intestinal disaccharidase, and some blood metabolites, *Poult. Sci.,* 76, 369–380, 1997.
16. Bartges, J. W., Osborne, C. A., Felice, L. J., Koehler, L. A., Ulrich, L. K., Bird, K. A., and Chen, M., *Am. J. Vet. Res.,* 5, 511–515, 1997.
17. Miller, D. M., Grover, T. A., Nayani, N., and Aust, S. D., Xanthine oxidase and iron dependent lipid peroxidation, *Arch. Biochem. Biophys.,* 301, 1–7, 1993.
18. Ross Breeders, *Ross Breeder Management Guide,* Ross Breeders, Huntsville, AL, 1996.
19. Shi, X., Flynn, D. C., Porter, D. W., Leonard, S. S., Vallyathan, V. V., and Castranova, V., Efficacy of taurine based compounds as hydroxyl radical scavengers in silica induced peroxidation, *Ann. Clin. Lab. Sci.,* 27, 365–376, 1997.
20. Monnier, V. M., Vishwanath, V., Frank, K. E., Elmets, C. A., Dauchot, P., and Kohn, P. R., Relationship between complications of type I diabetes mellitus and collagen-linked fluorescence, *N. Engl. J. Med.,* 314, 403–408, 1986.
21. Sell, D. R., Lapolla, A., Odetti, P., Fogarty, J., and Monnier, V. M., Pentosidine formation in skin correlates with severity of complications in individuals with long-standing IDDM, *Diabetes,* 41, 1286–1291, 1992.
22. Maekawa, T., Ratinasamy, K. I., Altman, Y. K., and Forbes, W. F., Changes in collagen with age. I. The extraction of acid soluble collagens from skin of mice, *Exp. Gerontol.,* 5, 177–186, 1970.
23. Iqbal, M., Kenney, P. B., and Klandorf, H., Age-related changes in meat tenderness and tissue pentosidine: effect of diet restriction and aminoguanidine in broiler breeder hens, *Poult. Sci.,* 78, 1328–1333, 1999.
24. Van Dyke, K., Cellular applications, in *Cellular Chemiluminescence,* Vol. 2, Van Dyke, K. and Castranova, V., Eds., CRC Press, Boca Raton, FL, 1987, 41–182.
25. Radi, R., Cosgrove, P., Beckman, J. S., and Freeman, B. A., Peroxynitrite-induced luminol chemiluminescence, *Biochem. J.,* 290, 51–57, 1993.
26. SAS Institute, SAT/STAT7 Users Guide: Statistics. Release 6.04, SAS Institute, Inc., Cary, NC, 1990.
27. Schweiger, F. F., Uehlein-Harrell, S., Von Hegel, G., and Weisner, H., Vitamin A (retional and retinyl esters), α-tocopherol and lipid levels in plasma of captive wild mammals and birds, *Zentralbl. Veterinarmed.,* A38, 35–42, 1991.
28. Klein, A. S., Joh, J. W., Rangan, W., Wong, D., and Bulkey, G. B., Allopurinol discrimination of antioxidant from enzyme inhibitory acitivities, *Free Radical Biol. Med.,* 21, 713–717, 1975.
29. Czarnecki, C. M., Olivero, D. K., and McVey, A. S., Plasma uric acid levels in ethanol fed turkey poults treated with allopurinol, *Comp. Biochem. Physiol.,* 86C; 67–71, 1987.
30. Morehouse, K. M., Flitter, W. D., and Mason, R. P., The enzymatic oxidation of desferal to a nitroxide free radical, *FEBS Lett.,* 222, 246–250, 1987.
31. Faure, M., Lissi, E. A., and Videla, L., Antioxidant capacity of allopurinol in biological systems, *Biochem. Int.,* 21, 357–366, 1990.
32. Czarnecki, C. M., Plasma uric acid levels in ethanol-fed turkey poults treated with allopurinol, *Comp. Biochem. Physiol.,* 186, 67–71, 1987.
33. Hande, K. R., Noone, R. M., and Stone, W. J., Severe allopurinol toxicity, *Am. J. Med.,* 74, 47–56, 1984.
34. Berlinger, W. G., Park, G. D., and Spector, R., The effect of dietary protein on the clearance of allopurinol and oxypurinol, *N. Engl. J. Med.,* 313, 771–776, 1985.
35. Appelbaum, S. J., Mayersohn, M., Dorr, R. T., and Perrier, D., Allopurinol kinetics and bioavailibility: intravenous, oral and rectal administration, *Cancer Chemother. Pharmacol.,* 8, 93–98, 1982.
36. Mehta, S., Drug disposition in children with protein energy malnutrition, *J. Pediatr. Gastroenterol. Nutr.,* 2, 407–417, 1983.
37. Fox, I. H. and Kelley, W. N., Uric acid and gout, in *The Kidney: Physiology and Pathophysiology,* Vol. 2, Seldin, D. W. and Giebisch, G., Eds., Raven Press, New York, 1985, 1747–1764.
38. Tappel, A. L., Unsaturated lipid oxidation catalyzed by hematin compounds, *J. Biol. Chem.,* 217, 721–733, 1985.
39. Aust, S. D., Morehouse, L. A. L., and Thomas, C. E., Role of metals in oxygen radical reactions. *J. Free Radical Biol. Med.,* 1, 3, 1985.
40. Ryan, T. P. and Aust, S. D., The role of iron in oxygen mediated toxicities, *Crit. Rev. Toxicol.,* 22, 119–141, 1992.

41. Grinbert, L. N., O'Brien, P. J., and Hakal, Z., The effects of heme-binding proteins on the peroxidative and catalatic activities of hemin, *Free Radical Biol. Med.,* 26, 214–219, 1999.
42. Masoro, E. J., Katz, M. S., and McMahan, C. A., Evidence for the glycation hypothesis of aging from the food restricted rodent model, *J. Gerontol.,* 44, B20–B22, 1989.
43. Sell, D. R., Lane, M. A., Johnson, W. A., Masoro, E. J., Mock, O. B., Reiser, K. M., Fogarty, J. F., Cutler, R. G., Ingram, D. K., Roth, G. S., and Monnier, V. M., Longevity and the genetic determination of collagen glycoxidation kinetics in mammalian senescence, *Proc. Natl. Acad. Sci. U.S.A.,* 93, 485–490, 1996.
44. Guitton, J. D., LePape, A., Sizaret, P. Y., and Muh, J. P., Influences of *in vitro* nonenzymatic glycosylation of type I collagen fibrillogenesis, *Biosci. Rep.,* 1, 945–954, 1981.
45. LePape, A., Guitton, J. D., and Muh, J. P., Distribution of non-enzymatically bound glucose *in vivo* and *in vitro* glycosylated Type I collagen molecules, *FEBS Lett.,* 170, 23–27, 1984.
46. Lyons, T. J., Bailie, D. E., Dyer, D. G., Dunn, J. A., and Baynes, J. W., Decrease in skin collagen glycation with improve glycemic control in patients with insulin-dependent diabetes mellitus, *J. Clin. Invest.,* 87, 1910–1915, 1991.
47. Monnier, V. M., Nonenzymatic glycosylation, the Maillard reaction and the aging process, *J. Gerontol. Biol. Sci.,* 45, B105–B111, 1990.
48. Beuchat, C. A. and Chong, C. R., Hyperglycemia in hummingbirds: Implications for hummingbird ecology and human health, *FASEB J.,* 11(3), A91, 1997.
49. Yu, B. P., Masoro, E. J., Murafa, I., Berfrand, H. Q., and Lynd, F. T., Life-span study of SPF Fisher 344 male rats fed *ad libitum* or restricted diets: longevity, growth, lean body mass and disease, *J. Gerontol. Biol. Sci.,* 37, 130, 1982.
50. Youngman, L. D., Park, J. Y., and Ames, B. N., Protein oxidation associated with aging is reduced by dietary restriction of protein or calories, *Proc. Natl. Acad. Sci. U.S.A.,* 89, 9112–9116, 1992.
51. Yu, B. P., Oxidative damage by free radicals and lipid peroxidation in aging, in *Free Radicals in Aging,* Yu, B. P., Ed., CRC Press, Boca Raton, FL, 1993, 57–88.
52. Ross, M. H., Length of life and caloric intake, *Am. J. Clin. Nutr.,* 25, 834–841, 1972.

30 Enhancement of Luminol-Dependent Peroxynitrite Luminescence in Dishes and Tubes from Various Macrophages: Rat Alveolar Macrophages Apparently Display a New Oxidative Mechanism

Knox Van Dyke, Michael Taylor, Paul McConnell, and Mark J. Reasor

CONTENTS

30.1 INTRODUCTION

Alveolar macrophages are a major portion of the immune protection of the lung. Whether the lung is attacked by virus, bacteria, fungi, parasites, or such foreign particles as silica, coal dust, or asbestos, alveolar macrophages are activated to kill or clear the unwanted foreign material. They act as a kind of vacuum cleaner and executioner of foreign matter in the lung so that oxygen exchange can occur in the small sacks known as alveoli. Oxygen diffuses through the alveolar walls into the blood and into the erythrocytes, which carry exchangeable oxygen to the tissues. The macrophages patrol the internal surface of the lung and keep it clean of foreign substances so that the lung can do its job and oxygenate the blood efficiently.

0-8493-0719-8/02/$0.00+$1.50

These macrophages can engulf and kill bacteria and other infectious organisms and the presence of foreign invaders (because of surface chemistry) trigger immune recognition that induces killing mechanisms like NO synthase II, which produces large amounts of nitric oxide (N·), and a second superoxide NO· is formed from oxygen because an electron is passed to oxygen via a membrane-bound NADPH oxidase. The two free radicals ($\cdot O_2^-$ and NO·) combine with high affinity and form peroxynitrite–OON=O–. This is a peroxide that is 1000 times more oxidative than equivalent amounts of hydrogen peroxide.[1] The Stratagene® enhancer produces a direct tenfold enhancement of luminol luminescence oxidized by peroxynitrite (formed from SIN-1 (linsidomine)). But quiescent cells produce as much as 100-fold stimulation of luminescence using the enhancer. In a nonstimulated state it does not seem likely that peroxynitrite would be much of a factor because in these cells NO synthase II has not been induced to form enough NO. Normally, these cells are in a nonstimulated state, but when foreign invaders enter their area, the macrophages are readily activated or induced to quickly (minutes to hours) produce large amounts of cytokines, chemokines, and nitric oxide and superoxide forming peroxynitrite in a sort of kill-on-demand response.

One of the ways we can measure the killing reaction of these cells is to measure cellular chemiluminescence of the major cells involved with killing, namely, macrophages and neutrophils. Usually when foreign invaders come into the lung, the macrophages attack the invaders first, and then large amounts of neutrophils invade the area, summoned by the chemokines released by macrophages. The oxidative activity is measured using a bystander molecule, i.e., luminol, which reacts with selective oxidants produced from either macrophages or neutrophils or a combination of the two. The main killing reaction from the macrophage that has been well characterized is peroxynitrite, as previously mentioned, and it is known that luminol does react with peroxynitrite and even more vigorously with its carbon dioxide derivative.[2] However, it is likely that other oxidative reactions are involved. The neutrophil seems to use at least two major mechanisms of oxidative killing. Both of these reactions produce light when they react with luminol. There is one from myeloperoxidase (MPO)/bleach[3] and a second from peroxynitrite formation via superoxide and nitric oxide.[2]

However, in the course of this chapter we will demonstrate that as inflammation occurs in the lung, initially macrophages are activated and then neutrophils come into the area. But when we use the Stratagene enhancer to stimulate the cellular light, bystander luminol apparently readily reacts with light from resting or quiescent macrophages. Macrophages do not have the MPO/bleach mechanism because this MPO enzyme is lost during early macrophage development from the monocyte. The quiescent macrophage should have little oxidative burst since it has not been induced by stimulation of the gene transcription factor NF-κb.

The purpose of this work was to characterize the enhancement of light that emanates from quiescent macrophages, U937 monocytes, and primary microglia. Whatever the actual mechanism of enhancement, there is apparently a product that is unique to the resting macrophage and it does not occur in the neutrophil since it is not involved with the enhancement of light. Furthermore, although peroxynitrite is formed in the activated macrophage, it seems to be an unlikely player in the stimulated (via enhancement) production of light in the resting or quiescent macrophage.

30.2 MATERIALS AND METHODS

All routine chemicals were obtained from Sigma (St. Louis, MO). However, the enhancer was a kind gift from Stratagene (La Jolla, CA), courtesy of Drs. Paul Kotturi and Peter Pingerelli.

30.2.1 Alveolar Macrophages from Rat Lungs

Alveolar macrophages were isolated from lungs of adult rats (250 to 300 g). Animals were anesthetized with sodium pentobarbital either intraperitoneally or intravenously. Animals were bled by cutting the renal artery or abdominal aorta. The diaphragm is slit, producing a pneumothorax. The trachea is cannulated with an 18-gauge needle, and lungs remain in the chest cavity for smaller animals.

For larger animals, the trachea is clamped, the heart and lungs are removed in total, and the trachea is cannulated with a tube of proper diameter ensuring a proper seal. The cannula is secured with a tight-fitting string. Alveolar macrophages are obtained from tracheal lavage by instillation and removal of a fluid, which is free of divalent cations and is isotonic. In general, enough fluid is instilled to expand the lungs fully. Care should be taken not to overexpand the lungs, which can cause leakage of fluid from the lung. The lung may be lavaged at least five times and the fluid collected together. Alveolar macrophages are concentrated by centrifugation (500 × *g* for 2 min). Cells that are collected in this manner are usually 90 to 95% alveolar macrophages. If the lavage fluid is warmed to 37°C rather than cold, and if the lungs are gently massaged during the lavage procedure, the yields are higher.

30.2.2 Chemiluminesence Assay via Tubes

The entire assay is accomplished in 3-ml round-bottom luminometer tubes that fit a Berthold LB9505 C 6 channel Luminometer. Each tube is filled to a total 500 μl volume with the following ingredients: 100 μl luminol, diluted from a stock of 1:10 in phosphate-buffered saline (PBS) buffer (the stock solution is made by adding 10 mg luminol (Sigma), dissolved in 1 ml dimethyl sulfoxide and then added to 99 ml PBS of pH = 7.4); 10 μl Stratagene enhancer; 100 μl phorbol myristate acetate (PMA) 2×10^{-5} *M* in PBS from a 10^{-3} *M* stock dissolved in dimethyl sulfoxide, which is added to start the reaction. The assay is followed for 20 min at 37°C in a Berthold Luminometer LB9505 C with KINB software on an IBM clone 386 computer. This system follows the generation of chemiluminescence in real time. The data can be quantitated by measuring the area under the light curve for the 20 min duration, i.e., integration by trapezoidal approximation via the KINB program. Lucigenin measured superoxide when added (1×10^{-6} *M*). L-NAME (10^{-3} *M*) was added to inhibit NO synthase. LNAC (10^{-5} *M*) and SoD superoxide dismutane (20 units) were added as inhibitors.

30.2.3 Microglial Primary Cell Culture

Primary cultures of microglia were prepared according to established procedures from several-day-old rats. Primary mixed brain cultures are prepared from Long–Evans rat pups and grown in 75-mm Primara flasks in 10 ml of DMEM–Ham's F10 1:1 supplemented with fetal calf serum (FCS) and 1% antimycotic–antibiotic solution as described by Grubinska. The cultures are maintained in a humidified 95% air/CO_2 atmosphere at 37°C. Briefly, the flasks are shaken at 37°C for 1 h at 200 rpm on a Braun rotary shaker, and the medium containing microglia cells is filtered through a 74-mm Nitex nylon screen to remove cell clumps. The cells are counted via hemocytometer/microscope, and 10^6 cells are seeded into 10 cm^2 petri dishes. After a 3-h incubation to allow for attachment of microglial cells, the medium containing nonspecific cells is removed and replaced with fresh DMED–F12/10% FCS. By morphological characteristics the cultures contain more than 95% microglial cells, and over 90% are positive for BS lectin staining. Cells were supplied by Dr. Greg Konat, Department of Anatomy, West Virginia University. γ-Interferon (human) was added 24 h prior to assay. The cells (10^6 per 35-mm dish) were adhered to clear polystyrene (Falcon). The cells were washed with physiological buffer (Hanks) and the medium discarded; 1 ml of medium without dye was added back. In addition, 500 μl of luminol solution (1 mg per 100 ml media) was added to the dish with 500 μl of 1×10^{-6} *M* PMA added last to trigger the oxidative burst. The assay for luminol luminescence was measured for 20 min in a Xylux Luminometer (Berthold, Maryville, TN) housed in a 33.7°C incubator. The reaction was followed in real time using software supplied from the manufacturer, Xylux.

30.2.4 Chemiluminesence in Dishes

U937 cells were obtained from American Type Culture Association (Bethesda, MD). They were grown in 75 cm^2 flasks in RPMI 1640 medium containing 10% FCS. The medium contained 2% antibiotic–antimycotic solution, which contained gentamicin, amoxicillin, and soluble amphotericin b. The cultures are seeded with 10^5 cells and grown in a 95% air/5% CO_2 atmosphere for several days. Cells are centrifuged to remove media and resuspended in fresh medium, prior to stimulation with γ-interferon and counted. Then 1000 units/ml γ-interferon (human or rat) is added to the culture, and cells are harvested in 20 h. Cells are centrifuged to remove media and resuspended in PBS without dye prior to assay. They are assayed in 35-mm dishes in a Xylux Luminometer. The entire unit is placed in a 33.7°C incubator. The assay is performed with a total volume of 2 ml with 10^6 cells in PBS (without dye), and each dish has 500 μl of luminol 0.1 mg% (1:10 dilution from stock) in PBS. The chemiluminescent (oxidative burst) reaction is started with PMA, 500 μl 2×10^{-5} *M*. If drugs are added, they are dissolved in PBS and are part of the 2.0 ml total volume. The Xylux Luminometer is linked to an IBM clone computer and the real-time analysis of light vs. time is followed using Microsoft-compatible software for Windows 98 supplied by Xylux.

30.3 RESULTS AND DISCUSSION

Figure 30.1 displays the chemical reaction of peroxynitrite formed from the degradation of SIN-1 with luminol to produce blue light at 425 nm (peak wavelength). This reaction is stimulated tenfold by Stratagene enhancer. It is surmised that there is a second reaction that emanates from resting or quiescent macrophages, which is greatly stimulated by the Stratagene enhancer.

In Figure 30.2 the number of rat neutrophils (PMNs) is plotted against the fold enhancement of luminescence by the Stratagene enhancer. It is clear that as the number of neutrophils increases, the fold enhancement hovers around zero. The second observation is that as fold enhancement increases from 20- to 100-fold, neutrophils are essentially absent. This means that fold enhancement correlates with macrophages, and the less the inflammation (as judged by the number of neutrophils present), the greater the fold enhancement. This clearly indicates that the more resting macrophages are producing the higher-fold enhacement by the Stratagene enhancer.

In Figure 30.3, no enhancer is added to the cells, and the luminescence from unenhanced cells vs. the number of neutrophils in a preparation is measured. There is an obvious correlation of nonenhancement related to the number of neutrophils. Therefore, we conclude that the unenhanced cells are neutrophils.

In Figure 30.4 initial luminescence is correlated inversely with the fold enhancement of luminesence in the presence of the Stratagene enhancer. The lower the initial luminol light, the higher the fold enhancement of light. Therefore, if the initial light is measured, we know that the higher-fold enhancement correlates with a low state of macrophage activation; i.e., the less the macrophage is activated, the greater the extent of enhancement.

In Figure 30.5, rat microglia isolated from the brain of young rats were developed as a primary culture on a 35-mm dish. 10^6 microglia are cultured on multiple 35-mm dishes in standard media and stimulated with 1000 units of γ-interferon for 24 h and assayed in a Xylux Luminometer built for the purpose at 33.7°C. Luminescence was followed for 20 min. Superoxide dismutase completely inhibits luminescence, whereas unenhanced light is a small fraction of the light and the NO inhibitor

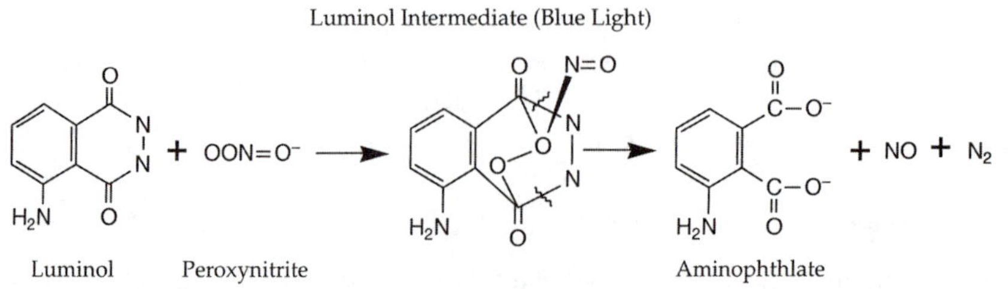

Luminol Intermediate (Blue Light)

Luminol + Peroxynitrite ($OON{=}O^-$) → Intermediate → Aminophthlate + NO + N_2

FIGURE 30.1 Chemical reaction of peroxynitrite.

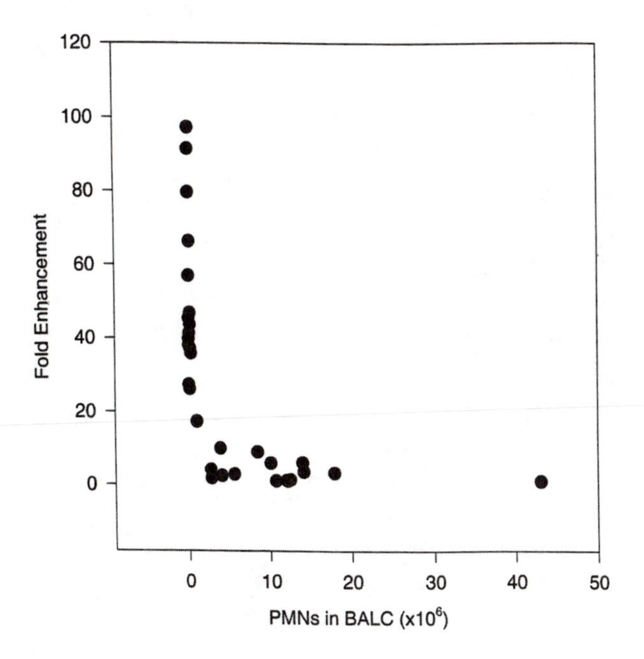

FIGURE 30.2 PMN number vs. fold enhancement.

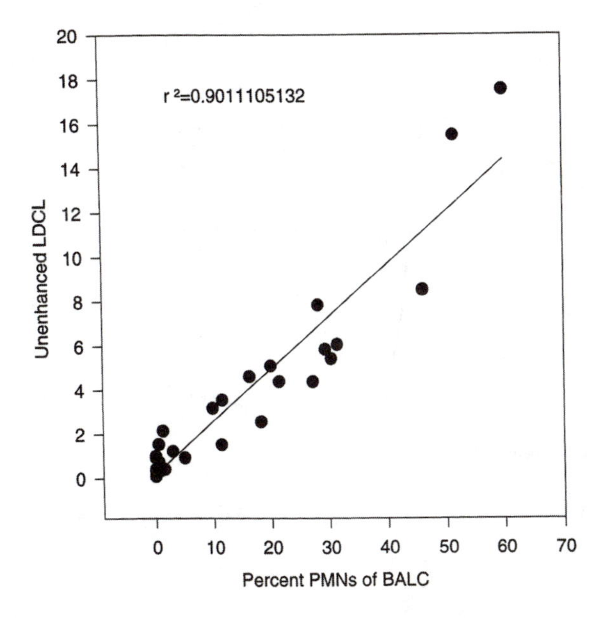

FIGURE 30.3 Unenhanced LDCL vs. percent PMNs.

L-NAME inhibited more than 90%. This indicates that the majority of enhanced luminol luminescence is probably due to peroxynitrite.

In Figure 30.6, unattached U937 monocytic cells, stimulated with γ-interferon overnight, are enhanced with Stratagene enhancer and are inhibited to a great extent by superoxide dismutase and to a lesser extent by L-NAME. This indicates that the luminescence of peroxynitrite can be enhanced. Luminescence from luminol and peroxynitrite can be enhanced with the Stratagene enhancer when generated from straight chemistry (SIN-1), from rat alveolar macrophages (no inflammation), from U937 activated monocytes, or from primary cultures of microglia from rat brain. In every case,

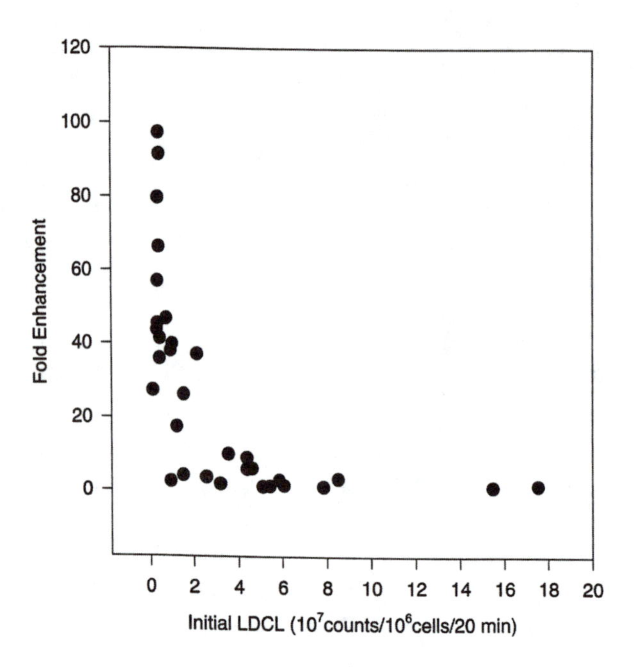

FIGURE 30.4 Enhancer studies: initial LDCL vs. fold enhancement.

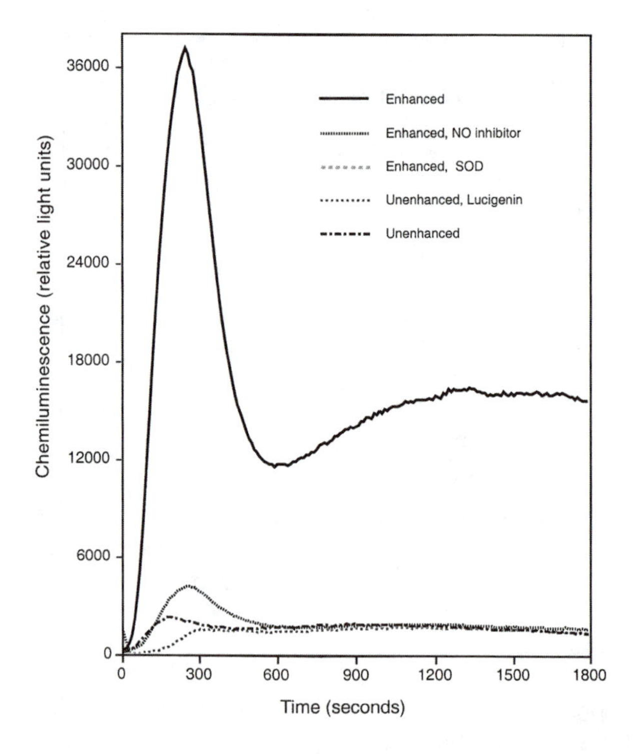

FIGURE 30.5 PMA-stimulated chemiluminescence from cultured microglia induced with interferon.

peroxynitrite-linked oxidation as detected using luminol luminescence is measured. However, in the case of the naive macrophages, a much greater stimulation of luminol-dependent luminescence is seen than can be accounted for by peroxynitrite alone. There must be another light-producing chemistry that is not related to peroxynitrite. In the resting rat alveolar macrophage, very little NO

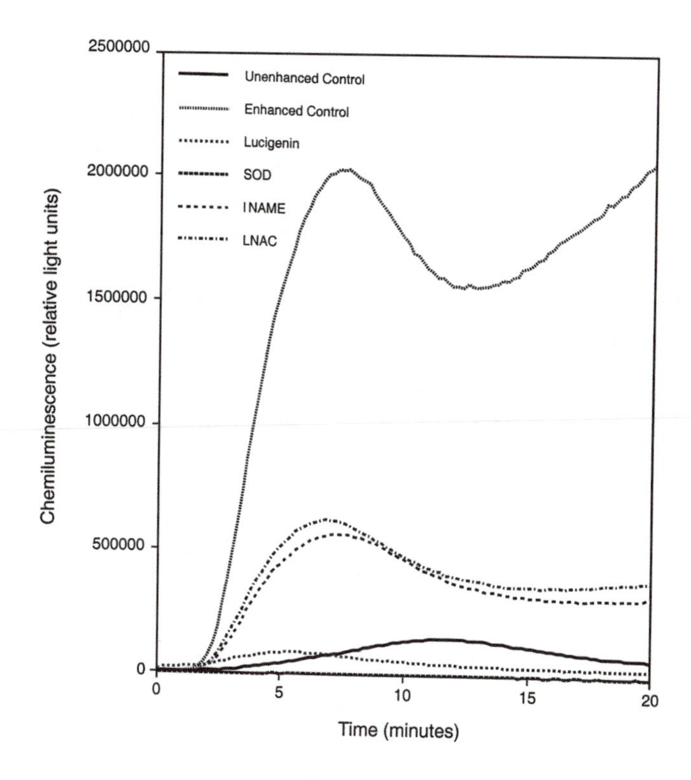

FIGURE 30.6 PMA-stimulated chemiluminescence from U937 monocytes 2 days postinduction with interferon.

is produced, and therefore the extreme enhancement of luminesence that occurs probably originates from an unrecognized chemical entity. Perhaps further studies will elucidate the chemical structure of the substance.

REFERENCES

1. Brunelli, L., Crow, J. P., and Beckman, J. S., The comparative toxicity of nitric oxide and peroxynitrite to *E. coli.*, *Arch. Biochem. Biophys.*, 316, 327–324, 1995.
2. Radi, R., Cosgrove, T. P., Beckman, J. S., and Freeman, B. A., Peroxynitrite-induced luminol chemiluminescence, *Biochem. J.*, 290, 51–57, 1993.
3. Brestel, E. P., Co-oxidation of luminol by hypochlorite and hydrogen peroxide: implications for neutrophil chemiluminescence, *Biochem. Biophys. Res. Commun.*, 126, 482–488, 1985.

31 A Search to Detect Antioxidants Using Luminescence Useful for Treating Oxidative Stress*

Knox Van Dyke, Paul McConnell, and Meir Sacks

CONTENTS

31.1 INTRODUCTION

Oxygen is essential for life. It is used by the body in metabolic reactions to oxidize or burn foodstuffs to supply energy. However, a variety of oxidative species, free radicals (containing an unpaired electron), and excited-state substances are formed. This would include singlet oxygen (excited states epsilon and delta), superoxide ($O_2^{\bar{\cdot}}$), hydrogen peroxide (HOOH), hydroxyl radical and certain metabolites, peroxynitrite ($OONO^-$), and hypochlorite (OCl^-). There are probably others. In fact it has been estimated that for every 10^{12} oxygen molecules that enter a cell, 1/100 damages protein and 1/200 damages DNA or RNA. We know lipids can be damaged as well via lipid peroxidation.[1] Some of this damage to DNA, RNA, protein, and lipid is repaired and some is not.

When the production of oxidants exceeds the amounts of antioxidants, oxidative stress occurs and the damage previously mentioned might occur.[2] If this condition is allowed to continue over a period of time without repair, diseases that involve inflammation, e.g., diabetes, cancer, neurodegeneration, arthritis, and even viral disease can occur.

What is oxidative stress and how do we prevent it from happening or, more to the point, how do we prevent the damage that is not repaired? First we must understand the two opposing forces that control the *yin* and *yang* of the situation. What constitutes an oxidant and an antioxidant (reductant)? An oxidant is any substance that loses an electron or hydrogen or gains an oxygen. Therefore, an antioxidant is a substance that gains an electron or hydrogen or loses oxygen. If the oxidant and antioxidant are balanced, we produce oxidative homeostasis. If the antioxidant is exessive, with enough reserve, your cells and organs are likely protected from oxidative damage.

* Reproduced in part from Van Dyke, K., *J. Biolumin. Chemilumin.*, 13, 339, 1998; and *Luminescence*, 15, 37, 2000. With permission of John Wiley & Sons.

0-8493-0719-8/02/$0.00+$1.50

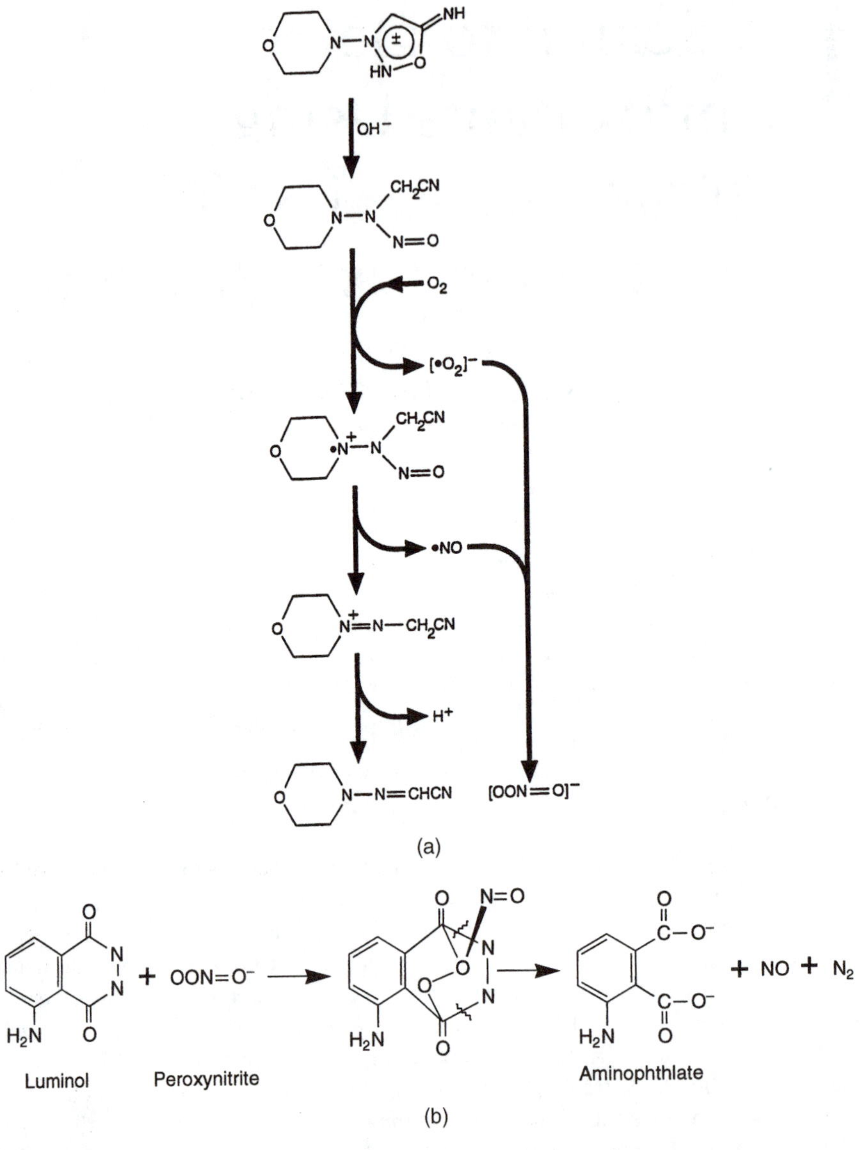

FIGURE 31.1 (a) Production of peroxynitrite from SIN-1 (linsidomine). (b) Possible reaction of luminol and peroxynitrite to produce luminescence. (From Van Dyke, K., *Luminescence,* 15, 37, 2000. With permission.)

Long-term oxidant deficit can lead to susceptability to disease. What of the possibility of antioxidative damage, can it occur? It appears very unlikely because oxidants are the aggressors chemically and antioxidants are the defense.

Luminescence can be used to detect antioxidants; because oxidation of a luminescent probe causes light to occur, inhibition of the production of light can be used as a method to detect antioxidants. In our initial work, the chemical SIN-1, which rearranges to produce the strong oxidant peroxynitrite, is used in the presence of luminol at body pH (7.4) to produce light (Figure 31.1a and b). If a drug interferes with the production of peroxynitrite or reduces peroxynitrite once it is formed, an inhibition of light production occurs. Real-time kinetics of the reaction of luminol and peroxynitrite reacting to produce light was measured with a Berthold Luminometer. The kinetics of luminesence can be seen moment to moment on the display monitor and printed once the reaction is ended, producing a hard copy of the complete kinetics. Data are obtained quickly with full or partial integration and stored to disk to be used at a later date. Reactions are done with a proper control containing all ingredients except for the presence of an inhibitor.

The body contains a variety of substances that are antioxidant in nature. Some of these are water-soluble, i.e., ascorbic acid (vitamin C), and are retained in the blood. Others are fat-soluble and are incorporated into membranes (vitamin E). There are enzymes, e.g., glutathione peroxidase or superoxide dismutase (SOD), which destroy oxidants, and some are proteins that are readily oxidized.

In the following narrative the details of antioxidant detection are explained.

31.2 MATERIALS AND METHODS

SIN-1 (linsidomine) was obtained from Dr. Karl Schönafinger of Casella AG (Frankfurt, Germany). It was dissolved in an aqueous solution of Hepes buffer 0.1 *M*, pH = 7.4, at a concentration of 3 mg/ml. When activated by heat, the SIN-1 rearranges to produce peroxynitrite, which reacts with luminol to produce blue light at 425 nm (see Figure 31.1a and b). Luminol (10 mg) was added to a 100-ml dry volumetric and dissolved with the addition of 1 ml dimethyl sulfoxide. Warm Hepes buffer was added to make a stock solution. When used for assay, the stock solution of luminol is diluted 1:10 with Hepes buffer. All other substances to be tested were dissolved in Hepes buffer as mentioned previously. All other components not mentioned per se were obtained from Sigma (St Louis, MO) (Figure 31.2). Lipton's black or green tea was purchased at a local foodstore; 15 mg of black tea was added to 50 ml hot (80°C) distilled water and allowed to sit for 20 min. Now at room temperature the aqueous extract was diluted in buffer for assay. A polyphenolic extract of green tea (Natural Brand) was used from a capsule (equivalent to four cups green tea). It was puchased from General Nutrition Center. The 315 mg of extract was emptied into 25 ml Hepes buffer, rapidly swirled, and allowed to stand at room temperature for 20 min. The supernatant fluid was decanted and used as the green tea extract.

31.3 PROCEDURE

All assays with SIN-1 were accomplished in a Berthold round-bottom, clear plastic 3-ml tubes with a total volume of 500 μl in Hepes buffer; 200 μl diluted luminol solution was added to 200 μl of SIN-1 with 100 μl of Hepes buffer as a control. This would generate a luminescent curve of the reaction without antioxidant. Then, various antioxidants dissolved in Hepes buffer would be used in place of the 100 μl of buffer solution. The SIN-1 was always kept in an ice bath when not in use. It was added last in the assay to start the production of light. A fresh tube of SIN-1 was prepared daily because it is unstable in warm solutions. The tubes were placed in a six-channel Berthold Luminometer at 37°C. The reaction was usually followed for 30 min. When the reaction is completed, the KINB software integrates the curve for the entire 30-min period. Graphs depict averages from three experiments with the associated error bars depicting standard error of the mean. Student's t test is used and significance set at $p = 0.5$, where appropriate.

Assays with peroxynitrite were performed by controlled-force injection of peroxynitrite into the luminometer tube rather than by using SIN-1 as the source of peroxynitrite. The conditions were 37°C and 1-min incubation, and injection of peroxynitrite was accomplished using a Hamilton gas-tite syringe, 2500 μl. A 1/50 volume is injected into the reaction mixture (50 μl). The peroxynitrite was purchased from Upstate Biotechnology (Lake Placid, NY). It is stabilized by being placed in a basic solution (0.3 *M* NaOH) and kept at −80°C prior to use. When used, it is thawed and diluted 1/1000 with distilled water. The original solution of peroxynitrite from Upstate is approximately 138 m*M*. Therefore, 50 μl of a 138-μM solution is injected in a tube containing luminol and buffer with or without antioxidant. The control was accomplished without antioxidant; the area under the curve was generated and compared with the area under the curve with a given amount of antioxidant using the KINB software as mentioned above. This reaction is activated as quickly as possible because the half-life of peroxynitrite is less than 1 s at neutral pH.

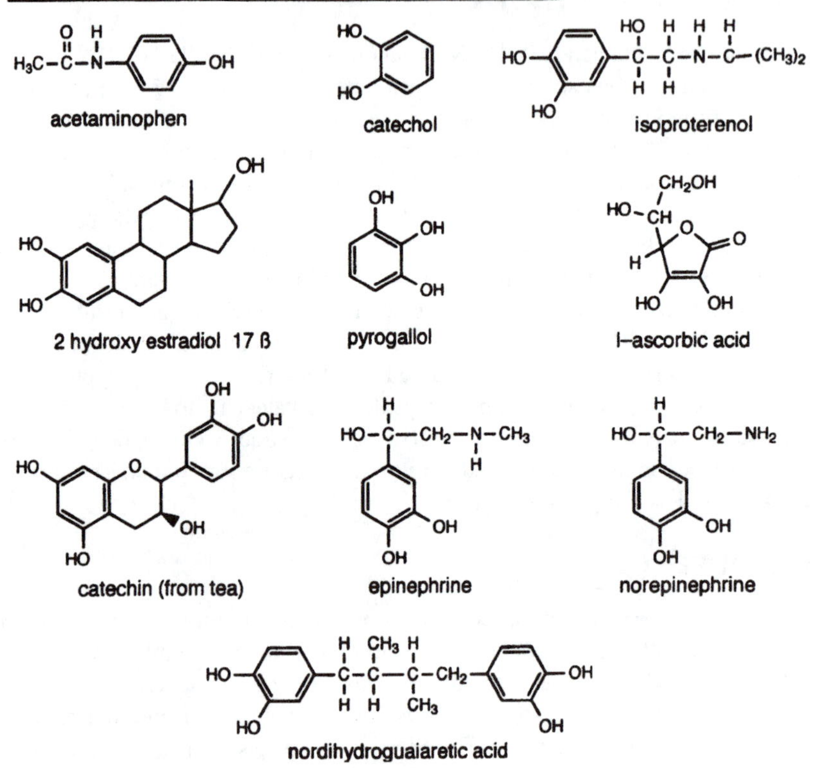

FIGURE 31.2 Chemical structures of SIN-1 and various mono, di, and triphenols used here. (From Van Dyke, K., *J. Biolumin. Chemilumin.,* 13, 339, 1998. With permission.)

31.4 RESULTS AND DISCUSSION

In Figure 31.1a the reaction of SIN-1 to produce peroxynitrite is shown. When SIN-1 is kept ice cold the compound is quite stable. But as soon as it is placed in a tube at 37°C, the chemical degradation to produce peroxynitrite occurs. Therefore, the SIN-1 is always kept cold until it is added to a tube and incubated at 37°C. In Figure 31.1b, the reaction between peroxynitrite and luminol is depicted where the luminol is attacked by the oxidant. This causes luminol to reach an excited state, which produces light at 425 nm. Then, it returns to ground state in an oxidized (spent) form. Thus, luminol is actually consumed in the reaction. The luminol is present in excess and never runs low during the time that the reaction occurs. It is clear from Figure 31.1a and b, that if a substance were added that interfered with any of the steps in the degradation of SIN-1 to peroxynitrite or reacted with the peroxynitrite in creating the excited state, it would be acting as an antioxidant. If the inhibiting compound could act as a chemical target, this would be another method to consume peroxynitrite. For example, certain phenolic substances (Figure 31.2) could

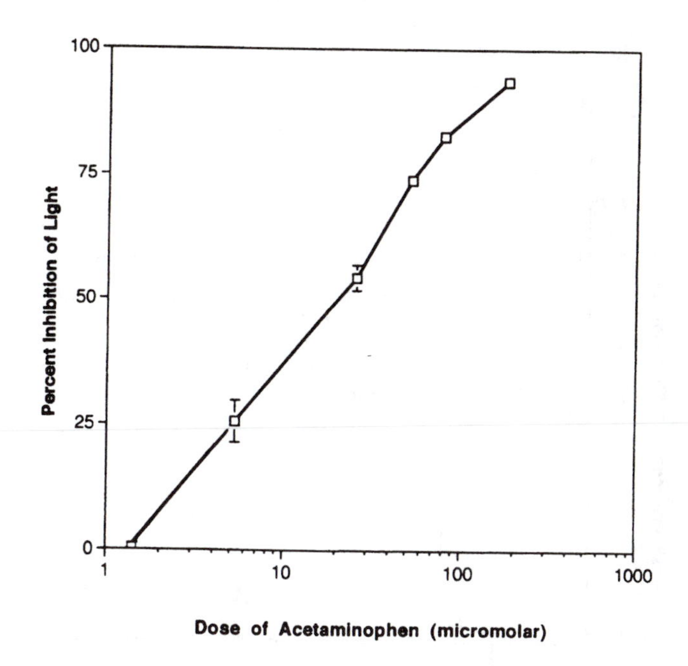

FIGURE 31.3 Dose response of acetaminophen vs. % inhibition of luminol light.

act as targets to be nitrated or oxidized by peroxynitrite, which would inactivate the peroxynitrite as an oxidant or nitrating reagent.

An example of target oxidation is acetaminophen. In Figure 31.3, this analgesic and antipyretic inhibits luminol luminescence activated by SIN-1, in a dose-responsive manner from 1.4 to greater than 100 μM. Figure 31.4 shows, whether acetaminophen is added early or late in the luminesence curve, it is highly inhibitory at 100 μM. It works very quickly to inhibit luminescence of luminol oxidized by peroxynitrite. In fact, one can actually observe the product turning yellow, possibly indicating nitration or at least oxidation of acetaminophen.

As shown in Figure 31.5, small doses of copper/zinc SOD can inhibit peroxynitrite-based luminescence very actively. It is known that SOD is nitrated and enzymic activity destroyed when it attacked by peroxynitrite. As shown in Figure 31.6, the catechol estrogen 2-hydroxy, 17β-estradiol inhibits in submicromolar amounts. Nordihydroguaiaretic acid is a potent inhibitor as shown in Figure 31.7 in similar doses as the estrogen. In Figure 31.8 SIN-1-driven luminescence is compared with several phenols. Catechol, epinephrine, isoproterenol, pyrogallol, and ascorbic acid are compared with the SIN-1 control without antioxidant. All of the compounds at 100 μM concentration inhibited luminescence except epinephrine and isoproterenol, which stimulated a small amount epinephrine and approximately threefold (isoproterenol). Certain phenols are known to have enhance the luminescent process, but it is not known why this occurs.[3] In Figure 31.9, the antioxidant activity of brewed black tea is inhibitory to the luminescence process at over 1000-fold dilution. Since there are many polyphenols in tea especially epigallocatechin gallate we know that these compounds are a perfect target for nitration by peroxynitrite. This is because phenols activate a benzene ring for aromatic substitution as part of the chemistry of aromatic substitution and they are ortho-para directing as well.[4] In the second series of experiments, we will explore green tea and some of its polyphenolic constituents, as shown in Figure 31.10.

In Figure 31.11, we used solubilized green tea extract as our source of polyphenols. This extract is dissolved at room temperature and diluted in buffer to free the polyphenols without degradation. It can be seen that the green tea extract produces inhibition in a dose-responsive manner from full strength to 1:10,000 dilution, when a peroxynitrite solution, shown in Figures 31.11 and 31.12, is

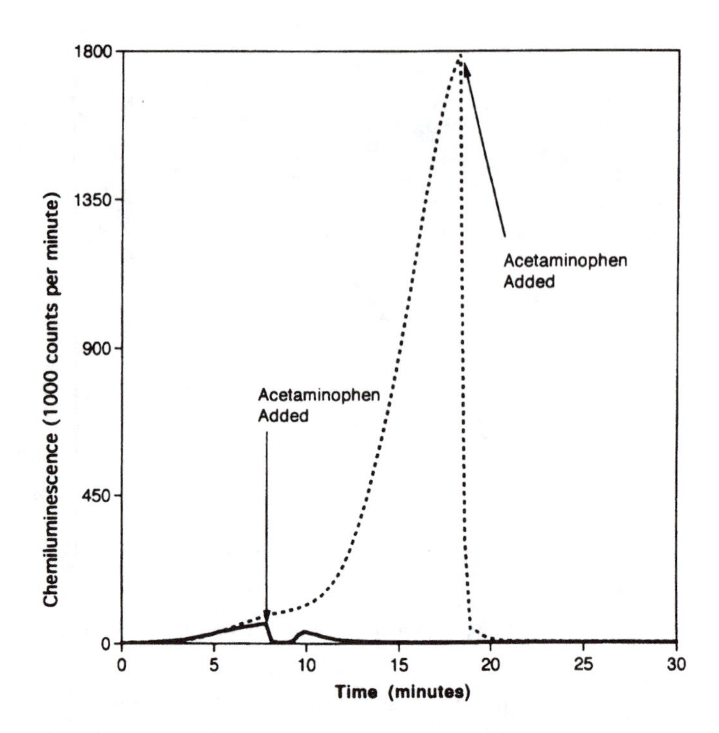

FIGURE 31.4 Time course of SIN-1 and luminol, 10^4 mol/l acetaminophen added at the beginning or at the maximum of the light of the emission. (From Van Dyke, K., *J. Biolumin. Chemilumin.,* 13, 339, 1998. With permission.)

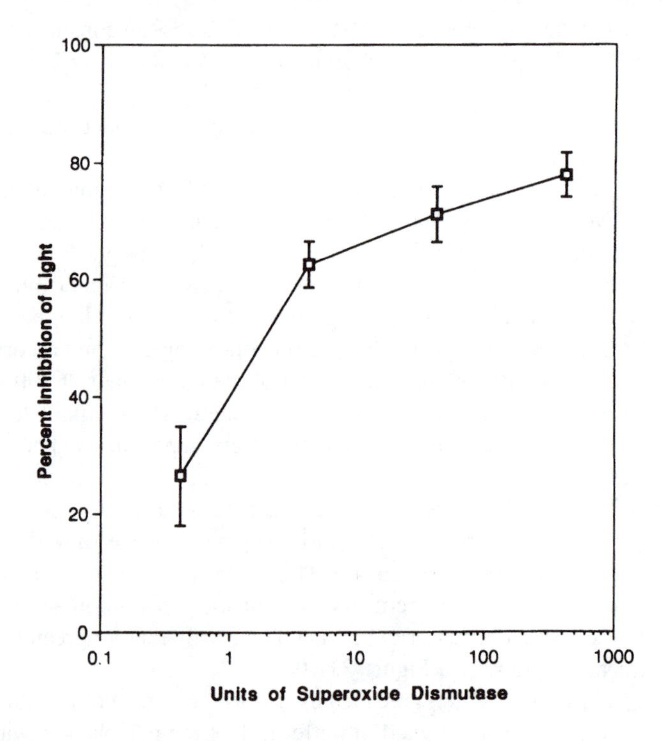

FIGURE 31.5 Superoxide inhibitory dose–response curve of SIN-1-activated luminol luminescence. (From Van Dyke, K., *J. Biolumin. Chemilumin.,* 13, 339, 1998. With permission.)

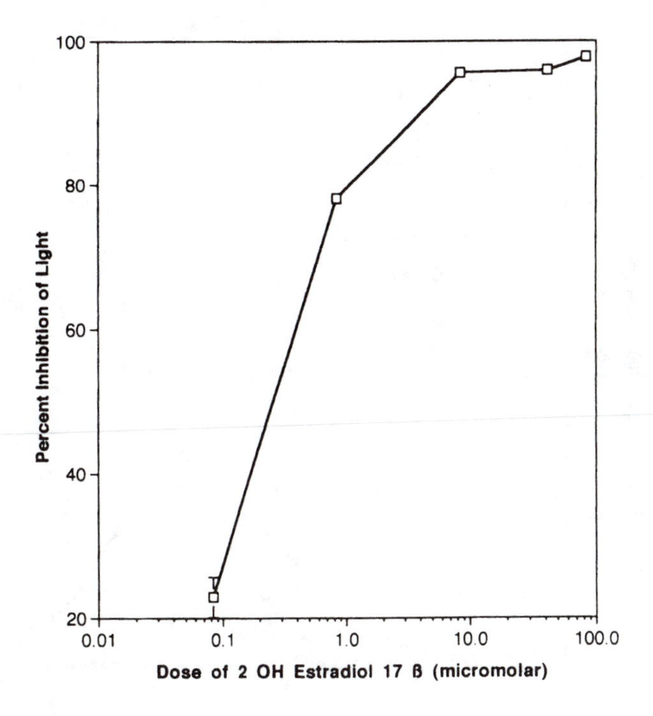

FIGURE 31.6 Dose response of 2-hydroxyestradiol of SIN-1-activated luminol luminescence. (From Van Dyke, K., *J. Biolumin. Chemilumin.,* 13, 339, 1998. With permission.)

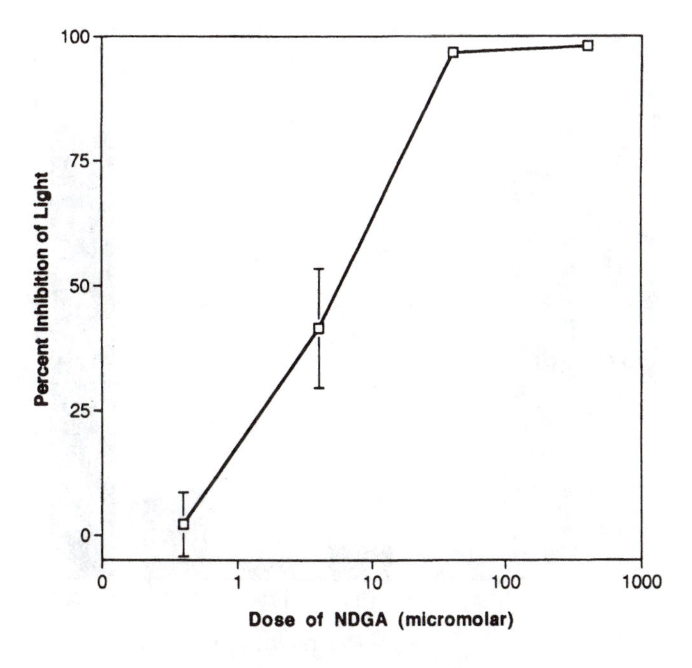

FIGURE 31.7 Dose response of nordihydroguaiaretic acid of SIN-1-activated luminol luminescence. (From Van Dyke, K., *J. Biolumin. Chemilumin.,* 13, 339, 1998. With permission.)

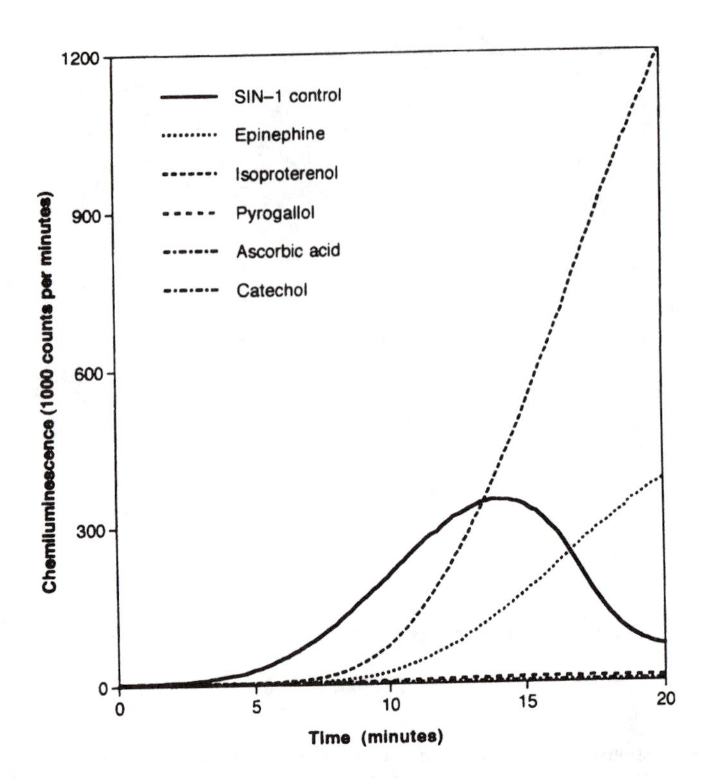

FIGURE 31.8 A comparison of catechol and other antioxidants against SIN-1-activated luminescence. Note early inhibition of luminescence with some catecholamines and then a stimulation. (From Van Dyke, K., *J. Biolumin. Chemilumin.*, 13, 339, 1998. With permission.)

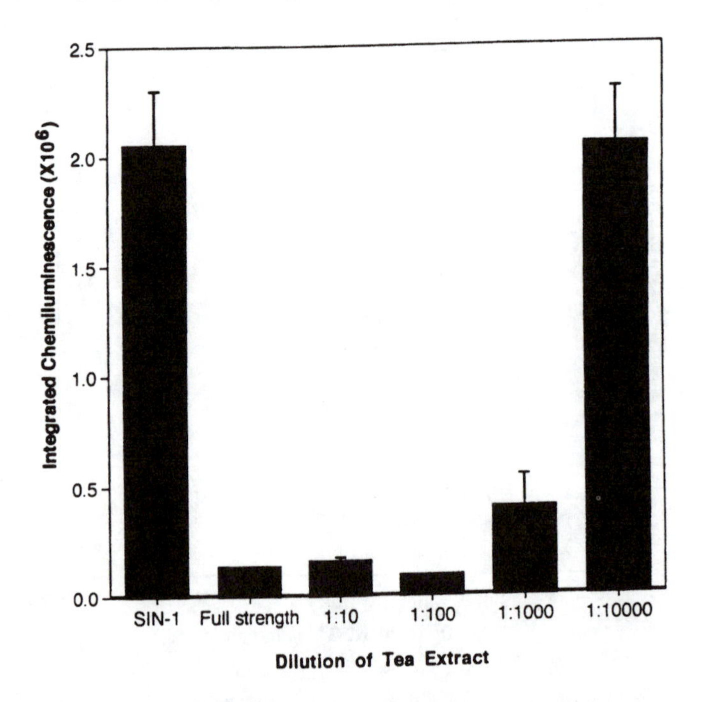

FIGURE 31.9 The inhibitory dose response of brewed black tea (orange pekoe and pekoe) against SIN-1-activated luminol light emission. (From Van Dyke, K., *J. Biolumin. Chemilumin.*, 13, 339, 1998. With permission.)

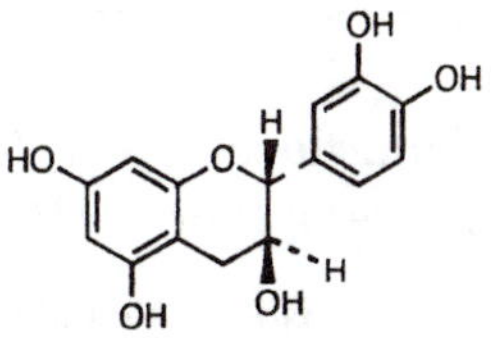

(+) Catechin

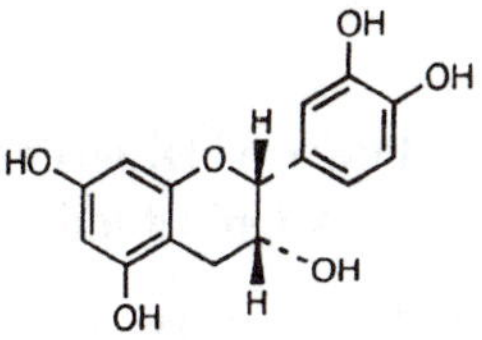

(−) Epicatechin

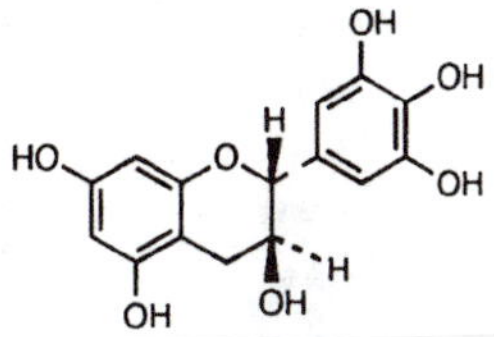

(+) Gallocatechin

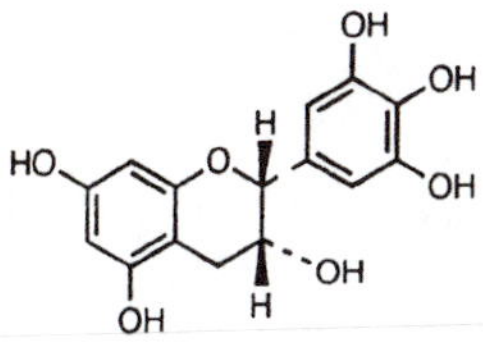

(−) Epigallocatechin

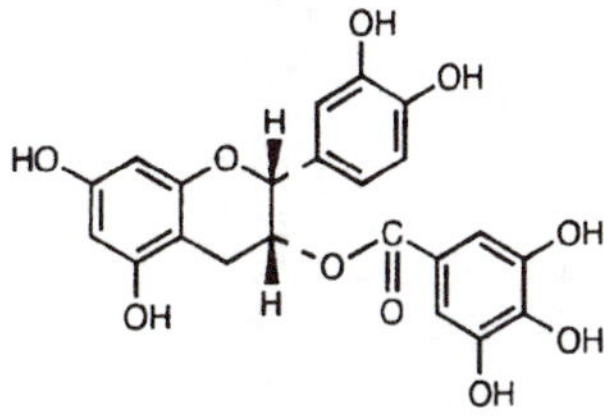

(−) Epicatechin gallate

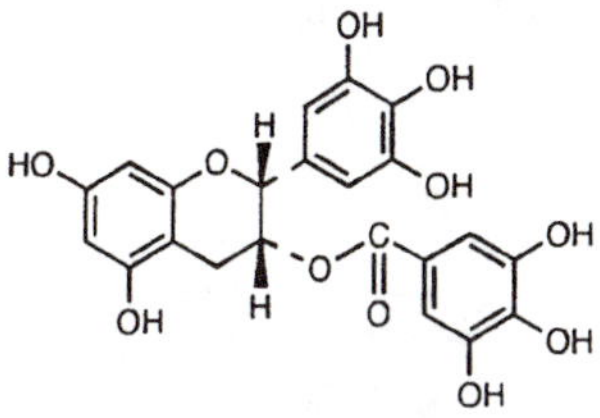

(−) Epigallocatechin gallate

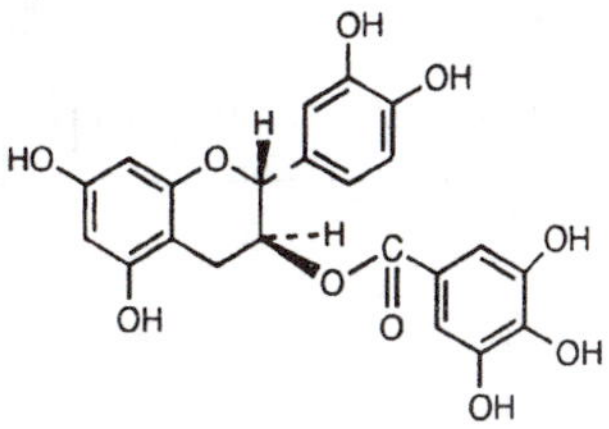

(+) Catechin gallate

FIGURE 31.10 Chemical structures of various polyphenol constituents of green tea. (From Van Dyke, K., *Luminescence,* 15, 37, 2000. With permission.)

used directly. This inhibition comes from a capsule that is aqueously extracted from approximately four cups of tea. As shown in Figure 31.12, SIN-1 was used as the source of peroxynitrite, and the same pattern of inhibition holds true that every dilution tested from 25 mg/ml to 2.5 μg/ml led to at least 50% inhibition. Although it is difficult to compare an extract against a brewed solution, it is clear that the polyphenolic substances in tea are very inhibitory to oxidation as judged by the inhibitory effect on luminescence. Although there are hundreds of different compounds in tea, by assaying some of the purified components that are present in high composition in teas it is clear that the polyphenols might be major contributors as antioxidants. For example (−) epicatechin is very inhibitory to luminescence caused both by SIN-1 and by peroxynitrite directly (Table 31.1). It appears to be more directly effective against peroxynitrite than against SIN-1, but the doses of oxidant and the kinetics of how the oxidant attacks luminol, hence causing oxidation that produces light, are not exactly the same. The same can be said of racemic (±) catechin: it is more effective against peroxynitrite than SIN-1 as the oxidant (see Table 31.1). Epigallocatechin gallate is inhibitory in the first two doses and then stimulates light at greater dilutions with both SIN-1 and peroxynitrite as the oxidant (see Table 31.1). The stimulation of light occurs with certain phenols but a mixed behavior probably stems from nitration or at least metabolism of the polyphenol. Epicatechin gallate is completely inhibitory to SIN-1 and

TABLE 31.1
Inhibitory Effect of Different Dilutions of Individual Green Tea Constituents When Added to Solution with Peroxynitrite SIN-1

	(–) Epigallocatechin Gallate		(±) Racemic Catechin	
Dose, μmol/l	**SIN-1**	**Peroxynitrite**	**SIN-1**	**Peroxynitrite**
452	100 ± 1	100 ± 1	100 ± 1	100 ± 1
45.2	100 ± 1	99 ± 1	100 ± 1	100 ± 1
4.52	–35	90 ± 10	97 ± 1	95 ± 1
0.452	–12	25 ± 4	47 ± 4	76 ± 5
0.0452	–10	12 ± 3	2 ± 3	53 ± 4
0.00452	n/a	n/a	–5	56 ± 8
	(–) Epicatechin		**(–) Epicatechin Gallate**	
Dose, μmol/l	**SIN-1**	**Peroxynitrite**	**SIN-1**	**Peroxynitrite**
452	100 ± 1	100 ± 1	98 ± 1	100 ± 1
45.2	100 ± 1	100 ± 1	97 ± 1	99 ± 1
4.52	95 ± 1	94 ± 3	84 ± 2	40 ± 10
0.452	47 ± 4	82 ± 4	68 ± 18	44 ± 8
0.0452	12 ± 8	78 ± 5	n/a	37 ± 10
0.00452	n/a	96 ± 5	n/a	27 ± 1

Results are expressed as percentage inhibition ±SEM. A negative symbol indicates a stimulation rather than an inhibition.

Source: Van Dyke, K., *Luminescence,* 15, 37, 2000. With permission.

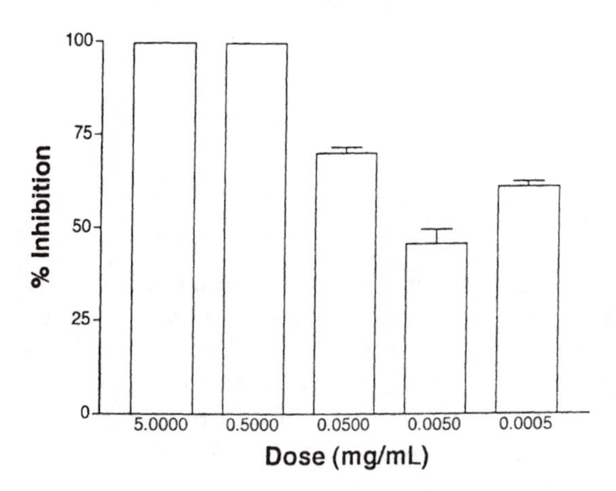

FIGURE 31.11 Dose response of green tea extract with SIN-1 and luminol. (From Van Dyke, K., *Luminescence,* 15, 37, 2000. With permission.)

peroxynitrite directly in the first two doses, but inhibition of light falls after the next three doses and greater inhibition occurs with SIN-1 than with peroxynitrite at the three lower doses.

A variety of *in vitro* and *in vivo* epidemiological studies point to the antitumor and antioxidant properties of tea (either green or black) or to its constituent polyphenols such as epigallocatechin gallate, a major component of tea.[5–11] Certainly tea has the correct compounds to protect against peroxynitrite-type oxidation.

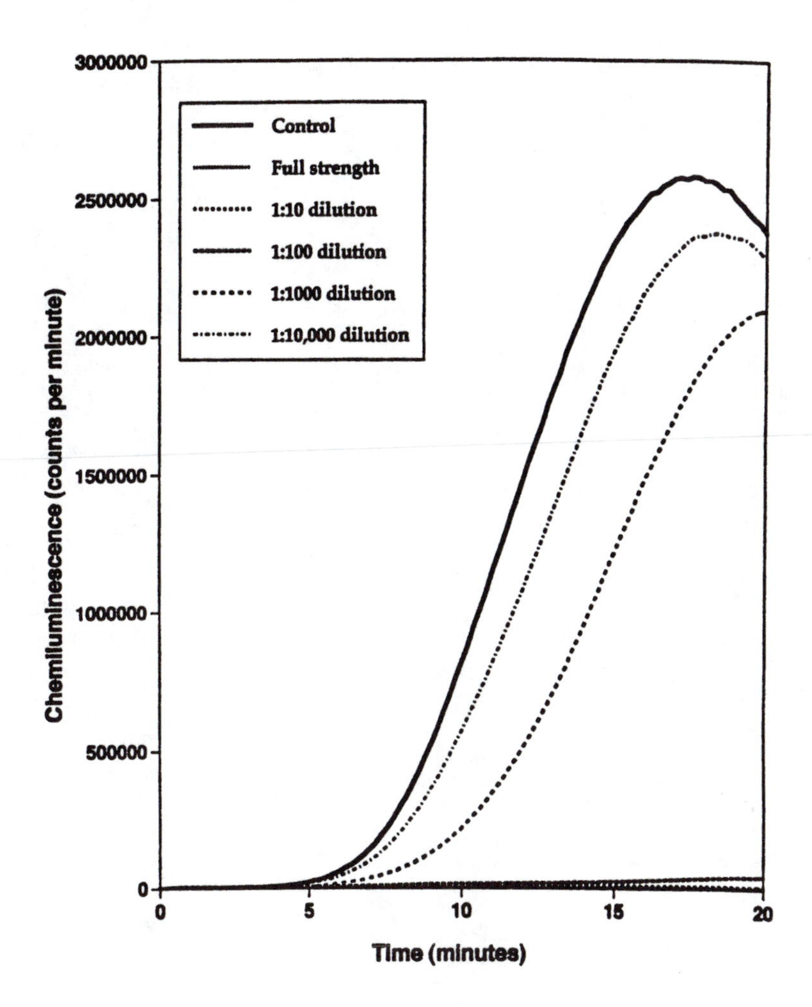

FIGURE 31.12 Kinetics of the dose response of green tea with SIN-1 and luminol.

ACKNOWLEDGMENTS

We gratefully acknowledge Dr. Karl Schönafinger of Casella AG, Frankfurt, Germany, for the gift of SIN-1. We are grateful to John Wiley & Sons Ltd., Chichester, U.K. for allowing us to republish material from our two publications with them.

REFERENCES

1. Van Dyke, K., A new screening method to detect water-soluble antioxidants: acetoaminophen (Tylenol) and other phenols react as antioxidants and destroy peroxynitrite-based luminol-dependent chemiluminescence, *J. Biolumin. Chemilumin.*, 13, 339–348, 1998.
2. Van Dyke, K., McConnell, P., and Marquardt, L., Green tea extract and its polyphenols markedly inhibit luminol-dependent chemiluminescence activated by peroxynitrite or SIN-1, *Luminescence,* 15, 37–43, 2000.
3. Coyle, P. M. Thorpe, G. H., Kricka, L. J., and Whitehead, T. P., Enhanced luminescence of horseradish peroxidase conjugates, *Ann. Clin. Biochem.*, 23, 42–48, 2986.
4 Morrison, R. T. and Boyd, R. N., *Organic Chemistry,* Allyn & Bacon, Boston, 284–301.

32 Luminol Luminescence Generated by a Novel Peroxynitrite Mechanism Is Inhibited by Polyphenolic Plant Extracts: Nitroglycerin Tablets and Superoxide Generate Peroxynitrite

Neena Agarwal and Knox Van Dyke

CONTENTS

32.1 INTRODUCTION

Nitroglycerin (NG) produces a vasodilator (nitric oxide) that relaxes most smooth muscles, including arteries and veins. Its use as a drug was invented by Sobrero in 1846, when he found that placing the oily substance under the tongue produced the effects of a severe headache, which resulted from vasodilation. In the following years, several scientists worked toward improving this drug. In 1847, Constantin Hering developed a better sublingual dosage form. T. Lauder Brunton, the renowned physician of Edinburgh, found amyl nitrite to relieve anginal pain within 30 to 60 s when administered by inhalation in 1857. Because the dosage was difficult to adjust, William Murrell used nitroglycerin to mimic the action of amyl nitrite and developed the use of sublingual nitroglycerin to relieve acute anginal attacks and as a prophylactic agent.[1]

Anginal pain is caused by the constriction of coronary blood vessels, which reduces the blood flow to the heart. Because of its rapid action and low cost, NG is a drug most useful as a coronary vessel vasodilator. It increases the amount of blood that flows to the heart and with it brings oxygen and relief of pain. As the blood vessels stretch, due to vasorelaxing effects of nitric oxide, NADPH oxidase is activated to produce superoxide.[2] Nitric oxide binds to guanyl cyclase and increases the

0-8493-0719-8/02/$0.00+$1.50

synthesis of cyclic GMP, which relaxes smooth muscle in blood vessels and in other tissues.[3] However, nitric oxide can also combine with superoxide to form peroxynitrite (OONO–). Peroxynitrite is a strong oxidizer that can attack a variety of biologically relevant targets, e.g., DNA, RNA, lipids, carbohydrates, and antioxidants, which protect against oxidative attack. As the antioxidants are depleted, the balance between antioxidants and oxidants is tipped toward oxidants, which are the damaging species. Therefore, the very relief of angina by nitric oxide can produce damage to the blood vessels that feed the heart with blood and nutrition. In addition, as NG is used to relieve multiple anginal attacks, the effect of continuous applications of NG produces a reduced coronary vasodilation termed tachyphylaxis. In part this is because, as more and more stretch occurs to the coronary arteries, more superoxide is produced. When it combines with nitric oxide and produces peroxynitrite, this diminishes the amount of nitric oxide available to produce vasorelaxation. Studies have shown that, by eliminating the superoxide before it can attack, the nitric oxide eliminates the tachyphylaxis or dwindling effect of NG.

32.2 METHODS

32.2.1 Reagents

In these experiments, we used several plant-derived extracts as the source of antioxidants. We obtained dried powders of pine bark and hawthorn berry from the Montiff (Beverly Hills, CA). Green tea extract (Natural Brand) was purchased at a local General Nutrition Center (GNC) store. Potassium superoxide was obtained from Aldrich Chemicals (Milwaukee, WI). Potassium superoxide (KOS) solution was made by taking 45 mg KOS and dissolving it in 5 ml of phosphate-buffered saline (PBS) buffer. It is kept on ice prior to use. The NG in 0.6-mg tablets was obtained from Parke, Davis & Co. (Detroit, MI).

A stock solution of Pine Bark Extract (PBE) was made by dissolving 30 mg PBE in 5 ml of PBS buffer. This is the full-strength substance, which is kept on ice and can be used for several hours. Then, serial dilutions were made by adding 4.5 ml of PBS and adding 500 μl of the full-strength solution, mixing, and taking a 500-μl aliquot of that solution into a tube with 4.5 ml PBS, etc. A stock solution of hawthorn berry extract (HBE) was dissolved using 30 mg HBE in 5 ml of PBS buffer. Serial dilutions were made as above. A final stock solution was made using 315 mg of Green Tea Extract (one capsule) dissolved in 5 ml of PBS buffer. This was the full-strength substance. The insoluble material settled and the supernatant solution was used as the full-strength stock. Then, three dilutions were made using the procedure described above.

The luminol was obtained from Sigma (St. Louis, MO). It was dissolved first in 1 ml DMSO and then diluted to produce a stock solution with 99 ml PBS at pH 7.4. The luminol solution was diluted to 1:10 with PBS, and covered with aluminum foil. It is then kept on ice until use.

32.2.2 Chemiluminescence Assay

After preparing these initial stock solutions the following volumes were added to a 3-ml Berthold round-bottomed, clear plastic test tube: 100 μl of luminol, 100 μl of each dilutions of PBE, 100 μl of KO_2, 200 μl of buffer alone. One tablet of the NG was added last to start the reaction by quickly swirling the contents for a few seconds in a luminometer tube (resulting in approximately a total volume of about 500 μl for each tube) and the mixture was placed in the instrument and assayed. There was a control solution made to function as a reference for the light produced by the combination of KO_2, NG, and luminol. All substances except the NG were kept on ice.

The tube was inserted into one of the six channels of a Berthold six-channel luminometer model LB9505C, for a period of 5 min with temperature control at 37°C. The assay was recorded and plotted in real time by an IBM clone computer running KINB software.

32.3 RESULTS AND DISCUSSION

There are a variety of antioxidants in herbs and plants that, when ingested, could provide important protection from strong oxidants, e.g., peroxynitrite and hypochlorite/hydrogen peroxide. It is important to supplement our major antioxidants, such as vitamins C and E, uric acid, and glutathione in the reduced state. These substances protect us against attack by oxidants, which can damage our important macromolecules including DNA, RNA, proteins, and lipids. Therefore, new nontoxic dietary antioxidants were explored that could supplement the usual fare of antioxdants.

Dose response was determined for the following extracts: PBE, HBE, and green tea at full strength and at 1:10, 1:100, and 1:1000 dilutions. The final concentration in the tube was actually one fifth of these concentrations. Using the NG and the superoxide together produced peroxynitrite, which reacts with luminol to produce light. The light is used as an easily determined end point to measure antioxidant effect or inhibition of light production in the presence of antioxidants. The antioxidants could inhibit the production of light by several mechanisms. First, they could react with the peroxynitrite and destroy it. Second, they could destroy one of the reactants, either superoxide or nitric oxide, before it makes peroxynitrite. Another mechanism is that if an antioxidant could react with peroxynitrite by acting as a target for nitration it could chemically destroy the peroxynitrite. A large amount of light was produced, about 2.5×10^9 by peroxynitrite and luminol reacting together. Based on the kinetic profile graph, the light appears to follow the dissolution of the NG tablet (see Figure 32.4 below). The extracts tested are known to have a variety of phenolic and polyphenolic substances that could inhibit luminol–peroxynitrite-based luminescence.

In Figure 32.1, PBE was the most potent of the extracts tested. This inhibition occurs in all dilutions tested. It can be seen that the undiluted, 1:50, and 1:500 dilutions had a percent inhibition of 99% with the final 1:5000 dilution having a percent inhibition of 68.9%. The dose–response graph shows that, compared with the control without antioxidants, the PBE sharply decreases the amount of chemically produced light. The undiluted solution of PBE, with only 10^5 light, produced a 10,000 times reduction in the amount of the light when compared with the light with 10^8 chemically produced light.

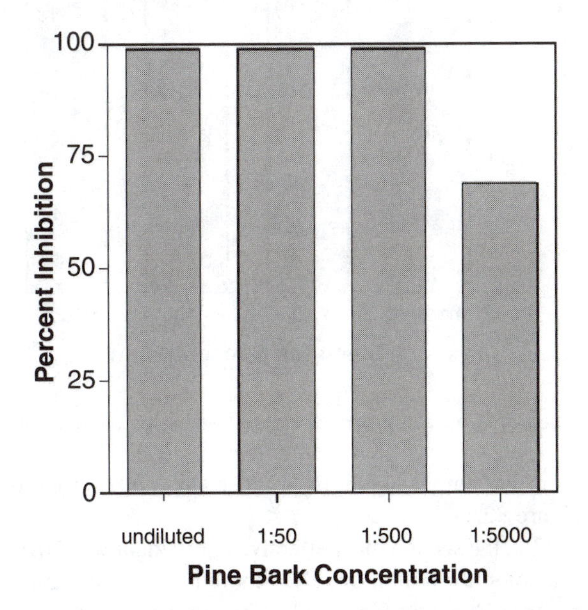

FIGURE 32.1 The percent inhibition for each dilution of the most potent antioxidant tested, PBE. The first three dilutions show a complete inhibition, whereas the third 1:5000 dilution inhibited tenfold compared with the control solution.

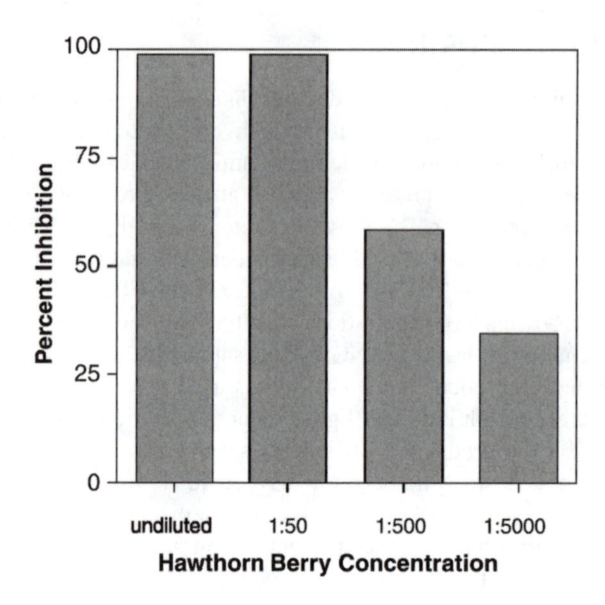

FIGURE 32.2 The second most effective antioxidant was HBE. The first two dilutions showed complete inhibition, and the other two also inhibited light when compared with the control without antioxidant.

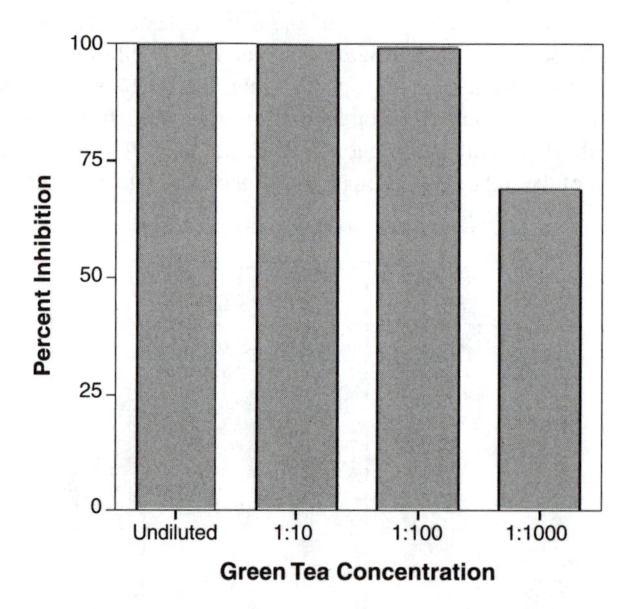

FIGURE 32.3 The inhibition of light with green tea extract used as an antioxidant.

Even in the final dilution, the amount of light produced is inhibited tenfold compared with the control solution (see Figure 32.1).

As shown in Figure 32.2, the second most effective antioxidant was HBE solution. Its undiluted and 1:50 dilution show a complete inhibition. The 1:500 dilution had a 58.2% and the 1:5000 had a 34.4% inhibition. Finally, as shown in Figure 32.3, the green tea with various dilutions: undiluted, 1:50, 1:500, and 1:5000 solutions produced major (98.8, 97.5, 89.8, and 59%) reduction of light, respectively.

Figure 32.4 displays the kinetics of the amount of light produced by each dilution of PBE over a specific period of time. Even at the lowest dilution, this antioxidant produces over 65% inhibition.

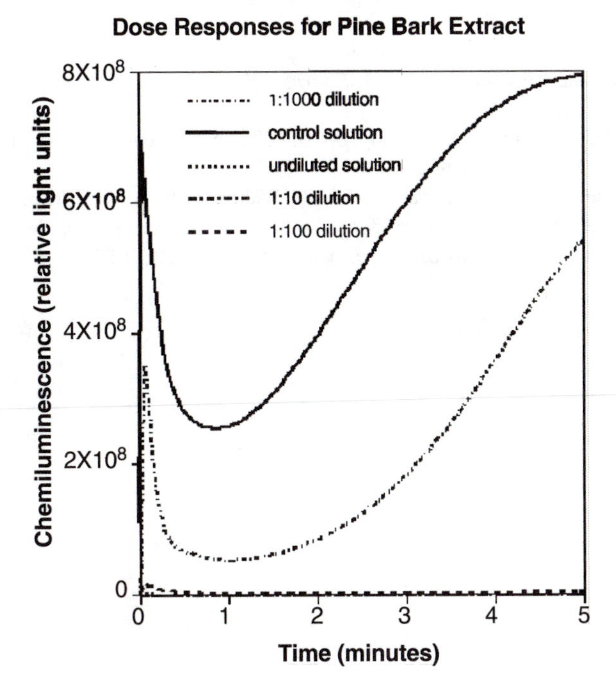

FIGURE 32.4 The kinetics of the amount of light produced by each dilution of PBE over a given amount of time. Each dose is compared to the control with no antioxidants. The initial peak is caused by the amount of NG available to react at the instant the first reading was taken. After that, a progressive increase can be seen as more NG begins to dissolve within the solution.

By directly or indirectly reacting with the peroxynitrite, the four extracts containing antioxidants mentioned above interfered with the oxidation of the luminol, to produce a reduced amount of light. It appears that peroxynitrite plays a major role in eliminating the body of unwanted parasites, viruses, and bacteria. However, these oxidants such as superoxide and peroxynitrite can produce a significant amount of toxicity in the body. Although they are necessary in maintaining the health of the body, these reactions can run out of control and cause severe inflammation. The NG, used for anginal pain, produces nitric oxide and, in the presence of superoxide, peroxynitrite, which is a toxic oxidant that produces inflammation. Therefore, using chemicals to prevent oxidants from running out of control, with antioxidants, makes sense. Of the three extracts chosen as antioxidants to conduct this experiment, PBE was the most effective. Daily use of this extract or the others in this study might prove to be beneficial for those using NG for anginal relief. Clearly, NG and other nitrates have been used to alleviate the pain of angina. But physicians are generally unaware that the substance causing vasodilation can also cause a toxic reaction to take place in the coronary vasculature. Therefore, it is imperative to develop chemical inactivation of peroxynitrite that accompanies vasodilation. Indeed, nitric oxide is needed, but peroxynitrite is not needed and should be prevented from occurring.

For years it has been noted that repeated or continuous use of NG produces a diminished vasodilatory effect on the coronary vessels. This is because as the vessel stretches from the dilatory effect of nitric oxide, the endothelial cells that line the inside of the coronary vessels produce superoxide $\cdot O_2^-$, which combines with available nitric oxide from the NG and diminishes the concentration of nitric oxide available to produce vasodilation. This is in part responsible for the diminishing effect of repeated doses of NG, termed tachyphylaxis. But by destroying the available superoxide or by destroying the peroxynitrite so formed, damage to the coronary vessels can be avoided or at least diminished.

REFERENCES

1. Hardman, J. G. and Limbard, L. E., *The Pharmacological Basis of Therapeutics,* 9th ed., McGraw-Hill, New York, 1996, 760–776.
2. Lowenstien, C. J., Dinerman, J. L., and Snyder, S. H., Nitric oxide: a physiological messenger, *Ann. Intern. Med.,* 120, 227–237, 1994.
3. Molina, C., Andersen, J. W., Papaport, R. M., Waldman, S. A., and Murad, F., Effect of *in vivo* nitroglycerin therapy on endothelium-dependent and -independent relaxation and cyclic GMP accumulation in rat aorta, *J. Cardiovasc. Pharmacol.,* 10, 371–378, 1987.

33 Antioxidant Activity in Grape and Other Fruit Extracts Inhibits Peroxynitrite-Dependent Oxidation (from SIN-1) as Measured by Luminescence

Knox Van Dyke, Candace L. Ogle, and Mark J. Reasor

CONTENTS

33.1 INTRODUCTION

A continuous and long-term deficiency in antioxidants is believed to be responsible for the etiology of many diseases,[1–3] including the neurodegenerative disorders Alzheimer's disease, Parkinson's disease, amyotrophic lateral sclerosis, Huntington's chorea, as well as a variety of inflammatory diseases such as atherosclerosis, diabetes, arthritis, and the aging process itself. Even diseases such as cancer and certain viral infections are exacerbated by a deficiency in antioxidants. Human papilloma virus (HPV) can cause genital warts, various types of cancer, and is a major risk factor in squamous cell carcinoma. The multiplication and possibly the transformation of HPV is inhibited by high levels of antioxidants, both *in vitro* and *in vivo*.[4,5] These diseases are linked to oxidative stress, a term coined by Seis,[6] which is an imbalance in the ratio of oxidants to antioxidants, where oxidants predominate in the body. Raising the level of different antioxidants can provide some protection from certain potent oxidants, e.g., peroxynitrite ($OONO^-$) or hypochlorite (OCl^-).

Oxidants like peroxynitrite and hypochlorite can cause mutation of DNA or RNA as well as damage to protein or lipids via direct chemical attack. Therefore, if taken over a lifetime and at high enough dosages, antioxidants may thwart the damage or mutation caused by such oxidants.

0-8493-0719-8/02/$0.00+$1.50

Certainly, vitamins C and E can be increased in most people without major toxicity; however, the largest natural contributor to the antioxidant load in the bloodstream is the purine metabolite uric acid. In most people it represents 75 to 90% of the total antioxidant level in the blood.[7] Uric acid exists in concentrations of 4 to 7 mg% in the blood of most people, a much higher concentration than other antioxidants.

In addition to the better-known endogenous antioxidants, it is recognized that phenols from dietary sources, especially polyphenols from the tea plant, can play an important role in this regard.[8,9] In the presence of extracts from black or green tea or their isolated polyphenols, the normally copious amount of blue light generated by the oxidative attack of peroxynitrite on luminol is markedly inhibited in a concentration-dependent manner at up to 1000-fold dilution of antioxidant (see Chapter 31). Similarly, studies concerning the link between red wine consumption and low incidence of cardiovascular disease, the so-called French paradox,[10] as well as the Zutphen Elderly Study relating to the health benefits available from other dietary sources of flavonoids,[11] have raised an interest in the identification of good dietary sources of antioxidants. In the present study, a simple chemiluminescence assay was employed to evaluate a variety of fruits and berries, as well as wine and fruit-derived beverages as potentially rich sources of antioxidants.

33.2 MATERIALS AND METHODS

33.2.1 Chemicals and Equipment

Linsidomine (SIN-1) was donated by Dr. Karl Schönafinger from Casella AG (Frankfurt, Germany). Luminol was purchased from Aldrich Chemical Co. (Milwaukee, WI). All fruit and wine were purchased from a local supermarket or the Forks of Cheat Winery (Morgantown, WV). Fruit samples were homogenized with a Ultra-Turrax Sonic Tissue Homogenizer. Luminescence was measured using a Berthold 9505 C, six-channel luminometer with Hamamatsu bialkali photomultiplier tubes. Standard 3-ml round-bottom Berthold tubes were used in all the assays.

33.2.2 Preparation of Fruit Extracts

33.2.2.1 Raisins

First, 10 g of Dole California Seedless regular dark and golden raisins were thoroughly washed with water and dried. They were brought to a final volume of 100 ml with 50% ethanol and soaked overnight at 4°C. They were then homogenized on ice and stirred at room temperature for 3 h and left to extract overnight at 4°C. The raisin homogenate was first filtered through a standard paper coffee filter and then through a Nalgene 0.2-μm Sterilization Filter Unit. The filtered extract was stored at –80°C.

33.2.2.2 Grapes and Fresh Berries

White, red, and black seedless grapes were washed thoroughly with water and dried. To keep the grape extraction concentrations as consistent with the raisin samples as possible, the average number of raisins contained in a 10-g sample was determined. Since there were 15 to 20 raisins in a 10-g sample, 15 grapes of all three varieties were weighed. It was determined that the average weight of 15 grapes is approximately 60 g. Therefore, all fresh samples were subsequently prepared using approximately 60 g of fruit. Because a final volume of 100 ml was found to be too thick, these samples were brought to a final volume of 200 ml with 50% ethanol. Thus, the grape and other fresh fruit samples are approximately half as concentrated as the raisins.

The fruit suspensions were homogenized on ice and stirred at room temperature for 3 h and then left to extract overnight at 4°C. The fruit homogenate was first filtered with an ordinary paper coffee filter, and then through a Nalgene 0.2-μm Sterilization Filter Unit. The filtered extract was stored at –80°C.

33.2.3 Luminescence Assay

All samples, including wine, raisin, grape, and berry extracts, were assayed for their ability to suppress oxidation using a chemiluminescence assay. SIN-1 produces approximately 1% peroxynitrite per minute. When exposed to this strong oxidant, luminol produces chemiluminescence measured by the luminometer. This is the 100% luminescence against which the extracts and wines were compared. First, 100 μl of sample at various dilutions in water was added to 200 μl 0.1 *M* phosphate-buffered saline (PBS), pH 7.4; 100 μl of luminol was dissolved in a small amount of DMSO, and this was added to PBS for a final concentration of 1 mg/ml; and 100 μl of SIN-1 (added last). The SIN-1 was made daily to a standard concentration of 20 mg/5 ml or 3.86 m*M*, and kept on ice. This mixture resulted in final concentrations of wine or extract of 1:5, 1:250, 1:500, and 1:2500 in a volume of 500 μl assay solution. Appropriate controls containing 50% ethanol instead of extract or wine were run simultaneously, and no inhibition of luminescence was observed.

All assays were repeated three to four times and data are shown as mean ± S.E.M.

33.3 RESULTS AND DISCUSSION

A simple chemiluminescence assay was used to measure the antioxidant activity of wines, grape juice, table grapes, and berries. Extracts from each fruit were prepared and assayed at various dilutions to examine the relationship between concentration and antioxidant activity in each sample. This is significant because beverages and fruit are diluted in the body and, therefore, antioxidant activity at a low concentration is desirable.

It was observed that the deeper-colored beverages exhibit the most antioxidant activity and antioxidant reserve upon dilution from 1:5 to 1:2500. This inhibition has little to do with color quenching by the beverage itself since at dilutions greater than 1:250 color was no longer observable. Nor was the alcoholic content of the wine an issue, because assays using only alcohol showed no antioxidant activity. The deeply colored wines, Chambourcin (aged in oak), Cabernet Sauvignon, Manischewitz, and Welch's grape juice were all powerful antioxidant fluids, even at great dilution (Figure 33.1). However, unlike the rest of the group, Manischewitz displayed stimulation rather than inhibition of SIN-1 light production with luminol at the highest dilution (1:2500). This phenomenon has been observed before by Kricka and co-workers,[12] who found that certain phenolic substances will actually stimulate light production from luminol at very low concentrations where the effect of the antioxidants has been diluted away.

Red rosé and blackberry wine showed moderately good antioxidant activity compared with the other wines, but our sample of white wine displayed little antioxidant activity. In fact, once this sample was diluted to 1:5 and or greater it actually showed marked stimulation of light (see Figure 33.1). This may indicate that antioxidant activity is as poor in lightly colored wines as it is good in deeply colored wines. The difference could be attributable, at least in part, to the difference in production techniques between red and white wines. White grapes are not squeezed as greatly during processing as are red grapes. Nor is white wine generally allowed to steep in its own seeds and skin for as long as darker wines. Because grape seeds and skins have a high antioxidant content,[13] it would not be surprising to find that exposure to these components contributed to the antioxidant content of the wine.

Like the wine in this study, the more deeply colored grape extracts displayed greater antioxidant activity and antioxidant reserve. As might be expected, the black grape extract had significantly more activity than the red or white grape samples. However, the red grape extract did not appear to be significantly better than the white grape (Figure 33.2). Because these were seedless table grapes and not wine grapes, the antioxidant content may be quite different from that of wine. Wine grapes are generally smaller and have much less juicy pulp than table grapes and, therefore, a much greater concentration of seed and skin.

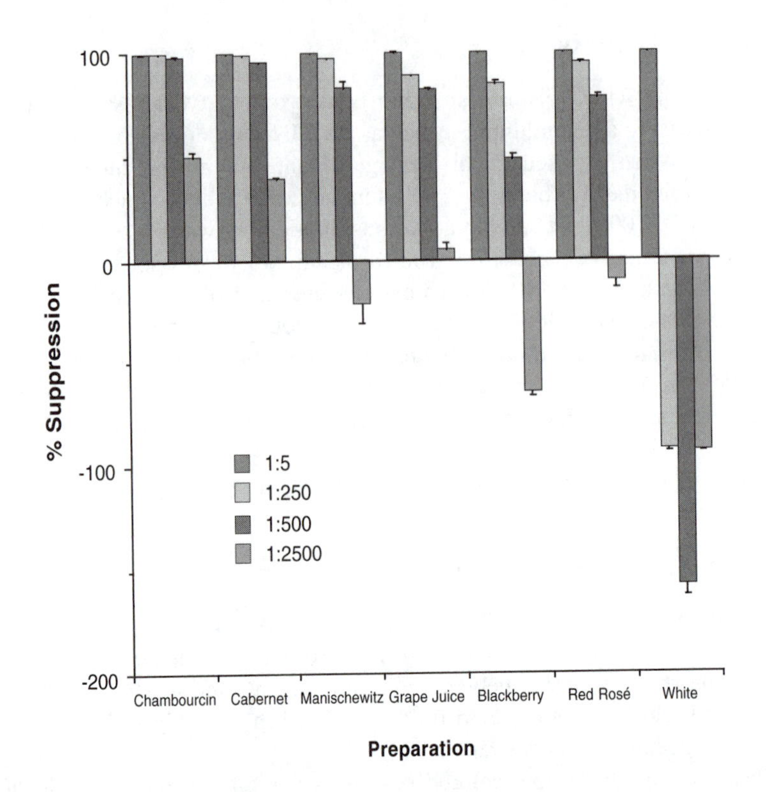

FIGURE 33.1 Suppression of SIN-1-induced chemiluminescence by grape-derived beverages.

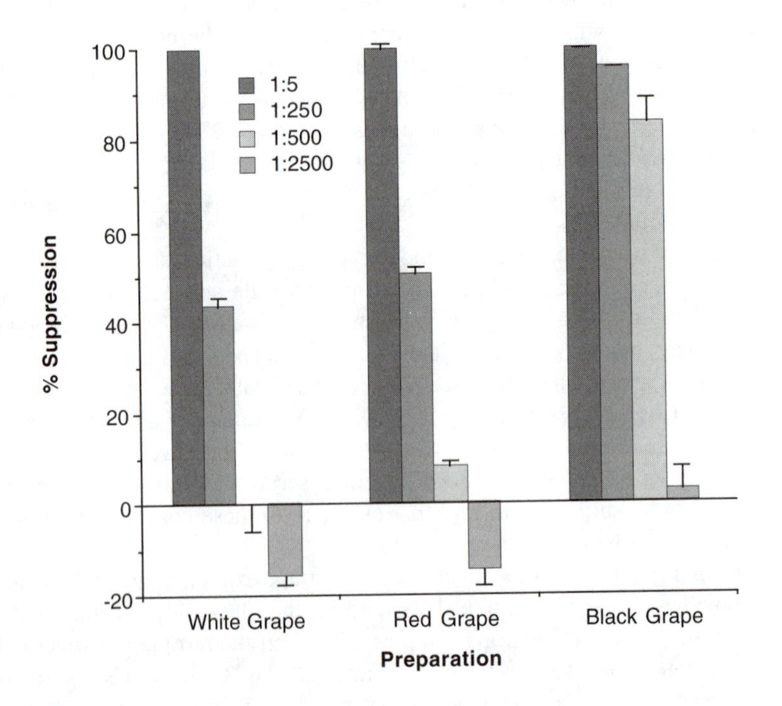

FIGURE 33.2 Suppression of SIN-1-induced chemiluminescence by grape extracts.

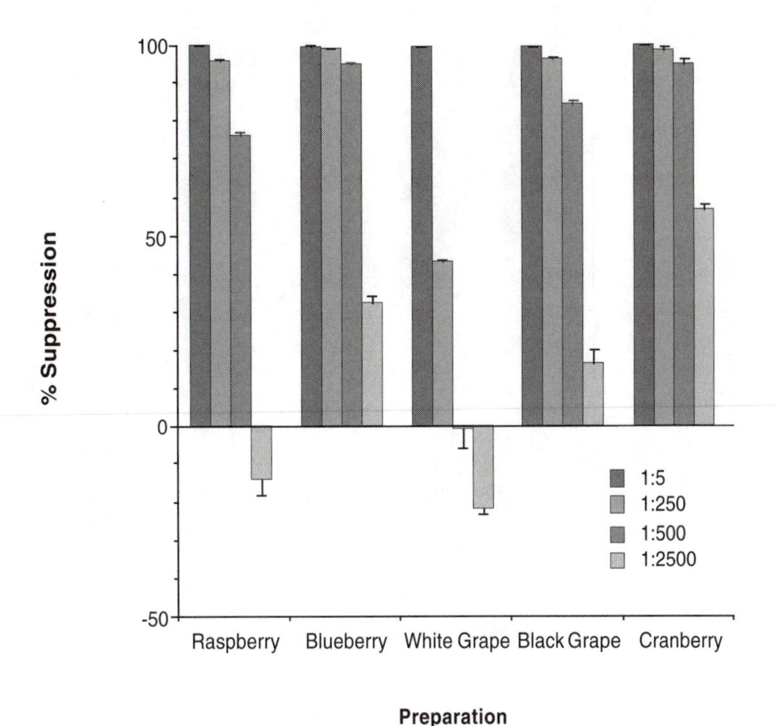

FIGURE 33.3 Suppression of SIN-1-induced chemiluminescence by berry extracts.

The relationship between color intensity and antioxidant activity becomes less apparent when comparing different berry extracts to grape extracts. Although both the cranberry and blueberry extracts contained even more active antioxidants than the black grape extract, the less intensely colored cranberry outperformed the deeply colored blueberry at the lowest concentration (Figure 33.3). At higher concentrations, the raspberry extract performed similarly to the black grape extract. However, like the white grape extract, it actually stimulated light production at the lowest concentration. It may be that the relationship between color and antioxidant activity is a reliable indicator only when comparing sources from the same plant species.

Even within a single plant species, preparation techniques might make antioxidant activity difficult to predict. For example, when comparing the activity of dark seedless raisins to golden seedless raisins, it would be reasonable to expect greater activity within the dark raisins. However, when extracts from these samples were actually compared, more activity was found in the golden raisins (Figure 33.4). The source of this unexpected outcome is not clear. Possibly the grapes from which dark raisins are made have less antioxidant content to begin with, or perhaps, some of the antioxidant content is lost in the drying process.

The composition of fresh grapes will vary according to the individual variety, but in general they contain many different phenolic compounds including simple phenols, phenolic acids, cinnamic acids, stilbenes, flavanoids, flavans, flavanols, and anthocyanins. Grapes also contain leukoanthcyanidins of different structures. There are a variety of phenolic compounds that can be attacked by the strong oxidants found in the body such as peroxynitrite and hypochlorite/peroxyhypochlorite. These phenols may be oxidized or the phenol or polyphenolic compound may be a bystander target for electrophilic substitution, namely, nitration (by peroxynitrite), or via chlorination from the hypochlorite/peroxyhypochlorite. In this way they can act as chemical targets that divert the oxidants from damaging the cellular constituents, e.g., DNA, RNA, protein, and unsaturated lipids.

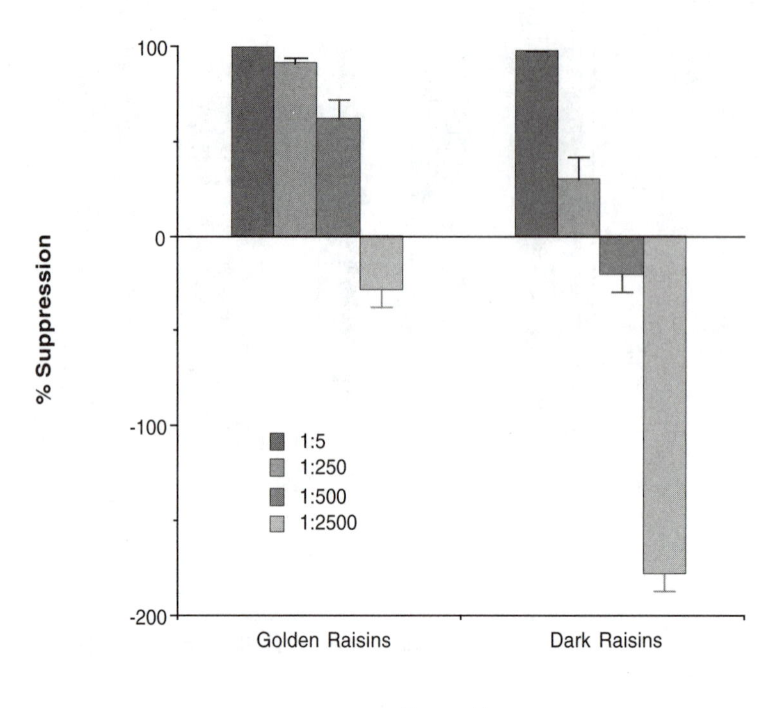

FIGURE 33.4 Suppression of SIN-1-induced chemiluminescence by raisin extracts.

Oxidation increases greatly during different disease states. Exposure to bright sunlight, radiation treatment, or even strenuous exercise can also cause oxidative stress. During major incidents of oxidative stress, raising the level of antioxidant-vitamins C and E and uric acid may have a protective effect, but in addition the dietary intake of fresh fruits and vegetables, as well as fruit-derived beverages can add to the total antioxidant load of the body. Since these incidents of stress may be difficult to predict or even perceive, it may be desirable to increase levels of antioxidants during an entire lifetime.

ACKNOWLEDGMENTS

The authors gratefully acknowledge the financial support of Dr. Robert D'Alessandri, Dean of the School of Medicine and Vice President of the Robert C. Byrd Health Sciences Center of West Virginia University, and the support of the Mylan Endowment and Dr. Robert E. Stitzel, Chairman of the Department of Pharmacology and Toxicology at WVU School of Medicine. We thank Dr. Karl Schönafinger of Casella AG, Frankfurt, Germany for supplying the SIN-1.

REFERENCES

1. Nuttall, S. L., Kendall, M. J., and Martin, U., Antioxidant therapy for the prevention of cardiovascular disease, *Quart. J. Med.*, 92, 239–244, 1999.
2. Das, D. K., Sato, M. et al., Cardioprotection of red wine: role of polyphenolic antioxidants, *Drugs Exp. Clin. Res.*, 25, 115–120, 1999.
3. Yamakoshi, J., Kataoka, S., Koga, T., and Ariga, T., Proanthocyanidin-rich extract from grape seeds attenuates the development of aortic atherosclerosis in cholesterol-fed rabbits, *Atherosclerosis*, 142, 139–149, 1999.

4. Kwasniewska, A., Tukendorf, A., and Semczuk, M., Frequency of HPV infection and the level of ascorbic acid in serum of women with cervix dysplasia, *Med. Dosw. Mikrobiol.*, 48(3–4), 183–188, 1996.
5. Giuliano, A. R., Papenfuss, M., Nour, M., Canfield, L. M., Schneider, A., and Hatch, K., Antioxidant nutrients: associations with persistent human papillomavirus infection, *Cancer Epidemiol. Biomarkers Prev.*, 611, 917–923, 1997.
6. Seis, H., Stahl, W., and Sundquist, A. R., Antioxidant functions of vitamins, in *Beyond Deficiency, New Views on the Function and Health Effects of Vitamins,* H. E. Suberlich and L. J. Machlin, Eds., The New York Academy of Sciences, New York, 1992, 7–20.
7. Scott, G. S. and Hooper, D. C., The role of uric acid in protection against peroxynitrite-mediated pathology, *Med. Hypotheses,* 56(1), 95–100, 2001.
8. Isemura, M., Sacki, K., Kimura, T., and Sazuka, M., Tea catechins and related polyphenols as anticancer agents, *Biofactors,* 13(1–4), 81–85, 2000.
9. Yang, C. S., Chung, J. Y., Yang, G. Y., Li, C., Meng, X., and Lee, M. J., Mechanisms of inhibition of carcinogenesis by tea, *Biofactors,* 13(1–4), 73–79, 2000.
10. Renaud, S. and DeLorgeril, M., Wine, alcohol, platelets, and the French paradox for coronary heart disease, *Lancet,* 339, 1523–1526, 1992.
11. Hertog, M. G. L., Hollman, P. C. H., Katan, M. B., and Krumhold, D., Estimation of daily intake of potentially anticarcinogenic flavonoids and their determinants in The Netherlands, *Nutr. Cancer,* 20, 21–29, 1993.
12. Thorpe, G. H., Kricka, L. J., Moseley, S. B., and Whitehead, T. P., Phenols as enhancers of the chemiluminescent horseradish perosedase-luminol-hydrogen peroxide reaction: application in luminescence-monitored enzyme immunoassays, *Clin. Chem.,* 31(8), 1335–1341, 1985.
13. Zhao, J., Wang, J., Chen, Y., and Agarwal, R., Anti-tumor-promoting activity of a polyphenolic fraction isolated from grape seeds in the mouse skin two-stage initiation-promotion protocol and identification of procyanidin B5-3′-gallate as the most effective antioxidant constituent, *Carcinogenesis,* 20, 1737–1745, 1999.

34 Antioxidant Actions of Acetaminophen Preventing Myocardial Injury Detected by Luminescence and Other Modalities

Gary F. Merrill

CONTENTS

Cardiovascular disease claimed 953,110 lives in the United States in 1997, the latest year for which complete data were available at the time of this writing. Cardiovascular disease was about 60% of total mentioned mortality, which means that of the more than 2,000,000 deaths from all causes, cardiovascular disease was listed as a primary or contributing cause on over 1,406,000 death

0-8493-0719-8/02/$0.00+$1.50

certificates in 1997. Since 1900, cardiovascular disease has been the No. 1 killer in the United States every year but 1918. More than 2600 Americans die each day of cardiovascular disease, *an average of one death every 33 s*. Cardiovascular disease claims more lives each year than the next seven leading causes of death combined (American Heart Association, *2000 Heart and Stroke Statistical Update*, Dallas, TX, American Heart Association, 1999).

Given such information, can anyone in good conscience conclude that we have spent enough money and time investigating this disease? In this chapter I wish to expose the reader to new data on an old compound. Acetaminophen, an active ingredient in many popular over-the-counter pain relievers, has been used in clinical medicine for more than 100 years.[1–3] Yet, because it was not investigated with enough vigor and attention in the experimental physiology laboratory, some have concluded that it lacks cardiovascular properties.[4] In the past two or three decades there has been growing interest in the phenomena of myocardial ischemia/reperfusion injury, myocardial stunning, and cardioprotection.[5–7] If these topics had attracted more attention in the first half of the past century, then perhaps investigation of the potential cardioprotective properties of acetaminophen would not have gone unnoticed for so long.

34.1 INTRODUCTION

Acetaminophen was introduced into Western medicine more than 100 years ago.[1–3] Its effects on relieving pain and lowering body temperature have been under investigation for several decades.[1–3] It is the active ingredient in Tylenol, the drug of choice for the relief of pain in patients in U.S. hospitals (according to McNeil Consumer Healthcare, Fort Washington, PA). Polls show that a rather substantial number of American households use Tylenol and other acetaminophen-containing products to avert cardiovascular events.[8] Increased use of acetaminophen by our aging population, without data to support or guide such use, is potentially dangerous. It reveals the need for further investigation of the previously undescribed cardiovascular actions and properties of acetaminophen in a wide range of clinically significant arenas.

Chemically, acetaminophen is a phenol.[1–3,9,10] Many phenols have important antioxidant properties.[1–3,9–11] Considering the increasing international interest in the redox state of cells during ischemia and reperfusion, and the central role that damaging oxidants such as hydroxyl radical and peroxynitrite seem to play, it is surprising that no one has paid attention to acetaminophen and the heart. This oversight might be explained, at least in part, on the basis of broad-sweeping generalities that acetaminophen has no effects on the mammalian cardiovascular system,[4,12] and by its lack of anti-inflammatory potential.[4] There are those who have questioned the efficacy of acetaminophen in the human cardiovascular system, but these reports contain limited, useful quantitative data, and are more-or-less anecdotal.[12] The only published report that the author has been able to find on the experimental use of acetaminophen in tissue ischemia and reperfusion is that of Nakamoto et al.[13] It is worth reviewing briefly here. These investigators studied the effects of acetaminophen on acute gastric mucosal injury caused by ischemia and reperfusion in rats. In anesthetized, instrumented animals, they clamped the celiac artery, causing reduced blood flow to the stomach for 30 min, and then released the clamp and reperfused the organ for 60 min. Animals were sacrificed, and the total area of gastric erosions was measured in vehicle- and acetaminophen-treated rats. Acetaminophen significantly reduced the area of erosions. It also inhibited the increase in lipid peroxide levels caused by ischemia/reperfusion, and the increase in lipid peroxidation caused by hydroxyl radicals. The authors concluded that acetaminophen protected against injury caused by ischemia and reperfusion by blocking membrane damage caused by hydroxyl radicals. Unfortunately, acetaminophen was administered in doses of 300 and 500 mg/kg 90 min before the onset of ischemia. Corresponding doses in humans would be lethal, or would at least produce circulating plasma concentrations well in excess of 300 μg/ml (concentrations known to produce severe, life-threatening liver damage).[1–4] Moreover, these investigators did not administer acetaminophen after the onset of ischemia to determine its efficacy. Thus, the application of their data to the human condition is limited. Nonetheless, the report of Nakamota et al. helped encourage the author and his co-workers to address the question of the utility of acetaminophen in myocardial ischemia/reperfusion injury. The rest of the chapter presents a discussion of our results.

34.2 MATERIALS AND METHODS

34.2.1 EXPERIMENTAL PREPARATION

The experiments reported in this chapter were performed exclusively in isolated, perfused hearts extracted from guinea pigs (Langendorff preparation). They were taken in about equal numbers from male and female young adults weighing approximately 400 ± 30 g (Hartley strain; see Table 34.1). On any given day, experiments were randomized to eliminate time-dependent effects, and acetaminophen-treated hearts were matched on a daily basis with vehicle-treated hearts. Vehicle was physiological salt solution, in this case Krebs–Henseleit buffer.

The guinea pig heart preparation developed by Bunger et al.[14,15] and Schrader et al.[16] (Figure 34.1) was used because it has *in vivo*-like properties (it autoregulates coronary blood flow, responds to

TABLE 34.1
Characteristics of Guinea Pigs and Hearts Used to Set Standards for This Investigation

	Body Weight (g)		Ventricular Weight (g)		V/B (g/100 g)	
	M	F	M	F	M	F
Sample size (n)	28	28	26	27	23	27
Mean	404	383*	1.14	1.08*	0.28	0.28
s.e.m.	10	10	0.04	0.03	0.005	0.006

Data are means plus or minus 1 s.e.m. M, male; F, female; V, ventricular weight; B, body weight; V/B, ventricular to body weight ratio; *, $P < 0.05$ relative to corresponding value for males.

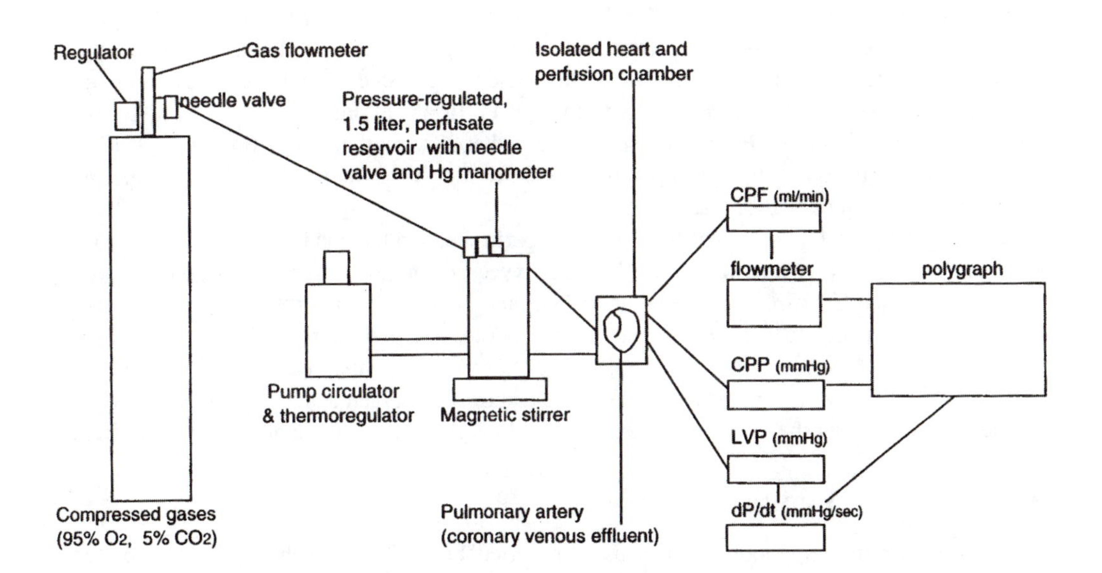

FIGURE 34.1 Experimental setup including isolated, perfused heart, and recording apparatus. O_2, oxygen in the gas phase; CO_2, carbon dioxide in the gas phase; CPF, coronary perfusate flow; CPP, coronary perfusion pressure; LVP, left ventricular pressures; *dP/dt*, differentiation of left ventricular pressures.

changes in oxygen supply and demand like the heart does *in vivo*, etc.). Few isolated, rodent heart preparations currently being used in experimental research can make this claim (including murine heart preparations that so many laboratories are scrambling to develop to meet the interests of genomics), and none is instrumented *in situ* like the Bunger et al. preparation. *In situ* instrumentation of isolated, perfused hearts is a critical quality of such preparations because it minimizes the chances of embolizing the aorta/coronary vasculature with air. Most investigators use hearts that are extracted and quickly plunged into a container of chilled physiological salt solution. Once hearts stop beating they are then cannulated for perfusion. These investigators generally are more concerned about the speed of arresting the hearts than with the prospects of rendering the hearts physiologically useless because of air embolization.

With our methodology, the animal is euthanized and a bilateral thoracotomy is quickly performed. The pericardium is removed and the heart arrested by superfusion with chilled Krebs–Henseleit solution. Once the heart is arrested, the aorta and pulmonary artery are cannulated (PE240 cannulae) and the aorta is perfused retrogradely at a low rate of flow (approximately 1 to 2 ml/min/g). The heart is then extracted, and transferred to the Langendorff apparatus. During the transfer (2 to 3 s) the ventricles are gently compressed between the index finger and the thumb. This expresses perfusate antegradely into the aortic cannula where a mixed perfusate/blood meniscus can be seen as the cannulated aorta is attached to the Langendorff perfusion system. Visualization of the perfusate/blood meniscus ensures freedom from air bubbles. The hearts are then reperfused incrementally (e.g., 2, 4, 6 ml/min/g) at 3- to 4-min intervals during the next 12 to 15 min. This minimizes the chances of reperfusion damage including ventricular fibrillation.

As long as we use guinea pigs of the standard 375 ± 25 g, ventricular wet weights at the end of our experiments are reliably 1.0 ± 0.2 g. This is a standard the author has used since graduate school in the early 1970s. It enables us to normalize coronary perfusate flow rate per gram of ventricular tissue, and experimental results during the past several decades have been uniformly consistent.

34.2.2 Experimental Perfusate

The physiological salt solution used in the studies described in this chapter was a modified Krebs–Henseleit buffer of the following composition (m*M*): KH_2PO_4 1.2, KCl 4.7, $NaHCO_3$ 24.9, NaCl 127.4, $MgSO_4 \cdot 7H_2O$ 1.2, $CaCl_2$ 2.5. It was fortified with 10 m*M* glucose, 2.0 m*M* pyruvate, and 200 μU/ml insulin. Perfusate was equilibrated with a gas mixture containing 95% O_2 and 5% CO_2. After several minutes of equilibration, this yielded perfusate partial pressures (mmHg) for these gases of 600 ± 50 (PO_2) and 35 ± 5 (PCO_2). Perfusate pH at these partial pressures was routinely 7.40 ± 0.02 (units). Perfusate was maintained at 38°C with a circulating pump/heater and water-jacketed perfusion system. Temperatures were routinely verified using miniature thermistor probes (Physitemp, model BAT-12, Clifton, NJ).

Knowing the partial pressures of oxygen in the perfusate, and the solubility of oxygen in physiological salt solutions, one can easily calculate the oxygen contents of arterial and coronary venous perfusate (CaO_2 and CvO_2). From these data and coronary perfusate flow rates, one can calculate the rate of oxygen delivery to the perfused myocardium (i.e., oxygen supply), the rate at which the myocardium extracts oxygen from the arterial supply (a-vO_2, or percent extraction), and the rate of oxygen consumption by the myocardium (MVO_2, also known as oxygen demand). These are invaluable variables for quantifying the effects of an experimental intervention on the mammalian myocardium.

34.2.3 Perfusion Modality

In cardiovascular investigations such as these reported here, there are three kinds of perfusion modalities from which an investigator can choose. Natural flow and natural pressure would be chosen if the investigation was being conducted *in vivo* and the investigator simply wanted to observe the consequences on the systemic circulatory system of a particular intervention (e.g., drug administration, exposure to hypoxia, etc.). Controlled pressure would be used and flow measured if the investigator wanted to study the ability of an organ to autoregulate its blood flow. This perfusion modality would be used *in vivo* and *ex vivo*. Flow would be controlled and pressure measured if the investigator wanted to administer a drug and deliver it to an organ at a known, constant rate.

With our isolated, perfused heart preparation, we have used both controlled pressure (while measuring flow) and controlled flow (while measuring pressure). In the experiments reported here, we used only controlled flow. This was initially chosen so we could have confidence in our calculations of the rates at which we were delivering acetaminophen to the myocardium. Later, we simply added acetaminophen to the perfusate reservoir to yield known circulating concentrations in the coronary perfusate. In these cases we still perfused the hearts at controlled flow rates because initial data were collected using this perfusion modality. Experiments reported in this chapter were conducted using a controlled coronary perfusion flow rate of 7 ml/min (approximately 6 to 7 ml/min/g). This rate of perfusion yielded coronary perfusion pressures of 40 to 60 mmHg (measured 2 to 3 cm upstream to the cannulated heart).

34.2.4 Experimental Protocols

34.2.4.1 Cardiac and Coronary Vascular Effects of Acetaminophen

In an initial group of hearts we studied doses of acetaminophen of 50, 250, and 1000 μg administered as bolus injections (i.c.) to (1) determine if acetaminophen affects the myocardium/coronary circulation and (2) to select a concentration that could be used in subsequent experiments with ischemia and reperfusion. All hearts were exposed to both acetaminophen and the corresponding volumes of vehicle. Data were collected when changes in monitored variables were evident, and only when such changes were in the steady state (Figure 34.2). We focused only on changes in coronary perfusate flow rate, coronary perfusion pressure, calculated coronary vascular resistance, and ventricular mechanics (e.g., left ventricular developed pressure, $\pm dP/dt_{max}$). From these experiments we selected concentrations of 0.1, 0.35, and 0.6 mmol/l (final concentrations in the perfusate) to use in subsequent experiments with ischemia/reperfusion.

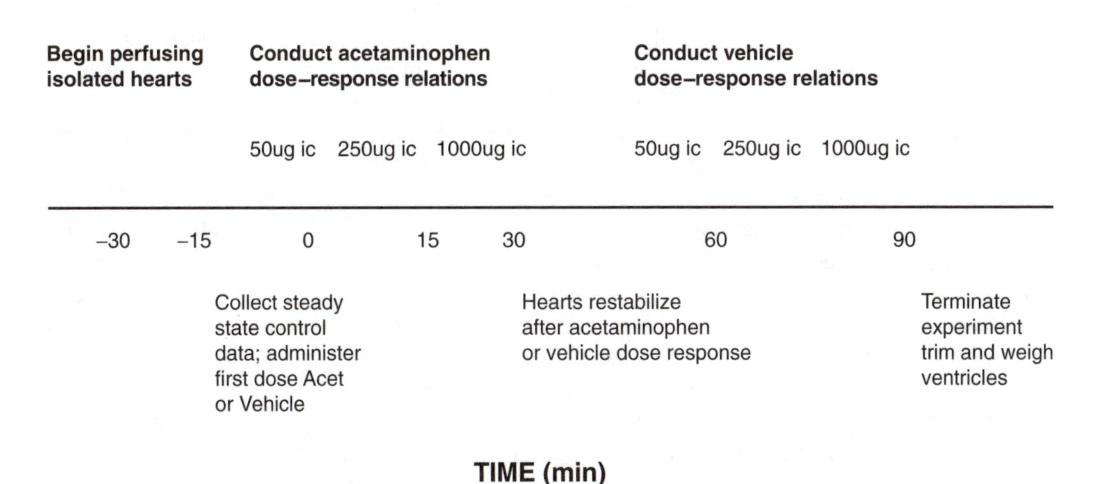

FIGURE 34.2 Experimental timeline. Time (30 min) was allowed for monitored variables to achieve their steady-state values. This timeline reflects experiments using bolus injections of acetaminophen; however, similar timelines were also used during ischemia and reperfusion. μg, micrograms; i.c., intracoronary route of administration.

34.2.4.2 Low-Flow Global Myocardial Ischemia and Reperfusion

To study the effects of acetaminophen on myocardial ischemia and reperfusion, we chose to use low-flow ischemia as opposed to zero-flow ischemia. In zero-flow ischemia, little or no coronary venous circulation exits the ischemic organ. Thus, it is impossible to collect coronary venous effluent perfusate (blood) under such conditions. If we wanted to know what was happening to myocardial oxygen extraction, myocardial oxygen consumption, etc. during ischemia, we could not collect such information because it is dependent on the collection of coronary venous effluent samples of blood or perfusate during the period of ischemia. Also, if we wanted to study the rates of release of a particular metabolite from the ischemic myocardium during the period of ischemia, we could not because of this same limitation. The rate of release of substance X from any tissue during a given period of time is calculated as follows: release of $X = ([X]_{cv} - [X]_{ca})(\text{CPF})$, where $[X]_{cv}$ is the concentration of substance X in the coronary venous perfusate, $[X]_{ca}$ is the concentration of substance X in the coronary arterial perfusate (these two measures must be based on arterial and coronary venous samples that were collected simultaneously), and (CPF) is the coronary perfusate flow rate at the time of collections of arterial and venous samples of X. This and related information cannot be collected using zero-flow ischemia, and thus is dependent on the investigator's ability to obtain simultaneously withdrawn samples of venous and arterial perfusate. This was our reason for using low-flow ischemia.

After hearts were extracted, instrumented (which included placing a Krebs–Henseleit-filled latex rubber balloon in the left ventricle for measuring left ventricular pressure and differentiating it, and placing a catheter in the trunk of the pulmonary artery for collecting coronary venous perfusate samples), and perfused at 7 ml/min for approximately 15 min, spontaneous heart rates were determined, pacing electrodes were attached, and hearts were paced at controlled rates equal to spontaneous rate plus approximately 10% (frequencies of approximately 4 to 5 Hz, delays of 0.01 ms, durations of 5.0 ms, voltages of 5 to 7 V). They were allowed another 15 min to adjust to this intervention, and for recorded variables to achieve a new steady state. A set of baseline, control data was collected at this time and was designated *predrug*. Subsequently, hearts were randomly divided into two treatment groups, those administered acetaminophen (0.35 mmol/l, final concentration in the perfusate reservoir) and those administered acetaminophen vehicle (vehicle, Krebs–Henseleit perfusate, in equal volume). Hearts were then allowed 20 min exposure to either acetaminophen or vehicle and a second set of data was collected and designated *postdrug*. At this point, coronary perfusate flow rate was reduced from 7 to 1 ml/min (manual adjustment of roller pump interposed in the perfusion circuit) for 20 min. During this period of low-flow, global myocardial ischemia, data were collected at 5-min intervals; however, only those collected at 20 min of ischemia are reported here. These were designated *ischemia*. After 20 min of ischemia, the roller pump was restored to its pre-ischemia rate, and the myocardium was reperfused for an additional 40 min. During reperfusion, data were collected at 10-min intervals. Only data collected at 40 min of reperfusion are reported here. These were designated *reperfusion*. Thus, this experimental protocol consisted of collecting data at four time periods in hearts treated with acetaminophen or its vehicle, then made ischemic and subsequently reperfused. Although not all data are reported here, our overall objective in these experiments was to compare morphological, mechanical, electrical, and metabolic variables in the two groups of hearts to determine if acetaminophen possesses cardioprotective properties that have not been reported previously. Morphological data included electronmicroscopic examination of the ischemic and reperfused myocardium. Mechanical data included measures of left ventricular function including left ventricular end diastolic pressure (LVPd, mmHg), left ventricular peak systolic pressure (LVPs, mmHg), left ventricular developed pressure (LVPD, mmHg), left ventricular pressure/rate product (LVPD × heart rate), and the rates of rise and decline of left ventricular pressure ($\pm dP/dt_{max}$, mmHg/s, a measure of contractility, as left ventricular end diastolic volume was controlled and therefore unable to change). Electrical data included measures of ventricular arrhythmias including incidence of ventricular fibrillation (VF), ventricular tachycardia (VT), ventricular salvos (VS), and ventricular premature beats (VPB) (Lambeth Conventions Guidelines,[17]). Finally, metabolic data included measures of oxygen extraction, oxygen consumption, and changes in perfusate gases, pH, and base excess.

34.2.4.3 Acetaminophen and Hydroxyl Radicals

This protocol was designed to determine if acetaminophen could affect the well-known burst of hydroxyl radicals that is observed in the early minutes of reperfusion following ischemia of the mammalian myocardium.[18–20] Acetaminophen is a phenol and has well-established antioxidant properties.[3,10] Unfortunately, these properties have never been tested in the mammalian myocardium, nor during cardiovascular disease including myocardial ischemia and reperfusion. Hydroxyl radical is a damaging oxygen species that is injurious to cell membranes/subcellular constituents in a variety of pathophysiological states, including ischemia/reperfusion.[21–23] Interest in this oxidant and cardiovascular disease has been mounting in recent years.[18–23] This provided the impetus for our hypothesis that acetaminophen might be cardioprotective if it combats the production/release of damaging oxygen species.

A protocol similar to that described above was used in this experiment. Predrug, postdrug, ischemia, and reperfusion data were collected in acetaminophen- and vehicle-treated hearts during low-flow, global, myocardial ischemia and reperfusion. During ischemia samples were collected at 10 and 20 min, and during reperfusion at 1, 5, 10, and 40 min. Only coronary venous effluent samples (0.5 to 1.0 ml) were collected. These were collected in prechilled vials, and were kept on dry ice until the end of the experiment when they were stored at –80°C until they could be analyzed.

Samples were analyzed after the methods of Powell and colleagues.[24,25] Briefly, analysis included collecting predrug data, followed immediately by addition of sodium salicylic acid to the perfusate reservoir (0.2 mmol/l). After hearts had been exposed to this agent for 10 min, either acetaminophen (0.35 mmol/l) or vehicle were added to the reservoir and hearts were exposed to each for 20 min. Subsequently, hearts were made ischemia for 20 min followed by 40 min of reperfusion. Coronary venous effluent samples (0.5 to 1.0 ml) were collected in prechilled vials at the following times: predrug, 10 min postexposure to salicylic acid, 20 min postexposure to acetaminophen, 20 min ischemia, and 2, 5, 10, and 40 min reperfusion. Samples were kept on dry ice until the end of the experiment and were then stored at –80°C until processed.

Because hydroxyl radical is unstable, it has a brief half-life. It combines with phenols yielding hydroxylated derivatives. Our molecular trap produced a more stable, measurable form of hydroxyl radical, 2,5-dihydroxybenzoic acid (2,5-DHBA), which was detected according to the methods of Powell and colleagues.[24,25] Briefly, 60 μl of filtered venous effluent perfusate was injected into an high-performance liquid chromatography (HPLC) system (Perkin Elmer, LC series 401, 20 μl loop) equipped with a coulemetric detector (Coulchem, model 5100, ESA). Detection parameters were electrode no. 1 oxidation potential, +0.4 V; electrode no. 2 reduction potential, –0.25 V; guard potential, –0.30 V. Chromatograms were monitored for approximately 10 to 12 min each, with 2,5-DHBA curves appearing around 2 min.

34.2.4.4 Acetaminophen and Peroxynitrite Anion

This experimental protocol was based, in part, on the published reports of Van Dyke et al.[10,11] Hearts were treated similarly to those described for hydroxyl radical, except salicylate was not present. Acetaminophen was administered in a perfusate concentration of 0.35 mmol/l, and vehicle was administered in an equal volume. Venous effluent samples (0.5 to 1.0 ml) were collected during predrug, postdrug, and ischemic conditions, as well as at 2, 5, 10, and 40 min reperfusion. Samples were treated with 100 μl luminol (final concentration 0.6 mmol/l), 100 μl SIN-1 (final concentration 5.8 mmol/l), and 300 μl physiological buffer solution. These were mixed then pipetted into 3-ml round-bottomed luminometer tubes and immediately placed in a luminometer (model LB9505C, six-channel Berthold). Samples were analyzed for 20 min each, and the light generated was acquired, plotted, and integrated with an IBM clone computer running KINB software. The assay was reported as counts per minute (cpm) integrated over the 20 min period. We also measured areas under the light curves, as well as peak heights and time to peak heights.

34.2.5 Statistical Analysis

All data were analyzed using analysis of variance for repeated measures. *A priori* tests (e.g., Tukey's w-procedure, Fisher's LSD) were used to compare means. Statistically significant differences were established at the conventional $P < 0.05$, and data are presented as means ± 1 s.e.m. (standard error of the mean).

34.3 RESULTS

34.3.1 Cardiac and Coronary Circulatory Effects of Acetaminophen

Table 34.2 summarizes the effects of the highest dose of acetaminophen on perfusate gases, pH, oxygen extraction, and oxygen consumption. Only oxygen consumption was significantly affected by acetaminophen, and only when perfusion pressure was controlled.

Doses of 250 and 1000 μg acetaminophen caused modest but significant positive inotropy whether flow or pressure was controlled (Table 34.3). For example, $\pm dP/dt_{max}$ increased about 50% at the highest dose when pressure was controlled, and about 20% when flow was controlled. Left ventricular developed pressures increased similarly. Whether coronary perfusate flow rate or coronary perfusion pressure was controlled, there were no significant effects of acetaminophen vehicle on cardiac and coronary circulatory variables.

When coronary perfusion pressure was controlled, acetaminophen at 250 and 1000 μg caused concentration-dependent, statistically significant coronary vasoconstriction ($P < 0.05$). For example, coronary flow decreased from about 6 to 7 ml/min to 4 to 5 ml/min in response to 1000 μg acetaminophen. Coronary vascular resistance increased correspondingly from about 8 or 9 to 11 mmHg/ml/min. These results were corroborated when coronary flow was controlled and coronary perfusion pressure was measured. Under these conditions, coronary perfusion pressure rose from 49 ± 5 to 52 ± 5 mmHg ($P < 0.05$) in response to the highest dose of acetaminophen. Calculated coronary vascular resistance increased correspondingly (Table 34.4).

Acetaminophen vehicle did not affect coronary perfusate flow rate, coronary perfusion pressure, or ventricular mechanics.

34.3.2 Acetaminophen and Cardiac Function/Structure during Myocardial Ischemia and Reperfusion

Only 0.35 mmol/l acetaminophen was used in this protocol. In general, hearts treated with acetaminophen appeared to perform better than those treated with vehicle. For example, both $\pm dP/dt_{max}$ and developed left ventricular pressure recovered 100% by 40 min reperfusion in the presence of acetaminophen. Vehicle-treated hearts recovered about 70% ($P < 0.05$) at the same time. During ischemia, $\pm dP/dt_{max}$ was significantly greater in the presence of acetaminophen. Developed left ventricular pressure did not differ between the two groups during ischemia.

Ultrastructurally, the acetaminophen-treated myocardium appeared to be preserved during reperfusion when compared with the vehicle-treated myocardium. For example, contraction bands, swollen mitochondria, and disrupted sarcolemmal membranes were omnipresent in vehicle-treated hearts during reperfusion. These were not as evident with acetaminophen.

TABLE 34.2
Influence of Acetaminophen (1000 μg i.c.) and Vehicle on Perfusate Gases, pH, Oxygen Extraction, and Oxygen Consumption in Isolated Guinea Pig Hearts Perfused at Constant Pressure (top) and Constant Flow (bottom)

	pHa (units)		pHv (units)		$PaCO_2$ (mmHg)		$PvCO_2$ (mmHg)		PaO_2 (mmHg)		PvO_2 (mmHg)		$a\text{-}vO_2$ ($\mu l/ml$)		MVO_2 ($\mu l/ml$)	
	ACE	VEH	ACE	VEH	ACE	VEH	ACE	VEH	ACE	VEH	ACE	VEH	ACE	VEH	ACE	VEH
Constant pressure	7.47	7.46	7.33	7.33	31	31	44	41	581	580	136	143	10	10	51*	73
(mmHg)	0.01	0.01	0.01	0.01	1	1	1	1	13	11	16	13	0.5	0.3	5	5
Constant flow	7.45	7.46	7.33	7.34	30	30	41	43	567	586	145	143	9	10	68	68
(ml/min/g)	0.01	0.01	0.01	0.01	1	3	2	2	11	10	19	20	0.4	0.4	4	4

Data are means (first and third rows) plus or minus 1 s.e.m. (second and fourth rows) ($n = 10$). pHa, pH in arterial perfusate; pHv, pH in venous perfusate; $PaCO_2$, partial pressure of carbon dioxide in arterial perfusate; $PvCO_2$, partial pressure of carbon dioxide in venous perfusate; PaO_2. partial pressure of oxygen in arterial perfusate; PvO_2, partial pressure of oxygen in venous perfusate; $a\text{-}vO_2$, difference in oxygen contents in arterial (a) and venous (v) perfusate (i.e., oxygen extraction); MVO_2, myocardial oxygen consumption; ACE, acetaminophen (1000 μg i.c.); VEH, vehicle for acetaminophen; *, $P < 0.05$ relative to corresponding value for VEH.

TABLE 34.3
Influence of Acetaminophen (μg i.c.) and Its Vehicle on Left Ventricular Function in the Isolated Guinea Pig Heart Perfused at Constant Pressure (top) or Constant Flow (bottom)

	LVPd (mmHg)				LVPs (mmHg)				+*dP*/*dt* (mmHg/s)				*dP*/*dt* (mmHg/s)			
	C	50	250	1000	C	50	250	1000	C	50	250	1000	C	50	250	1000
VEH	4	4	4	4	59	59	59	61	666	666	672	681	500	500	513	516
s.e.m.	0.3	0.3	0.3	0.3	6	6	6	7	54	54	47	57	30	26	26	23
ACET	4	4	4	4	59	60	67*	90*	672	697	772*	1119*	509	528	613*	722*
s.e.m.	0.3	0.3	0.3	0.3	6	5	5	6	45	50	47	128	24	26	27	78
VEH	2	2	2	2	56	56	56	58	689	689	689	697	527	527	527	527
s.e.m.	0.6	0.6	0.6	0.6	3	3	2	3	46	46	46	48	43	43	43	43
ACET	2	2	2	2	56	56	59*	65*	690	692	731*	794*	530	530	561*	613*
s.e.m.	0.6	0.6	0.6	0.6	3	3	2	4	46	46	49	55	43	43	46	49

Data are means plus or minus one standard error of the mean (s.e.m.; $n = 10$ ea). C, control; LVPd, left ventricular late diastolic pressure; LVPs, left ventricular peak systolic pressure; +*dP*/*dt*, first derivative of ventricular systolic pressure; *dP*/*dt*, first derivative of ventricular diastolic pressure; VEH, vehicle; ACET, acetaminophen; *, $P < 0.05$ relative to corresponding control value.

TABLE 34.4
Influence of Acetaminophen (µg i.c.) and Its Vehicle on Coronary Circulation in the Isolated Guinea Pig Heart Perfused at Constant Pressure (top) or Constant Flow (bottom)

	CPP (mmHg)				CPF (ml/min/g)				CVR (mmHg/ml/min/g)				HR (bpm)			
	C	50	250	1000	C	50	250	1000	C	50	250	1000	C	50	250	1000
VEH	50	50	50	50	6.3	6.3	6.3	6.3	7.8	7.8	7.8	7.8	203	201	203	204
s.e.m.	5	4	4	4	0.3	0.3	0.4	0.3	0.5	0.5	0.5	0.5	14	15	15	13
ACET	50	50	50	50	6.5	6.4	5.9	4.8	7.6	7.6	8.5*	10.6*	206	203	202	203
s.e.m.	4	4	4	4	0.3	0.3	0.3	0.4	0.4	0.4	0.5	0.7	14	15	15	13
VEH	49	49	49	49	6.7	6.7	6.7	6.7	7.4	7.4	7.4	7.4	204	204	204	204
s.e.m.	5	5	5	5	0.3	0.3	0.3	0.3	0.7	0.7	0.7	0.7	10	10	10	10
ACET	49	49	49	52*	6.7	6.7	6.7	6.7	7.4	7.4	7.4	7.8*	204	204	204	204
s.e.m.	5	5	5	5	0.3	0.3	0.3	0.3	0.7	0.7	0.7	0.8	10	10	10	10

Data are means plus or minus one standard error of the mean (s.e.m., $n = 10$). C, control; CPP, coronary perfusion pressure; CPF, coronary perfusate flow; CVR, calculated coronary vascular resistance; HR, heart rate; *, $P < 0.05$ relative to corresponding control values.

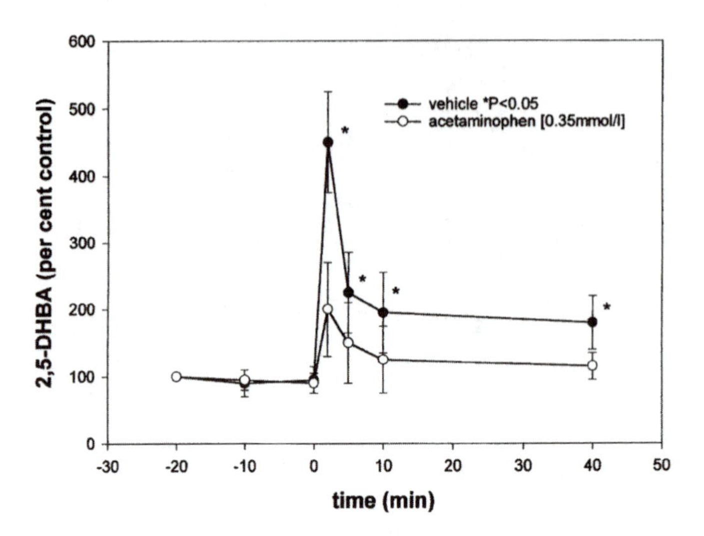

FIGURE 34.3 Changes in coronary venous effluent concentrations of 2,5-dihydroxybenzoic acid (2,5-DHBA), as a function of time of perfusion, in the absence and presence of 0.35 mmol/l acetaminophen (therapeutic dose) expressed as a percentage of the baseline control values. Note: 20 min (10 min postadministration of salicylate); −10 min (20 min postadministration of acetaminophen); 0 min (after 20 min ischemia); 0 to 40 min (during reperfusion). (From Merrill, G. et al., *Amer. J. Physiol.*, 280, H2631–H2638, 2001. With permission.)

34.3.3 Acetaminophen and Hydroxyl Radicals

Metabolically, vehicle- and acetaminophen-treated hearts did not differ significantly. Perfusate gases and pH were similar throughout the experiment in both groups of hearts. Changes in vehicle-treated hearts were parallel to changes in acetaminophen-treated hearts. For example, the pH of coronary venous effluent decreased significantly from about 7.20 ± 0.01 to 6.90 ± 0.01 in both groups between postdrug and ischemia.

In vehicle-treated hearts hydroxyl radicals increased to about $400 \pm 50\%$ of their corresponding predrug/postdrug control values after 2 min of reperfusion. In the presence of 0.35 mmol/l acetaminophen this burst was reduced to approximately $200 \pm 25\%$ ($P < 0.05$) of its corresponding control value. By 40 min reperfusion 2,5-DHBA concentrations in acetaminophen-treated hearts were not significantly different from their corresponding predrug/postdrug values. In vehicle-treated hearts 2,5-DHBA concentrations after 40 min of reperfusion were still modestly, but significantly elevated above their corresponding predrug/postdrug values. Thus, acetaminophen significantly attenuated the production/release of hydroxyl radicals during reperfusion (Figure 34.3).

34.3.4 Acetaminophen and Peroxynitrite Anion

The production of blue light (chemiluminescence) seen in vehicle-treated hearts was abolished during ischemia and significantly attenuated during reperfusion in the presence of 0.35 mmol/l acetaminophen. By 40 min of reperfusion chemiluminescence was evident in acetaminophen-treated hearts but was still significantly less than that found in vehicle-treated hearts. For example, during the early minutes of reperfusion, the peak cpm for blue light were $6.8 \pm 0.9 \times 10^{-6}$ and $1.4 \pm 0.1 \times 10^{-6}$ ($P < 0.05$), respectively, for vehicle- and acetaminophen-treated hearts. These differences persisted through the first 10 min of reperfusion, but diminished thereafter. These results were corroborated by the integrated cpm, areas beneath the respective chemiluminescence curves, and times for the curves to achieve maximal/half-maximal values (Figure 34.4).

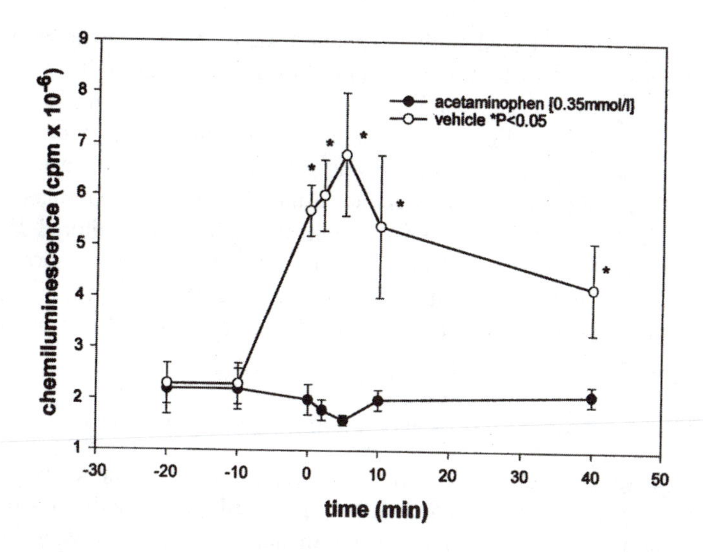

FIGURE 34.4 Luminol-mediated chemiluminescence in the absence (vehicle) and presence of 0.35 mmol/l of acetaminophen (therapeutic dose). In the presence of a donor of nitric oxide (NO), 3-morpholinosydnominine (SIN-1), NO, and superoxide combine to produce peroxynitrite. Peroxynitrite oxidizes luminol to produce blue light (chemiluminescence). Note: –20 min (predrug baseline conditions); –10 min (20 min postadministration of acetaminophen); 0, 2, 5, and 10 min (reperfusion after 20 min of ischemia). (From Merrill, G. et al., *Amer. J. Physiol.*, 280, H2631, 2001. With permission.)

34.4 DISCUSSION

34.4.1 Choice of Langendorff Preparation

We chose to do these initial experiments in isolated, perfused rodent hearts for several reasons. First, we are familiar with these preparations and techniques.[26–28] Second, rodents are relatively inexpensive to purchase and maintain. Third, we wanted to avoid the potentially confounding effects that can occur using *in vivo* preparations where whole blood is the organ/tissue perfusate. And finally, we wanted to eliminate intervening neurogenic effects. We chose the Bunger et al.[14–16] Langendorff preparation because it has *in vivo*-like properties, e.g., it can autoregulate coronary flow, and because phenomena such as "preconditioning" (often studied in isolated rat and rabbit hearts) was not of interest to us in this investigation.

34.4.2 Effects of Acetaminophen on the Heart

To the author's knowledge, ours are the first experiments aimed at determining if acetaminophen has direct effects on the mammalian myocardium and coronary circulation. One can find references in the literature to the lack of effects of acetaminophen in the mammalian cardiovascular system,[12] but such studies were not physiological, and were not designed to answer this question. Moreover, such reports were based more on clinical experience than on scientific methodology, and are more or less anecdotal. Still, these and related reports have led to broad assertions that acetaminophen has no effects in the human cardiovascular system.[4] Thus, they are responsible, in part, for delaying/averting a more thorough investigation of the cardiovascular properties of acetaminophen.

Two of our main concerns were to avoid concentrations of acetaminophen that are known to be toxic in humans (>300 μg/ml; 2.1 mmol/l), i.e., to stay within what we are calling the safe-to-therapeutic range of concentrations (e.g., 20 to 120 μg/ml; 0.1 to 0.8 mmol/l), and to identify effects that were not complicated or obscured by bloodborne elements (cell, platelets, macromolecules) and neurohumorals (an impossibility when using whole blood perfusates, and when studying the heart *in vivo* or *in situ*). In our initial studies we administered bolus injections of acetaminophen into the perfusion lines approximately 30 to 40 cm upstream to the cannulated hearts (dead space volume of approximately 30 ml). This was done when hearts were perfused at controlled flow rates (6 to 7 ml/min). Under such conditions, exact circulating concentrations of agents can be calculated with great precision. Bolus injections of 50, 250, and 1000 μg were routinely used. Only the highest doses caused statistically significant, reproducible results that included coronary vasoconstriction and cardiac positive inotropy. These results were replicated when we monitored coronary flow and controlled coronary perfusion pressure. Thus, we can safely conclude that in the perfused Langendorff guinea pig heart preparation, acetaminophen boluses of 250 to 1000 μg produced modest but statistically significant coronary vasoconstriction and increments in cardiac contractility.

34.4.3 Effects of Acetaminophen during Ischemia and Reperfusion

Ultrastructurally and functionally the reperfused myocardium appeared to be preserved in the presence of acetaminophen. Swollen mitochondria, myofibrillar contraction bands, and disrupted membranes were not as evident in acetaminophen-treated hearts as they were in vehicle-treated hearts. Acetaminophen-treated hearts developed greater left ventricular pressures during ischemia and reperfusion than vehicle-treated hearts. Contractility was also significantly greater in the former hearts. Ultrastructurally, such a clear difference was not evident during ischemia. Thus, whatever the mechanism of cardioprotection, it was not as evident morphologically during ischemia as it was during reperfusion.

Mechanistically, acetaminophen appeared to attenuate the production/release of both hydroxyl radicals and peroxynitrite during ischemia and reperfusion. Both of these agents have been implicated in the tissue damage caused by ischemia/reperfusion injury.[20,24,29–31] Thus, the actions of acetaminophen probably included an antioxidant action against the tissue-damaging effects of hydroxyl radical and peroxynitrite anion. If these and other agents accumulated in sufficient quantities in the ischemic interstitium and other compartments to damage sarcolemmal and other membranes, it is reasonable to assume that the phenolic, antioxidant properties of acetaminophen antagonized such actions. This is the most logical explanation for the observed ultrastructural effects of acetaminophen. The acetaminophen-mediated preservation of ultrastructure would in turn help explain the preservation/protection of mechanical function, i.e., contractility and developed pressure. These studies are ongoing in our laboratory, and we have additional preliminary evidence that the antioxidant actions of acetaminophen might also protect the electrical stability of the myocardium; i.e., might block the arrhythmogenic effects of ischemia/reperfusion. More work is needed to sort out this possibility. Still, these results reveal cardioprotective properties that have not been reported previously.

34.4.4 Administration of Acetaminophen during Ischemia

This chapter has focused on results of experiments obtained when acetaminophen was administered 20 min prior to the onset of 20 min global, low-flow myocardial ischemia. We have additional, as yet incompletely analyzed data, suggesting that administering acetaminophen in the same concentration (0.35 mmol/l) midway through the period of ischemia is also effective in providing cardioprotection during reperfusion. Comparing these two sets of results reveals the need for a more complete, temporal evaluation of the cardioprotective actions of acetaminophen. One of our next projects is to administer the compound chronically for several days (weeks) to *intact* guinea pigs, then to compare responses in isolated hearts from treated and untreated animals. We are currently looking at the efficacy of acetaminophen in hearts that have been challenged by means other than ischemia/reperfusion (e.g., hypoxia/reoxygenation, barbiturate-induced cardiac depression). Results are encouraging but too preliminary to discuss here.

34.5 SUMMARY AND CONCLUSIONS

In isolated, perfused rodent hearts, acetaminophen has modest, but statistically significant, concentration-dependent cardiac and coronary vascular properties. It displays cardioprotection, both ultrastructurally and functionally, in the reperfused, stunned myocardium. One mechanism of protection appears to involve antioxidant actions against both hydroxyl radical and peroxynitrite. More work is needed to confirm these results, and to extend them to include regional myocardial ischemia/reperfusion, chronic vs. acute administration of acetaminophen, and experiments in whole animals. Collectively, results from such investigations should guide us to the future design of experiments in cardiac patients.

REFERENCES

1. Ameer, J. and Greenblatt, J., Pharmacological review of paracetamol (acetaminophen), *Ann. Intern. Med.,* 87, 202, 1977.
2. Fairbrothers, J. E., Paracetamol: a comprehensive description, in *Analytical Profiles of Drug Substances,* K. Florey, Ed., Vol. 3, Academic Press, New York, 1974, 1–109.
3. Prescott, L. F., *Paracetamol (acetaminophen): A Critical Bibliographic Review,* Taylor & Francis, London, 1996.
4. Gilman, A. G., Rall, T. W., Nies, A. S., and Taylor, P., Goodman and Gilman's *The Pharmacological Basis of Therapeutics,* Pergamon Press, New York, 1990.
5. Bolli, R., Jeroudi, M. O., Patel, B. S., Aruoma, O. I., Halliwell, B., Lai, E. K., and McKay, P. B., Marked reduction of free radical generation and contractile dysfunction by antioxidant therapy begun at the time of reperfusion: evidence that myocardial stunning is a manifestation of reperfusion injury, *Circ. Res.,* 65, 607, 1989.
6. Cave, A. C., Collis, C. S., Downey, J. M., and Hearse, D. J., Improved functional recovery by ischaemic preconditioning is not mediated by adenosine in globally ischaemic isolated rat heart, *Cardiovasc. Res.,* 27, 663, 1993.
7. Ferdinandy, P., Danial, H., Ambrus, I., Rothery, R. A., and Schultz, R., Peroxynitrite is a major contributor to cytokine-induced myocardial contractile failure, *Circ. Res.,* 87, 241, 2000.
8. National Family Opinion Organization, Center for Cardiovascular Education, Inc., New Providence, NJ, available at http:/www.heartinfo.com/features/analgesics/index,htm., 1996–2000.
9. Dinis, T. C. P., Madeira, V. M. C., and Almeida, L. M., Action of phenolic derivatives (acetaminophen, salicylate, and 5-aminosalicylate) as inhibitors of membrane lipid peroxidation and as peroxyl radical scavengers, *Arch. Biochem. Biophys.,* 315, 161, 1994.
10. Van Dyke, K., Sacks, M., and Qazi, N., A new screening method to detect water-soluble antioxidants: acetaminophen and other phenols react as antioxidants and destroy peroxynitrite-based luminol-dependent chemiluminescence, *J. Biolumin. Chemilumin.,* 13, 339, 1998.
11. Van Dyke, K., McConnell, P., and Marquardt, L., Green tea extract and its polyphenols markedly inhibit luminol-dependent chemiluminescence activated by peroxynitrite or SIN-1, *Luminescence,* 15, 37, 2000.
12. Brent, J., New ways at looking at an old molecule, *Clin. Toxicol.,* 34, 149, 1996.
13. Nakamoto, K., Kamisaki, Y., Wada, K., Kawasaki, H., and Itoh, T., Protective effect of acetaminophen against acute gastric mucosal lesions induced by ischemia-reperfusion in the rat, *Pharmacology,* 54, 203, 1997.
14. Bunger, R., Haddy, F. J., and Gerlach, E., Coronary responses to dilating substances and competitive inhibition by theophylline in the isolated, perfused guinea pig heart, *Pfluegers Arch.,* 358, 212, 1975.
15. Bunger, R., Haddy, F. J., Querengasser, A., and Gerlach, E., An isolated guinea pig heart preparation with *in vivo*-like features, *Pfluegers Arch.,* 353, 317, 1975.
16. Schrader, J., Haddy, F. J., and Gerlach, E., Release of adenosine, inosine, and hypoxanthine from the isolated guinea pig heart during hypoxia, flow-autoregulation, and reactive hyperemia, *Pfluegers Arch.,* 369, 1, 1977.
17. Walker, M. J. A. et al., The Lambeth Conventions: guidelines for the study of arrhythmias in ischemia, infarction, and reperfusion, *Cardiovasc. Res.,* 22, 447, 1988.

18. Grootveld, M. and Halliwell, B., Aromatic hydroxylation as a potential measure of hydroxyl radical formation *in vivo*, *Biochem. J.,* 237, 499, 1986.
19. Das, K. D., Cordis, G. A., Rao, P. S., Liu, Z., and Maity, S., High-performance liquid chromatographic detection of hydroxylated benzoic acids as an indirect measure of hydroxyl radical in heart: its possible link with the myocardial reperfusion injury, *J. Chromatogr.,* 19, 209, 1992.
20. McHugh, N. A., Merrill, G. F., and Powell, S. R., Estrogen diminishes postischemic production of hydroxyl radical, *Am. J. Physiol.,* 274, H1950, 1998.
21. Kuppusamy, P. and Zweier, J. L., Xanthine oxidase reduces hydrogen peroxide to hydroxyl radical, *J. Biol. Chem.,* 264, 9880, 1989.
22. Bolli, R. and McKay, P. B., Use of spin traps in intact animals undergoing myocardial ischemia and reperfusion: a new approach to assessing the role of oxygen radicals in myocardial "stunning," *Free Radical Res. Commun.,* 9, 163, 1990.
23. Liu, X., Tosaki, A., Engelman, R. M., and Das, D. K., Salicylate reduces ventricular dysfunction and arrhythmias during reperfusion in isolated rat hearts, *J. Cardiovasc. Pharmacol.,* 19, 209, 1992.
24. Powell, S. R. and Hall, D., Use of salicylate as a probe for hydroxyl radical formation in isolated ischemic rat hearts, *Free Radical Biol. Med.,* 9, 133, 1990.
25. Powell, S. R. and Tortolani, A. J., Recent advances in the role of reactive oxygen intermediates in ischemic injury, *J. Surg. Res.,* 53, 417, 1992.
26. Merrill, G. F., Haddy, F. J., and Dabney, J. M., Adenosine, theophylline, and perfusate pH in the isolated, perfused guinea pig heart, *Circ. Res.,* 42, 225, 1978.
27. Young, M. A. and Merrill, G. F., Differential effects of adenosine and hypoxia on potassium-induced dilation in the isolated, perfused guinea pig heart, *Blood Vessels,* 19, 292, 1982.
28. Wei, H. M., Kang, Y. H., and Merrill, G. F., Coronary vasodilation during global myocardial hypoxia: effects of adenosine deaminase, *Am. J. Physiol.,* 254, H1004, 1988.
29. Liu, P., Hock, C. E., Nagele, R., and Wong, P. Y. K., Formation of nitric oxide, superoxide, and peroxynitrite in myocardial ischemia-reperfusion injury in rats, *Am. J. Physiol.,* 272, H2327, 1997.
30. Schulz, R., Dodge, K. L., Lopaschuk, G. D., and Clanachan, A. S., Peroxynitrite impairs cardiac contractile function by decreasing cardiac efficiency, *Am. J. Physiol.,* 272, H1212, 1997.
31. Yasmin, W., Strynadka, K. D., and Schulz, R., Generation of peroxynitrite contributes to ischemia-reperfusion injury in isolated rat hearts, *Cardiovasc. Res.,* 33, 422, 1997.

Reference added in proof:

Merrill, G., McConnell, P., Van Dyke, K., and Powell, S., Coronary and myocardial effects of acetaminophen: Protection during ischemia-reperfusion, *Am. J. Physiol.,* 280, H2631, 2001.

35 Luciferase Luminescence Can Be Used to Assess Oxidative Damage to Plasmid DNA and Its Prevention by Selected Fruit Extracts

Knox Van Dyke, Candace L. Ogle, and Mark J. Reasor

CONTENTS

35.1 INTRODUCTION

There have been a number of studies that indicate that peroxynitrite, a strong oxidizing chemical, causes nicking to plasmids or genomic DNA.[1–3] Such nicks are thought to be in the sugar-phosphate backbone of the DNA structure.[4] Peroxynitrite has been shown to cause breaks in chromosomal DNA when cells are exposed it.[5] However, if damage is too extensive to the DNA, the cellular repair mechanisms might not reverse the damage. A mutation might continue or the death of a cell might develop via activation of the PARS mechanism, which has been studied extensively by Szabo and associates.[6]

We have been interested in developing a strategy that might be able to prevent the actions of peroxynitrite or its production by annihilating one or both of the products that form this reactive molecule, namely, the free radicals superoxide and nitric oxide. We developed an assay using the drug linsidomine (SIN-1), which generates both necessary radicals, and because of their high affinity for each other, produces peroxynitrite anion or ($OONO^-$). The peroxynitrite so produced reacts with luminol, which generates blue light indicating the oxidation of luminol. This light, and indirectly

0-8493-0719-8/02/$0.00+$1.50

the level of peroxynitrite, can be quantified using a luminometer. The light can be inhibited or prevented from occurring by adding various polyphenols of plant origin, e.g., catechins from tea. In a second scenario, peroxynitrite may later react with carbon dioxide, which changes its reactivity toward various polyphenolic targets from oxidation to nitration of those compounds.[7] We found a variety of different polyphenolic substances in green tea that are very useful in inhibiting the formation of peroxynitrite.[8]

Therefore, we thought that these same compounds would be useful in preventing the damage to DNA caused by peroxynitrite itself or even with other strong oxidants such as hypochlorite. Further, we produced extracts of a variety of common fruits which were tested for their use in preventing extensive damage to plasmid DNA. We used an assay that would produce a simple but useful end point in determining whether DNA could be repaired or whether the damage to DNA was so extensive that repair was not possible. In the assay, a plasmid containing a luciferase reporter gene is incubated with competent *Escherichia coli* in the presence or absence of fruit extract. If the plasmid infects the *E. coli*, luciferase can be expressed, which can be easily assayed via its production of light. This can only happen if the plasmid DNA is not extensively damaged or, if damaged slightly, can be repaired by the bacterial repair mechanisms. These results were compared to experiments that quantified the protection against nicking of plasmid DNA by peroxynitrite afforded by these extracts. The application of this research is that perhaps we could use a simple nutritional extract to prevent dangerous mutations that might persist without repair.

35.2 MATERIALS AND METHODS

35.2.1 CHEMICALS

Plasmid DNA was purchased from Aldevron (Fargo, ND, e-mail, DNA@aldevron.com). DNA was purchased in 10-mg batches and supplied as a sterile and buffered solution at a concentration of 1 mg/ml buffer. The particular plasmid used was the G-Wiz™- luciferase high-expression plasmid, which contains a kanamycin resistance marker as well as the luciferase gene. The plasmid uses a cytomegalovirus infectious mechanism that infects both Gram-negative bacteria, e.g., *E. coli*, and mammalian cells. The plasmid has more than 6300 base pairs with a variety of restriction enzyme cutting sites available for possible gene insertions and deletions. A plasmid gene map displaying only the major genes relevant to this project is presented in Figure 35.1. Peroxynitrite was purchased from Upstate Biotechnology (Lake Placid, NY, e-mail, techserv@upstatebiotech.com). The original concentration of peroxynitrite is 138 m*M*. The solution contains 0.3 *M* NaCl and 0.3 *M* NaOH. Peroxynitrite is stored frozen at –80°C until use.

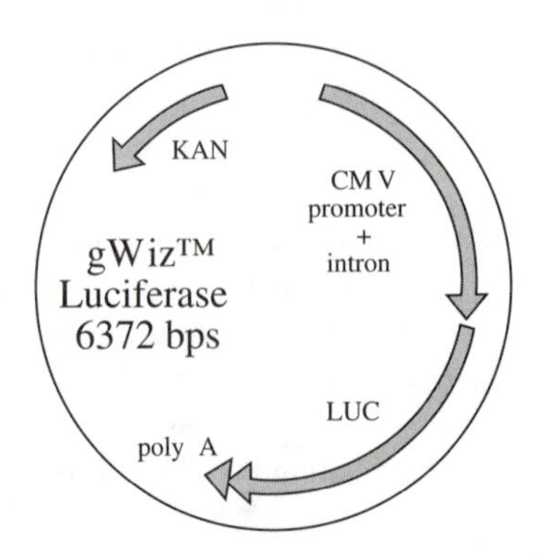

FIGURE 35.1 A incomplete gene map of the G-Wiz luciferase plasmid with a kanamycin resistance marker.

35.2.2 Extraction of Fruit and Incubation with Plasmids

Extraction of various fruits has been previously described (see Chapter 33). Briefly, the different fruits were homogenized in a solution of 50% ethyl alcohol and water using a Tekmar Tissuemizer to release their contents. They were then filtered and frozen directly, and in some cases, freeze-dried. The preparations were stored at –80°C.

35.2.3 Nicking Plasmid DNA with Peroxynitrite

The plasmid (50 μg) in Tris-EDTA buffer was incubated with peroxynitrite (final concentration of 6.9 m*M*) in the absence or presence of dilutions of the various fruit extracts. The samples were incubated at 37°C for 1 h to allow time for peroxynitrite-induced damage to occur to the DNA.

35.2.4 Electrophoresis of Plasmid DNA

The DNA samples were subjected to gel electrophoresis for 35 min using 0.8% agarose E gel cassettes (prestained with ethidium) and an E gel power supply. Skipping the first and last of 12 wells, 20 μl of each sample was loaded into each well. Upon completion of electrophoresis, the DNA was viewed under ultraviolet light to cause the ethidium-stained DNA to fluoresce. Nicked or broken DNA has a relaxed morphology and therefore travels a shorter distance from the origin than the native supercoiled plasmid. According to Aldevron, less than 5% of the plasmid will originate in the relaxed or nicked form. By using a FluoroChem imaging device for stained plasmid DNA the fluorescent spots of DNA were quantified using AlphaEase™ software 2D densitometry.

35.2.5 FluorChem Fluorescent Imaging

Using the FluoroChem imaging device for stained plasmid DNA, we quantified fluorescent spots with the ultraviolet box set on ethidium settings for the agarose gels using AlphaEase software 2D densitometry. The instrumentation is available from the Alpha Innotech Corporation (San Leandro, CA). The densitometry was performed using the software and percent protection calculated by comparing the density of the ethidium-based fluorescent signal from the nicked DNA to the density of the signal from the non-nicked plasmid DNA.

35.2.6 Infection of Competent *E. coli* with Plasmids to Test for Infectivity or Peroxynitrite-Induced Damage to Plasmid DNA

Escherichia coli strain DH1 α was made competent for infection with the G-Wiz plasmid DNA using protocol SOP #6 from Massey Cancer Center. Briefly, *E. coli* is cultured in Lacto Bacillus broth to the early log phase, which produces a turbidity of 0.2 to 0.3 o.d. units at 600 nm in 4 h. The culture is placed on ice for 15 min. The solution is centrifuged in 50-ml tubes at 3000 rpm at 2°C for 10 min. The supernatant solution is discarded and the pellet is resuspended in 100 m*M* $CaCl_2$, using half the original culture volume. This solution is incubated on ice for 10 min and centrifuged as before to produce a pellet. The resulting supernatant is again completely discarded. For every 20 ml of the original culture solution 1 ml of 100 m*M* $CaCl_2$ and 0.43 ml of 1:1 glycerol:sterile water is added to the cell pellet. The *E. coli* is resuspended in this solution, split into 0.5-ml aliquots, and frozen at –80°C.

Competent *E. coli* are thawed on ice for 10 min. DMSO (1.5 μl) is added and mixed gently. Plasmid DNA (100 to 500 ng) that had been incubated with peroxynitrite with or without fruit extracts (at a 1:2.5 dilution) is added. The mixture in small tubes is incubated on ice for 30 min. The tubes with the mixture of competent bacteria and plasmids is heat-shocked at 42°C for 90 s. The mixture is chilled on ice for 5 min. The entire tube of cells is added to 20 ml Luria broth without any antibiotics and incubated for 1 h at 37°C in a CO_2 incubator with shaking to allow antibiotic resistance to be expressed. Kanamycin is added at a final concentration of 5 μg/ml final solution of LB broth. The 50 ml of solution is incubated for 1 to 2 days to produce enough bacteria to test for the luciferase gene product.

35.2.7 Assay for Luciferase Gene Product

Bacterial solution (40 ml, with or without infected plasmids) is placed in a 50-ml centrifuge tube and centrifuged at 3000 rpm at 2°C for 10 min. The supernatant solution is decanted. A solution (200 μl) of 1% Triton X-100 with DTT and Tris-EDTA 0.1 *M* at pH = 8 is added and the lysed mixture sonicated briefly. This solution is added to 200 μl of a luciferin/ATP glow solution purchased from Promega (Madison, WI). The assay solution is placed in 3-ml round-bottom Berthold cuvettes and assayed for 50 s at room temperature in a LB9505 C Berthold Luminometer. The integrated light signal is assessed using KINB software. The blank signal when no luciferase is present is usually in the order of 1.2×10^3 counts and when luciferase is present the signal is 1.7×10^7 counts, which is more than a 10,000-fold difference. The luminescence end point produces a clear difference. Either there is no luciferase activity in the bacteria or there is good luciferase activity. Therefore, when luciferase activity appears in a preparation, it means that the plasmid infected the *E. coli* and was intact or repaired enough by the *E. coli* to be functional. If there is no activity, it means that the plasmid could not infect and be repaired by the *E. coli* DNA-repair system. This approach evaluates whether the antioxidants from the different fruit extracts are effective in preventing the damage to the DNA to the extent the antioxidants and/or *E. coli* repair system could produce an infective and genetically active plasmid.

35.3 RESULTS AND DISCUSSION

A representative experiment evaluating the actions of the fruit extracts is shown in Figure 35.2. The electrophoretic pattern of the control plasmid g-Wiz luciferase, which is in an intact supercoiled state, can be seen as almost a single band at the far right. The peroxynitrite-nicked plasmid is to the left of the plasmid control and it migrates just above the control plasmid. The effects of the raspberry extract at various dilutions, 1:5, 1:10, 1:25, and 1:50 are seen in order to the left of the peroxynitrite-nicked plasmid. The raspberry extract is somewhat protective against nicking at all dilutions but the 1:10 seems to be the most protective, with 1:5 and 1:25 dilutions appearing fairly similar, and the 1:50 dilution giving the least protection. The blueberry extract followed a more logical concentration–response curve. The most protective was the 1:5 extract and with further dilutions of 1:10, 1:25, or 1:50 the protection of the plasmid DNA from peroxynitrite-based nicking was progressively diminished. These trends occurred the same way with multiple experiments.

In Figure 35.3 is displayed all the fruit extracts at different dilutions with error bars representing standard error of the mean from three to five separate experiments. Extracts of raspberry, blueberry, black grape, dark raisin, golden raisin, and dried cranberry were all effective in conveying protection from oxidative nicking caused by peroxynitrite. Somewhat less effective were extracts of white grape and cranberry. Both raspberry and cranberry displayed less protection at the highest dose and became more active at the next dilution. It is interesting that dried cranberry (craisins) were more potent than cranberry that was not dried. The drying may cause polyphenols to polymerize or antioxidants may have been added to preserve the craisins. It is not clear why the raspberry and blueberry extracts were less protective at the most concentrated dilution than at the next greater dilution. When one uses crude extracts, there is probably a mixture of pro-oxidants and antioxidants and as dilution occurs one of them is more effective than the other.

Table 35.1 presents the effect of the various extracts on the infection of the plasmids into *E. coli* and the expression of the gene product, luciferase. Infection can only occur when the kanamycin resistance contained in the plasmid is expressed by the *E. coli*. Therefore, it can be seen that only black grape, golden raisin, blueberries, and dried cranberry are effective at protecting the plasmid

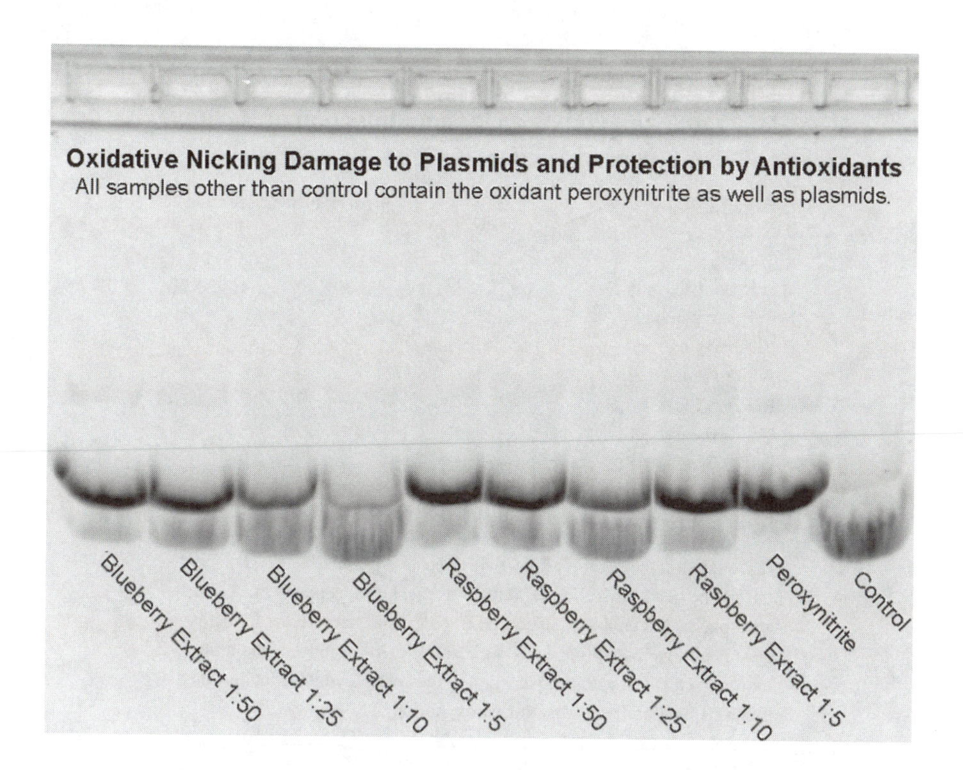

FIGURE 35.2 Peroxynitrite-induced oxidative nicking of plasmids and protection by fruit extracts. Representative electrophoretic separation of native supercoiled plasmid DNA (control, far right lane) from nicked DNA caused by peroxynitrite (next lane to the left). The effects of increasing dilutions of blueberry and raspberry extracts are shown. See details for conditions in Materials and Methods.

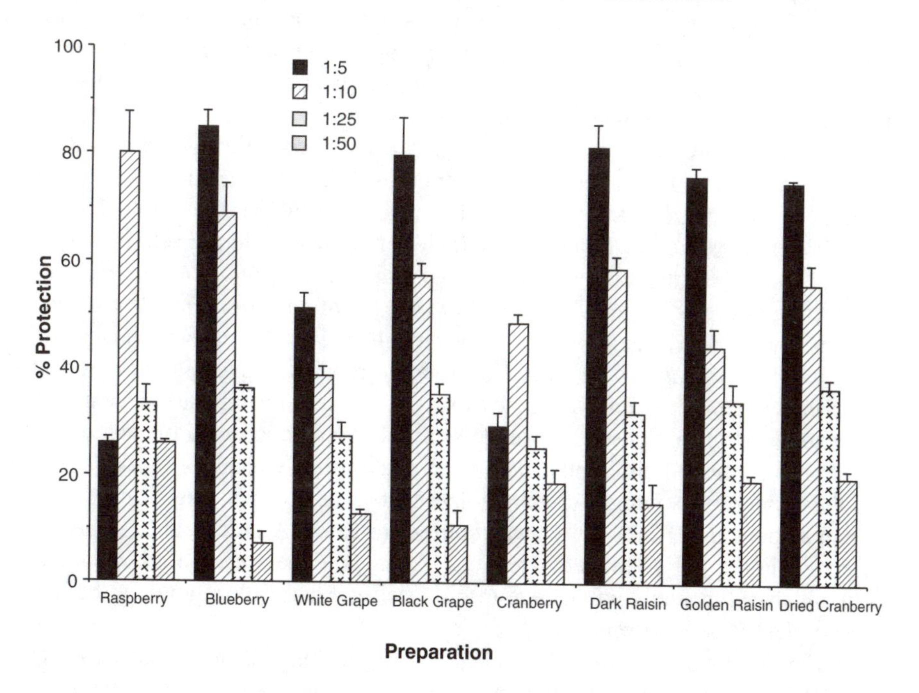

FIGURE 35.3 Percent protection of plasmids from peroxynitrite-induced oxidative nicking by extracts of various fruits.

TABLE 35.1
Effect of Fruit Extracts on Plasmid Infection of *E. coli* and Expression of Luciferase Gene

Preparation	Presence of Luciferase
Control	+
Peroxynitrite only	–
Raspberry extract	+
Blueberry extract	+
White grape extract	–
Black grape extract	+
Cranberry extract	–
Dark raisin extract	–
Golden raisin extract	+
Dried cranberry extract	+

Note: The undamaged plasmid will infect the permissive *E. coli* and impart the gene for luciferase to its host (displayed as positive). A damaged plasmid is unable to infect the *E. coli* and therefore the luciferase gene is not expressed (displayed as negative). All samples contain plasmids and $OONO^-$, with the exception of the control, which contains only plasmids, and no fruit extract or peroxynitrite.

from oxidation sufficiently to produce kanamycin-resistant bacteria, which produce the luciferase activity. The positive control, plasmid without peroxynitrite, is active as expected. The negative control without extract but with plasmids and peroxynitrite grows no bacteria and has no luciferase activity. All other fruit extracts were ineffective in protecting the plasmid from infecting the permissive *E. coli*. The damage inflicted on the plasmid DNA by peroxynitrite may have been totally or partially inhibited by the extracts since infection of the bacteria by the plasmid produced luciferase activity. Since genetic repair mechanisms exist in the *E. coli*, the bacterial repair mechanism may have repaired the small amount of damage left and luciferase activity was expressed. In the case of the extracts that did not allow luciferase activity to occur, the peroxynitrite-based damage was too extensive to repair. It should be noted that the extract of cranberries was ineffective in protecting the plasmid DNA from peroxynitrite attack and produced no luciferase activity. However, dried cranberries were effective in protecting DNA sufficiently so that luciferase activity was expressed. Why would this occur? To dry cranberries to create craisins, the cranberries are placed in heated air. This concentrates the polyphenols that could likely polymerize or metabolize the compounds to make them more effective antioxidants, which might prevent extensive damage to DNA by peroxynitrite, possibly enough so that the *E. coli* could repair the plasmid sufficiently to obtain expression of the luciferase. By analyzing the extract of craisins and comparing them to the extract from cranberries, it might be possible to understand how protection could occur from the same fruit that had been processed differently. The same relationship existed for black grapes and dark raisins, but the opposite was found for white grapes and golden raisins. The exact reason the fresh and dried preparations differ in activity awaits further research. A precedent has already been set by the difference in the polyphenols of green vs. black tea. Green tea is not heated compared to black tea in which polyphenols polymerize to produce the black color; however, both teas come from the same plant, *Camellia sinensis*.

Peroxynitrite causes damage to DNA other than by nicking. It has been demonstrated that peroxynitrite oxidizes or nitrates guanines producing 8-oxoguanines or 8-nitroguanine.[9] Furthermore, these investigators showed that consecutive guanines are readily attacked but certain isolated guanines were oxidized or nitrated even more readily, creating mutational hot spots. A mutational hot spot can be defined as the genetic entity mathematically most likely to produce a mutation. Therefore, the main effects of peroxynitrite are to nick the DNA backbone or oxidize or nitrate the guanines or all of these. To summarize, there are three levels of complexity at which antioxidant/cellular repair mechanisms might occur. First, the antioxidants could prevent the precursors of peroxynitrite (superoxide and nitric oxide) from being generated or from combining to form peroxynitrite. Second, antioxidants could interact with peroxynitrite once it is formed. An example of this would be the effect of uric acid, which reacts with peroxynitrite and forms a variety of compounds including allantoin and a nitrated imidazole.[10] This type of annihilation reaction with an antioxidant could prevent DNA from being damaged or mutated by peroxynitrite. This might include both genomic and mitochondrial DNA. Third, if the DNA was not damaged very much, the repair mechanisms of a cell might repair the minimal damage and the DNA might remain intact, and in the case of plasmids, DNA might become infective under the conditions used in our study. In our work it appears that of the various fruit extracts tested only extracts of blueberry, black grape, golden raisins, and craisins could prevent peroxynitrite damage to plasmid DNA sufficiently to allow infectivity and luciferase production. In contrast, all extracts provided at least partial protection against plasmid nicking by peroxynitrite as evaluated electrophoretically. It is apparent that under the conditions of this study, the effects of the fruit extracts on native plasmid do not predict the functional consequences of the DNA damage.

It is likely that the bacterial repair mechanisms played a role in whether the plasmid would infect and express luciferase activity. Matsumoto et al.[11] have recently demonstrated that Nth DNA glycosylase is involved with the removal of 8-oxoguanines/guanine mispairs in DNA. This would probably be the enzyme used by *E. coli* to repair peroxynitrite-mediated DNA damage and might include 8-nitroguanine, which would likely be repaired by the same DNA glycosylase.

In conclusion, two approaches have been utilized to study the ability of extracts of a variety of fruits to protect plasmid DNA from peroxynitrite-induced damage. Although all the extracts provided some degree of protection against nicking of the DNA, not all were effective in providing enough protection to allow infection of *E. coli* by the plasmid and gene expression to occur. Nevertheless, monitoring infection and gene expression by luciferase luminescence has proved to be a relatively simple method to assess the ability of fruit extracts to inhibit oxidative damage of DNA by peroxynitrite. It should prove equally effective in studying other food products and their individual components.

REFERENCES

1. Ohsima, H., Yoshie, Y., Aurivl, S., and Gilbert, I., Antioxidant and prooxidant action of flavonoids: effects on DNA damage induced by nitric oxide, peroxynitrite, and nitroxyl anion free radical, *Biol. Med.,* 25(9), 1057–1065, 1998.
2. Salgo, M. G., Stone, K., Squadrito, G. L., Battista, J. R., and Pryor, W. A., Peroxynitrite causes nicks in plasmid p BR 322, *Biochem. Biophys. Res. Commun.,* 210, 1025–1030, 1995.
3. La Marr, W. A., Sandman, K. M., Reeve, J. N., and Dedon, P. C., Differential effects of DNA supercoiling on radical-mediated DNA strand breaks, *Chem. Res. Toxicol.,* 10, 118–122, 1997.
4. Yoshie, Y. and Ohshima, H., Synergistic induction of DNA strand breakage caused by nitric oxide with catecholamines: implications for neurodegenerative disease, *Chem. Res. Toxicol.,* 10(9), 1015–1022, 1997.
5. Seis, H. and De Groot, H., Role of reactive oxygen species in cell toxicity, *Toxicol. Lett.,* 64–65, 547–551, 1992.
6. Virag, L., Scott, G. S., Antal-Szalmas, P., O'Connor, M., Okashima, H., and Szabo, C., Requirement of intra-cellular calcium mobilization for peroxynitrite-induced poly (ADP-vibose) synthase activation and cytotoxicity, *Mol. Pharmacol.,* 56(4), 824–833, 1999.
7. Yermilov, V., Yoshie, Y., Rubio, J., and Ohshima, H., Effects of carbon dioxide/bicarbonate on induction of DNA single strand breaks caused by peroxynitrite, *FEBS Lett.,* 399, 67–70, 1996.

8. Van Dyke, K., McConnell, P., and Marquardt, L., Green tea extract and its polyphenols markedly inhibit luminol-dependent chemiluminescence activated by peroxynitrite or SIN-1, *Luminescence,* 15, 37–43, 2000.
9. Tretyakova, N. Y., Burney, S., Wishnok, J. S., Wogan, G. N., and Tannenbaurm, S. R., Peroxynitrite-induced DNA damage in the Sup F gene: correlation with mutational spectrum, *Mutat. Res.,* 447, 287–303, 2000.
10. Santos, C. X., Anjos, E. I., and Augusto, O., Uric acid oxidation by peroxynitrite: multiple reactions, free radical formation and amplification of liquid oxidation, *Arch. Biochem. Biophys.,* 372(2), 285–294, 1999.
11. Matsumoto, Y., Zhang, Q. M., Takao, M., Yasui, A., and Yonei, S., *E. coli* Nth and human HNTH-1DNA glycosylase are involved in removal of 8-oxoguanine from 8 oxoguanine/guanine mispairs in DNA, *Nucl. Acids Res.,* 29(9), 975–981, 2001.

Section VI

Luminescent Imaging

36 Detection Systems Optimized for Low-Light Chemiluminescence Imaging

Mark A. Christenson

CONTENTS

36.1 INTRODUCTION

The luciferase enzyme has provided a powerful tool to examine biological processes within living organisms. By exploiting the ability of this protein to catalyze light emission from a substrate, biological researchers can examine gene expression within a cell, tissue, or whole organism, and measure local ATP release or monitor protein–protein interaction in the cell. Since the stoichiometry of the reaction is one photon emitted per one substrate converted, the light levels produced can be extremely low under some experimental conditions. This chapter describes the current state of the art in imaging technology and the optimization of the imaging conditions required to detect and accurately measure extremely low light level chemiluminescence emission from a variety of biological model systems.

36.2 THE METHOD

The key to good low-light chemiluminescence imaging is to have the best possible signal-to-noise (S/N) ratio for a given number of emitted photons. To achieve the best S/N ratio, the measured signal should be maximized while the all the noise contributions during the measurement should be minimized.[1,2] We can isolate the approach into three different components; the generation of photons from the source, the collection of photons by the optics, and the measurement of photons by the detector. We can then define the signal per pixel (S) as follows:

$$S = \text{Photons emitted from a source} * \text{Collection Efficiency} * \text{Detection Efficiency} \quad (36.1)$$

0-8493-0719-8/02/$0.00+$1.50

Similarly, the total noise per pixel, N, can be separated into terms describing contributions from the source, the optics, and the camera:

$$N = (N_P^2 + N_B^2 + N_C^2)^{1/2} \quad (36.2)$$

where
N_P = photon shot noise
N_B = noise from the background light collected by the optics
N_C = noise from the camera

The S/N measurement will describe the quality of the data collected. The ideal condition for imaging is where the background photon noise and the camera noise are eliminated completely, leaving the inherent noise of the measured photon signal as the only source of noise. This is called "shot noise"-limited imaging and the S/N ratio can be described as follows:

$$\text{S/N (shot noise limited)} \approx \text{Signal}/N_P = \text{Signal}/\text{Signal}^{1/2} = \text{Signal}^{1/2} \quad (36.3)$$

As the signal increases, the S/N also increases in proportion to the square root of the measured signal. In practice, shot noise–limited imaging can be very difficult to achieve, but the discussions in this chapter will focus on optimizing the imaging setup to approach this limit. Given optimized conditions, the researcher will be able to see extremely low light chemiluminescent signals and accurately measure their magnitudes.

The source of photons for chemiluminescence imaging is the breakdown of the luciferin substrate that is catalyzed by the luciferase enzyme. Maximizing the number of photons emitted from the cells or tissue of interest is an important issue, requiring an understanding of the different luciferase enzymes, the levels of that enzyme within a cell or tissue, the ATP levels, the substrate concentration, and the local oxygen levels.[3] Methods of increasing the photon emission rate from targets will not be addressed in this chapter. Nevertheless, the experimentalist should understand the factors affecting the photon generation rate *in situ* because this will affect the overall signal dynamics.

To examine the efficiency of light collection, we need to consider each of the two basic configurations for imaging chemiluminescence—microscope-based imaging and macroscopic imaging. The former is used for high-resolution, single-cell experiments, whereas the latter is used primarily for bulk target imaging like tissues or whole plants. Let us consider each of these in turn.

36.3 MAXIMIZING LIGHT COLLECTION THROUGH THE MICROSCOPE

In chemiluminescence imaging, the microscope should be set up to eliminate all excess optical elements since every interface results in a net reduction in photon number. Every surface that is coated with an antireflection coating will still lose approximately 1% of the light, while uncoated optics can lose 5% or more. The optimization should include the removal of all dichroic beam splitters and filters from the emission pathway. The user should choose the optical path in the microscope that minimizes the number of optical elements prior to the camera port, for example, the bottom port that is now found on the newer-generation inverted microscopes.

The main function of the objective in chemiluminescence imaging is to collect the maximum amount of light possible from the sample. The key measure of the ability of an objective to collect light is its numerical aperture (NA). The NA is defined as follows:

$$\text{NA} = n_0 \sin\theta \quad (36.4)$$

where n_0 = refractive index of the medium between the lens and the sample and θ is the half angle of light cone collected by the lens through the medium.

The larger the NA of the objective, the more light is collected. In fact, the amount of light collected is proportional to the NA^2 so small changes in NA can have large effects on the results.

Another factor affecting the measurement is the magnification. For a tenfold increase in magnification, the signal will be spread out over a 100-fold larger area so the net intensity per pixel will be reduced by 100-fold. Therefore, one should use an objective that gives the highest NA-to-magnification ratio. In fact, the light-gathering power (LGP) of an objective can be described as follows:[4]

$$\text{LGP} = 10{,}000 * (\text{NA}/\text{Magnification})^2 \quad (36.5)$$

Therefore, if the specimen can be visualized at sufficiently high resolution at 40× magnification, it would be better to use the 40× objective, which has an NA of 1.3 instead of a 100× objective with an NA of 1.4. Since resolution in chemiluminescence imaging tends to be limited by the diffusion of the luciferase during the typical long exposures, the user would do well to use the lower-power objective in most cases.

It may also be possible to select for an objective that has the highest throughput at the peak emission of the chemiluminescent substrate to achieve superior results.[5] In addition, the user should carefully clean any dirt, oil, or other residue from the objectives according to the manufacturer's recommended procedures. The microscope should be set up to direct 100% of the light to the camera port when measurements are to begin.

36.4 MAXIMIZING LIGHT COLLECTION THROUGH A PHOTOGRAPHIC LENS

In macroscopic imaging, a camera is mounted over the target and the light is collected through a standard F-mount-type lens (for charge-coupled devices, or CCDs, with >1 in. diagonal) or a C-mount type lens (for CCDs with <1 in. diagonal). The lens should have the highest light collection efficiency possible, which is specified by the lens f-number. The f-number is defined as follows:[6]

$$\text{f-number} = \text{lens focal length}/\text{entrance pupil diameter} \quad (36.6)$$

and

$$\text{NA}_{\text{lens}} \approx (2 * \text{f-number})^{-1} \quad (36.7)$$

Because the NA_{lens} is inversely related to the f-number, then the best light collection will be accomplished using the lens with the smallest possible f-number. In the case of standard F-mount lenses, the Nikkor 50-mm focal distance lens (Nikon USA, Melville, NY) with an f-number of 1.2 is recommended. For a standard C-mount lens, the Universe Kogaku 25-mm focal distance lens (Universe Kogaku, Oyster Bay, NY) with an f-number of 0.95 is recommended.

The relative light-collecting ability of one lens can be compared with another by taking the inverse ratio of the f-numbers squared. As an example, comparing two lenses with f-numbers of 0.95 and 1.2 on the same target:

$$\text{Light Collection Ratio} = 1.2^2/0.95^2 = 1.6 \quad (36.8)$$

So the f/0.95 lens would collect 1.6 times as much light as the f/1.2 lens.

One trick that can be exploited to increase the light-gathering ability of the lens is to add a macro ring or an extender tube to the back of the lens. The effect of this action can be predicted from the relationship of focal length to image and object distances from the lens:[6]

$$1/\text{focal length} = (1/\text{object to lens}) + (1/\text{image to lens}) \quad (36.9)$$

Note: This thin lens formula is valid for thick lenses when 1° and 2° principal points are used as the internal lens reference points for measuring distance.

Adding the macro ring causes an increase in the image-to-lens distance. Since the focal length of the optic is fixed, this lens movement is compensated for by a reduction in the object-to-lens distance. This means that the object can be brought closer to the lens surface when a macro ring or extender tube is used. By bringing the target closer to the lens, there are two effects. First, the solid angle of light collected from a point source in the field of view is increased, enhancing sensitivity. Second, the object is magnified to a higher degree, which can be very useful for smaller targets. The higher magnification will spread out the signal, reducing the intensity at each pixel, but this can be compensated for by increasing the binning level on the CCD (see discussion below).

36.5 MAXIMIZING THE NUMBER OF DETECTED PHOTONS

Now that the experimental setup has been optimized to deliver the highest number of photons to the detection system, it would be senseless to throw away all this work with a low-sensitivity detector. The efficiency with which the camera measures a signal, the camera detection efficiency, can be described by the following formula:

$$\text{Detection Efficiency} = \text{QE} * \text{Pixel Area} * \text{Time} \quad (36.10)$$

This means that the total signal can now be described as:

$$S = \text{Photon Flux Density} * \text{QE} * \text{Pixel Area} * \text{Time} \quad (36.11)$$

where the photon flux density (PFD) is defined as the number of photons per unit area per unit time arriving at the detector.

The PFD is the combined photon emission rate from the source and the collection efficiency of the optics as discussed above. Once the photons arrive at the camera, the most dominant factor in determining the signal size is the quantum efficiency (QE) of the CCD. The QE is defined as the fraction of incoming photons that actually generate a signal in the CCD and this number is a function of the wavelength of light hitting the CCD. For visible wavelengths of light, the energy level of the photon is only sufficient to generate a single excited-state electron in the CCD so the stoichiometry of electrons out to photons in is 1:1. For very energetic photons, such as X rays, each detected photon can generate multiple excited-state electrons.

The transfer of the photon energy into an excited-state electron occurs within a region of the CCD called the "depletion zone." To create this zone, a defined voltage is applied to surface gate structures on the CCD. These gate structures define the pixel boundaries in one dimension, whereas buried implants define the pixel boundaries in the other dimension.[7] Although necessary to create the pixel and to move charge across the CCD, the overlying gate structures cause a reduction in the QE over a range of visible wavelengths as a result of a combination of reflection and absorption.

To compensate for this kind of loss, CCD manufacturers have introduced a series of innovations. One approach is to use a semitransparent gate structure that allows more of the short wavelengths to reach the depletion zone. Kodak has introduced a series of blue-enhanced CCDs that use indium-tin-oxide in place of silicon oxide for one of the gate lines on each pixel. This enhances the CCD QE throughout the 380- to 580-nm region of the spectrum (Figure 36.1).

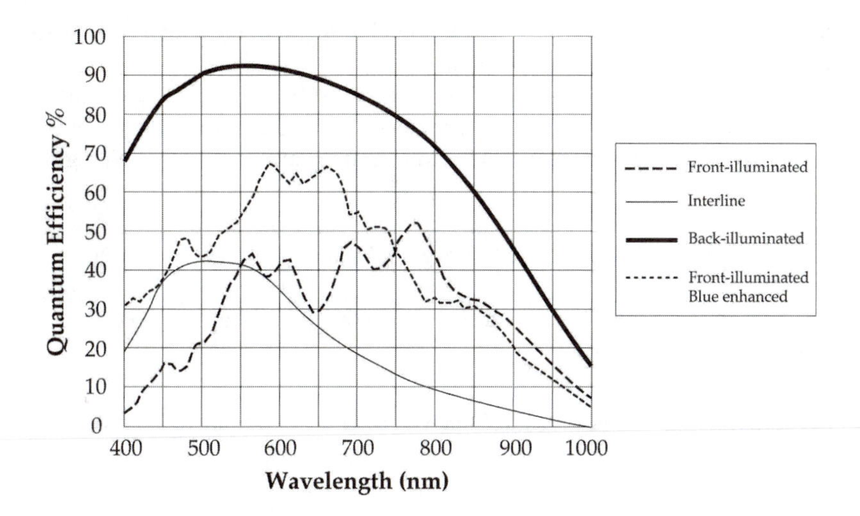

FIGURE 36.1 Quantum efficiency profiles of various CCD types. A comparison of the quantum efficiencies of several CCD types across the visible spectrum is provided. The quantum efficiency represents the number of photons that are converted to measurable signals in the CCD itself.

A unique approach employed in the Sony interline-type CCDs is the use of a photodiode-like sensor to collect the light, but a CCD-like element to store the signal. The net effect is very high QE, but a low area to collect the signal because of the need to place a light-impermeant mask over the storage region on each pixel. The low area of collection is partially compensated for by using a single or double microlens structure on each pixel to focus light onto the photosensitive region.

The most effective means of increasing the net QE is back-illumination, but it is most difficult to accomplish. Back-illuminated (BI) CCDs are produced by turning the CCD over and etching away the supporting silicon substrate to expose the depletion zone. This exposed surface is then layered with an antireflection coating optimized for the visible region of the spectrum. In this configuration, the BI CCD can often achieve quantum efficiencies in excess of 90% for light of approximately 600 nm wavelength.

Figure 36.1 shows the QE profiles of standard front-illuminated (FI), blue-enhanced front-illuminated (BE-FI), interline (IL), and back-illuminated (BI) CCDs. Comparing the QE profiles, it is clear that the BI CCDs have the highest QE of all the sensor types across the whole visible spectrum.

Because QE is a function of wavelength, the actual conversion efficiency of a real signal will depend on the intersection area of the emission wavelength profile of the chemiluminescent substrate and the QE curve of the CCD sensor. For luciferin, the emission maximum wavelength is 560 nm, where the BI CCD has an effective QE of 92%.

Once one has a detector with the optimal QE, the signal intensity will be a function of the cross-sectional area of the pixel: the larger the pixel, the larger the signal. In addition to collecting more light, larger pixels allow for a higher dynamic range. Since the sensor dynamic range is defined as the largest measurable signal divided by the readout noise specification, having a larger "bucket" to hold more charge allows for dynamic ranges that go from 4095 gray levels (12 bits) all the way to 65,565 gray levels (16 bits). There is a fundamental trade-off, however, in that increasing the pixel size reduces the spatial resolution that is achievable with the CCD.

Another approach that can be utilized to increase pixel area in a scientific-grade camera is the software-selectable on-chip binning feature. This allows the user to run the camera at a variety of resolutions and sensitivities. For example, collecting an image at 2×2 binning will yield four times the signal as compared with the unbinned state. Similarly, when 3×3 binning is selected, the user obtains nine times the unbinned signal. By using on-chip binning, the user also limits the noise in the signal by having one readout event per nine pixels.

The advantage of on-chip binning to S/N can be demonstrated as follows. Assume there are two units of signal per pixel and two units of noise per pixel, a condition that would not allow the user to collect meaningful data at full resolution. Now reading out a 3 × 3 pixel region yields the following:

$$\text{S/N (unbinned)} = \text{Signal/Noise} = 2/2 = 1$$

By reading out the 3 × 3 region unbinned and then combining the signal in software after the readout, we have the following S/N:

$$\text{S/N (software binning)} = \text{(Sum of the Signals)/(Sum of the squares of the noise terms)}^{1/2}$$

$$= 9 * 2/(9 * 2^2)^{1/2} = 3$$

If instead, we bin all the pixels in a 3 × 3 region we have only one readout event for all the accumulated signal and hence only one noise term to consider:

$$\text{S/N (binned } 3 \times 3) = (9 * 2)/2 = 9$$

So we can see that there is an increase in the S/N that is equal to the square root of the number of pixels binned when on-chip binning is used vs. software-based binning after full-resolution readout. In this example, some form of binning (hardware or software) would be required to see any significant measurement at all.

The other variable under the control of the user is the exposure time. For signals that are linear with time, an exposure that is two times longer will generate a signal that is two times larger since the CCD camera produces extremely linear data over its full 65,000 gray level range. When the signal itself is not linear, such as in the case where the enzyme activity plateaus early and then begins to decrease over time, the extended exposure will give a net increase in signal, but it will not be proportionally higher. For chemiluminescence imaging, there are a variety of luciferase-type molecules and each has a distinct activity-vs.-time curve and the user should measure the enzyme kinetics *in situ* if quantitative information is desired. Eventually, there will be a point of limited return from a longer imaging period due to drop off in the signal. This should be determined empirically from the preparation under study.

36.6 ELIMINATING NOISE SOURCES

As discussed above, the noise sources of a measurement include the photon shot noise, the background signal noise, and the camera noise. The photon shot noise is inherent to the light signal itself and is equal to the square root of the number of photons measured. So, for 100 measured photons, the associated shot noise would be ±10 while for 10,000 measured photons the associated shot noise would be ±100. This noise term cannot be altered and its contribution can only be minimized by measuring more photons. For many chemiluminescence experiments, there are a limited number of photons that can be measured.

The background signal arises from contaminating stray light entering the optical system. This signal can be eliminated if the contaminating light sources are constant in space and in time. However, the noise associated with this background signal cannot be removed from the data and serves to reduce the overall S/N_T in the measurement. The magnitude of the background noise equals the square root of the background signal. Minimizing this noise source is one of the most critical factors for successful detection of a faint chemiluminescence signal. To suppress this background light, the microscope must be insulated exceedingly well from stray light.

If all the light is directed to the camera port, there is no need to worry about light coming from the oculars. The main light leakage, therefore, comes via the objective during the integration time. To keep the contaminating light to a minimum, a black foam core material can be used to frame a black box around the microscope, with the seams sealed by a double layer of black tape. The setup also requires a door or access panel to allow the introduction of a sample to the microscope stage. Alternatively, a small, dark chamber can be constructed that covers the sample and blocks light entry into the objective of the inverted microscope.

In addition to the above measures, the room lights should be kept off and all major sources of light should be eliminated. This involves masking light leaks around doors and windows, turning off the computer monitor (or running the computer from an adjacent room) and using black tape to cover light-emitting diodes on equipment in the room. It is also a good idea to use dark clothing that does not have lots of whiteners that can reflect stray light back toward the microscope.

It is easy to measure the background noise in the experimental setup by running a control, such as a dish of cells without the luciferin substrate. Starting with shorter exposures and lower binning ratios, light leaks can be measured and systematically eliminated. Once most of the obvious light leaks have been blocked, a measure of the background noise is made. The easiest way to measure this noise is to use a program that can perform statistics on an image and report back the standard deviation of the signal across the whole field of view. If the value of this standard deviation is close to that standard deviation of a bias image (exposure time set to 0 so no light hits the CCD), then the data should be close to the limit of detectability allowed by the digital camera (see discussion below on measuring noise sources).

For macroscopic imaging, the sample is usually placed into a black box environment where all extraneous photons are excluded by the light tightness of the chamber. In practice, light tightness is a relative term since we have found that some commercial boxes produced for higher-level chemiluminescence signals are not sufficiently dark for the very low light chemiluminescence signals that can be measured by a high-performance digital camera. These boxes can often be made "darker" by sealing the seams with black RTV compound (Dow Corning) or a wide strip of black electrical tape. Also, the camera mount must have an efficient O-ring-type seal (preferably a double O-ring) to prevent light leaking in from the top.

In addition to the light tightness, a dark box should have a wide door with easy sample access. There should be movable stage that can be adjusted with the door closed and the sample platform should be large enough to accommodate multiple target types including multiple petri dishes as well as whole plants and animals. One example of such a dark box is shown in Figure 36.2.

This now brings us to camera-derived noise terms. These include the camera electronic readout noise and the dark current noise in the CCD itself.

$$N_C = (N_R^2 + N_D^2)^{1/2} \tag{36.12}$$

where N_R = readout noise and N_D = dark current noise.

The readout noise arises from the requirement to amplify and measure the small amounts of charge that are generated in the CCD. This is the area where a camera manufacturer uses its expertise in electronic design to minimize the absolute noise level during readout. Often this involves the use of double-correlated sampling techniques, well-isolated electronic circuits, use of high-quality electronic components, and other careful design elements.

For scientific-grade BI CCDs, a major factor influencing the readout noise is the speed at which the analog to digital converter (ADC) is operated. As the amplifier on the CCD is driven faster, the readout noise increases. A tenfold increase in readout rate causes a doubling of the readout noise. Conversely, slowing down the readout rate decreases the noise. For optimal operation,

FIGURE 36.2 Dark box imaging system. An example of a dark box setup along with the LN-cooled VersArray 1300B camera that is used to collect chemiluminescence images. During operation, the camera is mounted on the top of the box via a light-tight seal. The box allows macroimaging of chemiluminescence via a movable stage that can be focused from the outside.

scientific-grade BI CCDs are typically run with a dual ADC configuration; the fast ADC (typically at a 1 megapixel per second rate) is used for setup and focusing, while the slower ADC is used for optimal data collection (typically at a 100 kilopixel per second rate). The lowering of readout noise with slower readout rates eventually reaches a plateau, so systems are rarely used with readout rates much slower than 50 kilopixels per second.

When using a CCD to integrate a signal over time, the user must be aware of the presence of a background signal that grows over time. This background signal is called the dark current and it arises from the thermal energy within the CCD sensor itself.[7] Because of defect structures within the CCD, the thermal energy of the CCD is sufficient to create electron–hole pairs within the pixel well, which results in a signal that is indistinguishable from the photon-derived signal. The rate of dark current accumulation is a function of temperature, so lowering the temperature of the CCD reduces the dark current rate (Figure 36.3). As a rule of thumb, the dark current is lowered by a factor of 2 for every 7° drop in CCD temperature. Some scientific-grade BI CCDs have a substantial dark current rate at room temperatures and high-performance cooling is required to lower the rate to acceptable levels.

For moderate suppression of dark current, a multistage thermoelectric Peltier-type device is used in conjunction with forced-air cooling. The Peltier device transfers heat from one side of the stack to the other side when the unit is electrically powered. This heat is then dispersed by air as it crosses a thermal reservoir or heat block. This type of cooling can produce cameras that run down to −50°C or more, depending on the size of the CCD being used.

For the ultimate in dark current suppression, the CCD can be lowered to −100°C or more using liquid nitrogen (LN)-based cooling mechanisms. In this type of camera, the LN is in thermal contact with a heat block that is bonded to the CCD. The LN pushes the temperature of the CCD downward and stabilization is achieved by adding a small heater to the heat block to keep the temperature

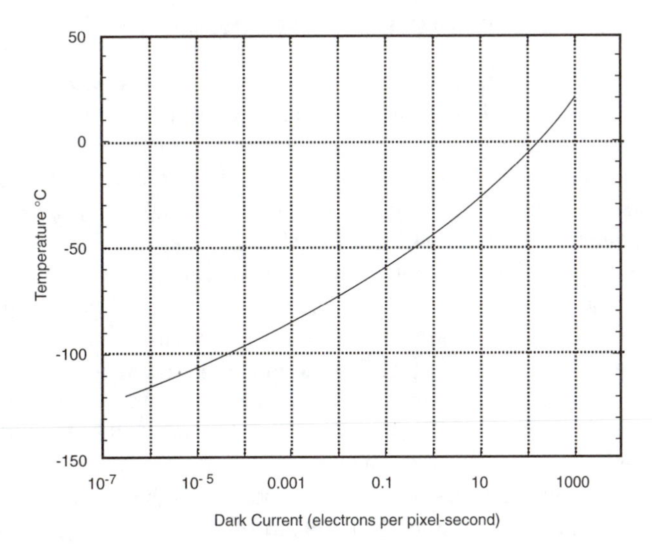

FIGURE 36.3 Dark current as a function of temperature. As the temperature of the CCD is lowered, the dark current rate continually decreases. To achieve the ultralow dark currents desired for chemiluminescence imaging, the CCD should be cooled to the −100°C level or lower.

regulated to within 50 millidegrees of the desired temperature. Under LN cooling, the BI CCDs can be kept to dark current rates that are totally negligible even for exposures of up to 20 min with moderate to high binning. This ultralow cooling is a hallmark of the highest-performance BI cameras that are used for extended integration on the CCD.

36.7 HOW TO MEASURE NOISE LEVELS

The procedure to measure noise levels is fairly straightforward. To measure the readout noise on a system, one can program the camera for a 0-s exposure and then collect three images. Subtract the second and third images on a pixel-by-pixel basis and perform a statistical analysis on the resulting image. The standard deviation of the values in the subtracted image should be divided by 1.414 (square root of 2) since we are deriving the data from a subtraction-based method. The net result will be the root mean square (rms) readout noise in counts.

Similarly, to measure noise due to stray light, collect three images over a time span similar to the desired exposure time, using the same binning as in the experimental measurements but with no sample present. Now subtract the second and third images and measure the standard deviation across the image. As above, divide this number by 1.414 and the result will be the rms noise due to background light and the readout noise combined.

To separate out the background noise from the readout noise, use the total noise formula (Equation 36.2) and the camera noise formula (Equation 36.12). Solving for the background noise (assuming that there is no dark noise due to deep cooling and that there is no signal present):

$$N_B = (N_T^2 - N_R^2)^{1/2} \tag{36.13}$$

where N_R = readout noise. By using this measurement, it is now possible to determine the level of background noise. Ideally, this noise should be reduced to the level of readout noise to ensure that the imaging system can measure the smallest possible chemiluminescent signal.

36.8 A NOTE ON COSMIC RAY–INDUCED SIGNALS

When measuring a very weak signal, obtaining the maximum S/N with a cooled CCD requires binning of the CCD and integration on the CCD for a period of time that can approach minutes. Both of these conditions (large area, large time), however, also increase the probability of detecting of background signals generated by cosmic rays. These background signals appear as very high local signals or spikes against a backdrop of constant low signal. They are caused by secondary particles that are created by cosmic rays entering the atmosphere.[1]

There is no way to prevent the cosmic events from appearing, but lowering the exposure time or reducing the binning will minimize the number of pixels affected by the spurious signal. This approach will limit the signal size and may not be the preferred route. Alternatively, these spikes can be removed by applying a median filter[8] to the image with a neighborhood size of around 5. This will result in some smoothing of the data in the image, but chemiluminescence images often do not have sharp features because of the diffuse distribution of the luciferase and the need to measure a signal over seconds to minutes, which tends to homogenize any localization of the signal that may be present. One other approach is to interpolate across the peaks to remove them. Ideally, this should be done with a two-dimensional interpolation to approximate the true background signal.

36.9 THE IDEAL CAMERA

The discussion above can now help us to craft the ideal system for chemiluminescence imaging. It should have the best collection optics available as measured by NA and magnification (microscope based) or by f-number (macro-imaging). The collection optics should be isolated from all extraneous light sources such that a control exposure under the standard experimental conditions produces noise in the image that is limited by the camera and not by the contaminating light; that is, the background noise should not be any higher than the camera readout noise.

To achieve the best possible signal conversion, the CCD should be of the BI type with peak QE values around 90% in the wavelengths of interest. To eliminate all dark current noise, the CCD should be cooled to −100°C or more. The camera should be equipped with dual ADCs to allow higher-speed operation for setup and focusing and slower-speed operation for low noise and optimal data collection.

The camera should also support on-chip binning capabilities to optimize signal vs. resolution for each experiment. The on-chip binning allows signal collection before readout and therefore optimizes the S/N of the measurement. The system should include easy-to-use software that controls all the features of the camera and presents the data in an easy-to-interpret fashion.

Combined with an environment that is appropriately shielded from light, a camera with these features will allow the experimentalist to image the most-light-starved chemiluminescence preparations, ranging from single cells under the microscope to whole organisms under macroscopic imaging conditions. For experimental conditions where the luciferase levels are much higher, it is sometimes possible to use camera systems which have a lower performance level. In these cases, however, it will be critically important to verify system performance on a realistic sample to ensure adequate sensitivity for the task at hand.

36.10 EXAMPLES OF CHEMILUMINESCENCE IMAGING

Bioluminescence resonance energy transfer (BRET) imaging can be used to measure the proximity of a luciferase–luciferin complex and a green fluorescent protein (GFP) molecule within a cell.[9] When the two protein partners are sufficiently close, the luciferase–luciferin complex donates its energy to the GFP molecule by nonradiative resonance energy transfer, resulting in GFP emission in the absence of illumination. The signals emitted from single cells are extremely low and can be virtually impossible to measure using most camera systems. The Princeton Instruments LN-cooled

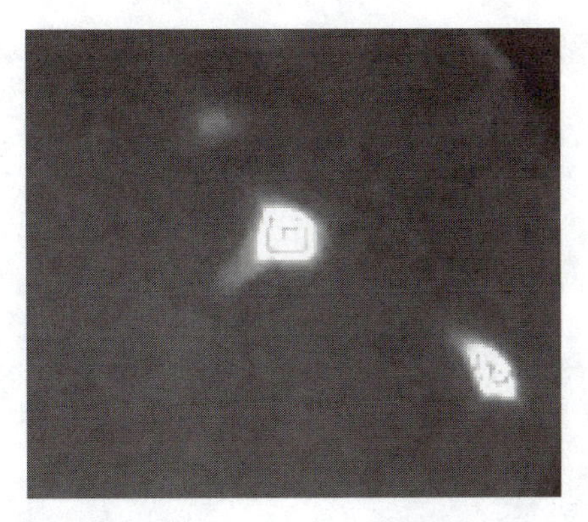

FIGURE 36.4 Example of single-cell BRET imaging. BRET imaged in single COS-7 cells, using *Renilla* luciferase and EYFP as reporters. Images were collected on an inverted microscope using a 40×, 1.3-NA objective. The luciferin was added 10 s before imaging. Exposures were 5 min in length with 6 × 6 binning. (Data are from the laboratory of Dave Piston, Vanderbilt University.)

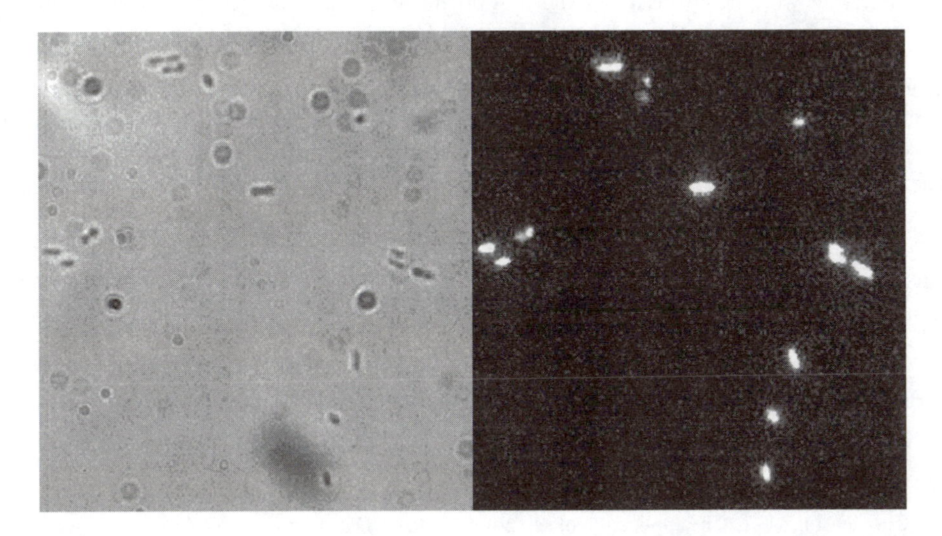

FIGURE 36.5 (Color figure follows p. 266.) Bacterial chemiluminescence imaged through the microscope. Natural bacterial chemiluminescence imaged through the microscope (100× lens, 1.3 NA) using a VersArray 1300B LN-cooled CCD camera system. The image on the left is a brightfield reference image and the image on the right is a bioluminescence image of *Vibrio harveyi* (2.5-min exposure). (Data were taken by Irina Mihalcescu in the laboratory of Stan Leibler, Princeton University.)

VersArray 1300B camera was used to collect the BRET images as shown in Figure 36.4 (Dave Piston laboratory, Vanderbilt University).

Another example of microscope-based chemiluminescence imaging is the characterization of endogenous luciferase activity in chemiluminescent bacteria. Individual bacteria can be imaged over time using the cryogenically cooled VersArray 1300B camera to monitor natural fluctuations in the luciferase under different growth or nutritional states. Figure 36.5 shows robust *Vibrio harveyi* chemiluminescence when examined on the single cell level (Stan Leibler laboratory, Princeton University).

One of the most common macroscopic imaging applications of chemiluminescence is the characterization of plant gene expression levels. The gene activities can be examined in response to a variety of factors such as light, pathogens, or phytohormones. Alternatively, bacterial chemiluminescence can be used as a tracer for infection of plant tissue. An example of this latter type

FIGURE 36.6 (Color figure follows p. 266.) Plant chemiluminescence image. Bacteria expressing luciferase were used to infect *Arabidopsis* plants and then whole plants were examined. The blue represents a very low signal, whereas the white represents a high signal. A brightfield reference image is shown on the left and the chemiluminescence image is shown on the right. (Data are from Dr. Jian-Min Zhou, Kansas State University.)

of imaging is shown in Figure 36.6, where *Arabidopsis* leaves were infected with different levels of bacterial pathogens expressing luciferase and imaged on in a black box environment using the LN-cooled VersArray 1300B system (Jian-Min Zhou laboratory, Kansas State University).

Clearly, the doors are now open to more creative and challenging experiments using the luciferase protein as a biomarker in living cells and whole organisms. As researchers become aware of the possibilities now open to them, we should begin to see a whole host of studies published demonstrating real-time changes in gene expression that were previously thought impossible to measure.

REFERENCES

1. Talmi, Y., Intensified array detectors, in *Charge-Transfer Devices in Spectroscopy,* Sweedler, J. W., Ratzlaff, K. L., and Denton, M. B., Eds., VCH Publishers, New York, 1994, chap. 5.
2. Christenson, M. A., The application of scientific-grade CCD cameras to biological imaging, in *Imaging Neurons,* Yuste, R., Lanni, F., and Konnerth, A., Eds., Cold Spring Harbor Laboratory Press, Cold Spring Harbor, NY, 2000, chap. 6.
3. Sala-Newby, G. B., Kendall, J. M., Jones, H. E., Taylor, K. M., Badminton, M. N., Llewellyn, D. H., and Campbell, A. K., Bioluminescent and chemiluminescent indicators for molecular signalling and function in living cells, in *Fluorescent and Luminescent Probes for Biological Activity,* Mason, W. T., Ed., Academic Press, New York, 1999, chap. 17.
4. Inoué, S., *Video Microscopy,* Plenum Press, New York, 1986, chap. 5.
5. Keller, H. E., Objective lenses for confocal microscopy, in *Handbook of Biological Confocal Microscopy,* 2nd ed., Pawley, J. B., Ed., Plenum Press, New York, 1995, chap. 7.
6. *RCA Electro-Optics Handbook,* RCA Solid State Division, Lancaster, PA, 1974, chap. 12.
7. Sims, G. R., Principles of charge-transfer devices, in *Charge-Transfer Devices in Spectroscopy,* Sweedler, J. W., Ratzlaff, K. L., and Denton, M. B., Eds., VCH Publishers, New York, 1994, chap. 2.
8. Russ, J., *The Image Processing Handbook,* 3rd ed., CRC Press, Boca Raton, FL, 1999, chap. 3.
9. Xu, Y., Piston, D. W., and Johnson, C. H., A bioluminescence resonance energy transfer (BRET) system: application to interacting circadian clock proteins, *Proc. Natl. Acad. Sci. U.S.A.,* 96, 114, 1999.

37 Applications of Bioluminescent and Chemiluminescent Imaging in Analytical Biotechnology

Aldo Roda, Massimo Guardigli, Patrizia Pasini, Monica Musiani, and Mario Baraldini

CONTENTS

37.1 INTRODUCTION

Chemiluminescence (CL) is the light emission due to a chemical reaction that produces electronically excited molecules, which decay to the ground state and emit photons. Bioluminescence (BL) is a type of CL that naturally occurs in living organisms and is also utilized *in vitro*.

The use of BL and CL as analytical tools offers several advantages over other detection principles that involve light, including high detectability, high selectivity, and wide dynamic range. The high detectability of BL/CL is partly due to the absence of the background signal present in absorption and fluorescence methods, which depends on photon emission from impurities and sample components, light scattering, etc. Sample components, which could hamper fluorescence or absorption measurements, usually do not produce light when the chemiluminescent reagents are added, making the BL/CL detection highly selective. Moreover, in suitable experimental conditions, the light output is related to the analyte concentration, thus allowing quantitative analysis to be performed. The absence of an excitation source for the light production and the wide range of response of phototransducers for light detection make the dynamic range of luminescent techniques very wide, so that samples can be measured across decades of concentrations without any dilution.

0-8493-0719-8/02/$0.00+$1.50

The measurement of BL and CL is typically performed using a luminometer. Luminometers are simple and relatively inexpensive instruments designed to measure the light output from samples, usually in the single tube or microtiter plate format. Low-light imaging devices represent a relatively recent advance in light detection technology. These devices, also known as luminographs, allow not only the measurement of the signal intensity at the single-photon level, but also the evaluation of the spatial distribution of the light emitted from the sample. Analytical applications of BL/CL imaging for the ultrasensitive quantitation and localization of analytes on a target surface are reported in this chapter. A wide range of applications on macrosamples and microsamples, by connecting the imaging device with an optical microscope, are described.

37.2 INSTRUMENTATION

The instrumentation for CL imaging includes an ultrasensitive image acquisition device, a suitable optical system, and appropriate software for image acquisition and analysis.[1]

The image acquisition device must be specifically designed for the detection of the weak CL emission. In the last few years high-performance charge-coupled device (CCD) cameras replaced previous devices based on intensified Vidicon tubes. Commercially available cooled, slow-scan, back-illuminated CCD cameras provide very high detection sensitivity because of their high quantum efficiency and low background noise and allow quantification of the emitted light at the single-photon level.[2–6] These devices are suitable for the measurement of static CL signals and do not require an image intensification step, which could negatively affect the image quality and the signal-to-noise ratio.[3,7–9] On the other hand, intensified CCD cameras are necessary for the real-time acquisition of dynamic CL signals.

The image acquisition device is connected to standard or custom camera optics and to a sample light-tight box. In this configuration it can be used for reading CL emissions from microtiter plates, target surfaces such as gels, thin-layer chromatography plates, and blot membranes,[10] or any kind of macroscopic samples. Commercially available low-light imaging devices are often provided with ultraviolet sources to detect fluorescence or a transilluminator to perform densitometric measurements. In this way fluorescence and colorimetric measurements can take advantage of the high sensitivity of the light detector and the easy acquisition and quantification of the signal.

A CCD camera (Figure 37.1) can be used in conjunction with an optical microscope to localize the light emission from tissues or cells, and to obtain semiquantitative information on the localization of the probed species.[9,11–17] The potential loss of light in the optical system should be minimized to achieve the maximum analytical detectability; therefore, the microscope should have a lens coupling system that is as simple as possible and objectives with the highest numerical aperture compatible with focal aberration and depth of field.

Factors that affect the light signal intensity, such as the sample geometry, should be taken into account in CL measurement procedures. When CL signals from a microtiter plate are measured by a luminograph, the sample geometry can cause relevant errors in the estimation of the CL intensities from the wells.[18] In the case of microsamples such as tissue sections, the presence of the CL cocktail as a liquid film over the sample can be responsible for internal refraction and reflection phenomena at the air–solution interface. This effect can be evaluated using model systems, such as CL enzymes chemically immobilized on suitable solid supports.[5,19] Evaluation and comparison of CL intensities may be difficult in samples with irregular surfaces.

The localization of the CL signal corresponding to the target analyte on a surface requires the following steps:

1. Sample live image is recorded as transmitted or reflected light image.
2. The luminescent signal is acquired using a suitable integration time, to obtain a satisfying signal-to-noise ratio.

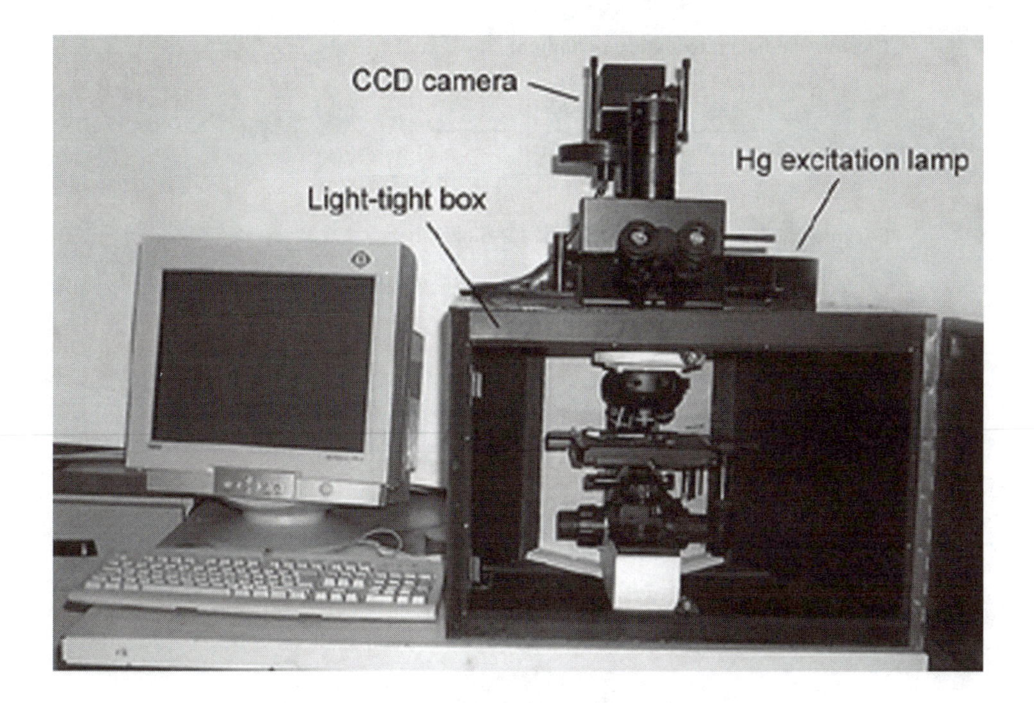

FIGURE 37.1 Conventional epifluorescence microscope modified for BL/CL imaging.

3. The CL image is converted to pseudocolors corresponding to the light intensity and overlapped to the live image, thus allowing the spatial distribution of the CL signal to be evaluated.

Image processing functions allow the reduction of the differences in the instrument response across the field of view (flat-field correction), the subtraction of the instrument background, and the enhancement of the image details. For quantitative analysis, the light emission from selected areas of the sample can be measured by summing the total number of photon fluxes from within those areas. The CL intensity is expressed as photons or relative light units per second per surface area unit (usually pixel or mm^2).

37.3 BIOLUMINESCENT AND CHEMILUMINESCENT REAGENTS

Many BL and CL reactions have been exploited in BL/CL imaging (Figure 37.2).

Some CL imaging applications make use of labeled biospecific probes (e.g., antibodies or gene probes). Labeling with CL enzymes is generally preferred to labeling with CL substrates because of the possibility of amplifying the luminescent signal in the excess of the enzyme substrate, thus increasing the label detectability. Horseradish peroxidase (HRP) and alkaline phosphatase (AP) are the most commonly used CL enzymes, even though other enzymes, such as glucose-6-phosphate dehydrogenase,[20,21] β-galactosidase,[22] and xanthine oxidase[23,24] have been used as CL labels. Both HRP and AP are suitable for the direct labeling of antibodies in immunochemical techniques or the indirect labeling of cDNA probes in hybridization techniques.

BL/CL reactions can be used not only for the detection of species directly involved in the luminescent reaction, but also for species that are involved in a reaction suitable for coupling to the luminescent one. For example, the luminol/H_2O_2/HRP system has been coupled with oxidase

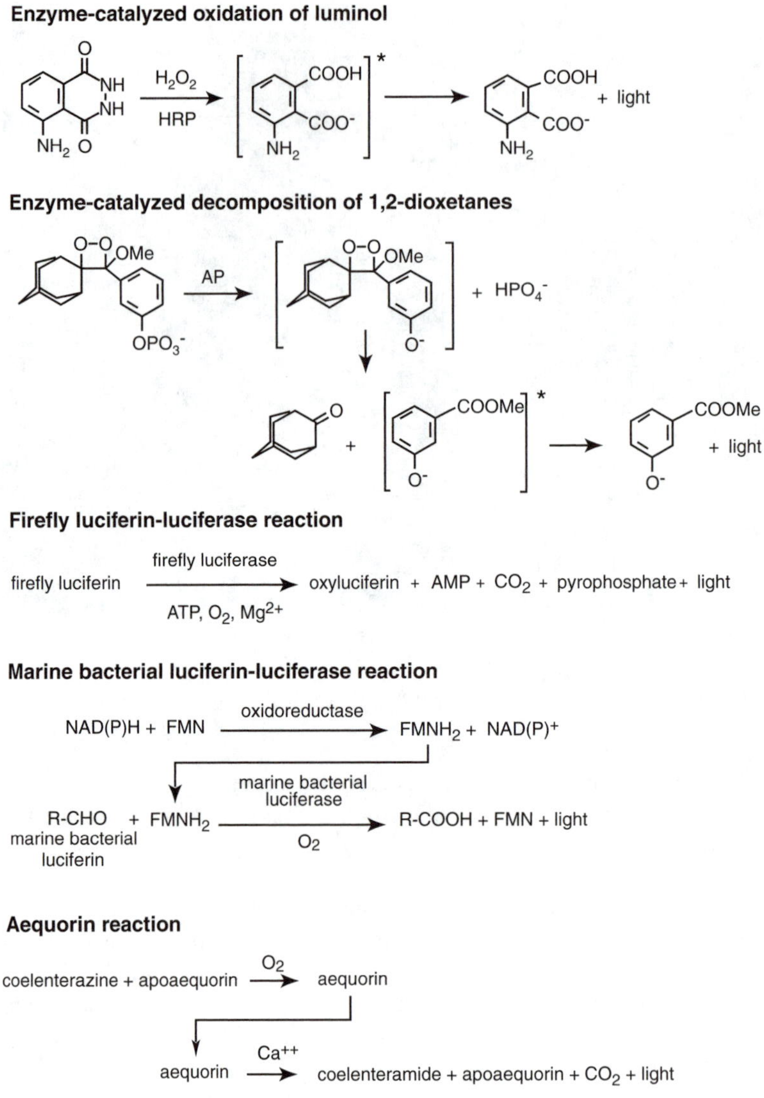

Enzyme-catalyzed oxidation of luminol

H_2O_2, HRP; COOH, COO^-, NH_2; + light

Enzyme-catalyzed decomposition of 1,2-dioxetanes

O-O, OMe, OPO_3^-; AP; + HPO_4^-; O, COOMe, O^-; + light

Firefly luciferin-luciferase reaction

firefly luciferin —(firefly luciferase; ATP, O_2, Mg^{2+})→ oxyluciferin + AMP + CO_2 + pyrophosphate + light

Marine bacterial luciferin-luciferase reaction

NAD(P)H + FMN —(oxidoreductase)→ $FMNH_2$ + $NAD(P)^+$

R-CHO (marine bacterial luciferin) + $FMNH_2$ —(marine bacterial luciferase; O_2)→ R-COOH + FMN + light

Aequorin reaction

coelenterazine + apoaequorin —(O_2)→ aequorin

aequorin —(Ca^{++})→ coelenteramide + apoaequorin + CO_2 + light

FIGURE 37.2 BL/CL reactions commonly used in bioanalytical imaging.

enzymes, and ATP-involving reactions (kinases) and NAD(P)H-producing or -consuming reactions (dehydrogenases) can be coupled with the firefly luciferin–luciferase and bacterial luciferin–luciferase systems, respectively.[25]

Different CL reagents for HRP and AP, which produce glow-type kinetics (i.e., the CL emission reaches a steady state) and permit the detection of very small amounts of enzyme (in the order of 10^{-18} to 10^{-21} mol), are commercially available. The steady-state light emission facilitates both the handling and measurement procedures and the standardization of the experimental conditions and quantitation of the labeled probes.

The commercially available CL substrates for HRP are usually based on luminol or luminol derivatives.[26,27] The HRP-catalyzed oxidation of luminol produces aminophthalate ion in the excited state that emits photons. These CL reagents also contain enhancers, such as phenols, naphthols, or amines, that increase the CL intensity and produce glow-type kinetics, with a steady-state light emission that is maintained for at least 10 to 15 min. Acridinium esters–based CL substrates for

HRP have also been synthesized.[28] The most efficient CL substrates for AP are based on adamantyl-1,2-dioxetane aryl phosphate derivatives,[29,30] which are dephosphorylated by AP to yield an unstable intermediate; the decomposition of this intermediate produces an excited-state aryl ester that emits light with characteristic steady-state kinetics. BL methods for the detection of AP labels (e.g., using D-luciferin-*O*-phosphate as a substrate) have been developed.[31–33]

BL systems are employed for imaging purposes, in particular the firefly and the bacterial luciferin–luciferase reactions. The well-known firefly luciferin–luciferase reaction, which requires adenosine triphosphate (ATP), magnesium ions, and oxygen, is characterized by a very high quantum yield[34] and luciferase can be used as a BL label.[35] Many different luciferins and mutant luciferases have been investigated to optimize the reaction performance.[36] BL bacterial systems involve bacterial luciferases that catalyze the oxidation of flavin mononucleotide in the reduced form ($FMNH_2$) and bacterial luciferin (a long-chain aldehyde) in the presence of oxygen. *In vivo*, the luciferase is coupled to an oxidoreductase that catalyzes the oxidation of nicotinamide adenine dinucleotide (phosphate) [NAD(P)H] leading to the formation of $FMNH_2$.[37] Aequorin-based BL systems have been developed using apoaequorin (the precursor of the photoprotein aequorin), which can be produced by recombinant DNA techniques. Apoaequorin is converted to aequorin by reaction with coelenterazine, and the flash light emission from aequorin is triggered by the presence of calcium ions.[38]

Most of the available BL/CL substrates can be used for imaging purposes, even if they are not specifically designed for this application. The half-life of the light-emitting species in the CL cocktail is a critical factor in adapting a CL system for imaging, especially in the case of imaging of microsamples.[39] In fact, to achieve precise and accurate localization of the enzyme-labeled probe on a target surface, the light emission should occur as near as possible to the site of the primary biospecific recognition. CL substrates should be characterized by very short-lived excited species to render diffusion of the emitting species negligible. The resolution could be improved by increasing the viscosity of the CL cocktail solution by adding substances like gelatin, glycerol, or polyvinylpyrrolidone at suitable concentrations. The problem related to the diffusion of the species involved in the CL reaction is even greater when the CL derives from a chain of enzymatic reactions. This situation could occur in the determination of an enzyme activity requiring two or more coupled reactions. In this case the diffusion of both the excited emitting species and the intermediate reagents must be taken into account.

Finally, it should be pointed out that the overall quality of a CL image depends not only on the absolute value of the CL signal but also on its signal-to-noise ratio. Therefore, the best results are not necessarily obtained with the CL reagents giving the maximum emission intensity but with those giving the best signal-to-noise ratio.[19]

37.4 APPLICATIONS ON MACROSAMPLES

37.4.1 FILTER MEMBRANES

CL labels have been widely used in filter membrane biospecific reactions (e.g., Southern, Northern, Western, and dot blot assays) with photographic detection, which has the limitation of not permitting a straightforward quantitation of the CL signal intensity. In recent years, there has been a growing interest in the use of CL imaging because of the possibility of performing a direct and rapid quantitative evaluation of the signal over a relatively wide dynamic range.[16] The analytical performance of CL imaging to detect immunological or genetic reactions is comparable or superior to that of systems using different principles, such as radioisotopes or color-producing substrates,[40,41] or even CL with photographic detection. Moreover, CL images can be permanently recorded for archiving, further elaboration, and exchange with other laboratories.

Nucleic acid hybridization techniques are able to detect viral genomes directly in clinical samples, allowing rapid and sensitive diagnosis of viral infections, especially for viruses that do not grow in cell cultures or have long replication cycles. In CL dot blot hybridization reactions the specimens are dotted on a membrane, hybridized with a specific gene probe labeled with a hapten, and then the hybrid is detected using an enzyme-conjugated antihapten antibody and a suitable CL substrate.[42–44] The analytical performance of different CL substrates for HRP or AP was evaluated in dot blot hybridization assays for the detection of B19 parvovirus DNA.[45] The assays used digoxigenin-labeled DNA probes that were immunoenzymatically revealed using antidigoxigenin Fab fragments conjugated with HRP or AP. The detection limits were between 0.5 and 2 pg of target homologous DNA using HRP as a label, and between 10 and 50 fg of DNA using AP as a label. Since the detection limits for colorimetric methods were about 5 pg and 100 fg when HRP and AP were used as labels, respectively, the CL method proved to be superior to colorimetry.

CL imaging with a cryogenically cooled CCD camera was used to detect DNA sequencing fragments covalently bound to a blotting membrane.[46] The direct detection of the DNA fragments using an AP–oligonucleotide conjugate probe was compared with the secondary detection of biotinylated oligonucleotides with an AP–streptavidin conjugate. It was found that the analytical sensitivity obtained using the direct AP–oligonucleotide conjugate as a hybridization probe is higher than that obtained using the secondary detection of biotinylated oligonucleotides, because of the nonspecific binding of the AP–streptavidin and the biotinylated hybridization probe.

A CL Western blot assay was developed to detect antibodies against linear epitopes of parvovirus B19 structural proteins VP1 and VP2 in human serum samples.[47] The CL assay combined the sensitivity of CL detection and the objective evaluation of the luminescent signal, measured with a videocamera-based, high-performance, low-light-level imaging luminograph. The potential for diagnostic purposes was evaluated using reference serum samples and clinical samples from patients with different clinical conditions and laboratory evaluations regarding B19 infection. The assay provided reproducible results, allowing a semiquantitative analysis of the presence of antibodies against VP1 and VP2, and results more sensitive than the colorimetric one.

A CL enzyme immunoassay for the rapid detection and quantification of *Lactobacillus brevis* contaminants in beer and pitching yeast was described.[48] The method was based on the immunometric reaction of *L. brevis* cells trapped on a membrane with a suitable HRP-labeled antibody and the subsequent CL detection of the bound antibody using a CCD camera and allowed the detection of *L. brevis* at the single-cell level.

37.4.2 Microtiter Plates

CL imaging can be used to measure CL emission from microtiter plates, even if CCD detectors possess a narrower dynamic range (two to three decades narrower) and a lower sensitivity (five to ten times lower) than conventional photomultiplier-based microtiter plate readers.[4] Although the photomultiplier-based readers should theoretically provide a lower detection limit, in practice the presence of a relatively high background signal due to the reagents and sample components often make the detection limits of the two devices comparable.

CL imaging offers some advantages over luminometric detection. In fact, imaging devices permit the simultaneous measurement of the luminescent signals from a whole 96- or 384-well microtiter plate (Figure 37.3a), and thus are more rapid than conventional luminometers that measure light output well by well or strip by strip. This is particularly important when the luminescent signal is stable for a short time and when a kinetic study must be carried out. Some commercially available luminographs allow the simultaneous measurement of up to four microtiter plates (corresponding to 4×384 microtiter wells) in a minute or less. This characteristic, together with the availability of CL reagents that rapidly reach the steady-state emission, makes CL imaging techniques appealing for the development of high-throughput screening (HTS) analytical methods. Such methods could be useful, for example, in the preliminary screening of the biological activity of compounds synthesized by combinatorial chemistry.

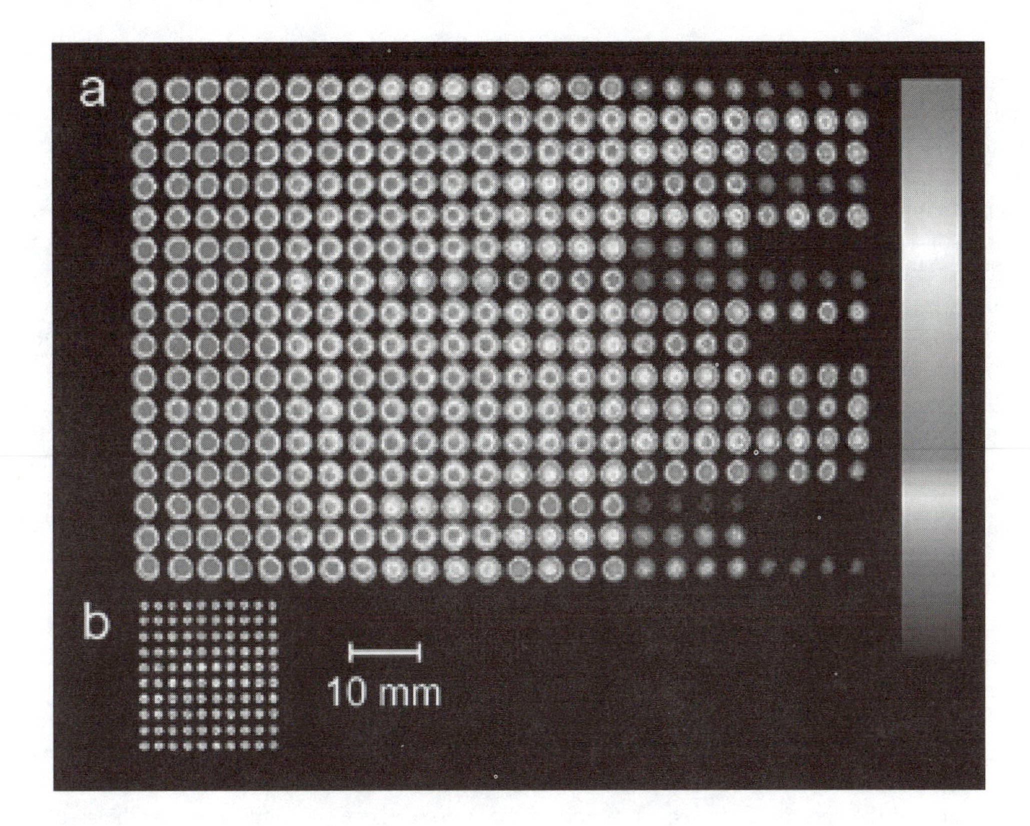

FIGURE 37.3 (Color figure follows p. 266.) Pseudocolored CL images of a 384-well microtiter plate (a) and a bidimensional array of HRP spots (spot diameter 1.0 mm) deposited on cellulose paper (b).

Another potential field of application for CL imaging is represented by analytical methods based on the monitoring of CL reaction kinetics to determine the activity and concentration of a given compound, rather than on a single light intensity measure. Standard luminometers are not suitable to record the emission kinetics across an entire microtiter plate, because of the relatively long time required to measure all the wells.

The use of a CL method in a 384-well microtiter-plate format for the evaluation of the antioxidant activity of natural compounds, referred to that of known antioxidants, was reported.[18,49] The assay relied on the modification of the kinetics of a CL luminol-based reaction in the presence of antioxidant compounds. The reaction exhibited a steady-state light output that was temporarily interrupted by the addition of antioxidants able to inactivate or block the formation of the luminol radicals driving the CL reaction. Light emission was restored after an interval of time that was directly proportional to the amount and activity of the antioxidant tested. The same 384-well microtiter plate analytical format was used for the determination of the activity of acetylcholinesterase (AChE) inhibitors.[18,50] AChE was coupled to choline oxidase and HRP through enzymatic reactions leading to a final light emission. Under suitable experimental conditions (i.e., in the excess of choline oxidase, HRP, and CL substrate for HRP), the intensity of the CL emission was proportional to the AChE activity. Therefore, the addition of AChE inhibitors to the system resulted in a reduction of the light output that was proportional to the enzyme inhibition. Both methods proved to be suitable for the analysis of the biological activity of a large number of samples (about 200 for each 384-well microtiter plate) in a short period of time. The results obtained for the AChE inhibitors were comparable with those obtained using conventional color-producing AChE substrates to evaluate the inhibitory activity. The high sensitivity of the CL detection allowed the analysis to be performed using very small volumes of reagents and samples.

A CL competitive immunoassay in a 384-well microtiter-plate format was developed for the detection of the β-agonist drug clenbuterol in bovine urine.[51] HRP was used as a label and the CL emission was recorded using either a conventional photomultiplier-based microtiter plate reader or a luminograph equipped with a sensitive, back-illuminated CCD camera. The assay matched the standard requirements of precision and accuracy, and the CL detection result was more sensitive than the conventional colorimetric one. In addition, the high detectability of the light signal allowed the use of a nonequilibrium method requiring only a few minutes of incubation, thus reducing the analysis time. The use of the 384-well microtiter-plate format allowed a fivefold reduction in reagent volume.

Daidzein concentration was measured in human serum samples by means of a CL competitive enzyme immunoassay developed in a 96-well microtiter-plate format, in which daidzein-specific antibodies were immobilized in the microtiter wells and a daidzein derivative coupled to HRP was used as a tracer.[52] The method exhibited a detection limit of 10 pg/well, which was lower than that of other developed immunoassays. The described CL immunoassay could represent a useful alternative to conventional chromatographic methods for the screening of daidzein content in large numbers of samples, as it is applied directly on serum specimens without any cleanup or extraction procedure.

Low-light imaging combined with the 96- or 384-well microtiter format could represent a very useful analytical tool in the development and application of luminescent recombinant cell-based biosensors. These biosensors are based on the ability of genetically engineered microorganisms (bacteria and yeast) or mammalian cells to emit visible light in response to specific substances. Such transformed cells are constructed by recombinant DNA techniques, using a plasmid vector containing a luminescent reporter gene under transcriptional control of a specific gene sequence, whose expression is regulated very precisely. The presence of the analyte induces the quantitative expression of the specific gene sequence and of the reporter gene, leading to the synthesis of the reporter protein that is detected by luminescent techniques. A number of luminescent recombinant cell-based biosensors have been developed for the rapid, sensitive, and selective detection of several inorganic and organic compounds.[53,54] Such biosensors are mainly used as first-level screening tools in the environmental monitoring of pollutants, thanks to the possibility of analyzing large numbers of samples in a short time (HTS assays). The use of low-light imaging could further improve the analytical throughput of the methods by allowing the simultaneous measurement of the light output from the microtiter wells.

A luminescence-based bacterial biosensor was used for the determination of mercury(II) in urine samples of subjects with dental amalgam restorations.[55] The system utilizes *Escherichia coli* as a host organism and firefly luciferase *luc* gene as a reporter under the control of the mercury-inducible promoter from the *mer* operon from transposon *Tn*21.[56] The biosensor showed a detection limit of 1.67×10^{-13} *M* Hg^{2+}, corresponding to 4.0×10^{-18} mol/well, with a dynamic range of six decades. The method fulfilled all the requirements of precision, accuracy, and robustness, and the detection limit was adequate to the Hg^{2+} levels in urine specimens, thus representing a straightforward, rapid, and sensitive tool to analyze Hg^{2+} in biological and environmental samples. The system appeared suitable for automation and the development of high-throughput screening assays. The use of 384-well microtiter plates and a CCD-based detector could easily allow the screening of more than 5000 samples/h.

In the case of some organic compounds, their ability to bind specific nuclear receptors is exploited to construct recombinant cell-based biosensors. A CL whole-cell biosensor based on recombinant yeast cells was developed and optimized for environmental monitoring of estrogen-like compounds. A recombinant yeast strain (*Saccharomyces cerevisiae*) in which the DNA sequence of the human estrogen receptor was stably integrated into the main chromosome was used. The yeast cells also contained an expression plasmid carrying estrogen-responsive sequences and the reporter gene *lac-Z*, encoding the enzyme β-galactosidase.[57] Upon binding an active ligand, the ligand-occupied receptor interacted with the estrogen-responsive elements, thus causing the expression of the reporter gene and the synthesis of the reporter enzyme whose activity was detected by a chemiluminescent 1,2-dioxetane-β-D-galactopyranoside substrate. The CL detection proved to be much more rapid than the conventional colorimetric one, allowing a detection limit of 10 pmol/l 17β-estradiol to be achieved after incubating the cells with the analyte for 24 h compared with the at least 72 h required with colorimetric detection. The applicability of the CL biosensor to the analysis of natural water samples was preliminarily checked by assessing samples from influent and effluent of an activated sludge sewage treatment plant.[58] The use of a CCD-based low-light imaging detector could reduce the measurement step, thus making the system even more suitable for environmental monitoring of estrogen-like compounds in surface waters and wastewaters.

37.4.3 Microarrays

CL imaging allowed the development of multiarray devices in which several different biospecific reagents are immobilized in an array arrangement on a suitable surface. This sensor design permitted parallel analyses to be performed on the same sample, allowing the simultaneous determination of different analytes and using reagent volumes much smaller than those needed to perform separate assays for each analyte of interest. Quantitation of the sample analyte content is performed by using imaging techniques: the position of the CL signal in the array and its intensity are related to the identity and the concentration of the analyte, respectively.

A parallel affinity sensor array based on CL labels for the detection of environmental contaminants in water was recently developed.[59] The required biospecific reagents (antibodies or haptens) were immobilized on a glass slide, thus achieving a biochip with an array of up to 1600 spots. The glass slide was inserted in a flow cell in which the immunoreactions took place and, after addition of the CL substrate, the luminescent signal was acquired by means of a CCD imaging device. The biochip was used to perform the simultaneous immunometric detection and quantitation of triazine, trinitrotoluene, and 2,4-dichlorophenoxyacetic acid (2,4-D) in water samples, both in the direct (immobilized antibody) and indirect (immobilized hapten) immunoassay formats. Both formats proved to be able to detect and quantitate the analytes, offering the possibility of many regeneration cycles of the biochip. This device could thus represent a potent analytical tool for performing multiple and simultaneous analysis on the same sample.

The development and optimization of a CL enzyme-linked immunosorbent assay (ELISA) for the simultaneous determination of 2,4-D in multiple samples, using innovative solid supports represented by gold-coated surfaces and glass capillaries with CCD imaging detection of the CL signal, were reported.[60] The authors observed that such solid supports offered advantages with respect to other supports used in the development of multiarray devices but the analytical sensitivity of the multiarray assay was still lower than that of the single sample assay.

37.4.4 Miscellaneous

An intensified CCD camera was used to evaluate the spatial distribution of the weak CL signal deriving from the slow auto-oxidation reactions and the process of water penetration into cereal food products.[61] The imaging technique showed the dynamics and the emission patterns of these processes. The assessment of the effect of antioxidants, as well as of radical promoters and scavengers, on the CL suggested that it derived, at least in part, from oxidative radical reactions.

CL imaging was used to count colonies of naturally luminescent or engineered bacteria in culture media[62] or to macroscopically visualize foci of virus infection in cell culture plates, after incubation with a CL enzyme-labeled specific antibody.[63] Imaging techniques permitted computer-assisted evaluation of parameters such as number, size, and luminescence intensity of colonies and virus foci, thus allowing a more accurate assessment of bacterial growth, extent of infection, and drug effects.

In recent years there has been a growing interest in the development of innovative techniques for monitoring either infection states or tumor growth in living animals. BL imaging offers a sensitive, quantitative, noninvasive, and real-time method to detect the disease progression and the efficacy of therapeutic agents. In addition, *in vivo* continuous detection on the same animal reduces the interexperiment variability. *Staphylococcus aureus* strains were transformed with plasmid DNA containing a genetically modified *Photorabdus luminescens lux* operon to obtain bacterial strains that are highly bioluminescent and do not require exogenous substrates.[64] These strains were used to monitor *S. aureus* infections in living mice by BL and to evaluate the effect of treatment with antimicrobial agents. A similar approach was used to develop a BL *E. coli* strain.[65] These BL bacteria permitted study of the correlation between BL levels and viable cell counts, both *in vitro* and *in vivo*, in the presence or absence of antimicrobial agents. In all cases, BL data correlated well with those obtained with conventional procedures (i.e., cell cultures), and the noninvasive approach produced results significantly faster and required a reduced number of animals. A constitutively expressed BL reporter gene encoding firefly luciferase was introduced into the chromosomes of the human cervical carcinoma cell line HeLa.[66] Small numbers of these labeled cells were transferred into immunodeficient mice and their distribution and growth were monitored by means of the light produced by the transformed cells and transmitted through the animals' tissues. With this methodology, both conventional chemotherapies and an immune cell therapy that uses cytokine-induced killer T cells were evaluated.

BL/CL imaging also represents a useful tool for evaluating the distribution of biomolecules immobilized on a given solid support, thus enabling to test the performance of the immobilization procedure. This approach was utilized to assess the immobilization of a protein on the gold surface of a quartz microbalance electrode.[67] The quartz microbalance allows the easy determination of the amount of protein on the electrode, but does not provide any information on its distribution on the electrode surface. By using suitable labeled proteins, CL imaging could allow comparison of different immobilization techniques in terms of both amount of immobilized protein and distribution on the solid support. In a recent paper, the performance of a simple and inexpensive device for the microdeposition of biomolecules on paper or plastic membranes, which was obtained by adapting a commercial ink-jet printer, was evaluated by means of CL imaging.[68] The system was used to obtain mono- and bidimensional arrays of spots containing HRP on cellulose paper and the quantity of HRP in each spot was measured by means of a CL reaction and a CCD-based low-light imaging luminograph (Figure 37.3b). This device proved to be rapid, reproducible, and easy-to-use, thus providing a suitable tool for the development of microarray-based bioanalytical microsystems.

37.5 APPLICATIONS ON MICROSAMPLES

BL/CL imaging coupled to optical microscopy represents a very useful tool for enzyme, metabolite, antigen, and nucleic acid localization and quantitation,[11,12] particularly when high detectability is required. The analytical performance of CL microscopy imaging in terms of detectability, precision, accuracy, and spatial resolution was evaluated. The system allowed for the detection of 400 amol of enzymes such as HRP or AP, with a spatial resolution as low as 1 μm and very low background. The analyte concentration in cell or tissue samples could be quantified using a calibration curve obtained by immobilizing different amounts of analyte on a target surface such as activated oxirane acrylic beads or nylon net.[5,19] This technique can be applied to any kind of specimen such as living cells, fixed cells, tissue cryosections, and paraffin-embedded sections. The resolving power of the CL image, obtained using a videocamera connected with an optical microscope, makes possible the localization of the CL signal in a tissue section or within a single cell, which is particularly suitable for analysis at subcellular level.[13,15,69]

37.5.1 Enzyme and Metabolite Localization

The spatial distribution and concentration of metabolites such as ATP, glucose, and lactate in normal tissue, tumors, and multicellular tumor spheroids used as a model system were determined. The method was based on single or coupled enzymatic reactions linking the metabolite of interest to firefly or bacterial luciferase, with consequent light emission. Structure-related distributions of the analytes were found in tumors, in contrast to more homogeneous distribution patterns in normal tissues.[17,70]

A method for the visualization of nitric oxide (NO) formation in cell cultures and living tissue was developed by using the luminol/H_2O_2 reaction to detect NO and a microscope coupled to a photon-counting camera to measure the light signal. This method has the potential for real-time imaging of NO formation, with high spatial and temporal resolution.[71]

A novel CL dynamic imaging method was developed to monitor ATP release from living cells using the firefly luciferin–luciferase reaction to detect ATP and an intensified CCD camera to measure and localize the photon emission. The assay had linear response over three orders of magnitude and a detection limit of 10^{-8} *M* ATP at millisecond exposure times.[72]

Low-light imaging allowed the localization and quantitation of calcium ions in the cytosol and in various compartments of living cells by using natural and targeted recombinant aequorins. Aequorin-based calcium imaging could be the method of choice for exploratory studies, since it is extremely sensitive, can detect a broad range of calcium concentrations, and allows for continuous recording during long periods of time; however, spatial resolution is lower than that obtained with fluorescence methods.[73,74]

Endogenous alkaline phosphatase activity was detected in rabbit intestine cryosections and localized in the epithelial cells with good resolution and sensitivity, and low aspecific signal (Figure 37.4). CL detection proved to be more sensitive than conventional histochemical colorimetric detection and also permitted a more accurate quantification of the enzyme.[5] Endogenous AChE activity was detected in rat coronal brain slices by using coupled enzymatic reactions terminating with light emission. The relative concentrations of the components of the CL reagent solution were optimized to avoid the diffusion of the light emission species in the surrounding solution, thus allowing a sharp localization of the CL signal with very low background emission. This imaging system could be a useful tool to study both the pathophysiological role of AChE distribution in the brain and the effect of *in vivo* administration of enzyme inhibitor drugs, providing a system that is more predictive than *in vitro* assay of inhibition activity.[50]

37.5.2 Immunohistochemical Assays

Immunohistochemistry (IHC) techniques provide important tools for the localization of specific antigens within individual cells. In CL IHC, the probes used are highly specific antibodies that bind to antigens such as proteins, enzymes, and viral or bacterial products. The bound specific antibody is revealed indirectly by species-specific or class-specific secondary antibodies conjugated to CL enzymes. The main advantage of using immunochemiluminescence techniques is that they allow more sensitive detection and permit quantification of tissue antigens.

A CL IHC method for the detection of epithelial components in thyroid tissue was developed. The CL detection, based on the use of HRP as a label and an enhanced CL reagent (ECL) as the substrate, proved to be more sensitive than the conventional chromogenic and fluorescent detection systems, with an adequate morphological resolution of the CL signal.[13]

A CL IHC technique was developed to localize interleukin 8 (IL-8) in gastric mucosa biopsy specimens. An increased IL-8 content was observed in epithelial cells in the presence of *Helicobacter pylori* infection (Figure 37.5), thus confirming earlier results achieved with an immunofluorescence technique.[39] The CL immunohistochemical localization of the neutrophil gelatinase associated lipocalin (NGAL), a protein thought to possess important immunomodulatory actions,

FIGURE 37.4 (Color figure follows p. 266.) CL localization of endogenous alkaline phosphatase in rabbit intestinal mucosa cryosection. Pseudocolor ruler on the right shows the relative light intensity.

in an *H. pylori* infected gastric mucosa cryosection is shown in Figure 37.6. The immunochemiluminescent approach could be used to localize discrete foci of pathogens using antibodies against conserved bacterial proteins and to identify bacterial phenotype using antisera against known strain-specific virulence determinants, thus allowing identification of bacterial phenotype *in situ* without microbial culture.

The immunohistochemical localization of von Willebrand factor in human tonsil tissue sections was performed using a HRP-labeled secondary antibody and colorimetric or CL substrates.[75] CL (Figure 37.7) allowed the rapid detection of the label, requiring a short incubation time (2 min) with the substrate, with a good spatial resolution of the light signal. Moreover, conventional chromogenic substrates for HRP, such as diaminobenzidine (DAB), are very toxic and require special handling and disposal procedures.

Immunocytochemical methods were developed to detect specific antibodies in sera from patients affected by different infectious diseases. P3-HR1 cells, which express Epstein–Barr virus–induced virus capsid antigens (VCA), were used for the search of specific human IgM to VCA in patients with infectious mononucleosis. After treatment of cells with serial dilutions of sera, HRP-conjugated anti-IgM antibody was added and detected with CL substrate.[5] Human embryo lung fibroblasts infected with a reference laboratory strain of Herpes simplex virus (HSV) type 2 were used to detect antibody to HSV type 2 in serum samples. After treatment of cells with serial dilutions of sera, HRP-labeled immunoglobulins to human IgG were added and detected with CL substrate.[50]

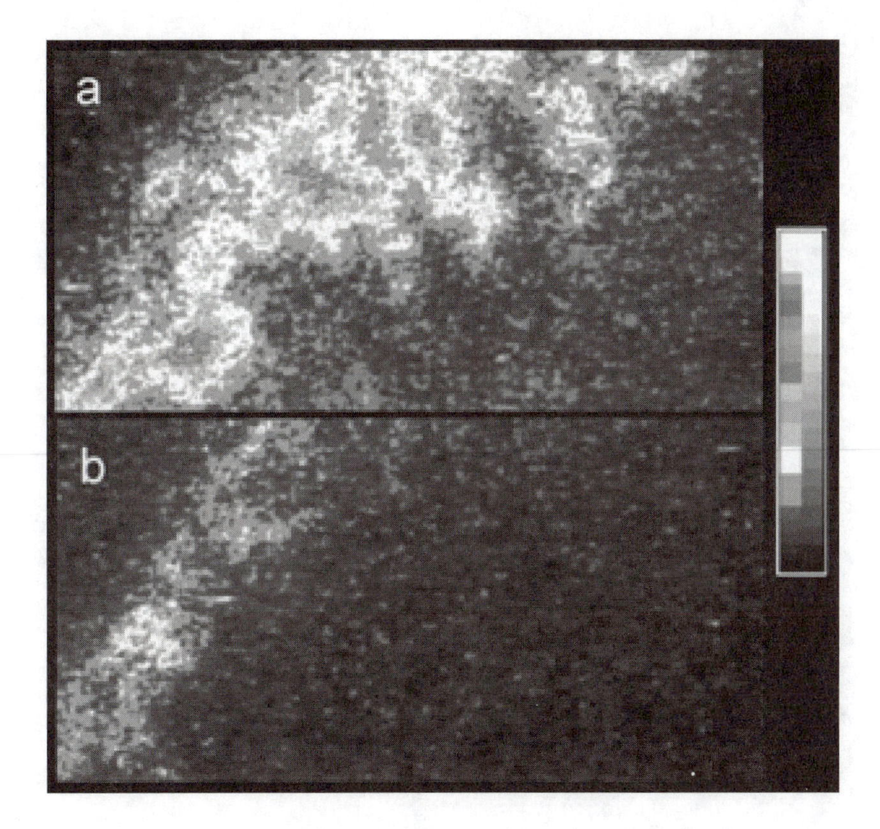

FIGURE 37.5 (Color figure follows p. 266.) Immunochemiluminescent localization of IL-8 in gastric mucosa cryosections of a patient infected with *H. pylori* (a) and an *H. pylori* negative control subject (b). The pseudocolor-processed CL signals are reported, with the pseudocolor ruler on the right showing the relative light intensity. (Courtesy of Dr. J. E. Crabtree, St. James's University Hospital, Leeds, U.K.)

37.5.3 *In Situ* Hybridization Assays

The *in situ* hybridization (ISH) technique is suitable for the localization of specific nucleic acids within individual cells with the preservation of cellular and tissue morphology, thus permitting the simultaneous assessment of the morphological alterations associated with the lesion. Moreover, the possibility of simultaneous detection of two or more CL probes can improve the diagnostic significance of such technique.[76] In recent years there has been a growing interest in the application of ISH for the rapid, specific, and reliable diagnosis of viral diseases, especially for those viruses that cannot be diagnosed by isolation procedures.

A CL ISH assay for the detection of human papillomavirus (HPV) DNA was developed, in which the hybridization reaction was performed using digoxigenin-, biotin-, or fluorescein-labeled probes. The hybrids were visualized using AP as the enzyme label and a highly sensitive 1,2-dioxetane phosphate as the CL substrate. This assay was applied to biopsy specimens from different pathologies associated with HPV, which had previously proved positive for HPV DNA by polymerase chain reaction (PCR). The analytical sensitivity was assessed using samples of HeLa and CaSki cell lines, whose content in HPV DNA is known (10 to 50 copies of HPV 18 DNA in HeLa cells and 400 to 600 copies of HPV 16 DNA in CaSki cells). The CL ISH assay proved sensitive and specific using digoxigenin-, biotin-, or fluorescein-labeled probes and provided an objective evaluation of the results. CL detection was more sensitive than colorimetric detection: clinical specimens that tested slightly positive at the colorimetric analysis were clearly positive with CL

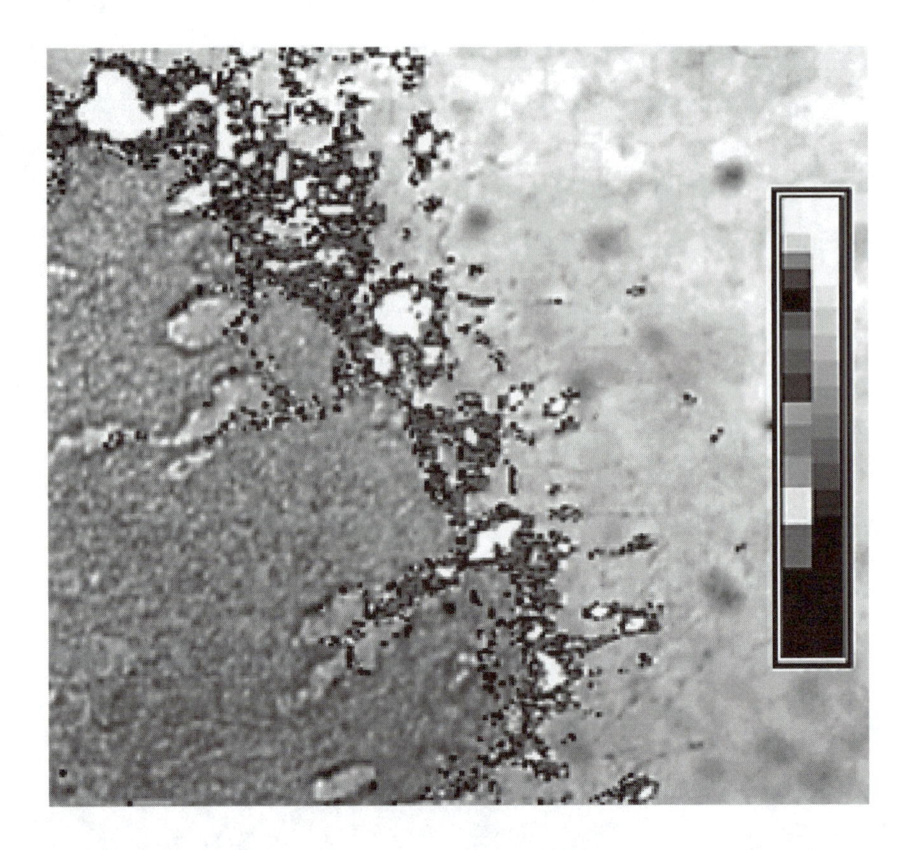

FIGURE 37.6 (Color figure follows p. 266.) Immunochemiluminescent localization of NGAL in *H. pylori*-infected gastric mucosa cryosection. Pseudocolor ruler on the right shows the relative light intensity. (Courtesy of Dr. J. E. Crabtree, St. James's University Hospital, Leeds, U.K.)

detection (Figure 37.8) and HeLa cells that tested negative in the colorimetric ISH assay proved to be positive in the CL assay.[77]

A different CL ISH assay for the detection of HPV DNA was developed, in which the hybridization reaction was performed using biotin-labeled probes detected by a two-step immunocytochemical reaction, using HRP-labeled secondary antibodies. The developed method was compared with a radioisotopic method using 35S-labeled secondary antibodies for autoradiography. The analysis of both human carcinoma cell lines containing known numbers of copies of HPV DNA (CaSki, HeLa, and SiHa cells) and biopsy specimens of human cervical preneoplastic and neoplastic lesions demonstrated that CL gave results comparable to those obtained with autoradiography. In addition, only 10 min of photon accumulation was required with the CL detection compared with the 3-week exposure necessary for 35S autoradiography.[78]

B19 parvovirus DNA was detected in bone marrow cells employing digoxigenin-labeled B19 DNA probes, antidigoxigenin Fab fragment conjugated with AP, and a 1,2-dioxetane phosphate CL substrate (Figure 37.9). The CL ISH was applied to samples that had been previously tested for B19 DNA content using ISH with colorimetric detection, dot blot hybridization, and nested PCR. The assay proved specific and showed an increased sensitivity in detecting B19 DNA when compared to ISH with colorimetric detection, being able to find a higher number of positive cells per 100 counted cells with highly statistical significant difference.[14] This CL ISH assay was applied for the prenatal diagnosis of parvovirus B19–induced hydrops fetalis.[79]

Cytomegalovirus (CMV) DNA in cultured CMV infected cells and in different clinical samples (tissue sections and cellular smears) was detected using digoxigenin-labeled probes, antidigoxigenin Fab fragments labeled with AP, and the 1,2-dioxetane phosphate CL substrate. The presence of

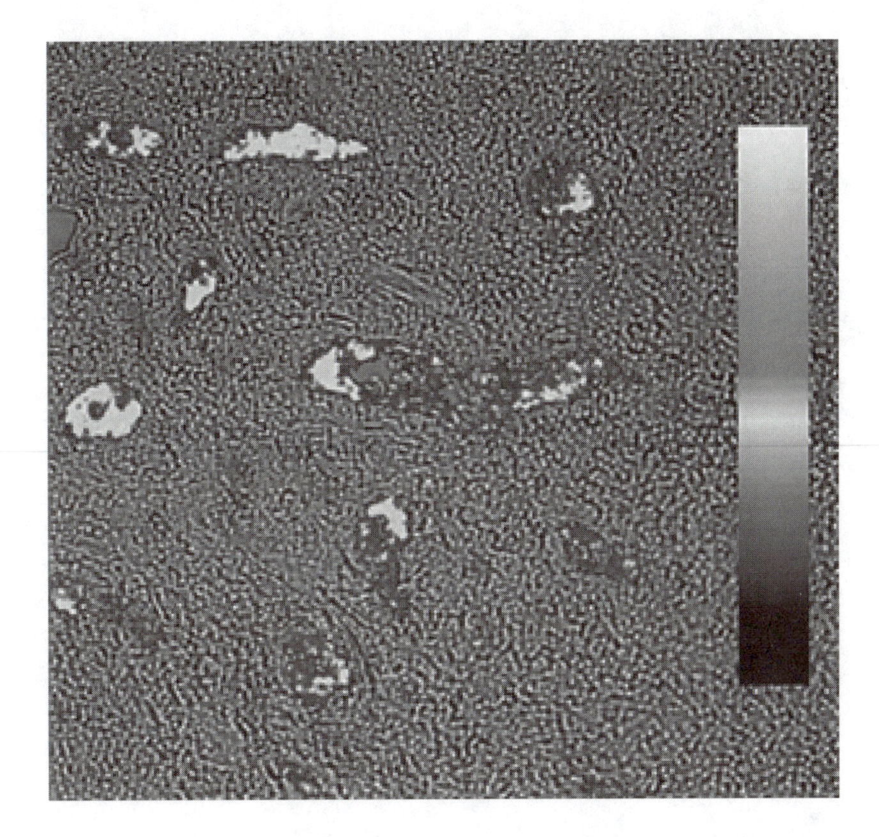

FIGURE 37.7 (Color figure follows p. 266.) Immunochemiluminescent localization of von Willebrand factor in endothelial cells of a paraffin-embedded section of human tonsil tissue. Pseudocolor ruler on the right shows the relative light intensity. (Courtesy of Dr. P. Chieco, Institute of Oncology "F. Addarii," Bologna, Italy.)

hybridized CMV DNA was observed in infected cells fixed at various times after infection, and it was possible to measure increasing light emission values thus following the CMV replication cycle. When the assay was performed on clinical samples from patients with acute CMV infections, CMV DNA was detected in all the positive samples tested, both in cellular samples and tissue sections.[15]

Herpes simplex virus (HSV) in human fibroblasts was detected using biotin-labeled HSV DNA probes, streptavidin–HRP complex, and ECL-enhanced substrate for HRP. The presence of HSV DNA was observed in cells infected with clinical samples known to contain the HSV fixed at 48-h postinfection, with a sharp topographical localization and a good preservation of cellular morphology.[50]

A double CL ISH assay for the simultaneous detection of two different viral DNAs (HSV and CMV DNAs) was developed utilizing both HRP and AP as reporter enzymes. A biotinylated HSV DNA probe and a digoxigenin-labeled CMV DNA probe were cohybridized with samples, and then the CL detection of the two probes was performed. The HSV DNA was revealed using a streptavidin–HRP complex amplified with biotinyl tyramide and a luminol-based CL substrate for HRP, whereas CMV DNA was detected by means of antidigoxigenin Fab fragments conjugated with AP and a dioxetane phosphate derivative as the CL substrate. Both HSV and CMV DNAs were found in infected cells with distinct localization and absence of cross reactions. In the double CL ISH, the enzymatic reaction of HRP is usually performed before that of AP. This is because the HRP–luminol reaction rapidly reaches a steady-state light output that is maintained for a relatively short time, whereas the kinetics of the AP–dioxetane reaction is slower and the CL emission lasts for a relatively long time. The sequential measure of the signals in double CL ISH represents a limitation with respect to other detection principles (colorimetry, fluorescence) that allow simultaneous double staining.[76]

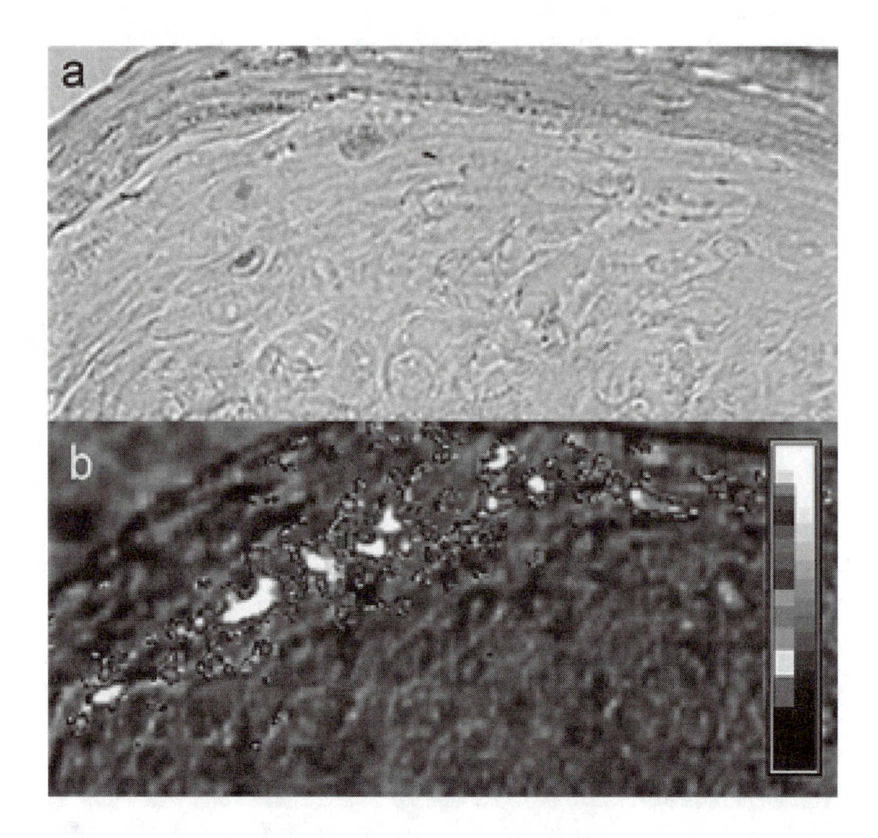

FIGURE 37.8 (Color figure follows p. 266.) ISH for the detection of human papillomavirus DNA in skin tissue cryosections with colorimetric (a) and CL (b) detection. Pseudocolor ruler on the right of b shows the relative light intensity.

37.6 CONCLUSIONS AND FUTURE PERSPECTIVES

BL/CL imaging proved to be a potent and versatile detection technique, thus representing a valuable bioanalytical tool suitable for a wide range of applications.

The major advantages of low-light imaging in the analysis of macrosamples are as follows: either multiple samples or multiple analytes in the same sample can be measured simultaneously using the microtiter plate and microarray formats, thus allowing the development of high-throughput screening assays; a quantitative and objective analysis can be performed in blot membrane format assays; the dynamics and spatial distribution of biological processes can be monitored *in vitro* and *in vivo*, thus permitting the real-time evaluation of disease progression and the efficacy of therapeutic agents. On the other hand, the main limitations of CCD-based imaging are the narrower dynamic range and the slightly lower sensitivity with respect to photomultiplier-tube-based detection devices.

BL/CL imaging coupled with optical microscopy allowed improvement in the detectability of labeled probes compared to colorimetric and fluorescence detection methods and paralleled the performance achieved with radioisotopic probes. This characteristic combined with the appropriate spatial resolution of the analytical signal makes BL/CL imaging a powerful diagnostic tool for the rapid and early diagnosis of infectious diseases and pathological conditions in general. Furthermore, once standardized, CL imaging is suitable for the straightforward, precise, and accurate quantitative detection of the labeled probes, which represents an important advance over conventional detection methods. The multistep procedure required to localize analytes would be facilitated by the use of an optical microscope equipped with a computer-assisted device able to locate precisely the same sample spots. The diagnostic power of such technique would be further increased by optimizing

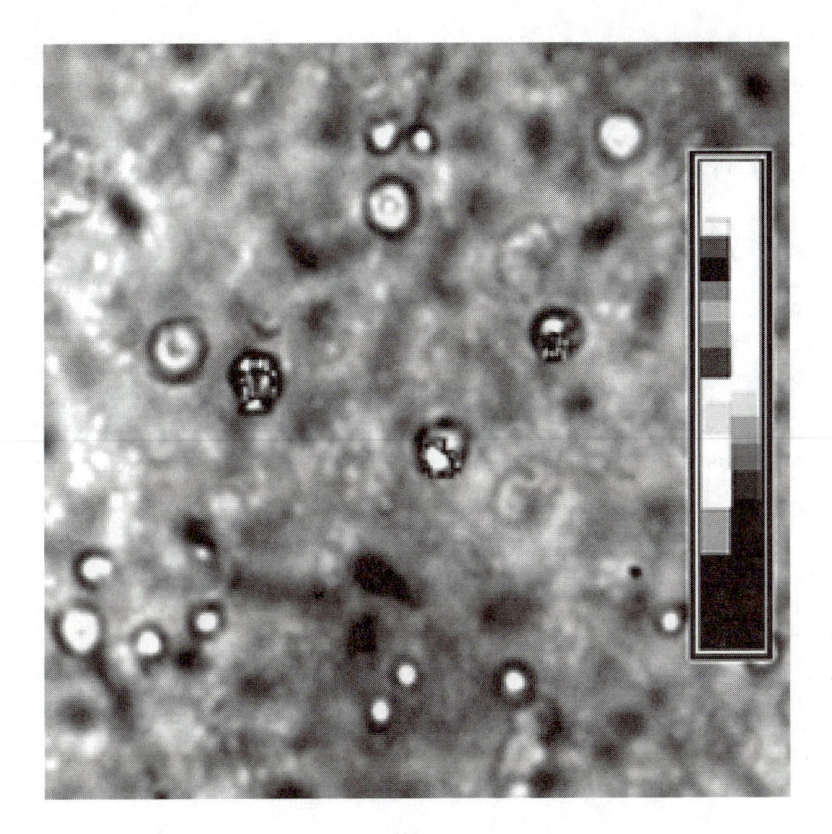

FIGURE 37.9 (Color figure follows p. 266.) Chemiluminescent ISH for the detection of parvovirus B19 DNA in bone marrow cells. Pseudocolor ruler on the right shows the relative light intensity.

the transmitted light image quality with appropriate staining of the samples, to better define the morphological structure.

An important recent trend in analytical chemistry and biotechnology has been toward micro-miniaturization using chip microfabrication technology. Miniaturized CL reactors, high-density arrays of reaction vessels, and microchip-based analyzers, which require very small amounts of samples and reagents, have been developed. BL and CL represent a convenient detection principle for these devices because of the high detectability of the luminescence signal. In conclusion, CL imaging is an important tool in the modern biotechnological and bioanalytical laboratory, at times complementary to other systems but, in many cases, unique.

REFERENCES

1. Stanley, P. E., Commercially available luminometers and low-light imaging devices, *Methods Enzymol.,* 305, 96, 2000.
2. Wick, R. A., Photon counting imaging: applications in biomedical research, *BioTechniques,* 7, 262, 1989.
3. Bräuer, R., Lübbe, B., Ochs, R., Helma, H., and Hofmann, J., Measuring luminescence with a low light-level imaging system using electronic light standards, in *Bioluminescence and Chemiluminescence: Status Report,* Szalay, A. A., Kricka, L. J., and Stanley, P. E., Eds., Wiley, Chichester, U.K., 1993, 13.
4. Hooper, C. E. and Ansorge, R. E., Quantitative photon imaging in the life sciences using intensified CCD cameras, in *Bioluminescence and Chemiluminescence: Current Status,* Stanley, P. E. and Kricka, L. J., Eds., Wiley, Chichester, U.K., 1991, 337.

5. Roda, A., Pasini, P., Musiani, M., Girotti, S., Baraldini, M., Carrea, G., and Suozzi, A., Chemiluminescent low-light imaging of biospecific reactions on macro- and microsamples using a videocamera-based luminograph, *Anal. Chem.,* 68, 1073, 1996.
6. Hooper, C. E., Ansorge, R. E., Browne, H. M., and Tomkins, P., CCD imaging of luciferase gene expression in single mammalian cells, *J. Biolumin. Chemilumin.,* 5, 123, 1990.
7. Nicolas, J. C., Applications of low-light imaging to life sciences, *J. Biolumin. Chemilumin.,* 9, 139, 1994.
8. Hooper, C. E., Ansorge, R. E., and Rushbrooke, J. G., Low-light imaging technology in the life sciences, *J. Biolumin. Chemilumin.,* 9, 113, 1994.
9. Hiraoka, Y., Sedat, J. W., and Agard, D. A., The use of a charge-coupled device for quantitative optical microscopy of biological structures, *Science,* 238, 36, 1987.
10. Roda, A., Pasini, P., Musiani, M., and Baraldini, M., Chemiluminescence imaging systems for the analysis of macrosamples: microtiter format, blot membrane, and whole organs, *Methods Enzymol.,* 305, 120, 2000.
11. Muller-Klieser, W., Walenta, S., Paschen, W., Kallinowski, F., and Vaupel, P., Metabolic imaging in microregions of tumors and normal tissues with bioluminescence and photon counting, *J. Natl. Cancer Inst.,* 80, 842, 1988.
12. Hawkins, E. and Cumming, R., Enhanced chemiluminescence for tissue antigen and cellular viral DNA detection, *J. Histochem. Cytochem.,* 38, 415, 1990.
13. Lorimier, P., Lamarcq, L., Labat-Moleur, F., Guillermet, C., Berthier, R., and Stoebner, P., Enhanced chemiluminescence: a high-sensitivity detection system for *in situ* hybridization and immunohistochemistry, *J. Histochem. Cytochem.,* 41, 1591, 1993.
14. Musiani, M., Roda, A., Zerbini, M. L., Gentilomi, G., Pasini, P., Gallinella, G., and Venturoli, S., Detection of parvovirus B19 DNA in bone marrow cells by chemiluminescence *in situ* hybridization, *J. Clin. Microbiol.,* 34, 1313, 1996.
15. Musiani, M., Roda, A., Zerbini, M., Pasini, P., Gentilomi, G., Gallinella, G., and Venturoli., S., Chemiluminescent *in situ* hybridization for the detection of cytomegalovirus DNA, *Am. J. Pathol.,* 148, 1105, 1996.
16. Martin, C. S. and Bronstein, I., Imaging of chemiluminescent signals with cooled CCD camera systems, *J. Biolumin. Chemilumin.,* 9, 145, 1994.
17. Muller-Klieser, W. and Walenta, S., Geographical mapping of metabolites in biological tissue with quantitative bioluminescence and single photon imaging, *Histochem. J.,* 25, 407, 1993.
18. Mirasoli, M., Pasini, P., Russo, C., Lotierzo, M., Valenti, P., Guardigli, M., and Roda, A., Chemiluminescence in high throughput screening for drug development on 384 well microtiter format, in *Bioluminescence and Chemiluminescence: Perspectives for the 21st Century,* Roda, A., Pazzagli, M., Kricka, L. J., and Stanley, P. E., Eds., Wiley, Chichester, U.K., 1999, 524.
19. Roda, A., Pasini, P., Baraldini, M., Musiani, M., Gentilomi, G., and Robert, C., Chemiluminescent imaging of enzyme-labeled probes using an optical microscope-videocamera luminograph, *Anal. Biochem.,* 257, 53, 1998.
20. Nicolas, J. C., Balaguer, P., Térouanne, B., Villebrun, M. A., and Boussioux, A. M., Detection of glucose 6-phosphate dehydrogenase by bioluminescence, in *Nonisotopic Probing, Blotting and Sequencing,* Kricka, L. J., Ed., Academic Press, San Diego, 1995, chap. 10.
21. Térouanne, B., Carrie, M. L., Nicolas, J. C., and Crastes de Paulet, A., Bioluminescent immunosorbent for rapid immunoassays, *Anal. Biochem.,* 154, 118, 1986.
22. Martin, C. S., Olesen, C. E. M., Liu, B., Voyta, J. C., Shumway, J. L., Juo, R. R., and Bronstein, I., Continuous sensitive detection of β-galactosidase with a novel chemiluminescent 1,2-dioxetane, in *Bioluminescence and Chemiluminescence: Molecular Reporting with Photons,* Hastings, J. W., Kricka, L. J., and Stanley, P. E., Eds., Wiley, Chichester, U.K., 1997, 525.
23. Baret, A., Detection of xanthine oxidase by enhanced chemiluminescence, in *Nonisotopic Probing, Blotting and Sequencing,* Kricka, L. J., Ed., Academic Press, San Diego, 1995, chap. 11.
24. Baret, A. and Fert, V., T4 and TSH ultrasensitive immunoassays using luminescent enhanced xanthine oxidase assay, *J. Biolumin. Chemilumin.,* 4, 149, 1989.
25. Campbell, A. K., *Chemiluminescence, Principles and Applications in Biology and Medicine,* VCH, Cambridge, 1988, chap. 5.
26. Thorpe, G. H. G. and Kricka, L. J., Enhanced chemiluminescent reactions catalyzed by horseradish peroxidase, *Methods Enzymol.,* 133, 331, 1986.

27. Thorpe, G. H. G. and Kricka, L. J., Enhanced chemiluminescence assays for horseradish peroxidase: characteristics and applications, in *Bioluminescence and Chemiluminescence: New Perspectives,* Schölmerich, J., Andreesen, R., Kapp, A., Ernst, M., and Woods, W. G., Eds., Wiley, Chichester, U.K., 1987, 199.
28. Akhavan-Tafti, H., DeSilva, F., Arghavani, Z., Eickholt, R. A., Handley, R. S., and Schaap, P. A., Lumigen® PS: new chemiluminescent substrates for the detection of horseradish peroxidase, in *Bioluminescence and Chemiluminescence,* Campbell, A. K., Kricka, L. J., and Stanley, P. E., Eds., Wiley, Chichester, U.K., 1994, 199.
29. Bronstein, I., Voyta, J. C., and Edwards, B., A comparison of chemiluminescent and colorimetric subtrates in a hepatitis B virus DNA hybridization assay, *Anal. Biochem.,* 180, 95, 1989.
30. Beck, S. and Koster, H., Applications of dioxetane chemiluminescent probes to molecular biology, *Anal. Chem.,* 62, 2258, 1990.
31. Hauber, R., Miska, W., Schleinkofer, L., and Geiger, R., New, sensitive, radioactive-free bioluminescence-enhanced detection system in protein blotting and nucleic acid hybridization, *J. Biolumin. Chemilumin.,* 4, 367, 1989.
32. Miska, W. and Geiger, R., Synthesis and characterization of luciferin derivatives for use in bioluminescence-enhanced enzyme immunoassays. New ultrasensitive detection systems for enzyme immunoassays, *J. Clin. Chem. Clin. Biochem.,* 25, 23, 1987.
33. Miska, W. and Geiger, R., Luciferin derivatives in bioluminescence-enhanced enzyme immunoassays, *J. Biolumin. Chemilumin.,* 4, 119, 1990.
34. DeLuca, M. and McElroy, W. D., Purification and properties of firefly luciferase, *Methods Enzymol.,* 57, 3, 1978.
35. Stults, N. L., Stocks, N. A., Cummings, R. D., Cormier, M. J., and Smith, D. F., Applications of recombinant bioluminescent proteins as probes for proteins and nucleic acids, in *Bioluminescence and Chemiluminescence: Current Status,* Stanley, P. E. and Kricka, L. J., Eds., Wiley, Chichester, U.K., 1991, 533.
36. Wood, K. V., Luc genes: introduction of colour into bioluminescence assay, *J. Biolumin. Chemilumin.,* 5, 107, 1990.
37. Hastings, J. W., Bacterial bioluminescence: an overview, *Methods Enzymol.,* 57, 125, 1978.
38. Actor, J. K., Nolte, F. S., and Smith, D. F., Aqualite®, the bioluminescent label for immunoassays, gene detection and molecular diagnostics, in *Bioluminescence and Chemiluminescence: Perspectives for the 21st Century,* Roda, A., Pazzagli, M., Kricka, L. J., and Stanley, P. E., Eds., Wiley, Chichester, U.K., 1999, 83.
39. Roda, A., Musiani, M., Pasini, P., Baraldini, M., and Crabtree, J. E., *In situ* hybridization and immunohistochemistry with enzyme-triggered chemiluminescent probes, *Methods Enzymol.,* 305, 577, 2000.
40. Bronstein, I., Voyta, J. C., Lazzari, K. G., Murphy, O., Edwards, B., and Kricka, L. J., Rapid and sensitive detection of DNA in Southern blots with chemiluminescence, *BioTechniques,* 8, 310, 1990.
41. Girotti, S., Musiani, M., Pasini, P., Ferri, E., Gallinella, G., Zerbini, M. L., Roda, A., Gentilomi, G., and Venturoli, S., Application of a low-light imaging device and chemiluminescent substrates for quantitative detection of viral DNA in hybridization reactions, *Clin. Chem.,* 41, 1693, 1995.
42. Chou, S., Newer methods for diagnosis of cytomegalovirus infection, *Rev. Infect. Dis.,* 12, 727, 1990.
43. Gentilomi, G., Musiani, M., Zerbini, M., Gallinella, G., Gibellini, D., and La Placa, M., A hybrido-immunocytochemical assay for the *in situ* detection of cytomegalovirus DNA using digoxigenin-labeled probes, *J. Immunol. Methods,* 125, 177, 1989.
44. Musiani, M., Zerbini, M., Gentilomi, G., Gallinella, G., Venturoli, S., Gibellini, D., and La Placa, M., Rapid detection of cytomegalovirus DNA in urine samples with a dot blot hybridization immunoenzymatic assay, *J. Clin. Microbiol.,* 28, 2101, 1990.
45. Girotti, S., Musiani, M., Ferri, E., Gallinella, G., Zerbini, M., Roda, A., Gentilomi, G., and Venturoli, S., Chemiluminescent immunoperoxidase assay for the dot blot hybridization detection of parvovirus B19 DNA using a low light imaging device, *Anal. Biochem.,* 236, 290, 1996.
46. Karger, A. E., Weiss, R., and Gesteland, R. F., Digital chemiluminescence imaging of DNA sequencing blots using a charge-coupled device camera, *Nucleic Acid Res.,* 20, 6657, 1992.
47. Manaresi, E., Pasini, P., Gallinella, G., Gentilomi, G., Venturoli, S., Roda, A., Zerbini, M., and Musiani, M., Chemiluminescence Western blot assay for the detection of immunity against parvovirus B19 VP1 and VP2 linear epitopes using a videocamera based luminograph, *J. Virol. Methods,* 81, 91, 1999.

48. Yasui, T. and Yoda, K., Imaging of *Lactobacillus brevis* single cells and microcolonies without a microscope by an ultrasensitive chemiluminescent enzyme immunoassay with a photon-counting television camera, *Appl. Environ. Microbiol.,* 63, 4528, 1997.
49. Speroni, E., Guerra, M. C., Rossetti, A., Pozzetti, L., Sapone, A., Paolini, M., Cantelli-Forti, G., Pasini, P., and Roda, A., Anti-oxidant activity of *Pueraria lobata* (Willd.) in the rat, *Phytother. Res.,* 10, S95, 1996.
50. Pasini, P., Musiani, M., Russo, C., Valenti, P., Aicardi, G., Crabtree, J. E., Baraldini, M., and Roda, A., Chemiluminescence imaging in bioanalysis, *J. Pharm. Biomed. Anal.,* 18, 555, 1998.
51. Roda, A., Manetta, A. C., Piazza, F., Simoni, P., and Lelli, R., A rapid and sensitive 384-microtiter wells format chemiluminescent enzyme immunoassay for clenbuterol, *Talanta,* 52, 311, 2000.
52. Bacigalupo, M. A., Ius, A., Simoni, P., Piazza, F., Setchell, K. D. R., and Roda, A., Analytical performance of luminescent immunoassays of different format for serum daidzein analysis, *Fresenius J. Anal. Chem.,* 370, 82, 2001.
53. Köhler, S., Belkin, S., and Schmid, R. D., Reporter gene bioassays in environmental analysis, *Fresenius J. Anal. Chem.,* 366, 769, 2000.
54. Ramanathan, S., Ensor, M., and Daunert, S., Bacterial biosensors for monitoring toxic metals, *Trends Biotechnol.,* 15, 500, 1997.
55. Roda, A., Pasini, P., Mirasoli, M., Guardigli, M., Russo, C., Musiani, M., and Baraldini, M., Sensitive determination of urinary mercury(II) by a bioluminescent transgenic bacteria-based biosensor, *Anal. Lett.,* 34, 29, 2001.
56. Virta, M., Lampinen, J., and Karp, M., A luminescence-based mercury biosensor, *Anal. Chem.,* 67, 667, 1995.
57. Routledge, E. J. and Sumpter, J. P., Estrogenic activity of surfactants and some of their degradation products assessed using a recombinant yeast screen, *Environ. Toxicol. Chem.,* 15, 241, 1996.
58. Pasini, P., Gentilomi, G., Guardigli, M., Baraldini, M., Musiani, M., and Roda, A., A chemiluminescent whole cell biosensor for assessing estrogenic activity, in *Bioluminescence and Chemiluminescence 2000,* Case, J. F., Herring, P. J., Robison, B. H., Haddock, S. H. D., Kricka, L. J., and Stanley, P. E., Eds., World Scientific Publishing Company, Singapore, in press.
59. Weller, M. G., Schuetz, A. J., Winklmair, M., and Niessner, R., Highly parallel affinity sensor for the detection of environmental contaminants in water, *Anal. Chim. Acta,* 393, 29, 1999.
60. Dzgoev, A., Mecklenburg, M., Xie, B., Miyabayashi, A., Larsson, P.-O., and Danielsson, B., Optimization of a charge coupled device imaging enzyme linked immuno sorbent assay and supports for the simultaneous determination of multiple 2,4-D samples, *Anal. Chim. Acta,* 347, 87, 1997.
61. Sltawinska, D. and Sltawinski, J., Chemiluminescence of cereal products III. Two-dimensional photocounting imaging of chemiluminescence, *J. Biolumin. Chemilumin.,* 13, 21, 1998.
62. Minko, I., Holloway, S. P., Nikaido, S., Carter, M., Odom, O. W., Johnson, C. H., and Herrin, D. L., Renilla luciferase as a vital reporter for chloroplast gene expression in Chlamydomonas, *Mol. Gen. Genet.,* 262, 421, 1999.
63. Heider, H. and Schroeder, C., Focus luminescence assay: macroscopically visualized foci of human cytomegalovirus and varicella zoster virus infection, *J. Virol. Methods,* 66, 311, 1997.
64. Francis, K. P., Joh, D., Bellinger-Kawahara, C., Hawkinson, M. J., Purchio, T. F., and Contag, P. R., Monitoring bioluminescent *Staphylococcus aureus* infections in living mice using a novel luxABCDE construct, *Infect. Immun.,* 68, 3594, 2000.
65. Rocchetta, H. L., Boylan, C. J., Foley, J. W., Iversen, P. W., LeTourneau, D. L., McMillian, C. L., Contag, P. R., Jenkins, D. E., and Parr, T. R., Jr., Validation of a noninvasive, real-time imaging technology using bioluminescent *Escherichia coli* in the neutropenic mouse thigh model of infection, *Antimicrob. Agents Chemother.,* 45, 129, 2001.
66. Sweeney, T. J., Mailänder, V., Tucker, A. A., Olomu, A. B., Zhang, W., Cao, Y., Negrin, R. S., and Contag, C. H., Visualizing the kinetics of tumor-cell clearance in living animals, *Proc. Natl. Acad. Sci. U.S.A.,* 96, 12,044, 1999.
67. Roda, A., Pasini, P., Musiani, M., Baraldini, M., Guardigli, M., Mirasoli, M., and Russo, C., Bioanalytical applications of chemiluminescent imaging, in *Chemiluminescence in Analytical Chemistry,* Baeyens, W. and García-Campaña, A. M., Eds., Marcel Dekker, New York, 2001, 473.
68. Roda, A., Guardigli, M., Russo, C., Pasini, P., and Baraldini, M., Protein microdeposition using a conventional ink-jet printer, *BioTechniques,* 28, 492, 2000.

69. Musiani, M., Pasini, P., Zerbini, M. L., Roda, A., Gentilomi, G., Gallinella, G., Venturoli, S., and Manaresi, E., Chemiluminescence: a sensitive detection system in *in situ* hybridization, *Histol. Histopathol.,* 13, 243, 1998.
70. Walenta, S., Doetsch, J., Mueller-Kieser, W., and Kunz-Schughart, L. A., Metabolic imaging in multicellular spheroids of oncogene-transfected fibroblasts, *J. Histochem. Cytochem.,* 48, 509, 2000.
71. Wiklund, N. P., Iversen, H. H., Leone, A. M., Cellek, S., Brundin, L., Gustafsson, L. E., and Moncada, S., Visualization of nitric oxide formation in cell cultures and living tissue, *Acta Physiol. Scand.,* 167, 161, 1999.
72. Wang, Z., Haydon, P. G., and Yeung, E. S., Direct observation of calcium-independent intercellular ATP signaling in astrocytes, *Anal. Chem.,* 72, 2001, 2000.
73. Creton, R., Kreiling, J. A., and Jaffe, L. F., Calcium imaging with chemiluminescence, *Microsc. Res. Tech.,* 46, 390, 1999.
74. Rutter, G. A., Burnett, P., Rizzuto, R., Brini, M., Murgia, M., Pozzan, T., Tavare, J. M., and Denton, R. M., Subcellular imaging of intramitochondrial Ca 21 with recombinant targeted aequorin. Significance for the regulation of pyruvate dehydrogenase activity, *Proc. Natl. Acad. Sci. U.S.A.,* 93, 5489, 1996.
75. Roda, A., Guardigli, M., Pasini, P., and Baraldini, M., Development of a time-resolved fluorescence microscope for imaging analysis, in *Bioluminescence and Chemiluminescence 2000,* Case, J. F., Herring, P. J., Robison, B. H., Haddock, S. H. D., Kricka, L. J., and Stanley, P. E., Eds., World Scientific Publishing Company, Singapore, in press.
76. Gentilomi, G., Musiani, M., Roda, A., Pasini, P., Zerbini, M., Gallinella, G., Baraldini, M., Venturoli, S., and Manaresi, E., Co-localization of two different viral genomes in the same sample by double-chemiluminescence *in situ* hybridization, *BioTechniques,* 23, 1076, 1997.
77. Musiani, M., Zerbini, M. L., Venturoli, S., Gentilomi, G., Gallinella, G., Manaresi, E., La Placa, M., D'Antuono, A., Roda, A., and Pasini, P., Sensitive chemiluminescence *in situ* hybridization for the detection of human papillomavirus genomes in biopsy specimens, *J. Histochem. Cytochem.,* 45, 729, 1997.
78. Lorimier, P., Lamarcq, L., Negoescu, A., Robert, C., Labat-Moleur, F., Gras-Chappuis, F., Durrant, I., and Brambilla, E., Comparison of ^{35}S and chemiluminescence for HPV *in situ* hybridization in carcinoma cell lines and on human cervical intraepithelial neoplasia, *J. Histochem. Cytochem.,* 44, 665, 1996.
79. Musiani, M., Pasini, P., Zerbini, M., Gentilomi, G., Roda, A., Gallinella, G., Manaresi, E., and Venturoli, S., Prenatal diagnosis of parvovirus B19-induced hydrops fetalis by chemiluminescence *in situ* hybridization, *J. Clin. Microbiol.,* 37, 2326, 1999.

38 Quantification of the α_3 Subunit of the Na^+/K^+-ATPase in Cerebellar Purkinje Neurons Using Luminescence

Peggy Biser, Jian-Qiang Kong, William Fleming, and David Taylor

CONTENTS

38.1 INTRODUCTION

The study of the postnatal development of the cerebellar Purkinje cells of the rat led to the observation that between 10 and 20 days an increase in the resting membrane potential occurs.[1] [^{3}H]-Ouabain binding in whole cerebellum correlated this hyperpolarization with an increased amount of the of the sodium (Na^+) pump.[1] The binding sites for ouabain, Na^+, and potassium (K^+), as well as the ATPase activity are found on the α subunit of the heterodimer, and the β subunit influences K^+ occlusion, regulates Na^+ and K^+ affinities of the α subunit, and is important in the delivery of the α subunit to the plasma membrane.[2] Western blots identified the α_3 subunit as the specific α subunit isoform involved and slot blots demonstrated a statistically significant difference in the amount of that subunit in 13- and 19-day-old animals.[3] Like the [^{3}H]-ouabain binding experiments that preceded them, however, Western and slot blots were necessarily limited to studies of the whole cerebellum.

0-8493-0719-8/02/$0.00+$1.50

The technique described in this chapter extended the study to the level of the various layers of the cerebellum, including the Purkinje cell layer. Thin tissue sections were labeled with antibodies specific for the α_3 subunit of the Na^+/K^+-ATPase. These primary antibodies were detected using fluorescently labeled second antibodies. The fluorescent light emitted by these tags produced images that were captured using a confocal microscope. Computer analysis of those images revealed that the electrophysiological data, obtained at the level of single Purkinje cells, was supported and extended by these methods, which were at both the tissue level and the level of the single cell.

38.2 EQUIPMENT AND REAGENTS

38.2.1 Animals

Sprague–Dawley rats, 13 and 19 days old, were sacrificed via decapitation in accordance with procedures approved by the Animal Care and Use Committee. All efforts were employed to minimize animal suffering.

38.2.2 Equipment

The cryostat used was a Jung Frigocut 2800E. The confocal microscope used was a Zeiss LSM 510.

38.2.3 Chemicals/Materials

The following chemicals and material supplies were obtained from Fisher Scientific (Pittsburgh, PA): reagent-grade paraformaldehyde; certified ACS-grade KH_2PO_4, Na_2HPO_4, and NaCl; Fisher*finest* premium cover glass (thickness #1); and Fisher*finest* premium frosted microscope slides. Picric acid in the form of the saturated aqueous solution was a product of Ricca Chemical Company (Arlington, TX). Fraction V bovine serum albumin (BSA) was obtained from Sigma/Aldrich (St. Louis, MO). O.C.T. Compound was obtained from VWR (Pittsburgh, PA). Fluoromount-G was a product of Southern Biotechnology Associates, Inc. (Birmingham, AL).

38.2.4 Antibodies

The antibody used against the α_3 subunit of the Na^+/K^+-ATPase was a monoclonal of mouse origin, XVIF9-G10, obtained from Affinity Bioreagents, Inc. (Golden, CO).[4] The second antibody was fluorescein (FITC)-conjugated affinity purified donkey anti-mouse IgG (H + L) with minimal cross-reactivity to bovine chicken, goat, guinea pig, syrian hamster, horse, human, rabbit, rat, and sheep serum proteins was obtained from Jackson Immuno Research Laboratories, Inc. (West Grove, PA).

38.3 PROCEDURES

Figure 38.1 gives an overview of the methodology used.

38.3.1 Tissue Fixation

1. 13- and 19-day-old rats from two litters were sacrificed on the same day as described.
2. The cerebellum was removed and placed in 10 ml of paraformaldehyde fixative buffer (2% w/v paraformaldehyde, 0.2% w/v picric acid, 0.024 *M* KH_2PO_4, and 0.12 *M* Na_2HPO_4, pH 7.3)[5] for 3 h at room temperature. In our laboratory this fixative was prepared in the following manner:
    a. A 1-l solution consisting of 3.30 g of KH_2PO_4 and 17.74 g of Na_2HPO_4 was prepared (solution A).

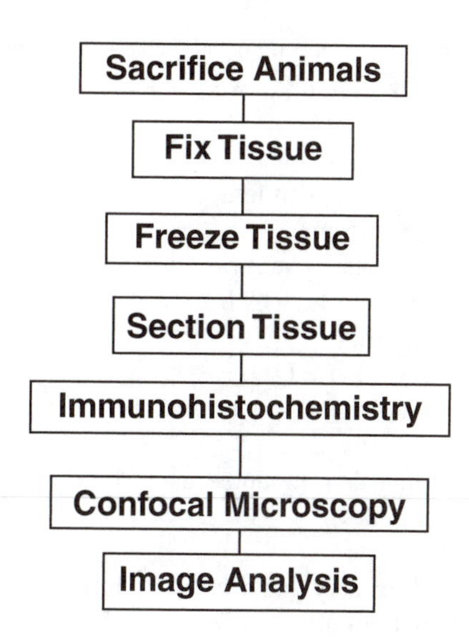

FIGURE 38.1 Overview of the procedure used.

   b. 20.0 g of paraformaldehyde was heated with 150 ml of saturated picric acid to 65°C with stirring. NaOH pellets were then added until the solution became a clear yellow (solution B).
   c. Solution A was added to solution B to approximately 800 ml. The pH was adjusted to 7.3 with concentrated HCl and then brought to a final volume of 1 l with solution B.
3. Tissues were rinsed three times in 10 ml of phosphate-buffered saline (PBS) (0.1 *M*, pH 7.8) and then stored in 10 ml PBS overnight at 4°C.[6]

Comments: The volume of fixative used was in excess of the recommended minimum of 20 times the tissue volume.[7] CH_3OH fixation followed by embedding with 30% sucrose as described by Peng et al.[8] has also provided good results for tissue fixation for immunohistochemistry in this laboratory.

38.3.2 Freezing

This technique was initially described by Dey et al.[6,9]

4. Pieces of cork slightly larger than the tissues to be frozen were wet down with PBS and then spotted with O.C.T. Compound.
5. Fixed tissues were coated with a slurry of O.C.T. Compound/PBS, attached to the prepared cork, and quick-frozen in isopentane chilled with $N_{2(l)}$.
6. Frozen tissues were stored at –70°C in airtight plastic bags until sectioned.

38.3.3 Immunohistochemistry

The procedure was a modification of a protocol from Upstate Biotechnology, Inc. The tissues from two 13-day-old animals and two 19-day-old animals were processed on a given day. All efforts were made to use identical assay conditions on sets of tissues processed on different days.

7. 16 μm sagittal sections were obtained using a cryostat and placed on subbed coverslips.
8. The sections were washed for 15 min (all washes took place at room temperature in PBS in Columbia jars).
9. Sections were blocked (this and all subsequent incubations took place in a humid box) with 8% BSA/PBS for 30 min at room temperature.
10. The sections were washed for 15 min.
11. Sections were incubated with 7.5 μg/ml primary antibody in 0.5% BSA/PBS overnight at 4°C, along with negative controls, which were incubated with vehicle only.
12. Sections were washed for 30 min.
13. FITC-conjugated donkey anti-mouse IgG (H + L) at a final dilution of 1:100 in PBS was applied to the sections, which were then incubated for 3 h in the dark.
14. The final 30-min wash was in the dark.
15. Coverslips were air-dried and then mounted using Fluoromount-G. Slides were stored at 4°C in the dark until viewed.

Comments: The concentrations of primary and secondary antibodies used were empirically determined to yield optimum staining with minimal background. The primary antibody has also yielded excellent results in guinea pig tissues in this laboratory.

38.3.4 Confocal Microscopy

16. Slides were viewed using a Zeiss LSM 510 confocal microscope. For the purpose of quantitation 200× images of 512 × 512 pixels were captured along with corresponding controls stained with second antibody only. Tissues were viewed under identical conditions.

38.3.5 Quantification of Images

Optimas® 6.2 imaging software was used to quantify 200× images.

17. The Purkinje cell layer: The soma and the nuclear region for each Purkinje neuron were circled and the mean scaled luminance of pixels (ArGV) was determined for the area within each circle (Figure 38.2). The ratio of these two values was then calculated for each cell. Between 5 and 21 cells were measured per animal and then averaged to give a mean value for that animal.
18. Granular Layer, Molecular Layer, and White Matter: A 9 × 9 μm square was placed at random within the molecular layer and the white matter. Within the granular layer the square was placed on regions that were stained. The mean scaled luminance of pixels (ArGV) was determined for the area within each square. In all, 11 squares were measured per layer per slide unless a given layer was not visible in the micrograph. These values were averaged for each animal and divided by the mean scaled luminance of pixels for the nuclei in the Purkinje cell layer for that animal.
19. Statistics: Data are reported as the mean ± SEM. All data were evaluated using *t*-tests with the exception of the granular layer and white matter adjusted optical density (OD) comparisons (see below), which, due to unequal variances, were evaluated using the Mann–Whitney rank sum test.

Comments: The question asked in this study was whether a difference in the levels of the α_3 subunit existed in the layers of the cerebellum in 13-day-old and 19-day-old rats. The measurement of luminance (gray values) and the comparison of those values to respective background luminance permitted the determination of a difference between age groups. Since luminance was not linear

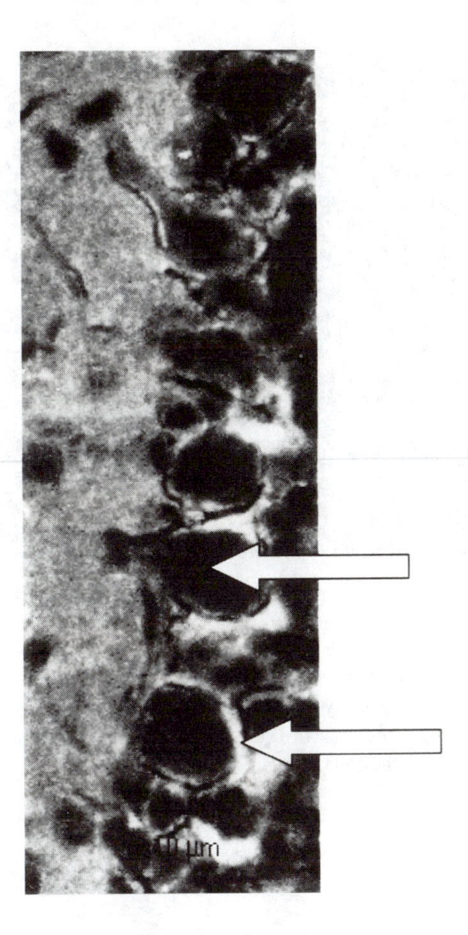

FIGURE 38.2 (Color figure follows p. 266.) 400× confocal micrograph (2048 × 2048 pixels) of cerebellar Purkinje cells from a 19-day-old animal stained with anti-α_3 subunit antibody XVIF9-G10. The upper arrow points to the nuclear region of a Purkinje cell. The lower arrow points to the Purkinje cell soma.

with respect to concentration, it did not allow the computation of a relative difference between age groups. For this purpose OD was calculated, which was linear with respect to concentration. The adjusted OD[10] was calculated using the equation:

$$OD_{\text{layer}} - OD_{\text{Purkinje cell nucleus}} = \log(ArGV_{\text{Purkinje cell nucleus}}/ArGV_{\text{layer}})$$

38.4 RESULTS

Figure 38.3 depicts the differences seen in the level of the expression of the α_3 subunit of the Na^+-pump. The differences in the Purkinje cell layer and in the white matter that are apparent upon visual inspection are supported by the results shown in Figures 38.3 and 38.4 and summarized in Table 38.1. In the Purkinje cell layer the ratio of the adjusted OD values of 19-day-old to 13-day-old rats is 1.59. In the white matter the ratio is 1.71. This indicates a relative increase of 59 and 71% in the expression of the α_3 subunit in 19-day-old rat cerebellar Purkinje cells and white matter vs. the same regions in 13-day-old rats. No difference in the distribution of the antigen in the molecular layers of 13- and 19-day-old animals was demonstrated.

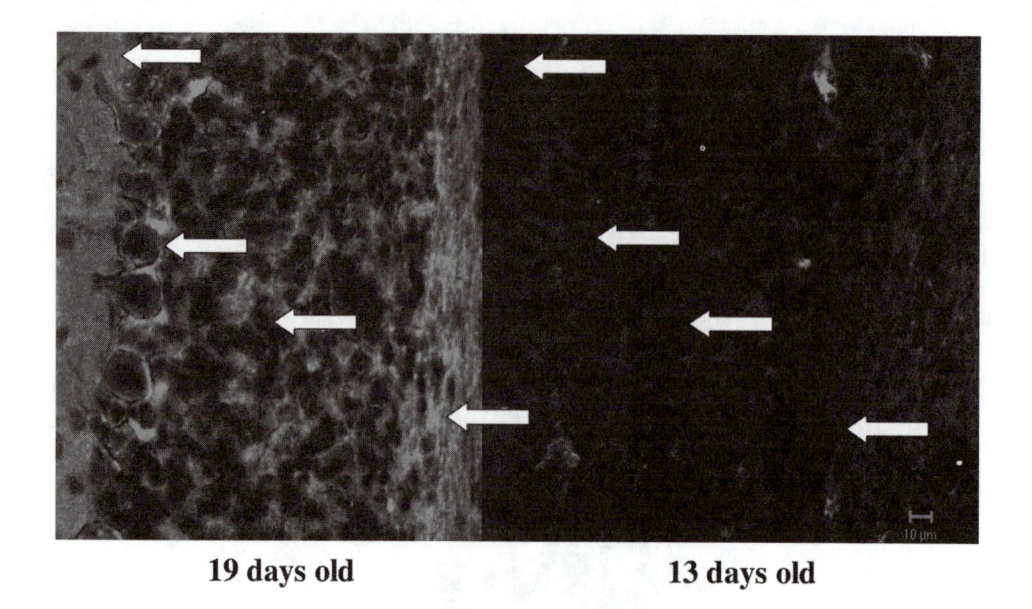

FIGURE 38.3 (Color figure follows p. 266.) 400× confocal micrographs (2048 × 2048 pixels each) of sagittal sections of the cerebella from 13- and 19-day-old animals stained with anti-α_3 subunit antibody XVIF9-G10. The arrows point to corresponding structures in both 13- and 19-day-old sections. From top to bottom the arrows point to: the molecular layer, the Purkinje cell layer, the granular cell layer, and the white matter.

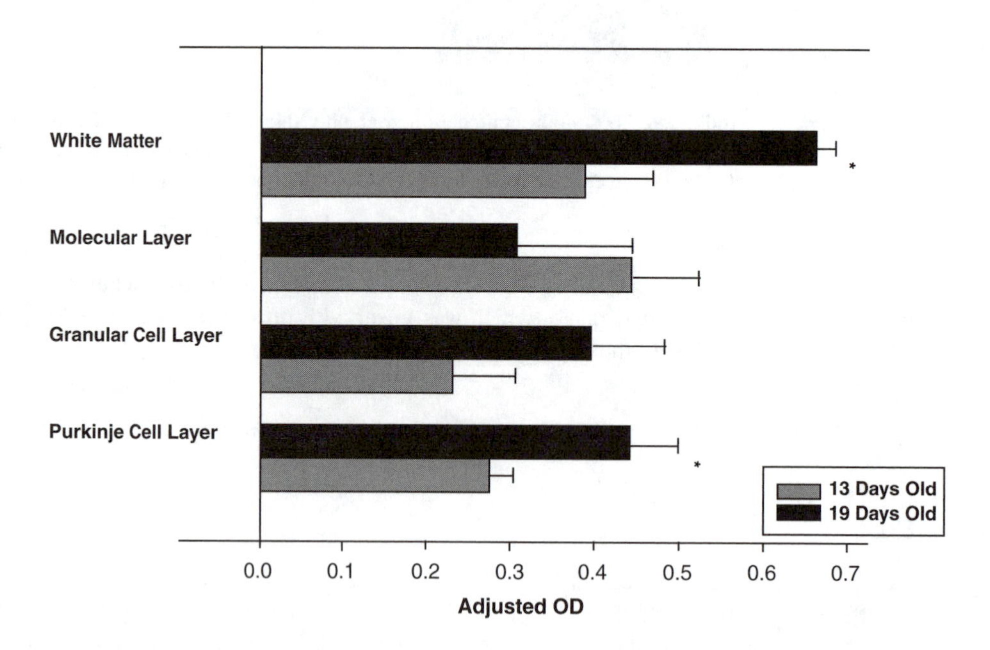

FIGURE 38.4 The graph shows the relative differences in the amount of the α_3 subunit of the Na^+/K^+-ATPase in the cerebellar layers of 13- and 19-day-old animals.

TABLE 38.1
Visually Apparent Differences in the Purkinje Cell Layer and in the White Matter

	Purkinje Cell Layer	Granular Cell Layer	Molecular Layer	White Matter
Ratio of ArGV	$p < 0.05$	No difference	No difference	$p < 0.01$
Adjusted OD	$p < 0.05$	No difference	No difference	$p < 0.03$

38.5 CONCLUSION

The combination of confocal microscopy and fluorescent immunohistochemistry has permitted this laboratory to identify differences in the localization of a specific isoform of the α subunit of the Na^+/K^+-pump at the tissue and cellular level within the rat cerebellum in animals 13 and 19 days of age. Coupled with computer-based image analysis and conversion of luminance to optical density values, these techniques have provided a means of identifying and quantifying relative differences in the distribution of the α_3 subunit between the animals of different age groups.

ACKNOWLEDGMENTS

Special thanks to Dr. Richard Dey, Brian Satterfield, Tanya Mulvey, Jeff Altemus, and Allison Reed. This work was supported in part by grants from the National Institutes of Drug Abuse.

REFERENCES

1. Molnar, L. R., Thayne, K. A., Fleming, W. W., and Taylor, D. A., The role of the sodium pump in the development regulation of membrane electrical properties of cerebellar Purkinje neurons of the rat, *Devl. Brain Res.,* 112, 287, 1999.
2. Blanco, G. and Mercer, R. W., Isozymes of the Na,K-ATPase: heterogeneity in structure, diversity in function, *Am. J. Physiol.,* 275 (Renal Physiol. 44), F633, 1998.
3. Biser, P., Thayne, K., Kong, J.-Q., Fleming, W., and Taylor, D., Developmental changes in the α_3 subunit of the Na^+, K^+ pump in rat cerebellar Purkinje neurons, *FASEB J.,* A139, 1999 (Abstr.).
4. Arystarkhova, E. and Sweadner, K. J., Isoform-specific monoclonal antibodies to Na,K-ATPase α subunits, *J. Biol. Chem.,* 271, 23,407, 1996.
5. Stefanini, M., de Martino, C., and Zamboni, L., Fixation of ejaculated spermatozoa for electron microscopy, *Nature,* 216, 173, 1967.
6. Dey, R. D., Shannon, W. A., Jr., and Said, S. I., Localization of VIP-immunoreactive nerves in airways and pulmonary vessels of dogs, cats, and human subjects, *Cell Tissue Res.,* 220, 231, 1981.
7. Høyer, P. E., Lyon, H., Møller, M., Prentø, P., van Deurs, B., Hasselager, E., and Anderson, A. P., Tissue Processing: III. fixation, general aspects, *Theory and Strategy in Histochemistry,* Lyon, H., Springer-Verlag, Berlin, 1991, chap. 12.
8. Peng, L., Martin-Vasallo, P., and Sweadner, K. J., Isoforms of Na,K-ATPase α and β subunits in the rat cerebellum and in granule cell cultures, *J. Neurosci.,* 17(10), 3488, 1997.
9. Dey, R. D., Hoffpauir, J., and Said, S. I., Co-localization of vasoactive intestinal peptide- and substance P-containing nerves in cat bronchi, *Neuroscience,* 24(1), 275, 1988.
10. Smolen, A. J., Image analytic techniques for quantification of immunohistochemical staining in the nervous system, in *Methods in Neurosciences,* Vol. 3, Conn, P. M., Ed., Academic Press, San Diego, 1990, chap. 11.

39 The Influence of Plates on Luminescence Measurement

Paul Hengster, Thomas Eberl, Walter Mark, Marialuise Kunc, and Wolfgang Steurer

CONTENTS

39.1 INTRODUCTION

Chemiluminescence (CL) today is widely used in molecular biology and serology to detect oxygen-free radicals and adenosine triphosphate.[1–3] CL is light emission as a result of a chemical reaction. This light is detected by a photomultiplier with a high dynamic range, which allows quantitative measurement in an extended linear range from zero to millions of counts. The biological systems so employed, however, may be influenced by various conditions, such as storage, protein content, temperature, pH value and concentration of the light-emitting substance like luminol or firefly luciferase.[4] Therefore, we used a chemically defined light standard for our measurements.

To cope with a large number of samples, CL is performed in microplates.[5–9] Various commercially available plates in white or black plastic material can be used in all CL counters. There are special plates with clear bottoms, suited for cell culturing and light microscopy, plates with smaller-diameter wells for the measurement of small volumes, and black plates with white inserts to reduce the cross talk of white plates. It is not known, however, whether all microtiter plates give the same results in CL counters.

39.2 MATERIALS AND METHODS

39.2.1 PLATES

The following plates were used in all experiments. There were three types of black white plates used:

1. One solid black plate (Dynatech MicroFluor®, Chantilly, VA);
2. A black plate with clear bottom suitable for light microscopy (Isoplate® 1450-571, Wallac, Turku, Finland);

0-8493-0719-8/02/$0.00+$1.50

3. A black plate with white inserts, which should combine the low cross talk of black plates with the higher sensitivity of white plates (Isoplate 1450-581, Wallac, Turku, Finland).

Regarding white plates, we tested five types: two different solid white plates, Polysorb 437842 (Nunc, Roskilde, Denmark) and 3917 (Corning Costar Corporation, Cambridge, MA). There were two plates with clear bottom available to us, the 1450-514 (Wallac) and the 3610 (Corning Costar Corporation). In addition we tested white plate with smaller-diameter wells designed for small volumes up to 100 μl (3693, Corning Costar Corporation). For background measurements plates were stored either in daylight or in the dark covered with aluminum foil.

39.2.2 Reagents

As the source of light we used the luminometric light standard with four different 1,2-dioxetanes that emit light upon decomposition (Lumigen, Inc., Southfield, MI). Four different standards with concentrations of 167, 944, 11,290, and 11,3900 relative light units in 100 μl are supplied. These standards emit light in a temperature-dependent fashion constantly over a time period of weeks. When stored in the dark, the emission of light is constant with a SD of 0.001 to 1.362. We used 10 μl of the most highly concentrated light standard per well in each experiment, except for determination of cross talk, when 100 μl was used.

39.2.3 Instrument and Measurements

The instrument used was the luminometer Lucy III (Anthos, Salzburg, Austria), which is combined with a temperature control unit. Plates were entered into the instrument in the dark. Measurements were performed at a constant temperature of 24°C eight times every 2.5 min. The first measurement was performed after a 5-min incubation time to avoid signals due to exposure to light before entering the plate into the counter.

39.2.4 Statistical Methods

No mathematical correction of counts was performed and results are expressed in kilocounts. Standard measurements were taken at least as quadruplicates. Results obtained were expressed as means of maximum values. The Excel program was used for analysis and graphics. Tests were performed at least twice. Cross talk was calculated as percent light detection in adjacent cells.

39.2.5 Experiments

In a first set of experiments, the background light emission of the plates was measured after storage of the plates at daylight or in the dark for 1 day. The measurement was performed for a 1-h period to establish a decrease in counts and to evaluate the steadiness of the background signals.

The second measurements dealt with the question of the sensitivity and cross talk of these various plates, by using the chemically defined standard as the light source in wells A1, A12, C3, C6, C9, E4, E7, E10, F6, H12.

The third question to be answered was the influence of volume on the signal detected. For this, 5 μl of the light standard was added to row 1, and together with 50, 100, 150, 200, and 300 μl of luminescence buffer into rows 3, 5, 7, 9, 11 with the exception of plate type 8 with a maximum volume of 150 μl (Figure 39.1).

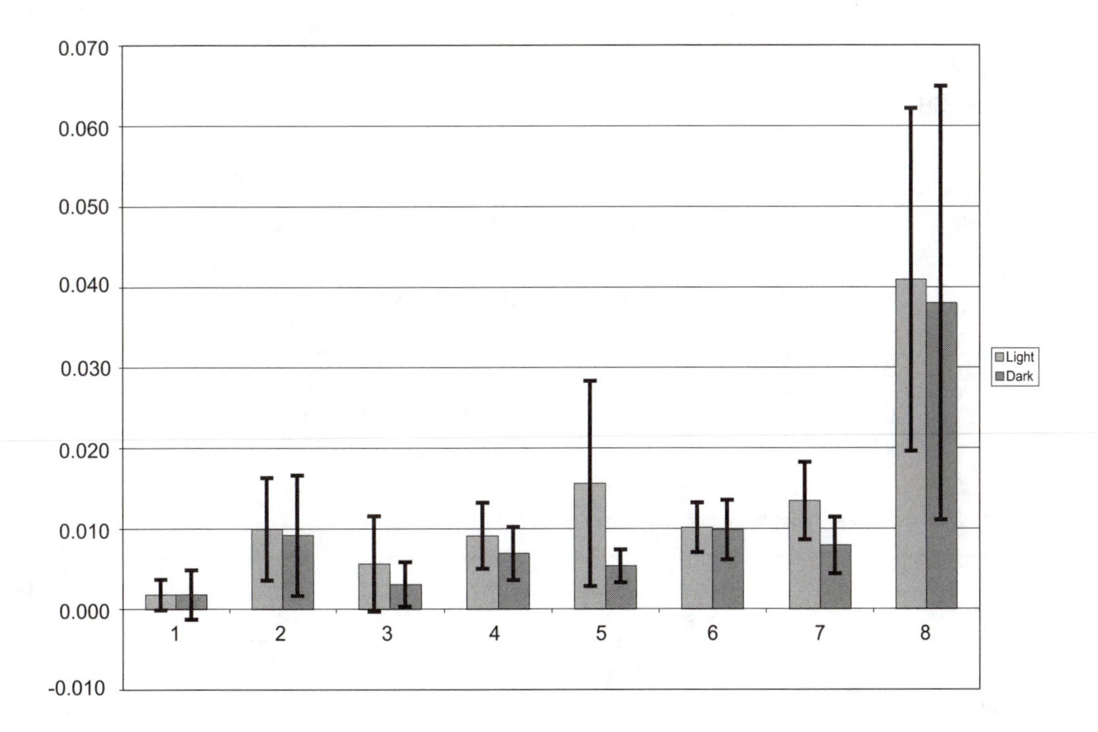

FIGURE 39.1 Autoluminescence of various plates after storage in daylight or in the dark. 1, Black plate Dynatech MicroFluor. 2, Black plate with clear bottom Isoplate 1450-571. 3, Black plate with white inserts Isoplate 1450-581. 4, White plate Polysorb 437842 Nunc. 5, White plate 3917 Corning Costar. 6, White plate clear bottom Wallac 1450-514. 7, White plate clear bottom 3610, Corning Costar. 8, White plate with small-diameter 3693, Corning Costar.

39.3 RESULTS

Background activity of black plates is slightly lower when compared with white plates and as much as 60% lower when stored in the dark. Only measurement of the plate for low volumes resulted in higher counts, up to 70 counts in the first measurement. Nevertheless, all values obtained with the other plates stayed below 30 counts (Figure 39.1). After the first 5 min a decrease of background can be observed but, interestingly, during the observation period of 1 h increases as well as decreases were observed (Figure 39.2). Sensitivity was found to be highest in plate 8 for low volumes. A high sensitivity was observed with white plates 5 and 3. White plate 6, designed with a clear bottom, had a lower sensitivity, which was even lower in plate 7 and white plate 4. The lowest counts were obtained with black plates (Table 39.1).

With regard to cross talk, black plates were superior to white plates. With white plate 5 and plate 8 for low volumes, highly satisfactory results were obtained. Only with plates having a clear bottom and with white plate 4, a cross talk of up to 0.7% was measured (see Table 39.1).

Volume had no significant influence on the number of counts detected (Figure 39.3).

39.4 DISCUSSION

CL has become a widely used routine diagnostic tool in research as well as in the clinical setting. To measure large numbers of samples, vials have been largely replaced with microplates.[5–9] Furthermore, some microplates are suitable for cell culturing with microscopic control. Amazingly, so far no systematic investigations have been conducted with an eye to comparing various commercially available microplates for their influence on CL.

TABLE 39.1
Sensitivity and Cross Talk

		Cross Talk			
Plate	Sensitivity	First Adjacent	Oblique Adjacent	Second Adjacent	Oblique Second Adjacent
1	9.2×10^3 (±SD 0.03)	3.63×10^{-4}	23.86×10^{-4}	2.37×10^{-4}	5.86×10^{-4}
2	1.0×10^4 (±SD 0.06)	2.04×10^{-3}	2.64×10^{-4}	2.31×10^{-4}	-1.7×10^{-4}
3	1.3×10^5 (±SD 25.96)	3.96×10^{-4}	1.43×10^{-4}	1.15×10^{-4}	5.43×10^{-5}
4	4.4×10^4 (±SD 28.62)	7.70×10^{-3}	4.01×10^{-3}	1.83×10^{-3}	6.90×10^{-3}
5	2.0×10^5 (±SD 29.45)	8.29×10^{-4}	3.43×10^{-4}	9.63×10^{-5}	9.61×10^{-5}
6	1.2×10^5 (±SD 3.53)	6.71×10^{-3}	2.31×10^{-3}	3.52×10^{-4}	2.16×10^{-4}
7	4.3×10^4 (±SD 0.38)	5.33×10^{-3}	2.18×10^{-3}	6.14×10^{-4}	4.62×10^{-4}
8	1.4×10^5 (±SD 51.50)	5.44×10^{-4}	3.19×10^{-4}	1.36×10^{-4}	9.01×10^{-5}

Straight and oblique adjacent in first and second order.

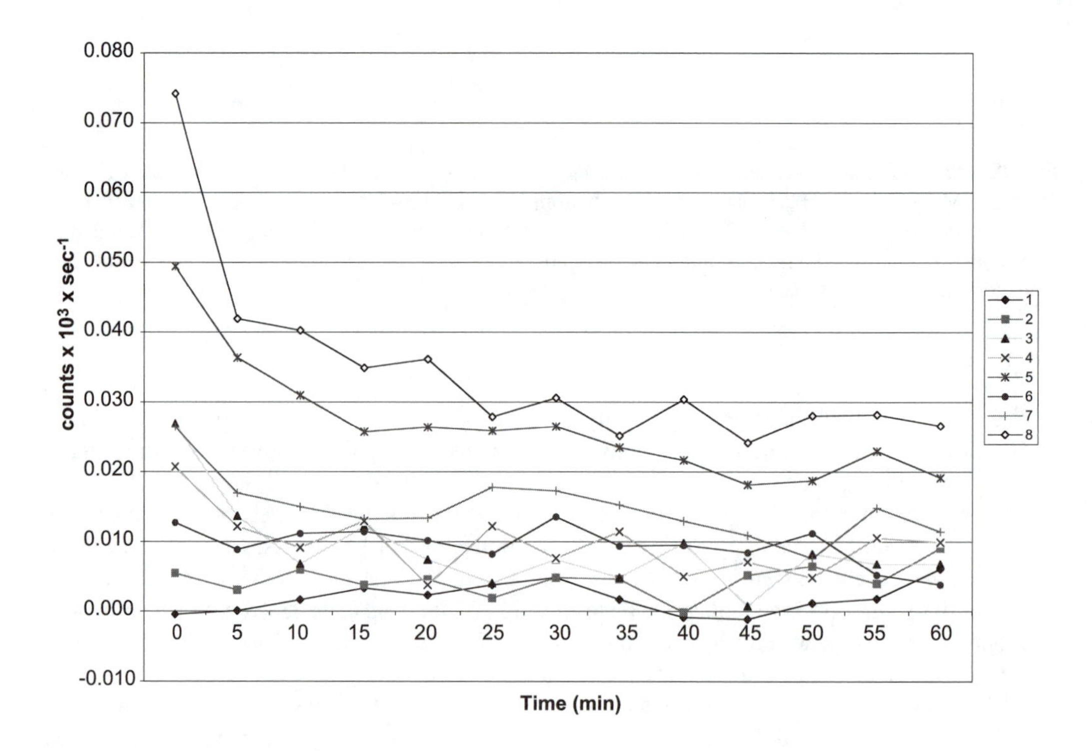

FIGURE 39.2 Vacillation of autoluminescence. Numbers of plates as in Figure 39.1.

The emission of light by the plates themselves is rather low even when stored in daylight and does not exceed 20 counts for black plates and 30 counts for most white plates. Only white plate 8 of one manufacturer shows values between 20 and 65 counts. Nevertheless, for measurements with low signals, the storage and selection of plates may be of concern. The emission of light is not constant and varies in a range of mean values between 3 and 40 counts. The high standard deviation as seen in Figure 39.1 is due to the low counts and the high variability.

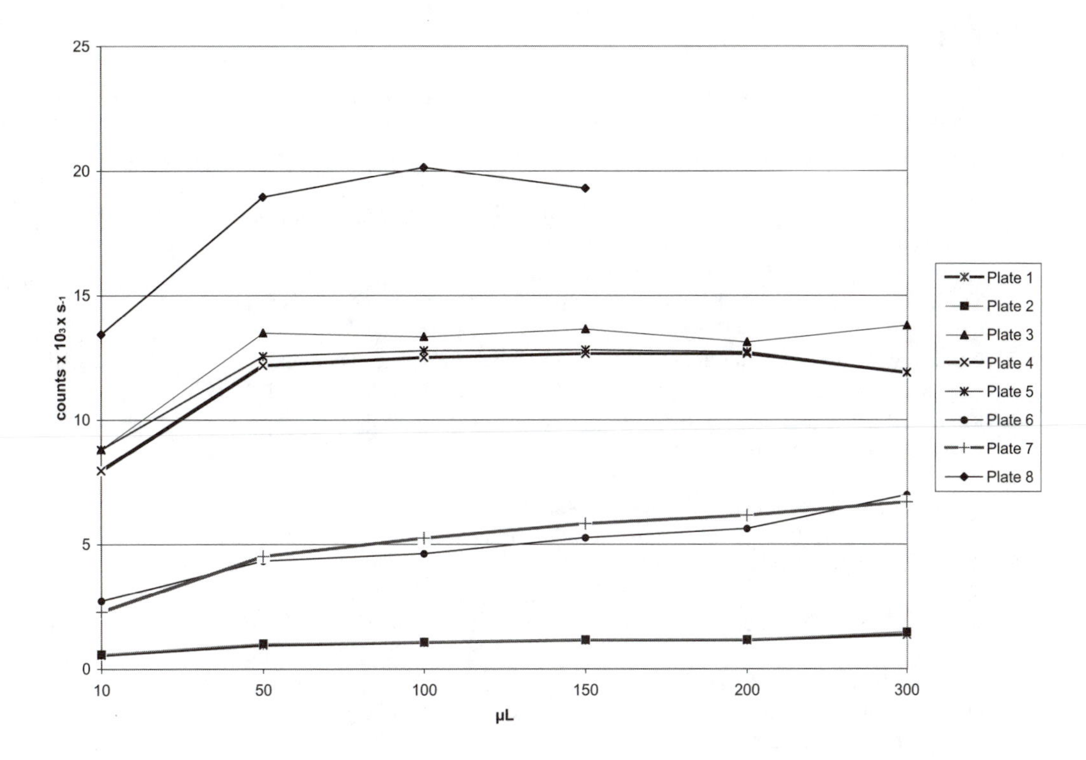

FIGURE 39.3 Influence of volume on CL counts. All wells contained the same amount of cells and had the same concentration of luminol and PMA.

Sensitivity in white plates was found to be higher than that of black plates, which can be explained by better light reflection and lesser absorption. Interestingly, marked differences between the material of different white plates were found to influence sensitivity.

Cross talk is a major concern of plates and has been reported to reach 30% in clear plates.[10] An increase in cross talk coincides with a decrease in distance between wells. It is further related to the material and the instrument.[10] Transparent material, especially that used in plates with clear bottoms, increases cross talk, whereas black material absorbs the light. All plates tested had clear bottom inserts for each well; nevertheless, cross talk was up to 0.7%. There are also older plates available (not tested) with an uninterrupted transparent bottom, which are not suitable for any measurement. Optimal results for sensitivity and cross talk were obtained with white plate 5. Counts for the determination of cross talk in plates 1, 2, 3, 4, and 8 were in the range of counts measured for autoluminescence. Plate 4, however, had a higher cross talk and a lower sensitivity than other white plates. Cross talk between diagonally adjacent wells was slightly less than in directly adjacent ones (Table 39.1). Since the instrument does not correct measured counts, negative values are conceivable if the signal is at the baseline deviation. In addition, it must be stated that instruments specifically designed for luminescence measurement are more appropriate than universally usable counters for sensitivity and cross talk.[10]

The yield of counts is only slightly influenced by sample volume. Although the same amount of standard in a high volume results in more counts as compared with a low volume (Figure 39.3). Liquids in the well form a convex surface, which functions as a dispersion lens; with small volumes the wall of the well reflects light toward the detector. In contrast to former reports, the influence of volume is less pronounced in a specialized luminometer rather than in a multiple-purpose measuring instrument.[10]

REFERENCES

1. Roda, A., Pazzagli, M., Kricka, L. J., and Stanley, P. E., *Bioluminescence and Chemiluminescence,* John Wiley & Sons, Chichester, U.K., 1999, 628 pp.
2. Kricka, L. J., Clinical and biochemical applications of luciferases and luciferins, *Anal. Biochem.,* 175, 14–21, 1988.
3. Heberer, M., Durig, M., Ganz, M., Harder, F., Zbinden, A. M., and Ernst, M., Measurement of chemiluminescence in diluted whole blood for the determination of granulocyte function in the perioperative period, *Helv. Chir. Acta,* 6, 273–280, 1983.
4. Trush, M. A., Wilson, M. E., and Van Dyke, K., The generation of chemiluminescence (CL) by phagocytic cells, *Methods Enzymol.,* 57, 462–494, 1978.
5. Cree, I. A., Blair, A. L., and Beck, J. S., Use of a microtitre plate chemiluminescence reader to study surface phagocytosis by human monocytes, *J. Biolumin. Chemilumin.,* 3, 71–74, 1989.
6. Blair, A. L., Cree, I. A., and Beck, J. S., Measurement of phagocyte chemiluminescence using a microtitre plate luminometer, *J. Biolumin. Chemilumin.,* 3, 67–70, 1989.
7. Blair, A. L., Cree, I. A., Beck, J. S., and Hastings, M. J., Measurement of phagocyte chemiluminescence in a microtitre plate format, *J. Immunol. Methods,* 112, 163–168, 1988.
8. Kournikakis, B. and Simpson, M., Optimization of a phagocyte microplate chemiluminescent assay, *J. Biolumin. Chemilumin.,* 10, 63–67, 1995.
9. Kaever, V., Robitzsch, J. T., Stangel, W., Schleinkofer, L., and Resch, K., Simultaneous detection of whole blood chemiluminescence in microtitre plates, *Eur. J. Clin. Chem. Clin. Biochem.,* 30, 209–216, 1992.
10. Hengster, P., Kunc, M., Linke, R., Eberl, T., Steurer, W., Öfner, D., Berthold, F., and Margreiter, R., Optimization of phagocyte chemiluminescence measurements using microplates and vials, *Luminescence,* 14, 91–98, 1999.

40 Whole-Body Bioluminescent Imaging for the Study of Animal Models of Human Bacterial Disease

Kevin P. Francis, Danny Joh, Stacy M. Burns, Christopher Gruber, Christopher H. Contag, and Pamela R. Contag

CONTENTS

40.1 INTRODUCTION

Animal model experiments involving bacterial diseases typically employ assays such as enumerating pathogenic cells by counting colony-forming units (CFU) or identifying and quantifying their proteins or nucleic acids. These assays are labor intensive and require the removal of tissues. Consequently, there is a loss of the contextual influences that were present in the living animal. Rapid *in vivo* assays that can reliably monitor etiological agents in experimental hosts, assess the extent of pathological changes, and reveal metabolic fluctuations noninvasively would greatly accelerate the analyses and enhance the predictive value of animal models. These assays will increase our understanding of pathogenic mechanisms and facilitate the development of new therapies and vaccines for human disease. A number of modalities are available for *in vivo* analyses. These include MRI (magnetic resonance imaging), PET (positron emission tomography), SPECT (single-photon emission computed tomography), ultrasound, and X-ray (CT; computed tomography), which are widely used in clinical settings to assess the extent of disease progression in a variety of conditions. However, these methods may require relatively expensive scanning devices that are not practical for use with most animal models. Furthermore, in the absence of significant pathological changes that lead to increased density, selective uptake of contrast dyes, incorporation of radioactive tracers, or other distinguishable features, these methodologies are not applicable.

0-8493-0719-8/02/$0.00+$1.50

We have developed an alternative *in vivo* imaging modality for the analyses of infectious diseases in animal models that is based on the detection of light from internal biological sources.[1–3] This modality was developed on the observation that light is transmitted through living tissue with efficiency suitable for use in monitoring both structure and function.[4–6] The application of light as a biological monitor currently has both routine clinical uses (e.g., pulse oximetry) and novel experimental applications in medicine.[5] In contrast to those methods that use external sources of light, the approach that we discuss here employs internally emitted biological sources of light to tag biological functions such as infection and gene expression that can then be detected externally.[1–3,7]

Internal sources of biological light can be generated using enzymes, such as luciferase, that produce bioluminescence from a substrate and an energy source. Luciferases have been used for decades as biological labels for monitoring gene expression[8] and for the determination of adenosine triphosphate (ATP) concentrations *in vitro*. Monitoring of luciferase activity *in vivo* was initially pioneered with the infectious disease model of gastrointestinal infection of mice by *Salmonella typhimurium*.[1] This early report indicated that the approach is feasible, and since it is quantitative and rapid, it provides more information in less time than conventional *ex vivo* assays. Since a large variety of bacterial species can be labeled with luciferase,[3,8–12] this approach can be generally applied in the field of microbiology. Moreover, since mammalian cells and transgenic animals can similarly be labeled, this method may also be applicable to the evaluation of genetic and cell-based therapies,[7] the study of gene expression in response to infection, and the developmental changes associated with susceptibility to infection.

40.2 TISSUE OPTICS AND SPECTROSCOPY

The depth that light penetrates tissue is wavelength dependent and is affected largely by hemoglobin absorption.[13,14] Shorter wavelengths of light (ultraviolet to blue) are absorbed by hemoglobin in mammalian tissue to a greater extent than are longer wavelengths of light (red to near infrared). At wavelengths of 700 nm or longer, absorption is minimal, yet scattering continues to affect the path that a given photon travels in tissue. Wavelengths of bioluminescent light have been described from 490 to 700 nm,[11,15] although the best-characterized bioluminescent reporters, those most often used in biological assays (bacterial, jellyfish, and firefly luciferases), are blue to yellow-green (490 to 560 nm). The ability to detect light through tissues is influenced by tissue depth and optical features, or opacity of the tissues. Therefore, detecting light emitted from tissues deep within a mammal or through relatively opaque organs, such as liver, is less efficient than detection through translucent structures or at more superficial sites (such as skin and bone).

Some knowledge of the depth of the signal and nature of the tissue between the signal and the detector is useful in the analysis of optically based reporters in living tissues. In addition, the consistency and intensity (as determined, in part, by the promoter strength) of the bioluminescent signal contribute to its detection and quantitation. Quantification of an internal signal is possible provided that the parameters of depth, intensity, and opacity are held constant in the model system. Thus, signals at specific tissue sites with a relatively constant physiology can be reliably assessed.[1,3,7]

There are a number of advantages to using an internal biological source of light over other optical signatures. A significant advantage of monitoring bioluminescence *in vivo* over fluorescence is that there are few, if any, sources of light in mammalian systems, and, therefore, background is minimal if not completely absent. The use of luminescent tags permits an integrated approach whereby the same label can be used *in vitro* and in correlative cell culture assays, and then used *in vivo* to test the predictions made *in vitro*. Moreover, once studies in the animal model are completed, the bioluminescent tag can be used to quantify the tagged organism or process in *ex vivo* biochemical assays. The results of the animal studies can thus be verified using the standard assays that measure protein levels, enzymatic activity, or amounts of nucleic acid. In these bacterial infectious disease models, the infecting pathogen can be readily distinguished from normal flora using the bioluminescent tag. For gastrointestinal models,[1] this feature is a tremendous asset, given the number of microorganisms that comprise the normal flora of the gut. In addition, since pathogens tend to adhere to tissues, *in vivo* monitoring of bioluminescence may provide more accurate quantitation than aspirates, washes, or homogenized tissues where adherent organisms pellet with debris or are not released from the tissues because the inhomogeniety of bacterial–tissue suspensions skews recovery and quantitation of pathogens.

40.3 LABELING OF BACTERIAL CELLS

lux operons from bioluminescent bacteria are ideal for labeling bacterial pathogens since they not only encode the luciferase genes but also the genes for the biosynthetic enzymes for synthesis of the substrate.[16,17] Use of these operons negates the need for an exogenous supply of substrate. Moreover, no deleterious effects of substrate biosynthesis on bacterial metabolism or virulence have been observed.[18] Among the *lux* operons from bioluminescent bacteria, those from *Photorhabdus luminescens* (previously known as *Xenorhabdus luminescens*)[19,20] appears to be ideally suited for use in mammalian animal models, given that mammalian body temperatures lie within the optimum temperature range for this enzyme.[17,21] This is in contrast to the low optimal temperature range for beetle luciferases (Luc) and other characterized bacterial luciferases (those from *Vibrio* spp.).

Five essential genes that are organized in an operon as *luxCDABE* are necessary for the synthesis of light in naturally occurring bioluminescent bacteria. Blue-green light is emitted from these bacteria with a peak at 490 nm as a result of a heterodimeric luciferase (encoded by *luxAB*) catalyzing the oxidation of reduced flavin mononucleotide ($FMNH_2$) and a long-chain fatty aldehyde (synthesized by a fatty acid reductase complex, encoded by *luxCDE*). Although a number of additional *lux* genes have been identified in bioluminescent bacteria, only *luxA–E* are essential for the biosynthesis of light.[17]

40.4 EXPRESSION OF THE *lux* OPERONS IN GRAM-NEGATIVE BACTERIA

Well-characterized vectors have been used for the expression of the *lux* operons in Gram-negative organisms. The original pUC-based plasmid, used to clone the *lux* operon from *P. luminescens* including the promoter region (vector designated pCGLS1)[16] is well suited for expressing luciferase from Gram-negative organisms such as *Escherichia coli* and *Salmonella*. Introduction of the *lux*-encoding vector into cells of these bacterial species can be carried out using standard methods of bacterial cell transformation for Gram-negative organisms. Alternatively, the *lux* operon can be stably integrated into the chromosome of a wide range of Gram-negative bacteria using *lux* transposons, such as Tn5 *luxCDABE*.[22]

Recently, using both *lux* transposons and homologous recombination, we have generated a wide range of stable, highly bioluminescent Gram-negative bacteria, including several strains of pathogenic *E. coli, Haemophilus influenza, Klebsiella pneumoniae*, and *Pseudomonas aeruginosa*, and different species of *Salmonella, Shigella*, and *Yersinia*. These bioluminescent bacteria are currently being tested in a number of different animal models of infection.

40.5 EXPRESSION OF THE *lux* OPERONS IN GRAM-POSITIVE BACTERIA

Although a light-encoding *lux* operon can easily be introduced into a variety of Gram-negative bacteria to confer a bioluminescent phenotype, because all identified species of naturally occurring marine and terrestrial bioluminescent bacteria are Gram-negative, the transformation of Gram-positive bacteria to a light phenotype has been problematic because of the differing genetics of these two bacterial groups. Bioluminescent Gram-positive bacteria such as *Staphylococcus aureus* and *Mycobacterium tuberculosis* have been constructed. However, these bacteria contain the firefly luciferase gene (*luc*) or variations of isolated bacterial *luxAB* luciferase genes, each of which require the addition of an exogenous substrate to allow bioluminescence. Moreover, most bioluminescent Gram-positive bacteria have been generated using bacterial luciferase genes that encode enzymes that are unstable at temperatures above 30°C. Although such bacteria are useful for environmental studies (e.g., the assessment of food products for contamination by such bacteria), *luxAB* constructs that only permit bioluminescence to occur below 30°C are of limited use for experimentation on pathogenicity carried out at 35°C and above *in vivo*. To address this problem, we have recently reengineered the entire *Photorhabdus luminescens luxCDABE* operon,[3] allowing Gram-positive infections to be monitored *in vivo* in live animal.

40.6 CORRELATIVE CELL CULTURE ASSAYS

Use of bioluminescent reporters in bacteria provides a rapid means of quantitation that is also amenable to use in high-throughput antibacterial drug screening. Assays for minimal inhibitory concentrations (MIC) of drugs can be rapidly performed on cultures of living bacteria and are particularly useful in the case of slow-growing bacteria such as *Mycobacterium*. More elaborate correlative cell culture assays, involving the interaction of bacterial cells with those of a mammalian host, can also be conducted to provide a rapid means for evaluating biological events for pathogenesis and gene expression prior to introduction into living animals. Contag et al.[1] described functional assays, *Salmonella* adherence and entry, that correlate cell–cell interactions to *in vivo* pathogenesis, where bioluminescence was used as a marker in these live cell assays. Similar assays have been described for many organisms, and the imaging can be performed on populations of cells using macroscopic detection, or on single living cells with microscopic detection.[23] Tagging bacterial pathogens with luciferase enables *in vitro, in vivo*, and *ex vivo* assays to be optimally integrated such that an analysis can be run full circle with a single reporter gene at its hub.

Rapidly evolving paradigms in drug development, including combinatorial chemistries, genomics, high-throughput *in vitro* screening, and chip technologies are generating hundreds or even thousands of potential lead compounds for a given disease that require efficacy testing in animal models. Drug screening in animals remains a significant bottleneck in the development of therapeutics and can be severely limiting. Noninvasive *in vivo* assays would accelerate the data acquisition, require fewer animals for drug testing, and produce significantly more information per protocol. The increased depth of understanding that could be obtained through noninvasive assays for pathogenic mechanisms and potential interventions would yield improved preclinical data and ultimately preservation of time, effort, and investments.

40.7 MONITORING BIOLUMINESCENT GRAM-NEGATIVE BACTERIA IN LIVE MICE

In the initial study by Contag et al.,[1] patterns of disease caused by three strains of bioluminescently labeled *Salmonella* were monitored over an 8-day time course in mice. Groups of mice were infected orally, the natural route of infection for mice or humans, with three strains of *Salmonella*: the wildtype, SL1344*lux*, the less invasive mutant, BJ66*lux*, or the less virulent strain, LB5000*lux*. Images were obtained daily using an intensified charge-coupled device (CCD) camera. At 1 to 2 days, the bioluminescent signal localized to a single focus in all infected animals. The distribution of bioluminescence did not spread in mice inoculated with the BJ66*lux*. This was in contrast to mice infected with the less virulent LB5000*lux*, where bioluminescence was not detected in any animal at 7 days. In the mice infected with the wild-type SL1344*lux*, bioluminescence was detected throughout the study period, with multiple foci of transmitted photons at 8 days. In one third of the animals infected with SL1344*lux*, transmitted photons were apparent over much of the abdominal

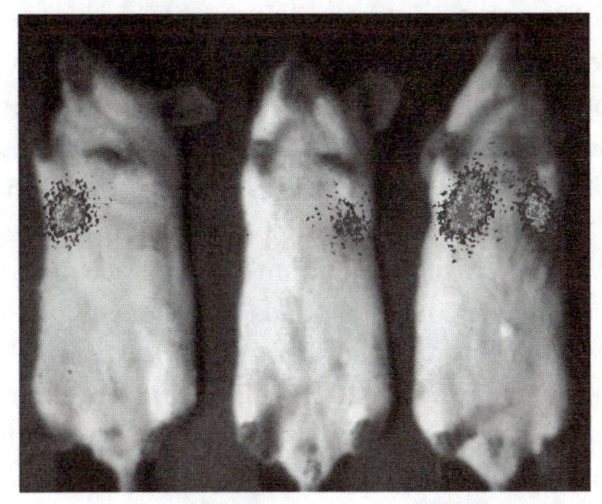

FIGURE 40.1 (Color figure folllows p. 266.) Murine models of bacterial pneumonia. Mice were inoculated by intranasal inoculation using a labeled strain of *P. aeruginosa*. Immediately after infection, the mice were imaged using an intensified CCD camera (model C2400-32, Hamamatsu, Japan).

area at 8 days, resembling the distribution of photons in a systemic infection following an intraperitoneal inoculation.[1]

The *lux* operon has also been used to generate labeled *Pseudomonas aeruginosa* and these organisms have been utilized in mouse lung infection models (Figure 40.1). In these models, variation in infection was observed within groups of mice that were inoculated in exactly the same manner. This illustrates how imaging can improve the data sets, in that the animals with bilateral infections may present with a different disease course than animals with unilateral infections. Also, if tissue samples are to be obtained from these animals, the images can serve as a guide for what is to be expected, and directed sampling may improve the data analyses.

In vivo imaging has been adapted for drug screening and drug development assays.[1,3,24] Recently, Rocchetta et al.[24] generated a highly bioluminescent strain of an *E. coli* clinical isolate, EC14, using a multicopy plasmid carrying the full *luxCDABE* operon. This bioluminescent reporter bacterium was used to study antimicrobial effects *in vitro* and *in vivo*, using the neutropenic-mouse thigh model of infection. Bioluminescence was monitored and measured *in vitro* and *in vivo*, and these results were compared to viable-cell determinations made using conventional plate counting methods. Statistical analysis demonstrated that in the presence or absence of antimicrobial agents (ceftazidine, tetracycline, or ciprofloxacin), a strong correlation (0.98) existed between bioluminescence levels and viable cell counts *in vitro* and *in vivo*. Moreover, this study showed that the ability to measure the same animals repeatedly reduced variability within the treatment experiments and allowed equal or greater confidence in determining treatment efficacy.

40.8 GENERATING AND *IN VIVO* MONITORING OF BIOLUMINESCENT GRAM-POSITIVE BACTERIA

Over the past decade, there has been a dramatic increase in the number of antibiotic-resistant strains of Gram-positive bacteria causing serious, life-threatening diseases in humans. Many strains of *Staphylococcus aureus*, which cause a variety of diseases ranging from pyoderma to toxic shock syndrome, are methicillin and gentamicin resistant, with some strains also showing limited resistance to vancomycin. Such bacteria are a major problem in nosocomial, or hospital-acquired, infections. *Streptococcus pneumoniae, M. tuberculosis*, and enterococci are also showing resistance to a wide range or, in some cases, all conventional antibiotics (e.g., MRSA and vancomycin-resistant enterococci). The occurrence of such multidrug-resistant bacteria is of increasing concern, since they represent a disease threat of epidemic proportions. Without the development of new antibiotics it is possible that, given time, such bacteria will be untreatable by conventional means. To combat such bacterial infections, novel and effective drugs will be needed. Therefore, new approaches for screening candidate antibiotics both *in vitro* and *in vivo* are essential to accelerate the development of new anti-infectives. The use of bioluminescent noninvasive imaging strategies to reveal the real-time effects of potential therapeutic agents on Gram-positive bacterial infections in animal models would greatly accelerate the analyses of compounds under development.

Recently, we have been able to construct a novel *Photorhabdus luminescens lux* operon, in which each gene (*luxA–E*) has been modified by introducing a Gram-positive ribosome-binding site (RBS).[3] This novel operon, when preceded by an appropriate promoter sequence on the broad host range vector, pMK4, allowed several highly bioluminescent *Staphylococcus aureus* strains to be generated. The minimum number of these *S. aureus* RN4220 *lux* cells detectable at 37°C using an intensified CCD camera was approximately 400 CFU. However using a more sensitive liquid nitrogen–cooled integrating CCD camera, we were able to detect as few as 80 CFU at 37°C.

A number of pathogenic strains of *S. aureus* were transformed with the plasmids pMK4 *luxABCDE* P1 and pMK4 *luxABCDE* P2 (shown to function as two of the brightest constructs), including RN6390, 8325-4, and a clinical isolate of MRSA. Figure 40.2 shows how bioluminescent *S. aureus* can be effectively monitored in living animals in real time and can be used to determine the efficacy of antibiotic therapy. Both *S. aureus* 8325-4 pMK4 *lux*ABCDE P1 and *S. aureus* 8325-4 pMK4 *luxABCDE* P2 produced significant bioluminescent signals in mice, allowing treatment with amoxicillin to be effectively assessed (only P1 ventral images shown). At 8 h (image not show), the infection in both groups of treated animals had begun to clear, as judged by a decrease in bioluminescence, and by 24 h no signal could be detected. In comparison, both groups of untreated mice had strong bioluminescent signals at 24 h, corresponding to an apparent poor state of health as indicated by ruffled fur and dragging of the hind quarters during movement.

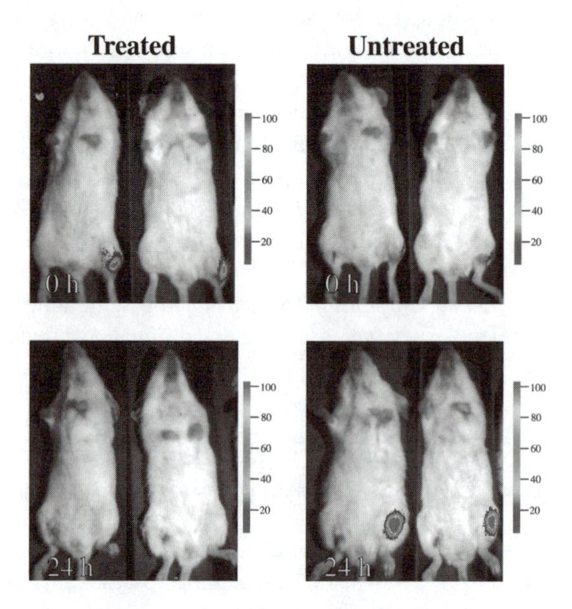

FIGURE 40.2 (Color figure follows p. 266.) Monitoring the effects of amoxicillin on bioluminescent *S. aureus* in mice. Shown are two treated and two untreated *S. aureus* 8325-4 pMK4 *luxABCDE* P1–infected mice, imaged ventrally for 5 min at 0 and 24 h postinfection using the intensified CCD camera.

Extraction of bacteria from the thigh muscles of each of the mice confirmed the bioluminescent data to be extremely accurate at predicting the number of viable bacteria in the tissue. For both strains of bacteria, *S. aureus* 8325-4 pMK4 *luxABCDE* P1 and *S. aureus* 8325-4 pMK4 *luxABCDE* P2, the number of CFU recovered from the thigh muscles of the untreated mice at 24 h postinfection was found to have increased approximately 25-fold over the inoculating dose. In comparison, the number of CFU (P1 and P2) in the thigh muscles of the antibiotic-treated mice at 24 h postinfection decreased at least a 1000-fold in CFU from that used to inoculate the mice. Moreover, by plating all of the recovered bacteria on media containing no antibiotics, the plasmid stability of both bioluminescent *S. aureus* strains could be assessed. *Staphylococcus aureus* 8325-4 pMK4 *luxABCDE* P2 was slightly more stable than *S. aureus* 8325-4 pMK4 *luxABCDE* P1, with 95% of P2 CFU retaining a high level of bioluminescence as opposed to 85% of P1 CFU when plated in the absence of selection.

40.9 SUMMARY AND CONCLUSIONS

Using both *lux* transposons and homologous recombination with *luxCDABE* cassettes, we have been able to generate a wide range of stable, highly bioluminescent Gram-negative bacteria with relatively little difficulty. The generation of highly bioluminescent Gram-positive bacteria has been much more laborious. In addition to *Staphylococcus. aureus*, we have used the modified *luxABCDE* operon to successfully transform strains of the Gram-positive bacterium *Streptococcus pneumoniae, S. pyogenes*, and *Listeria monocytogenes*. In each case, the Gram-positive bacteria carrying this modified *lux* operon are highly bioluminescent and can be monitored *in vivo* in animals. Furthermore, since plasmid loss in the absence of antibiotic selection has been shown to occur in animal models exceeding 48 h,[3] we have recently constructed a number of *luxABCDE* transposons. These have been successfully used for stable transformation and for monitoring gene regulation in a number of Gram-positive bacteria, in addition to monitoring long-term infections in live mice.[25]

The noninvasive monitoring of bioluminescent bacterial pathogens described here allows for experimental protocols that are significantly faster. Furthermore, because biostatistics are improved through collection of multiple data points in the same animal, the overall number of animals required is reduced. These two benefits should improve the capacity to relieve the bottleneck of animal studies that occur in drug development as high-throughput technologies increase the number of potential lead compounds that require animal evaluation. We believe that this technology also increases the quality of animal model data to provide more information for selecting the drug candidates that will be clinically successful.

ACKNOWLEDGMENTS

We would like to thank B. Nelson, B. Rice, and M. Cable (Bioimaging Section, Xenogen Corporation, Alameda, CA) for information pertaining to image analysis using the CCD cameras. Our work was supported, in part, by a grant from the National Institutes of Health number RO1 HD37543 (C.H.C) and unrestricted gifts from the Hess and Mary L. Johnson Research Funds (C.H.C.).

REFERENCES

1. Contag, C. H., Contag, P. R., Mullins, J. I., Spilman, S. D., Stevenson, D. K., and Benaron, D. A., Photonic detection of bacterial pathogens in living hosts, *Mol. Microbiol.,* 18, 593–603, 1995.
2. Contag, P. R., Olomu, I. N., Stevenson, D. K., and Contag, C. H., Bioluminescent indicators in living mammals, *Nat. Med.,* 4, 245–247, 1998.

3. Francis, K. P., Joh, D., Bellinger-Kawahara, C., Hawkinson, M. J., Purchio, T. F., and Contag, P. R., Monitoring bioluminescent *Staphylococcus aureus* infections in living mice using a novel *luxABCDE* construct, *Infect. Immun.,* 68, 3594–3600, 2000.
4. Jobsis, F. F., Noninvasive, infrared monitoring of cerebral and myocardial oxygen sufficiency and circulatory parameters, *Science*, 198, 1264–1267, 1977.
5. Benaron, D. A., Contag, P. R., and Contag, C. H., Imaging brain structure and function, infection and gene expression in the body using light, *Philos. Trans. R. Soc. Lond. B Biol. Sci.,* 352, 755–761, 1997.
6. Tromberg, B. J., Shah, N., Lanning, R., Cerussi, A., Espinoza, J., Pham, T., Svaasand, L., and Butler, J., Non-invasive *in vivo* characterization of breast tumors using photon migration spectroscopy, *Neoplasia,* 2, 26–40, 2000.
7. Contag, C. H., Spilman, S. D., Contag, P. R., Oshiro, M., Eames, B., Dennery, P., Stevenson, D. K., and Benaron, D. A., Visualizing gene expression in living mammals using a bioluminescent reporter, *Photochem. Photobiol.,* 66, 523–531, 1997.
8. Francis, K. P. and Gallagher, M. P., Light emission from a Mudlux transcriptional fusion in *Salmonella typhimurium* is stimulated by hydrogen peroxide and by interaction with the mouse macrophage cell line J774.2, *Infect. Immun.,* 61, 640–649, 1993.
9. Andrew, P. W. and Roberts, I. S., Construction of a bioluminescent mycobacterium and its use for assay of antimycobacterial agents, *J. Clin. Microbiol.,* 31, 2251–2254, 1993.
10. Prest, A. G., Winson, M. K., Hammond, J. R., and Stewart, G. S., The construction and application of a lux-based nitrate biosensor, *Lett. Appl. Microbiol.,* 24, 355–360, 1997.
11. Wilson, T. and Hastings, J. W., Bioluminescence, *Annu. Rev. Cell Dev. Biol.,* 14, 197–230, 1998.
12. Siragusa, G. R., Nawotka, K., Spilman, S. D., Contag, P. R., and Contag, C. H., Real-time monitoring of *Escherichia coli* O157:H7 adherence to beef carcass surface tissues with a bioluminescent reporter, *Appl. Environ. Microbiol.,* 65, 1738–1745, 1999.
13. Colin, M., Moritz, S., Schneider, H., Capeau, J., Coutelle, C., and Brahimi-Horn, M. C., Haemoglobin interferes with the *ex vivo* luciferase luminescence assay: consequence for detection of luciferase reporter gene expression *in vivo*, *Gene Ther.,* 7, 1333–1336, 2000.
14. Smith, A. D. and Trempe, J. P., Luminometric quantitation of photinus pyralis firefly luciferase and *Escherichia coli* beta-galactosidase in blood-contaminated organ lysates, *Anal. Biochem.,* 286, 164–172, 2000.
15. Hastings, J. W. and Nealson, K. H., Bacterial bioluminescence, *Annu. Rev. Microbiol.,* 31, 549–595, 1977.
16. Frackman, S., Anhalt, M., and Nealson, K. H., Cloning, organization, and expression of the bioluminescence genes of *Xenorhabdus luminescens, J. Bacteriol.,* 172, 5767–5773, 1990.
17. Meighen, E. A., Bacterial bioluminescence: organization, regulation, and application of the lux genes, *FASEB J.,* 7, 1016–1022, 1993.
18. Contag, C. H., Contag, P. R., Spilman, S. D., Stevenson, D. K., and Benaron, D. A., Photonic monitoring of infectious disease and gene regulation, in *Biomedical Optical Spectroscopy and Diagnostics,* E. Sevick-Muraca and D. Benaron, Eds., Optical Society of America, Washington, D.C., 1996, 220–224.
19. Rainey, F. A., Ehlers, R. U., and Stackebrandt, E., Inability of the polyphasic approach to systematics to determine the relatedness of the genera *Xenorhabdus* and *Photorhabdus*, *Int. J. Syst. Bacteriol.,* 45, 379–381, 1995.
20. Akhurst, R. J., Mourant, R. G., Baud, L., and Boemare, N. E., Phenotypic and DNA relatedness between nematode symbionts and clinical strains of the genus *Photorhabdus* (Enterobacteriaceae), *Int. J. Syst. Bacteriol.,* 46, 1034–1041, 1996.
21. Szittner, R. and Meighen, E., Nucleotide sequence, expression, and properties of luciferase coded by lux genes from a terrestrial bacterium, *J. Biol. Chem.,* 265, 16,581–16,587, 1990.
22. Winson, M. K., Swift, S., Hill, P. J., Sims, C. M., Griesmayr, G., Bycroft, B. W., Williams, P., and Stewart, G. S., Engineering the *luxCDABE* genes from *Photorhabdus luminescens* to provide a bioluminescent reporter for constitutive and promoter probe plasmids and mini-Tn5 constructs, *FEMS Microbiol. Lett.,* 163, 193–202, 1998.
23. Pettersson, J., Nordfelth, R., Dubinina, E., Bergman, T., Gustafsson, M., Magnusson, K. E., and Wolf-Watz, H., Modulation of virulence factor expression by pathogen target cell contact, *Science,* 273, 1231–1233, 1996.

24. Rocchetta, H. L., Boylan, C. J., Foley, J. W., Iversen, P. W., LeTourneau, D. L., McMillian, C. L., Contag, P. R., Jenkins, D. E., and Parr, T. R., Jr., Validation of a noninvasive, real-time imaging technology using bioluminescent *Escherichia coli* in the neutropenic mouse thigh model of infection, *Antimicrob. Agents Chemother.,* 45, 129–137, 2001.
25. Francis, K. P., Yu, J., Bellinger-Kawahara, C., Joh, D., Hawkinson, M. J., Xiao, G., Purchio, T. F., Caparon, M. G., Lipsitch, M., and Contag, P. R., Visualizing pneumococcal infections in the lungs of live mice using bioluminescent *Streptococcus pneumoniae* transformed with a novel Gram-positive *lux* transposon, *Infect. Immun.,* 69, 3350–3358, 2001.

Section VII

High-Throughput Screening

41 Resonance Energy Transfer as an Emerging Technique for Monitoring Protein–Protein Interactions *in Vivo*: BRET vs. FRET

Yao Xu, Akihito Kanauchi, David W. Piston, and Carl Hirschie Johnson

CONTENTS

41.1 INTRODUCTION

Protein–protein interactions are known to play an important role in a variety of biochemical systems. To date, thousands of protein–protein interactions have been identified by using the conventional two-hybrid system, but this method is limited in that the interaction must occur in the yeast nucleus. This means interactions that strictly depend upon cell-type-specific processing or compartmentalization will not be detected. Therefore, a number of new methods have been developed recently that rely on reconstitution of biochemical function *in vivo*, such as fluorescence resonance energy transfer (FRET), protein mass spectrometry, or evanescent wave.[1] Among those methods, the resonance energy transfer techniques have potential advantages for assaying protein–protein interactions in living cells and in real time. In this chapter, we will describe a recently developed resonance energy transfer method based on bioluminescence. This chapter is an update of a previously published review.[2]

0-8493-0719-8/02/$0.00+$1.50

41.2 FLUORESCENCE RESONANCE ENERGY TRANSFER (FRET)

FRET[3,4] is a well-established phenomenon that has been useful in cellular microscopy. When two fluorophores (the "donor" and the "acceptor") with overlapping emission/absorption spectra are within ~50 Å of one another and their transition dipoles are appropriately oriented, the donor fluorophore is able to transfer its excited-state energy to the acceptor fluorophore. Therefore, if appropriate fluorophores are linked to proteins that might interact with each other, the proximity of these candidate interactors could be measured by determining if fluorescence resonance energy is transferred from the donor to the acceptor. Thus, the presence or absence of FRET acts as a "molecular yardstick."

The discovery and development of green fluorescent protein (GFP) and its mutants made possible their use as FRET donors and acceptors.[5–11] Genetically fusing GFP derivatives to the candidate proteins enabled the detection of protein–protein proximity in real time in living cells of the organisms from which the proteins were originally obtained.[8,9] In those studies, blue fluorescent protein (BFP) was used as the donor fluorophore and GFP was the acceptor. As mentioned above, the efficiency of the resonance transfer depends upon the spectral overlap of the fluorophores, their relative orientation, as well as the distance between the donor and acceptor fluorophores. By targeting the fusion proteins to specific compartments, this FRET-based assay can also allow protein interactions to be observed within cellular compartments *in vivo*, as has been shown for mitochondria and nuclei.[8,9] However, because FRET demands that the donor fluorophore be excited by illumination, the practical usefulness of FRET can be limited because of the concomitant results of excitation: photobleaching, autofluorescence, and direct excitation of the acceptor fluorophore. Furthermore, some tissues might be easily damaged by the excitation light or might be photoresponsive (e.g., retina).

41.3 BIOLUMINESCENCE RESONANCE ENERGY TRANSFER (BRET)

Recently, we developed a bioluminescence resonance energy transfer (BRET) system for assaying protein–protein interactions that incorporates the attractive advantages of the FRET assay while avoiding the problems associated with fluorescence excitation.[12] In BRET, the donor fluorophore of the FRET pair is replaced by a luciferase, in which bioluminescence from the luciferase in the presence of a substrate excites the acceptor fluorophore through the same resonance energy transfer mechanisms as FRET.

The bioluminescent *Renilla* luciferase (RLUC; MW = 35 kDa) was originally chosen as the donor luciferase for BRET because its emission spectrum is similar to the cyan mutant of *Aequorea* GFP ($\lambda_{max} \approx 480$ nm), which has been shown to exhibit FRET with the acceptor fluorophore EYFP, which is a yellow-emitting GFP mutant.[7] The excitation peak of EYFP (513 nm) does not perfectly match to the emission peak of RLUC, but the emission spectrum of RLUC is sufficiently broad that it provides good excitation of EYFP. The spectral overlap between RLUC and EYFP is similar to that of EYFP and the enhanced cyan mutant of GFP, ECFP, which yields a critical Förster radius (R_0) for FRET of ~50 Å.[8] Thus, we would expect significant BRET between RLUC and EYFP, with an R_0 for BRET of ~50 Å. The fluorescence emission of EYFP is yellow, peaking at 527 nm, which is distinct from the RLUC emission peak. Furthermore, RLUC and EYFP do not naturally interact with each other. Fortunately, the substrate for RLUC, coelenterazine, is a hydrophobic molecule that is able to permeate cell membranes.

In the BRET assay of protein interactions, RLUC is genetically fused to one candidate protein, and EYFP is fused to another protein of interest that perhaps interacts with the first protein. If RLUC and EYFP are brought close enough for resonance energy transfer to occur, the bioluminescence energy generated by RLUC can be transferred to EYFP, which then emits yellow light (Figure 41.1). In the BRET assay for protein interaction, this resonance transfer can occur between RLUC/EYFP fusion proteins that interact. If there is no interaction between the two proteins of

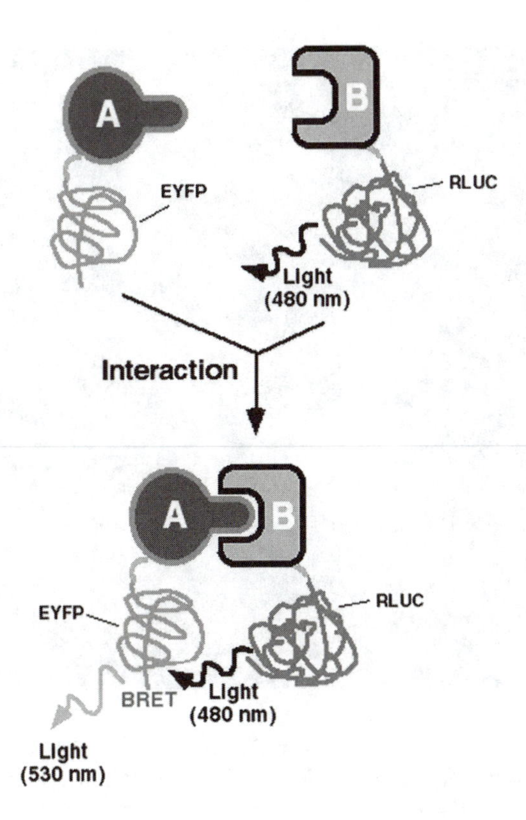

FIGURE 41.1 (Color figure follows p. 266.) A diagram of BRET for a protein–protein interaction assay. One protein of interest (B) is genetically fused to the donor luciferase RLUC, and the other candidate protein (A) is fused to the acceptor fluorophore EYFP. Interaction between the two fusion proteins can bring RLUC and EYFP close enough for BRET to occur, with an emission of longer-wavelength light.

interest, RLUC and EYFP will be too far apart for significant transfer and only the blue-emitting spectrum of RLUC will be detected. Thus, protein–protein interactions can be monitored both *in vivo* and *in vitro* by detecting the emission spectrum and quantifying the emission ratio at 530 nm/480 nm.

BRET between RLUC and EYFP was first demonstrated in control experiments in which RLUC was fused directly to EYFP through a linkage of 11 amino acids.[12] The luminescence profile of the *Escherichia coli* cells expressing the RLUC::EYFP fusion construct yielded a bimodal spectrum, with one peak centered at 480 nm (as for RLUC), and a new peak centered at 527 nm (as for EYFP fluorescence).[12] This result suggests that a significant proportion of the RLUC energy is transferred to EYFP and emitted at the characteristic wavelength of EYFP. We concluded that RLUC/EYFP could be an effective combination to apply in a protein–protein interaction assay.

41.4 APPLICATION OF BRET TO CLOCK PROTEINS

To test BRET as a protein–protein assay, we chose the proteins encoded by circadian (daily) clock genes from cyanobacteria and fused them to RLUC or EYFP, respectively. In cyanobacteria, the *kaiABC* gene cluster encodes three proteins, KaiA (MW = 32.6 kDa), KaiB (MW = 11.4 kDa), and KaiC (MW = 58 kDa), that are essential for circadian clock function.[13] Iwasaki et al.[14] have used the yeast two-hybrid and *in vitro* binding assays to discover that Kai proteins interact in various ways, such as formation of KaiB–KaiB homodimers. First, we tried the N-terminal fusions of KaiB to RLUC and to EYFP. The luminescence spectra of *E. coli* expressing these fusions showed a second peak in the cells expressing both RLUC::KaiB and EYFP::KaiB (Figure 41.2, right side). This spectrum is similar to that depicted for the fusion protein RLUC::EYFP.[12]

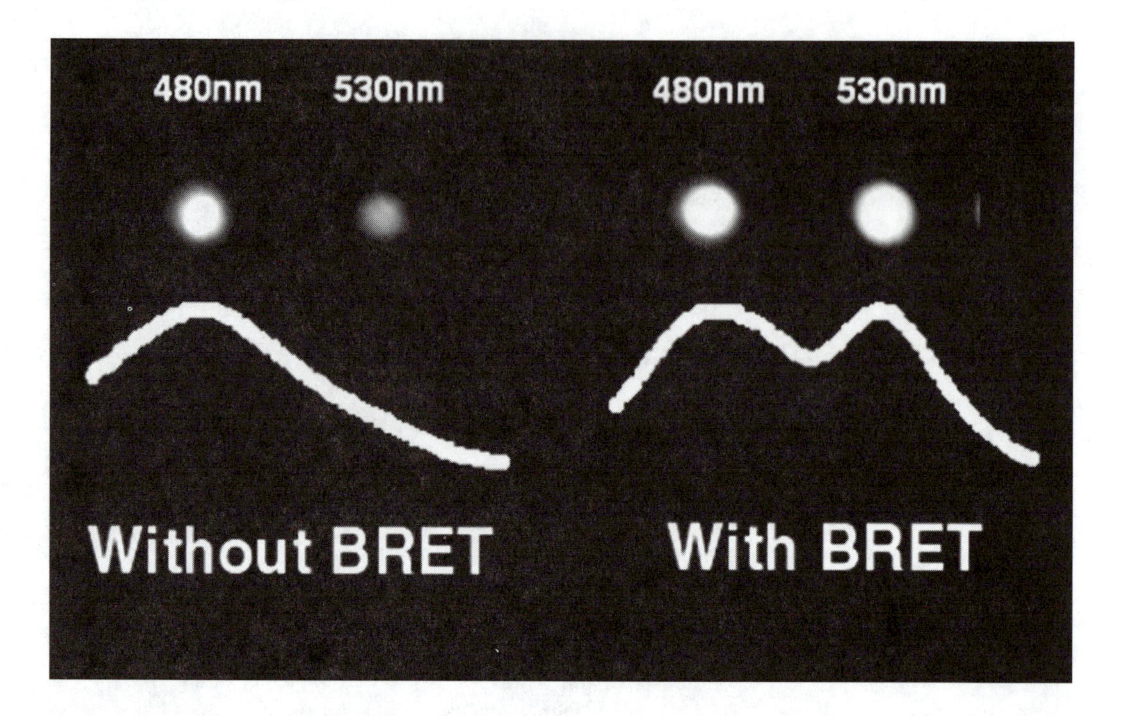

FIGURE 41.2 (Color figure follows p. 266.) Comparison of complete BRET spectra using a fluorescence spectrophotometer with camera images of *E. coli* cells. Top: Cultures imaged with a CCD camera through filters transmitting light of 480 or 530 nm from the transformed *E. coli* strains coexpressing fusion proteins exhibiting BRET on the right side (RLUC::KaiB and EYFP::KaiB) or fusion proteins that are not exhibiting BRET on the left side (RLUC::KaiB and EYFP::KaiA). Bottom: Luminescence emission spectra measured continuously from 440 to 580 nm for the same strains.

We further tested all possible combinations of KaiB fusions with RLUC or EYFP, including N- vs. N-, N- vs. C-, C- vs. N-, as well as C- vs. C-terminal fusions. All these combinations of the fusion proteins of KaiB showed BRET (unpublished data). KaiB interactions were also observed *in vitro* by BRET.[12] To demonstrate that this bimodal spectrum does not occur nonspecifically, we used KaiA as a control, in which EYFP was fused to a slightly truncated KaiA. The luminescence spectra of *E. coli* coexpressing EYFP::KaiA with RLUC::KaiB did not exhibit the second luminescence peak, indicating no interaction occurred between KaiA and KaiB. Our results, therefore, strongly suggest that interaction among KaiB molecules either in N-terminal or C-terminal fusions to the donor luciferase or the acceptor fluorophore has brought the RLUC and EYFP into close proximity such that energy transfer occurs for ~50% of the RLUC luminescence. Thus, BRET supports the data from the yeast two-hybrid assay,[14] demonstrating that the clock protein KaiB self-associates to form multimers.

In the experiments described above, the extent of BRET was determined by measuring emission spectra.[12] For applications such as microscopic imaging and high-throughput screening, it would be more convenient to measure the ratio of luminescence intensities at two fixed wavelengths, e.g., 480 and 530 nm. Ratio imaging has the advantage of automatically correcting for differences in overall levels of expression of RLUC and EYFP fusion proteins. The top portion of Figure 41.2 is the images of *E. coli* cultures expressing fusion proteins that either exhibit (right) or do not exhibit (left) BRET. These images of liquid *E. coli* cultures (5 μl cultures) were collected using a charge-coupled device (CCD) camera through interference bandpass filters centered at 480 and 530 nm, respectively. In the cultures coexpressing the interacting combination of RLUC::KaiB with EYFP::KaiB, the amount of light emitted at 480 and 530 nm is roughly equal, as would be predicted from the spectra depicted (Figure 41.2, right), In contrast, in the cultures containing a noninteracting combination of RLUC::KaiB with EYFP::KaiA, there is much less light emitted at 530 nm than at 480 nm (Figure 41.2, left). As we reported previously, the extent of BRET can be quantified according to the 530 nm/480 nm ratios of luminescence intensity in the image.[12] Thus, the 530 nm/480 nm ratios can apparently be used to evaluate BRET and thereby infer if protein–protein interaction has occurred.

41.5 BRET IN MAMMALIAN CELLS

The BRET technique has now been successfully extended to mammalian cells. Angers et al.[15] used BRET to demonstrate that human β2-adrenergic receptors form constitutive homodimers in HEK-293 cells. Treatment with the agonist isoproterenol increased the BRET signal, indicating that the agonist interacts with receptor dimers at the cell surface. In addition, Wang et al.[16,17] used BRET (they call it "LRET," but it is the same phenomenon) to demonstrate interaction between insulin-like growth factor II (IGF-II) and its binding protein, IGFBP-6, that are expressed by mammalian cells.

41.6 NEW TOOLS/APPLICATIONS FOR BRET

Very recently, several new tools for BRET have appeared that may prove useful. The first is that a RLUC, which is codon-optimized for mammalian expression, is now available from BioSignal ("hRluc"; www.BiosignalPackard.com). Transfection of the hRluc construct into mammalian cells results in significantly higher luminescence levels than with the original Rluc (Figure 41.3). The second new tool is the luciferase isolated from *Gaussia* that has an emission spectrum like that of RLUC but whose molecular weight is only 20 kDa (available from Prolume Ltd.; www.prolume.com).[18] Like RLUC, this luciferase uses coelenterazine as a substrate. By virtue of its smaller size, this luciferase may better allow native interactions without steric hinderance in fusion proteins. Transfection of the humanized version of *Gaussia* luciferase (hGluc) also allows strong luminescence signals in mammalian cells that are somewhat more stable over time than with hRluc (at least, in COS7 cells, Figure 41.3).

Another tool of potential advantage is a new fluorescent protein isolated from anthozoans.[19] These proteins, which are remote homologues of GFP, form a new group of fluorescent tags.

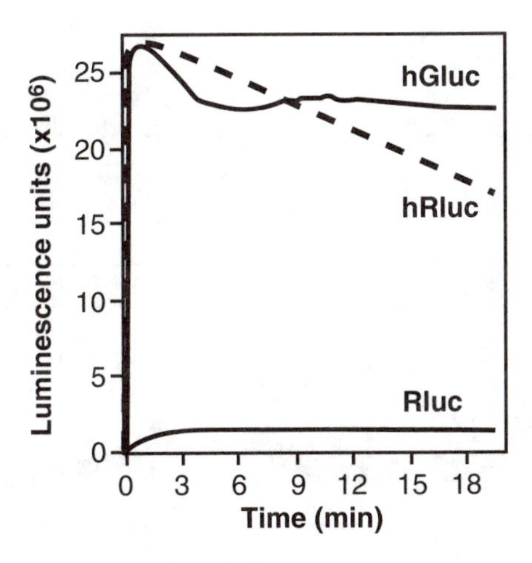

FIGURE 41.3 Comparison of the luminescence intensity of COS7 cells transfected with the genes encoding the original *Renilla* luciferase ("Rluc"), human-codon optimized Rluc ("hRluc"), or human-codon optimized *Gaussia* luciferase ("hGluc").

One of these proteins has a much longer wavelength than any other fluorescent protein yet isolated, with an excitation spectrum peaking at 558 nm and a sharp emission spectrum peaking at 583 nm. The excitation spectrum is broad enough that a luciferase-like RLUC might excite it. The advantage of this fluorescent protein is that its emission spectrum is sufficiently red-shifted that the separation between BRET and non-BRET luminescence is much greater than with YFP, and hence quantification of BRET could be more accurate. This red fluorescent protein is now available from Clontech as "DsRed," and it has been used in a FRET assay of protein–protein interactions in plants.[20] DsRed needs further development as a resonance tool, however, because it is a green fluorescent protein when first synthesized and matures to the red form over time. This means that it can undergo FRET with *itself* and its use could lead to misinterpretations of FRET/BRET signals.[21] Further, the natural tetramerization of DsRed makes its use in energy transfer studies problematic.[21] It is hoped a useful mutant form of DsRed can be developed that is naturally a monomer and synthesized immediately into a stable red fluorescent form.

New substrates for the luciferase are also available. In their study with mammalian cells, Angers et al.[15] used the coelenterazine analogue, "h coelenterazine," to increase the luminescence intensity. This coelenterazine analogue and others are available from Molecular Probes (www.probes.com). BioSignal markets another coelenterazine analogue in which the spectrum of emission is shifted to shorter wavelengths. When used with a GFP mutant adapted to this emission wavelength, this BRET pair results in a higher sensitivity and wider dynamic range. This system is now available under the trademark, BRET2. Finally, a new application of BRET is its use in a homogeneous noncompetitive immunoassay.[22] It is likely that many other applications will emerge as instrumentation as this assay becomes more common.[23]

41.7 A POTENTIAL SCREENING SYSTEM

Based on these data, we proposed[12] a relatively simple scheme for designing an *in vivo* library screening system for protein–protein interaction through BRET (Figure 41.4). By measuring the light emission collected through interference filters, the 530 nm/480 nm luminescence ratio of *E. coli*

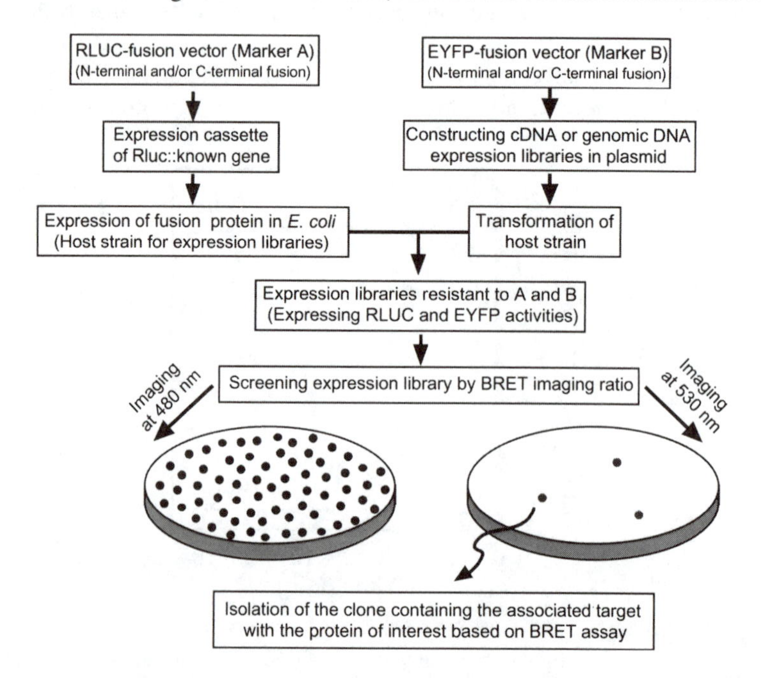

FIGURE 41.4 Schematic diagram of a screening system for protein–protein interaction using ratio or visual imaging of BRET.

(or yeast) colonies expressing a "bait" protein fused to RLUC and a library of "prey" molecules fused to EYFP (or vice versa) could be measured. It would be possible to screen colonies of bacteria or yeast on agar plates using a camera imaging system. On the other hand, a photomultiplier-based instrument designed to measure luminescence of liquid cultures in 96-well plates could be adapted to high-throughput BRET screening by insertion of switchable 480- or 530-nm filters in front of the photomultiplier tube. Colonies that show high light intensity (i.e., bright colonies) at 530 nm or exhibit an above-background ratio of the 530 nm/480 nm could be selected and the "prey" DNA sequence further characterized. Thus, an efficient BRET screening system could be practical by using an appropriate instrument.

41.8 ADVANTAGES OF RESONANCE ENERGY TRANSFER TECHNIQUES (BRET AND FRET)

Features of BRET or FRET techniques offer some attractive advantages over other current assays for protein–protein interactions, especially the yeast two-hybrid method, which is currently the most widely used. For example, BRET or FRET can be applied to determine whether the interaction changes with time because the measurement is noninvasive. BRET/FRET is suitable to assay the protein–protein interactions in different subcellular compartments or specific organelles of the native cells, as has already been shown to work for FRET.[8,9] In particular, the yeast nucleus may be a poor place for some compatible proteins to meet. This advantage of BRET/FRET could be especially useful in the case of interacting membrane proteins for which assays are limited with other traditional methods. BRET or FRET also may be used to reveal interactions that depend upon cell-type-specific post-translational modifications that do not occur in yeast and therefore cannot be assayed by the yeast-two hybrid method. By using cell-type-specific promoters and/or fusion to targeting sequences, the GFP-based BRET or FRET indicators can be observed specifically in the cell type and subcellular location of choice. Moreover, BRET/FRET assays could be adopted to monitor the dynamic processes of protein–protein interactions *in vivo*, such as intracellular signaling.

41.9 NO TECHNIQUE IS PERFECT

As with any technique, however, the resonance energy transfer methods have some limitations. For example, the efficiency of both BRET and FRET is dependent on proper orientation of the donor and acceptor dipoles. Conformational states of the fusion proteins may fix the dipoles into a geometry that is unfavorable for energy transfer. Further, because the fluorophore/luciferase tags are fused to the ends of the potentially interacting molecules, it is possible that some parts of the candidate molecules are interacting without allowing the fluorophore/luciferase tags to be close enough for energy transfer to occur. Consequently, two proteins might interact in a way that is blind to the FRET/BRET technique. In other words, a negative result with a resonance transfer technique does not prove noninteraction. In such a case, testing different combinations of N-terminal and C-terminal fusions in the BRET/FRET assays could help to determine the optimal orientation in which candidate proteins interact.

The luciferase/fluorescent protein tags that are fused to the candidate interacting proteins could interfere with the interaction by steric hinderance (this problem is true for the yeast two-hybrid assay as well). Therefore, the smaller the tags, the less likely will be the hindrance. This is a reason the *Gaussia* luciferase might prove to be superior to RLUC. These luciferase/fluorescent tags might cause inactive or incorrectly folded fusion proteins. For example, the bulkiness of the GFP (and its derivatives) cylinders (20 × 30 Å) have been shown to impede correct folding of fusion proteins.[11]

Another consideration in the use of GFP variants as fluorophore tags is that the slow kinetics of GFP turnover may hamper measuring the kinetics of interaction (whereas *Renilla* luciferase does not suffer these same disadvantages in turnover rate). New GFPs are available that have been engineered to be less stable (Clontech d2EGFP),[24] and the reengineering of BRET fluorophores to be less stable could be useful in temporal studies. Moreover, the pH sensitivity of some of the GFPs might restrict their application to subcellular areas with a higher pH value. However, this limitation can be overcome by utilizing the mutants that are less sensitive to pH.[25]

41.10 BRET VS. FRET

BRET has potential advantages over FRET because it does not require the use of excitation illumination. BRET should be superior for cells that are either photoresponsive (e.g., retina or any photoreceptive tissue) or damaged by the wavelength of light used to excite FRET. Cells that have significant autofluorescence would also be better assayed by BRET than by FRET. This is particularly true for highly autofluorescent tissue such as plants, but all cells are autofluorescent to a degree because of ubiquitous fluorescent molecules such as NADH, collagen, and flavins. Moreover, although photobleaching of the fluorophores can be a serious limitation of FRET, it is irrelevant to BRET. BRET assay requires a substrate for the luciferase, which in the case of RLUC and GLUC is coelenterazine. Coelenterazine is hydrophobic and can permeate all the cell types we have tested, including bacteria (*E. coli* and cyanobacteria), yeast, *Chlamydomonas*,[26] plant seedlings and calli,[27] and animal cells in culture (see Figure 41.3).[15]

In addition, FRET may be prone to complications due to simultaneous excitation of both donor and acceptor fluorophores. Specifically, even with monochromatic laser excitation, it is impossible with the current generation of fluorescent proteins to excite only the donor without exciting the acceptor fluorophore to some degree. In contrast, because BRET does not involve optical excitation, all the light emitted by the fluorophore must result from resonance transfer. Therefore, BRET is theoretically superior to FRET for quantifying resonance transfer. Related to this point is one of the most important advantages of BRET over FRET, namely, that the relative levels of expression of the donor and acceptor partners can be quantified independently: the donor by luminescence and the acceptor by fluorescence. This is difficult with FRET because the acceptor is generally excited to some extent by the excitation wavelength used to excite the donor. With BRET, measuring the fluorescence of the system gives the relative level of the acceptor (YFP/GFP fusion partner), and when coelenterazine is added, the total luminescence of the system measured in darkness gives the relative level of the donor (luciferase fusion partner). Knowledge of the relative levels of the fusion partners is crucial for comparing results from one experiment to the next.

The major limitation that BRET suffers in comparison to FRET is that the luminescence may often be too dim to measure accurately without a very sensitive light-measuring apparatus. With FRET, dim signals can be amplified by simply increasing the intensity or duration of excitation (possibly at the cost of light-induced damage to the cells), whereas with BRET the only option to improve low signal levels is to integrate the signal for a longer time. Nevertheless, Packard recently introduced a very sensitive new instrument designed to be capable of measuring BRET and FRET ratios (the "Fusion"), and other instruments are capable of BRET measurements.[23] Manufacturers are continuously developing improved instrumentation for measuring low light levels, and these improvements in technology will undoubtedly aid the further development of BRET assays of real-time protein–protein interactions in living organisms.

ACKNOWLEDGMENTS

This research was supported by the National Institute of Mental Health (MH 43836 and MH 01179 to C.H.J.), the National Institutes of Health (GM 59984 to C.H.J., DK534343 and CA86283 to D.W.P.), and the National Science Foundation (MCB-9874371 to C.H.J.); spectral data were acquired at the Vanderbilt Cell Imaging Shared Resource, supported in part by the NIH through the Vanderbilt Cancer Center (CA68485) and the Vanderbilt Diabetes Center (DK20593).

REFERENCES

1. Mendelsohn, A. R. and Brent, R., Protein interaction methods—toward an endgame, *Science,* 284, 1948, 1999.
2. Xu, Y., Piston, D. W., and Johnson, C. H., Resonance energy transfer as an emerging strategy for monitoring protein–protein interactions *in vivo*: BRET vs. FRET, *Spectrum,* 12, 9, 1999.
3. Wu, P. and Brand, L., Resonance energy transfer: methods and applications, *Anal. Biochem.,* 218, 1, 1994.
4. Clegg, R. M., Fluorescence resonance energy transfer, *Curr. Opin. Biotechnol.,* 6, 103, 1995.
5. Heim, R., Prasher, D. C., and Tsien, R. Y., Wavelength mutations and posttranslational autoxidation of green fluorescent protein, *Proc. Natl. Acad. Sci. U.S.A.,* 91, 12,501, 1994.
6. Heim, R. and Tsien, R. Y., Engineering green fluorescent protein for improved brightness, longer wavelengths and fluorescence resonance energy transfer, *Curr. Biol.,* 6, 178, 1996.
7. Miyawaki, A., Llopis, J., Heim, R., McCaffery, J. M., Adams, J. A., Ikura, M., and Tsien, R. Y., Fluorescent indicators for Ca^{2+} based on green fluorescent proteins and calmodulin, *Nature,* 388, 882, 1997.
8. Mahajan, N. P., Linder, K., Berry, G., Gordon, G. W., Heim, R., and Herman, B., Bcl-2 and Bax interactions in mitochondria probed with green fluorescent protein and fluorescence resonance energy transfer, *Nat. Biotechnol.,* 16, 547, 1998.
9. Periasamy, A. and Day, R. N., FRET imaging of Pit-1 protein interactions in living cells, *J. Biomed. Opt.,* 3, 154, 1998.
10. Xu, X., Gerard, A. L., Huang, B. C., Anderson, D. C., Payan, D. G., and Luo, Y., Detection of programmed cell death using fluorescence energy transfer, *Nucleic Acids Res.,* 26, 2034, 1998.
11. Gadella, T. W., Jr., van der Krogt, G. N., and Bissseling, T., GFP-based FRET microscopy in living plant cells, *Trends Plant Sci.,* 4, 287, 1999.
12. Xu, Y., Piston, D. W., and Johnson, C. H., A bioluminescence resonance energy transfer (BRET) system: application to interacting circadian clock proteins, *Proc. Natl. Acad. Sci. U.S.A.,* 96, 151, 1999.
13. Ishiura, M., Kutsuna, S., Aoki, S., Iwasaki, H., Andersson, C. R., Tanabe, A., Golden, S. S., Johnson, C. H., and Kondo, T., Expression of a gene cluster kaiABC as a circadian feedback process in cyanobacteria, *Science,* 281, 1519, 1998.
14. Iwasaki, H., Tanihuchi, Y., Ishiura, M., and Kondo, T., Physical interactions among circadian clock proteins KaiA, KaiB and KaiC in cyanobacteria, *EMBO J.,* 18, 1137, 1999.
15. Angers, S., Salahpour, A., Joly, E., Hilairet, S., Chelsky, D., Dennis, M., and Bouvier, M., Detection of beta 2-adrenergic receptor dimerization in living cells using bioluminescence resonance energy transfer (BRET), *Proc. Natl. Acad. Sci. U.S.A.,* 97, 3684, 2000.
16. Wang, Y., Wang, G., O'Kane, D. J., and Szalay, A. A., A study of protein-protein interactions in living cells using luminescence resonance energy transfer (LRET) from Renilla luciferase to Aequorea GFP, *Mol. Gen. Genet.,* 264, 578, 2001.
17. Wang, Y., Wang, G., O'Kane, D. J., and Szalay, A. A., The study of protein-protein interactions using chemiluminescence energy transfer, in Roda, A. P. M., Kricka, L. J., and Staley, P. E., Eds., *Bioluminescence and Chemiluminescence: Perspectives for the 21st Century,* Wiley, Chichester, U.K., 1999, 475.
18. Szent-Gyorgyi, C., Ballou, B. T., Dagnal, E., and Bryan, B., Cloning and characterization of new bioluminescent proteins, in *Biomedical Imaging: Reporters, Dyes, and Instrumentation: Proceedings of SPIE—The International Society for Optical Engineering,* 3600, 1999, 4–11.
19. Matz, V. M., Fradkov, A. F., Labas, Y. A., Savitsky, A. P., Zaraisky, A. G., Markelov, M. L., and Lukyanov, S. A., Fluorescent proteins from nonbioluminescent *Anthozoa* species, *Nat. Biotechnol.,* 17, 969, 1999.
20. Más, P., Devlin, P. F., Panda, S., and Kay, S. A., Functional interaction of phytochrome B and cryptochrome 2, *Nature,* 408, 207, 2000.
21. Baird, G. S., Zacharias, D. A., and Tsien, R. Y., Biochemistry, mutagenesis, and oligomerization of DsRed, a red fluorescent protein from coral, *Proc. Natl. Acad. Sci. U.S.A.,* 97, 11,984, 2000.
22. Arai, R., Nakagawa, H., Tsumoto, K., Mahoney, W., Kumagai, I., Ueda, H., and Nagamune, T., Demonstration of a homogeneous noncompetitive immunoassay based on bioluminescence resonance energy transfer, *Anal. Biochem.,* 289, 77, 2001.

23. Xu, Y., Piston, D., and Johnson, C. H., BRET assays for protein-protein interactions in living cells, in *Green Fluorescent Protein: Methods and Protocols,* Hicks, B. W., Ed., Humana Press, Totowa, NJ, in press.
24. Andersen, J. B., Sternberg, C., Poulsen, L. K., Bjorn, S. P., Givskov, M., and Molin, S., New unstable variants of green fluorescent protein for studies of transient gene expression in bacteria, *Appl. Environ. Microbiol.,* 64, 2240, 1998.
25. Miyawaki, A., Griesbeck, O., Heim, R., and Tsien, R. Y., Dynamic and quantitative Ca^{2+} measurements using improved cameleons, *Proc. Natl. Acad. Sci. U.S.A.,* 96, 2135–2140, 1999.
26. Minko, I., Holloway, S. P., Nikaido, S., Odom, O. W., Carter, M., Johnson, C. H., and Herrin, D. L., Renilla luciferase as a vital reporter for chloroplast gene expression in *Chlamydomonas*, *Mol. Gen. Genet.,* 262, 421, 1999.
27. Sai, J. and Johnson, C. H., Different circadian oscillators control Ca^{++} fluxes and *Lhcb* gene expression, *Proc. Natl. Acad. Sci. U.S.A.,* 96, 11,659, 1999.

42 BRET^2: Efficient Energy Transfer from *Renilla* Luciferase to GFP^2 to Measure Protein–Protein Interactions and Intracellular Signaling Events in Live Cells

Pierre Dionne, Mireille Caron, Anne Labonté, Kelly Carter-Allen, Benoit Houle, Erik Joly, Sean C. Taylor, and Luc Menard

CONTENTS

42.1 BIOLUMINESCENCE RESONANCE ENERGY TRANSFER

BRET^2 (bioluminescence resonance energy transfer) is a technology that allows the detection of protein–protein interactions and intracellular signaling events in live cells. This nondestructive assay is based on the transfer of resonant energy from a bioluminescent donor protein to a fluorescent acceptor protein using *Renilla* luciferase (Rluc) as the donor and a green fluorescent protein (GFP^2) as the acceptor molecule. This assay is analogous to fluorescence resonance energy transfer (FRET), but eliminates the need for an excitation light source and its associated problems (e.g., high background caused by autofluorescence). Rluc emits blue light at 395 nm in the presence of its substrate DeepBlueC™ coelenterazine. If the GFP^2 is in close proximity to Rluc, it absorbs the blue light energy and re-emits green light at 510 nm (Figure 42.1). The BRET signal is a ratiometric measurement of the amount of green light emitted by the GFP vs. the blue light emitted by Rluc (510/395 nm).

0-8493-0719-8/02/$0.00+$1.50

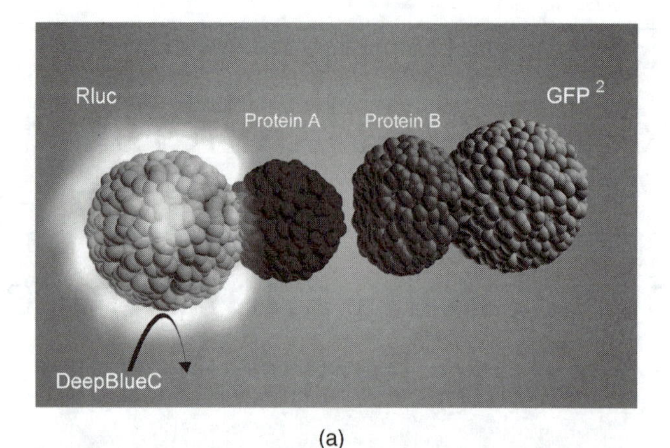

(a)

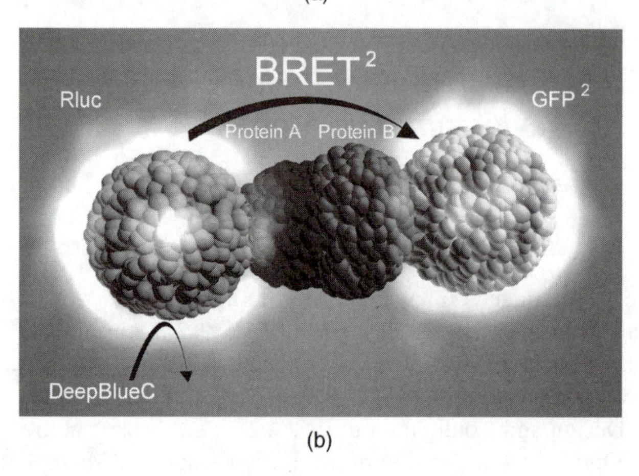

(b)

FIGURE 42.1 (Color figure follows p. 266.) Schematic representation of the $BRET^2$ principle. In the presence of its substrate DeepBlueC (a coelenterazine derivative), Rluc emits blue light (395 nm). When Rluc and GFP^2 are brought into close proximity by means of a biological interaction, e.g., interaction of proteins genetically fused to Rluc (Protein A) and GFP^2 (Protein B), blue light energy is transferred to GFP^2, which re-emits green light (510 nm). The BRET signal is calculated as a ratio of the signal at 510 nm vs. that at 395 nm.

42.2 UNDERSTANDING RESONANCE ENERGY TRANSFER

Resonance energy transfer is defined as the transfer of excited-state energy from a donor to an acceptor molecule. This phenomenon is highly dependent upon:

1. The overlap of the donor emission and the acceptor excitation spectra,
2. The quantum yield of the donor,
3. The distance between the pair, and
4. The orientation of their respective dipoles.

The typical effective distance between the donor and acceptor is 10 to 100 Å. This range correlates well with most biological interactions, making resonance energy transfer an excellent tool for monitoring macromolecular interactions. The energy transfer efficiency (E) may be directly related to the distance using:

$$E = \frac{R_0^6}{R_0^6 + r^6}$$

where r represents the distance between the donor and the acceptor dipoles and R_0, the Förster radius. R_0 is traditionally defined as the characteristic distance for a given donor–acceptor pair at which the efficiency of transfer is 50%. For most donor–acceptor pairs, R_0 varies from 20 to 50 Å and is defined by the equation:

$$R_0 = 9.79 \times 10^3 (k^2 n^{-4} \phi_d J)^{1/6} \qquad \text{(in Å)}$$

where k is a factor describing the relative orientation of the donor and acceptor dipoles, n is the refractive index of the medium, ϕ_d is the quantum yield of the donor in the absence of the acceptor, and J is the overlap integral (overlap between the donor emission and the acceptor excitation spectra).

The energy transfer rate is highly dependent on the distance between the donor and acceptor molecules since, as can be derived from the equation above, the energy transfer efficiency is inversely proportional to the sixth power of the distance between the donor and acceptor.

42.3 BRET^2 SPECTRAL PROPERTIES

In the original reports describing the use of BRET to monitor protein–protein interaction, Rluc was used as the donor, EYFP or Topaz was used as the acceptor, and coelenterazine *h* was used as the Rluc substrate.[1,2] Using this assay configuration, Rluc emits light at 475 to 480 nm and the GFPs emit light at 525 to 530 nm, giving a spectral resolution of 45 to 55 nm. Rluc emission spectrum is broad and substantially overlaps with GFP emission, which decreases the signal-to-noise ratio of the system. With BRET^2 the spectral resolution of the system has been increased to 105 nm. This was accomplished by the development of the new coelenterazine derivative DeepBlueC (DBC) that decreases the Rluc emission peak from 480 to 395 nm. A new GFP was developed to accommodate this new spectral requirement and GFP^2 was engineered to have a maximal absorption at 395 nm and maximal emission at 510 nm (Figure 42.2, Table 42.1).

As will be seen later in this chapter, the main advantage of the increase in spectral resolution in BRET^2 is an increased dynamic range of the assay, which offers improved sensitivity.

42.4 CONSIDERATIONS FOR DETECTION OF BRET^2 ASSAYS

Selection of a plate reader suitable for detection of BRET^2 assays requires two main considerations. The first decision can be made on whether or not the reader has the ability to use luminescence detection through two separate band-pass filters. As described earlier in this chapter, the BRET^2 emission contains two distinct peaks, the Rluc donor signal at 395 nm and the GFP^2 acceptor signal at 510 nm. The reader must be able to discriminate between these two signals. Most luminometers do not have filters for spectral discrimination. The second consideration, which is most obvious, will be to select the reader with the best sensitivity and lowest system noise. Given the above criteria, we found the Fusion™ Universal Microplate Analyzer (Figure 42.3; Packard Instrument Company, Downers Grove, IL) to be an excellent choice for detection of BRET^2 assays.

TABLE 42.1
Components and Emission Wavelengths of BRET vs. BRET^2

Generation	Coelenterazine	Donor	Acceptor	Wavelength (nm) (donor/acceptor)
BRET	WT, h	Rluc	EYFP, Topaz	480/530
BRET^2	DeepBlueC	Rluc, Rluc(h)	GFP^2	395/510

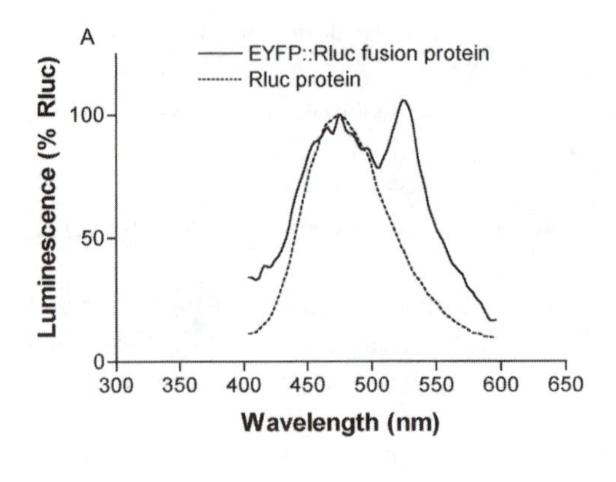

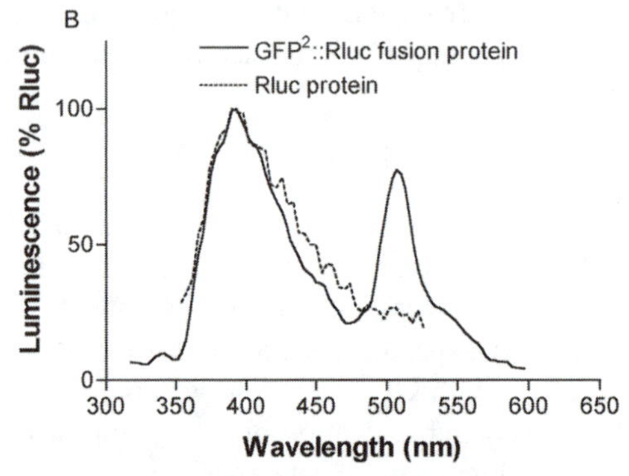

FIGURE 42.2 Emission scans of Rluc and GFP fusion proteins. Emission scans of CHO cells expressing either GFP^2::Rluc or EYFP::Rluc fusion proteins. Scans were generated using a Fluorolog-3 instrument. (Instruments S.A., NJ.)

42.5 PLATE READER DESIGN AND CONFIGURATION

Luminescence and fluorescence detection can be considered well established and routinely employed methods of detection in most research laboratories. The advent of high-throughput screening technology in drug discovery, as well as the desire to conserve reagents, has spurred innovation in microplate-based detection systems resulting in the large variety available today. Growing interest in functional cell-based assays in miniaturized microplate formats has elicited the market to respond with more sensitive instruments and non-isotopic assay reagents.

Instruments capable of multiple modes of detection are increasingly used for both basic research and assay development for high-throughput screening. Assay volumes have been scaled down from 96-well to 384-well and even 1536-well plates, which has become a key criterion for selecting new reagent technologies. In addition, the ability to measure fluorescence, absorbance, and luminescence assays on a single instrument gives the researcher a powerful and flexible tool. Combining fluorescence and luminescence in the same instrument enables the use of filters, typically only found in fluorometers, for the measurement of luminescence as produced by $BRET^2$ assay components.

The basic design of plate-reading systems can actually be described quite simply. Light emission from a sample in a well of an opaque microplate is directed to a detector, most commonly a highly sensitive photomultiplier tube (PMT). In some systems the light is collected using fiber optics positioned above the microplate well. Light loss within the fiber may cause a degradation of sensitivity, particularly if the positioning is not accurately controlled. The Fusion uses direct optics to detect luminescent signals from each well of the microplate. A filter wheel containing the specific interference filters for the Rluc donor emission (395 nm) and the GFP^2 acceptor emission (510 nm) is placed in the optical path to provide wavelength discrimination. The plate itself is placed into a light-tight chamber and optimally positioned beneath the optical pathway for gathering as much light as possible. In luminescence mode, the excitation pathway in the Fusion, which is required for fluorescence and absorbance modes, is disabled with the accompanying software. This is necessary to prevent photobleaching and cell damage from the intense illumination. A schematic representation of the system can be seen in Figure 42.4.

FIGURE 42.3 Fusion Universal Microplate Analyzer.

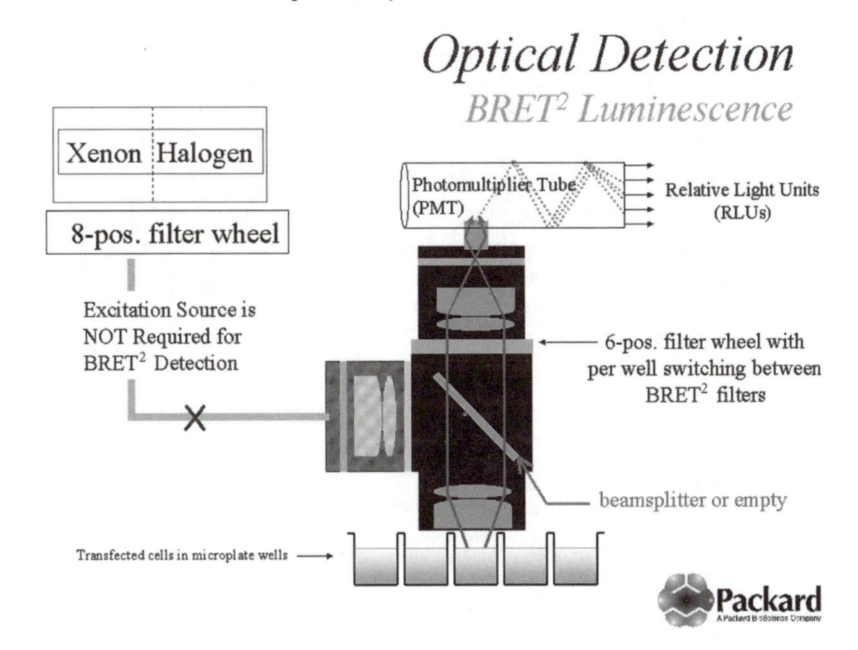

FIGURE 42.4 Schematic representation of direct optics for BRET² detection in the Fusion.

The design of the optics in the Fusion minimizes the amount of well-to-well cross talk. This is achieved through automated and accurate X–Y and Z-height positioning of the plate using the software parameters entered into the Plate Type Library. Many different manufacturers' plates can be found in this library, and the system allows for plates to be added as new developments are made. The patented design of collection and collimation of light from each well also contributes to an increase in sensitivity, particularly when compared with other multidetection plate readers.

42.6 SELECTING THE OPTIMAL CONFIGURATION OF THE FUSION UNIVERSAL MICROPLATE READER

Nearly all possible Fusion configurations are suitable for use in laboratories implementing $BRET^2$ assays. The exception is that instruments selected for TRF (time-resolved fluorescence) may limit the sensitivity of $BRET^2$ detection. In the optimized TRF configuration of the Fusion, the use of a red-sensitive PMT will slightly reduce the sensitivity of detection at blue and green emission wavelengths, and therefore may not be compatible with some $BRET^2$ assays. The basic TRF configuration with a blue sensitive PMT may be used for $BRET^2$ assays if careful optimization of the sensitivity is performed when setting up each $BRET^2$ assay. Table 42.2 illustrates the suitability of various Fusion detection modes for use with $BRET^2$.

It is important to select carefully the correct bandpass filters for detection of the $BRET^2$ emission signals. Packard Instrument Company provides $BRET^2$ optimized filters with center wavelengths at 410 nm with 80-nm bandpass and at 515 nm with 30-nm bandpass for use with the Fusion. These filters have been designed and tested with the $BRET^2$ reagents and are highly recommended. Use of filters other than those specified could produce different, perhaps even erroneous results (see Table 42.2 for optimum configuration).

TABLE 42.2
Compatibility of Fusion Detection Modes and Options with $BRET^2$ Assays

Description	Compatibility
AlphaScreen Detection Mode	++
TRF Detection Mode (blue PMT)	+
TRF Optimized Detection Mode (red PMT)	–
Fluorescence Polarization Mode	++
Absorbance Mode (visible range)	++
UV Absorbance Mode	+
Temperature control	++
Stacker/Bar code	++
1536-well Reading Option[a]	++
Red PMT	–
Blue PMT	++
Paired Emission Filter set: 410/80 nm and 515/30 nm	++

Legend: ++ highly suitable, + suitable with optimization, – not recommended.

[a] 1536-well measurement may not be suitable for very low signal luminescence assays.

With the flexibility of multiple detection modes, some instruments become complicated to use. However, the Fusion software automatically configures the instrument for plate positioning, filter selection, per-well filter switching, dual emission data collection, and deactivation of the excitation pathway when the user selects the dual-wavelength mode for BRET^2. User-definable or fully automated sensitivity optimization makes setting up the PMT high voltage and gain for optimal detection of the BRET^2 assay signal extremely easy to do. The combination of sensitive direct optics, low system backgrounds, and easy-to-use software makes performing BRET^2 assays on the Fusion simple and successful.

42.7 THE BRET^2 ASSAY TECHNOLOGY

BRET^2 assays are highly flexible and can be used to detect protein–protein interactions or to create biological sensors of intracellular signals. The only limitation of the technology is in the ability to express proteins of interest with the Rluc and GFP^2 tags. When working with new proteins, the experimental design used for cloning can be critical for the generation of optimal BRET^2 conditions. Two major consi-derations should be examined that reflect protein function and protein expression. First, the target proteins of interest should be tested in fusions with GFP^2 and Rluc at both the N- and C-terminus. Second, a variety of transfection conditions should be tested for each fusion protein. For example, cells can be cotransfected with a fixed amount of the Rluc DNA and increasing amounts of the GFP^2 fusion DNA partner. These combined considerations can have a major impact on the success of the BRET^2 assay.

42.8 APPLICATIONS OF BRET^2

Protein–protein interaction assays can be performed by genetically fusing Rluc and GFP^2 to biological partners that are expected to interact in a cell-based assay (Figure 42.5). Changes in the interaction, modulated, for example, by ligands or compounds, can be monitored by a change in the ratio of blue and green light emitted by the cells.

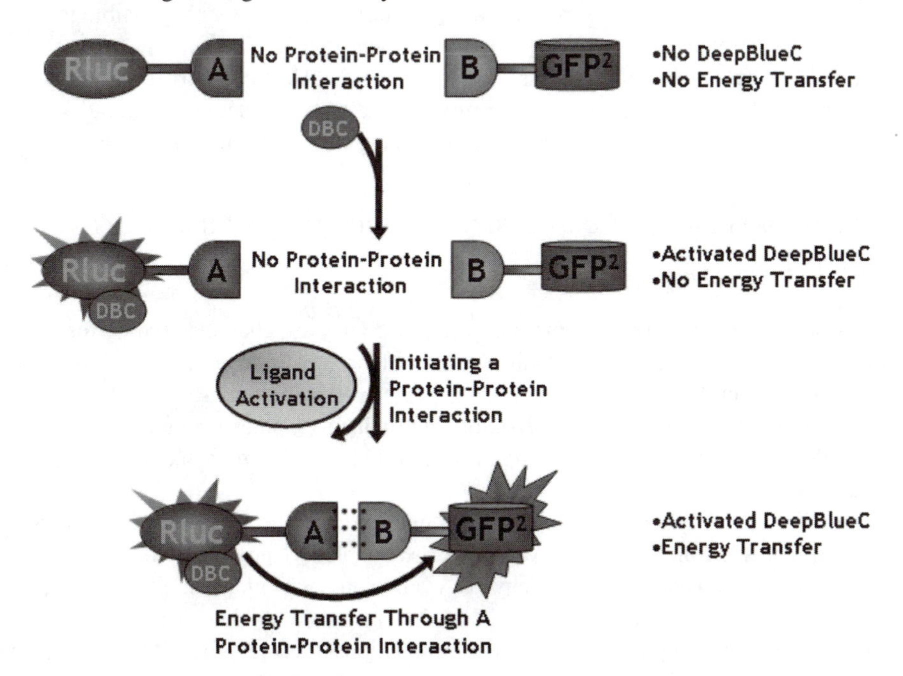

FIGURE 42.5 Protein–protein interaction BRET^2 configuration.

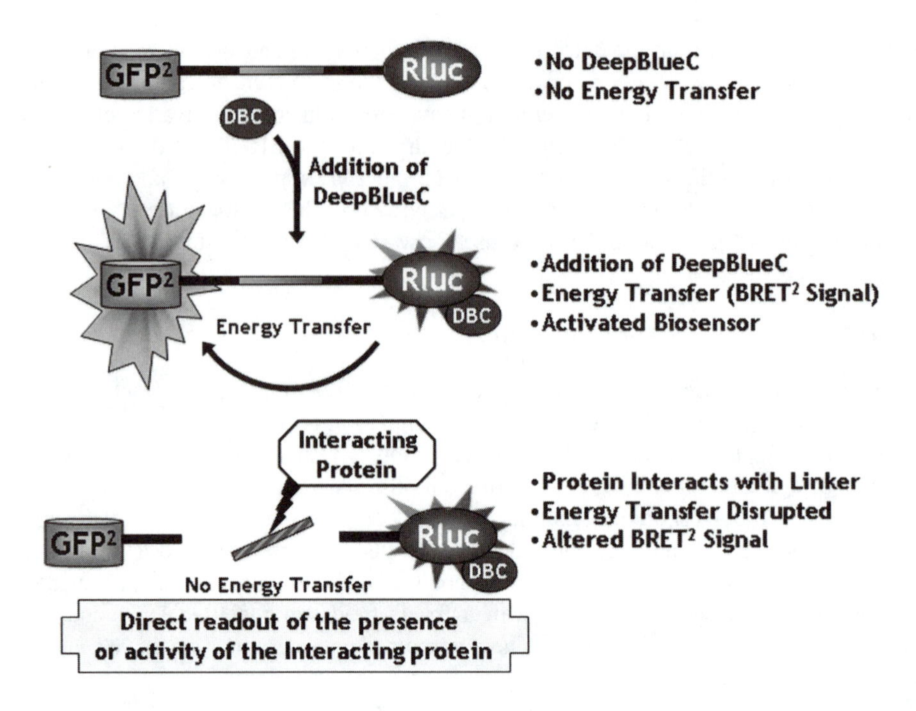

FIGURE 42.6 Sensor BRET^2 configuration.

The engineering of sensors involves the expression of a fusion protein in which Rluc and GFP^2 are linked together by a peptide sequence that can be modulated (Figure 42.6). This linker sequence could, for example, be a protease site. Upon activation, cleavage of the linker sequence will increase the distance between the donor and acceptor proteins, thus decreasing the energy transfer between Rluc and GFP^2 and result in a lower BRET^2 ratio. Other configurations are possible including the engineering of a kinase site or ion-sensitive linker sequences (e.g., Ca^{2+}).

An example of these two configurations is described below. The interaction between a G protein–coupled receptor (GPCR) and *β*-arrestin will be used to illustrate a protein–protein interaction assay, while the sensor configuration will be demonstrated using the apoptotic induction of caspase-3.

42.8.1 Study of G Protein–Coupled Receptor Activation in Intact Cells

42.8.1.1 GPCR Signaling

GPCRs represent a large family of membrane receptors. They are characterized by the presence of seven transmembrane domains and transduce extracellular signals from a wide variety of stimuli that include light, peptides, small molecules, and others (Watson and Arkinstall[3] and references therein). Because of their membrane localization and their involvement in numerous physiological processes, GPCRs are the targets of more than 50% of the current drugs and are intensely studied for drug discovery.

Signaling through GPCRs involves a cascade of events starting with the activation of the receptor by its ligand (agonist), followed by activation of the coupled G proteins, which in turn leads to activation of effector molecules (e.g., adenylate cyclase, ion channels, or phospholipase C) and changes in second messenger levels (reviewed in Neer[4]). In addition to the signal generation of GPCRs, homeostasis also requires proper signal termination. It has long been known that long-term exposure to an agonist leads to receptor desensitization. Part of this phenomenon occurs within seconds to minutes of exposure to the agonist, and is the result of sequestration and internalization of the receptors within the cells (Lefkowitz[5] and references therein). Rapid GPCR desensitization is

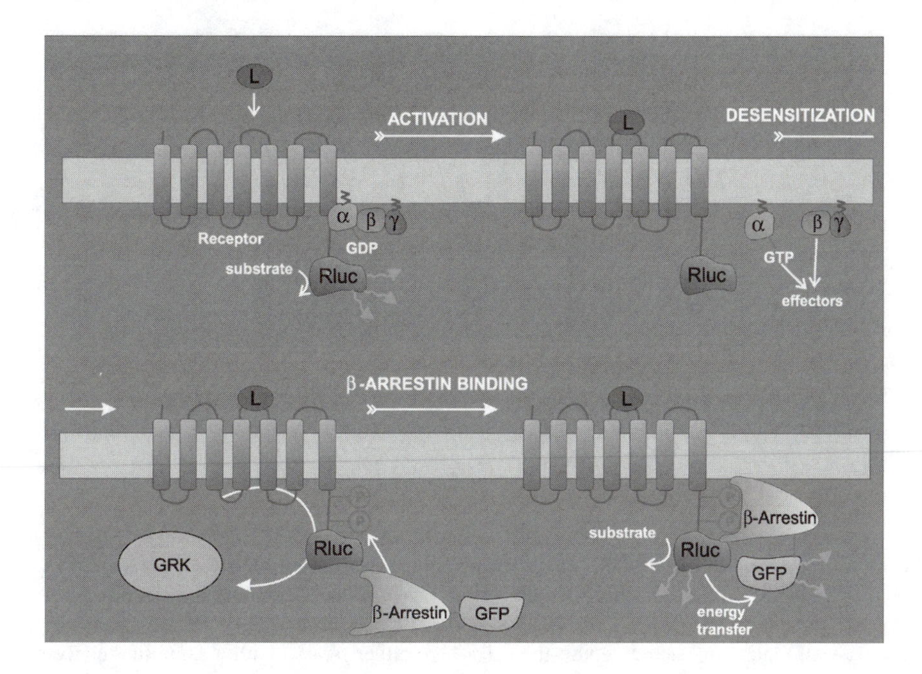

FIGURE 42.7 Schematic representation of GPCR/*β*-arrestin BRET^2 assay. (1) Binding of an agonist activates the receptor. (2) G protein subunits dissociate from the receptor and activate downstream effectors. (3) Receptor desensitization is initiated by the action of GRK, which phosphorylates the receptor. (4) *β*-Arrestin binds to phosphorylated receptor and contributes to receptor internalization.

a tightly regulated process involving specific proteins,[6] and the main signaling pathway takes place in two steps (Figure 42.7). First, the agonist-activated receptor is specifically recognized and phosphorylated by a GPCR kinase (GRK) at the phosphorylation sites on the C-terminal tail of the GPCR. This site-specific phosphorylation leads to the binding of cytosolic proteins called arrestins. The receptor–arrestin interaction prevents further activation of G proteins and targets the receptor to clathrin-coated pits, where receptor internalization occurs. Arrestins represent a family of cytosolic proteins of about 48 kDa. Of the six known members, arrestin 2 and arrestin 3 (also known as *β*-arrestin 1 and *β*-arrestin 2, respectively) have the widest tissue distribution. Although not all GPCRs internalize through clathrin-coated pits, it is estimated that the majority of GPCRs follow an arrestin-dependent internalization mechanism.

Since the majority of GPCRs interact with *β*-arrestin, monitoring the interaction between a receptor and arrestin could represent an attractive assay for GPCR activation, especially with regard to orphan receptors (receptors for which natural ligands have yet to be identified). Using the original BRET technology, Angers et al.[2] have developed an assay to detect the interaction between the chimeric *β*2 adrenergic receptor-*Renilla* luciferase and the *β*-arrestin 2-EYFP fusion protein and have shown that treatment with the agonist isoproterenol led to a dose-dependent increase of the BRET ratio. This assay has since been used to study other receptors, in a BRET^2 configuration.

Figure 42.8 shows the results obtained with four receptors, the vasopressin 2 receptor (V_2), and three adrenergic receptors, *β*1AR, *β*2AR, and *β*3AR. *β*3AR was used as a negative control since it has been shown not to interact with arrestin.[7] The results show that agonist treatment led to an increase of the BRET^2 ratio for V_2, *β*1, and *β*2 with no effect on *β*3 control.

Next, we performed cotransfections with various amounts of plasmid DNA from *β*2AR:Rluc and GFP^2:*β*-arrestin 2. To assess the effect of the amount of receptor DNA on expression levels, whole-cell binding assays were performed on transfected cells (Figure 42.9a). Total binding was determined using the lipophilic radiolabeled antagonist ^{125}I-pindolol, which labels receptors that

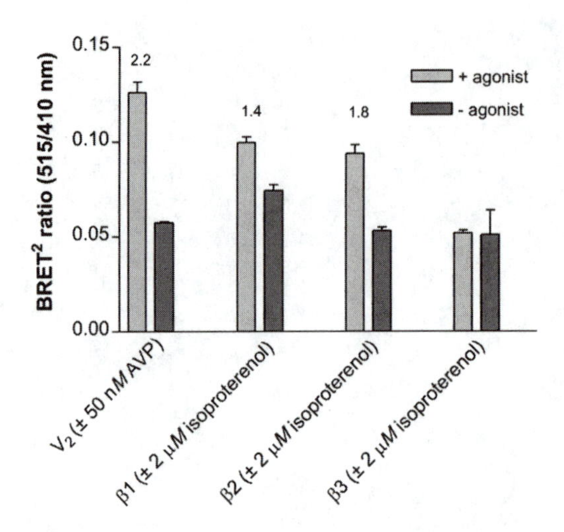

FIGURE 42.8 Agonist promoted GPCR/β-arrestin interaction measured using $BRET^2$. HEK 293T cells were transiently cotransfected with a construct encoding Rluc fused to the C-terminal portion of the receptor and a vector encoding the GFP^2:β-arrestin 2 fusion protein. At 48 h post-transfection, the cells were harvested and plated in a 96-well plate. For each receptor, the cells were either treated with a saturating concentration of agonist for 20 min or treated with buffer alone, and $BRET^2$ was determined using a Fusion Universal Microplate Analyzer. Values above bars represent signal-to-basal ratio (S/B).

are both intra- and extracellular. Nonspecific binding was assessed using either hydrophobic (cell permeant) *S*(−)-propanolol or hydrophilic CGP-12177. The results of Figure 42.9a show that receptor expression levels correlated with the amounts of β2AR:Rluc DNA used, and was estimated to be in the range of 1.7×10^5 (condition 1:40) to 2×10^6 (condition 25:25) sites per cell. Moreover, comparison of the results obtained with *S*(−)-propanolol vs. CGP-12177 shows that most of the receptors were localized at the cell membrane, which is an essential condition for performing a $BRET^2$/arrestin assay.

By using cells from the same transfections, the response to agonist was determined in a $BRET^2$/arrestin assay. Figure 42.9b shows that treatment with 10 μM of isoproterenol led to a $BRET^2$ increase for all conditions, yielding signal-to-basal (S/B) values of 1.44 to 1.62. The results of Figure 42.9b also show a basal $BRET^2$ ratio (noninduced) of 0.06 for condition 1:40. This value is slightly above the background $BRET^2$ ratio of 0.05 observed when Rluc is not in close proximity to GFP^2, and most likely reflects "basal" interactions resulting from overexpression of the GFP^2:β-arrestin 2 fusion protein. As expected, increasing β2AR:Rluc expression levels led to a gradual decrease in the stimulated $BRET^2$ ratios since overexpression of β2AR:Rluc leads to an increasing amount of fusion receptors that are not engaged in an interaction with GFP^2:β-arrestin 2 following stimulation with the agonist. Overall, these results highlight the importance of adequate expression levels for interacting partners in such an assay.

Using the β2 adrenergic receptor, the sensitivity of the $BRET^2$/arrestin assay was assessed. As shown in Figure 42.10a, a 20-min treatment of the cells with increasing concentrations of the agonist isoproterenol revealed a dose-dependent increase of the $BRET^2$ ratio, with an S/B of 1.8 at saturating concentrations, and an EC_{50} value in the nanomolar range (94 n*M*). In comparison, cells transfected with the wild-type receptor and assayed in a cAMP assay using the AlphaScreen™ system resulted in an EC_{50} of 10 n*M* (Figure 42.10b).

By using transiently transfected cells, three compounds showing affinity for the β2 adrenergic receptor—dichloroisoproterenol (DCI), alprenolol, and ICI 118-551—were tested for their capacity to antagonize the interaction of β-arrestin 2 with the β2AR elicited by treatment with 1 μM isoproterenol. Figure 42.11 shows that each compound could inhibit the response in a

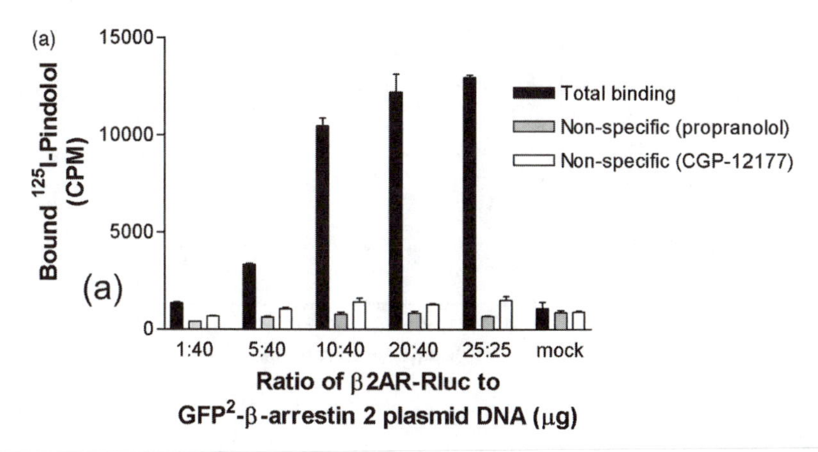

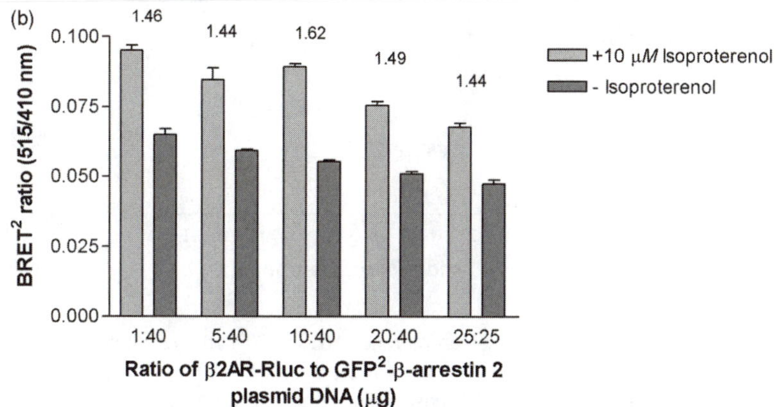

FIGURE 42.9 BRET^2 signal obtained using various ratios of GFP^2:*β*-arrestin 2 to *β*2AR:Rluc DNA amount. HEK 293T cells were transiently cotransfected at various ratios of *β*2AR:Rluc to GFP^2:*β*-arrestin 2 plasmid DNA amount. Cells were harvested 48 h post-transfection and assayed for either receptor level of expression (a) or receptor/arrestin interaction (b). (a) *Receptor level expression*: 10,000 cells were incubated for 4 h at 10°C with 0.59 n*M* ^{125}I-pindolol in the absence or presence of either 1 *μM S*(−)-propranolol or 0.3 *μM* CGP-12177. The amount of bound radioactivity was determined with a TopCount NXT instrument (Packard BioScience). (b) *Receptor/arrestin interaction*: The cells were harvested and plated in a 96-well plate. For each condition, the cells were either treated with 10 *μM* of the agonist isoproterenol for 20 min or treated with buffer alone, and BRET^2 was determined using a Fusion Universal Microplate Analyzer. Values above bars represent signal-to-basal ratio (S/B).

dose-dependent manner. The EC_{50} values were 3.1 n*M* for ICI 118-551, 4.1 n*M* for alprenolol, and 620 n*M* for DCI. For DCI and alprenolol, the sensitivity was very similar to that reported for cAMP assays and, most importantly, the ranking obtained with the BRET^2/arrestin assay was identical to that obtained in cAMP assays by others.[8] Altogether, these results confirmed that the BRET^2/arrestin assay provides pharmacologically relevant information and can be used to characterize compounds.

In conclusion, the BRET^2/arrestin assay is a live-cell assay that can be used for the functional characterization of GPCRs, yielding pharmacologically relevant data. Given the large number of GPCRs that undergo rapid desensitization through an arrestin-dependent pathway, this assay offers an attractive system for the screening of agonists specific to orphan receptors. Future work will aim at optimizing the system, namely, by generating cell lines stably expressing appropriate ratios of the binding partners. This will facilitate the use of the BRET^2/arrestin assay in a high-throughput screening mode.

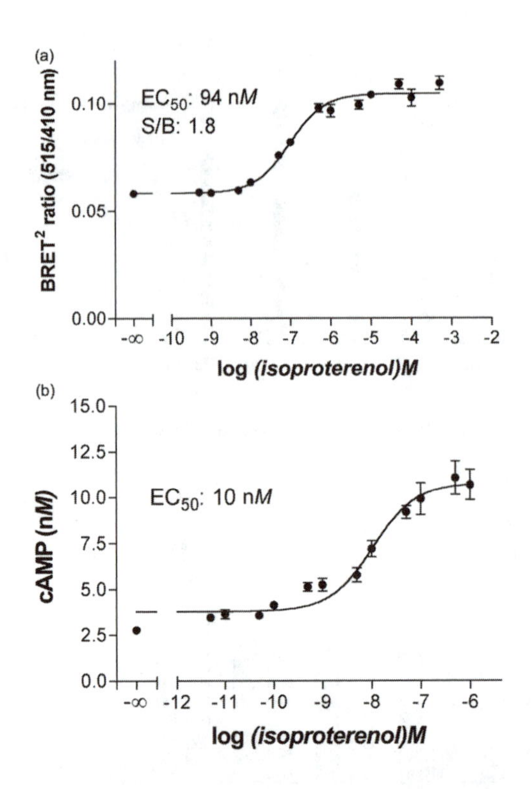

FIGURE 42.10 Dose–response curves using isoproterenol. HEK 293T cells were transiently transfected either with (a) a 10:40 ratio of β2AR:Rluc to GFP^2:β-arrestin 2 plasmid DNA amount or with (b) wild-type β2-adrenergic receptor. Cells were harvested 48 h post-transfection and dose–response curves were performed using either (a) the BRET^2 assay technology or (b) the AlphaScreen technology. (a) *BRET^2 dose–response curve*: Cells were plated in a 96-well plate and incubated for 20 min with increasing concentrations of isoproterenol. BRET^2 was determined using a Fusion Universal Microplate Analyzer. (b) *AlphaScreen dose–response curve*: Cells (not serum starved) were plated in a 96-well plate and were incubated with increasing concentrations of isoproterenol for 20 min. The cAMP levels were detected using a standard AlphaScreen protocol using an AlphaQuest Microplate Analyzer.

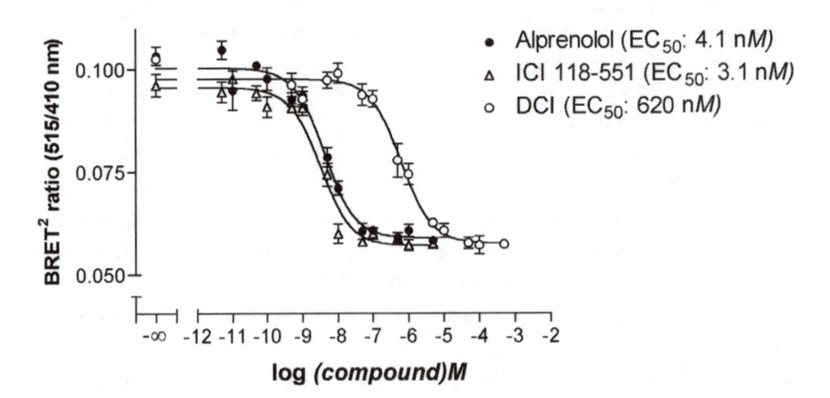

FIGURE 42.11 Dose–response curves for β2AR-specific compounds. HEK 293T cells were transiently co-transfected at a 10:40 DNA ratio of β2AR-Rluc:GFP^2-β-arrestin 2. At 48 h post-transfection, cells were harvested and plated in a 96-well plate and treated for 20 min with increasing concentrations of different β2AR antagonists in presence of 1 μM isoproterenol. BRET^2 was determined using a Fusion Universal Microplate Analyzer.

42.8.2 Caspase Assay

Apoptosis, or programmed cell death (PCD), is a regulated and structured process by which a cell self-destructs into membrane-encapsulated particles called apoptotic bodies that are rapidly phagocytosed and degraded. This mechanism allows for the elimination of cells that are not needed or that are potentially harmful for the organism such as cells that have been infected by a virus or tumor cells. PCD does not induce an inflammatory response, which differs from necrosis, the other type of cell death that is usually caused by a severe cell injury. Inappropriate PCD has been linked to diseases such as cancer and autoimmune or neurodegenerative diseases.

Many proteins are involved in the PCD pathway including the caspases, which are members of a large family of cysteine proteases.[9,10] The caspases are elements of a protease cascade ending in the activation of effectors responsible for the degradation of cellular proteins and genomic DNA.

The first line of caspases can be activated by either extracellular signals transduced by so-called death receptors such as FAS or TNF (caspase-8) or by intracellular signals such as the release of cytochrome *c* from mitochondria (caspase-9). Once activated, this first line of caspases activates downstream caspases (caspase-3, -4, -7, -9)[11] by proteolytic cleavage. These, in turn, activate other caspases or act on intracellular targets resulting in cell disassembly. Caspases recognize and cleave specific amino acid sequences usually composed of four amino acids[12] (Figure 42.12).

Apoptosis can also be triggered by chemical agents such as thapsigargin in HL-60 cells or staurosporine in HeLa cells. BRET^2 provides a convenient method to assess the effect of various conditions on the PCD pathway.

We show here that, by introducing a caspase-3 consensus sequence (DEVD in single amino acid letter code) in the linker region between GFP^2 and Rluc to give the GFP^2-DEVD-Rluc fusion construct, a decrease in the BRET^2 signal indicates induction of apoptosis in living cells (see Figure 42.12).

Figure 42.13 represents cellular morphology changes of HeLa cells undergoing apoptosis after treatment with staurosporine. Morphological changes in apoptotic HeLa cells were characterized

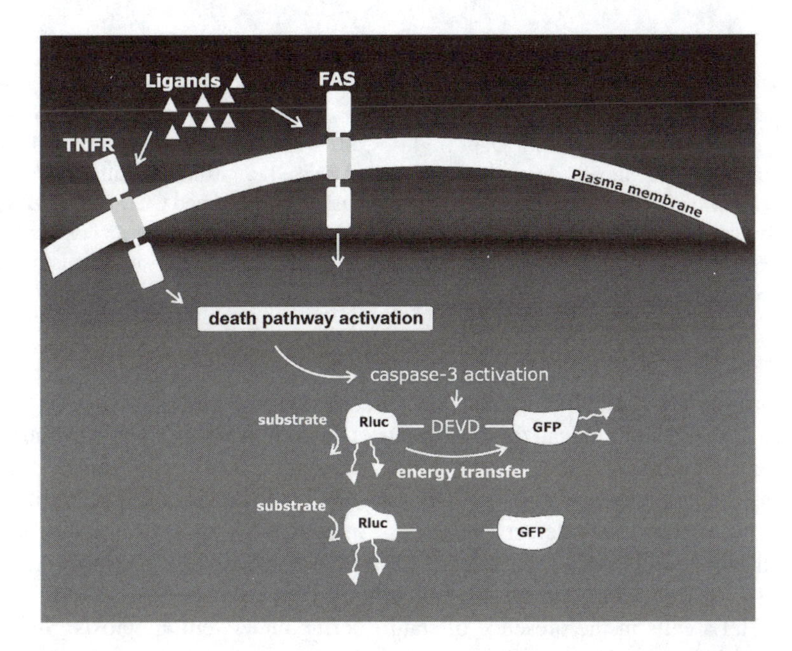

FIGURE 42.12 Schematic representation of apoptotic extracellular induction pathway. Agonists bind to the extracellular domain of death receptors which activate the intracellular domains initiating the death pathway. Cells transfected with the Rluc-DEVD-GFP^2 fusion protein can be directly assayed for initiation of apoptosis through caspase-3 production.

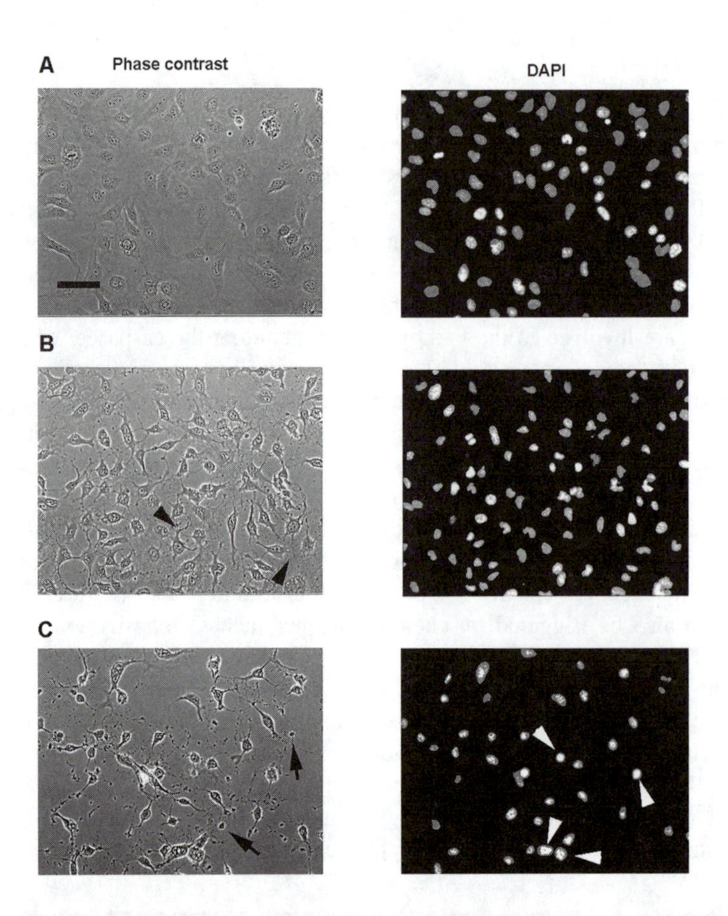

FIGURE 42.13 Morphological changes of apoptotic HeLa cells. At 24 h post-transfection, cells were harvested and seeded in six-well plates. Prior to seeding the cells, sterile coverslips were added to each well. Cells were allowed to attach to coverslips overnight. Apoptosis was induced with 1 μM staurosporine for 0, 1, and 5 h before staining the DNA with Dapi (right). Cells were finally fixed with frozen-cold methanol (–20°C) for 20 min and examined under an AxioSkop (Carl Zeiss, Inc.) using excitation and emission filters (Omega Optical, Inc.) for Dapi. Cells images were recorded using an intensified CCD camera (ZVS^{TM}3C75DE video camera system from Carl Zeiss, Inc.). Black arrowheads and arrows in phase contrast pictures (left) show membrane blebbing and apoptotic bodies, respectively. White arrowheads show cells in which nuclear DNA is condensed. Bar represents 50 μM.

using light (phase contrast) and fluorescence microscopy (Dapi staining). At time 0 (Panel A), cells appeared fully attached and elongated (phase contrast) with no sign of DNA condensation (Dapi). After 1 h of induction (Panel B), cells were rounded up and started to detach from the coverslip. Cell volume decreased and membrane blebbing was detectable in the majority of cells (phase contrast). DNA condensation (Dapi) was detectable but not in a majority of cells. At 5-h post-treatment (Panel C), cell numbers decreased significantly, and the presence of numerous apoptotic bodies was detected (phase contrast) with DNA condensation (Dapi) observed in almost all remaining cells. Clearly, HeLa cells in the presence of staurosporine underwent apoptosis. Similar morphological changes of HeLa cells undergoing apoptosis have been reported by others.[13]

Figure 42.14 represents the BRET^2 signal changes in HeLa cells transfected with either GFP^2:Rluc or GFP^2:DEVD:Rluc fusion construct after a 5-h apoptosis induction with staurosporine. Significant BRET^2 signal changes (0.31 unit) occur when cells transfected with the GFP^2:DEVD: Rluc construct were treated with staurosporine. The BRET^2 signal remained relatively stable in staurosporine-treated

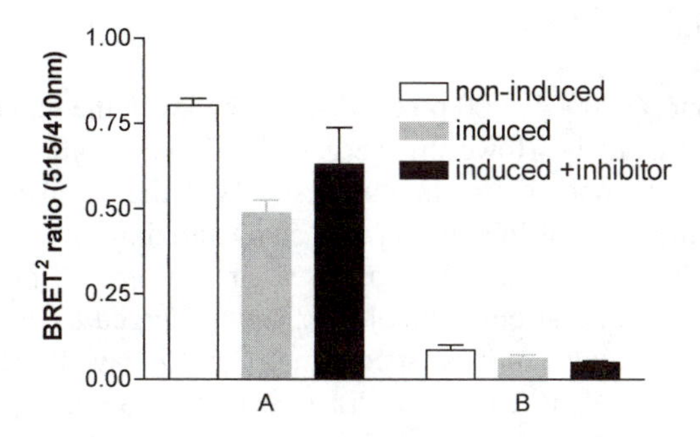

FIGURE 42.14 Staurosporine induction of apoptosis in HeLa cells. HeLa cells were transfected with a construct encoding either GFP^2::Rluc or GFP^2:DEVD:Rluc 100 mm dishes. At 24 h post-transfection, cells were harvested and distributed into a 96-well microplate. The following day, apoptosis was induced with 1 μM staurosporine for 5 h with or without the caspase-3 inhibitor I. The bioluminescence reaction was initiated with the addition of DeepBlueC. Light emission at 410 and 515 nm was detected using the Fusion Universal Microplate Analyzer in dual luminescence mode. Results represent the mean ± SD of quadruplicate wells taken 4 min after the addition of DeepBlueC.

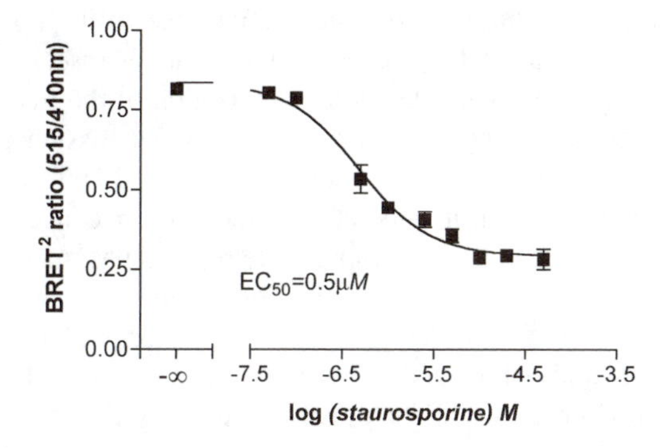

FIGURE 42.15 Staurosporine dependence of apoptosis induction. HeLa cells were transfected with pGFP^2:DEVD:Rluc vectors in 100-mm dishes. At 24 h post-transfection, cells were harvested and distributed into a 96-well microplate. The following day, apoptosis was induced with varying concentrations of staurosporine for 3 h. The bioluminescence reaction was started with the addition of DeepBlueC. Light emission at 410 nm and 515 nm was detected using the Fusion Universal Microplate Analyzer in dual luminescence mode. Results represent the mean ± SD of quadruplicate wells.

cells transfected with GFP^2:Rluc. These results suggest that the consensus protease sequence present in the GFP^2:DEVD:Rluc construct was cleaved by caspase.

Caspase-3 inhibitor I is a highly specific, potent, cell-permeable, and reversible inhibitor of caspase-3 as well as caspase-6, -7, -8, and –10. These data are supported using the caspase-3 inhibitor I,[14] which led to a decrease in BRET^2 signal for apoptotic cells transfected with GFP^2:DEVD:Rluc. Levels of Rluc luminescence for non-induced and apoptotic cells were equivalent suggesting that the Rluc was not recognized by caspase during the time period of the experiment (data not shown).

Figure 42.15 shows a dose–response relationship of staurosporine to the BRET^2 signal using cells transfected with the GFP^2:DEVD:Rluc construct. Clearly, staurosporine induced a dose-dependent decrease in the BRET^2 signal. From this curve, we calculated an EC_{50} of 0.5 μM of staurosporine. This value is similar to previously reported data[15] using a lysate-based peptide cleavage assay. The same group reported an EC_{50} ten times higher (5 μM) with FRET.

In conclusion, we have demonstrated that staurosporine-induced apoptosis in live cells can be easily detected using BRET^2. We showed that the apoptosis induction is time and dose dependent and can be blocked with caspase-specific inhibitor.

42.9 CONCLUSION

"There appear to be about 30,000–40,000 protein-coding genes in the human genome—only about twice as many as in worm or fly. However, the genes are more complex, with more alternative splicing generating a larger number of protein products."[16] With the initial draft of the human genome now available, the daunting task of identifying and characterizing the encoded proteins is now at hand with academic, government and industrial research facilities all playing a role in this burgeoning field of proteomics.[17,18] Various technologies have emerged to fulfill complementary roles in this complex task. High-throughput mass spectrometry and X-ray chrystallography have opened the door to rapidly identify and structurally characterize proteins. In combination with the time-honored tools of cell biology, these instrument-based techniques offer researchers the opportunity to map the proteome of organelles. However, these technologies do not provide simple methods for the precise identification of protein partners *in vivo*.

$BRET^2$ provides the complementary technical approach to organelle-based proteomics by unraveling the often-complex interactions between proteins localized to various organelles as well as those that traverse organelles as observed with secretory proteins. We have shown that $BRET^2$ allows the quantitation of protein–protein interactions and signaling events in live cells without the need for an excitation light source. This avoids the problems of FRET-based technology including cell damage, photobleaching, and high background signals due to the fluoresence of endogenous intracellular proteins. The Rluc and GFP^2 fusion proteins can themselves be used as markers to assure appropriate expression levels and subcellular localization of the fusion proteins being examined. Finally, the large spectral resolution between donor and emission peaks (see Figure 42.2) greatly improves the signal-to-noise ratio over traditionally used BRET and FRET technologies.

The advent of light-detection instruments such as the Fusion Universal Microplate Analyzer and highly sensitive camera-based high-throughput plate readers has made screening using $BRET^2$ in live cells (or organelles) possible. The combination of these technologies has created new opportunities for drug companies and academic research laboratories alike to find novel drug targets and interesting mechanisms and pathways inside the cell and at the cell surface *in vivo*. Limited only by the imagination of talented scientists, this technology opens new and exciting avenues for research and development.

REFERENCES

1. Xu, Y., Piston, D. W., and Johnson, C. H., A bioluminescence resonance energy transfer (BRET) system: application to interacting circadian clock proteins, *Proc. Natl. Acad. Sci. U.S.A.*, 96, 151–156, 1999.
2. Angers, S., Salahpour, A., Joly, E., Hilairet, S., Chelsky, D., Dennis, M., and Bouvier, M., Detection of β2-adrenergic receptor dimerization in living cells using bioluminescence resonance energy transfer (BRET), *Proc. Natl. Acad. Sci. U.S.A.*, 97, 3684–3689, 2000.
3. Watson, S. and Arkinstall, S., *The G-Protein Linked Receptor Facts Book*, Academic Press, San Diego, CA, 1994.
4. Neer, E. J., Heterotrimeric G proteins: organizers of transmembrane signals, *Cell*, 80, 249–257, 1995.
5. Lefkowitz, R. J., G protein-coupled receptors; III. New roles for receptor kinases and β-arrestins in receptor signaling and desensitization, *J. Biol. Chem.*, 273, 18,677–18,680, 1998.
6. Ferguson, S. G. S., Barak, L. S., Zhang, J., and Caron, M. G., G-protein-coupled receptor regulation: role of G-protein-coupled receptor kinases and arrestins, *Can. J. Physiol. Pharmacol.*, 74, 1095–1110, 1996.
7. Cao, W., Luttrell, L. M., Medvedev, A. V., Pierce, K. L., Daniel, K. W., Dixon, T. M., Lefkowitz, R. J., and Collins, S., Direct binding of activated c-src to the β_3-adrenergic receptor is required for MAP kinase activation, *J. Biol. Chem.*, 275, 38,131–38,134, 2000.
8. Chidiac, P., Hebert, T. E., Valiquette, M., Dennis, M., and Bouvier, M., Inverse agonist activity of β-adrenergic antagonists, *Mol. Pharmacol.*, 45, 490–499, 1994.

9. Nichelson, D. W. and Thornberry, N. A., Caspases: killer proteases, *Trends Biol. Sci.,* 22, 299–306, 1997.
10. Thornberry, N. A. and Lazebnik, Y., Caspases: enemies within, *Science,* 281, 1312–1316, 1998.
11. Villa, P., Kaufmann, S. H., and Earnshaw, W. C., Caspases and caspase inhibitors, *TIBS,* 22, 388–393, 1997.
12. Talanian, R. V., Quinlan, C., Trautz, S., Hackett, M. C., Mankovich, J. A., Banach, D., Ghayur, T., Brady, K. D., and Wong, W. W., Substrate specificities of caspase family proteases, *J. Biol. Chem.,* 272, 9677–9682, 1997.
13. Shinbrot, E., Spencer, C. M., and Kain, S. R., Morphological detection of plasma membrane changes during apoptosis using enhanced green fluorescent protein, *BioTechniques,* 26, 1065–1067, 1999.
14. Margolin, N., Raybuck, S. A., Wilson, K. P., Chen, W., Fox, T., Gu, Y., and Livingston, D. J., Substrate and inhibitor specificity of interleukin-1 beta-converting enzyme, *J. Biol. Chem.,* 272, 7223–7228, 1997.
15. Jones, J., Heim, R., Hare, E., Stack, J., and Pollok, B. A., Development and application of a GFP-FRET intracellular caspase assay for drug screening, *J. Biomol. Screening,* 5, 307–317, 2000.
16. Lander, E. S., Linton, L. M., Birren, B., Nusbaum, C., Zody, M. C., Baldwin, J., Devon, K., Dewar, K., Doyle, M., FitzHugh, W., Funke, R., Gage, D., Harris, K., Heaford, A., Howland, J., Kann, L., Lehoczky, J., LeVine, R., McEwan, P., McKernan, K., Meldrim, J., Mesirov, J. P., Miranda, C., Morris, W., Naylor, J., Raymond, C., Rosetti, M., Santos, R., Sheridan, A., Sougnez, C., Stange-Thomann, N., Stojanovic, N., Subramanian, A., Wyman, D., Rogers, J., Sulston, J., Ainscough, R., Beck, S., Bentley, D., Burton, J., Clee, C., Carter, N., Coulson, A., Deadman, R., Deloukas, P., Dunham, A., Dunham, I., Durbin, R., French, L., Grafham, D., Gregory, S., Hubbard, T., Humphray, S., Hunt, A., Jones, M., Lloyd, C., McMurray, A., Matthews, L., Mercer, S., Milne, S., Mullikin, J. C., Mungall, A., Plumb, R., Ross, M., Shownkeen, R., Sims, S., Waterston, R. H., Wilson, R. K., Hillier, L. W., McPherson, J. D., Marra, M. A., Mardis, E. R., Fulton, L. A., Chinwalla, A. T., Pepin, K. H., Gish, W. R., Chissoe, S. L., Wendl, M. C., Delehaunty, K. D., Miner, T. L., Delehaunty, A., Kramer, J. B., Cook, L. L., Fulton, R. S., Johnson, D. L., Minx, P. J., Clifton, S. W., Hawkins, T., Branscomb, E., Predki, P., Richardson, P., Wenning, S., Slezak, T., Doggett, N., Cheng, J. F., Olsen, A., Lucas, S., Elkin, C., Uberbacher, E., Frazier, M. et al., Initial sequencing and analysis of the human genome, *Nature,* 409, 860–921, 2001.
17. Blackstock, W. P. and Weir, M. P., Proteomics: quantitative and physical mapping of cellular proteins, *Trends Biotechnol.,* 17, 121–127, 1999.
18. Yates III, J. R., Mass spectrometry. From genomics to proteomics, *Trends Genet.,* 16, 5–8, 2000.

43 High-Throughput Screening of *Arabidopsis* Mutants with Deregulated Stress-Responsive Luciferase Gene Expression Using a CCD Camera

Byeong-ha Lee, Becky Stevenson, and Jian-Kang Zhu

CONTENTS

43.1 INTRODUCTION

Over the course of the past several years we have developed a high-throughput screening protocol to isolate stress signal transduction mutants in the plant *Arabidopsis thaliana* using luciferase imaging with a charge-coupled device (CCD) camera. This method involves using plants containing the firefly luciferase gene fused with different promoters from genes involved in cold temperature stress, osmotic stress, abscisic acid response, or in general stress perception. The system works by inducing the promoter–luciferase fusions by various environmental perturbations and recording the amount of light from luciferase-catalyzed reactions with a CCD camera. The level of luminescence is an indication of plant responsiveness to the treatment.

Using luciferase in this system has several advantages over other reporters. One is that the screening process is noninvasive to the plants. The fact that the luciferase enzyme is quickly degraded

0-8493-0719-8/02/$0.00+$1.50

within plants allows several experimental treatments of the same plant followed by CCD imaging. Additionally, the equipment required to detect luciferase expression has recently become more affordable, and currently there are several integrated systems available that are designed specifically for luciferase imaging. This makes it quick and easy for even a novice to set up the necessary equipment to start luciferase imaging.

43.2 EQUIPMENT

The CCD camera system as shown in Figure 43.1 is a product of Roper Scientific (Princeton, NJ). It consists of the camera itself, with a lens, a camera controller, a computer interface card, a Cryotiger® cryogenic cooler, a dark box, and a standard IBM-compatible computer running WinView/32 software provided with the camera.

The camera resolution is a function of the number of pixels. This model has 1300 × 1340 pixels. This high resolution makes it possible to image large numbers of small plants and also to

FIGURE 43.1 Camera system with dark box.

identify more clearly which part of the sample the luciferase signal is coming from. The lens on the camera is a standard f-mount Nikon lens. Since the f-mount is one of the most common types for 35 mm cameras, there are many types of lenses that are available, and one can be found that is tailored to each sample and image size. To detect luciferase expression, the CCD chip inside the camera must be cooled. This model uses the Cryotiger cryogenic cooler, a compressed gas system. It is able to cool the CCD chip inside the camera to –100°C. It is important to cool the CCD chip because the warmer the chip is, the higher the noise and the more likely the luciferase signal will be lost in the noise. There are thermoelectrically cooled cameras, which be cooled to approximately –40°C, that are still useful for luciferase detection. These thermoelectrically cooled models are slightly more affordable, although we have found the sensitivity of the cryogenically cooled model to be a benefit when studying genes that are not highly expressed.

The camera is fixed to the top of a dark box into which the sample to be imaged is placed. Inside the box there is a movable stage and a light to be used to take a brightfield image if one is required for sample orientation. The box is designed to be light tight when used in a room with normal light conditions. This eliminates the need for employing the camera in a darkroom.

The camera controller is an ST-133 model. It controls the temperature as well as the shutter of the camera. It also contains the analog-to-digital converter that is required for the computer to be able to interpret the camera data output. The computer interface card is a PCI serial card that is installed in the computer and connected to the camera controller.

The computer requirements for this system are operating system Windows 95 or newer, with 32 megabytes of RAM, a VGA monitor with at least 256 colors and at least 512 kilobytes of memory, and a Microsoft two button compatible mouse. A large hard drive and/or a CD writer are recommended because of the large numbers of image files that can be produced, and the fact that each image is approximately 530 kb. WinView is the software that comes with the camera. This software allows the user to control exposure time, the speed of the analog-to-digital converter, and pixel binning. Pixel binning is basically adding the signal from a group of pixels and treating them as one pixel. This has the effect of increasing the signal while reducing the time of exposure. There is a loss of resolution when pixel binning is used, but with this chip size of 1300 × 1340 the loss of resolution has no effect on this screening method until the range of about 6 × 6 binning.

43.3 SCREENING PROCESS

43.3.1 Generating the Bioluminescent Plants

We employed the luciferase reporter system to study stress signal transduction in plants. Because of a scarcity of morphological phenotypes for stress mutants, altered expression of stress-responsive genes can be used in screening for stress mutants. Luciferase was chosen in this case because it has several advantages over other frequently used reporters in plant biology such as β-glucuronidase (GUS) and green fluorescent protein (GFP).

A stress-responsive promoter fused with luciferase can be introduced into a plant to produce a bioluminescent plant. Several stress-responsive promoters are well characterized such as *RD29A* (also known as *COR78* or *LTI78*), *DREB1A* (*CBF3*), *DREB1B* (*CBF1*), *DREB1C* (*CBF2*), and *DREB2A*.[1,2] Here the application of *RD29A::LUC* lines will be described in detail.

Expression of the *RD29A* gene is induced by cold, abscicic acid (ABA, a plant hormone), osmotic stress, and drought.[1] The promoter of *RD29A* spanning from –650 to –1 was obtained by polymerase chain reaction (PCR) with *Arabidopsis* genomic DNA and two primers: 5′-TCGGGATCCGGTGAATTAAGAGGAGAGAGGAGG-3′ and 5′-GACAAGCTTTGAGTAAAA-CAGAGGAGGGTCTCAC-3′. This promoter fragment was inserted into a plant transformation vector containing the firefly luciferase coding region to produce the *RD29A::LUC* vector.[3] The vector was introduced into *Arabidopsis* (ecotype C24) via the root transformation method with

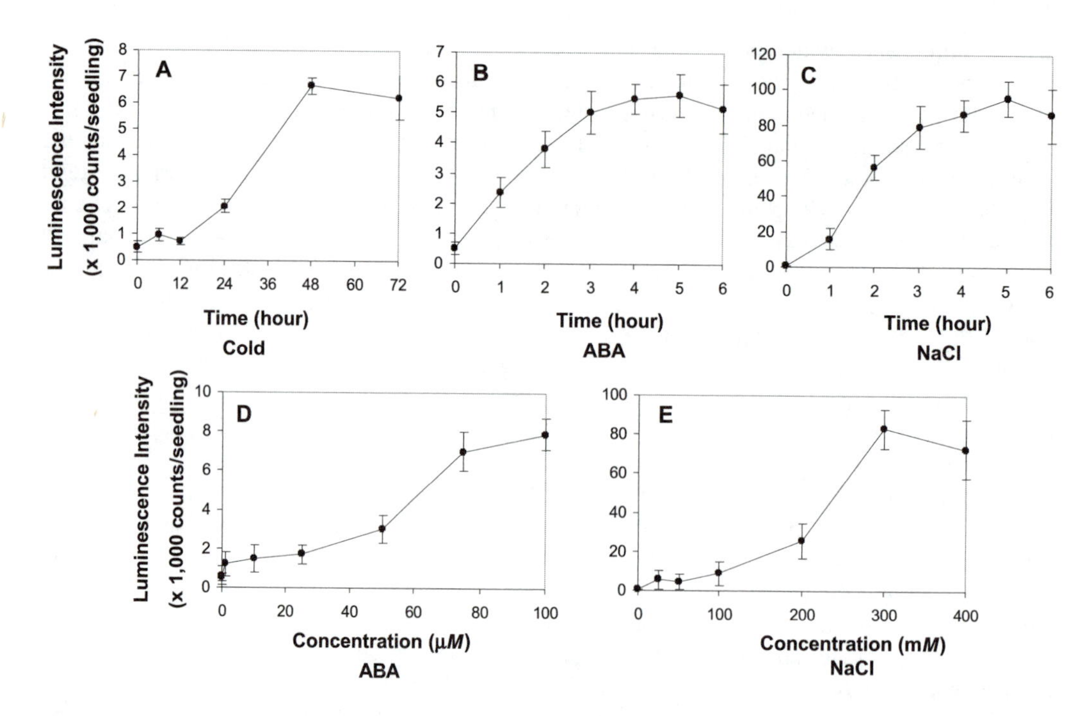

FIGURE 43.2 *RD29A::LUC* expression in response to cold, ABA, and NaCl treatment. (A) Time course of *RD29A::LUC* expression at 0°C; (B) time course of *RD29A::LUC* expression after 100 μM ABA treatment; (C) time course of *RD29A::LUC* expression after 300 m*M* NaCl treatment; (D) *RD29A::LUC* expression 3 h after ABA treatment; (E) *RD29A::LUC* expression 5 h after NaCl treatment.

Agrobacterium tumefaciens.[4] The plant line homozygous for the *RD29A::LUC* transgene was selected in the T2 generation (the second generation after transformation).

Bioluminescence of selected *RD29A::LUC* lines was tested and characterized under various stress conditions and at different time points, as shown in Figure 43.2.

43.3.2 Mutagenesis

We took a genetic approach to dissect stress-signaling pathways in plant by generating mutants showing altered stress gene expression (*RD29A* in this case) and searching for the mutated genes. There are several mutagens to induce mutations in plants; EMS (ethyl methanesulfonate), fast neutron, foreign DNA such as T-DNA and transposon, and so on. Mutagensis methods can be found elsewhere.[5–7] We used EMS and T-DNA as mutagens to mutagenize *RD29A::LUC Arabidopsis*. We generated about 300,000 EMS-mutagenized mutants and 50,000 T-DNA insertion lines. Recently, our T-DNA-mutagenized plants are publicly released and available at ABRC (*Arabidopsis* Biological Resource Center, Ohio State University). T-DNA mutants have advantages over EMS mutants in that it is easier to clone the genes responsible for the mutation. Because the DNA sequence is known for the T-DNA insert, the T-DNA can be used as a "tag." TAIL-PCR,[8] inverse PCR,[9] PCR-walking,[10] plasmid rescue,[11] and even genomic library screening with a probe from T-DNA have been successfully applied to clone the disrupted plant genes. For T-DNA mutants, the pSKI015 activation tagging vector was used [11] and ten T2 individuals were pooled to make one pool.

43.3.3 Plant Handling

Seeds were dispensed in eppendorf tubes (about 50,000 seeds/g) and were surface-sterilized with commercial bleach supplemented with 0.01% Triton X-100 for 5 to 10 min, and then washed with sterile water four to five times. One droplet of 0.3% low-melting agarose can be added into the sterilized seeds for easier handling during plating. Seeds were plated onto 0.6% agar medium with 3% sucrose (pH 5.7) in 150 × 15 mm petri dishes with a transfer pipette. As many as 500 seeds can be plated per 150 × 15 mm round plate. After being kept at 4°C for 2 to 4 days to break the seed dormancy, the plates were placed at 22°C under continuous light for germination and growth. Approximately 1-week-old seedlings were used for luminescence imaging. When needed, seedlings on agar medium were transferred to soil, and then allowed to grow in a growth chamber with cycles of 16-h light at 22°C and 8-h dark at 18°C.

43.3.4 Stress Treatments

Based on characterization of bioluminescence from *RD29A::LUC* plants (wild-type), 1-day cold treatment for cold stress (0°C), 3-h incubation under light after 100 *μM* ABA spray for ABA treatment, and 5-h incubation under light after 300 m*M* NaCl application for salt stress were chosen. Because of the short half-life of the luciferase enzyme (about 3 h) and nontoxicity of the marker, the same seedlings can be used repeatedly for several different stress applications.

43.3.4.1 No Stress

Constitutive expression of *RD29A::LUC* should be detected without stress treatment. One-week-old seedlings were first subjected to luminescence imaging without stress. The perturbations in the environment during manipulation may cause some luminescence induction. Therefore, care should be taken when screening for constitutive luciferase-expressing mutants.

43.3.4.2 Cold Stress

After imaging of seedlings without stress, the plates were placed at 0°C for 1 day. In our conditions, sometimes the 0°C 1-day incubation was not sufficient to induce strong luminescence. In such cases, more prolonged incubation such as 2 days at 0°C resulted in better luminescence images. Since the enzyme activity is reduced in cold conditions, the plates were occasionally warmed at room temperature for as long as 30 min. This usually gives higher luminescence signals.

After luminescence imaging, the plates were placed at 22°C under continuous light for at least 24 h to allow the luminescence signal to disappear.

43.3.4.3 ABA Treatment

After incubation of the plates under continuous light, 100 *μM* ABA was sprayed onto the seedlings to sufficiently wet each seedling. Before ABA treatment, the luminescence images of plates can be taken to ensure that no luminescence signal remains from the cold treatment. 100 *μM* ABA was prepared by diluting with sterile water from the stock solution of 10 m*M* (±)-*cis,trans*-abscisic acid (Sigma Chemical Co., St. Louis, MO) in ethanol. The stock solution should be stored at –20°C and the working solution can be stored at 4°C.

After ABA-sprayed plates were incubated at 22°C under continuous light for 3 h, the luminescence images were taken. After imaging, the plates were, again, placed at 22°C under continuous light.

43.3.4.4 NaCl Treatment

Seedlings on plates were carefully transferred onto filter paper saturated with nutrient solution supplemented with 300 m*M* NaCl. After 5 h, the images were taken. It should be noted that recovering the putative mutant after NaCl treatment might be difficult due to the severity of NaCl stress. In addition, the putative mutants previously marked after cold or ABA treatment may be lost because seedlings may be floating and moving in the 300 m*M* NaCl solution. Therefore, the NaCl treatment was usually applied only during the secondary screening process with progeny from the putative mutants.

43.3.5 Luminescence Imaging

After each treatment, the plates were sprayed evenly with 1 m*M* luciferin. The 1 m*M* luciferin solution was freshly prepared from 100 m*M* luciferin (Promega Co., Madison, WI) stock solution. The stock solution prepared in sterile water was stored at −80°C in 100 μl aliquots and diluted with 0.01% Triton X-100 to make 10 ml of 1 m*M* luciferin. The working solution of luciferin was kept at 4°C in the dark during use. Luciferin-sprayed plates were kept in dark for 5 min for luciferase enzyme reaction and decay of autofluorescence from chlorophyll. During the 5-min incubation, a background image was taken with an empty plate. A background image is generally needed because CCD cameras have some internal noise. Therefore, subtraction of background signal from the raw luminescence image enhances the image qualities. After a 5-min incubation of the plates, the plates were placed under the CCD camera in the dark. To prevent chlorophyll autofluorescence from interfering with the luminescence image, the plates should not be exposed to light after luciferin is applied. Luminescence images were acquired with 5-min exposure. Some representative mutants are shown in Figure 43.3.

In our system, 5-min exposure was sufficient to detect luminescence emitted from seedlings. However, the luminescence intensity is dependent on the nature of promoter and/or the position of transgene in the plant genome. Indeed, another *RD29A::LUC* line in the Columbia ecotype shows lower luminescence intensity than that in C24 ecotype, thus requiring longer exposures. Lower luminescence intensities were also observed in the *DREB1A::LUC, DREB1B::LUC*, and *DREB1C::LUC* lines. After imaging, the plates were aligned with the images and the putative stress mutants showing

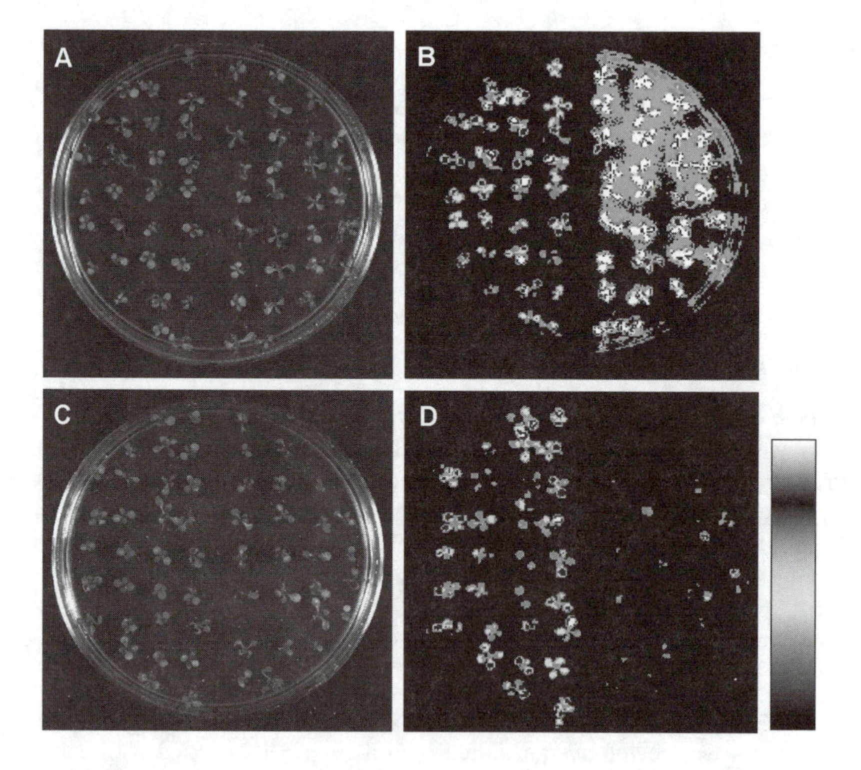

FIGURE 43.3 (Color figure follows p. 266.) Stress mutants showing altered RD29A::LUC expression. (A) The plate corresponding to luminescence image in B; (B) Mutants (right half) showing higher RD29A::LUC expression than wild-type (left half) after ABA treatment; (C) the plate corresponding to luminescence image in D; (D) Mutants (right half) showing lower RD29A::LUC expression than wild-type (left half) after ABA treatment. Right spectrum bar shows color changes depending on luminescence intensity; as intensity goes higher, color changes from black to white.

altered luminescence intensity—either higher or lower than wild-type (*RD29A::LUC* line)—were identified. For convenience in aligning with images, plates can be marked on the side with tape before luciferin application. After all stress imaging, all putative mutants were transferred to soil to produce seeds.

43.3.6 Secondary Screening

The progenies from the putative mutants were retested to confirm the mutant phenotypes. The plates were divided into eight to ten sections and each mutant progeny (about 30 seeds) was plated in each section. One section of wild-type plants should be included on each plate to compare with the mutants. If a mutant is real- and true-breeding, its progeny section will show all higher or lower intensity relative to the wild-type control. However, it should be noted that our T-DNA lines were generated with an activation tagging vector. Therefore, one may see a 3:1 segregation, if the mutation is dominant.

43.4 CONCLUSION

How cells perceive environmental signals and how the signals are transduced to activate adaptive responses have long been of interest to scientists. Molecular genetic approaches coupled with a chimeric transgene of the signal-inducible promoter fused with luciferase are valuable tools to study signal transduction. Here we have provided an example of luciferase imaging application to generate large numbers of stress-signaling mutants. This method can be applied to studies of any signaling pathway. However, because of the technical limitations of the CCD camera sensitivity, luminescence emitted from the plants should be strong enough to be detected. It is also helpful to have a basic knowledge of which treatment conditions are the most effective and how long after each treatment the luminescence signal is the highest before a large-scale screening is carried out.

REFERENCES

1. Yamaguchi-Shinozaki, K. and Shinozaki, K., Characterization of the expression of a desiccation-responsive rd29 gene of *Arabidopsis thaliana* and analysis of its promoter in transgenic plants, *Mol. Gen. Genet.,* 236, 331, 1993.
2. Liu, Q., Kasuga, M., Sakuma, Y., Abe, H., Miura, S., Yamaguchi-Shinozaki, K., and Shinozaki, K., Two transcription factors, DREB1 and DREB2, with an EREBP/AP2 DNA binding domain separate two cellular signal transduction pathways in drought- and low-temperature-responsive gene expression, respectively, in Arabidopsis, *Plant Cell,* 10, 1391, 1998.
3. Millar, A. J., Short, S. R., Chua, N. H., and Kay, S. A., A novel circadian phenotype based on firefly luciferase expression in transgenic plants, *Plant Cell,* 4, 1075, 1992.
4. Valvekens, D., Vanmontagu, M., and Vanlijsebettens, M., *Agrobacterium tumefaciens*-mediated transformation of *Arabidopsis thaliana* root explants by using kanamycin selection, *Proc. Natl. Acad. Sci. U.S.A.,* 85, 5536, 1988.
5. Bechtold, N. and Pelletier, G., *In planta Agrobacterium*-mediated transformation of adult *Arabidopsis thaliana* plants by vacuum infiltrantion, in *Arabidopsis Protocols,* Martinez-Zapater, J. and Salinas, J., Eds., Humana Press, Totowa, NJ, 1998, chap. 28.
6. Lightner, J. and Caspar, T., Seed mutagenesis of *Arabidopsis*, in *Arabidopsis Protocols,* Martinez-Zapater, J. and Salinas, J., Eds., Humana Press, Totowa, NJ, 1998, chap. 14.
7. Long, D. and Coupland, G., Transposon tagging with *Ac/Ds* in Arabidopsis, in *Arabidopsis Protocols,* Martinez-Zapater, J. and Salinas, J., Eds., Humana Press, Totowa, NJ, 1998, chap. 32.
8. Liu, Y. G., Mitsukawa, N., Oosumi, T., and Whittier, R. F., Efficient isolation and mapping of *Arabidopsis thaliana* T-DNA insert junctions by thermal asymmetric interlaced PCR, *Plant J.,* 8, 457, 1995.

9. Ochman, H., Gerber, A. S., and Hartl, D. L., Genetic applications of an inverse polymerase chain reaction, *Genetics,* 120, 621, 1988.
10. Siebert, P. D., Chenchik, A., Kellogg, D. E., Lukyanov, K. A., and Lukyanov, S. A., An improved PCR method for walking in uncloned genomic DNA, *Nucleic Acids Res.,* 23, 1087, 1995.
11. Weigel, D., Ahn, J. H., Blazquez, M. A., Borevitz, J. O., Christensen, S. K., Fankhauser, C., Ferrandiz, C., Kardailsky, I., Malancharuvil, E. J., Neff, M. M., Nguyen, J. T., Sato, S., Wang, Z. Y., Xia, Y., Dixon, R. A., Harrison, M. J., Lamb, C. J., Yanofsky, M. F., and Chory, J., Activation tagging in *Arabidopsis, Plant Physiol.,* 122, 1003, 2000.

44 NorthStar™ HTS Workstation: A CCD-Based Integrated Platform for High-Throughput Screening

Deborah M. Boldt-Houle, Yu-Xin Yan, Corinne E. M. Olesen, Anthony C. Chiulli, Brian J. D'Eon, Betty Liu, John C. Voyta, and Irena Bronstein

CONTENTS

44.1 INTRODUCTION

In the pursuit for new drug candidates, pharmaceutical companies test hundreds of targets against millions of potentially valuable compounds. To identify meaningful candidates as quickly as possible, pharmaceutical companies look for innovations that will accelerate the drug-screening process, while reducing nonspecific hits. To this end, high-throughput screening (HTS) instrumentation and assays that are robust, cost-effective, and amenable to automation are essential. The NorthStar™ HTS Workstation (Figure 44.1) is a fully integrated luminescent imaging system specifically developed to expedite lead discovery. Features include a cooled charge-coupled device (CCD) camera optimized to image highly sensitive luminescent assays; integrated liquid and plate handlers; standard and high-density microplate imaging formats; multiwavelength detection capability; easy-to-use, intuitive software; and the capacity to perform up to 500,000 assays per day. The use of proprietary chemiluminescence reagents combined with the NorthStar HTS Workstation

0-8493-0719-8/02/$0.00+$1.50

FIGURE 44.1 NorthStar™ HTS Workstation. (Courtesy of Applied Biosystems.)

further enhances the performance of HTS assays by providing an integrated system for highly sensitive detection with wide dynamic range and glow light emission kinetics for cell-based assays, and enzyme, protein, and nucleic acid quantitation. The total system approach of the NorthStar HTS Workstation provides pharmaceutical screeners a versatile solution for lead discovery.

44.2 INSTRUMENTATION

44.2.1 NorthStar HTS Workstation Optical System

The CCD camera is a liquid-cooled (–35°C) integrated thermoelectric (Peltier) device (Figure 44.2). The CCD chip dimensions are 20.5 × 16.4 mm, with a pixel density of 1280 × 1024. The large-aperture low-distortion Nikkor™ lens, 52 mm in diameter, has a 35-mm focal length, with F/1.4. A collimator with an integrated Fresnel field lens segregates light emanating from each well and minimizes cross talk. The Fresnel lens functions as a telecentric lens to image directly down into the sample well, minimizing distortions (see light path in Figure 44.2). The collimator/Fresnel assembly is microplate-format specific, and is easily changed or removed to permit imaging of flat membranes or chips. The read chamber under the collimator assembly features programmable temperature control to optimize experimental conditions.

44.2.2 Multiwavelength Detection

The motor-driven filter wheel contains multiple interchangeable band-pass filter positions, which enables software selection of various wavelength ranges and true multiplexing capabilities. Multiplex detection with the filter wheel enables concurrent measurement of multiple-wavelength signals, thereby increasing the data obtained from one sample.

44.2.3 Liquid and Plate Handling System

The NorthStar HTS system has 96- and 384-well reagent injection capability of 10 to 200 μl of reagent per well through 16 syringe pumps. A computer-controlled injector array enables rapid reagent additions for "flash" luminescence imaging. A Zymark extended-capacity Twister™ plate handler with a specialized intermediate plate platform provides automated rapid plate processing. The NorthStar HTS Workstation is compatible with other linear and rotary robotic systems,

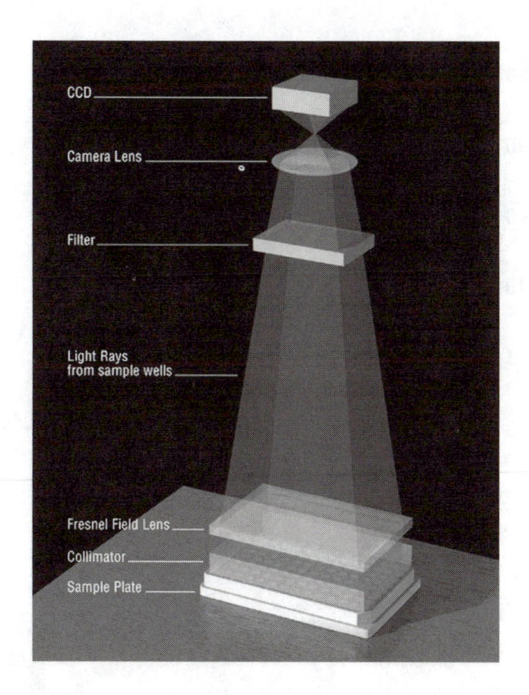

FIGURE 44.2 NorthStar HTS Workstation detection optics. (Courtesy of Applied Biosystems.)

and easy-to-use software allows the NorthStar workstation to function as a stand-alone detection platform or as part of an integrated high-throughput process system.

44.3 APPLICATIONS

A panel of sensitive, prepackaged, and robust chemiluminescent screening assays, as well as proprietary HTS technologies, is available from Applied Biosystems. Current screening assays accessible with the NorthStar workstation include reporter gene assays (e.g., Luc-Screen®, Gal-Screen®),[1] second messengers (cAMP-Screen™ assay),[2] detection of protein–protein interactions with a β-galactosidase enzyme fragment complementation technology,[3] specific mRNA expression (Xpress-Screen®), as well as other enzymes or biochemical cellular pathways. Standard experimental conditions for all the cell-based assays described below include the use of white/clear-bottom tissue culture-treated microplates; final reaction volumes of 200 μl (96-well), 50 μl (384-well), and 5 μl (1536-well); and imaging times of 1 min (96-well), 2 min (384-well), and 4 min (1536-well) on the NorthStar HTS Workstation.

44.3.1 Reporter Gene Assays

44.3.1.1 Luc-Screen Assay

Reporter gene assays are used in HTS to monitor the effect of compounds on gene expression levels. Examples of commonly used reporters include luciferase and β-galactosidase.[4–7] The Luc-Screen reporter gene assay system provides a homogeneous, extended glow light emission reaction for the quantitation of firefly luciferase directly in cultured cells. Following reagent addition, light emission reaches maximum within 10 to 20 min and has a half-life of 4 to 5 h.

To demonstrate the performance of the Luc-Screen assay in different microplate formats, serial dilutions of pGL3-transfected NIH/3T3 cells (ATCC, Rockville, MD) in DMEM/10% calf serum were seeded 24 h post-transfection in 96-, 384-, and 1536-well microplates and assayed 4 h after plating.

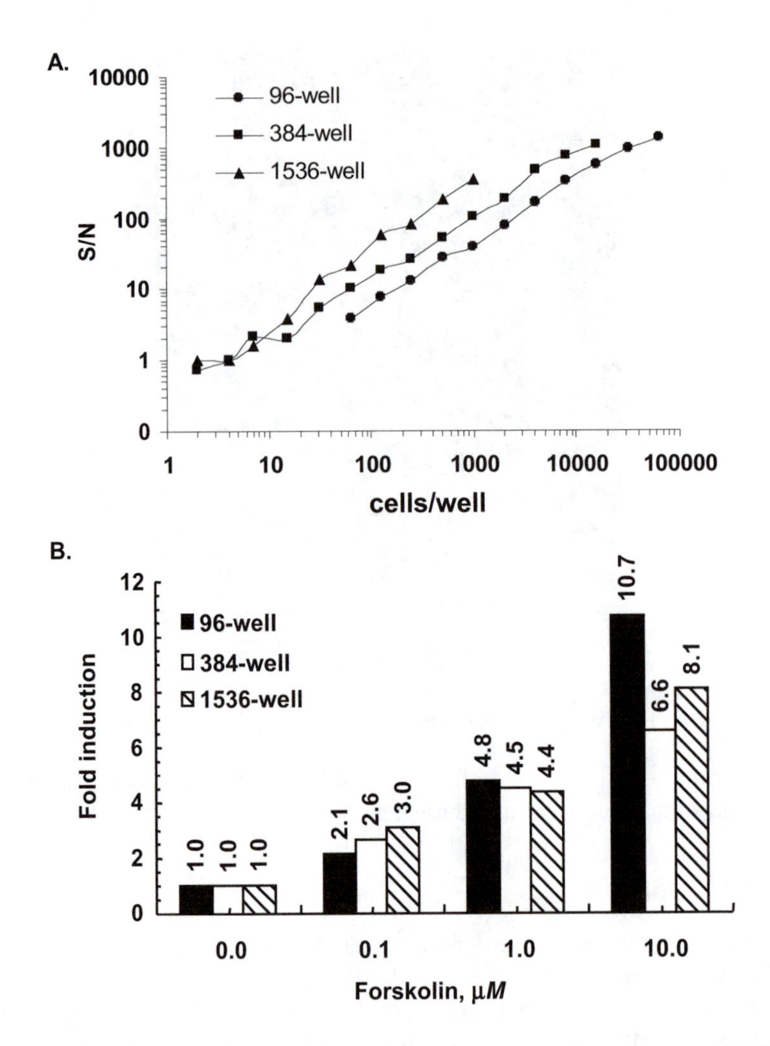

FIGURE 44.3 Luc-Screen Assay. (A) Detection of pGL3-transfected NIH/3T3 cells. NIH/3T3 cells were transfected with pGL3 and serially diluted into 96-, 384-, and 1536-well microplates. Luc-Screen reagents were added directly to cells in culture medium 24 h after transfection, and the signal was imaged on the NorthStar HTS Workstation 60 min later. S/N is the ratio of signal (S) from cells to reagent background (noise, N). (From Boldt-Houle, D., *Am. Lab.*, 32(3), 60, 2000. With permission.) (B) Forskolin induction of pCRE-Luc transfected NIH/3T3 cells. NIH/3T3 cells were transfected with pCRE-Luc and seeded into 96-, 384-, and 1536-well plates. Cells were incubated with varying concentrations of forskolin for 17 h. Luc-Screen reagents were added directly to cells in medium, and the signal was imaged on the NorthStar HTS Workstation. Forskolin-induced cAMP production is indicated for all concentrations and plate formats.

The plasmid pGL3 (Promega, Madison, WI) constitutively expresses firefly luciferase from an SV40 promoter/enhancer. As demonstrated in Figure 44.3A, each plate format shows linear detection over three orders of magnitude of cell concentration. Luciferase activity is detectable in as few as 20 cells in the 384- or 1536-well microplate format, and 100 cells in the 96-well format.

Detection of luciferase activity has also been demonstrated with forskolin-treated pCRE-Luc transfected cells. The plasmid pCRE-Luc (Stratagene, La Jolla, CA) is a cAMP-inducible firefly luciferase reporter vector. Exposure of transfected cells to forskolin, an activator of adenylate cyclase, increases the level of intracellular cAMP[8] and downstream activation of luciferase expression from the CRE promoter.[9] NIH/3T3 cells were transiently transfected with pCRE-Luc and seeded in DMEM/10% calf serum in 96-, 384-, or 1536-well plates. Cells were treated with varying amounts of forskolin (Calbiochem, San Diego, CA) for 17 h. As shown in Figure 44.3B, forskolin-induced cAMP levels via luciferase reporter readout are comparable among all plate formats.

44.3.1.2 Gal-Screen Assay

The Gal-Screen reporter gene assay system provides a homogeneous, glow light emission reaction for the quantitation of bacterial β-galactosidase reporter enzyme directly in mammalian or yeast cells. The system incorporates Galacton-*Star*® substrate, a luminescence enhancer, and cell lysis components in a single reagent, which is added directly to cells in culture medium. Light emission typically reaches maximum intensity within 60 to 90 min and exhibits glow kinetics for approximately 1 h.

To demonstrate the performance of the Gal-Screen assay on mammalian cells, serial dilutions of Ψ2BAGα cells (ATCC, Rockville, MD), which constitutively express bacterial β-galactosidase, were seeded in 96-, 384-, and 1536-well microplates in phenol red-free media. Gal-Screen Reaction Buffer A was added to the cells and the signal imaged 60 min later on the NorthStar HTS system. Figure 44.4A demonstrates highly sensitive detection of β-galactosidase activity: fewer than 100 cells are detected in the 1536- and 384-well formats, and 100 cells in the 96-well format, with a dynamic range of at least three orders of magnitude of cell concentration. The presence of phenol red in culture media reduces the overall signal intensity, but has little effect on detection sensitivity (data not shown).

The Gal-Screen assay is also optimized for yeast cells. Yeast cells that express β-galactosidase were serially diluted in YPD media in 96-, 384-, and 1536-well white microplates. An equal volume of Gal-Screen Reaction Buffer B was added to the yeast cell suspension, and the signal was imaged on the NorthStar HTS Workstation 60 min later. Figure 44.4B shows the detection curve of β-galactosidase activity in yeast cells in the 96-, 384-, and 1536-well plate formats. All formats show similar detection sensitivity curves over four orders of magnitude of cell concentration.

44.3.2 Second Messenger Assay

Another important class of HTS assays is the measurement of intracellular signaling molecules or second messengers.[10,11] Examples of second messenger targets include cyclic AMP (cAMP), Ca^{2+} ions, and inositol triphosphate.[2,11–14] The cAMP-Screen assay is a competitive immunoassay for ultrasensitive determination of cAMP levels.[2] The system incorporates a cell lysis solution, anti-cAMP antibody, CSPD® substrate for alkaline phosphatase (AP), cAMP–AP conjugate, and Sapphire-II™ enhancer. Glow light emission kinetics are obtained, which reach maximum intensity 30 min after substrate addition. The cAMP-Screen assay has been performed on the NorthStar HTS Workstation in both 96-well and 384-well microplate formats.

To demonstrate the performance of the cAMP-Screen assay on the NorthStar HTS Workstation, the cAMP standard control was diluted in Assay/Lysis Buffer and added to a 384-well assay plate. The cAMP–AP conjugate and anti-cAMP antibody were added and incubated for 1 h at ambient temperature. The assay plate was washed six times with Wash Buffer, CSPD/Sapphire-II RTU reagent was added and incubated for 30 min, and the plate was imaged on the NorthStar HTS Workstation and the TR717™ luminometer, a photomultiplier tube–based instrument.

Figure 44.5 demonstrates the performance of cAMP-Screen assay using either the NorthStar workstation or the TR717 luminometer. The results achieved are very similar between instruments, as can be seen by comparing the Z-factors at several different intervals on the curve. The Z-factor is a statistical parameter that takes into account the dynamic range of an assay as well as the variability of the sample and control data, and permits objective comparisons of assays or instruments.[15] An ideal assay has a Z value of 1, and a Z value within the range of 0.5 to 1 is considered excellent. The Z-factor values obtained on the NorthStar workstation indicate the superb data quality at cAMP concentrations near the sensitivity of the assay, and validate a second messenger quantitation assay on the NorthStar HTS Workstation.

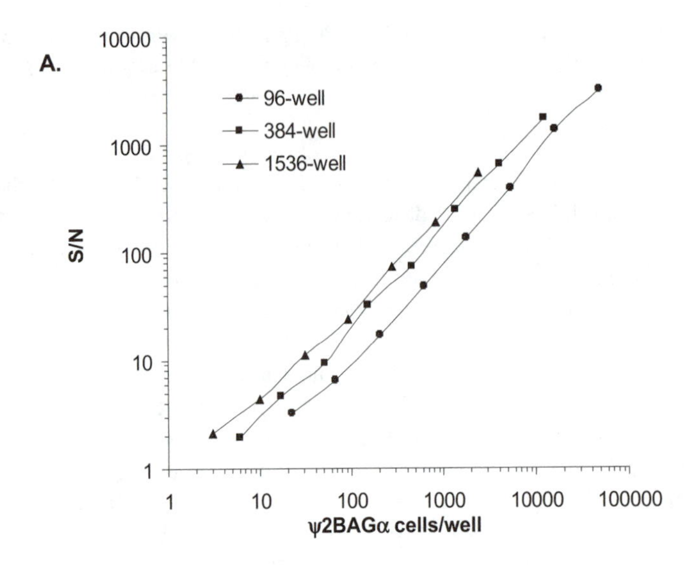

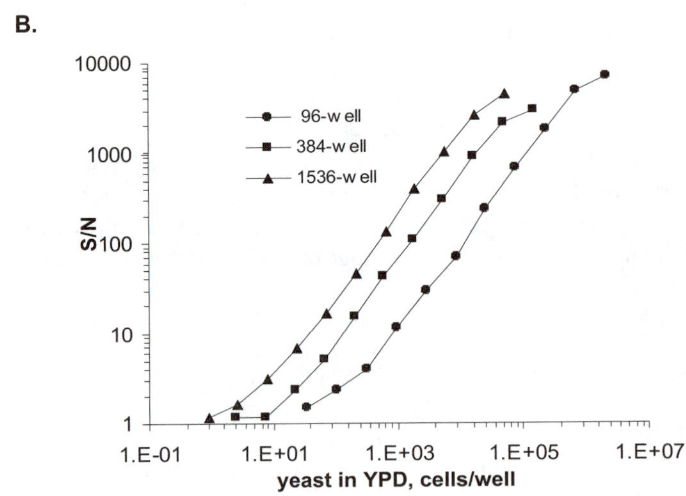

FIGURE 44.4 Gal-Screen Assay. (A) Detection of Ψ2BAGα cells. Ψ2BAGα cells were serially diluted in 96-, 384-, and 1536-well microplates in calf serum containing DMEM media without phenol red indicator. Gal-Screen reagent (Buffer A) was added directly to the cells in culture, and the plate was imaged 60 min after reagent addition. (From Boldt-Houle, D., *Am. Lab.*, 32(3), 60, 2000. With permission.) (B) Detection of β-galactosidase expressing yeast cells. Wild-type yeast cells transformed to express β-galactosidase were serially diluted in YPD media in 96-, 384-, and 1536-well microplates. Gal-Screen reagent (Buffer B) was added immediately to yeast cells in media, and the plate was imaged 60 min after plating. (From Boldt-Houle, D., *Am. Lab.*, 32(3), 60, 2000. With permission.)

44.4 EMERGING APPLICATIONS

The desire of pharmaceutical screening groups to develop multiplex detection assays, together with the availability of CCD imaging technology and instrumentation, has inspired the development of new 1,2-dioxetanes with green- and red-shifted emission spectra. (CCD cameras are more sensitive to green and red wavelengths.) Future applications on the NorthStar workstation will take advantage of its multiwavelength detection capabilities, enabling multiplexing in numerous applications, including immunoassays, nucleic acid detection, reporter enzyme quantitation, and microarrays.

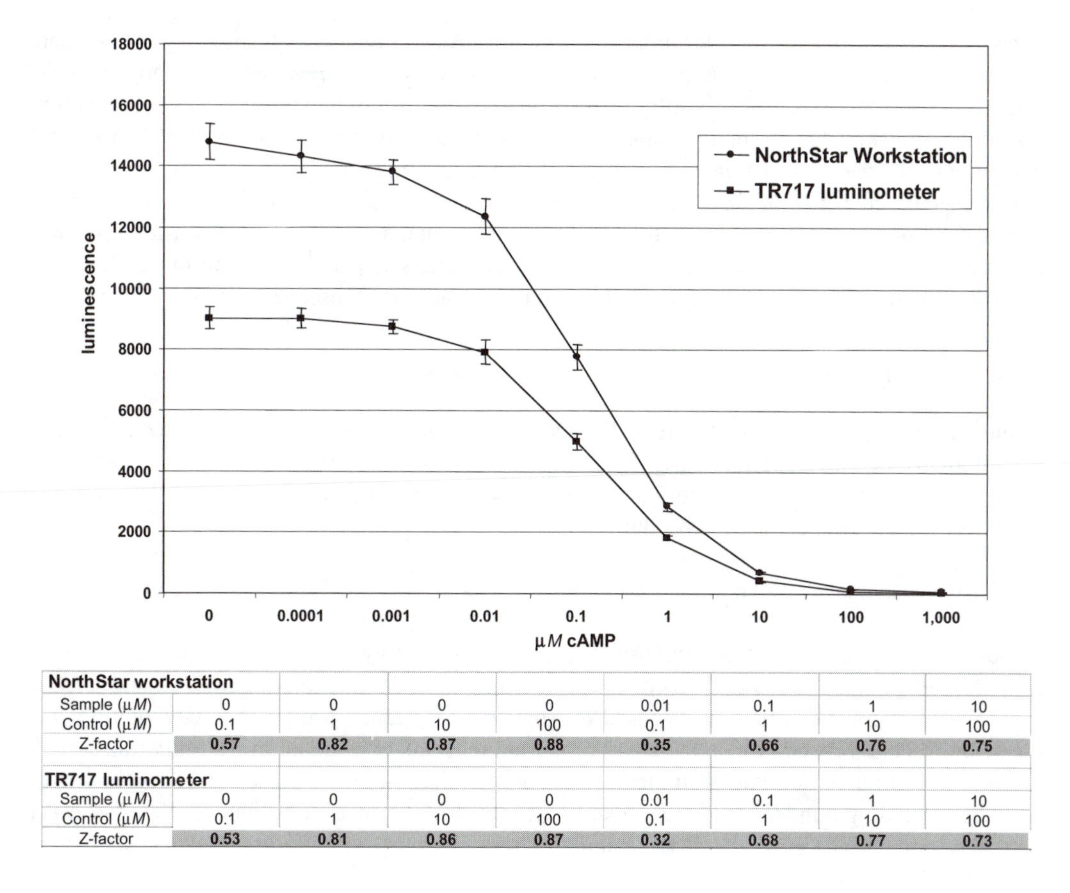

NorthStar workstation								
Sample (μ*M*)	0	0	0	0	0.01	0.1	1	10
Control (μ*M*)	0.1	1	10	100	0.1	1	10	100
Z-factor	**0.57**	**0.82**	**0.87**	**0.88**	**0.35**	**0.66**	**0.76**	**0.75**
TR717 luminometer								
Sample (μ*M*)	0	0	0	0	0.01	0.1	1	10
Control (μ*M*)	0.1	1	10	100	0.1	1	10	100
Z-factor	**0.53**	**0.81**	**0.86**	**0.87**	**0.32**	**0.68**	**0.77**	**0.73**

FIGURE 44.5 cAMP-Screen Assay. Comparison of the cAMP-Screen assay performance on the NorthStar workstation and the TR717 luminometer. cAMP standard was diluted into a 384-well plate and the results were read on both the NorthStar workstation and the TR717 luminometer. The luminescence data shown are the mean relative light units, with standard deviation, vs. the concentration (μM) of the cAMP standards. Z-factors were calculated for nine regions of the standard curve. The Z-factors show both the sensitivity and the dynamic range of the assay for the 384-well format.

The combination of chemiluminescent substrates such as a green-emitting 1,2-dioxetane AP substrate, BZPD[16] (550 nm emission), with a blue-emitting β-galactosidase substrate, Galacton-*Star* substrate (475 nm emission), will enable dual detection assays. Using BZPD and Galacton-*Star* substrate, 3.0×10^{-19} mol of AP can be detected in the presence of up to 6.4×10^{-16} mol of β-galactosidase, and 3.2×10^{-17} mol of β-galactosidase can be detected in the presence of up to 1.0×10^{-17} mol of AP.

44.5 SUMMARY

The NorthStar HTS Workstation is a valuable tool for HTS. It is a versatile integrated system, capable of imaging flash and glow luminescence assays, and is compatible with multiple microplate formats and can be interfaced with robotic handling. The Luc-Screen and Gal-Screen assays provide highly sensitive detection and rapid-throughput homogeneous cell-based reporter gene assays optimized for the NorthStar HTS Workstation. They are ideal HTS assays performed directly in cell culture media; the glow light emission kinetics provide flexibility in the time of reagent addition and measurement; and the high density capability enables conservation of reagents and sample. The cAMP-Screen assay is a highly sensitive second messenger assay, available in both the 96- and 384-well format and optimized for imaging on the NorthStar HTS system. All of these assays are easy to automate, providing seamless detection systems for reporter genes or second messengers, homogeneous or heterogeneous assay types. The NorthStar HTS Workstation offers flexibility with liquid-handling capability, multiwavelength imaging, and automation-friendly instrumentation, as well as true high throughput with rapid read times. As pharmaceutical screening assay trends continue to require miniaturization, higher throughput, homogeneous formats, and more biologically relevant cell-based assays, the NorthStar HTS system, with its high-density format and use of highly sensitive chemiluminescent substrates in specially designed HTS assays, offers unparalleled performance for high-throughput analyses and provides a versatile solution for these lead discovery goals.

ACKNOWLEDGMENTS

Applied Biosystems, CSPD, Gal-Screen, Galacton-*Star*, Luc-Screen, and Xpress-Screen are registered trademarks, and Applera, cAMP-Screen, NorthStar, Sapphire-II, and TR717 are trademarks of Applera Corporation or its subsidiaries in the U.S. and certain other countries. For Research Use Only. Not for use in diagnostic procedures.

Twister is a trademark of Zymark Corporation.

Intracistronic Complementation Analysis Screening Technology is patent pending. The use of Xpress-Screen® assay technology is covered by the methods claimed in pending U.S. and international patent applications that are licensed to Applied Biosystems from Digene Corporation. The use of this technology and the associated probes and reagents is personal to, and non-transferable by, the original purchaser, and is limited solely for the purchaser's own research for screening all analytes (excluding human papillomavirus of all types) for pharmaceutical and biopharmaceutical drug development and agricultural biotechnology research purposes. No use in diagnostic procedures is authorized.

REFERENCES

1. Boldt-Houle, D., Yan, S., Olesen, C., D'Eon, B., Lee, J., Liu, B., Bodziuch, U., Chiulli, A., Atwood, J., Gambini, M., Voyta, J., and Bronstein, I., A CCD-based integrated platform for accelerated high-throughput screening, *Am. Lab.*, 32(3), 60, 2000.
2. Chiulli, A., Trompeter, K., and Palmer, M., A novel high-throughput chemiluminescent assay for the measurement of cellular cAMP levels, *J. Biomol. Screening*, 5(4), 239, 2000.
3. Blakely, B. T., Rossi, F. M. V., Tillotson, B., Palmer, M., Estelles, A., and Blau, H. M., Epidermal growth factor receptor dimerization monitored in live cells, *Nat. Biotechnol.*, 18, 218, 2000.
4. Alam, J. and Cook, J. L., Reporter genes: application to the study of mammalian gene transcription, *Anal. Biochem.*, 188, 245, 1990.
5. Bronstein, I., Fortin, J., Stanley, P. E., Stewart, G. S. A. B., and Kricka, L. J., Chemiluminescent and bioluminescent reporter gene assays, *Anal. Chem.*, 219, 169, 1994.
6. Moreira, J. L., Wirth, M., Fitzek, M., and Hauser, H., Evaluation of reporter genes in mammalian cell lines, *Methods Mol. Cell. Biol.*, 3, 23, 1992.
7. Suto, C. M. and Ignar, D. M., Selection of an optimal reporter gene for cell-based high throughput screening assays, *J. Biomol. Screening*, 2, 7, 1997.
8. Seamon, K. B., Padgett, W., and Daly, J. W., Forskolin: unique diterpene activator of adenylate cyclase in membranes and in intact cells, *PNAS U.S.A.*, 78, 3363, 1981.
9. Montminy, M. R., Sevarino, K. A., Wagner, J. A., Mandel, G., and Goodman, R. H., Identification of a cyclic-AMP-responsive element within the rat somostatin gene, *PNAS U.S.A.*, 83, 6682, 1986.
10. Herz, J. M., Thomsen, W. J., and Yarbrough, G. G., Molecular approaches to receptors as targets for drug discovery, *Recept. Signal Transduct. Res.*, 17, 671, 1997.

11. Kariv, I., Stevens, M. E., Behrens, D. L., and Oldenburg, K. R., High throughput quantitation of cAMP production mediated by activation of seven transmembrane domain receptors, *J. Biomol. Screening,* 4, 27, 1999.
12. Berridge, M. J., Bootman, M. D., and Lipp, P., Calcium—a life and death signal, *Nature,* 395, 645, 1998.
13. Berridge, M. J. and Irvine, R. F., Inositol triphosphate, a novel second messenger in cellular signal transduction, *Nature,* 312, 315, 1984.
14. Karin, M., Signal transduction from the cell surface to the nucleus through phosphorylation of transcription factors, *Curr. Opin. Cell. Biol.,* 6, 415, 1994.
15. Zhang, J., Chung, T. D. Y., and Oldenburg, K. R., A simple statistical parameter for use in evaluation and validation of high throughput screening assays, *J. Biomol. Screening,* 4(2), 67, 1999.
16. Bronstein, I., Edwards, B., Voyta, J. C., Martin, C., Olesen, C., Ethier, C., and Lee, J. Y., Sensitive detection of bioanalytes with chemiluminesence, in *Book of Abstracts, 216th ACS National Meeting,* Abstr. 221, American Chemical Society, Washington, D.C., 1998.

45 Versatile, Fast Multilabel-Biochip-Reader Using a CCD Camera

Berthold Breitkopf, Fritz Berthold, and Knox Van Dyke

CONTENTS

45.1 INTRODUCTION AND HISTORY

Versatility in measuring light-based end points with different format conditions has long been an important and achievable goal for both instrument manufacturers and users. In this regard, it is important to be aware of the limitations placed on the physics of light measurements with regard to speed, sensitivity, reproducibility, and miniaturization. Certainly an instrument designed for high-throughput assay must make multiple measurements within an array of samples quickly and keep sample volume small so that reagents are judiciously used and placed within a quickly changeable platform. In addition, it is important to have solid physical and electronic construction so that the instrument will perform repetitively and reproducibly for long periods of time with low maintenance. Such instrumentation must be electronically linked for computerized analysis with software that is easy to use and that produces clear, real-time graphics. The computer software should be in a format that permits updating or possibly reprogramming to fit the needs of the user.

One method to meet many of these criteria is to image a luminescent, fluorescent, or color reaction on a mesoscale silicon-glass microstructure known as a chip. These chips can have micrometer-sized interconnecting channels and chambers, which presents a challenge to developing instrumentation with the characteristics described above. Kricka et al.[1] described the use of a microplate luminometer to measure luminescence from small volumes on a chip. However, detailed resolution was accomplished on a Berthold NightOwl backlit CCD imager.

This work demonstrates that measurements on such a small scale are possible because researchers were able to measure luminescence in 23 to 70 nl on a microchip. These chips are highly miniaturized devices that intersect with multiple technologies but that are particularly useful for

0-8493-0719-8/02/$0.00+$1.50

entities that need efficient screening systems for analytical purposes, e.g., the pharmaceutical industry, forensic analysis, clinical chemistry, and molecular biology/genetic or genomic assay. If we attempt to measure the polymorphic genetic variation of individuals to treat them with drugs that are custom-made for an individual's genetic makeup, we will need these high-throughput chip assays to perform the measurements. Gene screening to measure the quantitative effects of hormones, toxins, and inhibitors on DNA, RNA, and proteins is being done on a massive scale and will determine many of the new drugs in the future. Selective interference will be accomplished with biochemical and toxicological metabolites and pathways that are heretofore unrecognized.

This chapter describes the multiple different measurement capabilities using a unique and versatile new instrument: the Multilabel Biochip Reader utilizing a CCD camera (MLBR-CCD). This instrument package is highly adaptable to measure almost any light-based end point without purchase of additional instrumentation. Many of the instruments at present available have limited flexibility, measure a maximum of four fluorophores, and are probably too slow for many industrial applications. By adding a stacking mechanism, the MLBR-CCD can perform robotic measurements over a long period of time almost entirely without attendance. The MLBR-CCD can be easily adapted to measure visible or fluorescent light (using a combination laser or standard fluorescent lamp) or luminescence from a myriad of strongly luminescent reactions.

The instrument can isolate narrow light bands from broad wavelengths and measure multiple wavelengths quickly and sequentially, quasi-simultaneously as run on the bioluminescent resonance energy transfer (BRET), where protein–protein interaction causing energy transfer from a luminescent donor to a fluorescent acceptor protein as such system is needed (see Chapter 41). This technology uses luciferase from *Renilla* as the donor luminescent protein and green fluorescent protein as the acceptor molecule in an assay similar to FRET (Fluorescent Resonance Energy Transfer) but without the need of an excitation light source and the associated problems linked to the signal-to-noise ratio and fluorescent compounds interfering with the signal.

45.2 DESCRIPTION OF KEY FACTORS IN INSTRUMENT DESIGN

45.2.1 Speed of Analysis

The speed of the analysis is certainly a key factor in the development of a high-throughput assay system. Light from a lamp or laser is guided by a mirror to the microchip, the sample is illuminated, and a photomultipler measures the incident light. The light beam moves to the next sample and measurement starts anew as a sequential process. The time of scanner measurement of a chip is dependent on resolution, beam diameter, scanning speed, and integration time. In a sequential process a long integration time for each spot decreases overall speed of assay per plate too much. A different approach to this problem should be found.

One possibility would be to use highly energetic laser light to a small amount of time. But that can be photodestructive to a fluorescent tag. If a light source is used, it is helpful to have a multichromatic source of light if the end point is fluorescence because different wavelengths are needed for excitation for various fluorescent compounds. If narrow band-pass filters are used multiple different wavelengths associated with one or more fluorescent compounds can be measured. Therefore, laser light, which produces a single, narrow, highly intense wavelength of light, is not universally useful compared with white light, which contains all colors. However, one might be able to combine multiple laser sources to maintain the high intensity. Since highly energetic laser light causes photodestruction of fluorophores, laser light does not provide a simple answer. The common lasers are:

1. Argon-ion, 388 nm
2. Nickel cadmium, 325 nm

3. Green helium/neon, 543 nm
4. Red helium/neon, 630 nm

The helium lasers are often used with Cy3 and Cy5 cyanine dyes from Amersham. However, with their small Stokes' shift, where excitation and emission wavelengths overlap, the background noise is quite high. Better fluorescent dyes are available, but they may not be a good match for the wavelength of a given laser. Deep blue C (Packard) emits at 390 to 400 nm and has high Stokes' shift and has been used in BRET assays. However, this dye is not readily available because its use is proprietary. The idea behind this new concept is to use a standard halogen lamp. We chose a very sensitive CCD camera as the detector. A major advantage of a CCD camera is that it is sensitive to a much greater spectrum of wavelengths than a photomultiplier tube. We can measure all the area of the silicon chip at one time, which reduces the time of detection to a major extent because there is no switching from sample to sample as with the photomultiplier tube. Second, even though the initial signal is actually analog, it can be readily digitized, which enables data capture and manipulation by computer programs in real time or at a any time later after data is stored.

45.2.2 Optical System

The novel optical system consists of five different parts:

1. The light source
2. Filters
3. Illumination unit
4. Light trap
5. Detector

When measuring luminescence, the light source and filters are ignored and the detector reads the sample directly. The following will be directed toward an instrument configured to measure fluorescence.

45.2.2.1 Light Source

A standard halogen lamp is utilized that has an emission range from 340 to 700 nm with maximum emission in red wavelengths (600 to 700 nm). These lamps concentrate light in a small point at a set distance. The light-emitting portion of the lamp is quite small, but the energy driving the lamp is quite high possibly leading to a short lamp lifetime. Instrument manufacturers compensate by reducing the voltage to lengthen lamp lifetime, but because the energy of the emission spectrum is driven by the voltage, the lower end (less energetic) of the spectrum could be lost. This effect is greatest at dark blue wavelengths. The multilabel reader measures the excitation wavelengths after light is filtered and regulates it so that constant and sufficient light illuminates the sample without putting undue stress on the lamp.

45.2.2.2 Filters

The excitation of the fluorophore is selected by a narrow-pass interference filters which must be at 90° and perpendicular to the light beam. A slight deviation allows incorrect and possibly interfering wavelengths to pass. This incorrect light increases the background of the sample unacceptably. The optics of the Biochip reader are optimized, and only parallel light traverses the filter with only the correct wavelengths.

45.2.2.3 The Illumination Unit: The Ring Light

Once the filter selects the wavelength, it is guided by the fiber optic to a ring light. These lights have been previously used to illuminate flat fields, e.g., two-dimensional gels. The ring light is strategically placed so the Biochip is optimally illuminated. The lost light not striking the chip directly is back-reflected by a mirrored complex to increase the intensity of the light striking the chip (energy/mm^2).

45.2.2.4 Light Trap

The holder of the chip is a source of extraneous signal noise. The glass chip allows the light to penetrate through the chip and strike the background below the chip, thereby increasing background noise. If the background beneath the chip is absorbent but reflective, it can reconcentrate stray light at the detector, decreasing background and increasing signal. This can decrease noise by a factor of 100.

45.2.2.5 Detector/Camera

The camera used in this MLB-CCD is a slow-scan CCD type that is air-cooled to –70°C by a four-stage Peltier cooler (Figure 45.1) which reduces background noise to 4 electrons/pixel/h. The spectral range of this front-illuminated camera is 400 to 1100 nm with maximum quantum efficiency of approximately 40% at 650 nm. Recently, a liquid nitrogen–cooled CCD camera has become available with a background noise of 1 electron/pixel/h. It is unclear at this time whether this low a background would justify the problems of supplying liquid nitrogen for every assay.

With currently available printing arrayers, spot diameters of 100 μm can be accomplished. A resolution of 512 × 512 pixels on the detector chip produces a spot resolution of 10 × 10 pixels. For smaller spot diameters, detector chips with higher resolution can be used, but the readout time would be lengthened. In the case of luminometry where the lamp would be extinguished prior to assay, the chip luminescence could be multiply scanned and added to produce a clearer signal if necessary. This is a clear advantage of CCD cameras vs. photomultipliers, not to mention the ability to read an *X–Y* scan and simultaneously produce luminescence data of an entire field at once!

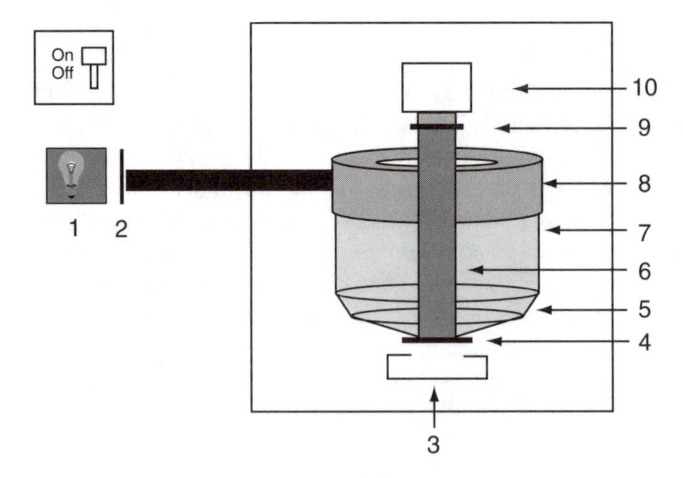

FIGURE 45.1 Schematics of a multilabel BioChip Reader. The hardware includes the lamp (1) with the filter (2). A ring light (8) illuminates the biochip (4) with the excitation light (7). The mirror optic (5) concentrates the light and makes it uniform. A light trap (3) reduces the background noise from the holder and the emission light (6) is separated from the excitation light by the emission filter (9) before the light is detected in the CCD camera (10).

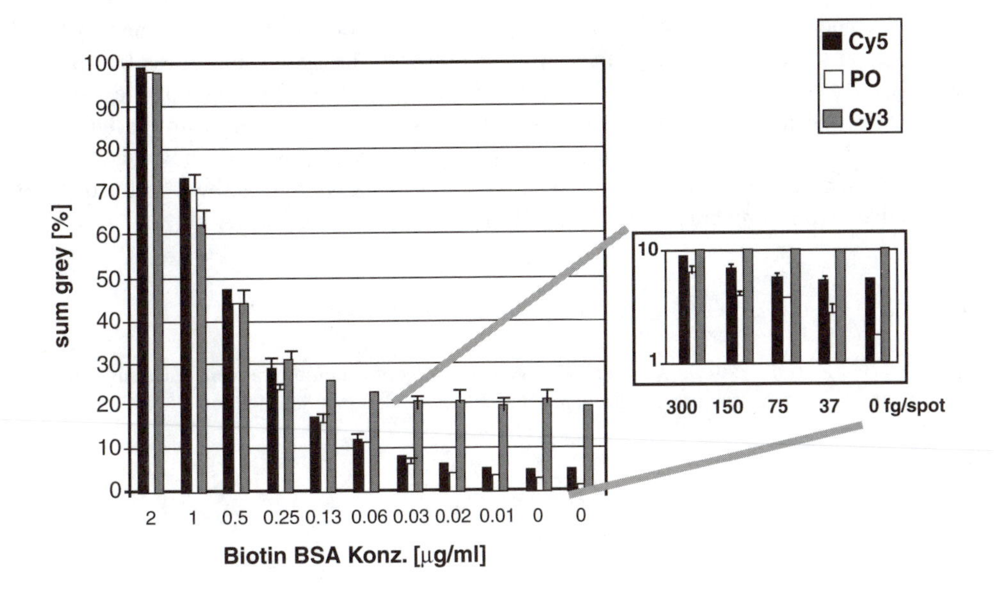

FIGURE 45.2 Sensitivity comparison of Cy3, Cy5, and peroxidase.

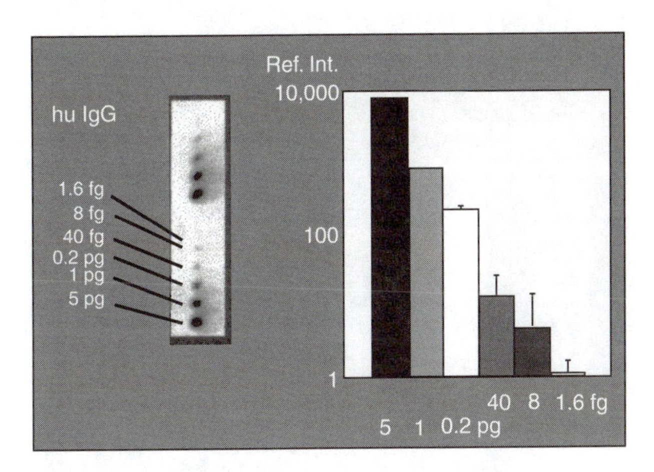

FIGURE 45.3 Chemiluminescence sensitivity <8 fg or 50 zeptomol.

Certainly this MultiLabel BioChip Reader will produce a versatility of measurement for fluorescence and luminometry at a competitive price for an instrument that will produce a high throughput of analysis.

Figure 45.2 shows data collected via the BioChip Reader with the fluorescent dyes Cy3, Cy5, and peroxidase conjugate. Figure 45.3 shows the exceptional sensitivity (50 zeptomol) for detecting biotin-labeled bovine serum albumin (BSA) spots from a volume of 10 nl. Less than 0.5 ml can be spotted, which measures 5000 molecules of IgG antibody.

45.2.3 Sensitivity

Biotin-labeled BSA spots with a volume of 10 nl and a diameter of approximately 250 μm have been analyzed in the BioChip Reader with streptavidin–Cy3 conjugate, streptavidin–Cy5 conjugate, and streptavidin–peroxidase conjugate. Figure 45.2 shows that the dectection limit for biotin-labeled BSA with Cy3 is approximately 0.25 to 0.50 μg/ml (5 pg/spot), Cy5-labeled streptavidin is 0.03 μg/ml

(0.3 pg/spot), and peroxidase-labeled streptavidin is 75 fg/spot. In Figure 45.3, human IgG has been spotted on the slide via the contact printing process. The spot volume was 0.5 nl and the diameter is 150 μm. IgG was detected at a volume less than 8 fg or 50 zeptomol—less than 5000 IgG molecules. Assuming that the polyclonal serum has only one binding with the IgG and that the spot field is 75,000 μm^2, the detection limit for this system is 0.07 molecules/μm^2 at an integration time of less than 4 min. The detection limit of the system is, by a factor of 5 to 10, more sensitive than a comparable, macroscopic chemiluminescence ELISA kit.

REFERENCE

1. Kricka, L. J., Ji, X., Nozaki, O., and Wilding, P., Imaging of chemiluminescent reactions in mesoscale silicon-glass microstructures, *J. Biolumin. Chemilumin.*, 9, 135–138, 1994.

Index

B

C

D

E

F

G

H

I

J

K

L

N

O

P

Q

R

S

T

U

V

W

X

Y

Z